Causal Inference and Machine Learning

Causal Inference and Machine Learning in Economics, Social, and Health Sciences bridges the gap between modern machine learning methods and the applied needs of economists, public health researchers, and social scientists. Designed with students and practitioners in mind, the book introduces machine learning through the lens of causal inference, offering a rigorous yet accessible roadmap for using data to answer real-world policy questions.

It combines econometric and machine learning methods such as penalized regressions, random forests, boosting, double machine learning, and the most up-to-date estimation methods for addressing selection on observables (e.g., matching, AIPW) and unobservables (e.g., instrumental variables, difference-in-differences, synthetic control). Readers learn how to estimate treatment effects, uncover heterogeneity, and work with high-dimensional data, while gaining clarity on assumptions, trade-offs, and limitations. The book also covers advanced and often underrepresented topics such as time series forecasting with machine learning methods, neural networks and deep learning, and core optimization algorithms like gradient descent. Each method is introduced with intuition, formal treatment, and applied examples from economics, health, labor, and development studies. It places special emphasis on transparency, identification, and interpretability.

Beyond introducing models, it provides step-by-step guidance from raw data to estimation, showing not just what works, but how and why—both methodologically and computationally. Unlike many texts that rely on pre-built software or assume deep technical knowledge, this book builds from foundational concepts such as estimation, error decomposition, and bias–variance trade-offs, then progresses to advanced machine learning approaches. Simulation-based pedagogy helps readers visualize model behavior under known conditions, enabling researchers and students alike to see how statistical tools perform across diverse empirical settings.

A distinctive feature of the book is its focus on when and how to use predictive versus causal models. Rather than treating them as separate tasks, it shows how each can inform the other. Practical insights, diagnostics, and examples guide readers in selecting appropriate tools based on research goals and data characteristics.

With its clear style, practical code in R, and integrated approach to prediction and causality, this book is an essential resource for applied researchers, students, and anyone using data to inform policy and decision-making.

KEY FEATURES

- Integrates causal inference with the latest econometric and machine learning methods to address real-world policy questions in economics, health, and the social sciences.
- Offers clear, detailed explanations and intuitive guidance—even for foundational concepts often overlooked in other sources—to build theoretical understanding and link econometric principles to application.
- Designed for applied researchers, students, and practitioners with limited technical background, with step-by-step instruction from raw data and basic code, including how both the methods and the underlying code function.
- Provides practical guidance on when and how to use predictive vs. causal models, highlighting their trade-offs and pitfalls to avoid, supported by real-world examples and simulation-based demonstrations.

Causal Inference and Machine Learning

In Economics, Social, and Health Sciences

Mutlu Yuksel and Yigit Aydede

CRC Press
Taylor & Francis Group
Boca Raton London New York

CRC Press is an imprint of the
Taylor & Francis Group, an **informa** business
A CHAPMAN & HALL BOOK

First edition published 2026
by CRC Press
2385 NW Executive Center Drive, Suite 320, Boca Raton FL 33431

and by CRC Press
4 Park Square, Milton Park, Abingdon, Oxon, OX14 4RN

CRC Press is an imprint of Taylor & Francis Group, LLC

ISBN: 978-1-032-82041-5 (hbk)
ISBN: 978-1-032-82491-8 (pbk)
ISBN: 978-1-003-50479-5 (ebk)

DOI: 10.1201/9781003504795

Typeset in CMR10 font
by KnowledgeWorks Global Ltd.

To Isik and Ege — thank you for your love, patience, and for reminding me daily what truly matters, even when I was lost in code.

To Mevlude, Eren, and Kaan — your love, laughter, and chaos have been the heartbeat behind every page of this book.

Contents

Preface

Why This Book, Why Now

This book is designed to fill a crucial gap in the intersection of econometrics, machine learning, and causal inference for applied economics, social and health scientists. While there are excellent texts in each of these areas, few books provide a unified, conceptually coherent, and practically grounded guide that connects classical statistical reasoning with modern machine learning tools—especially in the context of health, economics, and social science research.

Many existing resources on machine learning and causal inference are highly technical, often dense with notation, and filled with field-specific jargon. They may assume a deep familiarity with underlying concepts, reducing foundational explanations to just a few sentences or paragraphs. In contrast, we have deliberately chosen to explain even the most basic concepts in detail, especially in the initial chapters. If you are already comfortable with these topics, feel free to skip ahead. However, our experience has shown that many beginners lack clarity on these foundational ideas, and we believe a thorough understanding is essential for progressing to more advanced methods. Additionally, while most texts focus on using pre-built packages and discussing results, we go a step further. We often provide detailed insights into what these programs are doing behind the scenes—from raw data to algorithmic decisions. Whether you are exploring established methods or the latest approaches from the most recent papers of 2025, this book is designed to be a comprehensive learning companion.

Our motivation for writing this book stems from years of working at the intersection of econometrics and machine learning. We found that while machine learning provides powerful tools, its application in causal research is often misunderstood. Many researchers hesitate to adopt these methods due to concerns about interpretability and identification. With this book, we aim to bridge that gap—providing clear, practical guidance for applying machine learning in a way that complements traditional econometric approaches.

What sets this book apart is its layered approach to both teaching and understanding modeling. It builds from first principles—starting with foundational topics like estimation, model error, and bias–variance trade-offs—and walks the reader forward into high-dimensional predictive models, algorithmic tuning, and cutting-edge methods for causal inference using machine learning. Along the way, it maintains a careful balance between theoretical clarity and practical implementation, ensuring that each method is not only explained but also demonstrated through simulation, visualization, and applied examples. While the book provides a solid theoretical foundation, every concept is paired with applied examples, simulations, and case studies to ensure readers gain practical experience.

Simulation plays a central role throughout the book. For every concept—from ordinary least squares (OLS) bias to double machine learning—we show how estimators behave when the truth is known. This simulation-based pedagogy allows readers to build intuition, evaluate estimator performance, and visually understand identification failures and model misspecification. This is not commonly emphasized in econometrics texts, and it makes the learning process more transparent and engaging.

Another distinguishing feature is the book's integration of predictive and causal goals, rather than treating them as opposing paradigms. We show how predictive tools (e.g., penalized regression, trees, boosting) are used not just for forecasting, but also as core components of modern causal inference—particularly in estimating heterogeneous treatment effects, handling selection bias, and conducting robust policy evaluation. Rather than separating statistical modeling and machine learning into silos, we provide a framework that connects them, rooted in both theory and empirical needs.

We also place special emphasis on interpretability, identification, and transparency, especially for readers in economics, public health, and policy fields. Throughout, we draw attention to assumptions, estimation targets, and evaluation strategies. This helps bridge the gap between abstract modeling and applied credibility.

In short, this book is not just a technical manual or a high-level overview. It is a comprehensive, practical, and intellectually coherent guide for students, researchers, and practitioners who want to understand and apply modern modeling strategies with confidence, clarity, and rigor. By the end of this book, readers will be able to critically evaluate machine learning methods, apply them to empirical problems, and interpret results in a credible and transparent manner. Whether it's estimating the effect of a policy intervention or forecasting economic trends, readers will gain both the theoretical understanding and practical skills needed to conduct rigorous, data-driven research.

Structure of the Book

This book is structured to guide readers from core modeling ideas to advanced machine learning and causal inference techniques, always anchored in practical applications from economics, health, and the social sciences. We begin with fundamental statistical and causal reasoning, gradually progressing to high-dimensional models, algorithmic tools, and real-world policy applications. The emphasis throughout is on both understanding the mechanics of methods and implementing them thoughtfully in applied work.

The first few chapters introduce the building blocks of data-driven analysis—how we move from ideas to formal models, distinguish estimation from prediction, and build learning systems that generalize well. Core concepts such as correlation, regression, error decomposition, overfitting, and the bias–variance trade-off are developed with simulations and clear explanations. These early chapters form the conceptual foundation for later techniques.

The next group of chapters covers classical and modern estimation and prediction strategies. We begin with parametric and nonparametric estimators, then transition to topics like hyperparameter tuning, classification models, and model selection. This section includes an extended treatment of penalized regression methods—ridge, LASSO, elastic net, and adaptive LASSO—and ensemble methods such as bagging, random forests, and boosting. Each method is developed from the ground up with hands-on simulations and technical derivations.

From Chapter 16 onward, the book turns to causal inference. We begin with the Rubin causal model and the logic of counterfactual comparisons, followed by the design and analysis of randomized controlled trials, including regression adjustment, randomization inference, and covariate balance checks. We then cover causal inference under selection on observables, introducing regression adjustment, matching, and inverse probability weighting

(IPW). These chapters build toward doubly robust methods—particularly augmented IPW (AIPW)—that combine modeling and weighting strategies for more reliable estimation. All methods are demonstrated through intuitive illustrations and simulation-based exercises, and many include machine learning implementations, such as XGBoost-assisted matching or LASSO-based adjustment.

After establishing tools for dealing with observed confounding, we turn to models that handle selection on unobservables. Chapter 22 introduces instrumental variables (IV) in both the potential outcomes framework and classical linear models. We connect the logic of IV to randomization, explain how to assess instrument validity, and develop the core 2SLS estimator before extending it to high-dimensional and nonlinear settings. The chapter also introduces double machine learning for IV using LASSO and random forests, emphasizing the role of orthogonal scores, robustness to model misspecification, and finite sample performance. Simulations and algorithmic steps provide readers with the tools to implement DML-IV in applied research. We then explore how to estimate heterogeneous treatment effects using interaction models, causal forests, and meta-learners such as S-, T-, X-, R-, and DR-learners.

Chapters 26 and 27 introduce modern causal designs like difference-in-differences and regression discontinuity designs, including machine-learning extensions (e.g., DML-DiD, synthetic control, synthetic DiD). These chapters link traditional identification strategies with predictive methods.

The final part of the book focuses on advanced and emerging areas: causal inference for time series, neural networks and deep learning, matrix decomposition tools such as principal component analysis (PCA) and factor analysis, and core optimization algorithms such as gradient descent. Together, these topics expand the book's reach to high-dimensional, time-dependent, and unstructured data applications.

An online companion site is available at www.causalmlbook.com, where readers can access a digital version of the book together with an accompanying workbook. The workbook contains more than 300 chapter-specific conceptual and applied exercises, advanced real-world applications, and over 75 policy, business, and research case studies that apply prediction and causal methods across chapters. The site also hosts the datasets and R packages used in the book, along with interactive tools, teaching slides, Shiny applications, and AI-supported learning modules. Python versions of the R code and additional content are added on a regular basis. By organizing the book this way, we aim to support a wide range of readers—from those seeking clarity on core modeling logic to researchers implementing frontier methods. Throughout, examples, simulations, and applied illustrations reinforce each concept and provide a clear path from theory to practice.

Who Can Use This Book?

This book is designed for a wide range of readers who work with data to answer empirical questions—particularly in the fields of economics, public health, policy, and the social sciences. It is accessible to learners with a basic background in statistics or econometrics and is intended to support both classroom instruction and self-guided learning.

- *Undergraduate and Master's students* in economics, data science, public policy, or health disciplines will benefit from the step-by-step development of modeling concepts, supported by simulation and intuitive explanations.

- *Ph.D. students and early career researchers* can use this book to bridge the gap between traditional econometrics and modern machine learning techniques, especially for empirical work focused on prediction, estimation, or causal inference.
- *Instructors and course designers* can use the modular structure to build full-semester courses or topical modules—spanning estimation theory, machine learning, and causal inference with modern computational tools.
- *Industry professionals and government analysts* who apply data-driven decision-making in fields such as finance, healthcare, or policy evaluation will find practical insights for applying machine learning to solve real-world problems.
- *Readers with coding experience in R or Python* will find the simulations, examples, and implementations easy to follow, though prior programming is not strictly required to understand the concepts.

Whether you are building predictive models, estimating treatment effects, or exploring new forms of data like text or time series, this book provides a comprehensive and flexible resource to support rigorous, transparent, and effective empirical research.

Acknowledgments

We would like to extend our heartfelt gratitude to our loved ones for their constant support during the creation of this book. Their unwavering belief in our abilities and vision has been invaluable, and we could not have reached this milestone without them.

Yigit is deeply grateful for the sacrifices Isik has made and for her steadfast encouragement throughout the pursuit of this dream. He is also thankful for the opportunity to share his passion for learning with his son, Ege.

Mutlu would like to express his profound thanks to his wife, Mevlude, whose love, patience, and understanding have been a constant source of strength and inspiration. He also extends his gratitude to his sons, Eren and Kaan, whose laughter, curiosity, and boundless energy have fueled his determination to work harder and build a lasting legacy.

We're grateful to the colleagues and friends who read early drafts and offered thoughtful comments on specific chapters—your insights helped sharpen our arguments and improve the clarity of our writing. We also thank those who generously shared code, datasets, and lecture notes online; your openness made our work easier and this book possible.

Yigit Aydede
Mutlu Yuksel

About the Authors

Mutlu Yuksel is a Professor of Economics at Dalhousie University, Canada, and an applied microeconomist whose research spans labor, health, and development. His recent work applies machine learning and high-dimensional data to complex policy questions. He has received teaching awards and co-founded the ML Portal to support research and training in social and health policy.

Yigit Aydede is the Sobey Professor of Economics at Saint Mary's University, Canada, and an applied economist working at the intersection of econometrics, machine learning, and artificial intelligence (AI). He teaches data analytics and serves as Faculty in Residence at the Sobey School of Business and as an Affiliate Scientist at Nova Scotia Health. Aydede is also the co-founder of Novastorms.ai and the ML Portal, both focused on data-driven public policy and health research.

1

Introduction

This chapter introduces the major themes of the book and sets the foundation for how we think about using machine learning (ML) and econometrics to address empirical questions in economics, health, and the social sciences. We begin by distinguishing prediction from estimation, then explore applications of ML across domains, compare the roles of ML and econometrics, and outline core modeling frameworks including statistical versus ML paradigms, parametric and nonparametric methods, and predictive versus causal approaches. We also discuss model selection and introduce simulation as a tool for learning and evaluation. These ideas will be developed and applied throughout the book.

Let's start with a parable about how perspectives on problems can differ, much like varied views of the same truth.

A group of blind men heard that a strange animal had been brought to their town. None of them had seen or touched one before. Out of curiosity, they said: "We must inspect and understand it by touch, which we are capable of." So, they sought it out, and each of them touched a different part of the animal.

"An animal like a tree!" — said one, holding its leg.

"A wall!" — said another, feeling its broad side.

"A snake!" — said a third, grasping its trunk.

"A rope!" — said another, grabbing the tail.

"A fan!" — said one, feeling the ear.

"A spear!" — said the last, touching the tusk.

They were all describing the same animal—an elephant—but each had only part of the picture.

This parable mirrors how we approach ML and econometrics: distinct lenses on a shared pursuit of explanation through data. Econometricians debate the core of their field—frequentist or Bayesian, theory-driven or applied, parametric or nonparametric, micro or macro, prediction or causation—yet all wrestle with the same beast: understanding the world through data. ML, too, invites divergent definitions: a computer scientist may envision deep reinforcement learning, an economist sees LASSO for variable selection, and a statistician views it as optimized regression. Each grasps a piece of the truth, their perspectives as varied as touches on an elephant.

Contrary to its name, ML is not about autonomous "learning" in any cognitive sense. It is the process of constructing models using data-driven algorithms rooted in statistical principles, with the goal of improving predictive or inferential performance. These models—ranging from linear regressions to decision trees and neural networks—are built through algorithmic estimation and executed on modern computational systems.

DOI: 10.1201/9781003504795-1

Just as a child learns to distinguish cats from dogs by observing many examples, supervised learning uses labeled data to train algorithms that map inputs to outputs. These models power spam filters, voice assistants, and recommendation engines. In this book, however, we emphasize ML's role in empirical research—particularly in building better estimators, uncovering structure, and strengthening causal analysis.

Within economics, social and health sciences, ML is increasingly used to uncover causal relationships. Whereas prediction focuses on forecasting outcomes, causal inference explores how a variable, intervention, or treatment affects an outcome. Consider questions like: Does a scholarship program boost graduation rates? Does telemedicine reduce hospital readmissions? While ML has transformed domains like image recognition and language processing, our emphasis is on its contribution to applied data analysis—especially in improving estimation, modeling heterogeneity, and enabling flexible, interpretable models.

We structure this book around the estimation and inference challenges commonly encountered in applied research. A central conceptual anchor is the bias–variance trade-off, which provides a unifying framework for understanding model generalization, overfitting, and regularization.[1]

ML methods are generally categorized as:

- Supervised learning: learning from labeled data (e.g., regression, classification)
- Unsupervised learning: discovering patterns without labels (e.g., clustering, dimensionality reduction)
- Reinforcement learning: adapting through feedback over time
- Deep learning: modeling with multi-layered neural networks

Although these paradigms enable a broad range of applications—from robotics to genomics—our focus is on supervised learning, particularly regression and classification, with selected coverage of unsupervised methods such as clustering and dimensionality reduction. Our objective is to demonstrate how ML complements, rather than replaces, econometric approaches. We begin with foundational concepts, proceed through predictive modeling, and culminate with modern approaches to causal inference. The aim is to provide readers with a technically grounded, practically oriented framework for using ML in applied research.

1.1 Prediction vs. Estimation

In economics, health, and the social sciences, researchers routinely aim to identify patterns, quantify associations, generate forecasts and predictions, and estimate causal effects. Yet terms such as prediction, extrapolation, forecasting, and estimation are often used interchangeably—particularly in ML contexts. Despite their similarities, these concepts have distinct meanings and serve different analytical purposes.

In the simplest terms, prediction refers to using observed data to make accurate generalizations to unseen or future events. Related concepts include extrapolation, which extends patterns beyond the observed data range, and forecasting, which emphasizes time-indexed predictions based on historical trends—especially common in finance and macroeconomics. In ML, prediction typically involves estimating a function that maps input features (variables or covariates) to outcomes, with performance evaluated on out-of-sample (unseen) data.

On the other hand, social scientists and economists are often concerned with **estimation**, particularly in a causal framework. Estimation focuses on quantifying the relationship between variables, frequently aiming to recover population-level parameters from sample data, typically emphasizing causal inference. While prediction often focuses on accurately forecasting unit-level outcomes, estimation is concerned with understanding and quantifying relationships, such as determining the causal impact of an intervention or treatment.

Distinguishing between these objectives is critical for selecting appropriate methods and framing research questions. For instance, in clinical settings, accurately predicting patient outcomes may guide decision-making. In contrast, health economists are typically interested in estimating whether a treatment improves outcomes on average across a population.

However, predictive modeling in the social sciences presents unique challenges. Economic agents and institutions adapt in response to predicted outcomes, potentially altering the very patterns being forecast. This critique is formalized in the Lucas Critique, which states:

> Given that the structure of an econometric model consists of optimal decision rules of economic agents, and that optimal decision rules vary systematically with changes in the structure of series relevant to the decision maker, it follows that any change in policy will systematically alter the structure of econometric models. (Robert Lucas, 1976)

Unlike natural sciences—where identified relationships often remain stable under consistent conditions—social science contexts frequently involve changing relationships influenced by policy actions or behavioral responses, complicating predictive reliability. Nevertheless, well-calibrated predictions remain valuable. Effective predictions indicate deeper understanding and often require balancing model simplicity and flexibility.

Finally, real-world scenarios frequently demonstrate the practical importance of individual-level prediction versus average causal estimation. A parent may be more concerned with how an education policy affects their own child than the average student. A physician may prioritize personalized prognostics over population-level inference. Law enforcement agencies often rely on predictive models to allocate resources, reflecting the operational relevance of individual-level forecasting.

In sum, both prediction and estimation play foundational roles in applied empirical work. ML tends to associate prediction with learning a function that performs well on unseen data, while estimation emphasizes parameter identification and causal interpretation. Clearly articulating this distinction helps applied researchers align methodological choices with their analytical objectives.[2]

With this conceptual distinction in hand, we now turn to how ML methods are applied in practice—both to improve predictive accuracy and to support causal inference across applied domains.

1.2 Applications of Machine Learning in Economics and Social Sciences

ML has increasingly reshaped empirical research in economics and related fields. While commonly associated with prediction, ML's utility extends well beyond forecasting. When integrated with econometric reasoning, ML improves data utilization, strengthens causal estimation, uncovers treatment heterogeneity, and supports more granular policy evaluation. This section highlights key areas where ML contributes meaningfully to applied research.

A major contribution of ML lies in its ability to extract structured variables from unconventional or unstructured sources. Text, images, and historical documents can now be transformed into quantitative inputs—such as sentiment scores, complexity indices, or classification tags—using tools from natural language processing (NLP) and computer vision. These capabilities have expanded the measurable dimensions of human behavior, institutions, and policy environments.

Although causal inference remains a core strength of traditional econometrics, ML complements these frameworks in powerful ways. Methods such as double machine learning and targeted regularization facilitate treatment effect estimation in high-dimensional settings. Within established designs—including instrumental variables (IV), difference-in-differences (DiD), regression discontinuity (RD), and synthetic control—ML aids in modeling counterfactuals, selecting control variables, and improving estimation robustness. These hybrid strategies improve both credibility and precision.

Heterogeneity in treatment effects is often central to policy analysis. ML methods such as causal forests allow researchers to detect and interpret effect heterogeneity across subgroups—for instance, identifying whether low-income students gain more from education subsidies, or whether specific patient populations benefit differentially from medical interventions. These insights support more targeted and equitable policy design.

In labor and health economics, ML is used to process and analyze large-scale administrative datasets, detect latent patterns, and uncover interaction effects. In industrial organization, ML improves demand estimation by refining discrete choice models, such as logit and nested logit, and improving dynamic discrete choice frameworks. These advances yield better out-of-sample predictions and enable richer counterfactual simulations, supporting firm-level strategy and policy analysis.

In finance and macroeconomics, ML models are applied to tasks such as credit scoring, asset pricing, and economic forecasting. Boosting algorithms, deep learning architectures, and flexible time series methods enable the incorporation of diverse feature sets—from macroeconomic indicators to event-driven variables—to forecast inflation, output growth, and volatility. Central banks and financial institutions increasingly employ ML for risk monitoring, stress testing, and monetary policy analysis.

Social scientists also apply ML to study networks, communication, and behavioral dynamics. Clustering and graph-based algorithms reveal patterns in information diffusion, political polarization, and social influence. Urban and regional economists leverage ML to analyze satellite imagery, transportation flows, and housing data, informing the impact of infrastructure projects or zoning reforms. In environmental economics, ML tools facilitate remote sensing and the detection of deforestation, pollution, and climate vulnerability, aiding in the design of evidence-based environmental policies.

Despite their versatility, ML methods pose challenges. Model interpretability can be limited—especially with black-box algorithms—and results may be sensitive to data quality and algorithmic bias. These concerns are particularly acute in high-stakes or socially sensitive domains. As a result, ML applications in economics and the social sciences require theoretical grounding, domain expertise, and a commitment to transparency and reproducibility.

Ultimately, ML expands the empirical toolkit available to applied researchers. It enables new forms of measurement, improves causal inference, and deepens our understanding of effect heterogeneity. ML does not replace economic theory or classical econometrics; rather, it complements and strengthens them when applied thoughtfully. Throughout this book, we highlight empirical studies that demonstrate these methods in action—including recent applications in health, labor, education, finance, and development studies. Our aim is to illustrate how modern ML methods can be used to answer substantive, policy-relevant questions with rigor and clarity. These real-world applications also raise an important methodological question: when should researchers rely on ML, and when is econometric structure essential? That's the focus of the next section.

1.3 Machine Learning and Econometrics: Complementary Roles

ML is a powerful tool for analyzing data, but its usefulness depends on the research question. For tasks focused on prediction or classification, ML excels—it handles large datasets, captures complex nonlinear patterns, and adapts well to new data. Supervised learning algorithms like random forests, boosting, and neural networks are especially effective in forecasting outcomes such as disease progression, credit default, or stock prices. Unsupervised methods, including clustering and dimensionality reduction, can reveal hidden structure or groupings in data, offering exploratory insights.

However, when the objective is to understand causal relationships, explain mechanisms, or evaluate policy interventions, traditional econometric methods—such as regression, IV, DiD, and regression discontinuity designs (RDD)—are often more appropriate. These approaches are built around specific assumptions about counterfactuals, selection, and identification that ML algorithms do not address on their own. For example, estimating what would have happened to a treated group in the absence of treatment requires assumptions like unconfoundedness or valid instruments, which cannot be learned purely from the data.

Another challenge is interpretability. Many ML models function as "black boxes," offering limited visibility into how predictions are generated. This poses difficulties when the goal is not just to predict but to explain—particularly in policy contexts where transparency and accountability matter. Although explainable AI (XAI) techniques such as feature importance scores or SHAP values are improving interpretability, they still lack the clarity and inferential rigor of econometric models designed for hypothesis testing and causal analysis.

ML's reliance on historical data can also be a limitation. When facing rare or unprecedented events—such as financial crises, pandemics, or natural disasters—models trained on past data may fail. Overfitting is another common pitfall, particularly in flexible models trained on limited samples. This happens when a model captures noise rather than true patterns, performing well on training data but poorly on new observations. Regularization techniques help by penalizing model complexity, while cross-validation tests how well a model generalizes across different data subsets. These tools reflect a broader bias–variance trade-off: simpler models may underfit, missing important structure (high bias), while overly complex ones may overfit (high variance).

In practice, the choice between ML and econometrics is not either-or. They are most powerful when used together. For example, ML can support causal inference by selecting control variables in high-dimensional settings (e.g., using LASSO before DiD) or by modeling treatment effect heterogeneity with causal forests. Predictive models aid in feature engineering or understanding variable importance, while econometric tools provide the structure needed for credible causal conclusions.

As you move through this book, consider how prediction and causality serve different goals, and how aligning methods with your research question will lead to more credible and useful results. Later chapters will show how recent advances like double machine learning and generalized random forests integrate ML with causal inference frameworks. To understand how these approaches are implemented in practice, we now turn to the foundational modeling frameworks that shape empirical strategies.

1.4 Core Modeling Frameworks

In contemporary data analysis, modeling is not merely a technical task but a strategic decision that shapes how we extract insight from data. This chapter introduces key conceptual frameworks that guide empirical modeling across disciplines. We begin by contrasting statistical modeling and ML approaches, outlining their respective goals, assumptions, and practical implications. We then examine the distinction between parametric and nonparametric models, followed by a discussion of predictive versus causal modeling. Finally, we consider essential aspects of model selection, including the balance between complexity, interpretability, and generalization. Throughout this book, we use simulations to demonstrate concepts and evaluate model performance under known conditions. This strategy not only builds intuition but also makes the theoretical and practical implications of modeling choices more transparent.

1.4.1 Statistical vs. Machine Learning Paradigms

Modern empirical work often draws on two major modeling paradigms: statistical modeling and ML. While they are sometimes treated as interchangeable—especially under the umbrella of "statistical learning"—these approaches differ meaningfully in orientation, methodology, and objective. Understanding these differences is essential for aligning modeling choices with research goals—whether explanatory, predictive, or both.

Statistical modeling is primarily concerned with inference. These models are built on assumptions about the underlying data-generating process and are often motivated by theory or domain knowledge. They aim to identify structural relationships and provide interpretable estimates of parameters such as treatment effects or associations between covariates and outcomes. These models allow researchers to estimate effect sizes, test hypotheses, and draw generalizable conclusions—particularly in fields such as economics, health, and the social sciences, where inference, transparency, and theoretical coherence are highly valued.

ML, by contrast, is predominantly geared toward prediction. It uses flexible, data-driven algorithms that learn patterns from input-output pairs, often with minimal assumptions about functional form or distributional structure. The primary goal is to estimate a function

$\hat{f}(x)$ that accurately maps inputs to outputs, optimizing predictive accuracy on unseen data. Evaluation emphasizes performance metrics such as out-of-sample loss, cross-validation scores, and classification accuracy. These models excel in high-dimensional or nonlinear settings where traditional models struggle or fail to generalize.

Although both approaches can be applied to prediction and inference, they differ in prioritization. Statistical models emphasize interpretability and causal structure, while ML models prioritize flexibility and predictive performance, often at the expense of transparency. This leads to trade-offs in applied work, particularly when researchers must balance clarity of interpretation with predictive accuracy.

In many health and social science applications, the dominant tradition remains inferential statistical modeling. Researchers tend not to partition data into training and test sets but rather aim to specify models that reflect theoretical relationships. These models are used for hypothesis testing and estimation, often with a focus on identifying causal effects or quantifying uncertainty around parameter estimates. Typically, they are constructed with substantial input from domain knowledge and involve clear articulation of identification strategies—such as randomized assignment, instrumental variables, or fixed effects—to support valid inference.

In contrast, ML approaches are generally agnostic view of model structure. Their objective is not to model the data-generating process per se, but to capture empirical regularities that support accurate prediction or classification. As data sets grow in size and complexity, ML methods have proven to be especially valuable for uncovering structure in noisy environments and automating complex tasks.

These contrasting philosophies reflect not only methodological preferences but also disciplinary histories. Statistical modeling evolved from classical statistics and econometrics, emphasizing inference, hypothesis testing, and model transparency. ML emerged from computer science and engineering, with a focus on optimization, scalability, and algorithmic performance.

Leo Breiman's (2001) influential essay, "Statistical Modeling: The Two Cultures", formalized this division. In the "data modeling" culture—characteristic of traditional statistics—analysts begin with a stochastic model, estimate its parameters, and test hypotheses. In contrast, the "algorithmic modeling" culture treats the relationship between inputs and outputs as unknown and potentially complex, focusing on algorithms that achieve high predictive performance without assuming a specific model structure. Breiman's framing has had a lasting impact on how researchers think about complexity, generalization, and the role of theory in empirical work.

Many applied researchers now adopt hybrid strategies.[3] For example, ML methods such as LASSO or random forests, may be used for variable selection or to identify heterogeneity, followed by more structured statistical modeling for inference. This pragmatic blending allows researchers to leverage the strengths of both paradigms—exploratory flexibility and interpretive clarity—without fully committing to either extreme.

In practice, choosing between statistical modeling and ML—or combining them—depends on the research question, the nature of the data, and the importance of interpretability versus prediction accuracy. In applied settings, this often means using ML tools to explore high-dimensional spaces, identify patterns, or generate predictions, while relying on statistical models to interpret relationships, test hypotheses, or estimate causal effects.

Ultimately, the goal is not to declare one approach superior but to understand their respective strengths and limitations. By building a solid understanding of both paradigms and the trade-offs they entail, researchers are better equipped to select the right tools, interpret results with care, and contribute to meaningful, data-informed decision-making.

In sum, the central challenge in empirical modeling is not choosing between prediction and explanation, but aligning analytical strategies with well-defined research objectives. By understanding the assumptions, strengths, and limitations of different modeling frameworks, researchers can make informed methodological choices that improve both the rigor and relevance of their work.

The following sections examine key modeling distinctions—parametric versus nonparametric, predictive versus causal—and return to Breiman's "two cultures" in Chapter 7, where we revisit their connection to modern empirical practice.

1.4.2 Parametric and Nonparametric Models

Parametric and nonparametric models represent two foundational approaches to empirical modeling in statistics and ML. Each offers distinct advantages and limitations, and the choice between them depends on the research question, the structure of the data, and the plausibility of underlying assumptions.

Parametric models rely on explicit assumptions about the functional form of relationships among variables and the distribution of the data-generating process. These models are characterized by a finite set of parameters, estimated using methods such as least squares or maximum likelihood. Examples include linear regression, logistic regression, and polynomial regression. When their assumptions hold, parametric models tend to be efficient, interpretable, and computationally straightforward. However, misspecification—such as assuming linearity when the true relationship is nonlinear—can lead to biased estimates, invalid inferences, and poor predictive performance.

Nonparametric models, by contrast, impose minimal structure on the relationship between inputs and outcomes. Rather than specifying a functional form in advance, they allow the data to guide the shape of the function. This flexibility makes nonparametric methods valuable when the true relationship is unknown or inherently complex. Examples include k-nearest neighbors (k-NN), kernel density estimators, local polynomial regressions, decision trees, and rank-based statistics such as Spearman's correlation. These models are particularly useful for detecting nonlinearities, capturing interactions, and handling ordinal or categorical data without transformation.

In statistical terms, parametric modeling begins with an assumed form for an unknown function f, such as $f(x) = \beta_0 + \beta_1 x$. The model is then estimated using data to recover the parameters β_0, β_1. This structure promotes interpretability and parsimony and often requires less data for estimation. However, when the functional form is incorrect, the model can yield misleading conclusions.

Nonparametric methods, in contrast, place no constraints on the form of f. They aim to recover it directly from the data, enabling more flexible estimation of relationships. For instance, kernel regression estimates $m(x) = E[Y \mid X = x]$ without imposing linearity or a parametric distribution. This flexibility comes at a cost: these models typically require more data, are computationally intensive, and can overfit without appropriate tuning.

Tuning parameters—such as the bandwidth in kernel methods, the number of neighbors in k-NN, or the depth of a decision tree—play a central role in balancing bias and variance in nonparametric models. These settings control model complexity and directly affect generalization. Poor tuning can lead either to underfitting (overly smooth models) or overfitting (models too responsive to noise).

Another consideration is dimensionality. Nonparametric methods often struggle in high-dimensional settings due to the curse of dimensionality, where the data become sparse and local estimation becomes unstable. This limits their effectiveness unless combined with dimensionality reduction, regularization, or variable/feature selection strategies. These issues are revisited in detail in later chapters.

Between the two extremes lie semi-parametric models, which integrate structured parametric components with flexible, data-driven elements. A central example is the partially linear model which we will revisit throughout the causal inference chapters, particularly in the context of double machine learning and heterogeneity analysis. These models allow researchers to impose structure where theoretical relationships are well understood, while leaving other components unspecified and estimated nonparametrically. Semi-parametric methods provide a practical compromise, preserving interpretability where needed, while accommodating complexity where warranted.

These differences echo the broader contrast outlined by Leo Breiman (2001) in "Statistical Modeling: The Two Cultures", previously discussed. Parametric methods align with the data modeling tradition, emphasizing structure and interpretability. Nonparametric approaches reflect the algorithmic tradition, prioritizing flexibility and predictive performance. Increasingly, researchers blend both—using flexible models for exploration or variable selection and structured models for inference and explanation.

In Chapter 7, "Parametric Estimation—Basics", we return to classical estimation techniques and their relationship to the data modeling tradition. In Chapter 8, "Nonparametric Estimation—Basics", we focus on flexible approaches to estimating conditional expectation functions $E[Y \mid X = x]$, where no structural assumptions are imposed. These chapters build on the framework outlined here and explore how parametric and nonparametric perspectives interact in modern empirical work.

To summarize, parametric and nonparametric models represent complementary strategies for learning from data. Parametric models offer structure, interpretability, and efficiency but depend on strong assumptions. Nonparametric models provide flexibility and minimal prior restrictions but require careful tuning and larger datasets. Understanding the trade-offs between these approaches—and when to combine them—is central to effective, rigorous empirical analysis.

1.4.3 Predictive vs. Causal Thinking

Predictive and causal models reflect two fundamentally different goals in empirical analysis. While both rely on observed data, they differ in their estimands, assumptions, evaluation criteria, and implications for decision-making. Distinguishing between these objectives is essential for choosing appropriate methods and correctly interpreting results.

Predictive modeling aims to estimate the conditional expectation $E[Y \mid X]$—the expected outcome given observed covariates. The objective is to construct a function $\hat{f}(X)$ that minimizes prediction error on new or unseen data. Performance is typically assessed using out-of-sample metrics such as mean squared error, classification accuracy, or area under the curve (AUC). Predictive models are widely applied in domains such as marketing, finance, and healthcare, where the goal is to forecast outcomes, behaviors, or risks.

These models rely on empirical associations and do not attempt to establish causal relationships. While they can uncover valuable patterns, they offer no insight into the mechanisms generating those patterns. Tools commonly used for prediction include time series forecasting, regularized regression, decision trees, neural networks, and ensemble methods. However, these approaches are susceptible to overfitting—modeling noise instead of signal—especially in complex or high-dimensional settings. Techniques such as cross-validation and regularization are used to mitigate this risk and improve generalization.

Causal modeling, by contrast, is concerned with estimating the effect of interventions or exposures, often framed using the potential outcomes framework. A common causal estimand is the average treatment effect (ATE), defined as $E[Y(1) - Y(0)]$, where $Y(1)$ and $Y(0)$ represent the potential outcomes under treatment and control, respectively. Causal questions ask: *What is the expected change in outcome if we intervene on variable X?*

Identifying causal effects requires strong assumptions—such as unconfoundedness, exclusion restrictions, or monotonicity—and research designs that introduce or approximate exogenous variation. Common approaches include randomized controlled trials (RCTs), natural experiments, IV, DiD, RD, and matching techniques. These methods prioritize internal validity over predictive performance.

Importantly, a model with excellent predictive accuracy may fail to uncover causal effects, while a well-identified causal model may perform poorly at forecasting individual outcomes. The central distinction lies in the estimand and purpose: prediction seeks accuracy on future data; causality seeks to answer counterfactual questions.

These objectives also shape how data are used. Predictive modeling emphasizes performance on held-out samples, often via training-test splits or cross-validation. Causal inference focuses on isolating variation that mimics random assignment, often requiring the full sample or carefully defined subgroups to estimate credible treatment effects.

Recent advances increasingly bridge the gap between these paradigms. Methods such as causal forests and double machine learning integrate ML into causal estimation. These approaches use ML to model nuisance functions—like propensity scores or outcome regressions—while preserving valid inference under appropriate assumptions.

Ultimately, the appropriate modeling strategy depends on the research question. If the aim is to optimize decisions based on forecasts—e.g., predicting hospital readmissions or credit default risk—predictive modeling is appropriate. If the goal is to evaluate the effect of a policy, treatment, or intervention—e.g., estimating the impact of a job training program on earnings—causal inference is required. Being explicit about one's objective is critical to selecting the right tools and avoiding methodological or interpretive errors.

In summary, predictive and causal models serve distinct purposes. Predictive modeling seeks to minimize error in estimating $E[Y \mid X]$, relying on flexible algorithms and evaluated through out-of-sample performance. Causal modeling aims to identify counterfactual differences—such as $E[Y(1) - Y(0)]$—and requires careful attention to design, assumptions, and identification strategy. Understanding these conceptual and practical differences is essential for credible and relevant empirical analysis. All key concepts introduced here are explored in detail in later chapters.

1.4.4 Model Selection

Model selection is a critical step in empirical analysis. It directly influences the accuracy, interpretability, and policy relevance of results. The process involves choosing a model class, specifying its structure, selecting estimation techniques, and aligning these decisions with the underlying objective—whether prediction or causal inference. Each choice entails trade-offs between flexibility and interpretability, complexity and generalizability.

The first consideration typically involves the model family. As discussed earlier, parametric models rely on fixed functional forms and a finite set of parameters, while nonparametric models allow the data to determine the model's shape. For example, in predicting housing prices, a linear model may assume additive effects for square footage and location, whereas a tree-based model segments the feature space adaptively, capturing nonlinearities and interactions. The appropriate choice depends on whether the structural assumptions are defensible and whether interpretability or adaptability is more important.

Next, the researcher must specify the functional form. A linear model implies constant marginal effects, e.g., a fixed income gain per additional year of education. Adding polynomial or interaction terms introduces nonlinearities and captures diminishing or increasing returns. The key challenge is to enrich the model without overfitting. These specifications should be guided by exploratory analysis and theoretical expectations about the data-generating process.

After selecting a model structure, the next step is choosing an estimation method. In parametric settings, ordinary least squares (OLS) and maximum likelihood estimation (MLE) are standard. In high-dimensional contexts, regularization techniques such as Lasso or Ridge regression help control complexity by shrinking or selecting parameters. Nonparametric models typically require tuning hyperparameters—such as bandwidth, number of neighbors, or tree depth—using cross-validation to manage the bias–variance trade-off.

Model selection also depends heavily on the analytical goal. For prediction, the priority is out-of-sample accuracy, and flexible algorithms with high capacity may be preferred. For causal inference, the emphasis shifts to interpretability, alignment with identification assumptions, and consistency with the counterfactual framework. For example, estimating the effect of education on health may call for a theory-driven model with careful covariate selection, while predicting hospital readmissions may benefit from a data-driven ML model that prioritizes accuracy over interpretability.

Ultimately, model selection is not a purely technical process. It requires domain expertise, theoretical insight, and awareness of context. Researchers must balance statistical assumptions, computational cost, and the need for interpretability, all while remaining aligned with their research questions.

A full treatment of model selection—including evaluation criteria such as cross-validation, out-of-sample error, and model diagnostics—will be provided in Chapter 11 "Model Selection". There, we outline how to implement model comparison systematically and rigorously across different modeling contexts.

1.4.5 The Role of Simulation

Simulation is a foundational tool in modern empirical research. It allows researchers to study model behavior, explore uncertainty, and evaluate estimator performance in settings where analytical solutions are infeasible or the data-generating process is too complex to model directly.

By generating synthetic data under known conditions, simulation makes it possible to assess how different models perform relative to a defined ground truth. This is particularly valuable for illustrating key statistical concepts—such as bias, variance, overfitting, and generalization error—and for comparing competing methods in controlled environments. Throughout this book, we use simulation exercises to show how modeling decisions affect predictive accuracy, estimator properties, and robustness to assumptions.

Simulation is especially useful for evaluating the behavior of estimators in finite samples. Monte Carlo methods, for example, repeatedly sample from a known distribution to study properties like bias and mean squared error. These techniques also help visualize the impact of model complexity, hyperparameter choices, and identification failures—issues we revisit in later chapters on regularization, causal inference, and model selection.

Because simulation makes performance observable under known conditions, it supports not only validation and diagnostics but also pedagogical understanding. It helps connect theory to intuition, enabling researchers and students alike to see how statistical tools behave across different data environments.

1.5 Concluding Remarks

This chapter introduced the central themes of the book: the role of ML in applied empirical research, the distinctions between prediction and estimation, the integration of ML with econometric approaches, and the modeling frameworks that guide empirical practice. Along the way, we previewed key tools and concepts—such as causal inference, regularization, simulation, heterogeneity, parametric vs. nonparametric models, and model selection—that are explored in depth throughout the book. These ideas will reappear in both theoretical and applied contexts as we build from foundational principles toward practical implementations. In Chapter 2, we begin by formalizing the statistical and ML frameworks used to structure modeling decisions across empirical settings.

Notes

1. Other theoretical foundations—such as Vapnik–Chervonenkis theory, computational complexity, and Bayesian learning—are important in different contexts (e.g., deep learning, theoretical ML), but less central to the kinds of data and questions we focus on here.

2. For a deeper understanding of the distinctions between prediction and estimation, consider reviewing Bradley Efron's (2020) "Prediction, Estimation, and Attribution", Galit Shmueli's (2010) "To Explain or To Predict?", and Tetlock and Gardner's (2015) "Superforecasting: The Art and Science of Prediction."

3. Bayesian modeling provides yet another framework that unifies aspects of both traditions by formally incorporating uncertainty and prior beliefs into estimation. While powerful, Bayesian methods are beyond the scope of this book.

2

From Data to Causality

Imagine a college student, Alex, who occasionally gets sick after meals but can't figure out which food is causing it. Living in a time of growing awareness about food allergies, Alex is frustrated by recurring reactions and decides to investigate. To identify the cause of it, he starts logging everything he eats, noting down ingredients and tracking how he feels afterward—building a personal dataset.

Alex's method mirrors the early steps of a typical quantitative research process: collecting data, utilizing descriptive statistics, visualizing the data, finding correlations, and ultimately using methods to determine the causation of the allergy. He creates a daily log of his meals and reaction levels, laying the groundwork for systematic analysis. As the records grow, he begins visualizing the data to spot patterns and make sense of his experiences.

To better understand the connections, Alex uses a few simple but powerful data visualization tools:

- A Histogram of Reaction Intensity helps him see how often different severity levels occur. If most intense reactions (say, levels 8–10) cluster on garlic days, that's a clear signal worth investigating.

- A Bar Chart of Reactions by Food compares average reaction intensities across different ingredients like garlic, dairy, and wheat. If garlic stands out consistently, suspicion grows.

- A Time Series Line Graph maps reaction levels over time, helping him spot clusters of bad days and check if they align with specific meals.

These visuals make potential correlations easier to spot. For instance, the histogram allows Alex to see if high reaction intensities—say, levels 8 to 10—occur frequently after consuming certain foods. The bar chart helps him compare average reaction intensities across ingredients, making it clear if garlic consistently stands out. The time series graph lets him observe patterns over time, showing peaks in reaction intensity on specific dates, which he can then match with what he ate that day. Recognizing patterns and correlations visually through these tools gives Alex a strong foundation to hypothesize about the cause of his reactions—enabling him to test specific foods in isolation or move toward more advanced methods of analysis.

Investigating Correlations:

Observing Correlations: To quantify his experiences, Alex began rating each reaction on a 1 to 10 scale, with 10 being the most severe. As his data accumulated over several weeks, a clear pattern emerged—every time he consumed garlic, his reaction intensity consistently ranged between 8 and 10. In contrast, foods like dairy or wheat occasionally triggered mild reactions (around 3 or 4), but without consistency. This distinction helped Alex narrow down the most likely trigger.

Potential Confounding Factors: One day, Alex felt unwell after a meal that didn't include garlic but did include a milkshake. He began to wonder if dairy could be another potential allergen and started tracking dairy consumption more carefully. Over time, however, he

DOI: 10.1201/9781003504795-2

noticed that several dairy-rich days passed without any symptoms, suggesting that the earlier incident might have been coincidental or influenced by another variable—highlighting how potential confounders can cloud interpretation.

Strength of Correlation: With more data, the link between garlic and his symptoms became increasingly pronounced. In nearly every instance where garlic was consumed, a strong reaction followed. Statistically, this would be described as a strong positive correlation—garlic consumption and high reaction intensity consistently moved together more than any other food in his records.

Spurious Correlations: At one point, Alex observed that his reactions often occurred on weekends. At first glance, this seemed unrelated to diet. However, after some reflection, he realized weekends were when he typically dined out—meals more likely to include garlic. This was a textbook example of a spurious correlation: weekends weren't the cause, but they were associated with increased exposure to the actual culprit. Without careful consideration, such patterns can mislead conclusions.

Drawing Conclusions: Though correlation doesn't prove causation, the persistent and strong association between garlic and adverse reactions led Alex to a well-supported hypothesis—he may be allergic to garlic. The Mystery of Mia's Illness:

On certain weekends, Mia, Alex's girlfriend, also began to feel unwell. As she traced back her symptoms, she noticed a troubling pattern—her discomfort often coincided with the days she spent time at Alex's place. This raised concerns: was she developing a food allergy too? Had she unknowingly consumed garlic during shared meals?

Curious and cautious, Mia decided to track her symptoms alongside Alex's garlic consumption log. To her surprise, she discovered that she sometimes felt sick on days Alex ate garlic—even when she hadn't eaten any herself.

Spurious Correlation Revealed: Further investigation uncovered a crucial detail. Whenever Alex cooked with garlic at home, he would light a specific brand of aromatic candle to mask the lingering smell. It turned out Mia wasn't reacting to garlic at all, but to the fragrance of the candle. The apparent connection between her symptoms and Alex's garlic intake was a spurious correlation—the real trigger was environmental, not dietary.

This example highlights how easy it is to misattribute effects when variables move together by coincidence or through indirect associations. It underscores the importance of ruling out alternative explanations before concluding causation.

Alex's Exploration of the Effect of His Garlic Consumption on His Allergy Severity:

After discovering a strong correlation between garlic consumption and allergic reactions, Alex decided to go further. While the link was clear, he wanted to quantify how garlic affected his symptoms and whether other factors played a role. Was the reaction intensity influenced not just by garlic, but also by variables like his weight, the weather, or whether he ate indoors or outside?

Gathering Data: Over several weeks, Alex carefully recorded new variables alongside his usual food and reaction logs: the quantity of garlic in each meal, his daily weight, peak temperature of the day,and whether he dined indoors or outdoors.

To better understand the relationships, he applied an Ordinary Least Squares (OLS) regression, allowing him to estimate how each factor contributed to reaction severity while holding others constant. The results were revealing. The coefficient on garlic quantity was positive, confirming that more garlic led to stronger allergic responses. Interestingly, on days when his weight was slightly higher, the reactions were milder—suggesting his body might have had more capacity to buffer allergens. Higher temperatures were also linked to lower reaction intensity. Dining outside, however, was associated with more severe reactions. This initially seemed odd until Alex realized that eating outside often meant restaurants, where he had less control over ingredients and was more likely to encounter hidden garlic.

He recalled Mia once commenting that his reactions seemed worse on weekends. Reflecting on this, Alex saw the pattern: weekends were when he most often dined out, increasing his garlic exposure. The weekend itself wasn't the cause—just another case of spurious correlation, with garlic quietly driving the outcome behind the scenes.

Investigating Causation:

Having found strong correlations and accounted for confounding variables through OLS regression, Alex understood that correlation alone couldn't prove causation. While his descriptive analyses and OLS regression had helped isolate the effect of garlic alongside other variables—like his weight, temperature, and whether he ate indoors or out—it still couldn't fully rule out all alternative explanations. To confirm whether garlic truly triggered his allergic reactions, he needed more rigorous evidence—beyond what statistical controls could offer.

Then he remembered a news article he'd recently read. It mentioned certain foods that were structurally and chemically similar to garlic, which could also provoke allergic responses in sensitive individuals. This introduced a new layer of complexity: perhaps garlic wasn't the sole offender. Maybe related foods were amplifying his symptoms, or worse, confusing the picture entirely.

Determined to get a definitive answer, Alex turned to a more rigorous method—the experimental approach. As the gold standard for establishing causality, experimentation would allow him to control conditions and observe effects in a structured way. He planned to systematically introduce and remove garlic and garlic-like foods from his diet while keeping other factors constant.

To strengthen the reliability of the experiment, Alex asked three friends, who had no known food allergies, to participate as a control subject. By comparing their reactions under identical conditions, he could better isolate the specific effect of garlic.

For a week, they ate standardized meals, differing only in the inclusion of garlic and similar foods. All of his friends consistently showed no symptoms, even after consuming garlic. Meanwhile, Alex continued to experience reactions—especially on the days garlic was part of his meal. The contrast was clear and consistent.

His friends' reaction-free experience, even when eating foods chemically similar to garlic, helped Alex rule out those other candidates. It became evident that garlic alone was responsible for his symptoms. Including a control like three similar people not only confirmed the role of garlic but also helped eliminate the possibility that other related foods or environmental triggers were to blame.

Causation Established:

With repeated, consistent results and other potential causes ruled out, Alex concluded with confidence that garlic was not just associated with his allergic reactions—it was the cause. In scientific terms, he had successfully moved from correlation to causation.

This journey mirrors the broader scientific process: careful observation, systematic data collection, identification of confounding variables, regression analysis, and finally, controlled experimentation to isolate causal effects. It's a model that reflects how researchers in fields like statistics and machine learning work to move beyond pattern recognition toward valid inference.

2.1 Qualitative and Quantitative Research Methods

In research, two primary methodologies are widely used: qualitative and quantitative. Qualitative research involves collecting nonnumerical data through tools such as focus groups, unstructured or in-depth interviews, and document analysis to identify specific patterns or themes. In social sciences, a qualitative study might explore how individuals perceive the returns to education or the barriers women face in accessing formal employment, drawing on in-depth interviews to understand the lived experiences. Another example could be interviewing households in low-income communities to understand decision-making around child labor and schooling.

Quantitative research, in contrast, emphasizes measurable, numerical data and the use of structured tools such as surveys, structured interviews, and administrative records. It often relies on statistical, mathematical, or computational analysis. For example, an economist might estimate the causal effect of education on income using national survey data or evaluate how changes in minimum wage laws affect employment rates across different demographic groups. These studies frequently involve large datasets and use statistical techniques to establish correlations or causal effects.

Quantitative research can also use existing datasets in combination with computational methods to make predictions or simulate policy impacts. For instance, researchers might use household survey data to forecast future income distributions under alternative tax policies. In the health field, a quantitative study might assess the effect of a new health insurance program on access to care by comparing health outcomes across treated and untreated groups. Such studies rely on structured observations or measurements to analyze questions about a specific sample population, using approaches such as descriptive, quasi-experimental, or experimental designs.

While both approaches contribute to knowledge, their underlying processes differ. Qualitative research typically follows an inductive logic, aiming to build theories or generate hypotheses based on patterns observed in the data. Quantitative research, on the other hand, uses a deductive approach, testing existing hypotheses or theoretical models using empirical data.

The nature of data in these two approaches also diverges. Qualitative research is inherently subjective and seeks to understand conditions from the perspective of participants. For example, a sociologist might use interviews to investigate how small business owners in rural areas interpret new financial regulations. Quantitative research is more objective, often aiming to measure the impact of specific interventions—such as estimating how a cash transfer program affects household spending—based on observable, quantifiable outcomes.

Qualitative data is text-based and offers rich, in-depth insight from a limited number of cases, while quantitative data is number-based, often covering broader populations but with less depth per case. Analysis in qualitative research is typically interpretive, using techniques like coding and thematic analysis. Quantitative research, by contrast, applies statistical tests to assess significance, effect sizes, or predictive power, allowing for more standardized conclusions.

In terms of data collection, qualitative research tends to use unstructured or semi-structured formats, enabling open-ended responses. Quantitative studies use fixed-response formats with predefined scales or categories. While qualitative findings are often less generalizable due to their case-specific depth, quantitative results are typically more generalizable due to their broader coverage.

Though this discussion focuses on qualitative and quantitative methods as distinct approaches, many researchers employ mixed methods—combining both to enrich analysis. For instance, a researcher may use qualitative interviews to design a survey instrument and then conduct a quantitative evaluation of an impact of the program.

In summary, while qualitative and quantitative research serve different purposes, both offer valuable tools for understanding economic and social phenomena. They differ in their methods, processes, data types, and scope of generalizability, each with unique strengths depending on the research question. In this book, we focus on quantitative methods.[1]

2.2 Quantitative Research Methods

This section provides a general perspective on research methods commonly used in health, economics, and the social sciences. Research begins with a clear, answerable question. While we may not always find definitive answers with the data and tools available, we can often generate insights that improve our understanding of how the world works. A good research question is usually connected to a hypothesis—something we observe in the world or expect based on theory. If a theory predicts how something operates, we should, in principle, be able to test it using data.

To do this, researchers collect or access data, explore patterns, make assumptions, and analyze the data accordingly. The conclusions may reflect associations, correlations, or causal relationships, and the goal is often to explain observed phenomena, make predictions, or support decision-making. These core ideas were introduced at the beginning of the book.

Each field tends to rely on different methodological traditions. In health research, a major emphasis is placed on evidence-based methods.[2] In *Patient Care under Uncertainty*, Charles Manski explains that whether studying treatment response or personalized risk, the objective is similar: probabilistic prediction of patient outcomes given patient attributes. Econometricians refer to this as regression analysis; psychologists may call it statistical or actuarial prediction; computer scientists describe it as machine learning or AI; business researchers often use the term predictive analytics.

Health researchers typically begin with descriptive analysis to identify associations. When the goal is to estimate causal effects—such as whether treatment X causes outcome Y—they rely heavily on randomized controlled trials (RCTs), often referred to as the "gold standard." These methods also support prescriptive decisions, such as choosing between surveillance and aggressive treatment, and draw on both clinical judgment and empirical evidence.

RCTs are widely valued for their internal validity—that is, the credibility of findings for the sample studied. Yet, evidence-based research faces important limitations. Stronger identifying assumptions may allow more decisive conclusions but can reduce credibility. This trade-off has led to a preference for methods with fewer and more transparent assumptions—especially randomized experiments. The idea is that when treatment is randomly assigned, comparisons between treated and control groups are more likely to reflect true causal effects. Variants of this approach appear in other fields under different names: A/B testing in machine learning and randomized field experiments in economics.

Still, randomized trials have their own challenges. Trial samples may differ from the broader population, estimates are often statistically imprecise due to small sample sizes, and volunteer-based participation may not reflect how real-world patients respond. As a result, predictions based on trial data may be fragile, especially when uncertainty is not properly addressed. These concerns have prompted criticism, including from Deaton and Cartwright (2018), who argue that RCTs are sometimes misunderstood or overused, and that the quality of causal evidence depends as much on thoughtful design and interpretation as on randomization itself.

Despite these issues, researchers in health fields often give more weight to an RCT with 200 participants than an observational study with 200,000. This preference is often justified using Donald Campbell's framework of internal and external validity (Campbell and Stanley, 1963). Internal validity refers to the credibility of findings within the sample studied—similar to in-sample analysis in machine learning. External validity is about whether those findings generalize to other populations—akin to out-of-sample prediction. Since the 1960s, Campbell's argument that internal validity should take precedence has supported the primacy of experimental methods over observational ones, regardless of sample size.

Yet, credible inferences still require that the sample be representative of the target population. Without representativeness, even well-identified estimates may be biased or misleading. This issue is especially salient in observational studies, where selection into treatment or sampling is often non-random. We will return to this point when discussing sampling and population inference—key challenges in all empirical research.

Economics, in contrast to health research, places greater value on observational studies, especially those using large, representative datasets. Here too, Campbell's framework applies. The best observational studies are often those that most closely approximate randomized experiments while leveraging the broader generalizability that comes with representative data. As Banerjee and Duflo (2009) emphasize, high-quality observational and quasi-experimental studies—particularly when embedded in real-world settings—can yield policy-relevant insights even without full randomization.

These methodological tensions also reflect differences in research goals. Shmueli (2010) distinguishes between explanatory modeling, which aims to understand causality, and predictive modeling, which seeks to forecast future outcomes. Both goals are central to research across disciplines, but each places different demands on data, assumptions, and evaluation criteria. For instance, a model with high predictive power may offer little insight into causal mechanisms, and vice versa.

Finally, economics and statistics often adopt a pragmatic mindset toward modeling. As Box (1976) famously said, "All models are wrong, but some are useful." Milton Friedman (1953) made a similar point, arguing that unrealistic assumptions are not necessarily problematic—what matters is whether the resulting conclusions approximate reality. Cartwright (2007) builds on this perspective, arguing that the usefulness of a model lies not just in its formal properties but in how well it aligns with real-world structures. This philosophical view continues to shape how researchers evaluate the trade-offs between internal validity, external validity, and model realism.

2.3 Data and Visualization

Researchers in fields such as health, economics, business, and the social sciences rely heavily on data to understand both dynamic (time-varying) and static (time-invariant) phenomena. The primary forms of data used in empirical research are cross-sectional, time series, and panel (or longitudinal) data.

Cross-sectional data captures observations from multiple subjects—such as individuals, households, firms, or countries—at a single point in time or over a short period. It offers a snapshot of various attributes, allowing researchers to examine variation across units but not over time. **Time series data**, in contrast, consists of repeated observations on a single unit or aggregate at regular intervals. This structure is particularly useful for identifying temporal patterns and forecasting outcomes. **Panel data** combines features of both types of data: it tracks the same units over multiple time periods, enabling researchers to examine both within-unit changes and between-unit differences, while controlling for unobserved heterogeneity through techniques like fixed effects.

Researchers use a variety of data sources to assemble these datasets. These include survey data (e.g., household or labor force surveys), administrative records (e.g., healthcare billing data, school enrollment), experimental data (e.g., field trials), and increasingly, digital sources such as web-scraped data, sensor-generated data, or mobile app records.

Once data is obtained, the research process follows a structured workflow. The first step is to explore the raw data to understand its structure and content. Next comes data cleaning which includes identifying and correcting errors, handling missing values through deletion or imputation, and removing or adjusting outliers. Researchers must also consider ethical and legal responsibilities when working with sensitive data—such as protecting privacy, ensuring informed consent, and complying with data protection regulations. Clean data is then ready for analysis and visualization, which help uncover structure, identify patterns, and support interpretation.

Visualization is a foundational tool in this process. It facilitates both exploratory analysis and clear communication of results. Effective visualizations can reveal trends, anomalies, clusters, or relationships that may not be obvious in raw numbers. Common static techniques include histograms, barplots, boxplots, scatterplots, and heatmaps.

- *Histograms* display the frequency distribution of continuous or discrete variables. Each bar represents a range of values (a bin), and the height shows the count within that range—for example, the number of people whose income falls within $10,000 intervals.
- *Barplots* are typically used with categorical variables to compare means or medians across groups—such as average test scores by classroom or median house prices by city.
- *Boxplots* visualize the spread and skewness of a variable and highlight potential outliers. They show the interquartile range (middle 50%), the median, and the extremes of the data. These are especially helpful for comparing distributions across multiple categories.
- *Scatterplots* are useful when plotting two continuous variables, with each point representing a single observation. These plots help identify correlations or nonlinear relationships—for instance, between education level and wages.

- *Heatmaps* represent the intensity or magnitude of values using color gradients in a matrix layout. They are particularly effective for visualizing correlation matrices or geospatial distributions, such as temperature or crime intensity.

More advanced visualizations expand analytical possibilities. **Line graphs** are ideal for time series data to track changes over time. **Area charts** show how different components contribute to a total across periods. **Network diagrams** are used to represent relationships between entities, such as in social network analysis or biological systems.

Interactive visualizations have become increasingly popular, especially in web-based reporting. Tools such as dashboards, interactive maps, and sliders allow users to manipulate parameters and observe how patterns or results change in real time. These approaches are particularly effective in exploratory work, dashboards for policymakers, and educational tools.

Combining static and interactive visualizations deepens researchers' understanding of their data and strengthens communication of results. Together, these tools support better diagnostics, model development, and data-driven decisions.

Building on this foundation, the next section will introduce the concept of **correlation**, which quantifies the strength and direction of relationships between variables—a key step in both exploratory analysis and model building.

Throughout this book, we demonstrate how to implement prediction and causal models using cross-sectional data first, followed by panel data, and then time series data. Each type will be matched with appropriate methods and practical examples in R.

R offers a powerful and flexible environment for data analysis and visualization. Using base R and packages from the tidyverse—such as `readr`, `readxl`, `dplyr`, and `ggplot2`—you can access, clean, and summarize data efficiently. Tidy data principles are central to these tools. For visualization, `ggplot2` enables the creation of histograms, barplots, boxplots, scatterplots, and more, with extensive layering and customization. For time series and panel data, packages such as `tsibble`, `zoo`, and `plm` offer functions for reshaping, modeling, and visualizing data over time. For interactive visuals, `plotly`, `leaflet`, and `shiny` allow users to build dynamic, web-based visual tools. These packages together make R a reproducible, extensible, and transparent platform for applied research.

2.4 Correlation

As we build from data visualization into analytical methods, a natural next step is understanding how variables are related. This brings us to one of the most widely used concepts in data analysis: correlation.

The phrase "correlation does not imply causation" is often cited in discussions of statistical relationships to emphasize an important caution in data interpretation. To fully appreciate what this means, we must first understand the concept of correlation and its role in evaluating connections between two variables.

Correlation is a statistical measure that quantifies the degree of association between two variables, typically describing how closely they follow a linear relationship. The term *association* is broader than *correlation*, as it encompasses any kind of relationship between variables—linear or nonlinear. In practice, correlation plays a key role in many fields, including finance, healthcare, and marketing, where understanding relationships between variables guides decisions and strategies. Interestingly, the word *correlation* combines "co-" (meaning together) and "relation," reflecting its function in identifying how quantities move in connection with one another.

There are three types of correlation: positive, negative, and zero. A **positive correlation** indicates that both variables move in the same direction. For example, as advertising expenditure increases, sales revenue may also rise, suggesting a positive correlation. A **negative correlation** means the variables move in opposite directions, such as when increased hours of television watching are associated with lower academic performance. **Zero correlation** implies no discernible relationship—changes in one variable are not systematically associated with changes in the other.

Correlation analysis is the statistical method used to examine whether and how strongly pairs of variables are related. It helps researchers quantify how two variables move together or apart, and whether that movement is consistent enough to be considered meaningful. By measuring the strength and nature of associations, correlation analysis can guide further exploration or prediction, though it must be interpreted carefully to avoid incorrect conclusions about causality.

One of the most common tools in correlation analysis is the **Pearson correlation coefficient**. This coefficient quantifies the strength and direction of a linear relationship between two numerical variables. It ranges from -1 to 1: values close to -1 indicate a strong negative linear relationship, values near 1 suggest a strong positive relationship, and values near 0 imply little to no linear association. For example, a Pearson correlation of 0.8 between study hours and test scores suggests that, in general, more study time is associated with higher scores.

The Pearson correlation coefficient, usually denoted as r, can be calculated using two equivalent approaches. The first is:

$$r = \frac{\sum(x_i - \overline{x})(y_i - \overline{y})}{\sqrt{\sum(x_i - \overline{x})^2 \sum(y_i - \overline{y})^2}} \tag{2.1}$$

This formula computes r as the ratio of the sum of the product of deviations from the mean for both variables to the geometric mean of the sums of squared deviations.

The second approach uses the covariance and standard deviations of x and y:

$$r = \frac{\text{cov}(x, y)}{s_x s_y} \tag{2.2}$$

where $\text{cov}(x, y)$ is the covariance between x and y, and s_x and s_y are the standard deviations of x and y, respectively. This formula emphasizes that correlation is a standardized measure of covariance, allowing for comparisons across different datasets or variables. Here, the **covariance** between x and y is defined as:

$$\text{cov}(x, y) = \frac{1}{n-1} \sum(x_i - \overline{x})(y_i - \overline{y}) \tag{2.3}$$

The **standard deviations** of x and y are:

$$s_x = \sqrt{\frac{1}{n-1} \sum(x_i - \overline{x})^2} \quad \text{and} \quad s_y = \sqrt{\frac{1}{n-1} \sum(y_i - \overline{y})^2} \tag{2.4}$$

In these formulas, x_i and y_i represent individual values of the two variables, $\overline{x}$ and $\overline{y}$ are their means, and the summations run over all n observations. These expressions show that r normalizes the covariance by dividing it by the product of the standard deviations, thereby measuring how much the two variables covary in proportion to their individual variability.

Covariance itself tells us the direction of a linear relationship between two variables, such as whether they increase together or one increases while the other decreases. However, covariance does not indicate the strength of that relationship because its magnitude depends on the units and scale of the variables. To overcome this limitation, we use the correlation coefficient which standardizes covariance.

Correlation is essentially the covariance between two variables divided by the product of their standard deviations. This standardization results in a dimensionless number that always falls between -1 and 1, allowing meaningful comparisons across datasets regardless of the original scale of the variables. This makes correlation a powerful and versatile tool for exploring relationships in data.

In the following simulation, we demonstrate how scaling affects covariance but not correlation, illustrating an important distinction between these two concepts. We begin by generating two numerical variables, each with 20 observations, and then scale them by multiplying each by 1,000. This operation significantly changes the magnitude of the covariance, showing that covariance is sensitive to the scale of the variables. However, when we calculate the correlation coefficients—using both definitions described earlier—we find that the values remain unchanged. This example highlights the robustness of correlation as a standardized, unit-free measure of association.

```
# Set seed for reproducibility
set.seed(0)

# Generate two variables each with 20 observations, all less than 10
x <- runif(20, min = 0, max = 10)
y <- runif(20, min = 0, max = 10)

# Scaling the data by multiplying each element by 1000
x_scaled <- x * 1000
y_scaled <- y * 1000

# Print the original and scaled variables
#print(x)
#print(y)
#print(x_scaled)
#print(y_scaled)

# Calculate the covariance between x and y
covariance_original <- cov(x, y)
print(paste("Original covariance:", covariance_original))
```

```
## [1] "Original covariance: -1.26490786248817"
```

```
# Calculate the covariance between x_scaled and y_scaled
```

```
covariance_scaled <- cov(x_scaled, y_scaled)
print(paste("Covariance after scaling:", covariance_scaled))
```

```
## [1] "Covariance after scaling: -1264907.86248817"
```

```
# Compute standard deviations for both datasets
sd_x <- sd(x)
sd_y <- sd(y)
#print(paste("Standard deviation of x:", sd_x))
#print(paste("Standard deviation of y:", sd_y))

# Standardize the variables
x_standardized <- (x - mean(x)) / sd_x
y_standardized <- (y - mean(y)) / sd_y
#print(paste("Standardized variable x:", x_standardized))
#print(paste("Standardized variable y:", y_standardized))

# Calculate the covariance between the standardized x and y
#which give us the correlation coefficient:
correlation_original <- cov(x_standardized, y_standardized)
print(paste("Correlation after standardizing:", correlation_original))
```

```
## [1] "Correlation after standardizing: -0.152795399233766"
```

```
# Equivalent formula to calculate the correlation between the original x and y
correlation_original2 <- cov(x, y) / (sd_x * sd_y)
print(paste("Correlation after standardizing2:", correlation_original2))
```

```
## [1] "Correlation after standardizing2: -0.152795399233766"
```

```
# Compute standard deviations for both scaled datasets
sd_x_scaled <- sd(x_scaled)
sd_y_scaled <- sd(y_scaled)
# Standardize the variables
x_scaled_standardized <- (x_scaled - mean(x_scaled)) / sd_x_scaled
y_scaled_standardized <- (y_scaled - mean(y_scaled)) / sd_y_scaled

# Calculate the correlation between the scaled x and y
correlation_scaled <- cov(x_scaled, y_scaled)/(sd_x_scaled * sd_y_scaled)
print(paste("Correlation after scaling:", correlation_scaled))
```

```
## [1] "Correlation after scaling: -0.152795399233766"
```

```
# Equivalent formula to calculate the correlation between the scaled x and y
correlation_scaled2 <- cov(x_scaled_standardized, y_scaled_standardized)
print(paste("Correlation after scaling:", correlation_scaled2))
```

```
## [1] "Correlation after scaling: -0.152795399233766"
```

TABLE 2.1 Summary Statistics for Anscombe's Quartet

Statistics	Group I	Group II	Group III	Group IV
Mean X	9.000	9.000	9.000	9.000
Mean Y	7.501	7.501	7.500	7.501
Variance X	11.000	11.000	11.000	11.000
Variance Y	4.127	4.128	4.123	4.123
Covariance	5.501	5.500	5.497	5.499
Correlation	0.816	0.816	0.816	0.817
Intercept	3.000	3.001	3.002	3.002
Slope	0.500	0.500	0.500	0.500

Having seen how correlation remains consistent regardless of scale, it's worth revisiting the importance of data visualization, as emphasized in the previous subsection. A compelling illustration of why visual analysis is essential is provided by the Anscombe Quartet, a well-known example that underscores the limitations of relying solely on numerical summaries.

The Anscombe Quartet (1973) consists of four datasets with nearly identical statistical properties: they share the same means, variances, correlation coefficients, and linear regression lines.

Yet when plotted, the datasets reveal dramatically different patterns. This example powerfully demonstrates that statistical summaries can be misleading unless paired with visual inspection.

For instance, Group I in the Anscombe dataset displays a typical linear relationship, where a straight-line model is appropriate. Group II, despite having the same regression line, shows a curved, nonlinear pattern, making a linear model unsuitable. Group III includes an outlier that heavily influences the regression, masking the real structure of the data. Group IV features most data points clustered at the same x-value, with a single influential point dictating the entire regression line. These differences become clear only through visualization and cannot be detected from summary statistics alone.

This example reinforces a key message: visualizing data is essential before interpreting statistical metrics such as correlation or regression outputs. It also prepares us to address another common misconception—namely, that a correlation coefficient of zero implies independence between variables.

A correlation of zero simply means that there is no linear relationship between the variables. It does not rule out the presence of a nonlinear association. To illustrate this, we can construct a simple nonlinear relationship, such as defining one variable as a quadratic function of another. Even though the two variables are clearly related, the correlation coefficient may be close to zero, misleadingly suggesting no relationship at all.

```
set.seed(42)  # Set seed for reproducibility
x <- -10:10   # Generate x as a sequence of values
y <- 9 + 2 * x^2  # Define y as a non-linear function of x

# Plot x vs y to visualize the non-linear relationship
plot(
  x, y,
```

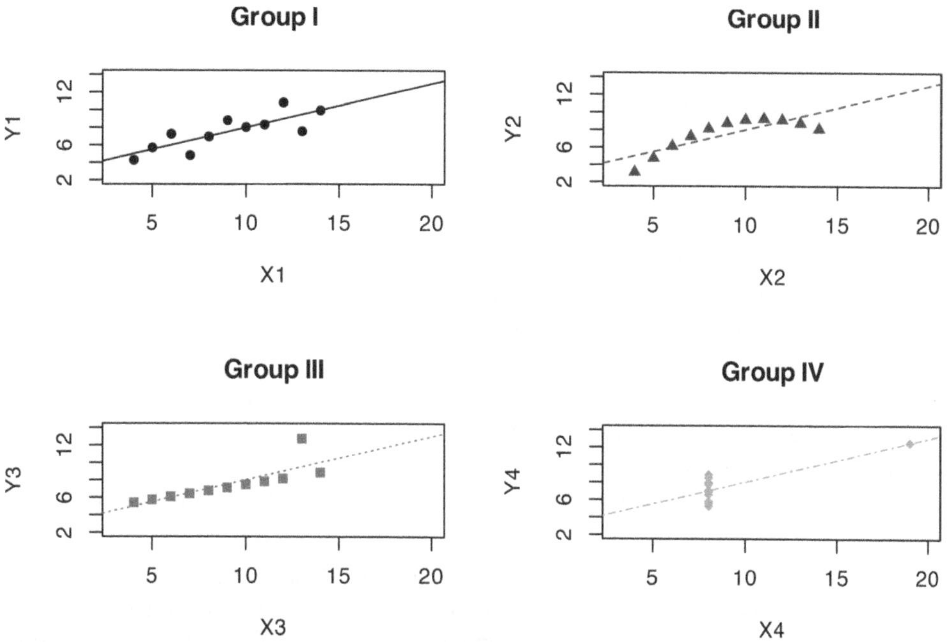

FIGURE 2.1 Anscombe's Quartet: Four datasets with identical statistics.

```
  main = "Non-linear Relationship: 9 + 2*x^2",
  xlab = "x", ylab = "y",
  pch = 19, col = "black"
)
```

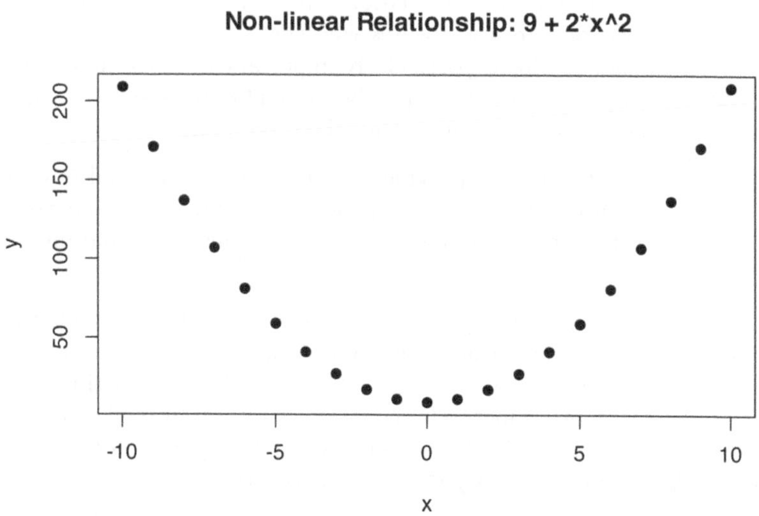

FIGURE 2.2 Nonlinear relationship between x and y.

```
# Calculate and print the correlation between x and y
correlation <- cor(x, y)
```

```
print(paste(
  "Correlation between x and y:",
  formatC(correlation, format = "f", digits = 3)
))
```

```
## [1] "Correlation between x and y: -0.000"
```

This limitation highlights the need for alternative correlation measures when relationships are nonlinear or more complex. Beyond the Pearson correlation, several other methods exist that can better capture these associations.

Partial correlation is one such method. It measures the relationship between two variables while holding one or more additional variables constant. This helps isolate the direct connection between two variables, free from confounding influences. For example, when examining the relationship between education and income, controlling for age and work experience through partial correlation provides a more accurate picture of the direct effect of education.

Conditional correlation is similar but often involves conditioning on categorical variables. It allows researchers to explore how the relationship between two continuous variables might change across different groups or segments. This method is especially helpful when relationships differ across subpopulations—for instance, analyzing how the correlation between exercise and blood pressure may vary by gender or age group.

Spatial correlation is relevant when the data include geographic or locational information. It assesses how values measured in nearby locations relate to each other. For example, pollution levels in one region may correlate with those in adjacent areas. Spatial correlation helps identify clustering effects or diffusion patterns, commonly used in environmental studies and urban planning.

Dynamic correlation focuses on how the relationship between variables evolves over time. In time series or panel data contexts, it captures whether variables move together consistently or if their relationship shifts. This is particularly important in economics and finance, where variables like interest rates and inflation may be more closely related in some periods than others. Dynamic correlation offers insights into these temporal changes.

Each of these correlation techniques provides unique insights, tailored to different types of data structures and research objectives. By choosing the appropriate correlation method, researchers can better understand the underlying connections and avoid misleading interpretations.

In summary, correlation analysis is a powerful and flexible tool for exploring the relationships between variables. Pearson correlation is widely used for measuring linear associations, but when data relationships are influenced by other variables, change over time, or are shaped by geography or nonlinearity, other methods such as partial, conditional, spatial, and dynamic correlation are more appropriate. Understanding and applying these tools enables researchers to gain deeper insights into complex patterns of association.

Importantly, while correlation can help describe how variables move together, it does not explain why. A strong correlation between two variables does not imply that one causes the other. Confounding variables, reverse causality, or coincidence may all be responsible. This is why the phrase "correlation does not imply causation" remains so central in empirical research. Drawing conclusions about causal effects requires additional tools and careful design. The next sections will explore the principles, strategies, and statistical methods for establishing causal relationships in data.

2.4.1 Correlation and Regression

Having explored how correlation helps us understand the strength and direction of relationships between variables, we now turn to a related but distinct concept: regression. While both correlation and regression examine relationships, they differ in purpose, interpretation, and application.

Correlation and **regression** are two foundational statistical methods used to analyze how variables relate to one another. They share similarities: both often assume a linear relationship between variables and involve coefficients that quantify these associations. These methods help illustrate how changes in one variable may relate to changes in another.

However, it is important to recognize their key differences. **Correlation** is primarily a **descriptive** technique. It measures the strength and direction of an association between two variables without attempting to model or predict outcomes. The correlation coefficient, which ranges from -1 to 1, summarizes how tightly two variables move together. A coefficient of 1 indicates a perfect positive relationship, -1 a perfect negative relationship, and 0 indicates no linear association. Correlation is especially useful when we want to summarize associations—such as between smoking and lung cancer—without making assumptions about causality or directionality.

In contrast, regression is a method used to explain and quantify the impact of one or more variables on another. It not only assesses the relationship between variables but also builds a model that estimates how changes in one variable influence another. This makes regression especially useful for forecasting and for evaluating the effect of an explanatory variable on an outcome. In econometrics, the term often refers to models that aim to estimate causal or structural relationships, under specific assumptions about the data-generating process.

Regression can accommodate both continuous and categorical variables. For instance, one might predict a student's GPA using hours of study (a continuous variable) and major (a categorical variable). The analysis produces coefficients that indicate the direction and magnitude of the effects, alongside measures like standard errors and p-values that help assess precision and statistical significance.

In summary, while both correlation and regression examine relationships, they serve distinct purposes. Correlation describes the strength and direction of an association but does not imply causality or predict outcomes. Regression goes further by modeling that relationship and allowing us to make predictions or infer potential effects. Understanding when and how to use each method is essential for aligning statistical tools with research objectives. The next section presents a slightly more technical discussion of regression, focusing on how we interpret coefficients, evaluate model fit, and think carefully about what it means to estimate the "effect" of X on Y.

2.5 Effect of X on Y/ Regression

Economists and social scientists typically work with observational data drawn from representative samples. Their primary goal is not prediction but understanding how changes in one variable, x, are associated with changes in another, y, on average. When they estimate a regression model, they aim to uncover a systematic relationship between these variables—specifically, how the average value of y changes with x, holding other factors constant. This relationship is captured by the conditional expectation function (CEF), denoted $E(y \mid x)$[3] which summarizes how y behaves in expectation across values of x.

Ordinary least squares (OLS) regression does not recover the full distribution of outcomes or individual-level variation; instead, it approximates the CEF linearly. That is, it finds the line that best summarizes how average y varies with x.[4] The simple linear regression model is:

$$y_i = \beta_0 + \beta_1 x_i + u_i \tag{2.5}$$

Here, β_1 represents the change in the mean of y associated with a one-unit change in x, assuming the error term u_i is not systematically related to x_i.[5]

For example, if x is years of education and y is income, the slope β_1 tells us how average income changes with an additional year of education—not how income changes for any specific individual. The key point is that regression is about relationships between means, not precise individual-level predictions. The variation of actual observations around the regression line—due to unmeasured or idiosyncratic factors—is captured by the error term u_i, but the model focuses on summarizing how $E(y \mid x)$ behaves.

This emphasis stands in contrast to machine learning (ML), which is primarily focused on prediction. ML algorithms aim to minimize individual-level prediction error and typically do not assume any specific functional form. Instead, they search for flexible patterns in the data that yield the most accurate predictions for y, given the input features x. Methods like LASSO or random forests often outperform OLS in predictive accuracy but do not yield easily interpretable summaries like regression coefficients. As Victor Chernozhukov noted in a 2013 NBER presentation, LASSO can approximate the CEF better than OLS for certain datasets, in terms of predictive fit—but its goal is predictive accuracy, not interpretability.

Bruce Hansen, in his wonderful 2022 book *Econometrics*, emphasizes that even when the true relationship between x and y is nonlinear, OLS still gives the best linear approximation to the CEF.[6] This is particularly important in economics and social sciences, where researchers seek to explain average effects and mechanisms rather than to maximize predictive performance.

Regression is a foundational topic in any introductory econometrics course. The goal here is not to reteach the mechanics, but to clarify the intuitive difference between estimating effects using regression—based on how average outcomes vary with inputs—and predicting outcomes as done in ML. In short, regression helps answer questions like "How does income change on average with more education?" while ML helps answer "What is this specific person's likely income given their education, experience, and other features?". Recognizing this distinction is crucial when choosing the appropriate method for a research question—whether the goal is understanding average effects or predicting specific outcomes.

2.5.1 How Can We Estimate the Population Parameters, β_0 and β_1?

In regression analysis, we aim to understand how an outcome variable y is related to one or more explanatory variables $x_1, x_2, \ldots, x_k$ in the population. Since we rarely observe the entire population, we rely on sample data to estimate these relationships. The population model we have in mind is:

$$y_i = \beta_0 + \beta_1 x_{i1} + \beta_2 x_{i2} + \ldots + \beta_k x_{ik} + \epsilon_i \tag{2.6}$$

Here, $\beta_0, \beta_1, \ldots, \beta_k$ are the population parameters we want to estimate. Each β_j reflects how the average value of y changes when x_j increases by one unit, holding other factors constant. The term ϵ_i represents all other influences on y_i that are not captured by the included regressors.

To estimate these parameters using a sample of data, we make several key assumptions: that the error term has mean zero $E(\epsilon_i) = 0$, and that it is uncorrelated with each explanatory variable $\text{Cov}(x_{ij}, \epsilon_i) = 0$.[7] Under these conditions, we can use the method of OLS to estimate the parameters in a way that is both unbiased and consistent.

OLS works by finding the line—or in the case of multiple regressors, the hyperplane—that minimizes the sum of squared residuals, which are the differences between the observed outcomes and the predicted outcomes from our model. This gives us coefficient estimates that make the predictions as close as possible to the actual data points, on average. Mathematically, we solve:

$$\min_{\hat{\beta}_0,\ldots,\hat{\beta}_k} \sum_{i=1}^{n} \left(y_i - (\hat{\beta}_0 + \hat{\beta}_1 x_{i1} + \ldots + \hat{\beta}_k x_{ik}) \right)^2 \tag{2.7}$$

To simplify and generalize the estimation, we often rewrite the model using matrix notation:

$$Y = X\beta + \epsilon \tag{2.8}$$

In this expression: Y is an $n \times 1$ vector of outcomes, X is an $n \times (k+1)$ matrix of regressors (including a column of 1s for the intercept), β is a $(k+1) \times 1$ vector of unknown coefficients, ϵ is the $n \times 1$ vector of errors.

The OLS estimator is:

$$\hat{\beta} = (X'X)^{-1}X'Y \tag{2.9}$$

This formula gives us a set of estimates $\hat{\beta}_j$ that minimize the average squared distance between the model's predictions and the observed outcomes. Under the standard Gauss-Markov assumptions, this estimator is BLUE—the best linear unbiased estimator.[8]

OLS also satisfies a key condition: the residuals it produces are, on average, uncorrelated with the regressors:

$$\frac{1}{n} \sum_{i=1}^{n} x_i (y_i - x_i'\hat{\beta}) = 0 \tag{2.10}$$

This condition reinforces the idea that the regression line is the best fit, in terms of capturing average patterns in the data.

2.5.2 From Estimation to Prediction

Once we have estimated the coefficients, we can use the model to predict outcomes. For any observation i, we calculate the fitted value:

$$\hat{y}_i = \hat{\beta}_0 + \hat{\beta}_1 x_{i1} + \ldots + \hat{\beta}_k x_{ik} \tag{2.11}$$

This fitted value represents the model's prediction for y_i, given the values of the regressors and the estimated parameters. The residual, defined as:

$$\hat{\epsilon}_i = y_i - \hat{y}_i \tag{2.12}$$

measures how far the model's prediction is from the actual outcome. These residuals are useful for assessing model fit and diagnosing potential violations of assumptions.

In matrix terms, we write the vector of fitted values and residuals as:

$$\hat{Y} = X\hat{\beta}, \quad \hat{\epsilon} = Y - \hat{Y} \tag{2.13}$$

This matrix form is particularly beneficial in multiple regression models, clarifying the relationship between multiple predictors and the response variable. Detailed examination of these relationships will continue in Chapter 4 and 5.

Note: While predicting values of y is one use of regression, in econometrics our main focus is often different: we care about understanding how y changes on average when one of the explanatory variables x_j changes, holding other variables constant—that is, we aim to estimate *ceteris paribus effects.* The regression coefficients β_j represent these average effects in the population, and we use sample estimates $\hat{\beta}_j$ to learn about them.

These estimates come from minimizing the sum of squared residuals using OLS, but it's important to keep in mind that different random samples would give different estimates. That's why we treat $\hat{\beta}_j$ as a random variable—its value depends on the sample we happened to draw. Despite this variability, the OLS estimate $\hat{\beta}_j$ serves as our best guess of the true effect β_j under certain conditions.

For an estimate to be unbiased, the sampling process must be such that the average estimate across repeated samples equals the true parameter. This holds when several assumptions are satisfied: the model must be linear in parameters, the data must come from a random sample, there must be variation in the regressor, the error term must have zero conditional mean, and the variance of errors should be constant (or appropriately adjusted for heteroskedasticity).

In addition to unbiasedness, we also want consistency, which ensures that as the sample size increases, the estimate $\hat{\beta}_j$ converges to the true β_j. This is particularly useful because in practice we usually only have one sample—but if it's large and the assumptions hold, we can treat $\hat{\beta}_j$ as a close approximation to β_j.

Finally, as the sample size becomes very large, the asymptotic distribution of $\hat{\beta}_j$ approaches a normal distribution. This property, known as asymptotic normality, allows us to use standard statistical tools—like confidence intervals and hypothesis tests—even when the finite-sample distribution is unknown.

Of course, all of these desirable properties depend on assumptions that may not hold in real-world data. In applied work, it is often difficult or even impossible to guarantee that the model is correctly specified or that errors are independently and identically distributed (i.i.d.). Issues such as omitted variable bias, measurement error, sample selection, endogeneity, or reverse causality can all threaten the validity of OLS estimates. In such cases, we should interpret our results as showing associations (or correlations) rather than causal effects. We'll return to these challenges and how to address them in more detail in Chapter 4.

2.6 Causal Effect

The phrase "correlation does not equal causation" is often used to caution against misinterpreting statistical associations. A more precise statement would be "correlation is not sufficient for causation," highlighting that the co-occurrence of two events does not necessarily mean one causes the other. This distinction is fundamental in research design in economics, social and health sciences, especially in the early stages of planning and data collection.

While correlation can uncover patterns and associations between variables, it falls short of determining whether one variable directly influences another. Understanding causation is crucial for effective decision-making, where interventions often depend on knowing both the direction and magnitude of causal relationships. For example, a strong correlation between economic growth and infrastructure investment does not confirm causation; it could result from reverse causality or unobserved confounding factors. Causal inference methods are essential to disentangle these complexities and derive actionable insights.

The goal of causal inference often goes beyond identifying relationships or correlations—it seeks to understand the underlying causes and their implications. By unraveling these causes, researchers can inform decisions, predict consequences, and take actions to alter relationships. For example, understanding the effects of specific actions, policies, interventions or treatments enables researchers to evaluate their potential impact on outcomes. Developing hypotheses about causal effects typically relies on prior observations, personal experiences, and theoretical insights. Researchers in economics, health, and social sciences frequently apply quantitative methods to analyze empirical data and evaluate the effects of interventions.

In the process of causal inference, researchers must carefully decide how treatments are assigned to sample units or individuals. When direct random assignment is not feasible, detailed information about covariates that might influence outcomes becomes essential. This proactive approach to experimental design is critical for generating valid causal inferences. Correlation alone, while useful for identifying relationships, cannot establish causality without rigorously controlled conditions. This shift from interpreting preexisting data to intentionally designing studies marks a fundamental departure from typical statistical and ML tasks, emphasizing the importance of planning in causal analysis.

A common challenge in causal inference is understanding the distinction between structural and reduced-form models. Structural models aim to uncover the underlying mechanisms driving causal relationships, relying on explicit theoretical assumptions. These models are particularly useful for policy simulation and counterfactual analysis, as they allow researchers to predict the effects of hypothetical interventions. However, their reliance on strong assumptions can limit their applicability when those assumptions are violated.

In contrast, reduced-form models focus on estimating causal effects with minimal assumptions, often leveraging natural experiments or observational data. These models emphasize credibility over mechanism, making them valuable when the primary goal is to identify causal effects rather than explain the processes behind them. Both approaches have their strengths, and choosing the right one depends on the research question and the data available.

Randomization is a cornerstone of causal inference, ensuring that treatment and control groups are statistically comparable by eliminating confounding variables. This framework provides a clear pathway for identifying causal effects since any differences in outcomes can be attributed directly to the treatment. However, randomization is not always feasible due to ethical concerns, logistical barriers, or high costs, which often make randomized controlled trials (RCTs) impractical in real-world settings.

In the absence of randomization, researchers rely on observational data and quasi-experimental methods. These approaches demand careful consideration of confounding factors to make robust causal inferences. Techniques such as regression adjustment, propensity score matching, and instrumental variables are designed to address specific challenges posed by observational data. Each method designed to address specific challenges in observational data by making assumptions about the data and the relationships between variables.

At the heart of these methods lies the concept of counterfactuals—hypothetical scenarios describing what would have happened to the same unit if it had not received the treatment or intervention. Counterfactual analysis is central to estimating causal effects but faces the inherent challenge that only one potential outcome can be observed for each unit, leaving the other as missing data. Overcoming this limitation requires strong assumptions, rigorous research designs, and increasingly advanced computational methods. While unit-level causal effects are unobservable, researchers focus on estimating average treatment effects (ATE), which provide insights into the expected difference in outcomes across a population.

Frameworks such as the potential outcomes model (Neyman, 1923; Rubin, 1975) and tools such as directed acyclic graphs (DAGs) are essential for defining and estimating causal effects. These frameworks help formalize causal questions and guide the selection of appropriate methods. The causal effect of an intervention or treatment is typically expressed as the difference between the potential outcomes for a unit: $Y_i(1)$ (the outcome if the unit receives treatment) and $Y_i(0)$ (the outcome if the unit does not receive treatment). Since a unit cannot exist simultaneously in both treated and untreated states, causal inference is limited by the fundamental problem of causal inference (Holland, 1986), which asserts that only one potential outcome can ever be observed. As Holland succinctly stated, "No causation without manipulation."

Counterfactuals, which represent the outcomes that would have occurred under an alternative treatment or policy, are inherently unobservable. Rubin has described causal inference as a "missing data problem" (Ding and Li, 2018), where researchers attempt to approximate the unobservable counterfactuals using various methods. Randomized experiments like RCTs or A/B tests remain the gold standard for identifying causal relationships because they ensure comparability between groups. However, when randomization is not feasible, researchers turn to observational and quasi-experimental methods to estimate causal effects.

In these cases, researchers confront the challenges posed by selection on observables and selection on unobservables. Selection on observables assumes that all factors influencing the assignment of treatment and the outcome are measurable and can be controlled for, enabling methods like propensity score matching or regression adjustment to produce valid causal estimates. In contrast, selection on unobservables addresses hidden biases that cannot be directly measured, using techniques such as instrumental variables, difference-in-differences, and regression discontinuity designs. These methods attempt to replicate the comparability achieved through randomization by creating conditions that mimic it.

Each of these approaches comes with its own set of assumptions and limitations. For instance, selection on observables relies on the assumption that all confounders are measurable and included in the analysis, while the validity of instrumental variables depends on identifying instruments that affect the outcome only through the treatment. Similarly, difference-in-differences assumes that, in the absence of treatment, treated and untreated groups would have followed parallel trends over time. Understanding these nuances is critical for drawing credible causal inferences and ensuring the validity of conclusions.

ML has emerged as a valuable tool in causal inference by offering the flexibility to model high-dimensional covariates. However, standard ML methods designed for prediction do not automatically produce valid causal inferences. Predictive models often focus on minimizing errors, such as regression or classification loss, but fail to address the complexities of causal identification, which can lead to misleading results, such as spurious correlations or unaccounted confounders. To overcome these challenges, causal ML techniques, such as double/debiased ML, causal forests, and methods that integrate causal graphs, have been developed. These approaches combine the predictive power of ML with the rigorous frameworks of causal inference, enabling researchers to uncover cause-and-effect relationships in complex, high-dimensional data settings.

By adapting ML algorithms to handle causal identification, researchers can address critical challenges in modern causal analysis. Techniques such as regularization and feature/variable selection (e.g., LASSO, elastic net) help manage large number of covariates, while other models like random forests and gradient boosting capture intricate interactions within the data. Furthermore, ML facilitates the estimation of heterogeneous treatment effects by directly modeling subgroup-level effects, offering insights into which populations or subgroups benefit most from an intervention. This integration of ML and causal inference is particularly valuable in fields like economics, health, and social sciences, where understanding causal relationships is essential for evaluating policies and designing interventions. We will cover all these topics, following an introduction to the foundational statistical and ML concepts necessary to understand these methods and their related applications.

Notes

1. Please read [this guide on writing about quantitative research](https://libguides.usc.edu/writingguide/quantitative).

2. The main source for the health section is Charles Manski, "Patient Care under Uncertainty", Princeton University Press, 2019.

3. The CEF gives the average value of y for each value of x. For example, if $x = 12$ years of schooling, the CEF tells us the average income for people with 12 years of schooling. It does not describe individual outcomes, only expected (mean) outcomes.

4. Formally, OLS provides the best linear approximation to the CEF: it selects coefficients β_0 and β_1 to minimize the average squared distance between the actual values of y_i and the predicted values $\hat{y}_i = \beta_0 + \beta_1 x_i$. Even if the true relationship between x and y is nonlinear, this line will be the best linear summary.

5. This is the zero conditional mean assumption: $E(u_i \mid x_i) = 0$. It ensures that, on average, the omitted factors (captured in u_i) are unrelated to x_i, so the estimated β_1 reflects the true average effect.

6. This makes regression valuable for understanding average relationships, even when we suspect more complex structures. It does not recover the full CEF, but it gives a useful and interpretable summary.

7. This ensures that the regressors are exogenous and not systematically related to the omitted influences captured in the error term.

8. The Gauss–Markov assumptions include linearity in parameters, exogeneity (zero conditional mean), no perfect multicollinearity among regressors, random sampling, and constant error variance (homoskedasticity).

3

Learning Systems

Imagine a young and curious person named Alex who occasionally feels sick after eating but isn't sure which specific food is to blame. He struggles with recurring allergic reactions and, driven by a need to identify the cause, sets out on a personal investigation. Wanting to understand his vulnerability, Alex begins to look for patterns and clues.

Data Collection: Each time Alex eats, he writes down everything he consumed that day. He also notes how he felt afterward.

Pattern Recognition: After several weeks of careful tracking, a pattern emerges. Every time he eats food containing garlic, he feels sick within a few hours. On days when he avoids garlic, he generally feels fine.

Making Predictions: Based on this pattern, Alex suspects that garlic might be causing his discomfort. To test this idea, he avoids garlic for a few days and monitors his health. Then, keeping all other foods the same, he reintroduces garlic on another day to see if the symptoms return.

Validation: On the garlic-free days, Alex feels completely fine. But after eating garlic again, the familiar symptoms come back. This strengthens his belief that garlic is the trigger.

Updating the Model: Wanting to be thorough, Alex tests other ingredients like onions and shallots on separate days. Since he experiences no negative reactions, he concludes that the issue seems specific to garlic.

In this example, Alex is essentially behaving like a simple machine learning model. He:

- Collects data (his meals and symptoms)
- Looks for patterns
- Makes predictions based on those patterns
- Validates predictions against actual occurrences.
- Adjust the predictive model considering new or contradictory data.

While Alex's learning process strongly suggests that garlic triggers his symptoms, it's important to recognize the limitations of his informal "model." Just as in any learning model, prediction errors could arise from multiple factors. For instance, there might be times when he consumes garlic but doesn't get sick because of variations in the quantity consumed, or the form in which it's ingested (raw versus cooked). There could also be external factors, like the combination of garlic with other foods, that influence his reaction. It's also possible that, on some days, other confounding variables like stress or a different underlying illness might mask or exaggerate his garlic-induced symptoms. Thus, while Alex feels confident in his findings, he understands that real-life scenarios can introduce unpredictability, making it essential to continually refine and reassess his conclusions.

This personal example helps to illustrate how learning from experience—through observing patterns and testing hypotheses—is the core idea behind machine learning. Machines follow a similar process, though typically with much more data and formalized steps.

DOI: 10.1201/9781003504795-3

Machine learning, which comes primarily from computer science, also focuses on interpreting data, similar to what we discussed in statistical learning in the previous chapter. However, its main emphasis often lies in making accurate predictions or decisions, even if it doesn't always explain the underlying reasons. You can think of it like an engineer who builds a tool that works reliably, without needing to fully understand every detail of its internal mechanics.

So why is machine learning sometimes referred to as "statistical learning"? Many machine learning techniques are rooted in statistics, and as both fields evolved, their methods and goals started to overlap significantly. Today, the boundary between them is often blurred. For most people, statistical learning can be thought of as a close relative of machine learning. Both aim to learn from data, but they may prioritize different things—statistical learning often seeks to explain, while machine learning tends to optimize prediction. At its core, machine learning is about teaching computers to make decisions or predictions from data, rather than giving them step-by-step instructions.

The goal in machine learning is to build a model with specific settings or parameters that can make accurate predictions on new, unseen data. Later in this book, we'll walk through the three central steps: prediction, training, and hyperparameter tuning.

Everything begins with data—whether it's images, text, economic indicators, political polls, public health records, or employment figures. This data is organized and used as training input for a machine learning model. Generally, more data leads to better model performance.

Once the data is prepared, the process follows three key phases:

Training (or Parameter Estimation) Phase: Here, the model learns from the training data. There are two main approaches: one searches for the best model by optimizing a chosen quality measure (producing a point estimate), and the other is Bayesian inference, which is beyond the scope of this book. Regardless of the method, the aim is the same—use numerical techniques to estimate parameters that best fit the data.

Hyperparameter Tuning (or Model Selection) Phase: In this step, we evaluate different models and their hyperparameters—settings that aren't directly estimated from the data—to choose the one that performs best on validation or test data. The ultimate goal is to select a model that generalizes well to data it hasn't seen.

Prediction (or Inference) Phase: This occurs when a trained model is applied to test data it hasn't seen before. At this stage, the structure of the model and settings are already fixed. The task is simply to use the model to generate predictions for new inputs.

A quick clarification: parameters are learned during training—these are internal values like weights in a regression or a neural network. Hyperparameters, on the other hand, are set before training begins, such as the learning rate or the maximum depth of a decision tree. Hyperparameter tuning helps find the best combination of these settings for optimal performance.

Once a model is trained, tuned, and validated, it can be used in a wide range of practical applications. In machine learning, "learning systems" are algorithms or models that improve over time as they are exposed to more data. They aren't explicitly programmed for each task but instead adapt by identifying patterns and making decisions based on input.

These systems serve different purposes, depending on the goals of the user or organization:

- The **Descriptive** role focuses on summarizing and understanding patterns in existing data. This is useful for uncovering historical trends and insights.
- The **Predictive** role goes further, using current data to forecast future outcomes.
- The **Prescriptive** role is the most action-oriented. It not only predicts but also recommends decisions or actions based on the analysis.

For example, industrial economists might use machine learning to suggest optimal pricing strategies, allocate resources efficiently, or make informed hiring decisions—drawing from data such as consumer demand, competitor pricing, product performance, labor market trends, and skill profiles. In healthcare, machine learning models can help identify cost-effective treatments, optimize hospital operations, target preventive interventions for at-risk populations, inform drug pricing, or guide health insurance premium setting—using data on patient outcomes, resource availability, public health patterns, drug efficacy, and insurance claims.

Now that we've introduced the key stages—prediction, training, and hyperparameter tuning—and how machine learning models are used across different sectors, the next subsection will walk through, step by step, how to find, train, and evaluate models that generalize well to unseen data.

3.1 Learning Systems

A machine learning system is a dynamic framework often used in practical machine learning projects. These systems typically follow an iterative cycle involving several key phases. It begins with Data Collection, where datasets are created, maintained, and continuously updated. This step is foundational, as the quality and relevance of the data greatly affect outcomes.

Next is the Experimentation phase. Here, the collected data is explored in depth, hypotheses are developed and tested, and suitable models are identified and evaluated. This stage also includes building training and prediction pipelines. Once a robust model is developed, the deployment phase puts it into use as a part of a real product or system. However, deployment is not the final step. The Operations phase monitors and updates the model, ensuring it adapts to changing patterns and continues to perform reliably.

While all of these phases—data collection, experimentation, deployment, and operations—are a part of a full machine learning system, it's important to note that they often require different skill sets and are typically handled by professionals from different domains. For example, deployment and operations are more commonly led by engineers or data infrastructure teams, particularly in industry settings. As economists, social scientists, and health researchers, our primary involvement tends to center around data preparation, model development, and evaluation. Still, understanding the full pipeline is valuable for collaboration and communication in interdisciplinary environments.

In this section, we introduce how machine learning systems are structured in practice, the typical steps in building predictive models, and the core theoretical ideas behind learning from data. We also raise key questions about what makes a model good for prediction versus explanation.

Machine learning systems use algorithms and models that improve over time as they process more data. They do not require detailed programming for every task but instead learn patterns and make decisions based on input data. Our focus in this book is on machine learning models, which are the central components of these systems. A model combines data with code to produce predictions or decisions. How do we find, train, and ensure that our models perform well on unseen data?

Here is a step-by-step overview:

Splitting the data: We begin by dividing the dataset into two main parts: a **training set** and a **testing set**.

Use training data wisely: Typically, about 80% of the data is used for training. Within this training portion, we split it again into two parts: **estimation data** (used to train the models) and **validation data** (used to evaluate model performance).

Keep testing data untouched: The testing data is set aside and not used during training or model selection. It will be used only at the final stage to assess how well the chosen model performs on unseen data.

Choose possible models: Before training, we select a few model candidates.
– For **parametric models**, we specify a functional form for the model, excluding the parameters to be estimated.
– For **nonparametric models**, we select which features/variables to include and decide on the values for any tuning parameters.

Train the model: Each candidate model is trained (fitted) using the **estimation data**, which is usually the larger portion of the training data.

Check and compare models: After training, we evaluate each model using the **validation data**, which the model hasn't seen before. This helps estimate how well each model generalizes.

Pick the best model: We select the model that performs best on the validation data.

Final training: We retrain the chosen model using all of the training data—estimation and validation combined—to make full use of available information.

Test: Finally, we use the untouched **testing data** to estimate how well the trained model generalizes to new data.

It is essential to distinguish between the roles of validation and testing data: validation data guides model selection, while testing data estimates final model performance.

This overview provides a plain-language summary of the typical machine learning workflow. In the sections that follow, we'll explore these steps using standard technical terminology. If any terms or ideas feel unfamiliar, don't worry—we'll explain each of them in more depth as the book continues.

1. The learner has a sample of observations. This is an arbitrary (random) set of objects or instances each of which has a set of features ($\mathbf{X}$ - features vector) and labels/outcomes (y). We call this sequence of pairs as a training set: $S = ((\mathbf{X}_1, y_1) \ldots (\mathbf{X}_m, y_m))$.

2. We ask the learner to produce a prediction rule (a predictor or a classifier model), $\hat{f}(x)$, so that we can use it to predict the outcome of new domain points (observations/instances).

3. We assume that the training dataset S is generated by a data-generating model (DGM) or some "correct" labeling function, $f(x)$. The "true" prediction function, $f(x)$, is unknown to the learner. The learner tries to approximate $f(x)$ using the sample S.

4. The learner will come up with a prediction rule, $\hat{f}(x)$, by using S, which will be different than $f(x)$. The quality of the prediction rule is measured by a loss function, $L_{(S,f)}(\hat{f})$, which quantifies the difference between the true function $f(x)$ and the learner's prediction function $\hat{f}(x)$. This is known as the generalization error or risk.

5. The goal of the algorithm is to find $\hat{f}(x)$ that minimizes the error with respect to the unknown $f(x)$. The key point here is that, since the learner does not know $f(x)$, it cannot calculate **the loss function**. However, it calculates the training error also called as the **empirical error** or the **empirical risk**, which is a function that defines the difference between $\hat{f}(x)$ and y_i.

6. Hence, the learning process can be defined as coming up with a predictor $\hat{f}(x)$ that minimizes the empirical error. This process is called **empirical risk minimization** (ERM).

7. Now the question becomes what sort of conditions would lead to bad or good ERM?

If we use the training data (in-sample data points) to minimize the empirical risk, the process can lead to $L_{(S,f)}(\hat{f}) = 0$. This problem is called **overfitting** and the only way to rectify it is to restrict the number of features in the learning model. The common way to do this is to "train" the model over a subsection of the data ("seen" or in-sample data points) and apply ERM by using the test data ("unseen" or out-sample data points). Since this process restricts the learning model by limiting the number of features in it, this procedure is also called **inductive bias** in the process of learning.

There are always two "universes" in a statistical analysis: the population and the sample. The population is usually unknown or inaccessible to us. We consider the sample as a random subset of the population. Whatever the statistical analysis we apply almost always uses that sample dataset, which could be very large or very small. Although the sample we have is randomly drawn from the population, it may not always be representative of the population. There is always some risk that the sampled data happens to be very unrepresentative of the population. Intuitively, the sample is a window through which we have partial information about the population. We use the sample to **estimate** an unknown parameter of the population, which is the main task of **inferential statistics**. Or, we use the sample to develop a **prediction rule** to predict unknown population outcomes.

When the outcome is numeric (non-binary), we often use a loss function such as the mean squared error (MSE) to assess the quality of a predictor or an estimator.[1] We use the term $\hat{f}(x)$ to refer to either. But can a good estimator also be a good predictor? In the next chapter, we will see that an estimator with a small estimation error can still perform better in prediction. Why does that happen? We will explore this question in detail in the next chapter.

In the next sections, we will carefully examine the difference between estimation and prediction, and explore under what conditions models generalize well to unseen data.

Note

1. For binary outcomes, classification errors—another type of loss—become relevant. We will cover those in Chapter 10.

4

Error

Many inquiries in economics, social sciences, and health sciences are centered around gaining insights about a population. "Population" in this context refers to the complete set of individuals, events, or objects of interest in a study, encompassing all members or occurrences that meet specified criteria. The process of learning from this population can broadly be divided into two categories: estimation and hypothesis testing. Estimation involves determining the value of various population parameters, including the average (mean), variance, standard deviation, median, proportion, range, skewness, and kurtosis. Additionally, estimation also encompasses the calculation of correlation, covariance, and model coefficients. These are crucial for quantifying the relationships between different variables in a model. On the other hand, hypothesis testing is a statistical method used to assess the strength of evidence against a null hypothesis, typically comparing it with an alternative hypothesis. It involves statistically determining whether the observed data deviates significantly from what would be expected under the null hypothesis, thereby providing insights into the likelihood of the occurrence of a specific event or characteristic within the population. This process is fundamental in validating or challenging existing theories and assumptions about the characteristics and behaviors of the the population. While hypothesis testing and estimation provide foundational methods for understanding populations, it is equally vital to consider the role of error in these processes, as it significantly influences the accuracy and reliability of our statistical and machine learning models.

In statistics and machine learning, **error** refers to the discrepancy between a calculated, estimated, or predicted value and the true or actual value. This encompasses estimation error, where estimations deviate from reality due to model limitations or assumptions; measurement errors, arising from inaccuracies in data collection; sampling errors, due to discrepancies between a sample and the overall population; and Type I and II errors in hypothesis testing, which are false positives and false negatives, respectively. In statistics, errors are often part of the model itself, such as the error term in a linear regression model, representing the part of the data that can't be explained by the estimation model. Sampling error arises because the sample may not perfectly represent attributes of the population, leading to differences between the true unknown parameters of a population and the observed sample statistics. In machine learning, prediction error is the difference between predicted values and the actual values of a model. Prediction error is crucial for evaluating model performance, encompassing two types: training error, which occurs on the training dataset and reflects learning accuracy of the mode; and generalization error (or test error), which is observed on new, unseen data, indicating the effectiveness of the model in applying its learning to different scenarios. Minimizing both types of error is key to developing accurate and reliable machine learning models. Model error is the discrepancy between the assumptions of the model and the real-world data it is trying to represent. Model error can occur due to incorrect model assumptions, oversimplification, or failure to capture complex relationships in the data. Algorithmic error arises due to the limitations or biases in the algorithms used for data processing and analysis. Additionally, errors in statistical and machine learning models are often categorized into irreducible and reducible errors. Irreducible error, inherent in

DOI: 10.1201/9781003504795-4

any dataset, is the noise that cannot be eliminated by any model, while reducible error is due to the model's inability to capture the true underlying pattern and can be minimized with better modeling techniques. Understanding and minimizing these errors, if possible, is crucial for effective and ethical decision-making based on data analytics, statistics, and machine learning. Throughout this book, we will explore all these various types of errors fundamental to statistics and machine learning. In this chapter, we'll focus on estimation error, a cornerstone of statistical analysis, and prediction error, particularly in in-sample or training data, which forms the basis of machine learning.

Assume you're interested in determining the average years of schooling among residents in your city and also understanding the effect of schooling on income. Collecting this data directly from every individual who constitutes the population of the city is often impractical. A more feasible approach is to gather data from a subset of people, say those passing through a main street, and use this as a sample to estimate the average years of schooling (using method of moments estimators) and estimating the effect of education on income (using ordinary least squares (OLS) to estimate coefficient). For simplicity, assume the subset of people you gather information, your sample, is random and representative of the population. The estimated average, derived from your sample, i.e. realized data, represents an attempt to infer a population parameter, in this case, the mean years of schooling. Using the data from your sample, you can find the the correlation between education and income as well as analyze how schooling effects income by estimating the coefficient in an ordinary linear regression model (with the assumptions we will discuss below). In this specific example, you can estimate the average years of schooling, as well as the correlation between education and income, and the regression coefficient of education on income.

The process of deriving any of these method of moment and coefficient estimates involves selecting an estimator—a statistical method or rule for estimating the unknown population parameter. While the estimator remains unchanged regardless of the sample obtained, and ideally needs to be specified before any sampling carried out, the specific estimate varies with each different sample collected. The ideal scenario is when the estimate from your sample aligns perfectly with the true, but unknown, population parameter. Since the actual parameter is never truly known, statistical properties are used to justify the assumption that your estimate nearly equal to this parameter. Main assumption/requirement of this process is obtaining a representative sample and identifying an unbiased estimator. Even among unbiased estimators, questions arise about which is the "best." This leads us to the concept of estimation error. For instance, the difference between the estimated average schooling and the actual average schooling of the population, or the coefficient which indicates the return of schooling on income. Estimation error is a key aspect of statistical analysis in trying to approximate a population parameter.

Similarly, when using the gathered data to predict an individual's or a group's years of schooling, the process will inevitably include some level of prediction error. This error reflects the discrepancy between the predicted value and the actual outcome. It is easy to assume that we would encounter prediction error primarily in models with limited variables, like those predicting a single parameter such as mean of years of schooling from a sample. However, even in models with large datasets includes numerous variables, prediction error remains an inherent challenge, as we will see in the following example.

Let's consider a different scenario to more effectively illustrate prediction error. Imagine you are developing a model to predict an individual's income based on factors like education level, profession, work experience, and any other factors you think might impact income. You use your sample data to train this predictive model. Initially, the model makes predictions about the income of these individuals within the training data. There will be a difference between the predicted income and the actual income for the individuals in your sample. This difference is known as in-sample prediction error and reflects accuracy of the model in predicting outcomes using the data it was trained on. Despite sophistication of the model, it cannot perfectly account for every variable influencing future income, such as changes in the state of the economy, individual career preferences, misreporting, or factors that are difficult to quantify numerically, like ability or chance. Moreover, when you apply this model to predict the income of a new set of individuals, outside the training dataset, there will likely be some discrepancy between predictions of the model and the actual incomes of these new individuals. This discrepancy, known as out-of-sample prediction error, is critical in assessing how well the model generalizes to new, unseen data. Both in-sample and out-of-sample prediction errors are crucial for understanding effectiveness of the model. Minimizing these prediction errors, especially the out-of-sample error, is a major focus in the field of machine learning, where the emphasis is often on the ability of the model to accurately predict outcomes for new, unseen data.

Recognizing the shift from traditional statistical methods, which concentrate on precise parameter estimation, we will now transition to discussing the concepts of estimation error and mean squared error (MSE). Following this, we will discuss the prediction error and mean squared prediction error (MSPE), highlighting their importance in the context of statistical and machine learning techniques.

4.1 Estimation Error

In this section, we show the concept of estimation error, using a simulation to understand how different estimators can be applied to estimate population parameters like the average years of schooling. We explore various estimators, examine their unbiasedness, and discuss the importance of efficiency and MSE in choosing the best estimator for accurate population parameter estimation. This simulation also highlights the balance between unbiasedness and variance in statistical estimation.

Our main objective is to estimate an unknown population parameter, symbolized as θ. This parameter, θ, could be the simple mean of a variable (or feature), μ_x, or a slope coefficient, β_1, from OLS estimation (for instance, the OLS regression coefficient of education on income). We use estimator to estimate θ. Our data comes from a random sample, yet, it may not perfectly represent the entire population, potentially due to sampling or measurement errors, or some other reasons. As a result, our estimate, denoted as $\hat{\theta}$, might not exactly match the true parameter, θ. The difference between both is known as estimation error (i.e., error $= \hat{\theta} - \theta$). Also, note that when we say we have random samples, we imply that these samples are independent and identically distributed from our population as well.

Let's simulate the scenario we discussed above, aiming to estimate the average years of schooling of residents in our city. Since collecting data from everyone is impractical, we decide to gather 10 representative samples instead. In this simulation, people report 9 years of education if they've completed only compulsory schooling, 12 years if they graduated from high school, or 16 years if they're college graduates. We'll assume that reported years of schooling can be any discrete number between 9 and 16. Each of our 10 different representative samples will consist of 5,000 individuals.

As previously discussed, estimating a parameter, denoted as θ, requires choosing an estimator. Estimator is a statistical method or rule used to estimate an unknown population parameter.

For estimating population mean, μ_x, in this simulation, which represents our specific parameter θ, we can use one of three different estimators.

First, we could use the average of years of schooling (sample mean) for everyone who reported it as an estimator. The average is a straightforward and commonly used estimator because it utilizes all available data points.

$$\bar{X} = \frac{1}{n}\sum_{i=1}^{n} x_i$$

or alternatively, we can just take the half of the first person's and last person's years of schooling. This estimator is simpler and less computationally intensive. It takes into account the variability in the data by considering the extremes (the first and last data points). Although it doesn't use all data points, it can be a quick and effective estimator when the data is expected to be uniformly. distributed.

$$\hat{X} = 0.5x_1 + 0.5x_n$$

or as a third alternative, we can just use weighted average of first person and the last person's years of schooling. This estimator allows for a more nuanced estimation than the equal weighting in the previous method. By adjusting the weights (0.25 for the first person and 0.75 for the last person), this estimator can be tuned to reflect certain assumptions about the distribution of the data, potentially providing a more accurate estimate under specific conditions. As a note, you can find unbiased estimator when you assign any positive values for weights as long as the sum is 1.

$$\tilde{X} = 0.25x_1 + 0.75x_2$$

In addition to these three estimators, other alternatives for estimating average years of schooling in our simulation may include using the **mode of the sample**, which identifies the most frequently reported years of schooling. Another option is the **trimmed mean**, which calculates the average after excluding extreme values at both ends of the spectrum, reducing the impact of outliers. Lastly, selecting a **single value** at random from the sample as the estimator. While this last method might seems overly simplistic and is subject to high variance, it could be useful in scenarios where a quick, single data-driven decision point is needed.

Therefore, we need to define **what makes an estimator the "best" among others**.

Let's say we have drawn various random samples that represent the population. For each sample, we find a single value for each parameter, such as the mean or variance of the sample. All these single means of samples will generate a distribution, thus forming the sampling distribution of sample means, or sampling distribution of sample variances. The sampling distribution is a statistical term that refers to the distribution of a parameter (like a mean, variance, or coefficient) across a large number of samples drawn from a specific population. In simpler terms, it represents the likelihood of the various outcomes (probability distribution) of an estimator calculated from multiple random samples of a population. Analyzing this distribution for a parameter estimate, using any of the estimators mentioned above, aids in developing the main criteria that guide us in choosing the "best" estimator. Initially, we need to discuss the expected means and variances of this distribution, which help establish certain principles leading to the development of the main criteria.

The first and most important criteria should be that the expected value, i.e., the mean, of all estimates obtained from various random samples, should be equal to the unknown population parameter, θ. An estimator that satisfies this condition is referred to as an **unbiased estimator** , (i.e., $\text{Bias}(\hat{\theta}) = \mathbb{E}[\hat{\theta}] - \theta = 0$ so $\mathbb{E}[\hat{\theta}] = \theta$. While an unbiased estimator doesn't ensure that the estimate from a particular sample will exactly match the population parameter, it implies that averaging the estimates from repeatedly drawn random samples, each calculated using the same estimator, will closely approximate the actual population parameter. We showed the therotical proof in the last section. It is also important to remember that this concept is theoretical, as in practical scenarios we typically work with a single random sample and rely on asymptotic (large sample) properties, which we will explore later.

Numerous econometrics textbooks, including ours in the last section of this chapter, do provide algebraic proofs to demonstrate that estimators for statistical parameters, including the average, variance, correlation, and OLS regression coefficients (both for single-variable and multiple-variable models), are unbiased under certain conditions. These proofs typically rely on assumptions such as linearity, independence, and normality of error terms, among others. However, in this case, we will demonstrate that all three estimators are unbiased through simulation. This approach can offer a clearer understanding of the concept, as visualizing it step-by-step in a simulation often makes it easier to grasp. Through this simulation, one can visually and numerically verify the fundamental statistical properties of independently and identically distributed (i.i.d) and unbiased estimators.

```
# Here is our population
populationX <- c(9,10,11,12,13,14,15,16)
#Let's have a containers to have repeated samples (5000)
samples <- matrix(0, 5000, 10)
colnames(samples) <- c("X1", "X2", "X3", "X4", "X5", "X6", "X7", "X8", "X9", "X10")
# Let's have samples (with replacement always)
set.seed(123)
for (i in 1:nrow(samples)) {
  samples[i,] <- sample(populationX, 10, replace = TRUE)
}
head(samples)
```

```
##      X1 X2 X3 X4 X5 X6 X7 X8 X9 X10
## [1,] 15 15 11 14 11 10 10 14 11  13
## [2,] 12 14 14  9 10 11 16 13 11  11
## [3,]  9 12  9  9 13 11 16 10 15  10
## [4,]  9 14 11 12 14  9 11 15 13  12
## [5,] 15 16 10 13 15  9  9 10 15  11
## [6,] 12 13 15 13 11 16 14  9 10  13
```

We define a population where years of schooling range from 9 to 16. Imagine this as a population where an equal number of individuals have each year of schooling, from 9 years up to 16. Consequently, the average years of schooling in this population, denoted as μ_x, is known to be 12.5. This average is calculated as the mean of the range of schooling years (9 to 16).

We generate 10 random samples, each consisting of 5,000 observations. These observations are created by randomly selecting from the defined population. Below we display the first six rows out of the 5,000 samples, where each row represents a different sample and each column represents a data point within that sample.

We can test the following points:

1. Is X independently and identically distributed (i.i.d)?

We need to verify that the mean and variance are consistent across different data points within the samples. In another words, an identical distribution requires $\mathrm{E}(x_1) = \mathrm{E}(x_2) = \ldots = \mathrm{E}(x_{10})$ and $\mathrm{Var}(x_1) = \mathrm{Var}(x_2) = \ldots = \mathrm{Var}(x_{10})$. In addition, independence is further confirmed by ensuring that the correlation (Corr) between any two different data points is zero, $\mathrm{Corr}(x_i, x_j) = 0$ where $i \neq j$.

2. Are the three estimators, $\bar{X}$, $\hat{X}$, and $\tilde{X}$, unbiased?

An estimator is considered unbiased if its expected value—the average of the values of the parameters obtained from different samples—equals the true population mean, μ_x , which in this case is 12.5. This value is known in simulation but unknown in reality.

Let's see:

```
# Check if E(x_1)=E(x_2)=...=E(x_10), rounded to 2 decimal places
colMeans <- round(colMeans(samples), 2)
print(colMeans)
```

```
##    X1    X2    X3    X4    X5    X6    X7    X8    X9   X10
## 12.48 12.51 12.48 12.57 12.54 12.51 12.45 12.50 12.51 12.45
```

```
# Check if Var(x_1)=Var(x_2)=...=Var(x_10), rounded to 2 decimal places
variances <- apply(samples, 2, var)
variances <- round(variances, 2)
print(variances)
```

```
##   X1   X2   X3   X4   X5   X6   X7   X8   X9  X10
## 5.22 5.17 5.28 5.30 5.18 5.31 5.21 5.20 5.27 5.31
```

```
# Check correlation, rounded to 2 decimal places
correlations <- cor(samples)
correlations <- round(correlations, 2)
print(correlations)
```

```
##        X1    X2    X3    X4    X5    X6    X7    X8    X9   X10
## X1   1.00  0.02  0.00  0.00  0.01  0.00 -0.01  0.00  0.00 -0.02
## X2   0.02  1.00  0.01  0.02  0.00  0.00  0.01 -0.01  0.01 -0.02
## X3   0.00  0.01  1.00  0.01  0.01  0.00 -0.02  0.02 -0.01  0.00
## X4   0.00  0.02  0.01  1.00 -0.02  0.00  0.00  0.02  0.01 -0.01
## X5   0.01  0.00  0.01 -0.02  1.00 -0.01  0.00  0.01  0.00  0.01
## X6   0.00  0.00  0.00  0.00 -0.01  1.00  0.01  0.02 -0.02 -0.01
## X7  -0.01  0.01 -0.02  0.00  0.00  0.01  1.00 -0.01  0.01  0.00
## X8   0.00 -0.01  0.02  0.02  0.01  0.02 -0.01  1.00  0.01  0.06
## X9   0.00  0.01 -0.01  0.01  0.00 -0.02  0.01  0.01  1.00  0.01
## X10 -0.02 -0.02  0.00 -0.01  0.01 -0.01  0.00  0.06  0.01  1.00
```

Note that if you use only unique set of samples, you can get exact results using following commands:

```
uniqsam <- unique(samples)
colMeans(uniqsam)
apply(uniqsam, 2, var)
cor(uniqsam)
```

The observed expected value (mean) and variance of each random sample are nearly equal. It is worth noting that increasing the number of observations in each sample from 5,000 to a larger number would likely result in these means and variances becoming even more similar. Additionally, the correlations between each sample are nearly zero. Thus, we can conclude that the condition of i.i.d is satisfied.

The next step involves determining if all three estimators are unbiased. For this, we apply each estimator to the random samples to estimate the population parameter. The code below is used to calculate the average value for a variable across multiple samples and then computes the overall average of these averages for each of the three estimators separately.

```
# First Xbar : sample mean
X_bar <- rep(0, nrow(samples)) #Container to have all Xbars
for(i in 1:nrow(samples)){
  X_bar[i] <- sum(samples[i,])/ncol(samples)
}
EX_bar <- sum(X_bar)/length(X_bar)
EX_bar
```

```
## [1] 12.49894
```

```
# Xhat: the half of the first person's and last person's years of schooling
X_hat <- rep(0, nrow(samples))
for(i in 1:nrow(samples)){
  X_hat[i] <- 0.5*samples[i,1] + 0.5*samples[i,10]
}
EX_hat <- sum(X_hat)/length(X_hat)
EX_hat
```

```
## [1] 12.466
```

```
# Xtilde: weighted average of first person and the last person's years of schooling
X_tilde <- rep(0, nrow(samples))
for(i in 1:nrow(samples)){
  X_tilde[i] <- 0.25*samples[i,1] + 0.75*samples[i,2]
}
EX_tilde <- sum(X_tilde)/length(X_tilde)
EX_tilde
```

```
## [1] 12.503
```

We can conclude all these three estimators are unbiased as $\mathbf{E}(\bar{X}) \approx \mathbf{E}(\hat{X}) \approx \mathbf{E}(\tilde{X}) \approx \mu_x \approx$ 12.5.

Increasing the number of observations in each sample, as well as the number of random samples, tends to bring these expected values closer to 12.5, the known population mean. However, it is important to note that these sample averages are not exactly the same as the population average. The discrepancy between the estimated value (from the sample) and the actual value (from the population) is known as error. Ideally, we aim for this error to be zero. As the number of observations in our sample approaches the size of the entire population, this error tends to diminish. Since we can never fully ascertain the exact characteristics of the entire population, we operate under the assumption that this error gets closer to zero as our sample size increases.

4.2 Efficiency

Up to this point, we have demonstrated that all three estimators provide unbiased estimates. This means that unbiasedness cannot be the sole criterion for determining the "best" estimator. We seek an estimator that closely approximates the population parameter with a higher likelihood, making the second criterion the choice of a relatively efficient estimator. In other words, the probability density function of the estimator should be concentrated around the true unknown population parameter, indicating that the estimator is **efficient**.

Before discussing efficiency, it is important to remind the difference between sample mean and variance, and sampling mean and sampling variance.

Let's say we have drawn various random samples that represent the population. For each sample, we find a single value for each parameter, such as the mean or variance of the sample. We previously defined a **sample mean** as $\bar{X} = \frac{1}{n}\sum_{i=1}^{n} x_i = \mu_x$, and showed as unbiased estimator of unknown population mean, μ_x. The formula $\hat{\sigma_X}^2 = \frac{1}{n-1}\sum_{i=1}^{n}(x_i - \bar{X})^2$ is the estimator for the unknown population variance $Var(X) = \sigma_x^2$, which is also unbiased (we show this in the last section). This is known as the **sample variance**, calculated by averaging the squared differences of each observation from the sample mean. We use n–1 in the denominator instead of n to provide an unbiased estimate of the unknown population variance, σ_x^2.

All these single sample parameters generate a distribution. One of the most important parameters we analyze is the mean. The collection of these means, when derived from a sufficient number of random samples, forms what is known as the sampling distribution of the sample mean. This distribution is a probability distribution of all possible means that could be obtained from sets of samples of the same size drawn from the same population. It plays a crucial role in statistical hypothesis testing and in constructing confidence intervals. The sampling distribution of the sample means has a mean and variance; hence, the mean of the sampling distribution of sample means is called the **sampling mean**, and the variance of the sampling distribution of sample means is called the **sampling variance**.

The sampling distribution refers to the distribution of a parameter (like a mean, variance, or coefficient) across many samples drawn from a specific population. This distribution of outcomes from an estimator features a sampling mean (which is used to check unbiasedness) and a sampling variance (which is used to check efficiency). The standard deviation of the sampling distribution, also known as the standard error, is the square root of the sampling variance and decreases as the sample size increases.

When each random sample with a mean μ_x and variance σ_X^2 is denoted as X_i, then the sampling mean is $E(\bar{X}) = \frac{1}{n}\sum_{i=1}^{n}\bar{X}_i = \mu_x$, and the sampling variance is $Var(\bar{X}) = \frac{1}{n}\sum_{i=1}^{n}(\bar{X}_i - \mu_x)^2 = \frac{\sigma_x^2}{n}$. (check the derivation in the last section).

In summary, when we have various random samples drawn from a population with mean μ_x and variance $Var(X) = \sigma_x^2$, the sampling mean mirrors the population mean. However, the sampling variance equals the population variance, σ_x^2, divided by the sample size, n. Therefore, as the sample size increases, the sampling variance approaches zero which is called *consistency.*

Generally, the variance of an estimator tends to decrease as the sample size increases (Law of Large Numbers). However, we cannot claim one estimator is more efficient than another solely based on a smaller variance if the variances are calculated from different sample sizes. When comparing two unbiased estimators of a parameter, the one with the smaller variance is considered relatively more efficient. Among all unbiased estimators, the one with the smallest variance is deemed the "best." If an estimator is linear, unbiased, and has the smallest variance among all unbiased linear estimators for a given dataset then it is called the best linear unbiased estimator (BLUE).

The term "relative efficiency" should be used when comparing different estimators that utilize the same information, meaning they are based on the same data and the same sample size. It is not applicable when comparing variances of the same estimator derived from different sample sizes.

Therefore, *the unbiased estimator with the smallest variance is the best estimate for unknown population parameter.* However, it is important to note that while an unbiased estimator may be more efficient than another, this doesn't guarantee it will always provide a more accurate estimate. It simply means that it is more likely to be accurate than the less efficient one. (check Figure 4.1).

Let's examine our simulation to determine which of the three unbiased estimators has the smallest variance.

```
var(X_bar)
```

```
## [1] 0.5385286
```

```
var(X_hat)
```

```
## [1] 2.590462
```

```
var(X_tilde)
```

```
## [1] 3.27012
```

As seen comparing variances, the $\bar{X}$, the sample mean, has the smallest variance. We showed the sample average is the most efficient of the all unbiased estimators.

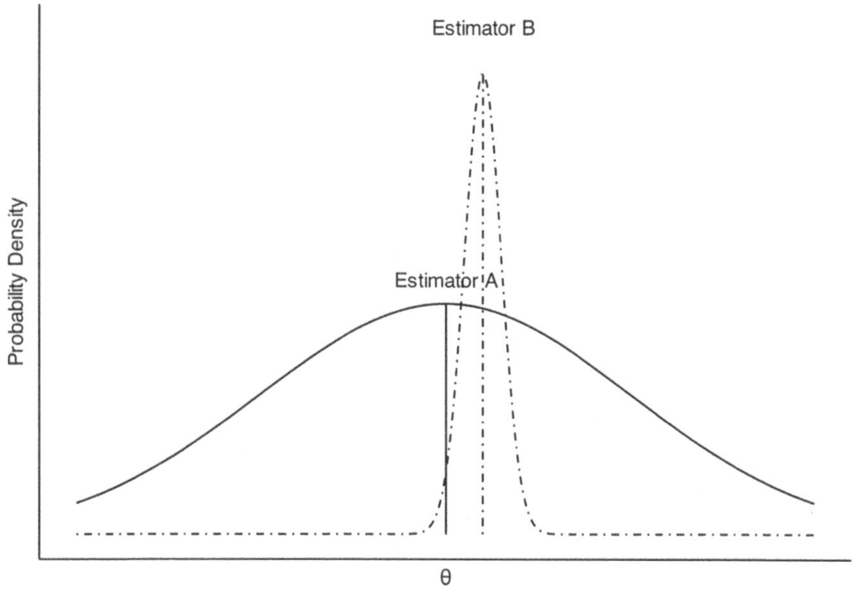

FIGURE 4.1 Which estimator is preferred?

4.3 Mean Square Error

The MSE is a fundamental measure for comparing the efficiency of different estimators. It encapsulates both the variance of the estimator and any bias it might introduce. This dual consideration is critical because it addresses the trade-off between precision and accuracy, making MSE a comprehensive criterion for estimator comparison.

Before proceeding, it is important to clarify the terminology used in this context. While some sources may use MSE interchangeably for both estimators and predictors, we will distinguish between these terms to avoid confusion. In this text, MSE refers exclusively to the performance measure of the estimator. In the following section, we will introduce the mean squared prediction error (MSPE), which, although conceptually and mathematically distinct from MSE, is sometimes conflated with it in other sources. By maintaining this distinction, we aim to enhance clarity and precision in our discussion.

How do we evaluate and compare different estimators when they are not all unbiased?

An unbiased estimator with high variance isn't always preferable, simply because it is unbiased, compared to a biased estimator with low variance. When comparing estimators, particularly when not all are unbiased, the choice isn't straightforward. For example, we might have two estimators for the same population characteristic: one is unbiased but has high variance, while the other is biased but with lower variance. The choice depends on our requirements. In applied microeconomics and social sciences, we often opt for an unbiased estimator when estimation errors are not a major concern. This is because we assume that estimation errors, on average, cancel each other out. This assumption is based on the error term having an expected value of zero ($\mathrm{E}[\epsilon] = 0$) and a variance of σ^2.

Which estimator is preferred? Estimator A is unbiased with high variance, while Estimator B has smaller variance but is biased. In Figure 4.1, you can see that from a random sample representing the population, you are more likely to find an estimate for θ that is closer to the real θ using Estimator B, even though it has some bias. We will demonstrate this in the next chapter with further simulation.

In cases where large errors are intolerable, an estimator with lower variance may be chosen, even if it introduces a small bias. For example, in weather forecasting, slightly biased temperature predictions might be favored if they consistently deliver more reliable results than unbiased models, which may fluctuate wildly from day to day. This reliability is crucial for industries relying on stable weather predictions, such as agriculture or aviation, where planning based on inaccurate forecasts can lead to severe operational disruptions. In medical diagnostics, tests that are slightly biased but have lower variance are sometimes preferred, especially in preliminary screening contexts. A diagnostic test for a disease might be calibrated to have a slight bias towards false positives. This means it may indicate the disease is present when it is not, but ensures that almost all actual cases are detected. In these scenarios, the high cost of errors necessitates the use of estimators that minimize variance, even at the expense of some bias. In other words, selecting an estimator often hinges on the cost associated with an error relative to its size. This cost is quantified by a *loss function*. In economics, social and health sciences, a commonly used loss function is the MSE.

The MSE of an estimator $\hat{\theta}$ quantifies the average of the squares of the errors—that is, the average squared difference between the estimated values and the true value θ. It is formally defined as the expected value of the square of the difference between the estimator $\hat{\theta}$ and the true parameter θ:

$$\text{MSE}(\hat{\theta}) = \mathbb{E}\left[(\hat{\theta} - \theta)^2\right] \tag{4.1}$$

This parameter θ could represent various statistical measures depending on the context, such as the mean or variance of a variable (or feature) μ_x, or a slope coefficient β from OLS regression.

In the realm of OLS regression, which is frequently encountered in econometrics and other disciplines, the MSE is crucial for estimating coefficients. Consider the simple linear regression model, $y = \beta_0 + \beta_1 X + \epsilon$. Here, y is the dependent variable, X represents the independent variable(s), β_0 and β_1 are coefficients to be estimated, and ϵ is the error term, typically assumed to have a mean of zero and constant variance, normally distributed (optional).

The goal in OLS regression is to minimize the MSE, which for the residuals ϵ_i is given by:

$$\text{MSE} = \frac{1}{n}\sum_{i=1}^{n}(\epsilon_i)^2 = \frac{1}{n}\sum_{i=1}^{n}(y_i - \hat{y}_i)^2 \tag{4.2}$$

where $\hat{y}_i = \hat{\beta}_0 + \hat{\beta}_1 X_i$ represents the predicted value of y based on the regression line. The coefficients β_0 and β_1 are estimated such that this MSE is minimized. It is important to note that in OLS, we technically minimize the residual sum of squares, $\text{RSS} = \sum_{i=1}^{n}(y_i - \hat{y}_i)^2$, which is the sum of the squared residuals; however, minimizing RSS is equivalent to minimizing MSE. Here, a residual is defined as the difference between the observed value and the predicted value, expressed as $y_i - \hat{y}_i$.

MSE serves as a fundamental measure for comparing the efficiency of different estimators because it encapsulates both the variance of the estimator and any bias it might introduce. This dual consideration is critical because it addresses the trade-off between precision and accuracy, making MSE a comprehensive criterion for estimator comparison.

The MSE can be decomposed into variance and bias as follows:

$$\text{MSE}(\hat{\theta}) = \mathbb{E}_{\hat{\theta}}\left[(\hat{\theta} - \theta)^2\right] = \left[\text{bias}(\hat{\theta})\right]^2 + \text{Var}(\hat{\theta}) \tag{4.3}$$

For the error term in OLS regression, this formulation also applies:

$$\text{MSE}(\hat{\epsilon}) = \mathbb{E}_{\hat{\epsilon}}\left[(\hat{\epsilon} - \epsilon)^2\right] = \left[\text{bias}(\hat{\epsilon})\right]^2 + \text{Var}(\hat{\epsilon}) \tag{4.4}$$

where $\hat{\epsilon}$ represents the estimated error term and ϵ the true error term. Minimizing the MSE of the error term is essential for achieving the best model fit, striving for zero bias and minimum variance.

Moreover, the decomposition of the MSE for the coefficient $\hat{\beta}_1$ is:

$$\text{MSE}(\hat{\beta}_1) = \mathbb{E}[(\hat{\beta}_1 - \beta_1)^2] = \text{Bias}^2(\hat{\beta}_1) + \text{Var}(\hat{\beta}_1) \tag{4.5}$$

Here, an unbiased estimator with minimum variance is ideal. In most applied work and basic econometrics texts, the methods presented are designed to yield unbiased estimators, so it is typically assumed that the final estimates are unbiased. As a result, MSE is equated with variance, and the square root of that variance—the standard error—is used to assess precision. The full decomposition of MSE into bias squared and variance is usually not emphasized at the introductory level.

As we conclude this section, it is important to note the other usages of the term MSE in statistics.

In regression analysis, MSE may refer to the unbiased estimate of error variance, which is calculated as the residual sum of squares divided by the degrees of freedom. This usage differs from the broader definition previously discussed. Specifically, the unbiased estimate of the variance of the unobserved errors is obtained by dividing the sum of the squared residuals by the degrees of freedom, df = n - p - 1, where n is the number of observations in the sample, and p is the number of parameters estimated in the model, excluding the intercept. This formula provides an unbiased estimate of the variance of the unobserved errors, and it is commonly referred to as the MSE.

In the context of analysis of variance (ANOVA), which is a statistical method used to analyze differences among group means in a sample, *MSE* has a specific usage and concept. The Error term in ANOVA represents the "variability within groups" or "unexplained random error." The primary goal of ANOVA is to identify whether any significant differences exist between the means of three or more unrelated groups. The method divides the observed aggregate variability found within a data set into two parts: variability between groups and variability within groups. Total Sum of Squares (SST) measures the total variance in the data and is calculated as the sum of the squared differences from the overall mean. Total sum of squares quantifies the variance caused by differences between group means, highlighting the impact of the independent variable on the dependent variable. It is calculated by summing the squared differences between each group mean and the grand mean (overall mean), each multiplied by the number of observations in that group. Sum of squares within (SSW), also known as the error term, measures the variance within each group and represents the sum of squares of deviations of each observation from their respective group mean.

These components are related through the formula: SST = SSB + SSW . From these sums of squares, ANOVA calculates the mean squares. The mean square between (MSB) quantifies the average variance between the groups and is calculated as MSB = SSB/k–1 , where k is the number of groups. The MSE, also known as the mean square within (MSW), measures the average the squares of the errors (the residuals) within the groups and is computed as MSE = SSW/n–k, where n is the total number of observations. MSE indicates the average variance within each group about their respective group means. A smaller MSE indicates that the data points within each group are closer to their group mean, suggesting less variability within the groups. The ratio of MSB to MSE forms the F-statistic, which is compared against a critical value from an F-distribution to determine the statistical significance of the observed differences among group means. A significant F-statistic indicates that the differences between the group means are statistically significant, suggesting that the independent variable has an effect on the dependent variable. See Gelman (2005) for a detailed discussion; full citation provided in footnote.[1]

4.3.1 From One Random Sample to Population Parameters

Before we move on to the next section where we will discuss the MSPE, in this section we want to address a critical issue. In practice, we often have only one sample at our disposal, unlike the multiple random samples used in the simulation above or depicted in Figure 4.1. The figure illustrates estimators A and B, utilizing multiple sample-derived estimates to analyze the population parameter θ, including calculating the estimated mean, variance, distribution, and ultimately visualizing these estimators. This leads us to a pivotal question: how can we estimate a population parameter θ using only a single random sample of the population?

The objective is to utilize the data within the sample to infer the value of a parameter in the population. "Inferring" involves deducing or concluding information from evidence and reasoning. **Statistical inference** encompasses the theory, methods, and practice of forming judgments about the parameters of a population and the reliability of statistical relationships, typically based on random sampling.[2] In essence, statistical inference is the process of drawing conclusions from data obtained from a subset of a larger population. It is important to note that no study can generalize universally. As an analyst, it is your responsibility to clarify the applicability of your results for making inferences and where they might not. This requires a clear description of the characteristics of the sample when interpreting results and making inferences.

To begin, we must collect and verify a random sample from the population of interest. Randomness ensures that every member of the population has an equal chance of being selected, making the sample representative. From this single sample, we calculate sample statistics, such as the sample mean, variance, standard deviation, or regression coefficients. These serve as estimates for the corresponding population parameters.Through statistical inference, specifically through estimation and hypothesis testing, we use the parameters estimated from this random sample to make educated guesses about the population parameters.

A key aspect to consider is the asymptotic or large sample properties of the estimator. The estimator used should ideally exhibit consistency. Consistency is the asymptotic property of the estimator related to its ability to converge to the true population parameter as the sample size increases. For an estimator to be considered consistent, it must fulfill two key conditions: it must be unbiased (or its bias approaches zero as the sample size increases), and its variance must diminish to zero as the sample size grows. Demonstrating that an estimator is consistent involves showing that it converges in probability to the true parameter value as the sample size increases to infinity. This concept is indeed closely related to the Law of Large Numbers (LLN).

Consistency suggests that as the sample size grows, the distribution of an estimator becomes more focused around the true parameter value. However, this concept doesn't provide any information about the shape of the distribution at a given sample size. To conduct interval estimation and hypothesis testing, we rely on the central limit theorem (CLT) to approximate the distribution of the estimator, allowing for an assumption of an asymptotically normal distribution. The CLT provides a description of the sampling distribution: by giving us information about an estimator (in hypothetical repeated sampling), it decreases the uncertainty of the estimation since now we can calculate how close the statistic is to the parameter.

Next, interval estimation involves calculating a confidence interval that is likely to contain the population parameter. For instance, a 95% confidence interval for the population mean suggests that, if you were to take 100 different samples and compute 100 confidence intervals, about 95 of them would be expected to contain the actual population mean. This approach, which is based on the CLT, allows us to infer population characteristics from a sample with a known level of confidence.[3] While it's often said that a sample size of 30 is "large enough" for the CLT to apply, this is just a rough rule of thumb—not a strict cutoff. The actual sample size needed for the sampling distribution to approximate normality depends on the underlying distribution's shape and variability.

The CLT is a foundational concept in statistics, explaining how the distribution of sample means becomes approximately normal as the sample size increases, regardless of the distribution of the population. This theorem applies when the random variables are independent and identically distributed, and it is most effective when the sample size is large. The significance of CLT lies in its allowance for the use of normal distribution assumptions in statistical methods, even when the underlying population is not normally distributed. Regression analysis, especially ordinary least squares, often relies on the assumption that the error term has normal distribution. This assumption is justified by considering error term is actually the sum of many independent error terms. Even if these individual errors are not normally distributed, the CLT allows us to approximate the sum of error terms as a normal distribution. This approximation is a crucial aspect of statistical inference in regression analysis.

With this, we want to note that there are several common misconceptions about the CLT, as highlighted by various studies and even in some widely-used textbooks. These misconceptions include the belief that the theorem applies to any random sampling of variables. However, the CLT specifically relates to the means (or sums) of independent and identically distributed random variables obtained through repeated sampling. Another misconception is that the theorem guarantees the emergence of a normal distribution for large samples of any random variable. In reality, what the sampling does is asymptotically reproduce the properties of the population distribution, not necessarily lead to a normal distribution in all cases. This distinction is critical for accurately applying and interpreting the CLT in statistical analysis.

In other words, if the number of observations in each sample large enough, $\bar{X} \sim N(\mu_x, \sigma_x^2/n)$ or when population variance is not known $\bar{X} \sim \mathcal{T}\left(\mu, S^2\right)$ where S is the standard deviation of the sample and $\mathcal{T}$ is the Student- t-distribution.

Why is this important? Because it works like a magic: with only one representative sample, we can generalize the results for the population. We will not cover the details of interval estimation here, but by knowing $\bar{X}$ and the sample variance S, we can have the following interval for the μ_x:

$$\left(\bar{x}-t^* \frac{s}{\sqrt{n}}, \bar{x}+t^* \frac{s}{\sqrt{n}}\right)$$

where t^*, the critical values in t-distribution, are usually around 1.96 for samples more than 100 observations and for the 95% confidence level. This interval would be completely wrong or misleading if $\mathbb{E}(\bar{X}) \neq \mu_x$ and would be useless if it is very wide, which is caused by a large variance. That is the reason why we don't like large variances.

"Then, finally, here is how inference works, in one paragraph: we use sample statistics to estimate population parameters—i.e., the statistics we calculate based on random sample data act as statistical estimators for what we truly want to know, the unknown population parameters. We do that by the postulates of the Central Limit Theorem which describe the sampling distribution, the bridge between the statistics and the parameters. By the CLT, we have the sampling distribution as normal. Again, by the CLT, we can center the sampling distribution on the sample mean, and calculate the sampling distribution's standard error using the sample standard deviation. By applying the properties of the normal probability distribution to the sampling distribution, we then produce population estimates. Ta-da!" (excerpt from CLT[4]; full source in footnote).

In making these inferences, it is crucial to consider the assumptions underlying the statistical methods used. Different methods may require certain conditions about the population from which the sample is drawn, such as assuming the population has a normal distribution or that observations within the sample are independent of each other. The accuracy and reliability of the inferences drawn from a sample significantly depend on how well these assumptions are met. By carefully considering these assumptions and applying appropriate statistical inference techniques, we can extend our findings from a single random sample to make broader conclusions about the entire population.

Reminder:

Assuming a true linear model $y = X\beta + \varepsilon$, estimate $\hat{\beta}$ and prediction $\hat{y} = X\hat{\beta}$. One can define, with $\|.\|$ the mean square error norm for example:

- Estimation error: $\|\beta - \hat{\beta}\|$
- Prediction error (residual): $\|y - \hat{y}\| = \|X(\beta - \hat{\beta})\|$ (note this definition omits the part related to the error term)

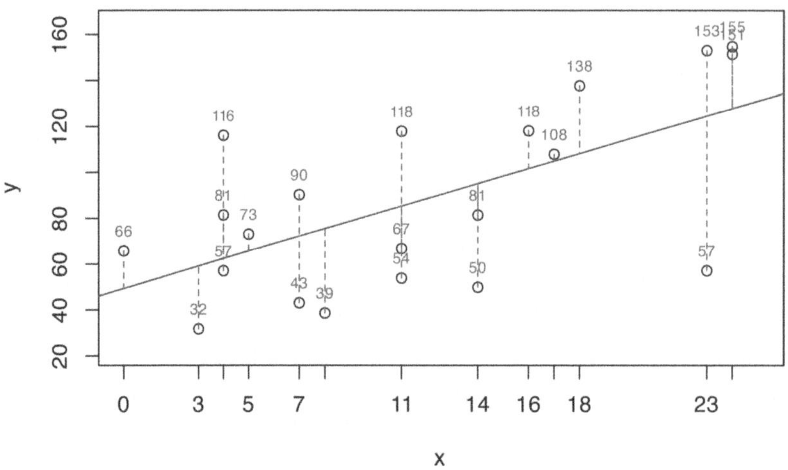

FIGURE 4.2 Observed values and residuals from a linear model.

Assuming a true linear model $y = X\beta + \varepsilon$, we estimate $\hat{\beta}_i$. The Gauss–Markov theorem states that if your linear regression model satisfies the first six classical assumptions, then

OLS regression produces unbiased estimates that have the smallest variance of all possible linear estimators, i.e. OLS is BLUE. For a detailed review of OLS properties and derivations, see the technical section at the end of this chapter.

OLS Assumption 1: The regression model is linear in parameters (the coefficients) and correctly specified.

OLS Assumption 2: The errors have mean zero.

OLS Assumption 3: All independent variables (regressors) are uncorrelated with the error term.

OLS Assumption 4: The regressors in X must all be linearly independent.

OLS Assumption 5: The error term has a constant variance (no heteroscedasticity).

OLS Assumption 6: No independent variable is a perfect linear function of other explanatory variables (no Multicolinearlity).

OLS Assumption 7: The error term is normally distributed (optional).

4.4 Prediction Error: MSPE

In the previous section, we defined MSE which is used exclusively to describe a performance measure of the estimator, and then decomposed between its variance and bias. While some sources might use MSE interchangeably for both estimators and predictors, we make a clear distinction between these terms to prevent confusion. We prefer the term MSPE in contexts specifically discussing prediction scenarios. We can define a predictor as a function mapping arbitrary inputs to a sample of values of some random variable. In this section, we aim to define the MSE specifically for prediction across various functions, thus use the term MSPE. However, in this chapter, we assume different prediction functions to find the best predictor function using the given sample dataset. This involves calculating the MSPE for each assumed function using the sample data, and then selecting the function with the lowest in-sample MSPE.

Our objective is the prediction of an outcome Y, which is typical in a supervised learning context where the outcome is known and is non-binary in a regression setup. We model the response variable Y as a function of features X, plus some random noise:

$$Y = f(X) + \varepsilon \tag{4.6}$$

We have a sample set (training set) consisting of data points $x_1, \ldots, x_n$ and corresponding real values y_i for each point. The data is assumed to be generated by a true function $f(x)$ with added noise ε, which has zero mean and a constant variance. Our goal is to find a predictive function $\hat{f}(x)$ that approximates $f(x)$ as closely as possible using a learning algorithm using the training dataset. But how do we define "close"? Commonly, this is done by minimizing the average squared error loss.

The loss function is $(Y - f(X))^2$, and the average squared loss function is the expected value of this loss, termed the Risk function:

$$\text{Risk function} = \mathrm{E}\left[(Y - f(X))^2\right] \tag{4.7}$$

Thus, our goal is to minimize the Risk function to "predict" Y using X. However, the true function $f(X)$ is unknown, so we aim to find a prediction function $\hat{f}(X)$, which is an estimate of unknown f using the data we have. This leads to an expected prediction error when predicting Y using $\hat{f}(X)$. Ultimately, our objective becomes minimizing the average square of this error, called as the **MSPE**:

$$\text{MSPE} = \text{E}\left[(Y - \hat{f}(X))^2\right] \tag{4.8}$$

The performance of this function is measured by how small the **mean squared prediction error** $(y - \hat{f}(x))^2$ is, not only for the training data but also for new, unseen data points. This model allows us to apply various supervised learning algorithms to find a function $\hat{f}$ that generalizes well beyond the training data, with its expected error decomposable into specific components based on unseen data points. However, due to the noise in y_i, there will always be some level of irreducible error in our predictions.

A good $\hat{f}(X)$ will exhibit a low MSPE. This error can be decomposed into two components: the reducible error (mean squared error), which is the expected squared error loss of estimating $f(X)$ using $\hat{f}(X)$ at a fixed point X, and the irreducible error, essentially the variance of Y given that $X = x$, representing noise that we aim not to learn.

Reducible error for a given $X = x$ is:

$$\text{MSE}(f(x), \hat{f}(x)) = \underbrace{(f(x) - \mathbb{E}[\hat{f}(x)])^2}_{\text{bias}^2(\hat{f}(x))} + \underbrace{\mathbb{E}\left[(\hat{f}(x) - \mathbb{E}[\hat{f}(x)])^2\right]}_{\text{var}(\hat{f}(x))} \tag{4.9}$$

The MSPE is:

$$\text{MSPE} = \text{E}\left[(Y - \hat{f}(X))^2\right] = \text{Bias}[\hat{f}(X)]^2 + \text{Var}[\hat{f}(X)] + \sigma^2 \tag{4.10}$$

where $\sigma^2 = \mathbb{E}[\varepsilon^2]$ represents the variance of the noise.

Moreover, the bias-squared and the variance of $\hat{f}$ is called *reducible error*. Hence, the MSPE can be written as:

$$\text{MSPE} = \text{Reducible Error} + \text{Irreducible Error} \tag{4.11}$$

We want to emphasize the difference between MSE and MSPE, and their decomposed forms in terms of their variances and biases. Even though the formulas for MSE for an estimator and MSE for a predictor are very similar, they serve distinct purposes. For MSE, bias and variance come from parameter estimation. For MSPE, bias and variance are derived from prediction functions. We try different prediction functions to find the best predictor function. In literature, finding an estimator is referred to as 'Point Estimation,' because θ is a point in a regular space. Conversely, determining the function f is described as 'Function Estimation,' since f represents a function within a functional space. A common source of confusion arises because MSE for both estimation and prediction is conceptually similar, leading to their interchangeable use in discussions about bias–variance decomposition, which we will discuss in the next chapter. The predictor with the smallest MSPE will be our choice among other alternative predictor functions. Yet, we have another concern that leads to over-fitting. We will discuss over-fitting in detail in Chapter 6. A detailed decomposition of the MSPE can be found in the technical point section at the end of this chapter.

Our job is to pick the best predictor, i.e., a predictor that will have the minimum MSPE among alternatives. In a perfect setting, we want a prediction function with zero bias and low variance to achieve the minimum MSPE. However, this never happens. Unlike an estimator, we can accept some bias as long as the MSPE is lower. More specifically, we can allow a predictor to have a bias if it reduces the variance more than the bias itself increases it. We will provide an example in Section 5.3.

Unlike estimations, this shows that, in predictions, we can achieve a reduction in MSPE by allowing a trade-off between variance and bias. We will discuss how we can achieve this in the next chapter. For instance, our predictor could be a constant, which, although it is a biased predictor, has zero variance. Or our predictor could be the mean of X as this predictor has zero bias but high variance. Or we could choose a predictor which has some bias and variance. We will show an example using these three predictors in the following simulation.

Let's follow the same simulation example from previous section. Our task is now different. We want to predict the next persons years of schooling using the data we have. Let's summarize some important facts about MSPE related to this simulation here:

1. Let x_0 be the value we want to predict, and let $\hat{f}$ be the predictor, which could be $\mathrm{E}(\bar{X})$, $\mathrm{E}(\hat{X})$, $\mathrm{E}(\tilde{X})$ from Section 4.1, or any other predictor.
2. We model x_0 as $x_0 = \mu_x + \varepsilon_0$, where $f = \mu_x$ represents the systematic component of x_0. Consequently, $\mathrm{E}[x_0] = f$, ensuring that $\mathrm{E}[\varepsilon_0] = 0$ because the random error ε_0 has an expected value of zero.
3. Since f is modeled as μ_x, and μ_x is a constant, the expected value of f is f itself: $\mathrm{E}[f] = f$ and specifically, $\mathrm{E}[\mu_x] = \mu_x$.
4. The variance of x_0 is calculated as follows:

$$\begin{aligned}\operatorname{Var}[x_0] &= \mathrm{E}\left[(x_0 - \mathrm{E}[x_0])^2\right] = \mathrm{E}\left[(x_0 - f)^2\right] = \mathrm{E}\left[(\mu_x + \varepsilon_0 - \mu_x)^2\right] \\ &= \mathrm{E}[\varepsilon_0^2] = \operatorname{Var}[\varepsilon_0] = \sigma^2 \end{aligned} \tag{4.12}$$

This illustrates that the variance of x_0 is entirely due to the variance of ε_0, given that $\mathrm{E}[\varepsilon_0] = 0$.

We want to **predict** the unobserved value of X rather than to estimate μ_x. Therefore, we need a **predictor**, not an **estimator**.

To answer these questions, we need to compare MSPEs or their square roots (RMSPE) as well. Note that we use MSPE here because our example involves predicting a continuous outcome, not a classification problem. In classification scenarios, where the outcome is binary, the loss function has a different algebraic structure than MSPE. The performance evaluation metrics and methods for classification problems will be discussed in Chapter 10 later.

As we know that most developed countries require to go to school between age 6 to 16 years old. As a first predictor, we may predict that the years of schooling for the next individual is 10 years. This is a very simple yet practical prediction function that may provide accurate predictions for some individuals. As a second predictor, we can use the average years of schooling in our data as a good predictor for the next individuals schooling level. Thus we have two different prediction functions. First one is a constant, 10, which has bias but zero variance. The other one is mean of our sample for each observation (average of each row), which has smaller bias and higher variance. For simplicity, we can use 1 sample which consist from 5,000 individuals in this simulation.

The two predictors are $\hat{f}_1 = 10$ and $\hat{f}_2 = \bar{X}$:

We will use the same example we worked with before. We sample from this "population" multiple times. Now the task is to use each sample and come up with a predictor (a prediction rule) to predict a number or multiple numbers drawn from the same population.

```
# Here is our population
populationX <- c(9,10,11,12,13,14,15,16)
#Let's have a containers to have repeated samples (2000)
Ms <- 5000
samples <- matrix(0, Ms, 10)
colnames(samples) <- c("X1", "X2", "X3", "X4", "X5", "X6", "X7", "X8", "X9", "X10")
# Let's have samples (with replacement always)
set.seed(123)
for (i in 1:nrow(samples)) {
  samples[i,] <- sample(populationX, 10, replace = TRUE)
}
head(samples)
```

```
##      X1 X2 X3 X4 X5 X6 X7 X8 X9 X10
## [1,] 15 15 11 14 11 10 10 14 11  13
## [2,] 12 14 14  9 10 11 16 13 11  11
## [3,]  9 12  9  9 13 11 16 10 15  10
## [4,]  9 14 11 12 14  9 11 15 13  12
## [5,] 15 16 10 13 15  9  9 10 15  11
## [6,] 12 13 15 13 11 16 14  9 10  13
```

As you see, this is the same sample with the previous simulation. You can change the data either setting different values in the seed or changing the sammle size, Ms. Now, let's use our predictors and find MSPEs:

```
# Container to record all predictions
predictions <- matrix(0, Ms, 2)
# fhat_1 = 10
for (i in 1:Ms) {
  predictions[i,1] <- 10
}
# fhat_2 - mean
for (i in 1:Ms) {
  predictions[i,2] <- sum(samples[i,])/length(samples[i,])
}
head(predictions)
```

```
##      [,1] [,2]
## [1,]   10 12.4
## [2,]   10 12.1
## [3,]   10 11.4
## [4,]   10 12.0
```

```
## [5,]   10 12.3
## [6,]   10 12.6
```

```
# MSPE
MSPE <- matrix(0, Ms, 2)
for (i in 1:Ms) {
  MSPE[i,1] <- mean((populationX-predictions[i,1])^2)
  MSPE[i,2] <- mean((populationX-predictions[i,2])^2)
}
head(MSPE)
```

```
##      [,1] [,2]
## [1,] 11.5 5.26
## [2,] 11.5 5.41
## [3,] 11.5 6.46
## [4,] 11.5 5.50
## [5,] 11.5 5.29
## [6,] 11.5 5.26
```

```
colMeans(MSPE)
```

```
## [1] 11.500000  5.788422
```

The MSPE of the unbiased $\hat{f}_2$ prediction function is the better as its MSPE is smaller than the other prediction function. What makes a good predictor? Is being unbiased predictor one of the required property? Would being a biased estimator make it automatically a bad predictor? Is having the smallest MSPE enough to be the best predictor? Lets check the following simulation to answer the last question.

Here is an example that illustrates MSPE using different polynomial prediction functions in a simulation.

In the simulation below, we generate a dataset with random x and y. We then create multiple prediction functions, extending up to a 10th-degree polynomial, to evaluate their MSPE.

Even though the "true" model is cubic, we can see that MSPE decreases as the degree of the polynomial increases. This is expected because higher-degree polynomials can better fit the data, resulting in lower prediction errors. However, this improvement comes at the cost of increased complexity, which can lead to overfitting, a topic that will be discussed in the next two chapters. The results highlight the trade-off between model complexity and prediction accuracy, emphasizing the importance of selecting an appropriate model that balances these factors. In practice, it is essential to evaluate the performance of different models using metrics like MSPE to identify the optimal model for a given dataset. This process helps ensure that the selected model provides accurate predictions while avoiding overfitting or underfitting.

What makes a good prediction function? Does being a biased prediction function automatically make it a bad predictor? In predictions, we can have a reduction in MSPE by allowing a trade-off between variance and bias. We will discuss this trade-off in the next chapter. Are complex prediction functions always better than simple ones? We will discuss this in the following chapter. We will also show these concepts by continuing to use this simulation.

```r
# Set seed for reproducibility
set.seed(123)
# Number of observations
# n <-5000 (when you increase N, irreducible error approximates to sigma^2)
  n <- 100
# population mean
mean <- 0
# population standard error (its square is the variance of the -unknown-error term)
sigma <- 8
#random X variable
X <- rnorm(n, 0, 1)
#x_1 is the first random sample (training sample)
x_1 <- X
# Generate lnw data using the provided cubic regression coefficients
b0 <- 1; b1 <- 2; b2 <- -2; b3 <- 3
Truefx <- b0 + b1*x_1 + b2*I(x_1^2)+ b3*(x_1^3)
RandomError <- rnorm(n, mean , sigma)
Y <- Truefx + RandomError
# Function to calculate MSPE
calculate_mspe <- function(Outcome, PredictedOutcome) {
  mean((Outcome - PredictedOutcome)^2)
}
# Function to calculate MSE (Reducible Error)
calculate_mse <- function(truefx, pred) {
  mean((truefx - pred)^2)
}
# Function to calculate Irreducible Error
calculate_IrrErr <- function(y, truefx) {
  mean((y - truefx)^2)
}
# Initialize storage for MSPE results
results <- data.frame(
  Degree = integer(),
  MSPE = numeric(),
  stringsAsFactors = FALSE
)
# Fit polynomial models of degrees 1 to 10 and calculate MSPE
PolyDegree <- 10
for (degree in 1:PolyDegree) {
  model <- lm(Y ~ poly(x_1, degree, raw = TRUE))
  predictions <- predict(model, newdata = data.frame(x_1 = x_1))

  mspe <- calculate_mspe(Y, predictions)
  mse <- calculate_mspe(Truefx, predictions)
  IrrError <- calculate_IrrErr(Y, Truefx)
  results <- rbind(results, data.frame(
    Degree = degree,
    MSPE = mspe,
    ReducibleError = mse,
```

```
    IrreducibleError = IrrError
  ))
}
# Print the results
print(results)
```

```
##    Degree     MSPE ReducibleError IrreducibleError
## 1       1 86.57600      23.336065         59.98584
## 2       2 79.56354      20.547115         59.98584
## 3       3 58.40780       1.578044         59.98584
## 4       4 57.70779       2.278051         59.98584
## 5       5 57.70180       2.284042         59.98584
## 6       6 57.59455       2.391297         59.98584
## 7       7 57.20530       2.780540         59.98584
## 8       8 55.74635       4.239495         59.98584
## 9       9 55.74594       4.239899         59.98584
## 10     10 55.19262       4.793224         59.98584
```

We have one sample, and we have ten different prediction functions. We can see that the MSPE decreases as the degree of the polynomial increases. This is expected because higher-degree polynomials can better fit the data, resulting in lower prediction errors. However, this improvement comes at the cost of increased complexity, which can lead to overfitting, a topic that will be discussed in the next two chapters. We can only calculate MSPE for the sample we have. This is calles emprical MSPE or training sample MSPE, or in-sample MSPE. We can not calculate the MSPE for the population.We can not also decompose MSPE into reducible and irreducible errors as we donot know the true function, and population parameters such as mean and variance. Our goal in to minimize MSPE by minimizing reducible error. However, we need to consider bias–variance tradeff, and complexity of the model which may lead the overfitting, as well as out-sample prediction error. We will discuss these in the next chapters.

4.5 Technical Points and Proofs

4.5.1 Unbiasedness of a Parameter

To estimate an unknown population parameter, symbolized as θ, our data comes from a random sample. We refer to $\hat{\theta}$ as the estimator of θ. The first and most important criteria choosing the "best" estimator should be that the expected value, i.e., the mean, of all estimates obtained from various random samples, should be equal to the unknown population parameter,θ. An estimator that satisfies this condition is referred to as an unbiased estimator , i.e.,

$$\text{Bias}(\hat{\theta}) = \mathbb{E}[\hat{\theta}] - \theta = 0 \quad \text{so} \quad \mathbb{E}[\hat{\theta}] = \theta$$

While an unbiased estimator does not guarantee that the estimate from a particular sample will exactly match the population parameter, it implies that averaging the estimates from repeatedly drawn random samples, each calculated using the same estimator, will closely approximate the actual population parameter. It is important to remember that this concept is theoretical, as in practical scenarios we typically work with a single random sample and rely on asymptotic (large sample) properties.

4.5.2 Unbiasedness of Sample Mean Estimator

We want to estimate population mean of a variable (feature). Its common notation to use μ_x instead of θ, using a sample mean $\bar{X}$ notation for an estimator to estimate this population mean.

Let's demonstrate the unbiasedness of the sample mean estimator, $\mathbb{E}[\bar{X}] = \mu_x$. The sample mean $\bar{X}$ is calculated as the average of n observations from a random sample. So

$$\hat{\theta} = \bar{X} = \frac{1}{n}\sum_{i=1}^{n} x_i$$

The expected value of the sample mean is:

$$\mathbb{E}[\bar{X}] = \mathbb{E}\left[\frac{1}{n}\sum_{i=1}^{n} x_i\right] = \mathbb{E}\left[\frac{1}{n}(x_1 + x_2 + \ldots + x_n)\right] = \frac{1}{n}(\mathbb{E}[x_1] + \mathbb{E}[x_2] + \ldots + \mathbb{E}[x_n])$$

Since the linearity of expectation allows the expectation operator to be distributed over addition.

Assuming each x_i is an independent and identically distributed sample from a population with mean μ_x, the expectation of each x_i is μ_x:

$\mathbb{E}[x_1] = \mathbb{E}[x_2] = \ldots = \mathbb{E}[x_n] = \mu_x$

Therefore, the equation simplifies to:

$$\mathbb{E}[\bar{X}] = \frac{1}{n}(\mathbb{E}[x_1] + \mathbb{E}[x_2] + \ldots + \mathbb{E}[x_n]) = \frac{1}{n}(n\mu_x) = \mu_x$$

This shows that $\bar{X}$, the sample mean, is an unbiased estimator of the population mean μ_x, as the expected value of $\bar{X}$ equals μ_x.

4.5.3 The Variance of the Sampling Distribution of Sample Means (Sampling Variance)

The sampling distribution refers to the distribution of a parameter (like a mean, variance, or coefficient) across many samples drawn from a specific population. This distribution of outcomes from an estimator has a sampling mean (which is used to check unbiasedness) and a variance, known as the estimator's sampling variance (which is used to check efficiency). The sampling variance of an estimator quantifies how much the estimates from different samples vary around the true population parameter. A smaller sampling variance indicates that the estimator is more precise and provides more consistent estimates across different samples.

The variance of the sample mean can be expressed using the definition of variance and the linearity of expectation:

$$\text{Var}(\bar{X}) = \text{Var}\left(\frac{1}{n}\sum_{i=1}^{n} X_i\right)$$

The variance of a constant times a random variable is the square of the constant times the variance of the random variable:

$$\text{Var}\left(\frac{1}{n}\sum_{i=1}^{n} X_i\right) = \frac{1}{n^2}\text{Var}\left(\sum_{i=1}^{n} X_i\right)$$

For independent random variables, the variance of their sum is the sum of their variances:

$$\text{Var}\left(\sum_{i=1}^{n} X_i\right) = \text{Var}(X_1) + \text{Var}(X_2) + \ldots + \text{Var}(X_n)$$

Since the x_i are i.i.d. with variance σ_x^2, we have:

$$\text{Var}(X_i) = \sigma_x^2 \quad \text{for all } i$$

Thus:

$$\text{Var}\left(\sum_{i=1}^{n} X_i\right) = \sigma_x^2 + \sigma_x^2 + \ldots + \sigma_x^2 = n\sigma_x^2$$

Substitute back into the variance formula for $\bar{X}$:

$$\text{Var}(\bar{X}) = \frac{1}{n^2} \cdot n\sigma_x^2 = \frac{\sigma_x^2}{n}$$

This derivation concludes that the variance of the sampling mean $\bar{X}$ is $\frac{\sigma_x^2}{n}$, highlighting how the variability of the sample mean decreases as the sample size n increases.

4.5.4 Unbiasedness of Sample Variance Estimator

We aim to estimate the population variance of a variable (or feature). It is common to use the notation σ_x^2 for the population variance (instead of θ), and $\hat{\sigma_X}^2$ as the notation for an estimator to estimate this population variance (instead of $\hat{\theta}$).

Let's demonstrate the unbiasedness of the sample variance estimator, where $\mathbb{E}[\hat{\sigma_X}^2] = \sigma_x^2$. The sample variance $\hat{\sigma_X}^2$ is calculated using the formula:

$$\hat{\sigma_X}^2 = \frac{1}{n-1}\sum_{i=1}^{n}(x_i - \bar{X})^2$$

where $\bar{X}$ is the sample mean. This formula corrects for the bias in the straightforward estimator of variance by dividing by $n-1$ instead of n, accounting for the degrees of freedom lost in estimating $\bar{X}$ from the sample. The expected value of the sample variance will be shown to be:

$$\begin{aligned}\mathbb{E}[\hat{\sigma_X}^2] &= \mathbb{E}\left[\frac{1}{n-1}\sum_{i=1}^{n}(x_i-\bar{X})^2\right] \\ &= \frac{1}{n-1}\mathbb{E}\left[\sum_{i=1}^{n}(x_i-\bar{X})^2\right] = \frac{1}{n-1}\sum_{i=1}^{n}\mathbb{E}[(x_i-\bar{X})^2] \\ &= \frac{1}{n-1}\sum_{i=1}^{n}\mathbb{E}[x_i^2-2x_i\bar{X}+\bar{X}^2] \\ &= \frac{1}{n-1}\sum_{i=1}^{n}\left(\mathbb{E}[x_i^2]-2\mathbb{E}[x_i\bar{X}]+\mathbb{E}[\bar{X}^2]\right) \stackrel{?}{=} \sigma_x^2\end{aligned}$$

Let's calculate each term in this part of the last equation ($\mathbb{E}[x_i^2] - 2\mathbb{E}[x_i\bar{X}] + \mathbb{E}[\bar{X}^2]$) separately:

Term 1: $\mathbb{E}[x_i^2] = \sigma_x^2 + \mu_x^2$ since x_i are i.i.d. with mean μ_x and variance σ_x^2, using variance decomposition formula (see below).

Term 2: $\mathbb{E}[x_i\bar{X}] = \mathbb{E}[x_i]\mathbb{E}[\bar{X}] = \mu_x^2$ since $\mathbb{E}[x_i] = \mu_x$ by definition and $\mathbb{E}[\bar{X}] = \mu_x$ as shown above in unbiasness of mean.

Term 3: $\mathbb{E}[\bar{X}^2] = \mathrm{Var}(\bar{X}) + \mathbb{E}[\bar{X}]^2 = \frac{\sigma_x^2}{n} + \mu_x^2$ since $\mathbb{E}[\bar{X}] = \mu_x$ and $\mathrm{Var}(\bar{X}) = \frac{\sigma_x^2}{n}$ (see above), and using variance decomposition formula (see below).

Substituting these into the expectation of the squared differences gives:

$\mathbb{E}[x_i^2] - 2\mathbb{E}[x_i\bar{X}] + \mathbb{E}[\bar{X}^2] = (\sigma_x^2 + \mu_x^2) - 2\mu_x^2 + \left(\frac{\sigma_x^2}{n} + \mu_x^2\right) = \sigma_x^2 - \frac{\sigma_x^2}{n}$

Thus, the expected value of the sample variance is:

$$\mathbb{E}[\sigma_X^2] = \frac{1}{n-1}\sum_{i=1}^{n}\mathbb{E}[(x_i-\bar{X})^2] = \frac{1}{n-1}n\left(\sigma_x^2 - \frac{\sigma_x^2}{n}\right) = \sigma_x^2$$

This proves that the sample variance formula is an unbiased estimator of the population variance σ_x^2, as the expected value of the sample variance equals the population variance.

Variance decomposition formula: The variance of a random variable x is defined as:

$$\mathrm{Var}(x) = \sigma_X^2 = \mathbb{E}[(x-\mu_x)^2] = \mathbb{E}[x^2 - 2x\mu_x + \mu_x^2] = \mathbb{E}[x^2] - 2\mu_x\mathbb{E}[x] + \mathbb{E}[\mu_x^2]$$

where μ_x is the expected value (mean) of x. Since the linearity of expectation allows the expectation operator to be distributed over addition.

As $\mathbb{E}[x] = \mu_x$ and μ_x is a constant, the expectation of a constant is the constant itself, and the expectation of a constant squared is also the constant squared: $\mathrm{Var}(x) = \mathbb{E}[x^2] - 2\mu_x^2 + \mu_x^2$ Thus, Variance decomposition formula is the following:

$$\mathrm{Var}(x) = \mathbb{E}[x^2] - \mu_x^2 \quad \text{or} \quad \mathrm{Var}(x) = \mathbb{E}[x^2] - \mathbb{E}[x^2]$$

Rearranging to solve for $\mathbb{E}[x^2]$ (to use in term 1 and term 3 above), we find:

$$\mathbb{E}[x^2] = \mathrm{Var}(x) + \mu_x^2$$

This equation demonstrates that the expected value of the square of the random variable x is equal to the variance of x plus the square of the mean of x, expressed as:

$$\mathbb{E}[x_i^2] = \sigma_x^2 + \mu_x^2$$

This relationship is fundamental in statistical analysis, particularly in understanding how the variance and mean of a distribution contribute to its second moment.

Proof of variance and bias of decomposition an estimator:

The MSE of an estimator $\hat{\theta}$ with respect to an unknown parameter θ is defined as:

$$\text{MSE}(\hat{\theta}) = \text{E}_\theta\left[(\hat{\theta} - \theta)^2\right] = \text{Var}(\hat{\theta}) + \text{Bias}(\hat{\theta})^2$$

The estimator $\hat{\theta}$ is derived as a sample statistic and is used to estimate some population parameter, then the expectation is with respect to the sampling distribution of the sample statistic.The MSE can be written as the sum of the variance of the estimator and the squared bias of the estimator.

The mean squared error of the estimator $\hat{\theta}$ is defined as the expected value of the square of the difference between $\hat{\theta}$ and the true parameter θ.

$$\text{MSE}(\hat{\theta}) = \text{E}\left[(\hat{\theta} - \theta)^2\right]$$

In the next step, we rewrite the squared term by adding and subtracting the expected value of $\hat{\theta}$, which helps in simplifying the analysis by separating the variance of the estimator from its bias.

$$= \text{E}\left[\left(\hat{\theta} - \text{E}[\hat{\theta}] + \text{E}[\hat{\theta}] - \theta\right)^2\right]$$

We apply the square of a sum formula. We obtain expected value of an equation which has three parts: the variance of the estimator, the middle term, and the square of the bias as the third term.

$$= \text{E}\left[\left(\hat{\theta} - \text{E}[\hat{\theta}]\right)^2 + 2\left(\hat{\theta} - \text{E}[\hat{\theta}]\right)\left(\text{E}[\hat{\theta}] - \theta\right) + \left(\text{E}[\hat{\theta}] - \theta\right)^2\right]$$

Next, we separate outside expectation. The linearity of expectation allow us to handle each component of the expanded square separately.

$$= \underbrace{\text{E}\left[\left(\hat{\theta} - \text{E}[\hat{\theta}]\right)^2\right]} + \underbrace{\text{E}\left[2\left(\hat{\theta} - \text{E}[\hat{\theta}]\right)\left(\text{E}[\hat{\theta}] - \theta\right)\right]} + \underbrace{\text{E}\left[\left(\text{E}[\hat{\theta}] - \theta\right)^2\right]}$$

The initial term represents the variance of the estimator $\hat{\theta}$. The middle term comprises two components. The first component is $[\hat{\theta} - \text{E}[\hat{\theta}]]$, which indicates the deviation of $\hat{\theta}$ from its expected value. The second component is $\text{E}[\hat{\theta}] - \theta$, which is identical to the term that appears in the third term and represents the bias of the estimator $\hat{\theta}$. Since this bias of an estimator is constant for any sample, it can be removed from the expectation operator in the middle term. Similarly, the initial expectation from the third term can be omitted, as the expected value of a constant squared is the constant itself.

We can simplify the expression by recognizing that the expected value of the deviation $\hat{\theta} - \text{E}[\hat{\theta}]$ is zero, leading to the final form of the MSE. The middle term simplifies to $2\left(\text{E}[\hat{\theta}] - \theta\right)\text{E}\left[\hat{\theta} - \text{E}[\hat{\theta}]\right] = 2\left(\text{E}[\hat{\theta}] - \theta\right) \cdot 0 = 0$

The final expression of the MSE is the sum of the variance of the estimator $\hat{\theta}$ and the square of the bias, highlighting the bias–variance trade-off.

$$= \mathrm{E}\left[\left(\hat{\theta} - \mathrm{E}[\hat{\theta}]\right)^2\right] + \left(\mathrm{E}[\hat{\theta}] - \theta\right)^2$$

which is equal to

$$= \mathrm{Var}(\hat{\theta}) + \mathrm{Bias}(\hat{\theta})^2$$

Concise proof of the MSE decomposition can also be demonstrated using the well-known identity for any random variable X, which states that:

$$\mathbb{E}\left(X^2\right) = \mathrm{Var}(X) + (\mathbb{E}(X))^2.$$

Applying this identity to the estimator $\hat{\theta}$ with respect to the true parameter θ, we set $X = \hat{\theta} - \theta$. Then, the MSE of $\hat{\theta}$ is given by:

$$\mathrm{MSE}(\hat{\theta}) = \mathbb{E}\left[(\hat{\theta} - \theta)^2\right].$$

Substituting X into the identity, we have:

$$\begin{aligned}\mathbb{E}\left[(\hat{\theta} - \theta)^2\right] &= \mathrm{Var}(\hat{\theta} - \theta) + (\mathbb{E}[\hat{\theta} - \theta])^2 \\ &= \mathrm{Var}(\hat{\theta}) + \mathrm{Bias}^2(\hat{\theta}),\end{aligned}$$

where the variance of the estimator $\mathrm{Var}(\hat{\theta})$ comes from the fact that the variance is invariant to shifts by a constant (in this case, θ), and $\mathrm{Bias}(\hat{\theta}) = \mathbb{E}[\hat{\theta}] - \theta$ by definition. Thus, the squared bias $\mathrm{Bias}^2(\hat{\theta})$ is the square of the expected deviation of $\hat{\theta}$ from θ. The final expression for the MSE shows that it is the sum of the variance of the estimator and the square of its bias.

In the case of unbiased estimators, the MSE simplifies to the variance of the estimator. Given that our criterion is to select only unbiased estimators, where $\mathbb{E}(\hat{\theta}) = \theta$, the MSE expression reduces to $Var(\hat{\theta})$. Consequently, evaluating the performance of alternative unbiased estimators by MSE effectively compares their variances, with the objective of selecting the estimator that offers the smallest variance. From our simulations in the previous section, consider the estimator $\hat{\theta} = \bar{X}$, where $\theta = \mu_x$ is the parameter we aim to estimate. This setup confirms that $\hat{\theta}$ is indeed unbiased, as $\mathbb{E}(\hat{\theta}) = \mu_x$, which aligns perfectly with our target θ. Thus, the efficiency of different unbiased estimators can be directly assessed by comparing their variances through MSE, guiding us toward the most precise estimator based on minimal variance.

OLS Review

OLS Regression Model

Consider the simple linear regression model:

$$Y_i = \beta_0 + \beta_1 X_i + \epsilon_i$$

where Y_i is the dependent variable, X_i is the independent variable, β_0 and β_1 are coefficients, and ϵ_i represents the *unknown* error term, assumed to have a mean of zero and constant variance.

Our objective is to find the values of β_0 and β_1 that minimize the MSE, which is given by:

$$\text{MSE} = \frac{1}{n}\sum_{i=1}^{n}(Y_i - \hat{Y}_i)^2$$

where $\hat{Y}_i = \hat{\beta}_0 + \hat{\beta}_1 X_i$ is the predicted value of Y based on the regression line. $Y_i = \hat{Y}_i + \hat{\epsilon}_i$ thus each value of Y_i is divided into fitted value, $\hat{Y}_i$, and residual, $\hat{\epsilon}_i = Y_i - \hat{Y}_i$.

The coefficients β_0 and β_1 are estimated such that the MSE is minimized. To find an estimator for unkown β_0 and β_1, which minimize MSE, we take its partial derivatives with respect to β_0 and β_1, set them to zero, and solve for β_0 and β_1. It is important to note that in OLS, a more common explanation for finding the estimates of β_0 which is $\hat{\beta}_0$and estimates of β_1 which is $\hat{\beta}_1$ involves the concept of a residual (or prediction error), defined as the difference between the observed value and the predicted value, expressed as $Y_i - \hat{Y}_i$. We minimize the RSS, given by

$$\text{RSS} = \sum_{i=1}^{n}(\hat{\epsilon}_i)^2 = \sum_{i=1}^{n}(Y_i - \hat{Y}_i)^2,$$

which is the sum of the squared residuals. However, as can be easily seen from the equations, minimizing RSS is equivalent to minimizing MSE.

Taking the partial derivative of MSE with respect to $\hat{\beta}_0$:

$$\frac{\partial}{\partial \hat{\beta}_0}\left(\frac{1}{n}\sum_{i=1}^{n}(Y_i - (\hat{\beta}_0 + \hat{\beta}_1 X_i))^2\right) = -\frac{2}{n}\sum_{i=1}^{n}(Y_i - \hat{\beta}_0 - \hat{\beta}_1 X_i)$$

Setting this to zero gives:

$$\sum_{i=1}^{n} Y_i - n\hat{\beta}_0 - \hat{\beta}_1 \sum_{i=1}^{n} X_i = 0$$

$$n\hat{\beta}_0 = \sum_{i=1}^{n} Y_i - \hat{\beta}_1 \sum_{i=1}^{n} X_i$$

$$\hat{\beta}_0 = \bar{Y} - \hat{\beta}_1 \bar{X}$$

where $\bar{Y}$ and $\bar{X}$ are the sample means of Y and X respectively.

Taking the partial derivative of MSE with respect to $\hat{\beta}_1$:

$$\frac{\partial}{\partial \hat{\beta}_1}\left(\frac{1}{n}\sum_{i=1}^{n}(Y_i - (\hat{\beta}_0 + \hat{\beta}_1 X_i))^2\right) = -\frac{2}{n}\sum_{i=1}^{n} X_i(Y_i - \hat{\beta}_0 - \hat{\beta}_1 X_i)$$

Setting this to zero gives:

$$\sum_{i=1}^{n} X_i Y_i - \hat{\beta}_0 \sum_{i=1}^{n} X_i - \hat{\beta}_1 \sum_{i=1}^{n} X_i^2 = 0$$

Substituting the expression for $\hat{\beta}_0$ we found earlier:

$$\sum_{i=1}^{n} X_i Y_i - (\bar{Y} - \hat{\beta}_1 \bar{X}) \sum_{i=1}^{n} X_i - \hat{\beta}_1 \sum_{i=1}^{n} X_i^2 = 0$$

$$\sum_{i=1}^{n} X_iY_i - \bar{Y}\sum_{i=1}^{n} X_i + \hat{\beta}_1\bar{X}\sum_{i=1}^{n} X_i - \hat{\beta}_1\sum_{i=1}^{n} X_i^2 = 0$$

$$\hat{\beta}_1 = \frac{\sum_{i=1}^{n} X_iY_i - \bar{Y}\sum_{i=1}^{n} X_i}{\sum_{i=1}^{n} X_i^2 - \bar{X}\sum_{i=1}^{n} X_i}$$

$$\hat{\beta}_1 = \frac{\sum_{i=1}^{n} X_iY_i - n\bar{X}\bar{Y}}{\sum_{i=1}^{n} X_i^2 - n\bar{X}^2}$$

This formula gives $\hat{\beta}_1$ as the slope of the regression line, which quantifies the relationship between X and Y. This formula also can be written as

$$\hat{\beta}_1 = \frac{\sum_{i=1}^{n}(X_i - \bar{X})(Y_i - \bar{Y})}{\sum_{i=1}^{n}(X_i - \bar{X})^2}$$

Note that: Expand the numerator and denominator of the alternative formula for $\hat{\beta}_1$ as follows to show they are equal with the initial formula for $\hat{\beta}_1$:

Recognize that:

$$\bar{X} = \frac{1}{n}\sum_{i=1}^{n} X_i \quad \text{and} \quad \bar{Y} = \frac{1}{n}\sum_{i=1}^{n} Y_i$$

$$\text{Numerator: } \sum_{i=1}^{n}(X_i - \bar{X})(Y_i - \bar{Y}) = \sum_{i=1}^{n}(X_iY_i - X_i\bar{Y} - \bar{X}Y_i + \bar{X}\bar{Y})$$

$$= \sum_{i=1}^{n} X_iY_i - \bar{Y}\sum_{i=1}^{n} X_i - \bar{X}\sum_{i=1}^{n} Y_i + n\bar{X}\bar{Y}$$

$$= \sum_{i=1}^{n} X_iY_i - n\bar{X}\bar{Y}$$

$$\text{Denominator: } \sum_{i=1}^{n}(X_i - \bar{X})^2 = \sum_{i=1}^{n}(X_i^2 - 2X_i\bar{X} + \bar{X}^2)$$

$$= \sum_{i=1}^{n} X_i^2 - 2\bar{X}\sum_{i=1}^{n} X_i + n\bar{X}^2$$

$$= \sum_{i=1}^{n} X_i^2 - n\bar{X}^2$$

Therefore, we confirm both formulas are equivalent.

Unbiasness of the OLS Estimator $\hat{\beta}_1$

The OLS estimator $\hat{\beta}_1$ for a simple linear regression model where $y = \beta_0 + \beta_1 X + \epsilon$ is given by:

$$\hat{\beta}_1 = \frac{\sum_{i=1}^{n}(X_i - \bar{X})(Y_i - \bar{Y})}{\sum_{i=1}^{n}(X_i - \bar{X})^2}$$

where $Y_i = \beta_0 + \beta_1 X_i + \epsilon_i$. Let us analyze the expectation $\mathbb{E}[\hat{\beta}_1]$.

Starting from the formula of $\hat{\beta}_1$, replace Y_i by its expression:

$$\hat{\beta}_1 = \frac{\sum_{i=1}^n (X_i - \bar{X})(\beta_0 + \beta_1 X_i + \epsilon_i - \bar{Y})}{\sum_{i=1}^n (X_i - \bar{X})^2}$$

We take the expectation of $\hat{\beta}_1$:

$$\mathbb{E}[\hat{\beta}_1] = \mathbb{E}\left[\frac{\sum_{i=1}^n (X_i - \bar{X})(\beta_0 + \beta_1 X_i + \epsilon_i - \bar{Y})}{\sum_{i=1}^n (X_i - \bar{X})^2}\right]$$

Applying the linearity of expectation and simplifying:

$$= \frac{\mathbb{E}\left[\sum_{i=1}^n (X_i - \bar{X})(\beta_0 + \beta_1 X_i + \epsilon_i - \bar{Y})\right]}{\sum_{i=1}^n (X_i - \bar{X})^2}$$

$$= \frac{\sum_{i=1}^n \mathbb{E}[(X_i - \bar{X})(\beta_0 + \beta_1 X_i + \epsilon_i - \bar{Y})]}{\sum_{i=1}^n (X_i - \bar{X})^2}$$

Since β_0 and β_1 are constants and ϵ_i are independent of X_i with zero mean, the expectations simplify:

$$= \frac{\sum_{i=1}^n (X_i - \bar{X})(\beta_0 + \beta_1 X_i - \beta_0 - \beta_1 \bar{X})}{\sum_{i=1}^n (X_i - \bar{X})^2}$$

$$= \frac{\beta_1 \sum_{i=1}^n (X_i - \bar{X})^2}{\sum_{i=1}^n (X_i - \bar{X})^2} = \beta_1$$

This shows that $\mathbb{E}[\hat{\beta}_1] = \beta_1$, proving that $\hat{\beta}_1$ is an unbiased estimator of β_1.

Variance of the OLS Estimator $\hat{\beta}_1$

Given the estimator $\hat{\beta}_1$ for a simple linear regression model:

$$\hat{\beta}_1 = \frac{\sum_{i=1}^n (X_i - \bar{X})(Y_i - \bar{Y})}{\sum_{i=1}^n (X_i - \bar{X})^2}$$

we want to find the variance of $\hat{\beta}_1$.

Expression for Variance

The variance of $\hat{\beta}_1$ is defined as: $\mathrm{Var}(\hat{\beta}_1) = \mathbb{E}[(\hat{\beta}_1 - \mathbb{E}[\hat{\beta}_1])^2]$

Calculation of Var($\hat{\beta}_1$)

Starting with the formula for $\hat{\beta}_1$, we know from earlier analysis that $\mathbb{E}[\hat{\beta}_1] = \beta_1$. Hence, we consider:

$$\hat{\beta}_1 - \mathbb{E}[\hat{\beta}_1] = \frac{\sum_{i=1}^n (X_i - \bar{X})(Y_i - \bar{Y})}{\sum_{i=1}^n (X_i - \bar{X})^2} - \beta_1$$

Since $Y_i = \beta_0 + \beta_1 X_i + \epsilon_i$, substituting Y_i gives:

$$= \frac{\sum_{i=1}^n (X_i - \bar{X})(\beta_0 + \beta_1 X_i + \epsilon_i - \bar{Y})}{\sum_{i=1}^n (X_i - \bar{X})^2} - \beta_1$$

$$= \frac{\sum_{i=1}^{n}(X_i - \bar{X})\epsilon_i}{\sum_{i=1}^{n}(X_i - \bar{X})^2}$$

we square it:

$$\left(\frac{\sum_{i=1}^{n}(X_i - \bar{X})\epsilon_i}{\sum_{i=1}^{n}(X_i - \bar{X})^2}\right)^2 = \frac{\left(\sum_{i=1}^{n}(X_i - \bar{X})\epsilon_i\right)^2}{\left(\sum_{i=1}^{n}(X_i - \bar{X})^2\right)^2}$$

Taking Expectations

Now, taking the expectation of the squared term:

$$\mathbb{E}\left[\frac{\left(\sum_{i=1}^{n}(X_i - \bar{X})\epsilon_i\right)^2}{\left(\sum_{i=1}^{n}(X_i - \bar{X})^2\right)^2}\right]$$

Applying the Independence of ϵ_i

Since the ϵ_i terms are independent and identically distributed (i.i.d.) with zero mean and variance σ^2, we can expand and simplify the expectation of the numerator:

$$\mathbb{E}\left[\left(\sum_{i=1}^{n}(X_i - \bar{X})\epsilon_i\right)^2\right] = \mathbb{E}\left[\sum_{i=1}^{n}\sum_{j=1}^{n}(X_i - \bar{X})(X_j - \bar{X})\epsilon_i\epsilon_j\right]$$

$$= \sum_{i=1}^{n}\sum_{j=1}^{n}(X_i - \bar{X})(X_j - \bar{X})\mathbb{E}[\epsilon_i\epsilon_j]$$

$$= \sum_{i=1}^{n}(X_i - \bar{X})^2\mathbb{E}[\epsilon_i^2] = \sum_{i=1}^{n}(X_i - \bar{X})^2\sigma^2$$

because $\mathbb{E}[\epsilon_i\epsilon_j] = 0$ for $i \neq j$ and $\mathbb{E}[\epsilon_i^2] = \sigma^2$.

Final Expression for Variance

The denominator of our original expression remains constant since it does not involve ϵ_i. Therefore, we simplify the expectation as:

$$\frac{\sum_{i=1}^{n}(X_i - \bar{X})^2\sigma^2}{\left(\sum_{i=1}^{n}(X_i - \bar{X})^2\right)^2} = \sigma^2\left(\sum_{i=1}^{n}(X_i - \bar{X})^2\right)^{-1}$$

This shows how the variance of $\hat{\beta}_1$ depends inversely on the sum of squared deviations from the mean of X, and directly on the variance of the error terms, σ^2.

Estimating the Variance of the Error Term σ^2

Estimation of σ^2

In the context of OLS regression, the variance of the error terms, σ^2, is unknown and must be estimated from the data. This estimate is typically obtained using the residual sum of squares (RSS) divided by the degrees of freedom, which in the simple linear regression model is $n - 2$ (subtracting the number of estimated parameters).

Residual Sum of Squares (RSS)

The residual for each observation is given by:

$$\epsilon_i = y_i - \hat{y}_i$$

where $\hat{y}_i = \hat{\beta}_0 + \hat{\beta}_1 X_i$ are the predicted values. The RSS is then:

$$\text{RSS} = \sum_{i=1}^{n} \epsilon_i^2 = \sum_{i=1}^{n} (y_i - \hat{\beta}_0 - \hat{\beta}_1 X_i)^2$$

The estimator for σ^2 is:

$$\hat{\sigma}^2 = \frac{\text{RSS}}{n-2}$$

This provides an unbiased estimate of the variance of the error terms.

Standard Error of $\hat{\beta}_1$

The standard error (SE) of $\hat{\beta}_1$ is the square root of the variance of $\hat{\beta}_1$, which we previously derived as:

$$\text{Var}(\hat{\beta}_1) = \sigma^2 \left(\sum_{i=1}^{n} (X_i - \bar{X})^2 \right)^{-1}$$

Substituting our estimate for σ^2, we have:

$$\text{Var}(\hat{\beta}_1) = \hat{\sigma}^2 \left(\sum_{i=1}^{n} (X_i - \bar{X})^2 \right)^{-1}$$

and for standard error of $\hat{\beta}_1$, we can use the formula:

$$\text{SE}(\hat{\beta}_1) = \sqrt{\hat{\sigma}^2 \left(\sum_{i=1}^{n} (X_i - \bar{X})^2 \right)^{-1}} = \sqrt{\frac{\text{RSS}}{n-2} \left(\sum_{i=1}^{n} (X_i - \bar{X})^2 \right)^{-1}}$$

This standard error is crucial for constructing confidence intervals and conducting hypothesis tests about the slope β_1.

Bias-Variance Decomposition for β_1

In a simple linear regression model $y = \beta_0 + \beta_1 X + \epsilon$, where ϵ is the normally distributed error term with zero mean and variance σ^2, we can analyze the estimator $\hat{\beta}_1$ for the coefficient β_1.

The OLS estimator for β_1 is given by (as shown above):

$$\hat{\beta}_1 = \frac{\sum_{i=1}^{n} (X_i - \bar{X})(Y_i - \bar{Y})}{\sum_{i=1}^{n} (X_i - \bar{X})^2}$$

where $\bar{X}$ and $\bar{Y}$ are the sample means of X and Y, respectively.

The expected value of $\hat{\beta}_1$ is:

$$\mathbb{E}[\hat{\beta}_1] = \mathbb{E}\left[\frac{\sum_{i=1}^{n}(X_i - \bar{X})(Y_i - \bar{Y})}{\sum_{i=1}^{n}(X_i - \bar{X})^2}\right]$$

Since $Y_i = \beta_0 + \beta_1 X_i + \epsilon_i$, we can substitute and simplify:

$$\mathbb{E}[\hat{\beta}_1] = \mathbb{E}\left[\frac{\sum_{i=1}^{n}(X_i - \bar{X})(\beta_0 + \beta_1 X_i + \epsilon_i - \bar{Y})}{\sum_{i=1}^{n}(X_i - \bar{X})^2}\right] = \beta_1$$

This shows that $\hat{\beta}_1$ is an *unbiased* estimator of β_1.(Check below for detailed calculations)

The *variance* of $\hat{\beta}_1$ is given by:

$$\text{Var}(\hat{\beta}_1) = \mathbb{E}[(\hat{\beta}_1 - \mathbb{E}[\hat{\beta}_1])^2] = \sigma^2\left(\sum_{i=1}^{n}(X_i - \bar{X})^2\right)^{-1} = \frac{\sigma^2}{(\sum_{i=1}^{n}(X_i - \bar{X})^2)}$$

This equation highlights that the variance of $\hat{\beta}_1$ depends inversely on the variability of X.

The MSE of $\hat{\beta}_1$ can be decomposed as follows:

$$\text{MSE}(\hat{\beta}_1) = \mathbb{E}[(\hat{\beta}_1 - \beta_1)^2] = \text{Bias}^2(\hat{\beta}_1) + \text{Var}(\hat{\beta}_1)$$

Given that $\hat{\beta}_1$ is unbiased:

$$\text{Bias}(\hat{\beta}_1) = \mathbb{E}[\hat{\beta}_1] - \beta_1 = 0$$

$$\text{MSE}(\hat{\beta}_1) = \text{Var}(\hat{\beta}_1) = \sigma^2\left(\sum_{i=1}^{n}(X_i - \bar{X})^2\right)^{-1}$$

This decomposition shows that the MSE of $\hat{\beta}_1$, in this case, is entirely due to its variance, reflecting the uncertainty due to the variability of X and the noise in the data.

Proof of variance and bias of decomposition a prediction function (MSPE)

We model the response variable Y as a function of features X, combined with some random noise:

$$Y = f(X) + \varepsilon$$

This formulation implies that to "predict" Y using features X means to find an approximation function $\hat{f}$ such that $\hat{f}(X)$ closely approximates Y.

The objective in predictive modeling is to minimize the Mean Square Prediction Error (MSPE), which quantifies how well the prediction function performs. The MSPE is defined as:

$$\text{MSPE} = \text{E}\left[(Y - \hat{f}(X))^2\right]$$

Given that Y can be expressed as $f(X) + \varepsilon$, we substitute Y in the MSPE formula to more explicitly consider the effects of the model and the noise:

$$\text{MSPE} = \text{E}\left[(f(X) + \varepsilon - \hat{f}(X))^2\right]$$

This representation allows us to decompose the error into parts directly attributable to the approximation of the model and the inherent noise in the data.

After now on, we will not use the subscript X for the sake of simplicity.

$$= \mathrm{E}\left[(f+\varepsilon-\hat{f})^2\right]$$

By adding and subtracting $\mathrm{E}[\hat{f}(X)]$, the MSPE can be rewritten as:

$$= \mathrm{E}\left[(f+\varepsilon-\hat{f}+\mathrm{E}[\hat{f}]-\mathrm{E}[\hat{f}])^2\right]$$

We can rewrite the same equation as:

$$= \mathrm{E}\left[(\underbrace{(f-\mathrm{E}[\hat{f}])}_{a}+\underbrace{\varepsilon}_{b}+\underbrace{(\mathrm{E}[\hat{f}]-\hat{f})}_{c})^2\right]$$

using $(a+b+c)^2 = a^2+b^2+c^2+2ab+2bc+2ca$, we will obtain

$$\begin{aligned} = \mathrm{E}\,[\,&\underbrace{(f-\mathrm{E}[\hat{f}])^2}_{a^2}+\underbrace{\varepsilon^2}_{b^2}+\underbrace{(\mathrm{E}[\hat{f}]-\hat{f})^2}_{c^2} \\ &+\underbrace{2(f-\mathrm{E}[\hat{f}])\varepsilon}_{2ab}+\underbrace{2\varepsilon(\mathrm{E}[\hat{f}]-\hat{f})}_{2bc}+\underbrace{2(\mathrm{E}[\hat{f}]-\hat{f})(f-\mathrm{E}[\hat{f}])}_{2ca}\,] \end{aligned}$$

Using linearity of expectation we can distribute the expectation across all terms, and taking 2's out as $E(cX) = cE(X)$ if c is constant, the MSPE can be decomposed into the following six terms:

$$\begin{aligned} = &\underbrace{\mathrm{E}\left[(f-\mathrm{E}[\hat{f}])^2\right]}_{1}+\underbrace{\mathrm{E}\left[\varepsilon^2\right]}_{2}+\underbrace{\mathrm{E}\left[(\mathrm{E}[\hat{f}]-\hat{f})^2\right]}_{3} \\ &+\underbrace{2\,\mathrm{E}\left[(f-\mathrm{E}[\hat{f}])\varepsilon\right]}_{4}+\underbrace{2\,\mathrm{E}\left[\varepsilon(\mathrm{E}[\hat{f}]-\hat{f})\right]}_{5}+\underbrace{2\,\mathrm{E}\left[(\mathrm{E}[\hat{f}]-\hat{f})(f-\mathrm{E}[\hat{f}])\right]}_{6} \end{aligned}$$

The simplification of this expression involves the following steps:

1. The first term, $\mathrm{E}\left[(f-\mathrm{E}[\hat{f}])^2\right]$, simplifies to $(f-\mathrm{E}[\hat{f}])^2$ is called bias(squared) since the expression $(f-\mathrm{E}[\hat{f}])$ (also appears in 4th and 6th term) is considered constant. This is due to f being the fixed true value and $\mathrm{E}[\hat{f}]$ representing its expected estimate. Both values remain invariant across different samples or realizations of the data, resulting in a constant difference when subtracted.
2. The second term $\mathrm{E}\left[\varepsilon^2\right]$ is σ^2, which is irreducible error or the variance of the noise.
3. The third term $\mathrm{E}\left[(\mathrm{E}[\hat{f}]-\hat{f})^2\right]$ is $(\mathrm{E}[\hat{f}]-\hat{f})^2$ which is variance of predictor, because $(\mathrm{E}[\hat{f}]-\hat{f})^2$ is a constant.

4. The fourth term $2\,\mathrm{E}[(f - \mathrm{E}[\hat{f}])\varepsilon]$ can be written as $2(f - \mathrm{E}[\hat{f}])\,\mathrm{E}[\varepsilon]$ because $(f - \mathrm{E}[\hat{f}])^2$ is a constant. As the expectation of ε is zero, this term simplifies to zero.
5. The fifth term $2\,\mathrm{E}[\varepsilon(\mathrm{E}[\hat{f}] - \hat{f})]$ can be written as $2\,\mathrm{E}[\varepsilon]\,\mathrm{E}[\mathrm{E}[\hat{f}] - \hat{f}]$. (Note that ε and $\hat{f}$ are independent). Since both the expectation of ε and second component are zero, this term also simplifies to zero.
6. The sixth term $2\,\mathrm{E}[(\mathrm{E}[\hat{f}] - \hat{f})(f - \mathrm{E}[\hat{f}])]$ also reduces to zero. Since $(f - \mathrm{E}[\hat{f}])^2$ is constant, we can rewrite this term as $2(f - \mathrm{E}[\hat{f}])\,\mathrm{E}[(\mathrm{E}[\hat{f}] - \hat{f})]$. Then, observing that $\mathrm{E}[\hat{f}]$ is the expected value of $\hat{f}$, we recognize that $\mathrm{E}[\mathrm{E}[\hat{f}] - \hat{f}] = \mathrm{E}[\hat{f}] - \mathrm{E}[\hat{f}] = 0$. Thus, the entire term simplifies to:

$2(f - \mathrm{E}[\hat{f}]) \cdot 0 = 0$

This result holds because the expectation of the deviation of $\hat{f}$ from its own expected value is zero, reflecting the property that the mean of deviations from the mean is always zero. Therefore, this term does not contribute to the MSPE.

Thus, we obtain the final simplified expression for the MSPE as follows:

$$\mathrm{MSPE} = (f - \mathrm{E}[\hat{f}])^2 + \mathrm{E}\left[\varepsilon^2\right] + \mathrm{E}\left[(\mathrm{E}[\hat{f}] - \hat{f})^2\right]$$

This result highlights the components of MSPE:

$$\mathrm{MSPE} = \mathrm{Bias}[\hat{f}]^2 + \mathrm{Var}[\hat{f}] + \sigma^2$$

where $\sigma^2 = \mathrm{E}[\varepsilon^2]$ represents the irreducible error or the variance of the noise, and the terms $\mathrm{Bias}[\hat{f}]^2$ and $\mathrm{Var}[\hat{f}]$ constitute the reducible error.

The MSPE is thus decomposed into:

$$\mathrm{MSPE} = \text{Reducible Error} + \text{Irreducible Error}$$

Another proof of bias–variance decomposition of MSPE[5]

We have a training set that includes data points and corresponding real values for each point. The data is assumed to be generated by a true function with added noise, which has zero mean and a specific variance. Our goal is to find a predictive function that approximates this true function as closely as possible using a learning algorithm and the training data. We evaluate the performance of this function by measuring how small the error is between the predicted values and the actual values, not only for the training data but also for new, unseen data points. However, due to the noise in the actual values, there will always be some level of irreducible error in our predictions. This approach allows us to use various supervised learning algorithms to find a function that generalizes well beyond the training data, with its expected error decompose into specific components based on unseen data points.

Suppose we have a training set consisting of a set of points $x_1, \ldots, x_n$ and real values y_i associated with each point x_i. We assume that the data is generated by a function $f(x)$ such as $y = f(x) + \varepsilon$, where the noise, ε, has zero mean and variance σ^2.

We want to find a function $\hat{f}(x;D)$, that approximates the true function $f(x)$ as well as possible, by means of some learning algorithm based on a training dataset (sample) $D = \{(x_1, y_1), \dots, (x_n, y_n)\}$. We make "as well as possible" precise by measuring the mean squared error between y and $\hat{f}(x;D)$: we want $(y - \hat{f}(x;D))^2$ to be minimal, both for $x_1, \dots, x_n$ and for points outside of our sample. Of course, we cannot hope to do so perfectly, since the y_i contain noise ε; this means we must be prepared to accept an irreducible error in any function we come up with.

Finding an $\hat{f}$ that generalizes to points outside of the training set can be done with any of the countless algorithms used for supervised learning. It turns out that whichever function $\hat{f}$ we select, we can decompose its expected error on an unseen sample x (i.e., conditional to x) as follows:

The derivation of the bias–variance decomposition for squared error proceeds as follows. For notational convenience, we abbreviate $\hat{f} = f(x;D)$ and we drop the D subscript on our expectation operators.

Let us write the MSE of our model:

$$\text{MSE} \triangleq \text{E}\left[(y - \hat{f})^2\right] = \text{E}\left[y^2 - 2y\hat{f} + \hat{f}^2\right] = \text{E}\left[y^2\right] - 2\text{E}[y\hat{f}] + \text{E}\left[\hat{f}^2\right]$$

First, since we model $y = f + \varepsilon$, we show that

$$\begin{aligned}
\text{E}\left[y^2\right] &= \text{E}\left[(f+\varepsilon)^2\right] \\
&= \text{E}\left[f^2\right] + 2\text{E}[f\varepsilon] + \text{E}\left[\varepsilon^2\right] && \text{by linearity of E} \\
&= f^2 + 2f\text{E}[\varepsilon] + \text{E}\left[\varepsilon^2\right] && \text{since } f \text{ does not depend on the data} \\
&= f^2 + 2f \cdot 0 + \sigma^2 && \text{since } \varepsilon \text{ has zero mean and variance } \sigma^2
\end{aligned}$$

Second,

$$\begin{aligned}
\text{E}[y\hat{f}] &= \text{E}[(f+\varepsilon)\hat{f}] && \text{by linearity of E} \\
&= \text{E}[f\hat{f}] + \text{E}[\varepsilon\hat{f}] && \text{since } \hat{f} \text{ and } \varepsilon \text{ are independent} \\
&= \text{E}[f\hat{f}] + \text{E}[\varepsilon]\text{E}[\hat{f}] && \text{since E}[\varepsilon] = 0 \\
&= f\text{E}[\hat{f}]
\end{aligned}$$

Last,

$\mathbb{E}[\hat{f}^2] = \text{Var}(\hat{f}) + \mathbb{E}[\hat{f}]^2 \text{since} \text{Var}(X) = \mathbb{E}[(X - \mathbb{E}[X])^2] = \mathbb{E}[X^2] - \mathbb{E}[X]^2$for any random variable X.

Eventually, we plug these three formulas in our previous derivation of MSE and thus show that:

$$\text{MSE} = f^2 + \sigma^2 - 2f\mathbb{E}[\hat{f}] + \text{Var}(\hat{f}) + \mathbb{E}[\hat{f}]^2$$

$$= (f - \mathbb{E}[\hat{f}])^2 + \sigma^2 + \text{Var}(\hat{f})$$

$$= \text{Bias}[\hat{f}]^2 + \sigma^2 + \text{Var}[\hat{f}]$$

Finally, MSE loss function (or negative log-likelihood) is obtained by taking the expectation value over $x \sim P$:

$$\text{MSE} = \mathbb{E}_x[\text{Bias}[\hat{f}(x;D)]^2 + \text{Var}[\hat{f}(x;D)] + \sigma^2]$$

Notes

1. See Gelman, A. (2005). Analysis of Variance—Why It Is More Important Than Ever. The Annals of Statistics, 33(1), 1–53. http://www.stat.columbia.edu/gelman/research/published/AOS259.pdf
2. Read [Statistical Inference](https://en.wikipedia.org/wiki/Statistical_inference)
3. https://en.wikipedia.org/wiki/Central_limit_theorem
4. https://pressbooks.bccampus.ca/simplestats/chapter/6-6-the-central-limit-theorem/
5. https://en.wikipedia.org/wiki/Bias%E2%80%93variance_tradeoff

5

Bias–Variance Trade-off

In this chapter, we explore the bias–variance trade-off and its critical role in model performance. In this chapter, we still assume prediction functions (i.e., we assume we know the prediction functions in simulations) to find mean square prediction error (MSPE) using the given sample dataset. We use single data set to calculate MSPE for that given data set (in-sample MSPE for a test data). In this chapter, we show calculating bias–variance trade-off for each assumed function using the sample data. Bias–variance trade-off, while complex, is reflected in everyday life situations, offering insights into its practical significance. We begin with the story of Baron, whose experiences in planning his biweekly work trips vividly illustrate this trade-off.

Baron aims to catch a flight for his biweekly work trips, with the gate closing at 8 AM. During his first year, he developed a routine to leave his house around 7:00 AM, planning to depart either 30 minutes before or after this time. However, he discovered that this approach sometimes led to him missing his flight. Even though his average or expected departure time from home was 7:00 AM, this process occasionally resulted in arrival times after the flight's gate closure, making his goal of catching the flight more error-prone. This approach represented a scenario with higher variance and lower bias, leading to costly errors concerning his airport arrival time and missed flights. In contrast, with his revised preparation routine, Baron aimed to leave around 7:00 AM but allowed a margin of 10 minutes either way. Consequently, his average departure time shifted to 7:05 AM, introducing a 5-minute bias. However, this new strategy led to more consistent success in catching his flight, reflecting higher bias but lower variance. This approach nearly eliminated all instances of missing the flight. This whole experience demonstrates the bias–variance trade-off. In the first case, a preparation process with higher variance and lower bias resulted in occasional flight misses. In the second case, adopting a process with reduced variance, even though it introduces a degree of bias, ultimately improves the overall result. Baron's experience illustrates that consistently opting for lower bias (even unbiased), is not always the best strategy to predict his arrival time, as a balance between bias and variance can lead to more effective real-world results.

As seen from Baron's narrative, the bias–variance trade-off is not an abstract statistical concept but a tangible factor in our everyday decision-making processes. Grasping this balance is key to understanding how to craft models that are not only accurate but also reliable across various situations. In the next chapter, we'll explore another critical concept in machine learning: overfitting. This concept is also very crucial for crafting precise and dependable models. We will explore how overfitting relates to and differs from the bias–variance trade-off.

We already discuss bias and variance of estimator and predictor, and decomposition of MSE for estimation and MSPE for prediction in the previous chapter. Now, lets discuss what is bias and variance and trade-off between them in predictive models. The bias–variance trade-off is a fundamental concept in statistical learning and machine learning, which deals with the problem of minimizing and balancing two sources of error that can affect the performance of a model. Bias occurs from oversimplifying the model, leading to underfitting

DOI: 10.1201/9781003504795-5

and missing relevant relations in the data. Variance occurs from overcomplicating the model, leading to overfitting and capturing random noise instead of the intended outputs. High bias results in poor performance on training data, while high variance causes poor generalization to new data. The goal is to find a balance, creating a model that is complex enough to capture true patterns but simple enough to generalize well. This balance is achieved through choosing appropriate model complexity, using techniques like cross-validation, applying regularization, and refining data or features.

Following is a formal discussion of bias–variance trade-off.

5.1 Bias–Variance Trade-off

In this section, we will discuss the formal definition of bias–variance trade-off. We will also discuss the MSPE and its components, including bias, variance, reducible error, and irreducible error. We will explore how these components interact to influence the overall prediction accuracy of a model.

To remind, our main task is prediction of an outcome, Y using the data (more accurately test/train data). To "predict" Y using features X, means to find some f which is close to Y. We assume that Y is some function of X plus some random noise.

$$Y = f(X) + \epsilon \tag{5.1}$$

where $\mathbb{E}[\epsilon] = 0$ and $\mathbb{V}[\epsilon] = \sigma^2$. In this formulation, we call $f(X)$ the signal and ϵ the noise, i.e irreducible error. Moreover, we can never know real $f(X)$. Thus our goal becomes to finding some $f(\hat{X})$ that is a good estimate of the regression function $f(X)$. There will be always difference between real $f(X)$ and $f(\hat{X})$. That difference is called reducible error. To be able to decrease reducibl error as well as to obtain a better prediction function, $f(\hat{X})$, which is an estimate of unknown, $f(X)$, we can use multiple sample data, which is random from population data, and estimate a function and find its expected value. In Chapter 5, we showed the MSPE and their decomposition. Using various samples we find different estimate of function , we attempt to decrease expected MSPE. We want to emphasize eventhough process looks like similar estimate unknown parameters using different representative samples to estimate parameter with each sample and then find the expected value of parameters. The main difference is instead of assuming same function to estimate parameters like in ordinary least square (OLS), Here we estimate different prediction function with each sample and try to estimate unknown function by calculating expected value of the estimated functions. We find a good estimate of $f(X)$ by reducing the expected mean square of error with each test data as much as possible (when Y is continuous. when it is binaty, we will use another method which we will cover in classification chapters in detail). To be able to have a variance of prediction functions, we need to use different samples to estimate the prediction function thus have different prediction functions. Below, we will show step by step the bias–variance trade-off for different polynomial degrees using the same sample data.

Thus, we can write MSPE for prediction function using one training sample (for test data, validation sample):

$$\text{MSPE}(Y - f(\hat{X}))^2) = (\text{Bias}[f(\hat{X})])^2 + \text{Var}[f(\hat{X})] + \sigma^2 \tag{5.2}$$

Then, MSPE (using multiple test data) can be written as:

$$\mathbb{E}[\text{MSPE}] = \mathbb{E}\left[(Y - \hat{f}(X))^2\right] = \underbrace{\mathbb{E}\left[(f(x) - \mathbb{E}[\hat{f}(x)])^2\right]}_{\text{bias}^2(\hat{f}(x))} + \underbrace{\mathbb{E}\left[(\hat{f}(x) - \mathbb{E}[\hat{f}(x)])^2\right]}_{\text{var}(\hat{f}(x))} + \sigma^2 \tag{5.3}$$

where $\sigma^2 = E[\varepsilon^2]$

More formally, we can write this as:

$$\text{MSPE}\left(Y, \hat{f}(X)\right) = \mathbb{E}_{X,Y,\mathcal{D}}\left[(Y - \hat{f}(X))^2\right] = \underbrace{\mathbb{E}_X\left[\text{bias}^2\left(\hat{f}(X)\right)\right] + \mathbb{E}_X\left[\text{var}\left(\hat{f}(X)\right)\right]}_{\text{reducible error}} + \sigma^2 \tag{5.4}$$

where Y is the outcome data we observe from the seen data and want to predict the ones from unseen data, $\hat{f}(X)$ is the prediction function, $\mathcal{D}$ is random sample from (X, Y), σ^2 is the irreducible error, $\text{bias}^2\left(\hat{f}(X)\right)$ is the squared bias of the prediction function, $\text{var}\left(\hat{f}(X)\right)$ is the variance of the prediction function, and reducible error is the sum of the squared bias and variance of the prediction function. The reducible error represents the error that can be minimized through model selection, while the irreducible error is the error that cannot be reduced due to the inherent noise in the data. The goal is to find a model that minimizes the reducible error while balancing the bias and variance of the prediction function. To simplify we donot use subscripts for the expected value of the functions.

The expected MSPE on the validation/training sample consists of the sum of the squared bias of the fit and the variance of both the fit and the error/noise term. The error term σ^2, also referred to as irreducible error or uncertainty, represents the variance of the target variable Y around its true mean $f(x)$. It is inherent in the problem and does not depend on the model or training data. When the data generation process is known, σ^2 can be determined. Alternatively, estimation of σ^2 is possible using the sample variance of y at duplicated (or nearby) inputs x.

Variance is the amount by which $\hat{f(X)}$ could change if we estimated it using different test/training data set. $\hat{f(X)}$ depends on the training data. (More complete notation would be $\hat{f}(X; Y_{train}, X_{train})$)If the $\hat{f(X)}$ is less complex/less flexible, then it is more likely to change if we use different samples to estimate it. However, if $\hat{f(X)}$ is more complex/more flexible function, then it is more likely to change between different test/training samples.

For instance, lets assume we have a data set (test or training data) and we want to "predict" Y using features X, thus estimate function $f(X)$. and lets assume we have 1-degree and 10-degree polynomial functions as a potential prediction functions. We say $\hat{f(X)} = \hat{\beta_0} + \hat{\beta_1}X$ is less complex than 10-degree polynomial $\hat{f(X)} = \hat{\beta_0} + \hat{\beta_1}X ... + \hat{\beta_{10}}X^{10}$ function. As, 10-degree polynomial function has more parameters $\beta_0, ..\beta_{10}$, it is more flexible. That also means it has high variance (as it has more parameters, all these parameters are more inclined to have different values in different training datasets). Thus, a prediction function has high variance if it can change substantially when we use different training samples to estimate $f(X)$. We can also say less flexible functions (functions with less parameters) have low variance. Low variance functions are less likely to change when we use different training sample or adding new data to the test sample. We will show all these with simulation in overfitting chapter as well.

Bias is the difference between the real! prediction function and expected estimated function. If the $f(\hat{X})$ is less flexible, then it is more likely to have higher bias. We can think this as real function (reality) is always more complex than the function approximates the reality. So, it is more prone to have higher error, i.e., more bias.

In the context of regression, parametric models are biased when the form of the model does not incorporate all the necessary variables, or the form of the relationship is too simple. For example, a parametric model assumes a linear relationship, but the true relationship is quadratic. In non-parametric models when the model provides too much smoothing.

There is a bias–variance trade-off. That is, often, the more bias in our estimation, the lesser the variance. Similarly, less variance is often accompanied by more bias. Flexible (i.e., complex) models tend to be unbiased, but highly variable. Simple models are often extremely biased, but have low variance.

So for us, to select a model that appropriately balances the tradeoff between bias and variance, and thus minimizes the reducible error, we need to select a model of the appropriate flexibility for the data. However, this selected model should not overfit the data as well which we will discuss in the next chapter.[1]

5.1.1 Simulation for Polynomial Regression

Remember that in real cases, we have data set which consist from outcome Y and features X. We can use different samples to estimate the prediction function, $f(\hat{X})$, and then find the expected value of the estimated functions. We can obtain these samples by splitting data using different methods which we will discuss in future. In this simulation, we will use the same sample data to estimate the prediction function for different polynomial degrees. We will use polynomial regression to estimate the function with different polynomial degrees ranging from 1 to 10. We will calculate the bias, variance, reducible error, and irreducible error for each polynomial degree. We will also calculate the expected in-sample MSPE for each polynomial degree. The simulation will provide insights into how the bias–variance trade-off influences the prediction accuracy of the model and the importance of selecting an appropriate model complexity. We want to remind the prediction function is polynomial function and we know the real function in this simulation. In real cases, we do not know the real prediction function.

First lets discuss data generation process. We start by defining a fixed set of input values, X, drawn from a normal distribution with mean 0 and standard deviation 1. These values are consistent across repeated samples to ensure comparability. Using these input values, we simulate a true function, f, which represents the underlying relationship between X and the dependent variable Y. The true function is specified as a cubic polynomial with coefficients $b_0 = 1$, $b_1 = 2$, $b_2 = -2$, and $b_3 = 3$. To remind, this true function is unknown in reality (outside simulation) and this is the prediction function we are trying to estimate. Additionally, random errors drawn from a normal distribution with mean 0 and standard deviation σ, which is 8 in this simulation, are added to the true function to generate the observed values of Y.

If we want to replicate the simulation using the new data and function that we described in Section 5.4, which demonstrates reducible and irreducible errors as well as MSPE, the results below will be observed. The variance is zero because we use the same sample (both X and Y) and only one function for prediction; therefore, there is no variance in the prediction function. The irreducible error is close to 60, but it can approach 64 if you increase n in the simulation program. These results can be achieved by changing the number of simulations, M, to 1 in the simulation program provided below. We have presented these results here because some sources use this simulation to illustrate the bias–variance trade-off. However, to estimate the prediction function more accurately using multiple datasets, we will need more than one function, which we discuss below.

```
# Function for X - fixed at repeated samples
xfunc1 <- function(n) {
  set.seed(123)
  x_1 <- rnorm(n, 0, 1)
  return(x_1)
}
# Defining a function for simulation
simmse <- function(M, n, sigma, poldeg) {
  x_1 <- xfunc1(n) # Repeated X's in each sample
  # Containers
  MSPE <- rep(0, M)
  Irrerr <- rep(0, M)
  yhat <- matrix(0, M, n)
  olscoef <- matrix(0, M, poldeg + 1)
  ymat <- matrix(0, M, n)
  xmat <- matrix(0, M, n) # to store x values
  f_values <- matrix(0, M, n) # to store f(x) values
  # Loop for samples
  for (i in 1:M) {
    b0 <- 1; b1 <- 2; b2 <- -2; b3 <- 3
    f <- b0 + b1 * x_1 + b2 * (x_1^2) + b3 * (x_1^3)
    RandomError <- rnorm(n, 0, sigma)
    y <- f + RandomError
    Irrerr[i] <- mean((y - f)^2)
    samp <- data.frame("y" = y, x_1)
    xmat[i, ] <- x_1
    f_values[i, ] <- f
    # Estimator
    ols <- lm(y ~ poly(x_1, degree = poldeg, raw = TRUE), samp)
    olscoef[i, ] <- ols$coefficients
    # Yhat's: This is equivalent to fhat for OLS
    yhat[i, ] <- predict(ols, newdata = samp)
    # MSPE - That is, residual sum of squares
    MSPE[i] <- mean((ols$residuals)^2)
    ymat[i, ] <- y
  }
  list(MSPE = MSPE, yhat = yhat, sigma = sigma,
       olscoef = olscoef, f = f, ymat = ymat, xmat = xmat,
       f_values = f_values, Irrerr = Irrerr)
}
# Running different fhat with different polynomial degrees
# Choose certain parameters: M, n, Maxpoldeg, sigma
```

TABLE 5.1 In-sample MSPE Decomposition of a Function

	Bias^2	Variance	Red.Err.	Irreduc.Err.	In-sample MSPEs
PolyDeg1	23.3361	0	23.3361	59.9858	86.5760
PolyDeg2	20.5471	0	20.5471	59.9858	79.5635
PolyDeg3	1.5780	0	1.5780	59.9858	58.4078
PolyDeg4	2.2781	0	2.2781	59.9858	57.7078
PolyDeg5	2.2840	0	2.2840	59.9858	57.7018
PolyDeg6	2.3913	0	2.3913	59.9858	57.5945
PolyDeg7	2.7805	0	2.7805	59.9858	57.2053
PolyDeg8	4.2395	0	4.2395	59.9858	55.7463
PolyDeg9	4.2399	0	4.2399	59.9858	55.7459
PolyDeg10	4.7932	0	4.7932	59.9858	55.1926

```r
M <- 1
n <- 100
Maxpoldeg <- 10
sigma <- 8
results <- lapply(1:Maxpoldeg, function(degree)
  simmse(M, n, sigma, degree))
# Table
tab <- matrix(0, Maxpoldeg, 5)
row.names(tab) <- paste0("PolyDeg", 1:Maxpoldeg)
colnames(tab) <- c("Bias^2", "Variance", "Red.Err.",
                   "Irreduc.Err.", "In-sample MSPEs")
f <- results[[1]]$f
fmat <- matrix(f, nrow = M, ncol = n, byrow = TRUE)
# Bias^2 = (mean(yhat) - f)^2 = mean((mean(fhat) - f)^2)
tab[, 1] <- sapply(results, function(res)
  mean((apply(res$yhat, 2, mean) - f)^2))
# Var(yhat) = var(fhat) = mean((fhat - mean(fhat))^2)
tab[, 2] <- sapply(results, function(res)
  mean(apply(res$yhat, 2, function(x)
    mean((x - mean(x))^2))))
# Reducible Error = mean((f - fhat)^2)
tab[, 3] <- sapply(results, function(res)
  mean(colMeans((fmat - res$yhat)^2)))
# Irreducible error = mean((y - f)^2)
tab[, 4] <- sapply(results, function(res) mean(res$Irrerr))
# MSPE = mean((y - yhat)^2)
tab[, 5] <- sapply(results, function(res)
  mean(colMeans((res$ymat - res$yhat)^2)))
knitr::kable(round(tab, 4), align = "c",
             caption = "In-Sample MSPE decomposition of a function")
```

To be able to estimate this unknown true function, we fit polynomial regression models of varying degrees (ranging from 1 to 10) to the simulated data. The polynomial regression models are estimated using OLS regression, with the degree of the polynomial specified as a parameter. The fitted models, denoted as $\hat{f}$, are used to predict the values of Y based on the input X. We calculate the MSPE for each model, which measures the average squared difference between the predicted values and the actual observed values. This provides an indication of the prediction accuracy of each model.

Using this setup, we generate 50 datasets, $\mathcal{D}$, with a sample size $n = 100$ and fit 10 models.

$$\hat{f}_1(x) = \hat{\beta}_0 + \hat{\beta}_1 x$$

$$\hat{f}_2(x) = \hat{\beta}_0 + \hat{\beta}_1 x + \hat{\beta}_2 x^2$$

$$\hat{f}_3(x) = \hat{\beta}_0 + \hat{\beta}_1 x + \hat{\beta}_2 x^2 + \hat{\beta}_3 x^3$$

$$\vdots$$

$$\hat{f}_{10}(x) = \hat{\beta}_0 + \hat{\beta}_1 x + \hat{\beta}_2 x^2 + \ldots + \hat{\beta}_{10} x^{10}$$

So, using dataset1, we estimate beta parameters for each polynomial degree. Then we use these beta parameters to estimate the prediction function for each polynomial degree. We repeat this process for 50 datasets. We calculate the expected value of the estimated prediction functions. We calculate the bias, variance, reducible error, and irreducible error for each polynomial degree. We also calculate the expected in-sample MSPE for each polynomial degree. We will explain this process step by step below.

```
# Function for X - fixed at repeated samples
xfunc2 <- function(n) {
  set.seed(123)
  x_1 <- rnorm(n, 0, 1)
  return(x_1)
}
# Defining a function for simulation
simmse <- function(M, n, sigma, poldeg) {
  x_1 <- xfunc2(n) # Repeated X's in each sample
  # Containers
  MSPE <- rep(0, M)
  Irrerr <- rep(0, M)
  yhat <- matrix(0, M, n)
  olscoef <- matrix(0, M, poldeg + 1)
  ymat <- matrix(0, M, n)
  xmat <- matrix(0, M, n) # to store x values
  f_values <- matrix(0, M, n) # to store f(x) values
  # Loop for samples
  for (i in 1:M) {
    b0 <- 1; b1 <- 2; b2 <- -2; b3 <- 3
    f <- b0 + b1 * x_1 + b2 * (x_1^2) + b3 * (x_1^3)
    RandomError <- rnorm(n, 0, sigma)
    y <- f + RandomError
    Irrerr[i] <- mean((y - f)^2)
    samp <- data.frame("y" = y, x_1)
```

```
    xmat[i, ] <- x_1
    f_values[i, ] <- f
    # Estimator
    ols <- lm(y ~ poly(x_1, degree = poldeg, raw = TRUE), samp)
    olscoef[i, ] <- ols$coefficients
    # Yhat's: This is equivalent to fhat for OLS
    yhat[i, ] <- predict(ols, newdata = samp)
    # MSPE - That is, residual sum of squares
    MSPE[i] <- mean((ols$residuals)^2)
    ymat[i, ] <- y
  }
  list(MSPE = MSPE, yhat = yhat, sigma = sigma, olscoef = olscoef, f = f,
       ymat = ymat, xmat = xmat, f_values = f_values, Irrerr = Irrerr)
}
# Running different fhat with different polynomial degrees
# Choose certain parameters: M, n, Maxpoldeg, sigma
M <- 50
n <- 100
Maxpoldeg <- 10
sigma <- 8
results <- lapply(1:Maxpoldeg, function(degree) simmse(M, n, sigma, degree))
# Table
tab <- matrix(0, Maxpoldeg, 5)
row.names(tab) <- paste0("PolyDeg", 1:Maxpoldeg)
colnames(tab) <- c("Bias^2", "Variance", "Red.Err.",
                   "Irreduc.Err.", "In-sample MSPEs")
f <- results[[1]]$f
fmat <- matrix(f, nrow = M, ncol = n, byrow = TRUE)
# Bias^2 = (mean(yhat) - f)^2 = mean((mean(fhat) - f)^2)
tab[, 1] <- sapply(results, function(res)
  mean((apply(res$yhat, 2, mean) - f)^2))
# Var(yhat) = var(fhat) = mean((fhat - mean(fhat))^2)
tab[, 2] <- sapply(results, function(res)
  mean(apply(res$yhat, 2, function(x) mean((x - mean(x))^2))))
# Reducible Error = mean((f - fhat)^2)
tab[, 3] <- sapply(results, function(res)
  mean(colMeans((fmat - res$yhat)^2)))
# Irreducible error = mean((y - f)^2)
tab[, 4] <- sapply(results, function(res) mean(res$Irrerr))
# MSPE = mean((y - yhat)^2)
tab[, 5] <- sapply(results, function(res)
  mean(colMeans((res$ymat - res$yhat)^2)))
knitr::kable(round(tab, 4), align = "c", caption = "InSample MSPE decomposition")
```

We explore the simulation designed to analyze the MSPE and its components. Initially, we generate a fixed set of predictor values, x, from a normal distribution with mean zero and standard deviation one, using a seed to ensure reproducibility. The primary simulation function takes parameters: number of simulations M, sample size n, standard deviation of the error term, σ, and polynomial degree *poldeg*. This function maintains consistency by using the same x values across simulations. Within this function, we initialize containers for

TABLE 5.2 In-sample MSPE Decomposition

	Bias^2	Variance	Red.Err.	Irreduc.Err.	In-sample MSPEs
PolyDeg1	22.4610	1.1039	23.5650	63.2363	87.7267
PolyDeg2	19.0787	1.6748	20.7535	63.2363	82.9245
PolyDeg3	0.1283	2.1724	2.3007	63.2363	60.9356
PolyDeg4	0.1292	2.9782	3.1075	63.2363	60.1289
PolyDeg5	0.1584	3.5336	3.6919	63.2363	59.5444
PolyDeg6	0.1692	4.0437	4.2129	63.2363	59.0234
PolyDeg7	0.1695	4.6262	4.7957	63.2363	58.4406
PolyDeg8	0.1730	5.3350	5.5080	63.2363	57.7283
PolyDeg9	0.2827	6.0106	6.2933	63.2363	56.9430
PolyDeg10	0.2907	6.7212	7.0118	63.2363	56.2245

storing results like MSPE, irreducible error, predicted values, OLS coefficients, and actual values of the dependent variable and predictor.

For each simulation, we define a true polynomial relationship between y and x, add random error to create observed y values, and calculate the irreducible error as the mean squared difference between observed and true y values. We fit an OLS regression to the data, store the coefficients, calculate the predicted values, and compute bias squared, variance, reducible and irreducible error to decompose the MSPE. We also compute the MSPE.

By analyzing simulation results, we find that higher-degree polynomial models tend to have lower MSPE due to their ability to better fit the data. However, this comes at the cost of increased model complexity and potential overfitting. Polynomial degree 3 exhibits the smallest reducible error, at 2.3007, suggesting it as the "true" prediction function. However, it is important to note that we cannot solely rely on reducible error for model validation, as it requires the true function, which is unobservable in practice. We can only calculate the MSPE with real data, and finding the lowest MSPE does not conclusively identify the true model because we must also consider the risk of overfitting, which will be discussed in the next chapter. As seen in this simulation, polynomial degree 10 has the lowest MSPE.

To visualize the results and illustrate each step in the simulation process, including how we calculate the MSPE and its components, we present figures from the initial three datasets (each containing 100 data points) for polynomial degrees 1 and 10. It is evident that as the degree of the polynomial prediction function increases, so does the variance of the function. This is noticeable in the increased "wiggly" appearance of the 10th-degree prediction function. As the complexity of the function increases—reflected by a higher number of parameters—the variance of the prediction function also rises. In the figures, gray lines represent polynomial degree 1 and 3, while darker gray corresponds to degree 10.[2]

Additionally, we generate plots that display the fitted polynomial regression lines for degrees 1, 3, and 10 separately across all 50 different datasets (training datasets). We plot the 50 fitted regression lines for polynomial degree 1 $\hat{f}$ in a single graph, along with the expected $\hat{f}$, which is the average of the 50 fitted regression lines. We do the same for ploynomial degree 3 and 10. You can visualize different polynomial functions by chooosing M number of data set, n number of observations in each dataset, standard deviation, or polynomial degree in command, results_deg1 <- simmse(50, 100, 8, 1), or changing betas or function f. From these figures, we can observe that when degree of polynomial is 1, variation in all prediction functions is lower than when polynomial degree is 10. As seen when polynomial degree is 10, different prediction functions obtained by different training sets have more variation, more

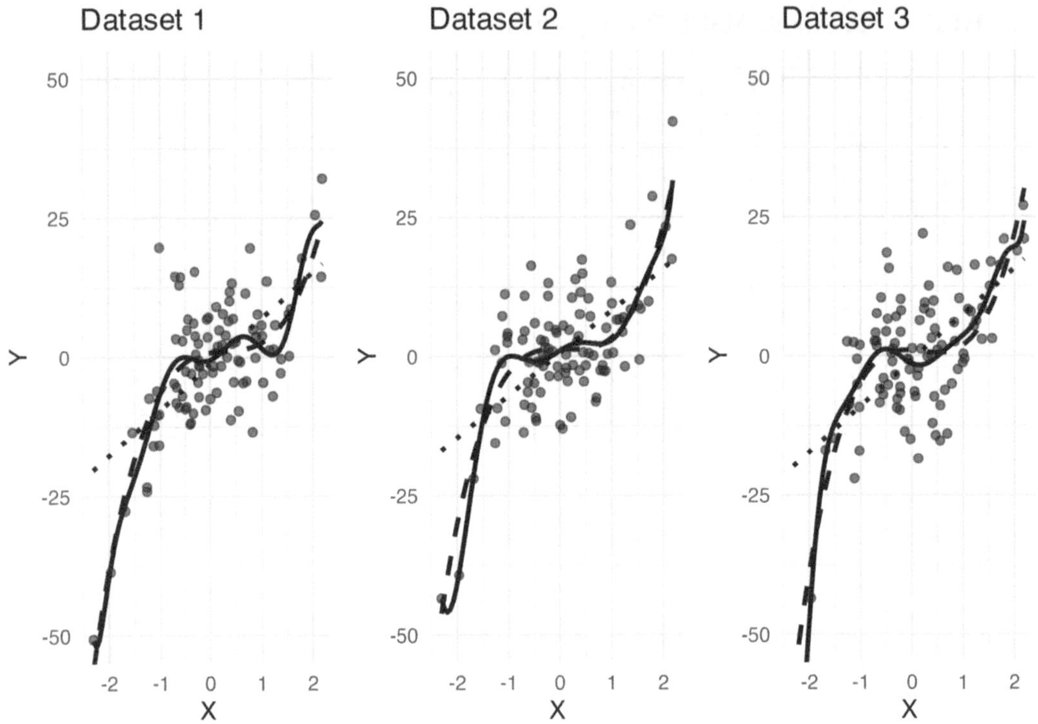

FIGURE 5.1 Fitted polynomials of degree 1, 3, and 10 across three datasets.

"wiggly" as complexity of function increases, i.e., has more parameters which may lead vary. The bold black lines in each figure indicates expected value these 50 different fhats which we will show how we calculate step by step below. Most sources do not provide this detailed explanation. This helps illustrate the trade-off between model complexity and prediction accuracy, emphasizing the importance of selecting an appropriate model that balances bias and variance.

To demonstrate how the simulation program operates, we will now explain the program step-by-step for a polynomial function of degree 3. Keep in mind that the simulation follows the same steps for polynomial degrees ranging from 1 to 10 (the maximum polynomial degree can be adjusted).

First, run the `simmse` function, starting from the `xfunc2` line at the top, up to selecting parameters in the snippet from the program above. Then, execute the simulation for a polynomial degree of 3 and calculate all steps by running the program snippets provided below.

To run the simulation, we set the following parameters: $M = 50$, $n = 100$, poldeg $= 3$, and $\sigma = 8$. These parameters define the number of simulations, the sample size, the degree of the polynomial used in the regression model, and the standard deviation of the error term, respectively. Additionally, we have a flag `show_all` to control the display of results which is set to `FALSE` by default.

After defining these parameters, we call the `simmse` function to perform the simulation. This function returns several results, including the predictor values (x), the true function values ($f(x)$), and the observed values (y). As we mentioned, we generated a fixed set of 100 input values, X, drawn from a normal distribution with mean 0 and standard deviation 1. These

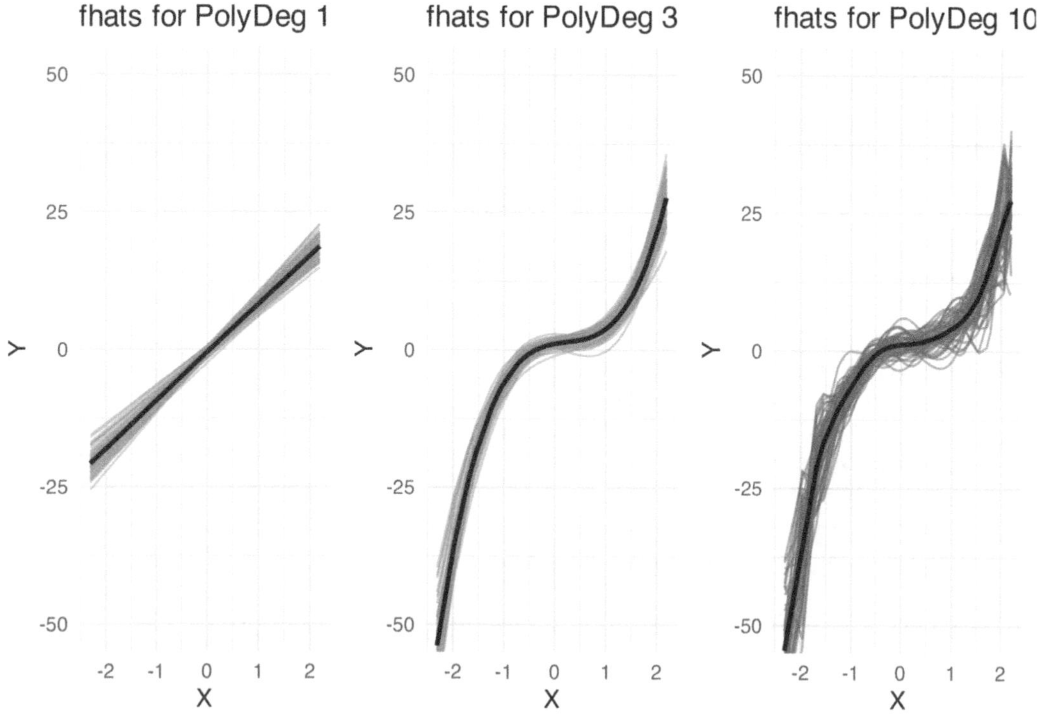

FIGURE 5.2 Estimated fhats from 50 samples using polynomial regression of degrees 1, 3, and 10. The black line shows the average fitted values. Gray lines reflect individual sample estimates.

values of X are consistent across repeated samples to ensure comparability in this simulation. However, we can use a different set of (X_i, Y_i) as training samples. Using these input values, we generated the true function, f, which represents the underlying relationship between X and the dependent variable Y. The true function is specified as $f_3(x) = 1 + 2x - 2x^2 + 3x^3$. To remind, this true function is unknown in reality (outside simulation) and this is the prediction function we are trying to estimate, and these values can be changed within the `simmse` function. Additionally, random errors drawn from a normal distribution with mean 0 and standard deviation σ, which is 8 in this simulation (which makes the irreducible error its square 64), are added to the true function to generate the observed values of Y_{sim1}.

For instance, $x_1 = -0.5604756$ then $f(x_1) = -1.2774088$ using the specified true function. Then we add the 1st value from the 100 random errors generated for simulation 1 (these error terms are not reported in this simulation) and obtain $y(x_1) = f(x_1) + \epsilon_{1_{\text{sim1}}} = -6.9606613$. We do this for all 100 values of X in simulation 1. Repeating this process 50 times for each simulated error term, we generated 50 different training datasets such as (X, Y_{sim1}),..., (X, Y_{sim50}). We then extract and round the x, $f(x)$, and y values from the simulation results for better readability. These values are organized into matrices for further analysis. Finally, we combine the x, $f(x)$, and y matrices into a single output matrix and print it below.

```
# Running the simulation
M <- 50
n <- 100
poldeg <- 3
```

```
sigma <- 8
show_all <- FALSE  # Set to TRUE to display all results

results <- simmse(M, n, sigma, poldeg)

# Extract x and f(x) values from each simulation
x_values <- round(results$xmat, 7)
f_values <- round(results$f_values, 7)
y_values <- round(results$ymat, 7)
# Create a matrix for (x, f(x)) values
if (show_all || n <= 5) {
  x_fx_matrix <- rbind(x_values[1,], f_values[1,])
  colnames(x_fx_matrix) <- paste0("X", 1:n)
  rownames(x_fx_matrix) <- c("X", "f(X)")
  # Add y_simulation rows
  y_simulation_rows <- y_values[1:M,]
  row.names(y_simulation_rows) <- c(paste0("y_sim", 1:M))
} else {
  x_fx_matrix <- rbind(
    x_values[1, c(1:3, n)],
    f_values[1, c(1:3, n)]
  )
  x_fx_matrix <- cbind(x_fx_matrix[,1:3], "...", x_fx_matrix[,4])
  colnames(x_fx_matrix) <- c(paste0("X", 1:3), "...", paste0("X", n))
  rownames(x_fx_matrix) <- c("X", "f(X)")
  # Add y_simulation rows
  y_simulation_rows <- rbind(
    y_values[1, c(1:3, n)],
    y_values[2, c(1:3, n)],
    y_values[3, c(1:3, n)],
    rep("...", 5),
    y_values[M, c(1:3, n)]
  )
  y_simulation_rows <- cbind(y_simulation_rows[,1:3], "...", y_simulation_rows[,4])
  colnames(y_simulation_rows) <- c(paste0("X", 1:3), "...", paste0("X", n))
  rownames(y_simulation_rows) <- c("y_sim1", "y_sim2", "y_sim3", "...",
                                   paste0("y_sim", M))
}
# Combine matrices for output
output_matrix <- rbind(x_fx_matrix, y_simulation_rows)
# Print the (x, f(x), y) matrix
knitr::kable(output_matrix, format = "latex", booktabs = TRUE, align = "c")
```

	X1	X2	X3	...	X100
X	−0.5604756	−0.2301775	1.5587083	...	−1.0264209
f(X)	−1.2774088	0.3970961	10.6192538	...	−6.4040475
y_sim1	−6.9606613	2.4521658	8.6457188	...	−15.8878882
y_sim2	16.313074	10.8963999	8.4980934	...	3.595269
y_sim3	−6.9993463	−5.6244156	3.1109442	...	−9.2403868
...	...	...	...	...	...
y_sim50	−5.2307998	9.4178438	1.4436574	...	−1.8972215

After obtaining 50 training datasets, we will discuss how each value of MSPE and its components is computed in the next program snippet. Let's continue to Using (X, Y_{sim1}) data, `simmsefunction` estimate cubic polynomial function, find and store the estimated coefficients, then calculate $\hat{f}_3(x) = \hat{\beta}_0 + \hat{\beta}_1 x + \hat{\beta}_2 x^2 + \hat{\beta}_3 x^3$ value using x_1 in simulation 1 data, X_1, then in simulation 1 data, x_2 in simulation 1 data, X_1, upto $\times 100$. Save it as a row in simulation 1 data, s_1, which is named as in simulation 1 data, y_{sim1} in the output table. This is done for all 50 training sample data. Thus, we have a data frame like following:

	x_1	x_2	$\cdots$	x_{100}
s_1	$\hat{f}(x_1)$	$\hat{f}(x_2)$	$\cdots$	$\hat{f}(x_{100})$
s_2	$\hat{f}(x_1)$	$\hat{f}(x_2)$	$\cdots$	$\hat{f}(x_{100})$
$\vdots$	$\vdots$	$\vdots$	$\ddots$	$\vdots$
s_{50}	$\hat{f}(x_1)$	$\hat{f}(x_2)$	$\cdots$	$\hat{f}(x_{100})$

Then, we calculate the expected value of the predicted $\hat{f}$ for all x values. Next, we determine the variance of the predicted $\hat{f}$ values for all x values. We also calculate the squared bias for all x values. To present the results, we create a data frame that includes the predicted values for each simulation, the sum of the predicted values, the expected predicted values, the variance, and the squared bias for the selected x values to calculate the MSPE and its components. For better readability, we add a column with ellipses ("...") to indicate the continuation of values between x_3 and x_n. We then print the data frame to display the results.

```
# Calculate expected fhat for all x values
expected_fhat_all <- apply(results$yhat, 2, mean)
# Calculate sum of fhats for all x values
sum_fhat_all <- colSums(results$yhat)
# Calculate variance of fhat for all x values
variance_fhat_all <- apply(results$yhat, 2,
                function(x) var(x) * (length(x) - 1) / length(x))
# Calculate squared bias for all x values
squared_bias_all <- (expected_fhat_all - results$f_values[1,])^2
# Select specific x values (x1, x2, x3, xn)
selected_indices <- c(1, 2, 3, n)
# Extract the specific values for display
expected_fhat <- expected_fhat_all[selected_indices]
sum_fhat <- sum_fhat_all[selected_indices]
variance_fhat <- variance_fhat_all[selected_indices]
squared_bias <- squared_bias_all[selected_indices]
# Add "..." column between X3 and Xn
```

```
add_dots <- function(mat) {
  dots <- matrix("...", nrow(mat), 1)
  cbind(mat[, 1:3], dots, mat[, 4])
}
# Create a data frame for output
output_matrix <- rbind(results$yhat[, selected_indices],
                       sum_fhat, expected_fhat, variance_fhat, squared_bias)
colnames(output_matrix) <- c("X1", "X2", "X3", paste0("X", n))
rownames(output_matrix) <- c(paste0("fhatsim_", 1:M), "Sum_fhat",
                          "Expected_fhat", "Variance_fhat", "Squared_Bias")
# Add dots column
output_matrix <- add_dots(output_matrix)
# Update column names to include the last X column name
colnames(output_matrix) <- c("X1", "X2", "X3", "...", paste0("X", n))
# Convert the numeric part of the matrix to numeric and round it
numeric_part <- output_matrix[, c(1:3, 5)]
numeric_part <- apply(numeric_part, 2, as.numeric)
numeric_part <- round(numeric_part, 5)
# Combine the numeric part with the "..." column
output_matrix <- cbind(numeric_part[, 1:3], "..." = output_matrix[, 4],
                       numeric_part[, 4])
# If-else structure for printing
if (show_all || n <= 5) {
  print(output_matrix, quote = FALSE)
} else {
  output_matrix_part <- rbind(
    output_matrix[1:3, ],
    rep("...", ncol(output_matrix)),
    output_matrix[M, ],
    output_matrix[(M+1):nrow(output_matrix), ]
  )
rownames(output_matrix_part) <- c(paste0("Simulation_", 1:3), "...",
paste0("Simulation_",M),rownames(output_matrix)[(M+1):nrow(output_matrix)])
knitr::kable(output_matrix_part, format = "latex", booktabs = TRUE, align = "c")
}
```

	X1	X2	X3	...	X100
Simulation_1	−1.41639	0.26594	8.23025	...	−6.936
Simulation_2	0.54238	1.29428	11.11243	...	−3.2481
Simulation_3	−0.11851	0.92696	9.36313	...	−5.12983
...	...	...	...	...	...
Simulation_50	−0.67598	1.03748	11.12986	...	−5.94959
Sum_fhat	−53.77726	29.44154	526.14357	...	−319.57438
Expected_fhat	−1.07555	0.58883	10.52287	...	−6.39149
Variance_fhat	1.06094	0.66369	3.05423	...	2.05494
Squared_Bias	0.04075	0.03676	0.00929	...	0.00016

Formally, let $\hat{f}_{X_1,i}$ represent the predicted $\hat{f}$ value at X_1 for the i-th simulation. The expected $\hat{f}$ value at X_1, denoted as $\mathbb{E}[\hat{f}_{X_1}]$, is calculated as:

$$\text{Expected } \hat{f} \text{ at } X_1: \quad \mathbb{E}[\hat{f}_{X_1}] = \frac{1}{M}\sum_{i=1}^{M} \hat{f}_{X_1,i} \tag{5.5}$$

Here is the output all 50 x_1s from all simulations (X1 column in simulation outtable above). Sum_hat is summation of these xs, and dividing this sum by 50 will give us Expected_fhat.

```
## fhat values at X1: -1.416392 0.5423814 -0.1185079 1.238812 -1.002696 -
1.728647 -1.624869 0.05110822 0.07516055 0.0766192 -1.034525 -0.5050097 -
1.735123 -0.5988579 -1.095943 -1.340411 -1.748233 -1.848451 -0.106242 -0.88527 -
1.412619 -2.880681 -1.754998 -2.231159 0.8346244 -2.680983 -2.629536 -0.2490797 -
0.4232302 -1.727644 0.03592634 -0.327935 -1.011624 0.07420027 -0.4940857 -
1.317213 -3.845779 -0.264057 0.5208008 -2.416152 -1.251696 -2.112036 -2.561129 -
1.392818 -1.065002 -1.547459 -1.362601 -1.980663 -0.8215615 -0.675977
```

```
## Expected fhat at X1: -1.075545
```

Formally, the variance of $\hat{f}$ at X_1, denoted as $\text{Var}(\hat{f}_{X_1})$, is calculated by averaging the squared differences between each predicted $\hat{f}$ value and the expected $\hat{f}$ value at X_1:

$$\text{Var}(\hat{f}_{X_1}) = \frac{1}{M}\sum_{i=1}^{M}(\hat{f}_{X_1,i} - \mathbb{E}[\hat{f}_{X_1}])^2 \tag{5.6}$$

These squared differences provide the foundation for calculating the variance of $\hat{f}$ at X_1. Displaying these values allows us to observe the individual contributions to the overall variance. For instance, first squared difference is $(-1.41639 - -1.07555)^2 = 0.1161766$.

$$\text{Squared differences at } X_1: \quad (\hat{f}_{X_1,i} - \mathbb{E}[\hat{f}_{X_1}])^2 \tag{5.7}$$

```
## Squared differences: 0.1161766 2.617687 0.9159205 5.35625 0.005307006 0.4265418
0.3017563 1.269348 1.324124 1.327483 0.00168266 0.3255109 0.4350434 0.2272309
0.0004160689 0.07015411 0.4525082 0.5973825 0.9395488 0.03620467 0.1136189
3.258514 0.4616558 1.335443 3.648748 2.577429 2.414887 0.6830453 0.425515
0.4252324 1.235369 0.5589212 0.004085916 1.321915 0.3380952 0.05840347 7.674195
0.6585131 2.548321 1.797226 0.03102923 1.074312 2.20696 0.1006622 0.0001111626
0.2227026 0.08240117 0.8192387 0.06450774 0.1596548
```

This variance provides a measure of the dispersion of the model's predictions, indicating how much the predicted values vary from the expected value. Displaying the calculated variance allows us to verify the consistency of the model's predictions at X_1.

$$\text{Variance of } \hat{f} \text{ at } X_1: \quad \text{Var}(\hat{f}_{X_1}) \tag{5.8}$$

```
## Variance of fhat at X1: 1.06094
```

Next, we calculate the squared bias for the predicted $\hat{f}$ at X_1. Formally, let $\mathbb{E}[\hat{f}_{X_1}]$ be the expected predicted value at X_1 and $f(X_1)$ be the true function value at X_1. The squared bias is given by:

$$\text{Squared bias at } X_1 = (\mathbb{E}[\hat{f}_{X_1}] - f(X_1))^2 \tag{5.9}$$

First, we extract the true function value $f(X_1)$ from the simulation results.

$$\text{True function value at } X_1: \quad f(X_1) \tag{5.10}$$

```
## f(X1) values: -1.277409
```

The expected $\hat{f}$ value at X_1, previously calculated as the mean of the predicted values across all simulations, is an essential part of the squared bias calculation.

$$\text{Expected } \hat{f} \text{ at } X_1 : \quad \mathbb{E}[\hat{f}_{X_1}] \tag{5.11}$$

```
## Expected fhat at X1: -1.075545
```

Displaying the squared bias allows us to directly observe the magnitude of this error. This measure helps in understanding the bias component of the model's error at X_1.

$$\text{Squared bias at } X_1 : \quad (\mathbb{E}[\hat{f}_{X_1}] - f(X_1))^2 \tag{5.12}$$

```
## Squared Bias at X1: 0.04074888
```

After calculating individual x values, we calculate aggregate measures for all x values for components of MSPE. As in this simulation we have 100 x values, we calculated 100 variance and squared bias.

First, we calculate the expected variance of the predicted $\hat{f}$ values for all x values. This is obtained by averaging the variances of the predicted $\hat{f}$ values across all x values. This expected variance provides a summary measure of the dispersion of the model's predictions.

$$\text{Expected variance of } \hat{f} \text{ for all } x \text{ values} = \frac{1}{n} \sum_{j=1}^{n} \text{Var}(\hat{f}_{x_j}) \tag{5.13}$$

Second, we calculate the expected squared bias for all x values. This is done by averaging the squared biases across all x values. The expected squared bias gives an overall measure of the systematic error in the model's predictions.

$$\text{Expected squared bias of } \hat{f} \text{ for all } x \text{ values} = \frac{1}{n} \sum_{j=1}^{n} (\mathbb{E}[\hat{f}_{x_j}] - f(x_j))^2 \tag{5.14}$$

Last, we compute the irreducible error, which is the average of the irreducible errors across all simulations. The irreducible error represents the portion of the total error that cannot be reduced by the model, as it is inherent in the data.

$$\text{Irreducible error} = \frac{1}{M} \sum_{i=1}^{M} \text{Irrerr}_i \tag{5.15}$$

```
## Expected Variance of fhat for polynomial degree 3 is: 2.172423

## Expected Squared Bias of fhat for polynomial degree 3 is: 0.1282653

## Irreducible error is: 63.23633
```

You can observe the same results as those in the row for PolyDeg3 from the main output table of the simulation, which presents results from PolyDeg1 to PolyDeg10. The main simulation performs this process for each polynomial degree in a loop. It is important to note that to decompose the MSPE as described above, we must use the true value of the function (as seen in the bias equation above). Since the true function is never known, we cannot calculate the components of the MSPE and obtain the MSPE by simply adding them. Instead, we must use the MSPE equation provided below. In our simulation, we calculated each MSPE for each training dataset and then found the expected value of these MSPEs. This approach is how MSPE is calculated with real datasets.

As we can calculate ($f(x_i)$ and $\hat{f}(x_i)$) for each value , we have a data frame:

	x_1	x_2	$\cdots$	x_{100}
s_1	$\left[f(x_1) - \hat{f}(x_1)\right]^2$	$\left[f(x_2) - \hat{f}(x_2)\right]^2$	$\cdots$	$\left[f(x_{100}) - \hat{f}(x_{100})\right]^2$
s_2	$\left[f(x_1) - \hat{f}(x_1)\right]^2$	$\left[f(x_2) - \hat{f}(x_2)\right]^2$	$\cdots$	$\left[f(x_{100}) - \hat{f}(x_{100})\right]^2$
$\vdots$	$\vdots$	$\vdots$	$\ddots$	$\vdots$
s_{50}	$\left[f(x_1) - \hat{f}(x_1)\right]^2$	$\left[f(x_2) - \hat{f}(x_2)\right]^2$	$\cdots$	$\left[f(x_{100}) - \hat{f}(x_{100})\right]^2$

Then using this data frame, we can calculate MSPE for each simulation and then find the expected value of these MSPEs. This is how MSPE is calculated in real datasets.

For each simulation i, the MSPE is computed as follows:

$$\text{MSPE}_i = \frac{1}{n}\sum_{j=1}^{n}(y_{ij} - \hat{y}_{ij})^2 \tag{5.16}$$

where y_{ij} represents the observed value and $\hat{y}_{ij}$ represents the predicted value for the j-th observation in the i-th simulation.

We calculate the error for x_i in simulation 1 by subtracting the predicted value from the true value and squaring the result as seen in data frame. We then calculate the errors for all values of x in simulation 1 and find their average. This average represents the MSPE for x_i in simulation 1. This process is repeated across all 50 simulations. Below are the MSPE values for the first three simulations.

```
## Simulation 1 MSPE: 58.4078
## Simulation 2 MSPE: 54.7409
## Simulation 3 MSPE: 66.193
```

The expected MSPE is the average of the MSPE values across all simulations. Formally, the expected MSPE is calculated as follows:

$$\text{MSPE} = \mathbb{E}\left[(y_i - \hat{f}(x_i))^2\right] \tag{5.17}$$

where y_i is the observed value and $\hat{f}(x_i)$ is the predicted value for the i-th observation.

This expected MSPE gives us a single value that summarizes the model's predictive performance over all simulations. It helps in understanding the average prediction error and serves as a benchmark for comparing different models or configurations.

```
## Expected MSPE: 60.9356
```

You can see these are the same results with those shown in the row for PolyDeg3 from the main output table of the simulation, which presents results from PolyDeg1 to PolyDeg10.

Before concluding this simulation, we will discuss its outcomes. From the main equation, it is evident that the MSPE can be decomposed into its components: bias, variance, and irreducible error. Ideally, adding these three components should yield the MSPE. However, this is not observed in the current simulation. For example, when the number of simulations, M, is increased to 500 and the number of x in each training dataset, denoted as n, is raised to 100,000; the MSPE is 64.0103 with a bias squared of 0.0025, a variance of 0.0024, and an irreducible error of 64.0123. If we further increase M to 2000, while maintaining n at 100,000; the MSPE adjusts to 63.9974 with a bias squared of 0, a variance of 0.002, and an irreducible error of 64.

These discrepancies arise due to the stochastic nature of the simulations and the finite sample sizes used. As M and n increase, the law of large numbers ensures that the estimated components (bias, variance, and irreducible error) converge to their true values. This convergence is why the calculated MSPE becomes more accurate and aligns more closely with the theoretical expectations as the number of simulations and the sample size increase. This trend is consistent across all polynomial degrees, illustrating the importance of large sample sizes and numerous simulations in achieving reliable statistical estimates.

5.2 Simulation for Population Parameter

Although the variance–bias trade-off conceptually seems intuitive, at least from a mathematical standpoint, another practical question arises: Is it possible to observe the components of the decomposed MSPE? In real-world data, observing these components is challenging as we do not know the true function and irreducible error. However, we can illustrate this concept through simulations.

For our analysis, we revisit the example discussed earlier. We consider years of schooling, varying between 9 and 16 years, as our variable of interest. From this "population," we repeatedly take random samples. The objective now is to utilize each sample to develop a predictor or a set of prediction rules. These rules aim to predict a number or a series of numbers (years of schooling in this simulation) that are also drawn from the same population. By doing so, we can effectively simulate and analyze the decomposition of the MSPE, providing a clearer understanding of its components and their interplay.

```
# Load necessary library
library(ggplot2)
M <- 5000 # Set the number of repeated samples
# Here is our population
populationX <- c(9,10,11,12,13,14,15,16)
# Generate repeated samples (M) with replacement
set.seed(123)
samples <- replicate(M, sample(populationX, 10, replace = TRUE))
samples <- t(samples)
head(samples)
```

```
##      [,1] [,2] [,3] [,4] [,5] [,6] [,7] [,8] [,9] [,10]
## [1,]   15   15   11   14   11   10   10   14   11    13
## [2,]   12   14   14    9   10   11   16   13   11    11
## [3,]    9   12    9    9   13   11   16   10   15    10
## [4,]    9   14   11   12   14    9   11   15   13    12
## [5,]   15   16   10   13   15    9    9   10   15    11
## [6,]   12   13   15   13   11   16   14    9   10    13
```

```
# Now, Let's use our predictors:
# Container to record all predictions
predictions <- matrix(0, nrow = M, ncol = 2)
colnames(predictions) <- c("fhat_1", "fhat_2")
# fhat_1 = 10
predictions[,1] <- 10
# fhat_2 - mean
predictions[,2] <- rowMeans(samples)
head(predictions)
```

```
##      fhat_1 fhat_2
## [1,]     10   12.4
## [2,]     10   12.1
## [3,]     10   11.4
## [4,]     10   12.0
## [5,]     10   12.3
## [6,]     10   12.6
```

```
# Now let's have our MSPE decomposition:
# MSPE
MSPE <- matrix(0, nrow = M, ncol = 2)
colnames(MSPE) <- c("fhat_1", "fhat_2")
for (i in 1:M) {
  MSPE[i,1] <- mean((populationX - predictions[i,1])^2)
  MSPE[i,2] <- mean((populationX - predictions[i,2])^2)
}
head(MSPE)
```

```
##      fhat_1 fhat_2
## [1,]   11.5   5.26
## [2,]   11.5   5.41
## [3,]   11.5   6.46
## [4,]   11.5   5.50
## [5,]   11.5   5.29
## [6,]   11.5   5.26
```

```
# Bias
bias <- colMeans(predictions) - mean(populationX)
bias_squared <- bias^2
# Variance (predictor)
var_fhat <- apply(predictions, 2, var)
# Variance (epsilon)
var_eps <- mean((populationX - mean(populationX))^2)
# Let's put them in a table:
VBtradeoff <- matrix(c(bias_squared, var_fhat, rep(var_eps, 2),
                       colMeans(MSPE)), nrow = 2, byrow = FALSE)
rownames(VBtradeoff) <- c("fhat_1", "fhat_2")
colnames(VBtradeoff) <- c("Bias^2", "Var(fhat)", "Var(eps)", "MSPE")
round(VBtradeoff, 3)
```

```
##            Bias^2 Var(fhat) Var(eps)    MSPE
## fhat_1    6.25     0.000     5.25 11.500
## fhat_2    0.00     0.539     5.25  5.788
```

This table clearly shows the decomposition of MSPE. The first column is the contribution to the MSPE from the bias, and the second column is the contribution from the variance of the predictor. These together make up the reducible error. The third column is the variance that comes from the data, the irreducible error. The last column is, of course, the total MSPE, and we can see that $\hat{f}_2$ is the better predictor because of its lower MSPE.[3]

5.3 Biased Estimator as a Predictor

Up to this point, we have shown through simulation that a prediction function with zero bias but high variance often produces better predictions than a prediction function with zero variance but high bias. However, we can potentially obtain an even better prediction function that has some bias and some variance. A better prediction function means a smaller MSPE. The key idea is that if the reduction in variance more than compensates for the increase in bias, then we have a better predictor.

To explore this, let's define a biased estimator of μ_x:

$$\hat{X}_{\text{biased}} = \hat{\mu}_x = \alpha \bar{X} \tag{5.18}$$

where $\bar{X}$ is the sample mean. The sample mean $\bar{X}$ is an unbiased estimator of μ_x, and the parameter α introduces bias. When α is 1, $\hat{\mu}_x$ becomes the unbiased sample mean.

The bias of the estimator $\hat{\mu}_x$ is given by:

$$\text{Bias}(\hat{\mu}_x) = \mathbb{E}[\hat{\mu}_x] - \mu_x = \alpha\mu_x - \mu_x = (\alpha - 1)\mu_x \tag{5.19}$$

The variance of the estimator $\hat{\mu}_x$ is:

$$\text{Var}(\hat{\mu}_x) = \text{Var}(\alpha\bar{X}) = \alpha^2 \,\text{Var}(\bar{X}) \tag{5.20}$$

Since the variance of the sample mean $\bar{X}$ is $\frac{\sigma_\varepsilon^2}{n}$ (check chapter 4.5), we have:

$$\text{Var}(\hat{\mu}_x) = \alpha^2 \frac{\sigma_\varepsilon^2}{n} \tag{5.21}$$

The MSPE and its bias–variance components:

$$\text{MSPE} = \mathbb{E}[(\hat{\mu}_x - \mu_x)^2] = \text{Bias}^2(\hat{\mu}_x) + \text{Var}(\hat{\mu}_x) + \sigma_\varepsilon^2 \tag{5.22}$$

where σ_ε^2 is the irreducible error.

First, calculate the bias squared:

$$\text{Bias}^2(\hat{\mu}_x) = [(\alpha - 1)\mu_x]^2 = (1 - \alpha)^2 \mu_x^2 \tag{5.23}$$

Next, calculate the variance:

$$\text{Var}(\hat{\mu}_x) = \alpha^2 \frac{\sigma_\varepsilon^2}{n} \tag{5.24}$$

Finally, the irreducible error is σ_ε^2. After combining these terms,we can shows that the MSPE for the biased estimator $\hat{\mu}_x = \alpha \bar{X}$ is:

$$\text{MSPE} = (1-\alpha)^2 \mu_x^2 + \alpha^2 \frac{\sigma_\varepsilon^2}{n} + \sigma_\varepsilon^2 = [(1-\alpha)\mu_x]^2 + \frac{1}{n}\alpha^2\sigma_\varepsilon^2 + \sigma_\varepsilon^2 \tag{5.25}$$

The final expression for the MSPE of the biased estimator $\hat{\mu}_x = \alpha \bar{X}$ combines the squared bias term $(1-\alpha)^2\mu_x^2$, the variance term $\frac{\alpha^2\sigma_\varepsilon^2}{n}$, and the irreducible error term σ_ε^2. By adjusting α, we can balance the trade-off between bias and variance to minimize the MSPE. This highlights that a small amount of bias can be beneficial if it significantly reduces the variance, leading to a lower MSPE and thus a better predictor.

Our first observation is that when α is one, the bias will be zero. Since it seems that MSPE is a convex function of α, we can search for α that minimizes MSPE. The value of α that minimizes the MSPE is:

$$\frac{\partial \text{MSPE}}{\partial \alpha} = 0 \rightarrow \quad \alpha = \frac{\mu_x^2}{\mu_x^2 + \sigma_\varepsilon^2/n} < 1 \tag{5.26}$$

Check end of the chapter for step-by-step derivation of the optimal value of α that minimizes the MSPE.

Using the same simulation sample above, lets calculate alpha and MSPE with this new biased prediction function, and compare all three MSPEs.

```
pred <-rep(0, Ms)
# The magnitude of bias
alpha <- (mean(populationX))^2/((mean(populationX)^2+var_eps/10))
alpha
```

```
## [1] 0.9966513
```

```
# Biased predictor
for (i in 1:Ms) {
  pred[i] <- alpha*predictions[i,2]
}
# Check if E(alpha*Xbar) = alpha*mu_x
mean(pred)
```

```
## [1] 12.45708
```

```
alpha*mean(populationX)
```

```
## [1] 12.45814
```

```r
# MSPE
MSPE_biased <- rep(0, Ms)
for (i in 1:Ms) {
  MSPE_biased[i] <- mean((populationX-pred[i])^2)
}
mean(MSPE_biased)
```

```
## [1] 5.786663
```

Let's add this predictor into our table:

```r
VBtradeoff <- matrix(0, 3, 4)
rownames(VBtradeoff) <- c("fhat_1", "fhat_2", "fhat_3")
colnames(VBtradeoff) <- c("Bias", "Var(fhat)", "Var(eps)", "MSPE")
VBtradeoff[1,1] <- bias1^2
VBtradeoff[2,1] <- bias2^2
VBtradeoff[3,1] <- (mean(populationX)-mean(pred))^2
VBtradeoff[1,2] <- var1
VBtradeoff[2,2] <- var2
VBtradeoff[3,2] <- var(pred)
VBtradeoff[1,3] <- var_eps
VBtradeoff[2,3] <- var_eps
VBtradeoff[3,3] <- var_eps
VBtradeoff[1,4] <- mean(MSPE[,1])
VBtradeoff[2,4] <- mean(MSPE[,2])
VBtradeoff[3,4] <- mean(MSPE_biased)
round(VBtradeoff, 3)
```

```
##         Bias Var(fhat) Var(eps)   MSPE
## fhat_1 6.250     0.000     5.25 11.500
## fhat_2 0.000     0.539     5.25  5.788
## fhat_3 0.002     0.535     5.25  5.787
```

As seen, increase in bias is lower than decrease in variance. The prediction function with some bias and variance is the best prediction function as it has the smallest MSPE. This example shows the difference between estimation and prediction for a simplest predictor, the mean of X.

In the next chapter, we will explore overfitting and how it relates to and differs from the bias–variance trade-off.

Note: To find the value of α that minimizes the MSPE, we take the derivative of the MSPE with respect to α, set it to zero, and solve for α.

The MSPE is given by:

$$\text{MSPE} = [(1-\alpha)\mu_x]^2 + \frac{1}{n}\alpha^2\sigma_\varepsilon^2 + \sigma_\varepsilon^2 \tag{5.27}$$

First, we take the derivative of MSPE with respect to α:

$$\frac{\partial \text{MSPE}}{\partial \alpha} = \frac{\partial}{\partial \alpha}\left[(1-\alpha)^2\mu_x^2 + \frac{1}{n}\alpha^2\sigma_\varepsilon^2 + \sigma_\varepsilon^2\right] \tag{5.28}$$

We compute the derivative term by term:

$$\frac{\partial \text{MSPE}}{\partial \alpha} = 2(1-\alpha)(-1)\mu_x^2 + 2\frac{1}{n}\alpha\sigma_\varepsilon^2 \tag{5.29}$$

Simplifying, we get:

$$\frac{\partial \text{MSPE}}{\partial \alpha} = -2(1-\alpha)\mu_x^2 + 2\frac{1}{n}\alpha\sigma_\varepsilon^2 \tag{5.30}$$

To find the minimum MSPE, we set the derivative equal to zero:

$$-2(1-\alpha)\mu_x^2 + 2\frac{1}{n}\alpha\sigma_\varepsilon^2 = 0 \tag{5.31}$$

Solving for α:

$$-2\mu_x^2 + 2\alpha\mu_x^2 + 2\frac{1}{n}\alpha\sigma_\varepsilon^2 = 0 \tag{5.32}$$

$$-\mu_x^2 + \alpha\mu_x^2 + \frac{1}{n}\alpha\sigma_\varepsilon^2 = 0 \tag{5.33}$$

$$-\mu_x^2 + \alpha\left(\mu_x^2 + \frac{\sigma_\varepsilon^2}{n}\right) = 0 \tag{5.34}$$

$$\alpha\left(\mu_x^2 + \frac{\sigma_\varepsilon^2}{n}\right) = \mu_x^2 \tag{5.35}$$

$$\alpha = \frac{\mu_x^2}{\mu_x^2 + \frac{\sigma_\varepsilon^2}{n}} \tag{5.36}$$

Thus, the value of α that minimizes the MSPE is:

$$\alpha = \frac{\mu_x^2}{\mu_x^2 + \frac{\sigma_\varepsilon^2}{n}} \tag{5.37}$$

Notes

1. Read [https://threadreaderapp.com/thread/1584515105374339073.html] and [https://www.simplilearn.com/tutorials/machine-learning-tutorial/bias-and-variance]

2. Polynomial degree is set in stat_smooth(...). The plots are generated using create_plot(results, degree, color), where the degree is 1, 3, or 10.

3. Other simulation examples are [1](https://blog.zenggyu.com/en/post/2018-03-11/understanding-the-bias–variance-decomposition-with-a-simulated-experiment/) , [2](https://www.r-bloggers.com/2019/06/simulating-the-bias–variance-tradeoff-in-r/), and [3](https://daviddalpiaz.github.io/r4sl/simulating-the-biasvariance-tradeoff.html)

6

Overfitting

Following our exploration of the bias–variance trade-off, in this chapter we explore the equally crucial concept of overfitting in machine learning.

Through the compelling stories of Baron's travel planning and Jennifer's exam preparation, we'll uncover how overfitting manifests in both everyday scenarios and complex algorithms, revealing its impact on decision-making and model accuracy. Let's extend Baron's story from the previous chapter, which not only illustrates the bias–variance trade-off but also seamlessly incorporates the concept of overfitting, providing a deeper understanding of its implications.

As we mentioned at the beginning of previous chapter, Baron initially developed a routine to leave his house around 7:00 AM, with a 30-minute window on either side to catch his bi-weekly flights. This approach, while seemingly flexible, had high variance. It is like a machine learning model that is not sufficiently trained or fine-tuned—it has a broad range of responses (leaving times from his home) but lacks precision (cathing flights).

Realizing the issue with his initial strategy, Baron decides to analyze past data more closely. He starts noting down the exact traffic conditions, weather patterns, and personal preparation times for each day of his travel. Based on this detailed analysis, he creates a very specific and complex routine, factoring in all these variables to predict the optimal departure time. This new strategy is akin to an overfitted model in machine learning – it is overly complex and tailored to past data, potentially capturing random error (like a one-time traffic jam or an unusual personal delay) rather than the overarching trend.

Initially, Baron's new strategy seems to work perfectly, as he catches his flights consistently. However, on a day with unexpected conditions (not previously encountered in his data, like a new road closure), his overly complex routine fails him, and he continue to miss some of his flights. This is similar to an overfitted model performing well on training data but poorly on new, unseen data.

Baron realizes that his overfitted strategy is not sustainable. He then revises his approach, aiming to leave around 7:05 AM with only a 10-minute window. This strategy introduces a slight bias (always leaving a bit later than the original 7:00 AM plan) but significantly reduces variance. It is like a well-balanced machine learning model that has been trained to generalize well, rather than just perform on specific past data.

With this balanced approach, Baron consistently catches his flights, demonstrating that a balance between bias and variance can lead to more effective real-world results. His experience underscores the importance of not overcomplicating models (or plans) based on past data, as real-world conditions often present new, unforeseen challenges. This balance is crucial in machine learning to create models that perform well both on historical data and in new, unpredictable situations.

After discussing Baron's experience, let's shift to another example that deepens our grasp on overfitting. We now turn to Jennifer, a student facing challenges in her study methods

DOI: 10.1201/9781003504795-6

for a history exam. Her story, paralleling Baron's, can illustrate the concept of overfitting in everyday situations and in machine learning models.

Jennifer gathers an array of study materials, including textbooks, class notes, and online resources. She compiles detailed notes on all the topics that might be covered in the exam (Data Collection). She initially tries to get a general understanding of all historical events and themes, but soon realizes it is too much to handle. As she studies, Jennifer notices that certain events and dates are emphasized more in her materials (Pattern Recognition). Then, Jennifer forms a hypothesis that the exam will primarily consist of questions about these specific events and dates she has identified as important (Making Predictions).

Jennifer faces a dilemma. She can either maintain a broad focus on history, which would result in a high bias but low variance in her study approach, potentially causing her to miss finer details. Alternatively, she can concentrate intensely on specific events and dates, which corresponds to a low bias but high variance approach, risking a lack of preparedness for broader questions. This scenario mirrors the bias–variance trade-off in machine learning, where the challenge lies in balancing the generalization of the model (a broad study scope) with the accuracy on specific details (focused study on certain events).

Jennifer opts to focus on specific details, thinking this will prepare her better for the exam. She memorizes dates and specific events, neglecting the broader themes. However, she soon realizes the potential pitfalls of this method – while it might prepare her well for certain types of questions, it could leave her ill-equipped for others that require a more comprehensive understanding of history. This is akin to a machine learning model that is overfitted - it performs exceptionally well on the training data (or the specific details Jennifer studied) but fails to generalize well to new data (or the broader exam questions).

On the exam day, Jennifer does well on questions about specific events but struggles with questions that require a broader understanding of history. This reflects the problem in machine learning where a model with low bias but high variance performs well on training data but poorly on unseen data. Jennifer's experience shows the downside of focusing too narrowly and not maintaining a balance between the specific details and the broader context.

Jennifer's study method serves as an effective analogy for machine learning. Her intense focus on memorizing specific historical details mirrors a model with low bias but high variance—performing well on familiar material (training data) but struggling with broader, unfamiliar questions (test data). This illustrates the bias–variance trade-off: effective learning, like effective modeling, requires balancing detailed understanding with the ability to generalize.

Overfitting is a phenomenon that occurs when a statistical model is trained too closely on the training dataset, capturing random error (noise, irreducible error) in the data rather than the underlying pattern. This results in a model that performs well on the training data (like Jennifer's success with specific questions) but poorly on new, unseen data (like the broader questions on the exam). A model that has been overfitted has poor predictive performance, as it overreacts to minor fluctuations in the training data. Overfitting is a common problem in statistics and machine learning. The more flexible (complex) a model is, the more likely it is to overfit the data. A good predictive model is able to learn the pattern from your data and then to generalize it on new data. Generalization, i.e., learning, is the ability of a model to perform well on unseen data. Overfitting occurs when a model does not generalize well. Underfitting occurs when a model is not flexible enough to capture the underlying relationship as well.

A classic example is fitting a polynomial to a set of data points. A low-degree polynomial (e.g., a line) may underfit by missing important curvature. A high-degree polynomial may perfectly pass through all training points but behave erratically outside that range, capturing noise rather than signal. We will simulate this example below. The same logic applies to complex models like decision tree with a large number of branches, which may memorize the training data instead of learning general rules.

All in all, the optimal model is the simplest one that achieves the right balance in the bias–variance trade-off and fits the data without overfitting. To detect overfitting, one should monitor both the training error (e.g., mean square prediction error (MSPE), as shown in simulations) and the test error during model training. As model complexity increases, the training error typically decreases. However, once the model begins to overfit—capturing noise rather than signal—the test error starts to rise.

In linear models, we assess model complexity using Mallows' C_p, the Akaike Information Criterion (AIC), or the Bayesian Information Criterion (BIC). The accuracy of these criteria depends on knowledge of the population model, which we will cover in Chapter 11. When data is relatively limited, cross-validation is a common approach for estimating prediction error while mitigating overfitting. The data is split into training and test sets; the model is trained on the former and evaluated on the latter. This process is repeated across different data partitions, and results are averaged to reduce variability. We discuss these model complexity scores and cross-validation techniques in related chapters.

Overfitting can be reduced by using a simpler model, collecting more data, applying regularization, or using early stopping. The simplest strategy is to select a model with just enough complexity to reflect the signal without capturing noise—for example, using a lower-degree polynomial instead of a high-degree one. More data can reduce variance and prevent overfitting when the model has too many parameters relative to the sample size. Regularization addresses overfitting by adding a penalty term to the loss function that discourages excessive complexity. In neural networks, early stopping monitors validation error during training and halts training once the error begins to rise. In time series models, overfitting can be mitigated through smoothing methods like moving averages or by removing seasonal trends before modeling and reintroducing them afterward.

Though often associated with prediction, overfitting also creates challenges in estimation, where the goal is unbiased and generalizable inference. In estimation contexts, overfitting arises when overly complex models capture sample noise rather than the true data-generating process. Even if such models appear unbiased in theory, they may yield unstable estimates that don't generalize. This reinforces the importance of managing the bias–variance trade-off to avoid fitting noise. In prediction settings, relying on in-sample MSPE can mask overfitting and inflate confidence in a model's performance. Similarly, in earlier chapters we evaluated estimation accuracy using in-sample MSPE, which can lead to misleading conclusions. To accurately evaluate model performance, MSPE should be calculated using out-of-sample data. We illustrate this distinction in the simulation below, comparing MSPE values from training and test data.

Figure 6.1 visualizes the relationship between model complexity and error for both training and test datasets. The red solid line represents the test error, which typically decreases initially as model complexity increases but eventually rises again due to overfitting. This reflects how a simple model underfits the data, leading to high test error, while an overly complex model fits noise in the training data, reducing generalization performance on unseen data. The blue dashed line shows the training error, which steadily decreases as model complexity increases, since more complex models can better capture patterns in the training data, including noise. The key takeaway from this plot is that while increasing complexity reduces training error, it eventually leads to a significant rise in test error due to overfitting,

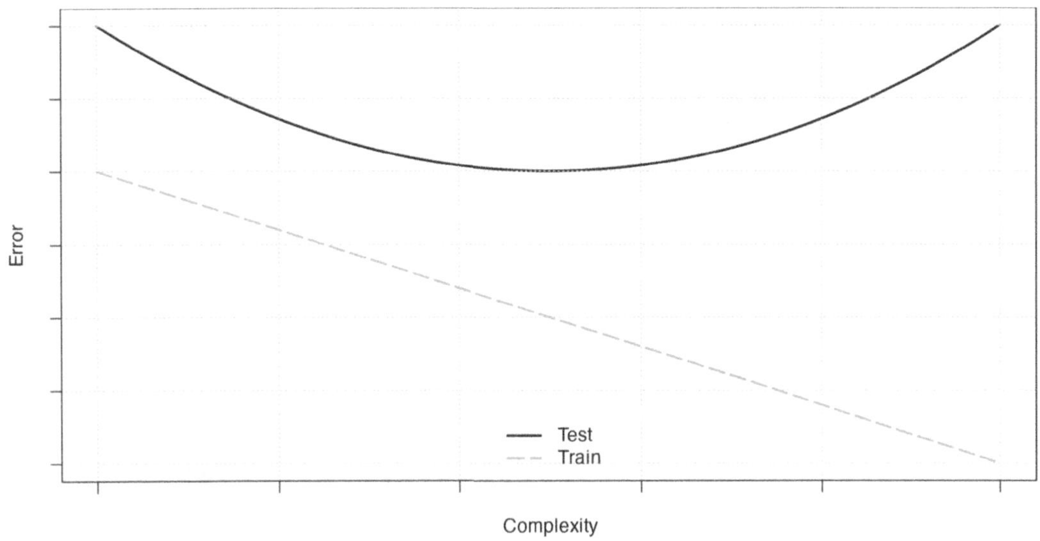

FIGURE 6.1 Error versus model complexity.

highlighting the importance of balancing model complexity to minimize generalization error. The absence of axis labels keeps the focus on the qualitative shape of the error curves, while the legend at the bottom clearly distinguishes between the behavior of the test and train errors.

Let's begin by presenting the ordinary least square (OLS) model fits for polynomial regression with degrees 1, 3, 10, and 30 in a figure. We will use the same simulation setup to maintain consistency. In the previous chapter, we computed the in-sample MSPE for polynomial degrees up to 10. Now, we will extend this analysis by calculating the out-sample MSPE and comparing it with the in-sample MSPE. The results will be presented in a figure, where we will examine how they align with the general relationship between error and model complexity shown in Figure 6.1.

In Figure 6.2, we visualize the fits of OLS models for polynomial degrees 1, 3, 10, and 30, as applied to the generated data. Each subplot represents fit of the model for a specific degree, allowing us to compare how the complexity of the polynomial impacts the model's ability to capture the underlying data-generating mechanism. As the degree increases, we observe how the model becomes increasingly flexible (complex), fitting the training data more closely. In particular, the degree 1 and 3 models provide smoother, simpler fits, while the degree 10 and 30 models capture more intricate patterns, and begin to overfit the noise in the data.

6.1 Out-sample MSPE

In Table 6.1, we compare the in-sample and out-sample MSPE for polynomial degrees ranging from 1 to 10. The in-sample MSPE values, starting from 87.7267 for degree 1 and decreasing to 56.2245 for degree 10, are the exact same ones calculated in the previous chapter. In contrast, the out-sample MSPE values are newly calculated, with the lowest out-sample error occurring at degree 3 (66.1662), after which it begins to rise, reaching 71.0181 at degree 10. This behavior illustrates how, as model complexity increases, the

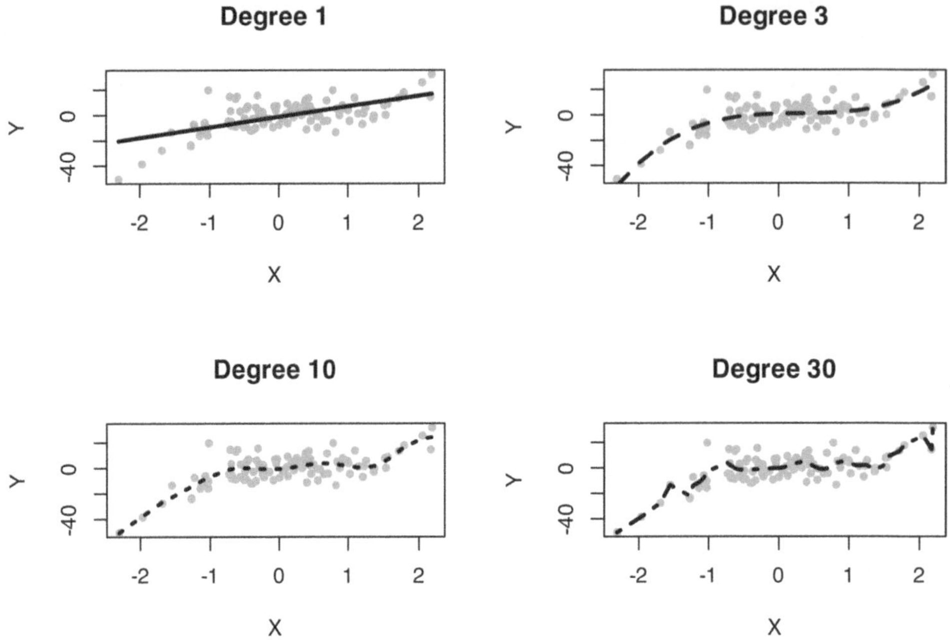

FIGURE 6.2 OLS fits for polynomial degrees 1, 3, 10, and 30.

TABLE 6.1 Comparison of In-sample and Out-sample MSPEs across Polynomial Degrees

Polynomial Degree	In-sample MSPE	Out-sample MSPE
1	87.7267	87.7241
2	82.9245	84.2082
3	60.9356	66.1662
4	60.1289	66.9232
5	59.5444	67.6132
6	59.0234	68.4805
7	58.4406	68.9593
8	57.7283	69.5755
9	56.9430	70.3915
10	56.2245	71.0181

in-sample MSPE continues to decrease as the model better fits the training data, while the out-sample MSPE initially decreases but eventually rises due to overfitting.

To select the "best" predictive model, we focus on the model with the lowest out-sample MSPE, rather than the in-sample MSPE. This is because the out-sample MSPE better reflects the ability of the model to generalize to unseen data, while the in-sample MSPE only measures how well the model fits the training data. In this case, the model with polynomial degree 3 offers the best predictive performance, as it achieves the lowest out-sample MSPE. Below, we will discuss how these out-sample MSPE values were calculated and the implications for model selection.

Table 6.1 presents a comparison of in-sample and out-of-sample MSPEs for polynomial degrees 1 through 10. It highlights the trade-off between model complexity and predictive performance, aligning with the U-shaped behavior illustrated in the earlier figure. The in-sample MSPE is shown in Figure 6.3 with a gray dashed line and triangle markers, while the out-sample MSPE is plotted using a black solid line and circle markers.

As the plot illustrates, the in-sample MSPE decreases steadily with increasing polynomial degree. This is expected, as more flexible models fit the training data more closely—capturing both the signal and some noise—which drives down the training error. By degree 10, the in-sample MSPE reaches its lowest point, reflecting an almost perfect fit to the training data.

In contrast, the out-sample MSPE follows a U-shaped curve. It initially decreases and reaches its minimum around degree 3, indicating that a degree 3 model best balances fit and generalization. Beyond that point, the out-sample MSPE rises again, suggesting that higher-degree models begin to overfit the training data and perform worse on unseen data.

This pattern exemplifies the bias–variance trade-off: low-degree models are too rigid (high bias), while high-degree models become too sensitive to noise (high variance). The degree 3 polynomial offers the optimal trade-off, minimizing out-sample MSPE and yielding the best predictive performance.

Ultimately, the simulation reinforces a central principle of statistical learning: the model that minimizes out-sample error—not in-sample error—is the most reliable for prediction. Choosing the right level of complexity is key to avoiding overfitting and achieving strong generalization.

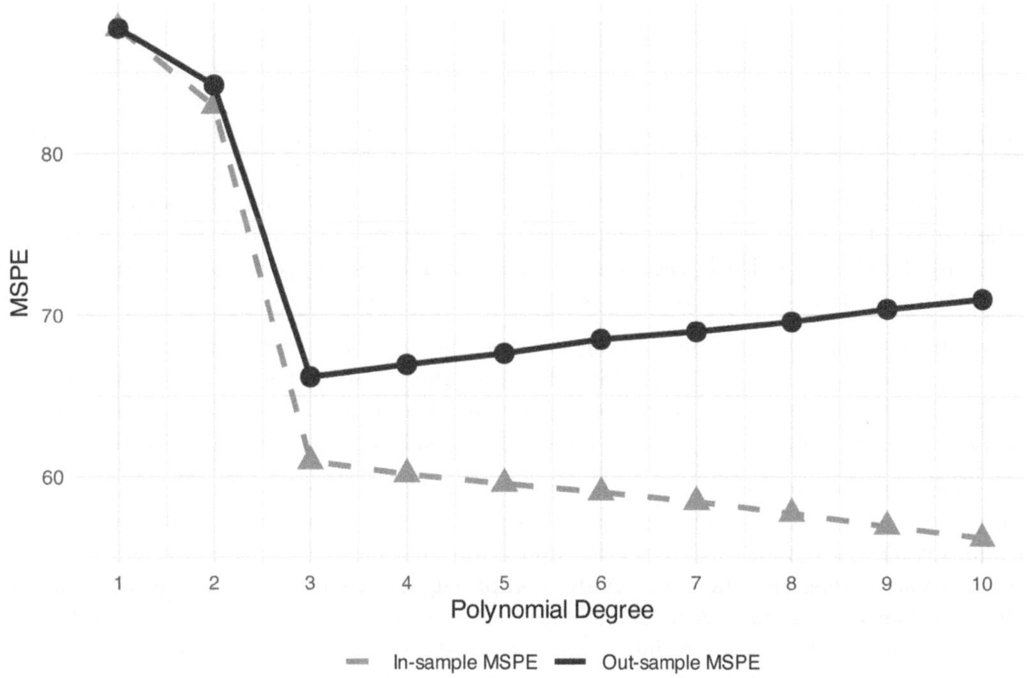

FIGURE 6.3 Comparison of in-sample and out-of-sample MSPEs across polynomial degrees.

In the previous chapter, we showed in detail how to calculate in-sample MSPE as well as its components such as bias squared, variance, and irreducible error. Although the process for calculating out-sample MSPE is very similar, there is a major difference. However, you can easily implement a similar simulation for out-sample MSPE, so we will not discuss every step in detail. Here is how out-sample MSPE is calculated, not only in this simulation but also for all real data applications.

To evaluate the generalization performance of the models, we calculate the out-sample MSPE. For this, we use a similar approach as with the in-sample MSPE but with one key difference: instead of calculating the error on the same data, i.e., the training data, used to fit the model, we use an independent test dataset. The out-sample MSPE gives us an indication of how well the fitted models, $\hat{f}$, generalize to new, unseen data.

In this setup, we generate 50 training datasets, $\mathcal{D}$, with a sample size of $n = 100$, and fit polynomial regression models of degrees ranging from 1 to 10 using OLS. These models are trained (estimate the model parameters, β) on the training data, and we calculate the prediction functions for each degree as follows:

$$\hat{f}_1(x) = \hat{\beta}_0 + \hat{\beta}_1 x$$

$$\hat{f}_2(x) = \hat{\beta}_0 + \hat{\beta}_1 x + \hat{\beta}_2 x^2$$

$$\hat{f}_3(x) = \hat{\beta}_0 + \hat{\beta}_1 x + \hat{\beta}_2 x^2 + \hat{\beta}_3 x^3$$

$$\vdots$$

$$\hat{f}_{10}(x) = \hat{\beta}_0 + \hat{\beta}_1 x + \hat{\beta}_2 x^2 + \ldots + \hat{\beta}_{10} x^{10}$$

For each dataset, we estimate the model parameters, β, based on the training data. These parameters are exactly the same β's we estimated in the simulation used to calculate the in-sample MSPE in the previous chapter. Then, using these estimated parameters, we predict the values of Y on a separate, independent test dataset. This allows us to calculate the *out-sample MSPE*, which is the average squared difference between the predicted values from the model and the actual observed values in the test data.

We repeat this process for 50 training and test datasets. For each polynomial degree, we compute the expected out-sample MSPE by averaging the results across the 50 test datasets. This allows us to assess how each model performs on new data, providing a clear picture of which degree of polynomial achieves the best trade-off between bias and variance.

To conclude, understanding and managing the bias–variance trade-off is essential in selecting a model that not only fits the training data well but also generalizes effectively to unseen data, ultimately minimizing overfitting and improving predictive accuracy. In the context of regression, selecting the right model complexity is crucial to balancing the trade-off between bias and variance. A model is biased when it oversimplifies the relationship between the variables, such as when a linear model is used to fit a quadratic relationship. On the other hand, a model becomes too variable when it overfits the data, incorporating too many variables or allowing for excessive flexibility, such as fitting a cubic model to a linear relationship. In nonparametric models, bias occurs when too much smoothing is applied, while variance increases when the model becomes overly "wiggly" and captures noise in the data.

Throughout this chapter, we have explored how increasing model complexity affects both train MSPE and test MSPE. As model complexity grows, the train MSPE decreases, since the model fits the training data more closely. However, the test MSPE follows a U-shaped curve, decreasing at first as the model captures the underlying patterns in the data, but eventually increasing as overfitting sets in. This behavior underscores the importance of selecting a model that balances bias and variance to minimize reducible error and perform well on unseen observations.

By understanding the bias–variance trade-off, we can fine-tune model complexity to avoid the extremes of underfitting and overfitting. Ultimately, the goal is to find the optimal point where the model is flexible enough to capture the true relationship in the data, but not so complex that it loses its ability to generalize to new data. This balance is key to building models that predict effectively and robustly.

6.2 Technical Points about Out-sample and In-sample MSPE

For linear regression with a Linear Population Model, we define MSPE over some data points, as we did in our simulation above, and rewrite it as follows:

$$\mathbf{MSPE}_{out} = \mathrm{E}\left[\frac{1}{n}\sum_{i=1}^{n}\left(y_i' - \hat{f}(x_i)\right)^2\right], \qquad \text{where } y_i' = f(x_i) + \varepsilon_i' \tag{6.1}$$

This type of MSPE is also called unconditional MSPE. Inside the brackets is the "prediction error" for a range of out-of-sample data points. The only difference here is that we distinguish y_i' as out-of-sample data points. Likewise, we define MSPE for in-sample data points y_i as:

$$\mathbf{MSPE}_{in} = \mathrm{E}\left[\frac{1}{n}\sum_{i=1}^{n}\left(y_i - \hat{f}(x_i)\right)^2\right], \qquad \text{where } y_i = f(x_i) + \varepsilon_i \tag{6.2}$$

Note that ε_i' and ε_i are independent but identically distributed. Moreover, y_i' and y_i have the same distribution.

The relationship between the out-sample MSPE and in-sample MSPE can be expressed as:

$$\mathbf{MSPE}_{out} = \mathbf{MSPE}_{in} + \frac{2}{n}\sigma^2(p+1) \tag{6.3}$$

The last term quantifies the overfitting, which reflects how the in-sample MSPE systematically underestimates the true MSPE, i.e., the out-sample MSPE. Additionally, the overfitting

1. **Grows** with the noise σ^2 in the data,
2. **Shrinks** with the sample size n,
3. **Grows** with the number of variables p.

Thus, as we discussed earlier, the overfitting problem worsens as p/n becomes larger. Minimizing the in-sample MSPE completely ignores overfitting, resulting in models that are too large and have very poor out-sample prediction accuracy.

Now, let's look closer at $\mathrm{E}\left[(y_i' - \hat{f}(x_i))^2\right]$. Using the definition of variance, we can express it as:

$$\begin{aligned}\mathrm{E}\left[(y_i' - \hat{f}(x_i))^2\right] &= \mathrm{Var}\left[y_i' - \hat{f}(x_i)\right] + \left(\mathrm{E}\left[y_i' - \hat{f}(x_i)\right]\right)^2 \\ &= \mathrm{Var}\left[y_i'\right] + \mathrm{Var}\left[\hat{f}(x_i)\right] - 2\,\mathrm{Cov}\left[y_i', \hat{f}(x_i)\right] + \left(\mathrm{E}\left[y_i'\right] - \mathrm{E}\left[\hat{f}(x_i)\right]\right)^2 \end{aligned} \tag{6.4}$$

Similarly, for in-sample points:

$$\begin{aligned}\mathrm{E}\left[(y_i - \hat{f}(x_i))^2\right] &= \mathrm{Var}\left[y_i - \hat{f}(x_i)\right] + \left(\mathrm{E}\left[y_i - \hat{f}(x_i)\right]\right)^2 \\ &= \mathrm{Var}\left[y_i\right] + \mathrm{Var}\left[\hat{f}(x_i)\right] - 2\,\mathrm{Cov}\left[y_i, \hat{f}(x_i)\right] + \left(\mathrm{E}\left[y_i\right] - \mathrm{E}\left[\hat{f}(x_i)\right]\right)^2 \end{aligned} \tag{6.5}$$

Recall our earlier discussion on variance-bias decomposition: when predicting out-sample data points, y_i' and $\hat{f}(x_i)$ are independent. Alternatively, ε_i' is independent of $\hat{f}(x_i)$. In other words, how we estimate $\hat{f}(x_i)$ is independent of y_i'. Hence, $\mathrm{Cov}\left[y_i', \hat{f}(x_i)\right] = 0$. However, it is important to understand that $\mathrm{Cov}\left[y_i, \hat{f}(x_i)\right] \neq 0$ for in-sample data. This is because $\hat{f}(x_i)$ is chosen to minimize the difference from y_i. Therefore, the estimator is not independent of in-sample points y_i; rather, they are used to estimate $\hat{f}(x_i)$. In fact, if $\hat{f}(x_i) = y_i$, the MSPE would be zero, and the correlation between $\hat{f}(x_i)$ and y_i would be 1.

Using the fact that $\mathrm{E}(y_i') = \mathrm{E}(y_i)$ and $\mathrm{Var}(y_i') = \mathrm{Var}(y_i)$, we can rewrite $\mathrm{E}\left[(y_i' - \hat{f}(x_i))^2\right]$ as:

$$\begin{aligned}\mathrm{E}\left[(y_i' - \hat{f}(x_i))^2\right] &= \mathrm{Var}[y_i] + \mathrm{Var}[\hat{f}(x_i)] + \left(\mathrm{E}[y_i] - \mathrm{E}[\hat{f}(x_i)]\right)^2 \\ &= \mathrm{E}\left[(y_i - \hat{f}(x_i))^2\right] + 2\,\mathrm{Cov}[y_i, \hat{f}(x_i)] \end{aligned} \tag{6.6}$$

Averaging over all data points:

$$\mathrm{E}\left[\frac{1}{n}\sum_{i=1}^{n}\left(y_i' - \hat{f}(x_i)\right)^2\right] = \mathrm{E}\left[\frac{1}{n}\sum_{i=1}^{n}\left(y_i - \hat{f}(x_i)\right)^2\right] + \frac{2}{n}\sum_{i=1}^{n}\mathrm{Cov}[y_i, \hat{f}(x_i)] \tag{6.7}$$

$$\frac{2}{n}\sum_{i=1}^{n}\mathrm{Cov}[y_i, \hat{f}(x_i)] = \frac{2}{n}\sigma^2(p+1) \tag{6.8}$$

Thus, we arrive at the relationship:

$$\mathbf{MSPE}_{out} = \mathbf{MSPE}_{in} + \frac{2}{n}\sigma^2(p+1) \tag{6.9}$$

7

Parametric Estimation: Basics

Statistical analysis relies on choosing appropriate models to interpret data and make inferences. This chapter focuses on parametric estimations which assume a specific data distribution and provide structured, interpretable insights. Widely used in applied econometrics, these models offer simplicity and efficiency but may lack flexibility when the data deviate from assumed patterns. We will explore the foundational concepts of parametric estimation, examine key models like the linear probability model (LPM) and logistic regression, and compare their applications, strengths, and limitations. This will build a comprehensive understanding before transitioning to the broader discussion of data modeling and algorithmic modeling in the next section. Following this, we will examine the differences between parametric and nonparametric models before exploring specific parametric methods like LPM and logistic regression.

7.1 The Dichotomy of Statistical Modeling

Statistical modeling can be broadly divided into data modeling and algorithmic modeling. Data modeling, rooted in parametric assumptions, focuses on understanding variable relationships and making causal inferences. Algorithmic modeling, often associated with machine learning, emphasizes predictive accuracy without strict distributional assumptions. While traditional econometrics has focused on causal inference, recent advancements in machine learning have expanded its toolkit, facilitating the use of both modeling approaches. The following sections will explore these distinctions further, providing a solid foundation for understanding how parametric and nonparametric models are applied in practice.

7.1.1 Data versus Algorithmic Approaches

In statistical modeling, a fundamental distinction exists between two approaches: data modeling and algorithmic modeling. These perspectives offer different philosophies on extracting insights from data and have significantly shaped the evolution of statistics. This chapter explores their contrast, drawing on insights from Leo Breiman's seminal work *Statistical Modeling: The Two Cultures* (Breiman, 2001) and subsequent reflections from leading statisticians (Gelman, 2021; Ogburn & Shpitser, 2021). Readers are encouraged to consult Breiman's paper and the *Observational Studies* special issue (2021) for a deeper understanding of this discussion.

The concept of the "two cultures" in statistical modeling echoes CP Snow's 1959 work on the divide between traditional and scientific cultures. In 2001, Breiman applied this notion to statistics, arguing that around 98% of statisticians adhered to the data modeling culture, which prioritizes interpretability through simplified models. He advocated for the broader

DOI: 10.1201/9781003504795-7

adoption of algorithmic models, particularly in cases where predictive accuracy is a primary objective.

Data modeling, or classical statistical modeling, assumes that the data follow a specific probabilistic structure, such as linear or logistic regression models. These models prioritize interpretability, often facilitating clear insights into variable relationships. Model validation typically relies on goodness-of-fit tests, and parametric assumptions provide a structured approach to inference. However, when data are high-dimensional or complex, strict reliance on parametric models may lead to oversimplified or biased conclusions.

Algorithmic modeling is often associated with machine learning, takes a more flexible approach by focusing on function estimation. Methods like neural networks, decision trees, and ensemble techniques emphasize predictive accuracy without relying heavily on assumptions about data distribution. While these models often excel in large-scale applications, they are typically less interpretable, posing challenges in fields where understanding the reasoning behind predictions is critical.

Breiman (2001) argued that the dominance of data modeling limited statisticians' engagement with real-world problems that could benefit from algorithmic methods. He highlighted scenarios where algorithmic models, particularly in fields like medical diagnosis and financial forecasting, had demonstrated superior predictive performance. In response, statisticians like Cox (2001) and Efron (2001) acknowledged the strengths of algorithmic models but cautioned against the potential loss of interpretability. They emphasized the importance of methodological diversity and the thoughtful application of both approaches.

Gelman (2021) further reflected on Breiman's perspective, critiquing his dismissal of Bayesian methods and advocating for their integration into modern statistical practice. He emphasized that algorithmic modeling could complement data modeling, especially in complex tasks such as hierarchical modeling and model checking. Similarly, Ogburn and Shpitser (2021) explored the role of algorithmic modeling in causal inference, arguing that machine learning can improve the estimation of the causal effect when used alongside traditional econometric methods.

In econometrics, the initial reaction to Breiman's paper was one of confusion, as the field was predominantly focused on causal inference rather than prediction. Econometricians were more concerned with estimating association and causal effects, leading to a divergence from Breiman's prediction-centered approach. Prominent statisticians such as Cox (2001), Efron (2001), and Gelman (2021) weighed in on Breiman's arguments providing thoughtful critiques and extensions of his viewpoints. Cox emphasized the importance of methodological diversity and cautioned against applying models mechanically. Efron, while praising Breiman's openness to new ideas, questioned the heavy emphasis on prediction in Breiman's work. In response, Breiman clarified that algorithmic modeling should not replace data modeling but rather complement it, particularly when tackling complex prediction problems where traditional methods struggle.

The evolving convergence of econometrics and machine learning illustrates the growing acceptance of algorithmic models in causal research. While econometricians historically prioritized causal effect estimation, the incorporation of machine learning techniques has expanded their toolkit. This shift allows for improved estimation of average treatment effects and treatment effects conditional on observable characteristics, as highlighted by recent applications in supply and demand estimation and returns to education (Athey & Imbens, 2019). Since then, the integration of these methods has continued to advance rapidly, contributing to more accurate and robust causal analysis.

Over the past two decades, Breiman's algorithmic perspective has gained significant traction. The rise of deep learning and machine learning, especially in prediction and classification tasks, has underscored the value of algorithmic approaches in a variety of scientific fields. Moreover, the integration of machine learning into econometric methods has expanded the field's focus to include causal discovery, creating a productive convergence between the two cultures. This development highlights the growing recognition that predictive accuracy and causal analysis can complement each other. For further reflections on the relevance of Breiman's work, see the commentaries in *Observational Studies* (2021).

In summary, the two cultures are not mutually exclusive. A balanced approach that leverages the strengths of both data and algorithmic modeling can improve research outcomes. By acknowledging the limitations and advantages of each, statisticians and econometricians can make more informed methodological choices. This perspective will remain essential as data complexities grow and the demand for accurate, interpretable insights continues to rise. Building on this discussion of data and algorithmic modeling, we now turn to another fundamental distinction in statistical analysis: parametric versus nonparametric models. This contrast further highlights how methodological choices are influenced by the balance between interpretability and predictive accuracy.

7.1.2 Parametric versus Nonparametric Models

Leo Breiman's distinction between data and algorithmic modeling has a nuanced relationship with parametric and nonparametric models. The data modeling approach, often associated with parametric models, operates under strict assumptions about the underlying data distribution. These models—such as linear regression, logistic regression, analysis of variance (ANOVA), and Poisson regression—rely on assumptions like normality or binomial distributions. Parametric models excel in their simplicity and interpretability, providing efficient ways to make inferences about the population, particularly when the data align well with their underlying assumptions. In these cases, parametric models offer clarity in estimating distribution parameters such as the mean and standard deviation. However, their reliability diminishes when the data deviate from these assumptions, potentially leading to biased or inaccurate predictions and conclusions.

On the other hand, the algorithmic modeling approach aligns more naturally with nonparametric models, which make no specific assumptions about the data distribution. Techniques like k-nearest neighbors, kernel density estimation, and decision trees such as classification and regression trees (CART) fall under this category. These models offer greater flexibility and adaptability, particularly in situations where the underlying distribution is unknown or does not meet the assumptions of parametric models. Nonparametric models are especially effective for handling ordinal or categorical data and are often used when parametric models are unsuitable. While they offer robustness and the ability to model complex, nonlinear relationships that parametric models struggle with, nonparametric models typically require larger datasets for effective learning and can be less efficient in terms of statistical power. This flexibility comes at the cost of interpretability, making nonparametric models less desirable in contexts where understanding the model's inner workings is critical.

It is important to note that the distinction between Breiman's two cultures and the choice between parametric and nonparametric models is not rigid. The data modeling approach, traditionally focused on inference and understanding the underlying structure of the data, can still incorporate nonparametric techniques when necessary. Conversely, the algorithmic approach, often associated with complex, black-box models, can utilize parametric models in contexts where simplicity or interpretability is valued. This fluidity highlights that the decision between parametric and nonparametric models should be guided by the characteristics of the data and the research objectives.

A balanced perspective also acknowledges the role of semi-parametric models, which combine elements of both parametric and nonparametric methods. For example, the Cox proportional hazards model is a widely used semi-parametric model in survival analysis. By incorporating both parametric assumptions and nonparametric components, semi-parametric models offer a compromise that provides interpretability while maintaining flexibility. Such models are particularly useful in fields where understanding the relationship between variables is crucial, but the underlying data distribution is complex or unknown.

Parametric models are well-suited for situations where the underlying data distribution is known and the focus is on estimating parameters within this framework. Their clarity and efficiency make them valuable tools in applications ranging from economics to the natural sciences. In contrast, nonparametric models offer a flexible alternative, particularly valuable when dealing with data that do not conform to specific distributional assumptions or when the distribution is unknown. These models have proven indispensable in fields such as machine learning, where handling high-dimensional, complex data requires adaptive techniques without the constraints of parametric assumptions.

Ultimately, the decision to use parametric or nonparametric models transcends the boundaries of Breiman's two cultures. It is shaped by the specific context, the nature of the data, and the goals of the analysis. As the lines between data modeling and algorithmic modeling continue to blur, so too does the distinction between parametric and nonparametric models. Both approaches provide valuable tools for understanding and predicting complex phenomena, and their ongoing interaction reflects the evolving nature of statistical modeling in modern data analysis.

As we continue exploring the intricacies of statistical modeling, the next section will focus on parametric estimations, where we will discuss two examples: linear probability models and logistic regression. Having already covered ordinary least square (OLS) in previous chapters, we will build on that foundation. Following this, we will turn our attention to nonparametric models, introducing flexible approaches that do not depend on specific distributional assumptions and highlighting their usefulness in addressing more complex data structures in the next chapter.

7.2 Parametric Estimations

In previous chapters, we covered the OLS method, simulating with polynomial functions to estimate the relationship between predictors and a continuous outcome. OLS is a regression method used in both data modeling and machine learning, where the goal is to estimate the parameters, β, of a known functional form that is linear in its coefficients. This aligns with the parametric data modeling approach discussed earlier, where we assume we know the form of the data-generating process.

However, when the outcome variable is binary, OLS can still be used but shifts into the realm of classification. In this case, the model is called the linear probability model (LPM). Like OLS, LPM assumes a linear relationship between the predictors X and the response Y, with β representing the parameters to be estimated. The key difference is that the outcome is now binary, taking values of 0 or 1. Despite this change, the functional form remains the same as in OLS.

Now, we will explore LPM, a simple model for binary classification, and demonstrate it using a simulation. This discussion will provide the groundwork for the next section, where we will cover logistic regression.

7.2.1 Linear Probability Model

The LPM is a straightforward method for predicting binary outcomes. It models the probability of an event occurring (e.g., high mpg or low mpg) as a linear function of the predictor variables:

$$\mathbb{P}(Y = 1|X) = X\beta \tag{7.1}$$

Here, β represents the coefficients estimated using least squares, and X contains the predictor variables. Similar to OLS, LPM uses the identity link function to directly relate the probability of $Y = 1$ to the linear combination of the predictors. However, one limitation of LPM is that it can predict probabilities outside the range of [0, 1], making it less suitable for classification in some cases. Nonetheless, it provides a simple and intuitive starting point for binary classification tasks.

We will demonstrate the LPM by simulating a dataset and predicting whether a vehicle has high or low fuel efficiency (mpg). First, we create a binary variable *mpg*, where 1 indicates high efficiency and 0 indicates low efficiency.

```
# Load necessary libraries
library(fueleconomy)
library(ggplot2)

# Load and prepare data
data(vehicles)
df <- as.data.frame(vehicles)

# Keep only complete cases
data <- df[complete.cases(df), ]

# Create binary mpg variable: 1 if hwy > mean(hwy), 0 otherwise
data$mpg <- ifelse(data$hwy > mean(data$hwy), 1, 0)

# Convert character variables to factors
for (i in 1:ncol(data)) {
  if(is.character(data[,i])) data[,i] <- as.factor(data[,i])
}

# Fit Linear Probability Model (LPM)
model_lpm <- lm(mpg ~ fuel + 0, data = data)
summary(model_lpm)
```

```
##
## Call:
```

```
## lm(formula = mpg ~ fuel + 0, data = data)
##
## Residuals:
##      Min       1Q   Median       3Q      Max
## -0.8571 -0.4832 -0.2694  0.5168  0.7306
##
## Coefficients:
##                                Estimate Std. Error t value Pr(>|t|)
## fuelCNG                        0.362069   0.065383   5.538 3.09e-08 ***
## fuelDiesel                     0.479405   0.016843  28.463  < 2e-16 ***
## fuelGasoline or E85            0.269415   0.015418  17.474  < 2e-16 ***
## fuelGasoline or natural gas    0.277778   0.117366   2.367   0.0180 *
## fuelGasoline or propane        0.000000   0.176049   0.000   1.0000
## fuelMidgrade                   0.302326   0.075935   3.981 6.87e-05 ***
## fuelPremium                    0.507717   0.005364  94.650  < 2e-16 ***
## fuelPremium and Electricity    1.000000   0.497942   2.008   0.0446 *
## fuelPremium Gas or Electricity 0.857143   0.188205   4.554 5.27e-06 ***
## fuelPremium or E85             0.500000   0.053081   9.420  < 2e-16 ***
## fuelRegular                    0.483221   0.003311 145.943  < 2e-16 ***
## fuelRegular Gas and Electricity 1.000000   0.176049   5.680 1.36e-08 ***
## ---
## Signif. codes:  0 '***' 0.001 '**' 0.01 '*' 0.05 '.' 0.1 ' ' 1
##
## Residual standard error: 0.4979 on 33370 degrees of freedom
## Multiple R-squared:  0.4862, Adjusted R-squared:  0.486
## F-statistic:  2631 on 12 and 33370 DF,  p-value: < 2.2e-16
```

This model predicts the probability of mpg = 1 (high fuel efficiency) based on the type of fuel used. In this case, the identity link function relates the probability directly to the linear function of the predictor variables.

We can examine the predicted probabilities and classification accuracy using a threshold of 0.5 for mpg = 1.

```
# Predicted probabilities for mpg = 1
mpg_hat <- fitted(model_lpm)

# Classify based on threshold of 0.5
mpg_pred <- ifelse(mpg_hat > 0.5, 1, 0)

# Summary of predicted probabilities
summary(mpg_hat)
```

```
##    Min. 1st Qu.  Median    Mean 3rd Qu.    Max.
##  0.0000  0.4832  0.4832  0.4824  0.5077  1.0000
```

```
# Table of predicted vs actual mpg values
table(Predicted = mpg_pred, Actual = data$mpg)
```

```
##          Actual
## Predicted     0     1
```

```
##           0 12993 11712
##           1  4287  4390
```

The predicted values above show how the LPM operates in practice. However, as seen, the predicted probabilities are not restricted between 0 and 1, which is one of the limitations of the LPM. We will address these issues by introducing a more suitable classification model in the next section.

Although LPM provides a simple approach to classification, its limitations make it unsuitable for more complex tasks. To overcome these issues, we will explore logistic regression in the following section, a more robust model that ensures predicted probabilities are always between 0 and 1.

7.2.2 Logistic Regression

After discussing the LPM in the previous section, which demonstrated a simple yet limited approach for binary classification, we now turn to a more robust method: logistic regression. While LPM assumes a linear relationship between predictors and the probability of the outcome, it can predict probabilities outside the [0, 1] range and lacks flexibility in modeling nonlinear relationships. This is where logistic regression becomes crucial. By using a nonlinear function, it overcomes the issues associated with LPM, particularly in cases where probabilities must be confined within valid bounds.

Logistic regression is widely used across various fields, including healthcare for predicting disease presence, finance for assessing credit risk, and marketing for customer churn prediction. Its ability to provide interpretable probabilities makes it a preferred choice in many real-world applications.

Logistic regression is also a special case of a generalized linear model (GLM), using the logistic function as its link function to model the relationship between predictors and a binary outcome. This connection to GLMs allows it to apply linear modeling principles while ensuring the output remains within the valid probability range.

We begin by introducing key notation for logistic regression, a method widely used in economics and statistics to model the probability of a binary outcome. Let Y denote a binary dependent variable, taking values $Y = 1$ (e.g., success, event occurs) or $Y = 0$ (e.g., failure, event does not occur), and let $\mathbf{X}$ represent a vector of explanatory variables (predictors). The goal is to estimate the conditional probability of $Y = 1$ given $\mathbf{X} = \mathbf{x}$:

$$p(\mathbf{x}) = P(Y = 1|\mathbf{X} = \mathbf{x}) \tag{7.2}$$

Since Y is binary, the probability of the complementary outcome, $Y = 0$, is:

$$P(Y = 0|\mathbf{X} = \mathbf{x}) = 1 - p(\mathbf{x}) \tag{7.3}$$

In logistic regression, we model this probability using the logistic function, also known as the sigmoid function, which maps a linear combination of predictors into the interval $[0, 1]$. The general form of the logistic function is:

$$p(x) = \frac{L}{1 + e^{-k(x - x_0)}} \tag{7.4}$$

Here, L denotes the upper asymptote (the maximum value of the function), k governs the steepness of the curve, and x_0 is the value of x at which the function reaches $L/2$, the midpoint of its range.

In statistical applications like logistic regression, we typically standardize this function. The standard logistic function, commonly employed in logistic regression, arises by setting $L = 1$ (to bound probabilities between 0 and 1), $k = 1$ (for a standard slope), and $x_0 = 0$ (centering the function at zero):

$$p(\mathbf{x}) = \frac{1}{1 + e^{-z}} \tag{7.5}$$

where $z = \beta_0 + \beta^\top f(x)$ is the linear predictor, with β_0 as the intercept and β as the vector of coefficients corresponding to $\mathbf{x}$.

This function, often denoted $\sigma(z)$, is also called the sigmoid or expit function (the inverse of the logit). It can be equivalently expressed as:

$$\sigma(z) = \frac{e^z}{1 + e^z} = \frac{1}{1 + e^{-z}} \tag{7.6}$$

In this context, z encapsulates the effect of the predictors on the log-odds of the outcome. For a simple case with one predictor x, we have $z = \beta_0 + \beta_1 x$. If $\beta_0 = 0$ and $\beta_1 = 1$, then $z = x$, reducing the expression to the standard form:

$$p(x) = \frac{1}{1 + e^{-x}} \tag{7.7}$$

The primary advantage of the logistic function is its ability to map any real-valued input to a value between 0 and 1, making it ideal for modeling probabilities. The curve's S-shape is particularly useful in binary classification tasks, as it reflects scenarios where the relationship between the input and the outcome probability is nonlinear. Near the midpoint, the probabilities change most rapidly, while the changes slow down as the values approach the asymptotes at 0 and 1. This characteristic is especially relevant in practical applications, such as predicting the probability of disease presence based on medical indicators or assessing loan default likelihood based on financial variables. In summary, the logistic function's bounded nature and flexible shape provide an effective way to model the probabilities of binary outcomes, making it a foundational component of logistic regression models.

While logistic regression resolves the probability range issue, it still relies on a parametric form, assuming a linear relationship between the predictors and the log-odds of the outcome. This can be a limitation when dealing with more complex data structures. In such cases, more flexible models like decision trees or neural networks may be preferred. Nevertheless, logistic regression remains a valuable tool when interpretability and clear probabilistic predictions are essential.

To better understand how the logistic function maps inputs to probabilities, we can visualize its characteristic S-shaped curve for different values of z:

As x moves toward large positive or negative values, the predicted probabilities approach 1 or 0, respectively. This nonlinear mapping is particularly useful in real-world scenarios where marginal effects diminish at extreme values. For instance, in predicting the effect of savings on home ownership, small changes in savings may have little effect at both low and high savings levels. The logistic function captures this behavior through its characteristic S-shaped curve.

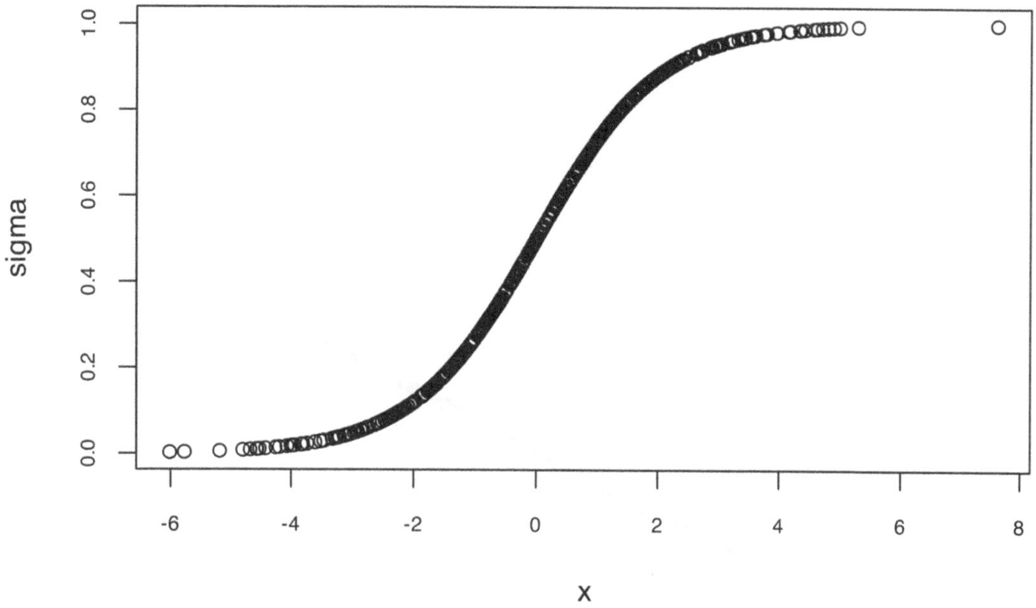

FIGURE 7.1 S-shaped curve of the logistic function.

In logistic regression, the probability of the outcome is modeled as:

$$p(x) = P[Y = 1|X = x] = \frac{1}{1 + e^{-(\beta_0 + \beta_1 x)}} \tag{7.8}$$

This equation highlights the nonlinearity of the relationship between the predictor and the probability of $Y = 1$, driven by the logistic function. While a small change in x near the middle of the curve may result in a significant change in probability, the effect diminishes as the curve flattens at the tails.

To further understand this, consider the scenario of predicting home ownership based on savings. At extremely low savings, the probability of owning a home remains close to zero, even with slight increases in savings. Similarly, once savings are sufficiently high, further increases have little effect, as the probability approaches one. This diminishing effect is a key characteristic of the nonlinearity of the logistic function.

The logistic function can also be expressed in terms of log-odds, which linearizes the relationship:

$$\ln\left(\frac{p(x)}{1 - p(x)}\right) = \beta_0 + \beta_1 x \tag{7.9}$$

Here, the term $\frac{p(x)}{1-p(x)}$ represents the odds of the outcome occurring, and the logarithm of the odds is referred to as the logit. This transformation is fundamental in logistic regression, allowing for a direct interpretation of the coefficients β_0 and β_1 in terms of their effect on the log-odds.

We can visualize this relationship by plotting the log-odds transformation of the predicted probabilities:

```
p_x <- sigma
Logit <- log(p_x / (1 - p_x))
plot(Logit ~ x, col = "black", cex.axis = 0.7)
```

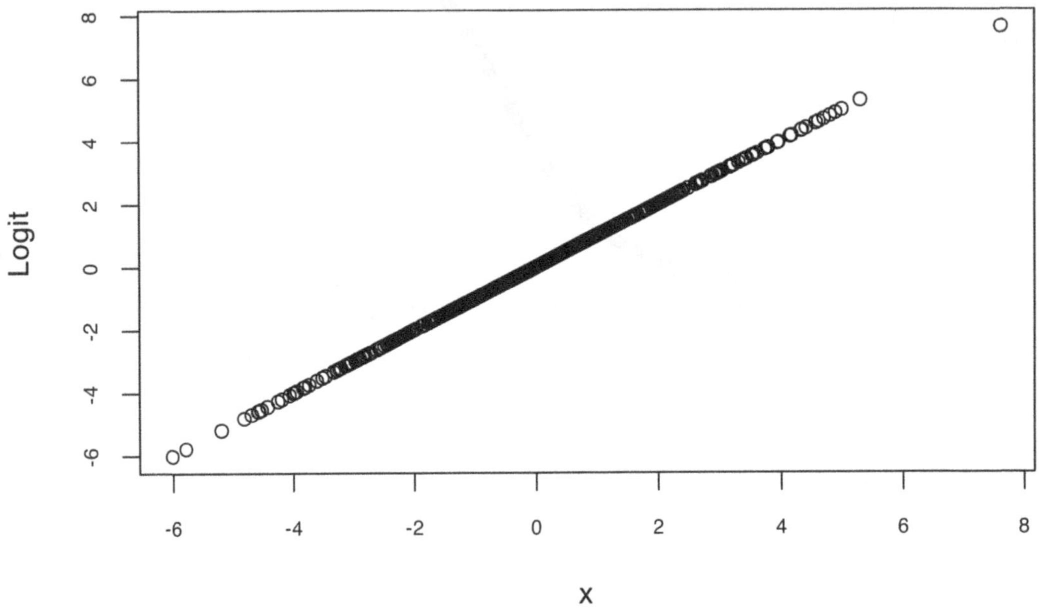

FIGURE 7.2 Log-odds transformation of predicted probabilities.

As expected, the log-odds form a linear relationship with x, confirming the underlying assumption of logistic regression. This linearity in the logit space allows logistic regression to model complex, nonlinear relationships in the probability space while maintaining interpretability and computational efficiency.

7.2.3 Linear vs. Logistic Probability Models

When deciding between linear and logistic models, the choice depends on the nature of the data and the desired output. The LPM is simple to implement but can predict probabilities outside the [0, 1] range and struggles with nonlinearity. Logistic regression solves these issues by using the logistic function, which maps predicted values within the [0, 1] range.

In many scenarios, researchers turn to logistic regression for binary outcomes. While the linear model may be suitable for certain cases, logistic regression is more appropriate when the relationship between the predictors and the probability of the outcome is nonlinear. To compare both models, consider the following representations:

$$P[Y = 1|\mathbf{X} = \mathbf{x}] = E[Y|\mathbf{X} = \mathbf{x}] = \beta_0 + \beta_1 x_1 + \beta_2 x_2 + \ldots + \beta_k x_k \tag{7.10}$$

$$\ln\left(\frac{P[Y = 1|\mathbf{X}]}{1 - P[Y = 1|\mathbf{X}]}\right) = \beta_0 + \beta_1 x_1 + \ldots + \beta_k x_k \tag{7.11}$$

In the linear model, P is a linear function of the predictors, while in the logistic model, the log-odds of P is linear. The logistic model requires maximum likelihood estimation (MLE), while LPM can be estimated using OLS.

The logistic function is defined as:

$$p(\mathbf{x}) = E[Y|\mathbf{X} = \mathbf{x}] = P[Y = 1|\mathbf{X} = \mathbf{x}] = \frac{1}{1 + e^{-(\beta_0 + \beta_1 x_1 + \cdots + \beta_{p-1} x_{p-1})}} \tag{7.12}$$

One key advantage of LPM is its interpretability. In a linear model, a coefficient like $\beta_2 = 0.05$ means a one-unit increase in x_2 raises the probability of $Y = 1$ by 5 percentage points. On the other hand, in logistic regression, $\beta_1 = 0.05$ means a one-unit increase in x_1 leads to a 0.05 increase in the log-odds, which is harder to interpret directly.

The logistic model is essential when it fits the data better, especially when probabilities are nonlinear. However, in many cases, both models give similar results, but the logistic model's estimates are more complex to interpret.

7.2.3.1 Simulation Example

Let's illustrate this with a simulation.

```
#Creating random data
set.seed(1)
n <- 500
x = rnorm(n)
z = -2 + 3 * x

#Probabilities are defined by a logistic function
p = 1 / (1 + exp(-z))

#Bernoulli distribution defines Y as 1 or 0
y = rbinom(n, size = 1, prob = p)

#Create the data
data <-  data.frame(y, x)
head(data)
```

```
##   y          x
## 1 0 -0.6264538
## 2 0  0.1836433
## 3 0 -0.8356286
## 4 0  1.5952808
## 5 0  0.3295078
## 6 0 -0.8204684
```

```
table(y)
```

```
## y
##   0   1
## 353 147
```

Next, fit the data using LPM:

```
lpm <- lm(y ~ x, data = data)
summary(lpm)
```

```
##
## Call:
## lm(formula = y ~ x, data = data)
##
## Residuals:
##      Min       1Q   Median       3Q      Max
## -0.76537 -0.25866 -0.08228  0.28686  0.82338
##
## Coefficients:
##             Estimate Std. Error t value Pr(>|t|)
## (Intercept)  0.28746    0.01567   18.34   <2e-16 ***
## x            0.28892    0.01550   18.64   <2e-16 ***
## ---
## Signif. codes:  0 '***' 0.001 '**' 0.01 '*' 0.05 '.' 0.1 ' ' 1
##
## Residual standard error: 0.3504 on 498 degrees of freedom
## Multiple R-squared:  0.411,  Adjusted R-squared:  0.4098
## F-statistic: 347.5 on 1 and 498 DF,  p-value: < 2.2e-16
```

```
#Plot the fitted probabilities against the true values
plot(x, p, col = "gray", cex.lab = 0.7, cex.axis = 0.8)
abline(lpm, col = "black")
legend("topleft", c("Estimated Probability by LPM", "True Probability"), lty = c(1, 1),
pch = c(NA, NA), lwd = 2, col = c("black", "gray"), cex = 0.7)
```

Now, fit a logistic regression:

```
logis <- glm(y ~ x, data = data, family = binomial)
summary(logis)
```

```
##
## Call:
## glm(formula = y ~ x, family = binomial, data = data)
##
## Coefficients:
##             Estimate Std. Error z value Pr(>|z|)
## (Intercept)  -1.8253     0.1867  -9.776   <2e-16 ***
## x             2.7809     0.2615  10.635   <2e-16 ***
## ---
## Signif. codes:  0 '***' 0.001 '**' 0.01 '*' 0.05 '.' 0.1 ' ' 1
##
## (Dispersion parameter for binomial family taken to be 1)
##
##     Null deviance: 605.69  on 499  degrees of freedom
## Residual deviance: 328.13  on 498  degrees of freedom
```

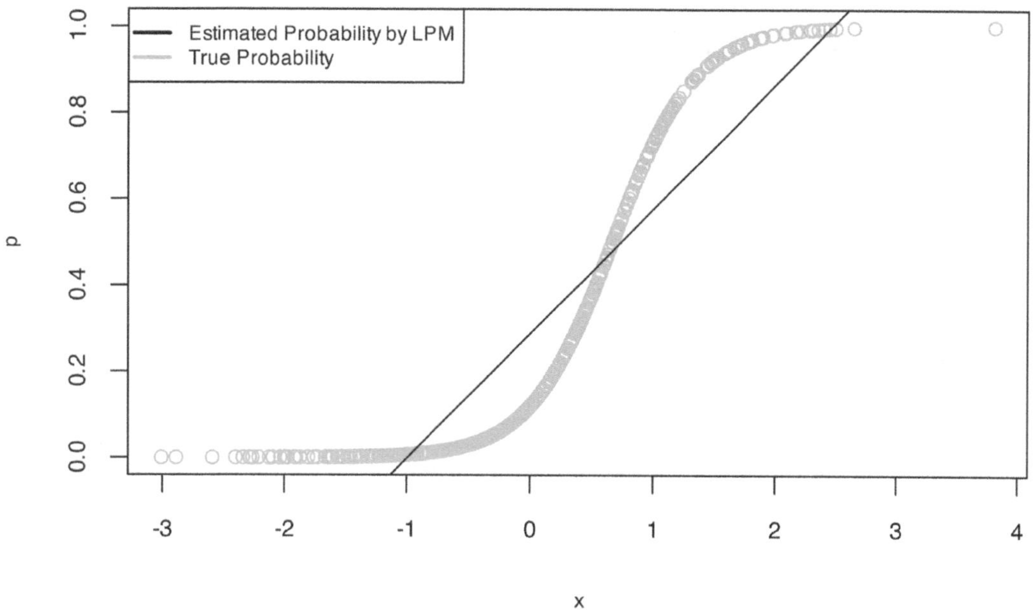

FIGURE 7.3 True vs. estimated probabilities from LPM.

```
## AIC: 332.13
##
## Number of Fisher Scoring iterations: 6
```

```
#Plot the fitted probabilities for logistic regression
plot(x, p, col = "gray", cex.lab = 0.8, cex.axis = 0.8)
curve(predict(logis, data.frame(x), type = "response"), add = TRUE, col = "black", lty = 2)
legend("topleft", c("Estimated Probability by GLM", "True Probability"), lty = c(1, 1),
pch = c(NA, NA), lwd = 2, col = c("black", "gray"), cex = 0.7)
```

As shown, logistic regression yields estimates consistent with the data-generating mechanism, whereas LPM tends to fit a straight line. The fitted logistic regression equation is:

$$\log\left(\frac{1-\hat{p}(x)}{\hat{p}(x)}\right) = -1.8253 + 2.7809x \tag{7.13}$$

This demonstrates the suitability of logistic regression for binary outcomes where nonlinear probability predictions are required.

Parametric models, with their clear interpretability and efficient estimation, remain central to econometric analysis. Techniques such as the LPM and logistic regression are invaluable tools for understanding binary outcomes and estimating relationships within structured data. Other common parametric models include Probit models, Poisson regression, and Tobit models, each designed to handle specific types of dependent variables such as binary, count, or censored data. These models offer interpretability and efficiency when the underlying assumptions about the data are met. However, in many real-world cases, these models are not without limitations, particularly when assumptions about the data distribution are violated. As we move into the next chapter, we will explore nonparametric models that relax many of these assumptions, offering greater flexibility in capturing complex relationships.

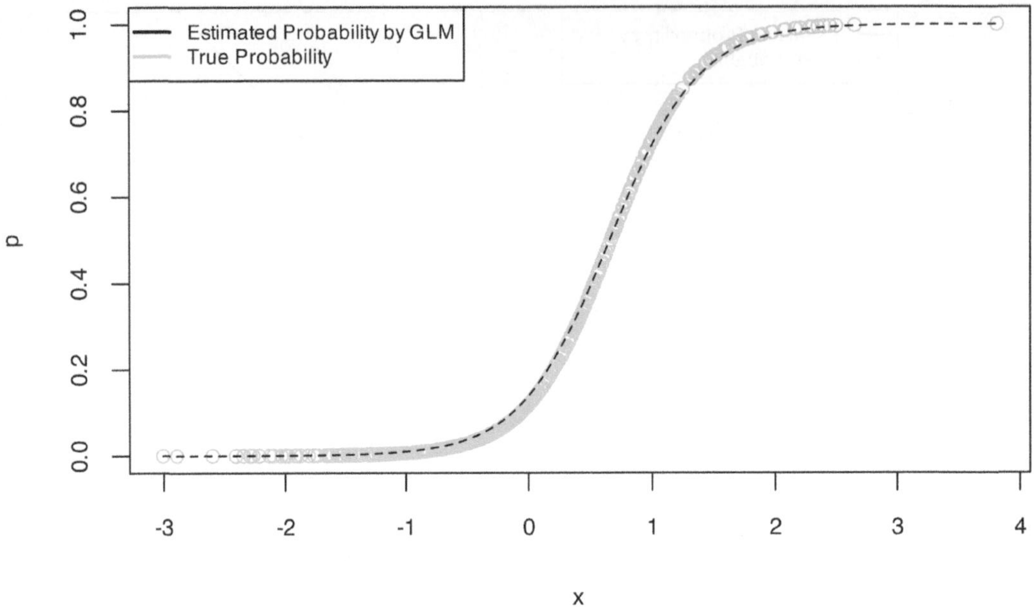

FIGURE 7.4 True vs. estimated probabilities from logistic regression.

By understanding both parametric and nonparametric approaches, you will be equipped to make informed decisions about which model best suits your analytical objectives. This comprehensive perspective will enable more robust and nuanced interpretations of data in applied research.

8

Nonparametric Estimations: Basics

In previous chapter, we explored the Dichotomy of Statistical Modeling with a focus on parametric vs. nonparametric models. Parametric models, as discussed, assume that the data-generating process follows a specific form, with a fixed number of parameters that we estimate. This assumption allows for simpler interpretations and computational efficiency, making parametric models a popular choice in both economics, social and health sciences, and machine learning. In the previous chapter, we concentrated on parametric estimations, such as the linear probability model and logistic regression, which rely on known functional forms and assumptions about the relationship between predictors and outcomes. These models allow us to find associations, estimate causal effects, and offer robust predictions and interpretations when the underlying assumptions are valid. However, they can be limiting when the true data-generating process is complex or unknown, which is often the case.

This chapter introduces nonparametric estimations, a more flexible approach that does not assume a specific functional form for the data. Nonparametric methods adapt to the structure of the data, allowing for more complex relationships to emerge without predefined assumptions. However, this flexibility comes at the cost of interpretability and computational efficiency, especially with larger datasets.

Nonparametric regressions form the foundation of many modern machine learning techniques due to their flexibility in modeling complex relationships without making strong assumptions about the underlying data structure. Unlike parametric models, which assume a specific functional form, nonparametric methods allow the data to guide the shape of the model, making them ideal for uncovering patterns in high-dimensional or nonlinear datasets. This flexibility is key in machine learning, where tasks often involve capturing intricate relationships between variables, such as predicting customer behavior or disease outcomes. Nonparametric methods such as kernel regression, decision trees, and k-nearest neighbors serve as the basis for more advanced algorithms like random forests, support vector machines, and deep learning, which further improve the predictive power while maintaining the ability to generalize well to unseen data. These methods are particularly powerful in situations where the data is too complex for simple parametric models, making nonparametric estimations a crucial building block for machine learning applications.

We will begin by discussing the basics of density estimation, a core concept in nonparametric statistics. From there, we will explore one of the most widely used nonparametric techniques: kernel density estimation and regression. These methods provide a foundation for understanding more advanced nonparametric models, making this chapter an essential starting point for those looking to extend beyond the constraints of parametric assumptions.

DOI: 10.1201/9781003504795-8

8.1 Density Estimation

In statistical modeling, understanding the distribution of a dataset is essential for many types of analysis. Parametric models typically assume that the data follow a known distribution, such as the normal distribution, with a few parameters to estimate. However, when we wish to avoid making strong assumptions about the underlying distribution, nonparametric density estimation offers a flexible approach to estimating the probability distribution of a random variable.

Let's assume that a sample of n observations, $y_1, \ldots, y_n$, is drawn from a parametric distribution $f(y;\theta)$. The joint density function for these independent and identically distributed (i.i.d.) observations is:

$$f(y;\theta) = \prod_{i=1}^{n} f(y_i;\theta) \tag{8.1}$$

To estimate θ, we maximize the likelihood or log-likelihood:

$$\ell(y;\theta) = \sum_{i=1}^{n} \log f(y_i;\theta) \tag{8.2}$$

This approach is called *parametric estimation*. If $f(y;\theta)$ is not the correct form, it leads to biased estimates.

To avoid such biases, we use *nonparametric estimation*, which estimates the density directly from the data without assuming a specific parametric form. Given a set of data points $X_1, X_2, \ldots, X_n$ drawn from some unknown probability distribution with a probability density function (PDF) $f(x)$, the goal of density estimation is to estimate $f(x)$ without assuming any particular form for it. In nonparametric methods, the shape of the density function is determined directly from the data, offering more flexibility compared to parametric methods. Here, y is used for parametric estimation, as it typically represents the dependent variable or outcomes in parametric estimations, and x is used in the context of nonparametric density estimation, where we are focused on the independent variable whose distribution we wish to estimate.

8.1.1 Histogram Density Estimation

One of the simplest forms of nonparametric density estimation is the histogram. The histogram groups the data into bins and counts the number of observations that fall into each bin. The density is then estimated by normalizing these counts by the total number of observations and the width of the bins. The histogram estimate of the density at a point x is given by:

$$\hat{f}(x) = \frac{1}{nh} \sum_{i=1}^{n} \mathbf{1}(x_i \in \text{Bin}(x)) \tag{8.3}$$

Where:

- $\mathbf{1}(x_i \in \text{Bin}(x))$ is an indicator function that is 1 if x_i is within the bin containing x, and 0 otherwise. Sum of this indicator function gives the number of observations in the bin containing x.

- h is the bin width.
- n is the total number of observations.

For instance, assume there are 2,000 total observations, $x = 3$, and the bin width $h = 1$. The bin containing $x = 3$ covers the interval $[x - h, x + h]$, which is $[3 - 1, 3 + 1] = [2, 4]$. If there are 100 observations in this bin, the estimated density $\hat{f}(3)$ would be: $\hat{f}(3) = \frac{1}{2000 \times 1} \times 100 = \frac{100}{2000} = 0.05$.

Thus, the estimated density at $x = 3$ is 0.05, meaning that approximately 5% of the total observations fall in the interval $[2, 4]$. While histograms are easy to understand, they suffer from limitations such as their dependence on bin width and location, which can lead to poor approximations of the underlying density.

8.1.2 Naive Estimator (Parzen Windows)

A more sophisticated approach to nonparametric density estimation is the *naive estimator* or *Parzen windows*. This method smooths the distribution by using a moving window (with intersections between bins) instead of fixed bins. The naive estimator is given by:

$$\hat{f}(x) = \frac{1}{nh} \sum_{i=1}^{n} I\left(x - \frac{h}{2} < x_i < x + \frac{h}{2}\right) \tag{8.4}$$

Where $I(\cdot)$ is an indicator function, and h is the bin width, which defines the range of the bin. While the naive estimator offers some improvement over histograms, it assigns equal weight to all points within a given window, regardless of their distance from the estimation point.

8.2 Kernel Density Estimation

Kernel density estimation (KDE) addresses the limitations of both histograms and the naive estimator by assigning weights to each data point based on their distance from the estimation point. Instead of using uniform weights, KDE smooths the distribution of data points by placing a "kernel" over each data point and then summing the contributions from all kernels to estimate the density. The kernel function, often a smooth, symmetric function such as a Gaussian, controls how much influence each data point has on the estimate of $f(x)$.

The kernel density estimator of $f(x)$ is given by:

$$\hat{f}(x; h) = \frac{1}{nh} \sum_{i=1}^{n} K\left(\frac{x - x_i}{h}\right) \tag{8.5}$$

Where:

- $K(\cdot)$ is the kernel function, typically a smooth, symmetric function such as the Gaussian, Epanechnikov, or Uniform kernel.
- h is the bandwidth parameter, controlling the smoothness of the estimate.
- n is the total number of data points.

The choice of bandwidth h is crucial in determining the quality of the density estimate. A small h leads to an overfit, with the estimate capturing too much noise, while a large h may oversmooth the data, losing important details of the distribution.

Another common notation is:

$$K_h(z) = \frac{1}{h} K\left(\frac{z}{h}\right), \tag{8.6}$$

so the KDE can be compactly written as:

$$\hat{f}(x; h) = \frac{1}{n} \sum_{i=1}^{n} K_h(x - X_i). \tag{8.7}$$

8.2.1 Common Kernels

- *Gaussian kernel*:

$$K(u) = \frac{1}{\sqrt{2\pi}} e^{-\frac{u^2}{2}} \tag{8.8}$$

- *Epanechnikov kernel*:

$$K(u) = \frac{3}{4}(1 - u^2) \quad \text{for} \quad |u| \leq 1 \tag{8.9}$$

- *Uniform kernel*:

$$K(u) = \frac{1}{2} \quad \text{for} \quad |u| \leq 1 \tag{8.10}$$

- There are others such as triangular kernel, quartic kernel, Biweight kernel, Triweight kernel, etc.

For instance, suppose we have 2,000 total observations, and we want to estimate the density at $x = 3$ with a bandwidth $h = 1$. The Epanechnikov kernel gives more weight to observations near $x = 3$ and less weight to those further away. If we have the following five data points: $x_1 = 2.5$, $x_2 = 2.7$, $x_3 = 3.1$, $x_4 = 3.5$, and $x_5 = 3.8$, we can compute the contribution of each point to the density estimate at $x = 3$ as follows:

$$\hat{f}(3) = \frac{1}{5 \times 1} \left[\frac{3}{4}(1 - (0.5)^2) + \frac{3}{4}(1 - (0.3)^2) + \frac{3}{4}(1 - (0.1)^2) + \frac{3}{4}(1 - (0.5)^2) + 0 \right]$$

Since the Epanechnikov kernel assigns a weight of 0 for any point more than 1 unit away from x, the observation at $x_5 = 3.8$ does not contribute to the estimate because $\left|\frac{3-3.8}{1}\right| = 0.8$ exceeds the support of kernel of $[-1, 1]$. The density estimate is then computed as $\hat{f}(3) = 0.51$. Thus, the estimated density at $x = 3$ using the Epanechnikov kernel with $h = 1$ is approximately 0.51. This example demonstrates how the Epanechnikov kernel assigns different weights to data points based on their proximity to the point of interest, leading to a smooth density estimate. Although the type of kernel function $K(u)$ influences the estimate, the choice of bandwidth h is far more important in determining the overall smoothness.

8.2.2 Simulation: KDE vs. Histogram

We can now simulate a density estimation example using both a histogram and a kernel density estimator. We will generate a set of random data points from a mixture of normal distributions and compare the estimated density using both the histogram and KDE methods.

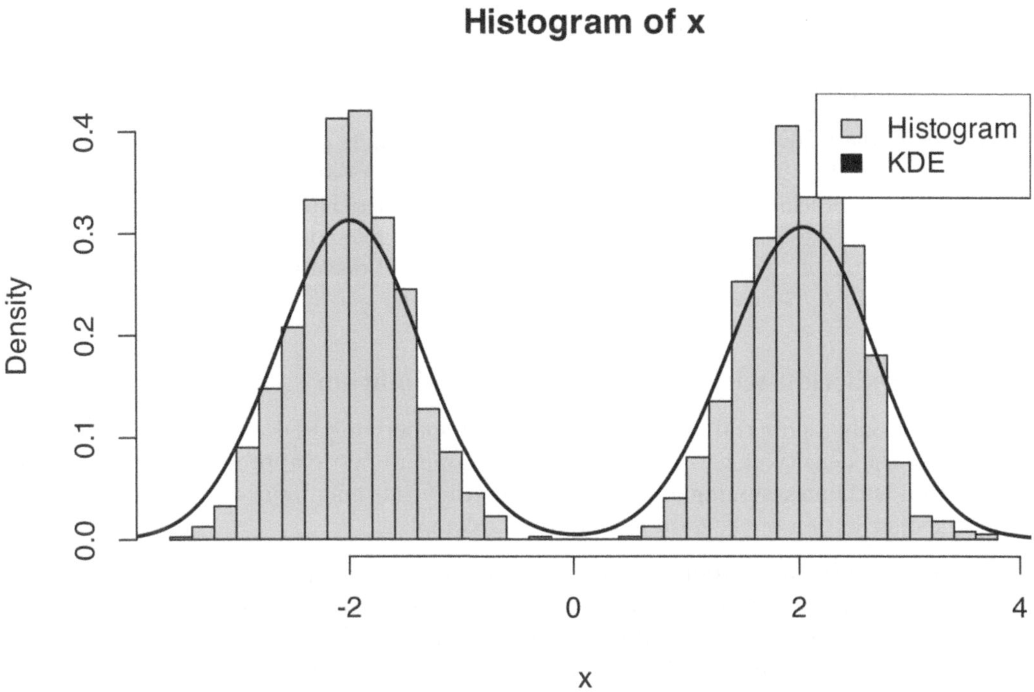

FIGURE 8.1 Histogram vs. kernel density estimation.

This will demonstrate how KDE offers a smoother and more flexible estimate compared to the histogram method.

In this simulation, we generated $n = 2000$ data points from a mixture of two normal distributions, one centered at -2 and the other at 2, both with a standard deviation of 0.5. The goal is to estimate the PDF using both a histogram and KDE.

The histogram groups the data into bins. For each bin, it counts the number of observations and normalizes by the total number of observations $n = 2000$ and the bin width h. For example, at $x = 3$, we check the bin containing $x = 3$. Let's assume the bin width $h = 0.2$. If there are 15 observations in the bin, the density estimate is computed as:

$$\hat{f}_{\text{hist}}(3) = \frac{1}{n \cdot h} \sum_{i=1}^{n} \mathbf{1}(x_i \in \text{Bin}(3)) = \frac{15}{2000 \cdot 0.2} = 0.0375$$

For every value of x, this process is repeated to estimate the density across all bins.

For the KDE, we use a kernel function (in this case, the Gaussian kernel). The kernel function places a smooth, continuous curve over each observation, and the density estimate at $x = 3$ is computed by summing the contributions from all observations weighted by their distance from $x = 3$.

$$\hat{f}_{\text{kde}}(3) = \frac{1}{2000 \times 1} \sum_{i=1}^{2000} \frac{1}{\sqrt{2\pi}} e^{-\frac{(3-x_i)^2}{2}} \approx 0.05$$

For each point x, both the histogram and KDE methods compute the density using the same principles described above, with KDE providing more flexibility due to the choice of kernel and bandwidth.

In this case, the KDE (shown by the line in Figure 8.1) offers a smoother estimate compared to the histogram, especially in regions where the data points are sparse. The choice of bandwidth h is crucial in determining the smoothness of the KDE. A small bandwidth leads to a more detailed estimate (but may overfit the data), while a large bandwidth smooths out the estimate (but may miss important features). We will discuss how to select appropriate hyperparameter h, bandwidth, in the next chapter.

8.2.3 Prediction with Kernel Density Estimation

KDE is not only a powerful tool for estimating the underlying PDF of a dataset but also for predicting the density at new data points. This nonparametric approach allows for flexible estimation without assuming a specific parametric form, making it especially useful when the true distribution of the data is unknown or complex.

In many real-world applications, we often encounter situations where we need to predict the likelihood or density of a new observation that hasn't been seen before. In finance, KDE can be used for estimating the probability of extreme returns for asset pricing. In healthcare, KDE helps in predicting the likelihood of disease occurrence based on patient data. In machine learning, it is useful for model calibration, anomaly detection, and probabilistic forecasting. In environmental science, KDE can predict rare events like natural disasters based on historical data. By avoiding strict parametric assumptions, KDE provides a more flexible and accurate method in cases where traditional models may fail to capture complex, multimodal distributions.

The process of KDE involves placing a smooth kernel (like a Gaussian) over each data point and then summing these contributions to estimate the density. The bandwidth h determines the smoothness of the resulting estimate, with smaller bandwidths capturing more details but risking overfitting, and larger bandwidths providing a smoother, more generalized estimate.

Once the density is estimated, we can also use it to predict the PDF for new data points. This can be crucial for making probabilistic forecasts in various domains, including risk management, anomaly detection, and machine learning model calibration. For example, instead of assuming a normal distribution for a dataset, KDE lets us model the true underlying distribution directly from the data, allowing for more accurate predictions.

How does the simulation calculate the prediction? The simulation below shows how we can estimate the density using KDE and then predict the PDF at a new point $x = 1.832$. We first generate a dataset of 100 observations, perform kernel density estimation, and create an interpolation function from the KDE result. Finally, we use this interpolation function to predict the density for the new data point.

Here is a simulation example that predicts the PDF for a new data point using KDE:

```
set.seed(123)
data <- rnorm(100)  # Generate 100 random normal values
# Perform kernel density estimation
kde_result <- density(data)
# Create an interpolation function from the density estimate
pdf_estimate <- approxfun(kde_result$x, kde_result$y)
```

```
# Predict the PDF at a new point (e.g., x = 1.832)
new_x <- 1.832
predicted_pdf <- pdf_estimate(new_x)
# Print the predicted PDF value
predicted_pdf
```

```
## [1] 0.09141111
```

In this simulation, first we generate 100 random normal values. The kernel density estimation process fits a smooth curve over the data points to estimate the underlying PDF. The `density()` function in R computes this kernel density estimate, and the `approxfun()` function is used to create a smooth interpolation function that can predict the PDF at any new point.

For example, we predict the PDF for $x = 1.832$, which yields a density value (i.e., the likelihood of observing a value around $x = 1.832$). This prediction is made without any parametric assumptions about the data distribution, allowing us to model more complex or multimodal distributions effectively. By leveraging the flexibility of KDE, we can avoid restrictive assumptions and create more accurate models for data that may not follow traditional distributional patterns.

8.2.4 Kernel Regression

Building on the concepts introduced in the chapter on density estimations, we now explore **kernel regression**, a nonparametric method for estimating the regression function $E[Y|X = x]$. While kernel density estimation focuses on estimating the probability distribution of a set of data points, kernel regression aims to estimate the conditional mean of the response variable Y given the predictor X, without assuming any specific functional form for the relationship between them. This is analogous to the OLS with one exogenous variable used in linear regression. The KDE can be viewed as a smooth, weighted average of observed values, where the weights are determined by the proximity of each observation to the point of interest. This flexibility allows kernel regression to capture nonlinear patterns in data that parametric models might miss, avoiding the risk of model misspecification.

In healthcare, it can be used to predict patient recovery times based on factors such as age and treatment type, offering a flexible alternative to linear models. Another example is predicting housing prices based on features like square footage, number of rooms, and distance to the city center. The Nadaraya–Watson estimator would weigh the influence of nearby houses (in terms of square footage) more heavily than distant ones.

The Nadaraya–Watson estimator is a common form of kernel regression, given by:

$$\hat{m}(x;h) == \sum_{i=1}^{n} \frac{K_h(x - X_i)}{\sum_{i=1}^{n} K_h(x - X_i)} Y_i = \sum_{i=1}^{n} W_i^0(x) Y_i, \tag{8.11}$$

Where:

- $W_i^0(x) = \frac{K_h(x-X_i)}{\sum_{i=1}^{n} K_h(x-X_i)}$ is the weight assigned to the ith data point in the regression estimate at x.
- $K_h(x - X_i)$ is the kernel function applied to the distance between x and the observed data points X_i. $K_h(x - X_i) = \frac{1}{h} K\left(\frac{x-X_i}{h}\right)$. The kernel function $K(\cdot)$ determines how much influence each data point has on the regression estimate at a particular x. Common kernel

functions include the Gaussian, Epanechnikov, and uniform kernels, whose equations were provided in the previous section.

- h is the bandwidth parameter which controls the smoothness of the regression curve. Smaller h results in a curve that fits more closely to the data (possibly overfitting), while larger h produces a smoother curve. The bandwidth h, which controls the width of the kernel, is crucial in determining the bias–variance trade-off in the model.
- Y_i is the observed value of the dependent variable corresponding to X_i.

The Nadaraya–Watson estimator can be seen as a weighted average of $Y_1, \ldots, Y_n$ by means of the set of weights $\{W_i(x)\}_{i=1}^n$ (they add to one). The Nadaraya–Watson estimator is a local mean of $Y_1, \ldots, Y_n$ about $X = x$.

For instance, let's calculate $\hat{m}(3)$ using the Nadaraya–Watson estimator with the following assumptions. Let's assume the bandwidth is $h = 0.5$, and we have three data points in our bandwidth around $x = 3$ from the sample. These data points are as follows: $x_1 = 2.5, y_1 = 6.25$, $x_2 = 3.2, y_2 = 10.24$, and $x_3 = 3.1, y_3 = 8.41$.

Kernel values in the Nadaraya–Watson estimator while using the Gaussian kernel function is given by:

$$K_h(x - X_i) = K\left(\frac{x - X_i}{h}\right) = \frac{1}{h}\frac{1}{\sqrt{2\pi}}e^{-\frac{(\frac{x-X_i}{h})^2}{2}} \tag{8.12}$$

We now calculate the kernel values for each data point:

For $x_1 = 2.5$: $K_h(3 - 2.5) = \frac{1}{0.5}\frac{1}{\sqrt{2\pi}}e^{-\frac{\left(\frac{3-2.5}{0.5}\right)^2}{2}} \approx 0.48394$

For $x_2 = 3.2$: $K_h(3 - 3.2) = \frac{1}{0.5}\frac{1}{\sqrt{2\pi}}e^{-\frac{\left(\frac{3-3.2}{0.5}\right)^2}{2}} \approx 0.73654$

For $x_3 = 3.1$: $K_h(3 - 3.1) = \frac{1}{0.5}\frac{1}{\sqrt{2\pi}}e^{-\frac{\left(\frac{3-3.1}{0.5}\right)^2}{2}} \approx 0.78208$

After calculating each values, we sum these kernel values and find $K_h(3 - 2.5) + K_h(3 - 3.2) + K_h(3 - 3.1) = 2.00256$.

Next, the weights for each data point are computed by dividing the kernel values $K_h(3 - x_i)$ by the sum of all the kernel values. The weights represent the influence of each data point in the final estimate of $\hat{m}(3)$.

$W_1(3) = \frac{K_h(3-2.5)}{K_h(3-2.5)+K_h(3-3.2)+K_h(3-3.1)} = \frac{0.48394}{2.00256} \approx 0.2417$

Similarly for $x_2 = 3.2$: $W_2(3) = 0.73654/2.00256 \approx 0.36788$ and for $x_3 = 3.1$: $W_2(3) = 0.78208/2.00256 \approx 0.39042$.

Finally, we compute $\hat{m}(3)$ by summing the weighted values of y_i for each data point:

$\hat{m}(3) = W_1(3)y_1 + W_2(3)y_2 + W_3(3)y_3 = 0.24170 \times 6.25 + 0.36788 \times 10.24 + 0.39042 \times 7.84 = 9.0314$

Thus, the estimated regression value at $x = 3$ using the Nadaraya–Watson estimator is $\hat{m}(3) \approx 9.0314$. The values x_1, y_1 the other two data points were generated from the function $y_i = x_i^2 + \epsilon_i(0, 0.2)$, thus the actual value of $m(3)$ is 9. Instead of using a parametric approach (for instance OLS), we have estimated $m(3)$ using the nonparametric Nadaraya–Watson estimator, demonstrating how nonparametric methods can provide flexible alternatives when the functional form is unknown.

8.2.5 Simulation: Kernel Regression Example

To illustrate kernel regression, we simulate data from a sine function with added noise and apply kernel regression to estimate the true underlying relationship:

$$Y = X^2 + \varepsilon \tag{8.13}$$

where $\varepsilon \sim N(0, 4)$. Below is the R code for fitting a kernel regression model using the `np` package and visualizing the results.

```
# Simulating data
set.seed(123)
n <- 50
X <- sort(runif(n, -10, 10))
Y <- X^2 + rnorm(n, 0, 4)
# Kernel regression using np package
library(np)
options(np.messages = FALSE)
model <- npreg(Y ~ X, regtype = "lc", bws = 0.5)
#regtype = "lc": Nadaraya-Watson estimator uses Gaussian kernel;
#bws = 0.1: This sets the bandwidth h to 0.1
#( can change with bwmethod = "cv.aic": Cross-validation for
#bandwidth selection); can include ckertype = "epanechnikov"
#or "triweight" for different kernel types)

# Plot the true function and estimated function
plot(X, Y, main="Kernel Regression", col="black", pch=20)
lines(X, X^2, col="black", lwd=2, lty=2)  # True function
lines(X, fitted(model), col="black", lwd=2)   # Kernel estimator
legend("topright", legend=c("True Function", "Kernel regression"),
       col=rep("black", 2), lty=c(2, 1), lwd=2, bty="n")
```


```
# Simulating data
set.seed(123)
n <- 50
X <- sort(runif(n, -10, 10))
Y <- X^2 + rnorm(n, 0, 4)
# Kernel regression using np package
library(np)
options(np.messages = FALSE)
model <- npreg(Y ~ X, regtype = "lc", bws = 0.5)
#regtype = "lc": Nadaraya-Watson estimator uses Gaussian kernel;
#bws = 0.1: This sets the bandwidth h to 0.1
#( can change with bwmethod = "cv.aic": Cross-validation for
#bandwidth selection); can include ckertype = "epanechnikov"
#or "triweight" for different kernel types)

# Plot the true function and estimated function
plot(X, Y, main="Kernel Regression", col="black", pch=20)
lines(X, X^2, col="black", lwd=2, lty=2)
```

```
lines(X, fitted(model), col="black", lwd=2, lty=1)
legend("topright", legend=c("True Function", "Kernel regression"),
       col=rep("black", 2), lty=c(2, 1), lwd=2, bty = "n")
```

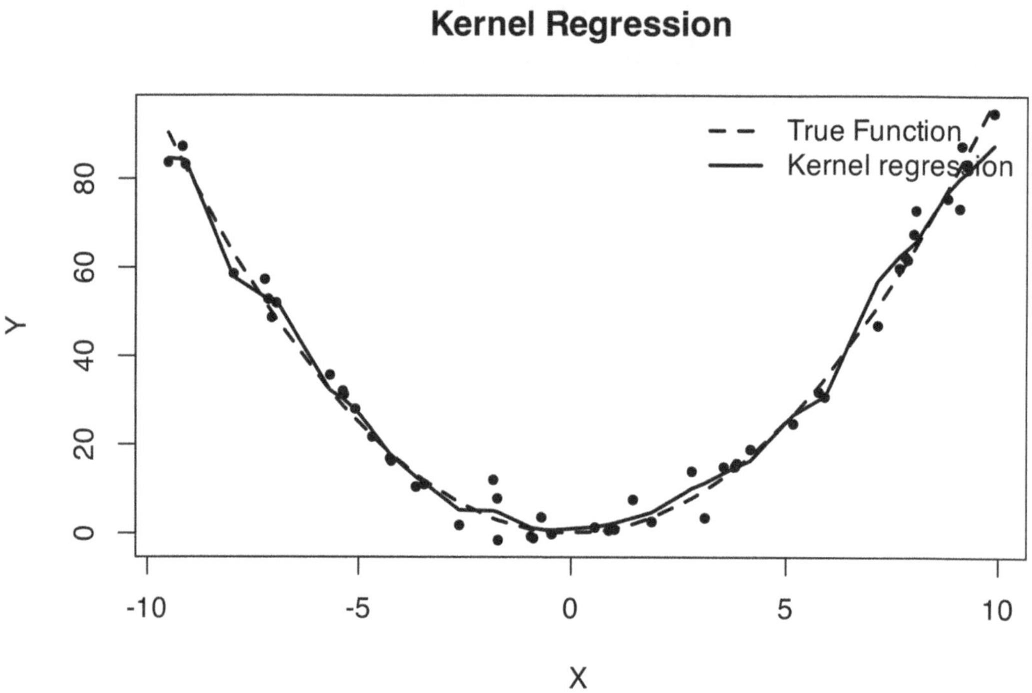

FIGURE 8.2 Kernel regression example.

In this example, we generated data where the true regression function is $\sin(X)$, and we used kernel regression to estimate it. The black dashed line represents the true function, and the solid line represents the kernel regression estimate. By adjusting the bandwidth parameter h, we can control the smoothness of the kernel regression fit. As seen in the Figure 8.2, kernel regression provides a flexible, nonparametric way to estimate the relationship between Y and X without making strong assumptions about the functional form of the relationship.

8.2.6 Multivariate and Additive Models

Although the Nadaraya–Watson estimator handles one-dimensional regression, kernel regression can be extended to handle multivariate data. For high-dimensional data, however, kernel regression suffers from the *curse of dimensionality*, as the number of required data points increases exponentially with the number of predictors.

To address this, additive models provide a flexible structure for modeling relationships in higher dimensions. The additive model for kernel regression is expressed as:

$$y = \beta_0 + m_1(x_1) + m_2(x_2) + \cdots + m_p(x_p) + \varepsilon \tag{8.14}$$

Here, the function $m_j(x_j)$ is estimated for each predictor separately, allowing the model to handle more predictors while reducing the complexity.[1]

Another widely used nonparametric regression method is locally estimated scatterplot smoothing (LOESS), also known as locally weighted regression. LOESS is a form of kernel regression. It is a nonparametric regression method that fits multiple weighted local regressions to subsets of data around each point. Unlike the Nadaraya–Watson estimator, which provides a local mean, LOESS fits a local linear or polynomial regression, thereby increasing its flexibility in capturing nonlinear relationships across the predictor space. We will use LOESS in the simulation in the data splitting section to demonstrate how to tune kernel regression, which will be covered in more detail in the next chapter.

Local polynomial estimators, in general, are a broader framework where a polynomial of a specified degree is fit locally to the data. The degree of the polynomial can vary (e.g., local constant, local linear, or local quadratic), offering different levels of smoothness in the regression function. The degree of the polynomials is also hyperparameter. LOESS is a specific case of this framework, often using local linear or quadratic fits with a tricube weight function to assign weights to nearby data points.

The *span* parameter in LOESS controls the size of the neighborhood around each point x_0, with smaller spans focusing more on local data points and larger spans leading to smoother fits. This makes LOESS highly flexible and particularly useful for smoothing scatterplots and capturing trends in data without assuming a global parametric form.

In summary, while local polynomial regression is a general technique, LOESS is a practical and popular method for local smoothing, providing flexibility in modeling complex relationships with options to control the level of smoothing through the span parameter.

8.2.7 Bandwidth Selection in Kernel Density Estimation and Kernel Regression

The choice of bandwidth h is critical in both KDE and kernel regression, as it controls the trade-off between bias and variance. In both contexts, a small bandwidth results in a highly flexible model that may overfit the data, while a large bandwidth produces a smoother model that may underfit the data. Although bandwidth plays a similar role in both methods, the criteria for selecting bandwidth may differ based on the specific task at hand—whether it is density estimation or regression.

In KDE, the bandwidth determines the smoothness of the estimated PDF. Choosing an appropriate bandwidth h is crucial because a bandwidth that is too small will result in overfitting (too much noise), while a bandwidth that is too large will oversmooth the estimate, potentially missing important features of the distribution. One common method for bandwidth selection is Silverman's Rule of Thumb, where the bandwidth is calculated using the formula: $h = 1.06\hat{\sigma}n^{-1/5}$ where $\hat{\sigma}$ is the standard deviation of the data and n is the number of data points. Cross-validation is another common method used to minimize the mean integrated squared error (MISE) by testing different bandwidths. The optimal bandwidth minimizes the prediction error by balancing bias and variance, ensuring generalizability to unseen data. Plug-in methods are also used in KDE to minimize the asymptotic mean integrated squared error (AMISE), which provides an approximation to MISE and becomes accurate as the sample size n increases. AMISE is easier to compute than MISE and is often employed for bandwidth selection in density estimation.

In kernel regression, the bandwidth selection process is similar but with some important distinctions. Kernel regression estimates the relationship between a dependent variable Y and an independent variable X. Here, the bandwidth again controls the smoothness of the estimated regression function $E[Y|X]$. A small bandwidth leads to a flexible model that may overfit, capturing too much noise, while a large bandwidth produces a smoother model that may miss important features in the data. Like KDE, bandwidth can be chosen using Silverman's Rule of Thumb: $h = 1.06\hat{\sigma}n^{-1/5}$ where $\hat{\sigma}$ is the standard deviation of the independent variable X, and n is the sample size. However, in kernel regression, the most common method of bandwidth selection is cross-validation. In this approach, the bandwidth is selected by minimizing the cross-validated mean squared error (MSE), which measures the accuracy of predictions. The data is split into training and validation sets, and the MSE is computed for different bandwidths to find the optimal one. This method directly minimizes the prediction error on unseen data and is widely used in practice.

Plug-in methods are also applicable in kernel regression, where they rely on estimating the second derivative of the true regression function to select the bandwidth that minimizes the integrated squared error (ISE). While MISE is used for density estimation in KDE, ISE is more appropriate for regression tasks because it focuses on minimizing the error in approximating the true regression function.

Although the principles of bandwidth selection are similar between KDE and kernel regression, the specific criteria for evaluating the performance of the models differ. In KDE, we aim to minimize MISE or AMISE, while in kernel regression, the focus is on minimizing MSE or ISE. Both approaches face a bias–variance trade-off: a small bandwidth reduces bias but increases variance, while a large bandwidth increases bias but reduces variance. These methods help balance the trade-off to achieve better generalization and prediction accuracy.

In higher-dimensional settings, both methods suffer from the curse of dimensionality. As the number of predictors increases, the data becomes sparse, and the number of required data points grows exponentially. Kernel regression, in particular, is sensitive to this issue, as the performance deteriorates in high dimensions. Practitioners often use additive models or methods like local polynomial regression (LOESS) to mitigate the curse of dimensionality.

We also want to mention the distinction between bandwidth and span in smoothing methods. Bandwidth is commonly used in kernel methods like kernel density estimation and kernel regression, controlling the width of the kernel and determining the influence of nearby points on the estimate. In contrast, span is specific to methods like LOESS, where it represents the proportion of data used for each local regression. Both parameters play a crucial role in balancing the trade-off between bias and variance, with smaller values yielding more detail and larger values providing smoother estimates.

In conclusion, bandwidth selection plays a crucial role in both KDE and kernel regression. Cross-validation is often the preferred method as it minimizes prediction error on unseen data. This approach, along with detailed implementation, will be covered in the next chapter.

8.3 Conclusion Remarks

In this chapter, we explored two important nonparametric methods: KDE and kernel regression. These methods offer flexibility by allowing the data to determine the shape of the model, avoiding strong assumptions about the underlying distribution or functional form. We began with kernel density estimation, where we used kernel functions to estimate the probability density function of a random variable, and highlighted the critical role of bandwidth selection in balancing the trade-off between bias and variance.

Building on the concepts from KDE, we introduced kernel regression, a nonparametric technique for estimating the regression function $E[Y|X = x]$. Kernel regression uses kernel weights to model the relationship between Y and X in a flexible way, making it suitable for capturing complex, nonlinear patterns in the data. We discussed various kernel functions, the Nadaraya–Watson estimator, and bandwidth selection methods such as cross-validation and plug-in estimators. Furthermore, we addressed the challenges of kernel regression, including the curse of dimensionality and computational costs, particularly when dealing with high-dimensional data. Both KDE and kernel regression are powerful tools for nonparametric analysis, but they require careful consideration of bandwidth selection and performance trade-offs.

Aside from KDE and kernel regression, several other nonparametric methods are commonly used in statistics and machine learning. One such method is **k-nearest neighbors (k-NN)**, which classifies an observation based on the majority label of its nearest neighbors. For instance, in a healthcare study, k-NN can be used to predict whether a patient has a certain disease based on the characteristics of similar patients. Nonparametric methods can be applied to both regression and classification tasks, depending on the data and the problem at hand. k-NN can function as both a classifier and a regression method. In classification, it assigns the majority label from the nearest neighbors, as in predicting whether a customer will make a purchase based on demographic similarities. In regression, k-NN averages the values of the nearest neighbors to predict a continuous outcome, such as estimating housing prices based on nearby properties.

Another popular technique is **decision trees**, which recursively split the data into subsets based on the most significant variables. In economics, decision trees can be used to predict household income based on demographic factors such as education, occupation, and location. Decision trees can be used for both classification and regression. In classification tasks, a decision tree predicts categorical outcomes, such as determining whether a student will drop out based on their grades and attendance. In regression, it predicts continuous outcomes, like estimating stock prices based on past performance and economic indicators.

Spline regression is another nonparametric method that fits piecewise polynomials to the data, ensuring smooth transitions between segments. This method is particularly useful in modeling relationships where different behaviors exist over different ranges of the predictor, such as predicting the effect of education on wages across different age groups in labor market studies.

Random forests, an ensemble of decision trees, are widely used for more robust predictions, such as predicting credit risk for loan applicants based on financial history and other factors. Random forests can be applied to both classification and regression problems. For instance, random forests can classify whether a bank customer will default on a loan (classification) or predict the actual loan amount that a customer is likely to default on (regression). These non-parametric methods offer flexible tools to capture complex patterns in data without relying on strict parametric assumptions. In the following chapters, we will explore all the mentioned nonparametric methods, in greater detail.

Hyperparameters play a critical role in nonparametric models as they directly affect the trade-off between bias and variance. In the next chapter, we will explore hyperparameter tuning techniques, such as cross-validation, to optimize these values and ensure the model generalizes well to new data.

In both density estimation and kernel regression, a key hyperparameter is the bandwidth h. The bandwidth controls the smoothness of the estimate by determining how much weight is given to nearby data points. In density estimation, a smaller bandwidth results in a more detailed but jagged estimate, while a larger bandwidth smooths the estimate, potentially at the cost of losing finer details. Similarly, in kernel regression, the bandwidth controls the flexibility of the regression function, balancing between overfitting and underfitting. A well-chosen bandwidth is crucial for the performance of both kernel density and regression estimations.

Note

1. Very detailed notes about kernel estimation and regression can be found at https://bookdown.org/egarpor/NP-UC3M/.

9

Hyperparameter Tuning

Building on the concepts explored in the previous chapter, where we introduced kernel density estimation and kernel regression, we now turn our attention to one of the most critical aspects of predictive modeling: hyperparameter tuning. As we discussed, hyperparameters such as bandwidths in kernel methods play key role in controlling the trade-off between bias and variance, and overfitting, thus significantly impacting the performance of the model. In kernel density estimation, bandwidth selection determines the smoothness of the estimated probability density function, while in kernel regression, it controls the flexibility of the regression curve.

Hyperparameters are model parameters set prior to the training process. Why is it important to tune these hyperparameters? The optimal values of hyperparameters are typically unknown ahead of time, making it essential to explore different values. Essentially, a model with different hyperparameters can be considered a distinct model, potentially leading to variations in performance. Thus, tuning is crucial as it directly impacts the effectiveness and accuracy of the model.

In general, each prediction method involves multiple tuning parameters—known as hyperparameters—which must be set before the learning process begins. These are external to the model and cannot be directly estimated from the data. For example, the bandwidth h in kernel regression, the number of neighbors k in k-nearest neighbors (kNN), the number of hidden layers in neural networks, the degree of polynomials in polynomial regression, and the penalty parameter λ in LASSO regression are all hyperparameters that require tuning in advance. This contrasts with model parameters, which are internal to a parametric model and can be estimated from the data once the hyperparameters are specified.

To find optimal hyperparameters, several methods are employed. While brute force approaches like grid search and randomized grid search are common, where hyperparameters are chosen randomly or systematically without considering previous outcomes, they can be inefficient for larger search spaces. Other methods like Bayesian optimization and genetic algorithms, which will not be covered in this book, use informative criteria to guide the search for optimal hyperparameters. These methods follow a sequential process, leveraging past decisions to inform future choices, and are generally more efficient than brute force methods, particularly for expansive search spaces.

One of the most important tools for hyperparameter tuning is cross-validation, particularly a version called k-fold cross-validation. This technique helps us choose hyperparameters that result in a model that performs well not just on the data it was trained on, but also on new, unseen data. In simple terms, k-fold cross-validation works by dividing the dataset into k equally sized parts (or 'folds'), using $k-1$ of them to train the model and the remaining fold to test it. This process repeats k times, each time using a different fold for testing, and the results are averaged to estimate the model's overall prediction accuracy. We'll explain this process in more detail in later chapters. But the main idea is to answer a key question:

DOI: 10.1201/9781003504795-9

how can we be confident that our estimated regression model will generalize beyond the sample data used to fit it? Ideally, we would test the model on a completely independent dataset—either by refitting the model to see if it gives similar results or by using it to predict outcomes and measuring the error, typically with mean squared prediction error. When such independent data aren't available, cross-validation offers a reliable alternative for estimating predictive performance, helping us strike the right balance between model complexity and generalization.

In addition to cross-validation, hyperparameter tuning often involves grid search, a straightforward yet powerful method where a predefined grid of potential hyperparameter values is systematically evaluated to identify the combination that minimizes prediction error. Grid search involves defining a range of possible values for each hyperparameter and testing every possible combination of these values. The performance of each combination is assessed through cross-validation, and the best combination is chosen based on the highest average performance across the cross-validation folds. This chapter will explore the rules and applications of setting up grid searches and how to evaluate combinations of hyperparameters across different machine learning algorithms. As with cross-validation, grid search plays a vital role in optimizing hyperparameters, ensuring that the final model achieves the right balance between bias and variance, which is crucial for producing reliable predictions.

It is also essential to understand the role of different data subsets during model building. Typically, data is divided into three sets: the training set for building the model, the validation set for tuning hyperparameters and selecting the model, and the test set for evaluating the final generalization performance of the model. This process helps in developing models that perform well not just on training data but also on unseen data, which we will discuss in the next section.

Throughout this chapter, we will explore methods for data splitting, cross-validation, and grid search, all of which are key to successful model building.

9.1 Training, Validation, and Test Datasets

In machine learning, one of the fundamental principles for building robust and generalizable models is the proper division of data into training, validation, and test sets. This process ensures that the model is not only fitted to known data but also evaluated on unseen data to assess its ability to generalize.

The training set is the portion of the dataset used to fit the model. It allows the model to learn patterns and relationships between the input features X and the target variable Y. During training, the parameters of the model are optimized to minimize a specific objective function, such as mean squared error for regression or cross-entropy for classification. However, relying solely on the training set can lead to overfitting, where the model performs well on the training data but fails to generalize to new data. We showed in the overfitting chapter how overfitting can occur with high polynomial function, i.e., when the model is too complex, capturing noise in the training data rather than the underlying patterns.

The validation set plays a crucial role in hyperparameter tuning, particularly in selecting the best hyperparameters such as bandwidth, h in kernel regression, k in k-NN, λ in LASSO. Unlike the training set, the validation set is not used to directly adjust model parameters; instead, it is used to evaluate the performance of the model during the training phase. This helps in selecting the best combination of hyperparameters to ensure that the model generalizes well. Key methods for tuning hyperparameters are cross-validation and bootstrapping.

In k-fold cross-validation, the dataset is split into k sections. In each iteration, one section is held out as the validation set, while the remaining sections serve as the training set. The model is trained on the training set, and its performance is evaluated on the validation set. This process is repeated k times, with each section being used as the validation set once. The aim of cross-validation is to minimize overfitting by ensuring that the model generalizes well to unseen data. Cross-validation is also often paired with grid search, where a predefined grid of possible hyperparameter values is evaluated to identify the combination that minimizes prediction error.

A typical hyperparameter tuning process could involve testing different values of a hyperparameter, like λ in ridge regression or bandwidth in kernel regression. For instance, we could establish a grid of potential λ values, split the data into k sections, and evaluate the performance of the model for each λ value. The model with the best performance across the validation sets is selected. This is crucial for preventing overfitting, as it allows the model to be evaluated on data it has not seen before.

The test set is used for the final evaluation of the predictive performance of the model after the training and validation phases. It serves as an unseen dataset that was not involved in model tuning, providing an unbiased estimate of how well the model is likely to perform in real-world applications. By evaluating the performance of the model on the test set, we can determine whether it has generalized beyond the training data. This step is essential because it ensures that the performance of the model is not overly optimistic and reflects its true predictive ability. Key performance metrics, such as root mean square error (RMSE) for regression or accuracy for classification, are calculated on the test set.

It is crucial to avoid data leakage during this process, as this can lead to overly optimistic performance estimates. Data leakage occurs when information from the test or validation sets influences the training process. One common source of leakage is applying preprocessing steps (like normalization or feature scaling) before splitting the data. To avoid this, preprocessing should always be performed independently on the training data, with the parameters derived from training applied to the validation and test sets.

A common confusion in practice arises between the validation and test sets, as they are sometimes used interchangeably. However, they serve distinct purposes. The validation set is used during the model selection process to fine-tune the hyperparameters, while the test set is reserved for evaluating the final predictive accuracy of the model. After the model has been tuned using the validation set, it is crucial to assess its performance on the test set, which simulates how the model will perform on new, unseen data.

To summarize the typical process in machine learning:

- *Training data* is used to build the model and estimate the parameters.
- *Validation data* is used to tune the hyperparameters and avoid overfitting.
- *Test data* is used to evaluate the final predictive performance of the model on unseen data.

This sequence ensures that the final model generalizes well to new data and is not simply overfitting to the training dataset. As the datasets grow in complexity and size, these methods become even more important to ensure robust model performance. Additionally, as we will see in the next sections, techniques like grid search and k-fold cross-validation help streamline the process of hyperparameter tuning, allowing us to optimize models more effectively.

Understanding the relationship between training, validation, and test data is foundational for modern machine learning. While the training set allows the model to learn, the validation set helps in fine-tuning, and the test set provides a final unbiased evaluation. In summary, the data-splitting process is fundamental to building and evaluating machine learning models. By dividing the data into training, validation, and test sets, we ensure that the model can learn, be optimized, and ultimately generalize to unseen data effectively. We will explore hyperparameter tuning in greater detail, including the processes of cross-validation and grid search, which help refine our models for better predictive performance, after discussing how to split data randomly.

9.2 Data Splitting

In fields like economics, social sciences, and health sciences, datasets are often limited due to the high cost of data collection. To address this, researchers commonly use data splitting and subsampling techniques to make the most of available data and to build models that can be reliably evaluated. These approaches allow for the creation of representative subsets, which support robust analysis and better estimates of model performance.

Data splitting is also a core practice in machine learning, where the goal is not only to explain relationships but also to predict accurately. Unlike traditional statistical models that often rely on the full dataset for estimation, machine learning models must be tested on data they haven't seen during training to evaluate how well they generalize. For example, in a dataset on housing prices and their predictors, using the entire dataset for training might result in overfitting—where the model memorizes patterns in the training data, including noise, and performs poorly on new data.

To avoid overfitting and to ensure that model performance reflects true predictive power, the dataset is typically divided into three parts. The training set is used to fit the model, allowing it to learn patterns from the data. The validation set is used during the training process to tune hyperparameters and monitor for overfitting—helping to strike the right balance between model complexity and generalization. Finally, the test set is kept completely separate and is used only for the final evaluation, providing an unbiased estimate of how well the model will perform on new, unseen data. This structured approach ensures that performance metrics are not simply measuring the model's ability to memorize the training data, but its ability to generalize to future cases.

To operationalize this process, several practical data splitting strategies are commonly used. These include:

Random splitting (or subsampling): Also called Monte Carlo cross-validation, this method involves randomly dividing the dataset into training and testing subsets. It's often used repeatedly to assess model stability, especially with limited data.

K-fold cross-validation: The data is split into k equal parts (folds). The model is trained k times, each time using a different fold as the test set and the rest for training. The results are averaged to estimate prediction accuracy. We will explain this technique in detail in later chapters.

Stratified sampling: Used when dealing with imbalanced classes, this ensures each subset contains approximately the same proportion of each class as the full dataset.

Rolling window cross-validation: Tailored for time-series data, this method uses sequential overlapping windows for training and testing, preserving the temporal order.

Blocked cross-validation: Also for time-series settings, it divides the dataset into blocks and performs training and testing on sequential blocks to prevent leakage across time.

Regardless of the specific technique used, maintaining the integrity of the test set throughout the modeling process is essential for valid performance evaluation. One of the most serious risks during data splitting is data leakage—when information from the test set, which should represent truly unseen data, inadvertently influences the training process. This undermines the integrity of model evaluation and can lead to overly optimistic performance metrics that don't generalize in practice.

There are two main forms of data leakage. The first is train-test contamination, which arises when preprocessing steps—such as feature selection, scaling, or transformation—are applied to the entire dataset before splitting. This allows information from the test set to seep into the training set, violating the principle that the test data must remain unseen (i.e., untouched) during model development. The second form is target leakage, which occurs when the model is given access to information that would not realistically be available at the time of prediction. A classic example is using future housing prices to predict current ones, which gives the model an unfair advantage and leads to misleading results.

To prevent data leakage, it is essential to split the data into training, validation, and test sets before performing any preprocessing. All transformations—such as variable generation, normalization, or feature engineering—should be carried out using only the training data. The same transformations are then applied to the validation and test sets using parameters learned from the training set. This disciplined approach ensures that the evaluation of model performance remains honest and reflects how the model would perform on truly unseen data. Cross-validation must also be implemented carefully—ensuring that transformations and selections are confined within each fold—to avoid leakage across folds.

Some modeling contexts, like time-series analysis, present additional challenges for data splitting. Random splitting can break this order and lead to misleading results. Instead, use methods like rolling window or blocked cross-validation. Time-series data often involve autocorrelation—dependence between values across time. To address this, one can stratify by time intervals or apply techniques such as Fourier or wavelet transforms to reduce temporal dependence.

There is no universal rule for how to split a dataset; the ideal strategy depends on the size of the dataset and the specifics of the problem. However, some general guidelines are commonly followed in practice. For many applications, allocating between 60% and 80% of the data for training—including both the training and validation sets—is typical, while the remaining 20% to 40% is set aside for final testing. When working with smaller datasets, a more detailed split such as 70% for training, 20% for validation, and 10% for testing may be more appropriate to ensure each phase of model development is adequately supported.

While splitting strategies help allocate available data, another tool used for statistical inference—though conceptually different—is bootstrapping. It's important to distinguish bootstrapping from data splitting. Bootstrapping is a resampling technique—not a splitting method. It involves repeatedly drawing samples with replacement from the dataset to create multiple resampled datasets, typically used to estimate uncertainty or variability in a statistic. Unlike data splitting, it doesn't separate the data into training and testing subsets.

To summarize, proper data splitting must be paired with disciplined data preparation practices. Before modeling, always visualize your data to identify trends, outliers, and dependencies that may inform your splitting strategy. Split your data before performing any transformations. Apply the same preprocessing steps to the validation and test sets as used on the training data. Keep all data subsets strictly separate throughout the modeling process to ensure a fair and unbiased assessment of your model.

9.2.1 How Can We Split the Data Randomly?

We can use the `sample()` function in R to randomly shuffle data or create random subsets for various tasks, such as splitting a dataset into training and validation sets. It allows for sampling with or without replacement, with uneven probabilities, data frame rows, lists, and vectors. This function is particularly useful in data modeling when we need to divide datasets into different sections, like in cross-validation, ensuring that the process remains unbiased and representative of the entire dataset.

Let's walk through an example where we randomly split a dataset into two equal parts—one for training and the other for validation. The dataset will be generated using a simple data-generating mechanism (DGM), where Y depends linearly on X with some added noise.

```
set.seed(1)
n <- 10000
X <- sort(runif(n) * 2 * pi)
# Create the response variable Y as sin(X) with added noise
Y <- sin(X) + rnorm(n)/4
data <- data.frame(Y, X)
random <- sample(n, n, replace = FALSE)
data <- data[random, ]

k <- 2  # Number of folds in cross-validation
# Divide the data into k equal folds
folds <- cut(seq(1, n), breaks = k, labels = FALSE)
for (i in 1:k) {
  val_idx <- which(folds == i)        # Indexes of validation data for the current fold
  train_data <- data[-val_idx, ]  # Training data (all except the current fold)
  val_data <- data[val_idx, ]     # Validation data (the current fold)
}
```

We create a dataset then we shuffle the dataset using sample(). This randomizes the order of the data points. Next, the dataset is split into k approximately equal segments. The first half ($i = 1$) is used for validation the model and the second half is reserved for training.

In real-world applications, we often split the data differently. For example, a 90% for training and 10% for validation split is common. You can achive that by changing the value of k to 10 and then using the all but i^{th} 9/10th of the data for training and the ith 1/10th for validation.

For five-fold or ten-fold cross-validation, which we will discuss next section, the data is divided into five or ten equal parts. In each iteration, one of these parts is designated as the validation set, while the other remaining parts are used as the training set. To modify the code for 5-fold cross-validation, you would need to divide the data into five segments and rotate through them as training and validation sets in each iteration.

How can we use splitting data randomly into two equal segments to evaluate the performance of a model?

We have prepared a dataset with a 50-50 split between training and validation sets using the first code snippet above (when $k = 2$).

To tune the span hyperparameter in a kernel regression model, we use the `loess()` function. We fit multiple kernel regression models with different span values using the training set, then predict the target variable (Y) on the validation set for each model. Finally, we calculate the root mean squared prediction error (RMSPE) for each model to assess its performance and determine the optimal span value.

```
# Train loess models with different spans on the training data
loe0 <- loess(Y ~ X, degree = 2, span = 0.02, data = train_data)  # Overfitting
loe1 <- loess(Y ~ X, degree = 2, span = 0.1, data = train_data)   # Moderate span
loe2 <- loess(Y ~ X, degree = 2, span = 1, data = train_data)     # Underfitting

# Predict on the validation set
fit0 <- predict(loe0, val_data$X)  # Predictions for small-span = 0.02
fit1 <- predict(loe1, val_data$X)  # Predictions for moderate-span = 0.1
fit2 <- predict(loe2, val_data$X)  # Predictions for large-span = 1
```

We have trained three different kernel regression models using the loess() function with varying span values: a small span (0.02), a moderate span (0.1), and a large span (1). These spans control the smoothness of the regression curve, with smaller spans leading to more flexible models and larger spans leading to smoother curves.

To evaluate the performance of these models, we calculate the RMSPE on the validation set, which measures the average squared difference between the predicted and actual values. Let's compute the RMSPEs for each model.

```
# Compute RMSPE (Root Mean Squared Prediction Error) for each model
rmspe0 <- sqrt(mean((val_data$Y - fit0)^2))  # RMSPE for span = 0.02
rmspe1 <- sqrt(mean((val_data$Y - fit1)^2))  # RMSPE for span = 0.1
rmspe2 <- sqrt(mean((val_data$Y - fit2)^2))  # RMSPE for span = 1

# Output the RMSPE values
c(paste("RMSPE with span = 0.02:", rmspe0),
  paste("RMSPE with span = 0.1:", rmspe1),
  paste("RMSPE with span = 1:", rmspe2))
```

```
## [1] "RMSPE with span = 0.02: 0.253191865839389"
## [2] "RMSPE with span = 0.1: 0.250909577867399"
## [3] "RMSPE with span = 1: 0.316932583440793"
```

The RMSPE is calculated as the square root of the average squared differences between the true values (val$Y) and the predicted values (fit0, fit1, fit2) for each span value. In this simulation, we tested three span values to fit the kernel regression models. The goal is to choose the span that minimizes RMSPE, indicating better predictive accuracy. The model with the lowest RMSPE offers the best performance.

However, our current approach has some limitations. We have used arbitrary span values, and the single random split may not accurately represent the generalization performance of the model. Additionally, comparing RMSPEs manually can be time-consuming and prone to error.

To address these limitations, we use grid search to systematically explore a wider range of span values and perform k-fold cross-validation to evaluate the model on multiple folds of the data. This will help us fine-tune the hyperparameter and ensure robustness of the model on unseen data.

Having discussed random data splitting, we now move on to k-fold cross-validation, a more structured and reliable method for evaluating models.

9.3 k-Fold Cross-Validation

One of the key challenges in building predictive models is ensuring that the model generalizes well to unseen data. A single random split between training and validation sets, as discussed in the previous section, can sometimes lead to unreliable estimates of model performance. This is because the performance of the model may vary depending on how the data is split. To address this issue, the process often begins with a full dataset that typically includes predictors (X) and the response (Y), which serves as the foundation for data splitting and model building.

The first step is dividing the dataset. A common approach is to allocate 60–80% of the full dataset to the training set, which is used to fit the models. The remaining 20–40% of the data is set aside as the validation set, which plays a crucial role in evaluating and fine-tuning model parameters. A better approach involves holding out a final test set, typically 20% of the full dataset, distinct from the training and validation sets, to provide an unbiased evaluation of the predictive performance of the final model. The training and validation sets should be split using 80% of the full dataset. This test set, often left untouched until the final stage, ensures that performance of the model is not influenced by prior tuning with the training and validation data.

Once the data is split, models are fitted to the training data, and their performance is assessed on the validation set using metrics such as RMSPE to compare predicted values against actual values. After selecting and tuning the best model based on validation performance, performance of the model is tested on the final test set, providing a more reliable measure of how well it can generalize to new, unseen data.

However, the aforementioned setup may be prone to variability in performance estimates due to the potential instability of a single data split, which can lead to overfitting or unreliable results. To mitigate this issue, a more robust technique called k-fold cross-validation is often employed. This method divides the training dataset into k equally sized folds, systematically training the model on k−1 folds while using the remaining fold for validation (never include final test data to this step). By repeating this process k times, each fold serves as a validation set once, and the final model performance is averaged across all iterations, offering a more reliable estimate of the true predictive power of the model. This approach not only helps in overcoming the variability inherent in a single random split but also ensures a more consistent evaluation of the ability of the model to generalize.

9.3.1 What is k-Fold Cross-Validation?

In k-fold cross-validation, the dataset (mostly, training data set) is split into k equally-sized (or almost equally-sized) subsets, or "folds." The model is then trained on $k - 1$ folds and validated on the remaining fold. This process is repeated k times, each time using a different fold as the validation set and the remaining $k - 1$ folds as the training set. The final performance metric is the average of the performance across all k iterations. We want to remind you that the final test data should be split from the raw data at the very beginning, before any further processing, and should not be included in the subsequent steps.

The steps for k-fold cross-validation are as follows:

1. **Split the data** into k folds of approximately equal size. (training and validation data).
2. **Train the model** on $k - 1$ folds.
3. **Validate the model** on the remaining fold and calculate the chosen performance metric (e.g., RMSPE, Accuracy, etc.).
4. **Repeat the process** k times, with each fold being used as the validation set once.
5. **Calculate the average performance** across all k folds to estimate the generalization performance of the model.

There are also alternative cross-validation methods such as **leave-one-out cross-validation (LOOCV)**, **leave-one-group-out**, and **time-series cross-validation**. These methods may be better suited to certain types of data and prediction tasks, which we will explore later.

9.3.2 Building a Simple k-Fold Cross-Validation Loop

Let's start with a simple example to calculate the means of a random variable X in both training and validation sets using ten-fold cross-validation. This example introduces how to structure k-fold cross-validation in practice.

```
set.seed(1)
n <- 10000
X <- sort(runif(n) * 2 * pi)
Y <- sin(X) + rnorm(n)/4
data <- data.frame(Y, X)
random <- sample(n, n, replace = FALSE)
data <- data[random, ]

k <- 5  # Number of folds in cross-validation
folds <- cut(seq(1, n), breaks = k, labels = FALSE)
spans <- c(0.02, 0.1, 1)  # Different span values for tuning
rmspe_matrix <- matrix(NA, nrow = k, ncol = length(spans))
for (i in 1:k) {
  val_idx <- which(folds == i)  # Indexes of validation data for the current fold
  train_data <- data[-val_idx, ]       # Training data (all except the current fold)
  val_data <- data[val_idx, ]          # Validation data (the current fold)
```

```
  for (j in 1:length(spans)) {
    loe_model <- loess(Y ~ X, degree = 2, span = spans[j], surface = "direct",
                       data = train_data)
    fit <- predict(loe_model, val_data$X)  # Predict using the current model
    # Calculate RMSPE for the current fold and span
    rmspe_matrix[i, j] <- sqrt(mean((val_data$Y - fit)^2, na.rm = TRUE))
  }
}
# Prepare a data frame to store RMSPE values for each fold and span
rmspe_table <- data.frame(matrix(NA, nrow = k + 1, ncol = length(spans)))
colnames(rmspe_table) <- paste("Span =", spans)  # Set column names
# Fill in the RMSPE values for each fold
for (j in 1:length(spans)) {
  rmspe_table[1:k, j] <- rmspe_matrix[, j]  # Fold-wise RMSPE values
}
# Add a row for the average RMSPE for each span across all folds
rmspe_table[k + 1, ] <- colMeans(rmspe_matrix, na.rm = TRUE)
rownames(rmspe_table) <- c(paste("Fold", 1:k), "Average")
# Print the RMSPE table
print(rmspe_table)
```

```
##           Span = 0.02 Span = 0.1  Span = 1
## Fold 1      0.2466543  0.2449180 0.3083954
## Fold 2      0.2453188  0.2430646 0.3056569
## Fold 3      0.2528307  0.2498545 0.3137374
## Fold 4      0.2554829  0.2536209 0.3195956
## Fold 5      0.2501773  0.2476371 0.3131323
## Average     0.2500928  0.2478190 0.3121035
```

This k-fold cross-validation method helps us evaluate different hyperparameters across multiple training-validation splits. By averaging the performance metrics, we can select the best-performing hyperparameter (e.g., the optimal `span`) that generalizes well to unseen data.

9.3.3 Advantages of k-Fold Cross-Validation

- *More reliable performance estimates*: Since each fold is used for both training and validation, the final performance estimate is based on multiple evaluations, reducing the variability that arises from a single train-test split.
- *Efficient use of data*: All data points are used for both training and validation, which is especially useful when the dataset is small.
- *Prevents overfitting*: By averaging the performance across multiple folds, k-fold cross-validation helps prevent the model from overfitting to a single validation set.

In the next sections, we will explore additional techniques for hyperparameter tuning, including more cross-validation grid search and Bootstrapped Grid Search.

9.4 Grid Search

One of the most common methods for hyperparameter tuning in machine learning models is grid search. This technique systematically explores a predefined set of hyperparameters by training the model on different combinations and selecting the one that yields the best performance. Often referred to as a parameter sweep, grid search involves an exhaustive search through a manually specified subset of the hyperparameter space of a given learning algorithm.

While grid search is considered one of the simplest optimization techniques, it remains powerful because it guarantees finding the most optimal solution by brute-forcing through all possible combinations of hyperparameters. It operates by creating a "grid" of potential parameter values and evaluating each combination to determine which set performs best. Each combination is tested, mostly using cross-validation, and the point on the grid that maximizes the average performance metric in cross-validation is selected as the optimal combination of hyperparameters. This exhaustive approach ensures that the best possible point within the hyperparameter space is identified. However, the main drawback of grid search is its inefficiency—it is slow because it requires evaluating every combination of parameters, which can be computationally expensive, especially when k-fold cross-validation is involved. For each point in the grid, the model must be trained and validated k times, making the process time-consuming and complex when dealing with large datasets or numerous hyperparameters.

Despite this, grid search remains a valuable method when the goal is to find the absolute best combination of hyperparameters. It guarantees that no possible combination is overlooked, making it ideal for scenarios where accuracy is the top priority and computational resources are available.

The algorithm follows a few key steps:

- Define hyperparameter grid, i.e., a range of values for each hyperparameter.
- Create a matrix of all possible combinations of these values.
- Each combination is then tested through an objective function that returns a score, ranking the combinations based on performance.
- In the context of machine learning, each combination is used to train and evaluate the model, with the results captured using performance metrics such as accuracy or mean squared error.
- After testing all possible combinations, the hyperparameters with the best performance is selected, and the model is retrained using this optimal set of hyperparameters for final predictions.

An alternative approach, which we will not discuss into in this book, is random search. Random search follows a similar concept but instead of evaluating every point on the grid, it tests a randomly selected subset of points. This makes random search faster and less computationally expensive, though potentially less accurate. The smaller the subset, the faster the search, but the likelihood of finding the optimal point decreases. Conversely, using a larger subset increases the accuracy of the search, bringing it closer to grid search in effectiveness. Random search is particularly useful when the hyperparameter space is large, and a fine-grained grid of values is impractical to explore exhaustively. While it may not find the absolute best combination, random search often identifies a good enough set of hyperparameters, making it a practical choice in situations where time and resources are limited.

In summary, grid search, though straightforward and computationally costly, is highly effective in hyperparameter tuning by ensuring that the best parameter combination is found, and it forms the foundation for fine-tuning machine learning models. We provide a simulation of this method below.

9.5 Simulation of Grid Search with Cross-Validation

In the previous example, we used three specific values for the `span` hyperparameter in the `loess()` function. While this was a simple demonstration, it highlights the key concept of grid search: evaluating different combinations of hyperparameters to find the one that minimizes prediction error. Additionally, hyperparameters like `degree` (which controls the degree of the polynomial used in `loess()`) were not included in our search, even though they could significantly impact the performance of the model.

Let's now perform a simple grid search using the `loess()` model with cross-validation. This will help us better understand the concept of tuning hyperparameters. In this example, we will search for the optimal combination of `span` and `degree`. We will use the same program snippet as before, but with one change: we will modify the `span` line to `span <- seq(from = 0.02, to = 0.1, by = 0.02)` to create a sequence of span values for tuning. As seen, the results for `span = 0.02` and `span = 0.1` are the same as in the previous example. You can modify the range by changing `to = 1` instead of `0.1` to observe the same results as when `span = 1`. Alternatively, you can experiment with different span values to explore how they affect performance of the model.

```
##            Span = 0.02 Span = 0.04 Span = 0.06 Span = 0.08 Span = 0.1
## Fold 1       0.2466543   0.2457323   0.2452745   0.2450727  0.2449180
## Fold 2       0.2453188   0.2441281   0.2435240   0.2431653  0.2430646
## Fold 3       0.2528307   0.2507143   0.2501278   0.2500375  0.2498545
## Fold 4       0.2554829   0.2542328   0.2539443   0.2537793  0.2536209
## Fold 5       0.2501773   0.2484649   0.2479637   0.2477718  0.2476371
## Average      0.2500928   0.2486545   0.2481668   0.2479653  0.2478190
```

After demonstrating the basic concept of grid search with cross-validation, let's simulate a more typical use case and further refine the approach in the final step. For our simulation, we generate a dataset and then split it into two parts: 80% for training data (used to fit the model and tune hyperparameters with cross-validation) and 20% for testing data (used to evaluate the performance of the final model). Once we have tuned the hyperparameters using cross-validation, it is essential to assess the performance of the model on the unseen test data. This ensures that the model generalizes well and has not overfit during the cross-validation process.

In this simulation, we introduce not only the span hyperparameter but also the degree hyperparameter for tuning. We continue using the loess model as our prediction model and employ tenfold cross-validation to evaluate the performance of the model on multiple folds of the training data. Our objective is to find the optimal combination of span and degree values for the specified prediction function (i.e., perform hyperparameter tuning using cross-validated grid search) that minimizes the RMSPE on the test set.

```
# Data generation
set.seed(1)
n <- 1000
X <- sort(runif(n) * 2 * pi)
Y <- sin(X) + rnorm(n) / 4
data <- data.frame(x = X, y = Y)

set.seed(321)  # Ensure reproducibility for the split
# Random shuffle data
random <- sample(n, n, replace = FALSE)
data <- data[random, ]

# Train - Test Split using cut command
indices <- cut(seq(1, n), breaks = c(0, 0.2 * n, n),
               labels = c("Test", "Train"), include.lowest = TRUE)
testset <- data[indices == "Test", ] #20% of data set aside for testing
trainset <- data[indices == "Train", ] #80% of data for training

# Grid of parameters
grid <- expand.grid(seq(from = 0.01, to = 1, by = 0.02), c(1, 2))
colnames(grid) <- c("span", "degree")

# Setting up cross-validation
k <- 10  # Number of folds in cross-validation
folds <- cut(seq(1, nrow(trainset)), breaks = k, labels = FALSE)

# Prepare to store optimal settings and RMSPE values
OPT <- vector("list", k)
rmspe_matrix <- matrix(NA, nrow = k, ncol = length(grid))

# Cross-validation and model fitting
for (i in 1:k) {
  val_idx <- which(folds == i)  # Indexes of validation data for the current fold
  train_data <- data[-val_idx, ]  # Training data (all except the current fold)
  val_data <- data[val_idx, ]  # Validation data (the current fold)

  # Vector to store RMSPE for each parameter set
  RMSPE <- numeric(nrow(grid))

  # Your prediction model; Loop through each parameter set in the grid
  for (s in 1:nrow(grid)) {
    model <- loess(
      y ~ x,
      control = loess.control(surface = "direct"),
      data = train_data,
      degree = grid$degree[s],
      span = grid$span[s]
    )
      fit <- predict(model, val_data$x)
```

```
    RMSPE[s] <- sqrt(mean((val_data$y - fit) ^ 2))
  }

  # Store the index of the best parameter set for this fold
  OPT[[i]] <- which(RMSPE == min(RMSPE), arr.ind = TRUE)
}

# Compile the best settings for each fold
opgrid <- grid[unlist(OPT), ]
colnames(opgrid) <- c("span", "degree")
rownames(opgrid) <- paste("Fold", 1:k, sep=" ")
# Choose the optimal hyperparameters
# Calculate mean of spans and modal degree;
library(raster)
opt_degree <- modal(opgrid[, "degree"])
opt_span <- mean(opgrid[, "span"])

# Print results
print(opgrid)
```

```
##         span degree
## Fold 1  0.71      2
## Fold 2  0.61      2
## Fold 3  0.11      2
## Fold 4  0.03      1
## Fold 5  0.07      2
## Fold 6  0.25      1
## Fold 7  0.21      1
## Fold 8  0.59      2
## Fold 9  0.53      2
## Fold 10 0.11      2
```

```
print(paste("Optimal mean span:", opt_span))
```

```
## [1] "Optimal mean span: 0.322"
```

```
print(paste("Optimal modal degree:", opt_degree))
```

```
## [1] "Optimal modal degree: 2"
```

We decided that the span will range between 0.01 and 1, increasing by 0.02, and the degree can be either 1 or 2. The grid of potential parameter values will include combinations such as (0.01, 1), (0.03, 1)...(1, 1), (0.01, 2), (0.03, 2)...(1, 2).

In the hyperparameter tuning process, using the training data (which is 80% of the original dataset, as 20% is reserved for the final test), the simulation will fit the model using the `loess()` function with each of these parameter combinations. The prediction function is applied one by one to fold-1 training data (90% of the training data, i.e., all but the first 10%). The model is then validated using fold-1 validation data (the first 10% of the training data). The combination of hyperparameters that minimizes the RMSPE for fold-1 is recorded.

This process is repeated for each fold, resulting in the best hyperparameters for each fold. Once all folds are processed, the optimal hyperparameters are determined by averaging the best span values across folds and finding the mode for the degree values.

Finally, the model, with these optimal hyperparameters, is evaluated using the final test data (the remaining 20% of the original dataset, which has been untouched up to this point).

```
# Use the test set for final evaluation
model <- loess(
  y ~ x,
  control = loess.control(surface = "direct"),
  degree = opt_degree,
  span = opt_span,
  data = trainset
)
fit <- predict(model, testset$x)
RMSPE_test <- sqrt(mean((testset$y - fit) ^ 2))
# Output the RMSPE on the test set
print(paste("RMSPE on Test Set:", RMSPE_test))
```

```
## [1] "RMSPE on Test Set: 0.260009376550889"
```

9.5.1 Repetition for Robust Evaluation

To further ensure the stability of the performance of our model, we can repeat the process of splitting the data into training and test sets multiple times. This approach helps account for any variability in performance due to the random train-test split and provides more reliable estimates of the generalization performance of the model. It is recommended as a best practice to evaluate the model on the test set multiple times, which allows for a more robust estimate of its ability to generalize. By repeating the evaluation, we can calculate the average performance and assess the variance in the performance of the model across different splits of the data. This method offers a more comprehensive understanding of the predictive power of the model and stability by repeatedly testing it 100 times.

```
# Initial settings and data preparation
set.seed(1)
n <- 1000
X <- sort(runif(n) * 2 * pi)
Y <- sin(X) + rnorm(n) / 4
data <- data.frame(x = X, y = Y)

# Grid of parameters
grid <- expand.grid(seq(from = 0.01, to = 1, by = 0.02), c(1, 2))
colnames(grid) <- c("span", "degree")

t = 100 # number of times we loop
RMSPE_test <- c() # container for 100 RMSPE's

for (l in 1:t) {
set.seed(321+l)  # Ensure reproducibility for the split
```

```
# Random shuffle data
random <- sample(n, n, replace = FALSE)
data <- data[random, ]

# Train - Test Split using cut command
indices <- cut(seq(1, n), breaks = c(0, 0.2 * n, n),
               labels = c("Test", "Train"), include.lowest = TRUE)
testset <- data[indices == "Test", ] #20% of data set aside for testing
trainset <- data[indices == "Train", ] #80% of data for training

# Setting up cross-validation
k <- 10  # Number of folds in cross-validation
folds <- cut(seq(1, nrow(trainset)), breaks = k, labels = FALSE)

# Prepare to store optimal settings and RMSPE values
OPT <- vector("list", k)
rmspe_matrix <- matrix(NA, nrow = k, ncol = length(grid))

# Cross-validation and model fitting
for (i in 1:k) {
  val_idx <- which(folds == i)  # Indexes of validation data for the current fold
  train_data <- data[-val_idx, ]  # Training data (all except the current fold)
  val_data <- data[val_idx, ] # Validation data (the current fold)
  # Vector to store RMSPE for each parameter set
  RMSPE <- numeric(nrow(grid))
  # Loop through each parameter set in the grid
  for (s in 1:nrow(grid)) {
    model <- loess(
      y ~ x,
      control = loess.control(surface = "direct"),
      data = train_data,
      degree = grid$degree[s],
      span = grid$span[s]
    )
      fit <- predict(model, val_data$x)
      RMSPE[s] <- sqrt(mean((val_data$y - fit) ^ 2))
  }
  # Store the index of the best parameter set for this fold
  OPT[[i]] <- which(RMSPE == min(RMSPE), arr.ind = TRUE)
}
# Compile the best settings for each fold
opgrid <- grid[unlist(OPT), ]
colnames(opgrid) <- c("span", "degree")
rownames(opgrid) <- paste("Fold", 1:k, sep=" ")
# Calculate mean of spans and modal degree
library(raster)
opt_degree <- modal(opgrid[, "degree"])
opt_span <- mean(opgrid[, "span"])
# **** Using the test set for final evaluation ****
```

```
model <- loess(
  y ~ x,
  control = loess.control(surface = "direct"),
  degree = opt_degree,
  span = opt_span,
  data = trainset
)
fit <- predict(model, testset$x)
RMSPE_test[l] <- sqrt(mean((testset$y - fit) ^ 2))
}
mean(RMSPE_test)
```

```
## [1] 0.2578508
```

```
var(RMSPE_test)
```

```
## [1] 0.0001485687
```

The model is built using a grid search over a range of values for the span and degree hyperparameters of a LOESS model, with cross-validation applied to optimize these hyperparameters, and evaluate it using test data. This process is identical to what was explained in detail earlier. By systematically looping through 100 iterations, we shuffle the initial data and split it into training and test sets. The training set (80% of the data) is used for model fitting and hyperparamer tuning, while the test set (20% of the data) is reserved for final evaluation, which remains untouched until each iteration's end.

By repeating the test set evaluation 100 times, this simulation evaluates not only the average performance (measured by the mean of the 100 RMSPEs) but also the variance of these RMSPEs obtained from different random splits of the data. The plot displays the RMSPE for each iteration, with the black line representing the average RMSPE across all iterations, offering a clear visual assessment of the performance and stability of the model .

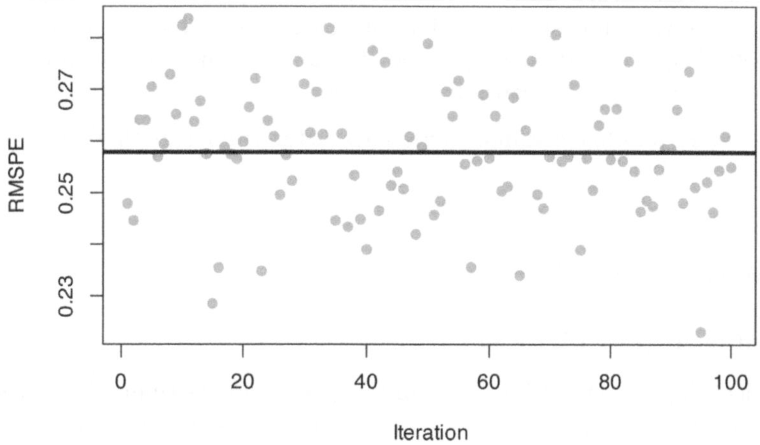

FIGURE 9.1 RMSPE for each iteration.

Moreover, this repetition for robust evaluation is particularly useful when comparing different prediction functions for the dataset using the same evaluation criteria. For continuous outcomes, this typically involves employing various regression-based prediction functions. It is common practice to compare models by evaluating their RMSPEs and selecting the one that delivers the best performance. This approach ensures a fair comparison and increases confidence in the robustness and reliability of the chosen model. The final results, which include both the average RMSPE and its variance, provide a comprehensive reflection of the overall performance of the model. This allows for a more stable and accurate estimate of the ability of the model to generalize to new, unseen data, ultimately ensuring the selection of the most reliable prediction function.

In this section, we introduced grid search and cross-validated grid search as essential techniques for hyperparameter tuning. Grid search systematically evaluates hyperparameter combinations, while cross-validation ensures robust model performance by reducing the risk of overfitting and accounting for variability in training data. Additionally, we highlighted the importance of testing on a separate test set and repeating the process to obtain stable performance estimates.

Most machine learning workflows do not require manually implementing hyperparameter tuning with cross-validated grid search, as demonstrated in the previous simulation. While building such a program helps to understand the process step by step, most practitioners rely on established R packages that streamline and automate this task. If you prefer, you can write similar programs yourself to explore the process, but using ready-made packages is often more easier. Here is a brief overview of some of the most commonly used R packages for hyperparameter tuning and cross-validated grid search.

`caret` is a popular package that simplifies the model training process by providing tools for data splitting, pre-processing, feature selection, and model tuning, including grid search with cross-validation, across a wide variety of algorithms. Another flexible package, `mlr3`, supports a broad range of machine learning models and provides automated hyperparameter tuning through grid search and cross-validation. For users familiar with the tidyverse, `tidymodels` offers a consistent framework for modeling and includes functions like `tune_grid()` to perform grid search with cross-validation. Other specialized packages include `xgboost`, which has built-in cross-validation for gradient boosting models and particularly popular for tree-based models, and `e1071`, often used for tuning support vector machines. Finally, packages like `glmnet` and `randomForest` are commonly used for penalized regression and random forests, respectively, with easy integration into `caret` for grid search and cross-validation. These packages help automate hyperparameter tuning and model evaluation, making them indispensable tools in machine learning workflows.

Now, let's explore another common hyperparameter tuning method: bootstrapped grid search.

9.6 Bootstrapped Grid Search

In machine learning and statistical modeling, hyperparameter tuning is a critical step that significantly affects model performance. Grid search is a common method for systematically exploring predefined hyperparameter values. However, traditional grid search may not fully capture the variability inherent in data, particularly when datasets are small or imbalanced. While grid search and cross-validated grid search are widely used, another powerful method for improving the robustness of hyperparameter selection is bootstrapped grid search.

Bootstrapped grid search combines grid search with bootstrap resampling techniques to create a more robust hyperparameter tuning process. Bootstrapping, a resampling technique where a dataset is sampled with replacement to create multiple new datasets (called bootstrap samples), allows us to estimate the variability of model performance across different hyperparameter settings. This leads to more generalizable models. In bootstrapped grid search, multiple versions of the training dataset are generated through resampling, and grid search is applied to each bootstrap sample. The performance of each hyperparameter combination is evaluated, and the results are aggregated across all samples. By selecting hyperparameters based on the overall performance, this method ensures that the chosen parameters are not overly influenced by the specific characteristics of a single dataset, providing a more stable and generalizable solution.

Bootstrapped grid search offers several advantages over traditional grid search. One key benefit is increased robustness: by averaging results over multiple resampled datasets, bootstrapped grid search provides a more reliable estimate of model performance and reduces the likelihood of overfitting to any one sample. This is especially important when working with smaller or with high variance. Another advantage is better handling of small datasets. Bootstrapping allows the creation of multiple training sets from the original dataset without losing valuable data. This is critical when data is scarce and we want to make the most out of every observation while still being able to perform model tuning and evaluation. Additionally, the process reduces variance in hyperparameter selection. Aggregating performance across multiple bootstrap samples leads to more stable and reliable hyperparameters, which can help improve the generalization ability of the model. This makes bootstrapped grid search a more dependable choice compared to methods that might overfit or select hyperparameters based on a single sample or validation set.

Bootstrapped grid search also provides confidence intervals for performance metrics, offering estimates of uncertainty in model evaluation. By reducing the risk of overfitting to a particular train-test split, bootstrapped grid search helps models generalize better to unseen data, making it a powerful technique for hyperparameter tuning, especially in cases where the dataset is small, has high variance, or imbalanced.

The bootstrapped grid search process can be broken down into the following steps:

- *Define hyperparameter grid:* Determine the hyperparameters to tune and their respective ranges.
- *Generate bootstrap samples:* Create B bootstrap samples from the original dataset by sampling with replacement.
- *Perform grid search:* For each bootstrap sample, apply grid search to tune the hyperparameters of the model (train and validate the model).
- *Aggregate results:* Aggregate the performance metrics from all bootstrap samples, and select the hyperparameters that perform best on average.
- *Evaluate on the test set:* Finally, evaluate the model with the selected hyperparameters on a held-out test set.

9.6.1 Simulation: Bootstrapped Grid Search in R

Here are the typical steps involved in implementing a bootstrapped grid search. First, generate bootstrap samples and choose the number of bootstrap iterations, typically ranging from 50 to 200. Next, select appropriate performance metrics based on the problem (e.g., RMSPE for regression, accuracy, or ROC-AUC for classification). Apply grid search within each bootstrap sample to find the optimal hyperparameters. Finally, train the model using the selected hyperparameters on the test set and evaluate its performance.

This specific simulation provides a comprehensive evaluation of a predictive model by repeatedly testing it 100 times. The model is built using a grid search over a range of values for the span and degree hyperparameters of a `loess()` model, with bootstrap sampling applied to optimize these hyperparameters, and finally evaluated using test data. We define the span to range between 0.01 and 1, increasing by 0.02, and the degree to be either 1 or 2.

In the hyperparameter tuning process, using the training data (which is 80% of the original dataset, with 20% reserved for the final test), the simulation will fit the model using the loess() function for each of these hyperparameter combinations. For each hyperparameter combination, we perform bootstrap resampling within the training data. Instead of dividing the training data into folds, we repeatedly sample with replacement to create bootstrap samples. The observations not included in the bootstrap sample form the out-of-bag (OOB) validation set. The model is trained on the bootstrap sample and evaluated on the OOB data, with the RMSPE recorded for each bootstrap iteration.

This process is repeated for 50 bootstrap samples per hyperparameter combination. The combination of hyperparameters that minimizes the average RMSPE across the 50 bootstrap samples is recorded. This procedure accounts for the variability in the training data and provides a more robust estimate of model performance compared to a single train-test split. Once all hyperparameter combinations have been evaluated through bootstrapping, the optimal hyperparameters are determined by selecting the combination with the lowest average RMSPE.

Finally, the model, with these optimal hyperparameters, is retrained using the entire training set and then evaluated using the final test data (the 20% of the original dataset set aside at the beginning). The test set RMSPE is calculated and stored for each iteration.

By systematically looping through 100 iterations, we shuffle the data each time and split it into training and test sets. The training set (80% of the data) is used for bootstrapped hyperparameter tuning, while the test set (20% of the data) remains untouched until the final evaluation step in each iteration. This repeated evaluation provides insights into the stability and generalization performance of the model over different random splits of the data.

```
# Simulating the data
n <- 1000
set.seed(1)
x <- sort(runif(n) * 2 * pi)
y <- sin(x) + rnorm(n) / 4
data <- data.frame(y = y, x = x)

# Hyperparameter grid
span_values <- seq(from = 0.01, to = 1, by = 0.02)
degree_values <- c(1, 2)
grid <- expand.grid(span = span_values, degree = degree_values)

t <- 100  # Number of repetitions
RMSPE_test <- numeric(t)  # Container for test RMSPEs

for (l in 1:t) {
  # Training-test split (80% training, 20% testing)
  set.seed(321 + l)
```

```
test_indices <- sample(nrow(data), size = 0.2 * nrow(data), replace = FALSE)
testset <- data[test_indices, ]
trainset <- data[-test_indices, ]

OPT <- numeric(nrow(grid))  # Container for validation RMSPEs

# Hyperparameter grid loop
for (s in 1:nrow(grid)) {
  B <- 10  # Number of bootstrap samples
  RMSPE_bootstrap <- numeric(B)  # Container for RMSPEs across bootstrap samples

  # Bootstrap sampling loop
  for (i in 1:B) {
    # Bootstrap sample from training data
    set.seed(1234 + i)
    boot_indices <- sample(nrow(trainset), size = nrow(trainset), replace = TRUE)
    bootstrap_data <- trainset[boot_indices, ]

    # Validation/Out-of-bag (OOB) data
    oob_indices <- setdiff(1:nrow(trainset), unique(boot_indices))
    validation_data <- trainset[oob_indices, ]

    # Check if OOB data is available
    if (nrow(validation_data) > 0) {
      # Fit model on bootstrap sample with current hyperparameters
      model <- loess(
        y ~ x,
        data = bootstrap_data,
        degree = grid$degree[s],
        span = grid$span[s],
        control = loess.control(surface = "direct")
      )

      # Predict on validation data
      predictions <- predict(model, newdata = validation_data$x)

      # Compute RMSPE for current bootstrap sample
      RMSPE_bootstrap[i] <- sqrt(mean((validation_data$y - predictions) ^ 2))
    } else {
      # If no OOB data, assign NA
      RMSPE_bootstrap[i] <- NA
    }
  }

  # Average validation RMSPE for current hyperparameter combination
  OPT[s] <- mean(RMSPE_bootstrap, na.rm = TRUE)
}

# Select hyperparameters with lowest average validation RMSPE
```

```
  best_index <- which.min(OPT)
  best_span <- grid$span[best_index]
  best_degree <- grid$degree[best_index]

  # Train model on entire training set with best hyperparameters
  best_model <- loess(
    y ~ x,
    data = trainset,
    degree = best_degree,
    span = best_span,
    control = loess.control(surface = "direct")
  )

  # Predict on test set
  test_predictions <- predict(best_model, newdata = testset$x)

  # Compute RMSPE on test set
  RMSPE_test[l] <- sqrt(mean((testset$y - test_predictions) ^ 2))
}

# Calculate mean and variance of test RMSPEs
mean_RMSPE_test <- mean(RMSPE_test)
var_RMSPE_test <- var(RMSPE_test)

# Output mean and variance
cat("Mean RMSPE on test set:", mean_RMSPE_test, "\n")
```

```
## Mean RMSPE on test set: 0.2584881
```

```
cat("Variance of RMSPE on test set:", var_RMSPE_test, "\n")
```

```
## Variance of RMSPE on test set: 0.0001447856
```

The plot displays the RMSPE for each iteration, with the black line representing the average RMSPE across all iterations, offering a clear visual assessment of the performance and stability of the model.

```
# Plot RMSPE_test over iterations
plot(RMSPE_test, col = "grey",
     xlab = "Iteration", ylab = "RMSPE")
abline(h = mean_RMSPE_test, col = "black", lwd = 3)
```

There are several alternatives and extensions to bootstrapped grid search that can improve both efficiency and effectiveness. One option is cross-validation can be applied after bootstrapped grid search, where bootstrapping is used for hyperparameter tuning, and cross-validation is employed afterward for model evaluation to test data, offering a more comprehensive assessment of model performance across different data splits. Another option is randomized search with bootstrapping, which randomly samples hyperparameter combinations instead of exhaustively searching the entire grid, reducing computational cost. Additional alternative method is Bayesian optimization, which uses probabilistic models

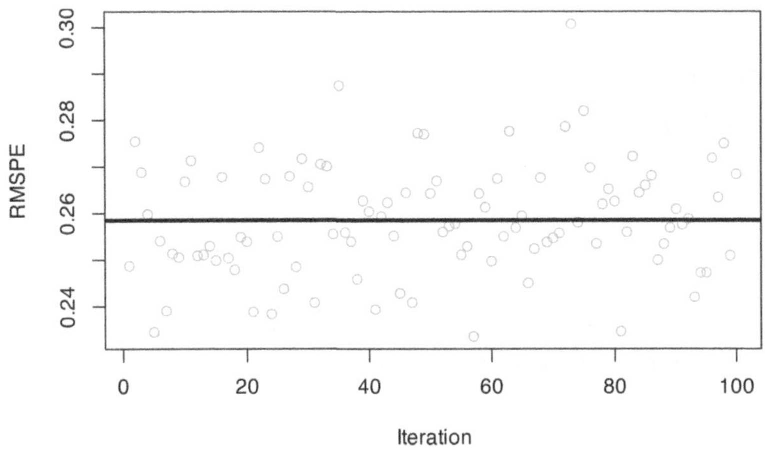

FIGURE 9.2 RMSPE on test set over iterations.

to guide the hyperparameter search and can incorporate bootstrap estimates to refine the search process.

Practitioners should consider the trade-offs involved in bootstrapped grid search and can leverage computational strategies like parallel processing to mitigate runtime concerns. This method is a valuable tool for hyperparameter tuning, also helps reduce the risks of overfitting and high variance, making it particularly useful in small-sample or imbalance datasets. While it introduces additional computational demands, the benefits in terms of reliability and performance often outweigh the costs.

The same packages used for cross-validation grid search can also be applied to bootstrapped grid search. `caret`, `mlr3`, and `tidymodels` all support both cross-validation and bootstrapping for hyperparameter tuning. Packages like `xgboost` and `e1071` remain useful for tuning models such as gradient boosting and support vector machines. Additionally, `glmnet` and `randomForest` can easily integrate with `caret` for bootstrapped grid search. For improved efficiency, `doParallel` can be employed to parallelize the process, reducing runtime.

10

Classification

Supervised learning encompasses two central tasks: regression and classification. Both aim to learn a mapping from input features (variables) to an outcome using labeled training data. What distinguishes them is the type of outcome being predicted. Regression models predict continuous outcomes such as prices, income, test scores, temperatures, electricity consumption, or hospital expenditures. Classification models, on the other hand, predict discrete categories like whether someone is employed or not, if an email is spam, whether a student will pass an exam, if a customer will make a purchase, or if a tumor is malignant. In education, regression might be used to model test scores based on study habits, while classification might flag students at risk of dropping out. In economics and the social sciences, regression helps forecast earnings or housing prices, while classification is used to predict defaults or categorize employment status. In healthcare, regression may predict recovery times or cholesterol levels, while classification supports decisions such as identifying disease presence or classifying high-risk patients.

While regression models typically optimize for metrics like mean squared error by fitting a continuous prediction surface, classification models focus on finding decision boundaries that best separate the classes and are evaluated using measures like accuracy, precision, recall, and the area under the ROC curve.

Common regression methods include linear regression, ridge and LASSO regression, support vector regression, decision trees, random forests, and boosting. Classification methods range from logistic model, K-nearest neighbors, decision trees, support vector machines, random forests, and neural networks.

This chapter focuses exclusively on classification. We introduce the general framework, discuss kNN in depth as an accessible starting point, evaluate model performance using metrics beyond accuracy, and walk through real applications using the MNIST dataset. We also illustrate how ROC curves, area under the curve (AUC), and threshold selection inform decision-making.

10.1 Classifying Handwritten Digits

Reading handwritten letters and numbers is no longer a challenging task for machines. For instance, post offices routinely use software that reads zip codes and sorts mail accordingly. Such applications rely heavily on classification algorithms. To explore how this works, we turn to a real dataset: MNIST, a widely used benchmark in machine learning for digit recognition. As described on Wikipedia:

DOI: 10.1201/9781003504795-10

The MNIST database (Modified National Institute of Standards and Technology database) is a large database of handwritten digits that is commonly used for training various image processing systems. The MNIST database contains 60,000 training images and 10,000 testing images. Half of the training set and half of the test set were taken from NIST's training dataset, while the other half of the training set and the other half of the test set were taken from NIST's testing dataset. There have been a number of scientific papers on attempts to achieve the lowest error rate; one paper, using a hierarchical system of convolutional neural networks, manages to get an error rate on the MNIST database of 0.23%.

These images are converted into $28 \times 28 = 784$ pixels and, for each pixel, there is a measure that scales the darkness in that pixel between 0 (white) and 255 (black). Hence, for each digitized image, we have an indicator variable Y between 0 and 9, and we have 784 variables that identifies each pixel in the digitized image. Let's download the data. (More details about the data[1] .

```
#loading the data
library(tidyverse)
library(dslabs)

#Download the data to your directory.  It's 21 MB!
mnist <- read_mnist()
save(mnist, file = "mnist.Rdata")
load("mnist.Rdata")
str(mnist)
```

```
## List of 2
##  $ train:List of 2
##   ..$ images: int [1:60000, 1:784] 0 0 0 0 0 0 0 0 0 0 ...
##   ..$ labels: int [1:60000] 5 0 4 1 9 2 1 3 1 4 ...
##  $ test :List of 2
##   ..$ images: int [1:10000, 1:784] 0 0 0 0 0 0 0 0 0 0 ...
##   ..$ labels: int [1:10000] 7 2 1 0 4 1 4 9 5 9 ...
```

For the train set, we have two nested sets: `images`, which contains all 784 features for 60,000 images. Hence, it's a 60000×784 matrix. And, `labels` contains the labels (from 0 to 9) for each image.

The digitizing can be understood from this image better:

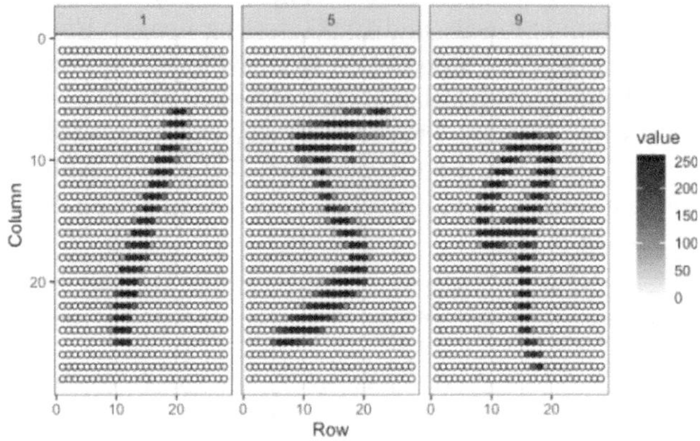

For each image, the pixels are features with a label that shows the true number between 0 and 9. This methods is called as "flattening", which is a technique that is used to convert multi-dimensional image into a one-dimension array (vector).

For now, we will use a smaller version of this data set given in the `dslabs` package, which is a random sample of 1,000 images (only for 2 and 7 digits), 800 in the training set and 200 in the test set, with only two features: the proportion of dark pixels that are in the upper left quadrant, `x_1`, and the lower right quadrant, `x_2`.

```
data("mnist_27")
str(mnist_27)
```

```
## List of 5
##  $ train      :'data.frame': 800 obs. of  3 variables:
##   ..$ y  : Factor w/ 2 levels "2","7": 1 2 1 1 2 1 2 2 2 1 ...
##   ..$ x_1: num [1:800] 0.0395 0.1607 0.0213 0.1358 0.3902 ...
##   ..$ x_2: num [1:800] 0.1842 0.0893 0.2766 0.2222 0.3659 ...
##  $ test       :'data.frame': 200 obs. of  3 variables:
##   ..$ y  : Factor w/ 2 levels "2","7": 1 2 2 2 2 1 1 1 1 2 ...
##   ..$ x_1: num [1:200] 0.148 0.283 0.29 0.195 0.218 ...
##   ..$ x_2: num [1:200] 0.261 0.348 0.435 0.115 0.397 ...
##  $ index_train: int [1:800] 40334 33996 3200 38360 36239 38816 8085 9098
##  15470 5096 ...
##  $ index_test : int [1:200] 46218 35939 23443 30466 2677 54248 5909 13402
##  11031 47308 ...
##  $ true_p     :'data.frame': 22500 obs. of  3 variables:
##   ..$ x_1: num [1:22500] 0 0.00352 0.00703 0.01055 0.01406 ...
##   ..$ x_2: num [1:22500] 0 0 0 0 0 0 0 0 0 0 ...
##   ..$ p  : num [1:22500] 0.703 0.711 0.719 0.727 0.734 ...
##   ..- attr(*, "out.attrs")=List of 2
##   .. ..$ dim     : Named int [1:2] 150 150
##   .. .. ..- attr(*, "names")= chr [1:2] "x_1" "x_2"
##   .. ..$ dimnames:List of 2
```

```
##    .. .. ..$ x_1: chr [1:150] "x_1=0.0000000" "x_1=0.0035155"
##   "x_1=0.0070310" "x_1=0.0105465" ...
##    .. .. ..$ x_2: chr [1:150] "x_2=0.000000000" "x_2=0.004101417"
##   "x_2=0.008202834"
 "x_2=0.012304251" ...
```

10.2 Linear Classifiers

To begin our exploration of classification methods, we start with linear classifiers. These models define a decision rule based on a linear combination of the input features, which results in a boundary—called a *hyperplane*—that separates observations into different categories. In simple two-dimensional problems, this boundary is just a line; in higher dimensions, it becomes a flat surface that divides the feature space. For example, in our reduced MNIST dataset containing only digits 2 and 7, each image is represented by just two features: the proportion of dark pixels in the upper-left and lower-right corners. A linear classifier attempts to draw a straight line between these two digit classes using these features, assigning new observations based on which side of the line they fall. We begin with one of the simplest linear classifiers: the linear probability model (LPM).

$$\Pr(Y = 1|X_1 = x_1, X_2 = x_2) = \beta_0 + \beta_1 x_1 + \beta_2 x_2 \tag{10.1}$$

```r
# LPM requires numerical 1 and 0
y10 = ifelse(mnist_27$train$y == 7, 1, 0)
train <- data.frame(mnist_27$train, y10)
plot(train$x_1, train$x_2, pch = ifelse(train$y10 == 1, 17, 1), col = "black", cex = 0.72)
```

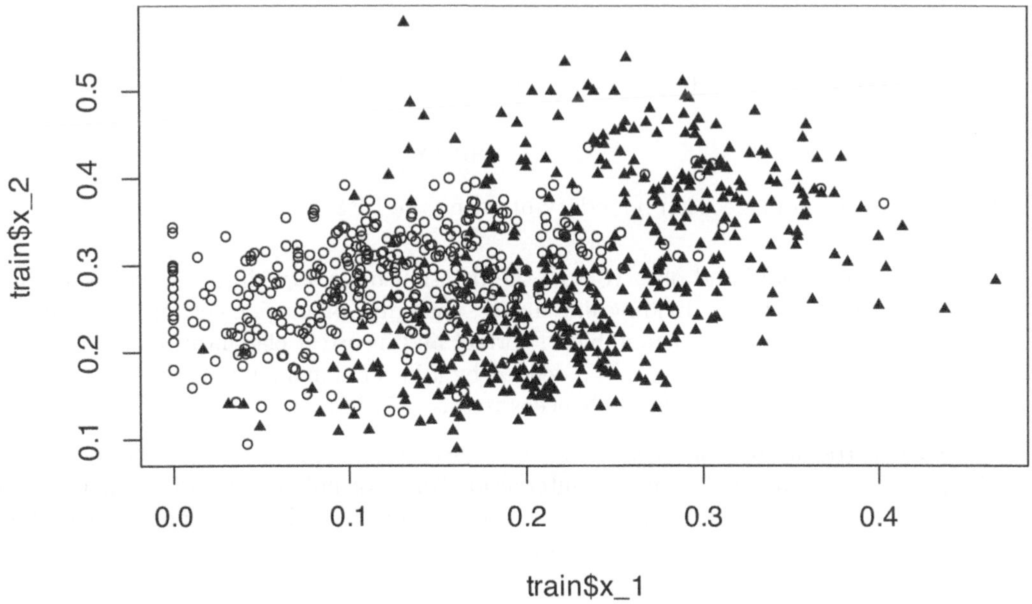

FIGURE 10.1 Training data: triangles for 7s and circles for 2s.

Here, the circles are 2 and triangles are 7. Note that if we use 0.5 as a decision rule such that it separates pairs (x_1, x_2) for which $\Pr(Y = 1|X_1 = x_1, X_2 = x_2) < 0.5$ then we can have a hyperplane as:

$$\hat{\beta}_0 + \hat{\beta}_1 x_1 + \hat{\beta}_2 x_2 = 0.5 \Longrightarrow x_2 = \left(0.5 - \hat{\beta}_0\right) / \hat{\beta}_2 - \hat{\beta}_1 / \hat{\beta}_2 x_1. \tag{10.2}$$

If we incorporate this into our plot for the train data:

```
model <- lm(y10 ~ x_1 + x_2, train)
tr <- 0.5
a <- tr - model$coefficients[1]
a <- a / model$coefficients[3]
b <- -model$coefficients[2] / model$coefficients[3]
plot(train$x_1, train$x_2, pch = ifelse(train$y10 == 1, 17, 1), col = "black", cex = 0.72)
abline(a, b, col = "black", lty = 2, lwd = 2.8)
```

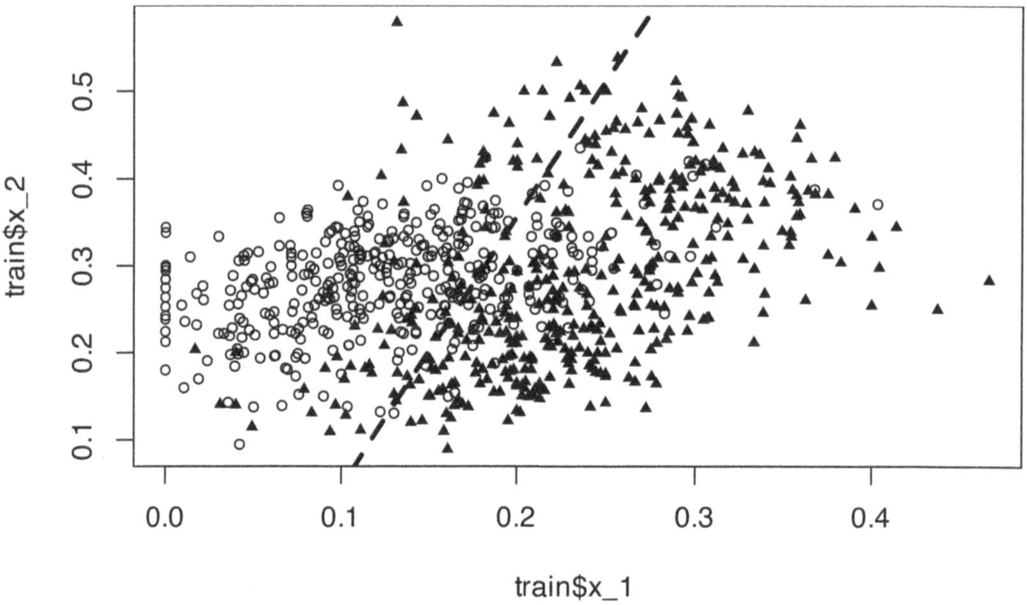

FIGURE 10.2 Linear decision boundary fitted by the LPM.

Play with the threshold and observe how the decision boundary shifts. As you vary this cutoff between 0 and 1, the dividing line moves, changing which side certain observations fall on and, in turn, altering the number of correct and incorrect classifications. This line is the estimated hyperplane, and since it's defined by a linear combination of the features, the resulting decision boundary is linear—hence why the LPM is referred to as a *linear classifier*.

But what if a straight line isn't flexible enough to separate the two classes effectively? Could adding nonlinear components, such as interaction terms or polynomial transformations of the features, lead to a better separation of 2s and 7s in our plot? Let's explore how changing our model in this way affects its performance.

```
model2 <- lm(y10 ~ x_1 + I(x_1 ^ 2) + x_2, train)
summary(model2)
```

```
##
## Call:
## lm(formula = y10 ~ x_1 + I(x_1^2) + x_2, data = train)
##
## Residuals:
##      Min       1Q   Median       3Q      Max
## -1.14744 -0.28816  0.03999  0.28431  1.06759
##
## Coefficients:
##             Estimate Std. Error t value Pr(>|t|)
## (Intercept)  0.09328    0.06571   1.419   0.1562
## x_1          4.81884    0.55310   8.712  < 2e-16 ***
## I(x_1^2)    -2.75520    1.40760  -1.957   0.0507 .
## x_2         -1.18864    0.17252  -6.890 1.14e-11 ***
## ---
## Signif. codes:  0 '***' 0.001 '**' 0.01 '*' 0.05 '.' 0.1 ' ' 1
##
## Residual standard error: 0.3891 on 796 degrees of freedom
## Multiple R-squared:  0.3956, Adjusted R-squared:  0.3933
## F-statistic: 173.7 on 3 and 796 DF,  p-value: < 2.2e-16
```

```
tr <- 0.5
s <- model2$coefficients
a = tr / s[3]
b = s[1] / s[3]
d = s[2] / s[3]
e = s[4] / s[3]
x22 = a - b - d * train$x_1 - e * (train$x_1 ^ 2)
plot(train$x_1, train$x_2, pch = ifelse(train$y10 == 1, 17, 1), col = "black", cex = 0.72)
lines(train$x_1[order(x22)], x22[order(x22)], col = "black", lty = 2, lwd = 2.8)
```

The coefficient of the polynomial term in our extended LPM is small and only marginally significant, and visually, the classification does not appear substantially better than the linear case. This shows that simply adding nonlinear terms doesn't always improve performance—especially when the relationship between features and class labels is not well captured by the chosen polynomial form.

Would logistic regression provide a better decision rule? Even without estimating the model, we can describe how it works. Logistic regression models the probability that $Y = 1$ using the logistic (sigmoid) function:

$$P(Y = 1 \mid x) = \frac{\exp(w_0 + \sum_i w_i x_i)}{1 + \exp(w_0 + \sum_i w_i x_i)}$$

This function smoothly maps any real-valued linear combination of predictors to a probability between 0 and 1. The complementary probability for class 0 is:

$$P(Y = 0 \mid x) = 1 - P(Y = 1 \mid x) = \frac{1}{1 + \exp(w_0 + \sum_i w_i x_i)}$$

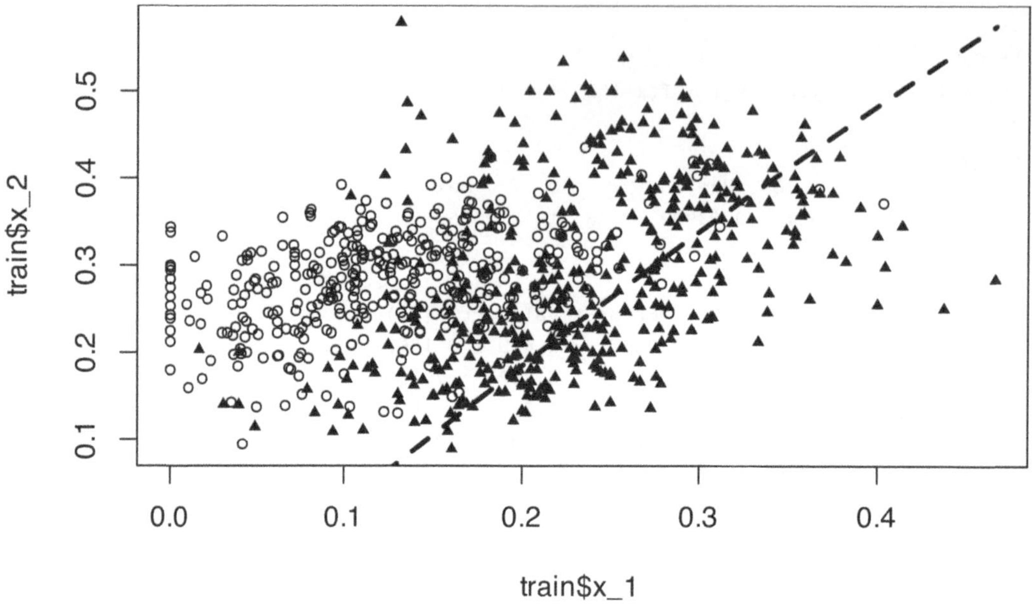

FIGURE 10.3 Linear decision boundary fitted using a polynomial LPM.

Taking the odds ratio of class 1 to class 0 yields:

$$\frac{P}{1-P} = \exp(w_0 + \sum_i w_i x_i)$$

We classify an observation as $Y = 1$ when this odds ratio exceeds 1, i.e., when $P > 0.5$. Taking the logarithm of both sides, the condition becomes:

$$w_0 + \sum_i w_i x_i > 0$$

So, the decision boundary is defined by the linear equation:

$$\hat{\beta}_0 + \hat{\beta}_1 x_1 + \hat{\beta}_2 x_2 = 0 \quad \Longrightarrow \quad x_2 = -\frac{\hat{\beta}_0}{\hat{\beta}_2} - \frac{\hat{\beta}_1}{\hat{\beta}_2} x_1$$

This confirms that the decision boundary in logistic regression, like in the LPM, is a straight line. Both models are therefore classified as *linear classifiers*, and they tend to work well only when the classes are approximately linearly separable.

But what if the actual relationship between the features and the class label is nonlinear? In such cases, linear classifiers may perform poorly. This motivates the use of more flexible, *nonparametric methods*, such as *k-nearest neighbors (kNN)*, which can adapt to more complex boundaries and capture subtle patterns in the data without imposing a specific functional form.

10.3 k-Nearest Neighbors

kNN is a simple yet powerful nonparametric method used for both classification and regression. Unlike models that assume a specific functional form, kNN makes predictions based on local patterns in the data. For classification, a new observation is assigned to the most common class among its k nearest neighbors in the feature space. For regression, the predicted value is the average of the outcomes of these neighbors. In essence, kNN estimates relationships in the data using a logic similar to bin smoothing—by relying on nearby points rather than fitting a global model.

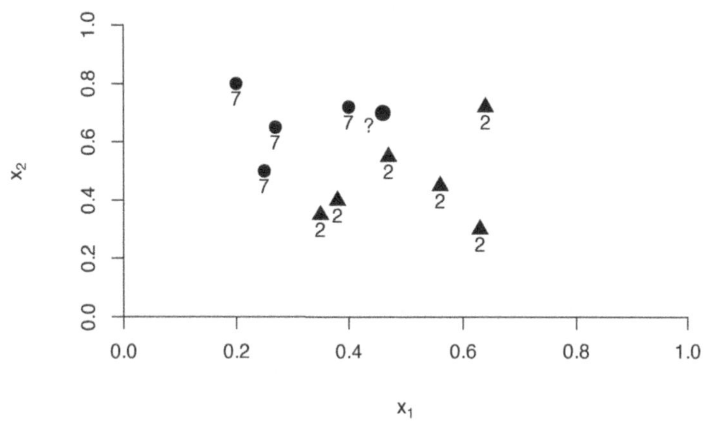

FIGURE 10.4 Example of kNN classification based on nearest neighbors.

Suppose we have to classify (identify) the "unkown" dot, labeled with ?, as 7 or 2. Since it is a nonparametric approach, we have to define bins. If the number of observations in bins set to 1 ($k = 1$), then we need to find one observation that is nearest to the unkown dot. How? Since we know to coordinates (x_1, x_2) of that unkown dot, we can find its nearest neighbors by some distance functions among all points (observations) in the data. A commonly used distance metric in kNN is the **Euclidean distance**, defined as:

$$d(x, x') = \sqrt{(x_1 - x'_1)^2 + \ldots + (x_n - x'_n)^2}$$

This gives the straight-line distance between two points. Depending on the nature of the data, alternative distance metrics may be more appropriate. For instance, *Manhattan distance* sums the absolute differences across features, *Chebyshev distance* considers the maximum difference along any single dimension, and *Hamming distance* is used for binary features. Since our features are continuous, Euclidean distance is a natural choice here.

To apply kNN, we compute this distance between the point we want to classify and every other observation in the dataset. For example, if we are comparing one red dot to ten other points, we must calculate ten distances. In practice, we often compute the full pairwise distance matrix for all n observations, resulting in an $n \times n$ symmetric distance matrix.

For example, for two-dimensional space, we can calculate the distances as follows:

```
x1 <- c(2, 2.1, 4, 4.3)
x2 <- c(3, 3.3, 5, 5.1)

EDistance <- function(x, y){
  dx <- matrix(0, length(x), length(x))
  dy <- matrix(0, length(x), length(x))

  for (i in 1:length(x)) {
    dx[i,] <- (x[i] - x)^2
    dy[i,] <- (y[i] - y)^2
    dd <- sqrt(dx^2 + dy^2)
  }
  return(dd)
}

EDistance(x1, x2)
```

```
##            [,1]       [,2]       [,3]       [,4]
## [1,] 0.00000000 0.09055385 5.65685425 6.88710389
## [2,] 0.09055385 0.00000000 4.62430535 5.82436263
## [3,] 5.65685425 4.62430535 0.00000000 0.09055385
## [4,] 6.88710389 5.82436263 0.09055385 0.00000000
```

```
plot(x1, x2, col = "black", lwd = 3)
#segments(x1[1], x2[1], x1[2:4], x2[2:4], col = "blue" )
#segments(x1[2], x2[2], x1[c(1, 3:4)], x2[c(1, 3:4)], col = "green" )
#segments(x1[3], x2[3], x1[c(1:2, 4)], x2[c(1:2, 4)], col = "orange" )
segments(x1[4], x2[4], x1[1:3], x2[1:3], col = "gray40" )
```

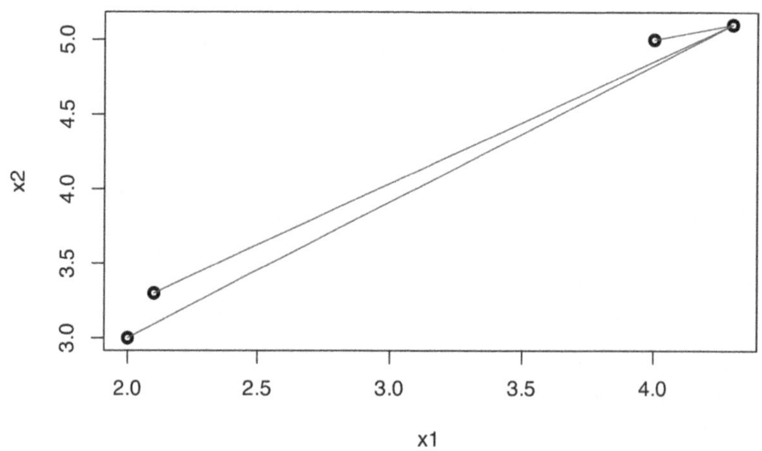

The matrix shows all distances for four points and, as we expect, it is symmetric. The green lines show the distance from the last point ($x = 4.3$, $y = 5.1$) to all other points. Using this matrix, we can easily find the k-NNs for any point.

When $k = 1$, the observation that has the shortest distance is going to be the one to predict what the unknown dot could be. This is shown in Figure 10.6:

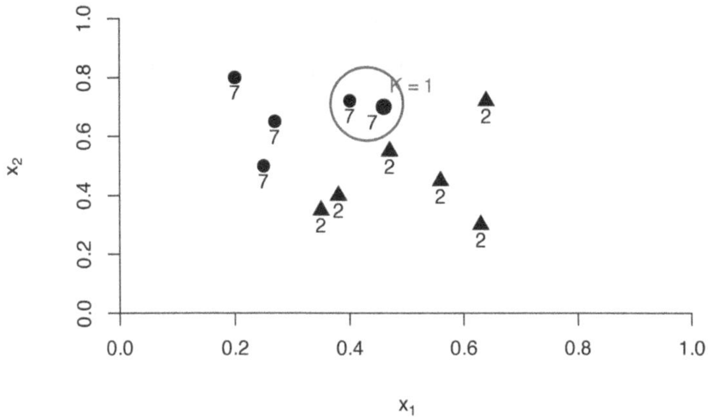

FIGURE 10.5 Example of kNN classification based on nearest neighbor.

If we increase k to 3, we classify the point based on a majority vote among the 3 nearest neighbors. Suppose two are 7s and one is a 2, then the unknown dot is classified as 7:

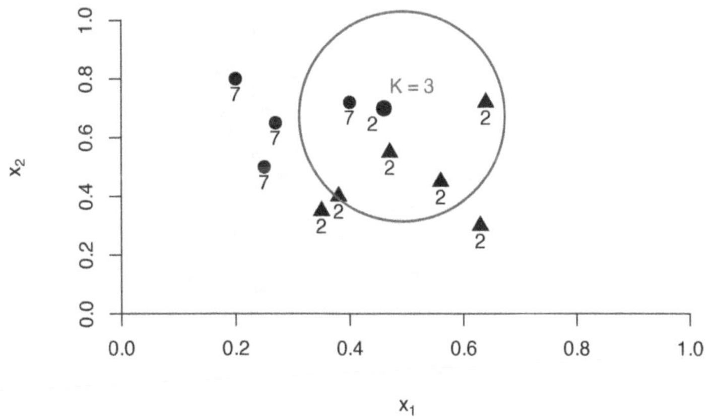

FIGURE 10.6 Example of kNN classification based on 3 nearest neighbors.

Formally, kNN estimates the probability that $Y = 1$ as:

$$\hat{P}_k(Y = 1|X = x) = \frac{1}{k} \sum_{i \in \mathcal{N}_k(x,D)} I(y_i = 1)$$

Based on this probability, we assign the class using a simple rule:

$$\hat{C}_k(x) = \begin{cases} 1 & \text{if } \hat{P}_k(Y = 1 \mid x) > 0.5 \\ 0 & \text{otherwise} \end{cases}$$

For instance, if $\hat{P}(Y = 7|x_1 = 4, x_2 = 3) = \frac{2}{3}$, we classify the unknown dot as 7.

The parameter k in k-NN is a **hyperparameter** that controls the smoothness of the decision boundary. A small k (e.g., $k = 1$) often leads to overfitting, as the model becomes highly sensitive to noise in the training data. On the other hand, a very large k results in oversmoothing, where the model averages over many observations and fails to capture relevant local patterns—leading to underfitting. Choosing the right k is crucial and is typically done through validation techniques that evaluate predictive accuracy.

Before diving into practical tuning, it's helpful to visualize what decision boundaries look like under different values of k. One way to do this is by examining *Voronoi diagrams*, which represent how the space is partitioned based on the nearest neighbors:

```
set.seed(1)
x1 <- runif(50)
x2 <- runif(50)

library(deldir)
tesselation <- deldir(x1, x2)
tiles <- tile.list(tesselation)

plot(tiles, pch = 19, close = TRUE,
     fillcol = rep("gray80", length(tiles)),
     xlim = c(-0.2, 1.1))
```

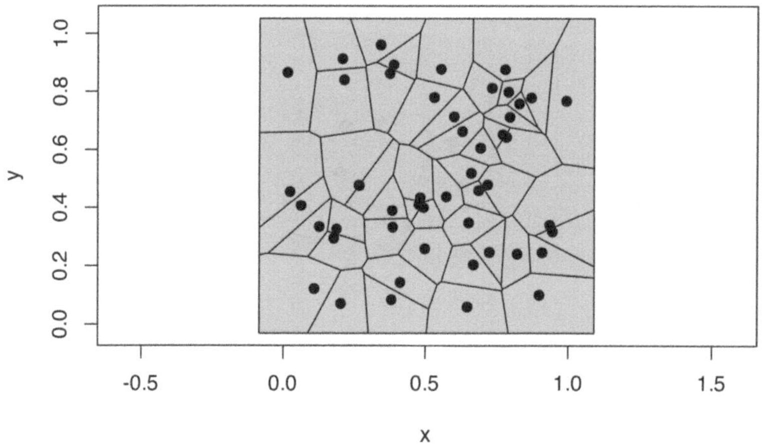

These Voronoi cells are polygons that surround each training point, with edges defined by the perpendicular bisectors between neighboring points. For 1-nearest neighbor classification, any new point is assigned to the same class as the training point whose cell it falls into. The decision boundary is formed by the borders between adjacent cells of different classes. In practice, plotting the actual boundary between classes involves overlaying class labels and merging same-colored regions, which reflect the union of Voronoi cells for each class.

To see how decision boundaries evolve with different values of k, we use the `knn3()` function from the *Caret* package on the `mnist_27` dataset. While we won't assess model accuracy in this visualization, the visual comparison of boundaries illustrates the trade-off between overfitting and oversmoothing.

```
library(tidyverse)
library(caret)
library(dslabs)
```

```
theme_set(theme_minimal(base_family = "", base_size = 11))

# With k = 2
model1 <- knn3(y ~ ., data = mnist_27$train, k = 2)

x_1 <- mnist_27$true_p$x_1
x_2 <- mnist_27$true_p$x_2
df <- data.frame(x_1, x_2)

p_hat <- predict(model1, df, type = "prob")
p_7 <- p_hat[, 2]
df <- data.frame(x_1, x_2, p_7)

p1 <- ggplot() +
  geom_point(data = mnist_27$train, aes(x = x_1, y = x_2, shape = factor(y)),
             size = 1, stroke = 0.72, color = "black") +
  stat_contour(data = df, aes(x = x_1, y = x_2, z = p_7),
               breaks = c(0.5), color = "black", linewidth = 1.2) +
  scale_shape_manual(values = c(16, 17))

plot(p1)
```

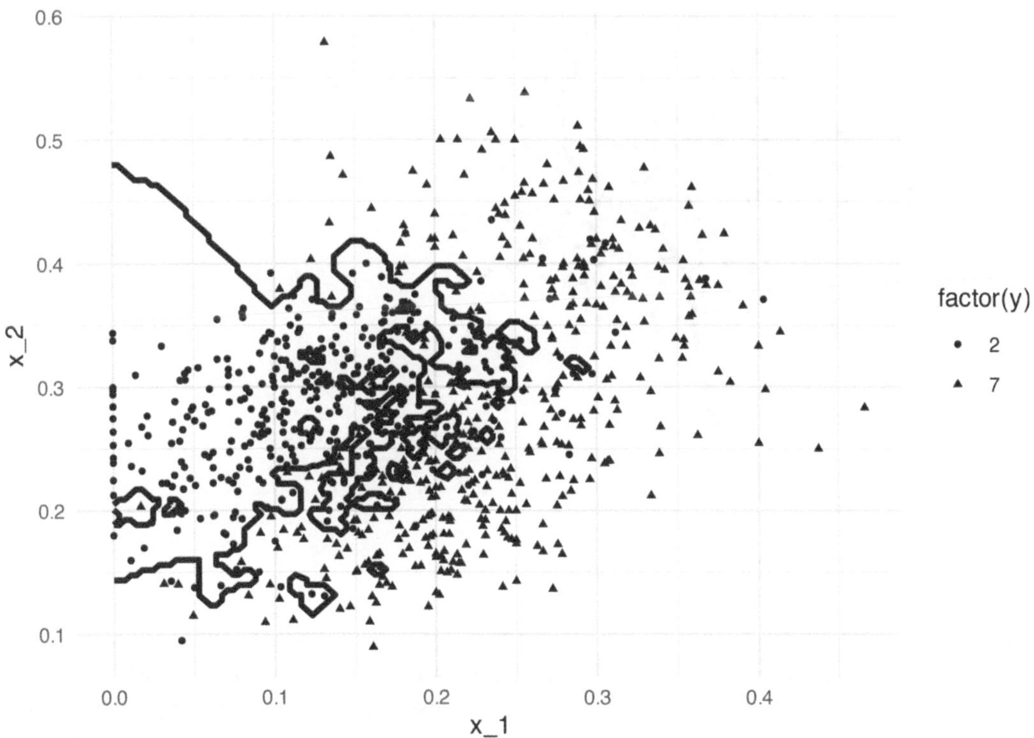

FIGURE 10.7 kNN decision boundary with k = 2.

```
# With k = 400
model2 <- knn3(y ~ ., data = mnist_27$train, k = 400)

p_hat <- predict(model2, df, type = "prob")
p_7 <- p_hat[, 2]
df <- data.frame(x_1, x_2, p_7)

p1 <- ggplot() +
  geom_point(data = mnist_27$train, aes(x = x_1, y = x_2, shape = factor(y)),
             size = 1, stroke = 0.72, color = "black") +
  stat_contour(data = df, aes(x = x_1, y = x_2, z = p_7),
               breaks = c(0.5), color = "black", linewidth = 1.2) +
  scale_shape_manual(values = c(16, 17))

plot(p1)
```

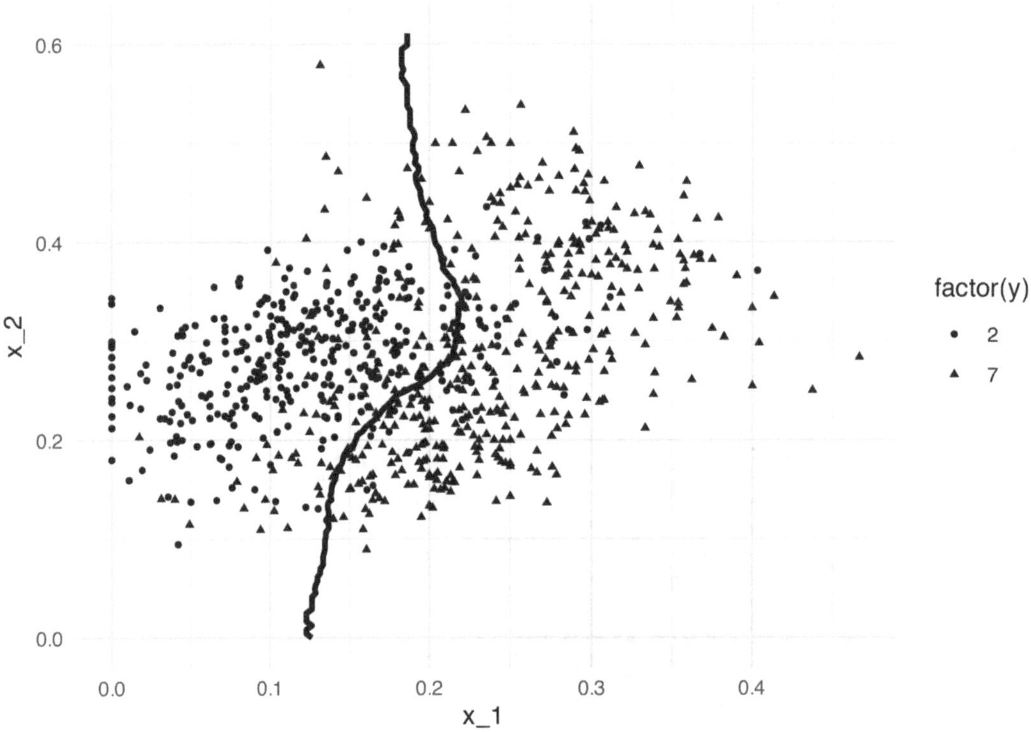

FIGURE 10.8 kNN decision boundary with k = 400.

As shown, the boundary with $k = 2$ is highly flexible, capturing local patterns but also overfitting noise. In contrast, the decision boundary with $k = 400$ is overly smooth, essentially collapsing to a linear boundary that ignores finer distinctions. These examples emphasize the importance of tuning k to balance bias and variance for better predictive performance.

So far we've relied on Euclidean distance to define "nearness," but kNN can use a variety of distance measures depending on the application:

Manhattan Distance (also known as L1 distance) is the sum of absolute differences

$$d(\mathbf{x}, \mathbf{y}) = |x_1 - x_2| + |y_1 - y_2| \tag{10.3}$$

```
# Manhattan Distance
MDistance <- function(x, y){
  n <- length(x)
  dd <- matrix(0, n, n)

  for (i in 1:n) {
    for (j in 1:n) {
      dd[i, j] <- abs(x[i] - x[j]) + abs(y[i] - y[j])
    }
  }
  return(dd)
}
```

The Chebyshev distance is the maximum absolute difference between the two points. The Chebyshev distance between two points $\mathbf{x} = (x_1, x_2)$ and $\mathbf{y} = (y_1, y_2)$ is given by:

$$d(\mathbf{x}, \mathbf{y}) = \max(|x_1 - x_2|, |y_1 - y_2|) \tag{10.4}$$

```
# Chebyshev Distance
CDistance <- function(x, y){
  n <- length(x)
  dd <- matrix(0, n, n)

  for (i in 1:n) {
    for (j in 1:n) {
      dd[i, j] <- max(abs(x[i] - x[j]), abs(y[i] - y[j]))
    }
  }
  return(dd)
}
```

10.3.1 Example: Finding Nearest Neighbors with $k = 2$

Let's now compute the distance matrix for 15 points in 2D space using the Euclidean distance function:

```
x1 <- c(2, 2.1, 4, 4.3, 1.5, 3.2, 6, 5.5, 7, 2.5, 4.5, 3.8, 6.2, 5, 7.3)
x2 <- c(3, 3.3, 5, 5.1, 2, 4.2, 6.5, 5.8, 7.2, 3.1, 4.7, 3.6, 6.1, 5.2, 7.6)

distanceMatrix <- EDistance(x1, x2)
print(distanceMatrix)
```

```
##              [,1]       [,2]       [,3]       [,4]     [,5]      [,6]
##  [1,] 0.00000000 0.09055385 5.65685425 6.88710389 1.030776 2.0364675
```

```
## [2,] 0.09055385 0.00000000 4.62430535 5.82436263 1.727918 1.4560907
## [3,] 5.65685425 4.62430535 0.00000000 0.09055385 10.957304 0.9050967
## [4,] 6.88710389 5.82436263 0.09055385 0.00000000 12.402326 1.4560907
## [5,] 1.03077641 1.72791782 10.95730350 12.40232639 0.000000 5.6371713
## [6,] 2.03646753 1.45609066 0.90509668 1.45609066 5.637171 0.0000000
## [7,] 20.15099253 18.33580377 4.58938994 3.49194788 28.637825 9.4577852
## [8,] 14.54400564 13.14138882 2.33925202 1.52108514 21.552578 5.8768784
## [9,] 30.59688873 28.42224833 10.21888448 8.52010563 40.573687 17.0150992
## [10,] 0.25019992 0.16492423 4.25377479 5.14758196 1.569745 1.3054501
## [11,] 6.88582602 6.08434056 0.26570661 0.16492423 11.582059 1.7083911
## [12,] 3.25993865 2.89140104 1.96040812 2.26384628 5.876878 0.5091169
## [13,] 20.08784956 18.54836111 4.98895781 3.74594447 27.758678 9.6970150
## [14,] 10.21888448 9.15205988 1.00079968 0.49010203 15.966217 3.3908111
## [15,] 35.16807785 32.75731521 12.81755437 10.95730350 45.990208 20.4012181
##              [,7]       [,8]       [,9]      [,10]      [,11]      [,12]
## [1,] 20.1509925 14.5440056 30.5968887 0.2501999 6.8858260 3.2599386
## [2,] 18.3358038 13.1413888 28.4222483 0.1649242 6.0843406 2.8914010
## [3,] 4.5893899 2.3392520 10.2188845 4.2537748 0.2657066 1.9604081
## [4,] 3.4919479 1.5210851 8.5201056 5.1475820 0.1649242 2.2638463
## [5,] 28.6378246 21.5525776 40.5736873 1.5697452 11.5820594 5.8768784
## [6,] 9.4577852 5.8768784 17.0150992 1.3054501 1.7083911 0.5091169
## [7,] 0.0000000 0.5500909 1.1135978 16.8432806 3.9446293 9.7032829
## [8,] 0.5500909 0.0000000 2.9839739 11.5820594 1.5697452 5.6371713
## [9,] 1.1135978 2.9839739 0.0000000 26.3180280 8.8388348 16.5172395
## [10,] 16.8432806 11.5820594 26.3180280 0.0000000 4.7490631 1.7083911
## [11,] 3.9446293 1.5697452 8.8388348 4.7490631 0.0000000 1.3054501
## [12,] 9.7032829 5.6371713 16.5172395 1.7083911 1.3054501 0.0000000
## [13,] 0.1649242 0.4981967 1.3688316 16.3834093 3.4919479 8.4994176
## [14,] 1.9636955 0.4382921 5.6568542 7.6492222 0.3535534 2.9372096
## [15,] 2.0785091 4.5820519 0.1835756 30.6741601 11.4975519 20.1509925
##             [,13]      [,14]      [,15]
## [1,] 20.0878496 10.2188845 35.1680779
## [2,] 18.5483611 9.1520599 32.7573152
## [3,] 4.9889578 1.0007997 12.8175544
## [4,] 3.7459445 0.4901020 10.9573035
## [5,] 27.7586779 15.9662175 45.9902077
## [6,] 9.6970150 3.3908111 20.4012181
## [7,] 0.1649242 1.9636955 2.0785091
## [8,] 0.4981967 0.4382921 4.5820519
## [9,] 1.3688316 5.6568542 0.1835756
## [10,] 16.3834093 7.6492222 30.6741601
## [11,] 3.4919479 0.3535534 11.4975519
## [12,] 8.4994176 2.9372096 20.1509925
## [13,] 0.0000000 1.6521804 2.5547211
## [14,] 1.6521804 0.0000000 7.8205946
## [15,] 2.5547211 7.8205946 0.0000000
```

The result is a symmetric matrix showing pairwise distances. From this, we can identify the two closest neighbors for each point. Here's a function that does exactly that:

```
findNN <- function(distanceMatrix, k) {
  n <- nrow(distanceMatrix)
  nearestNeighbors <- matrix(NA, n, k)

  for (i in 1:n) {
    sortedDistances <- sort(distanceMatrix[i, ], index.return = TRUE)
    nnIndices <- sortedDistances$ix[2:(k + 1)]
    nearestNeighbors[i,] <- nnIndices
  }
  return(nearestNeighbors)
}

nearestNeighbors <- findNN(distanceMatrix, 2)
print(nearestNeighbors)
```

```
##       [,1] [,2]
##  [1,]    2   10
##  [2,]    1   10
##  [3,]    4   11
##  [4,]    3   11
##  [5,]    1   10
##  [6,]   12    3
##  [7,]   13    8
##  [8,]   14   13
##  [9,]   15    7
## [10,]    2    1
## [11,]    4    3
## [12,]    6   11
## [13,]    7    8
## [14,]   11    8
## [15,]    9    7
```

Now that we understand how distance metrics and the value of k affect the structure of kNN, we turn to a more realistic example. We'll generate labeled and unlabeled data, and then use a simple kNN procedure to classify the new observations.

10.3.2 More Realistic Example

In this example, our task is to classify new, unlabeled data points (shown in cross) based on the labels of existing training data (triangle and circle points). We will implement a simple kNN classification procedure from scratch to demonstrate how it works in practice.

First, we generate the training dataset and the new points we want to classify:

```
n = 25

# Generate training data
set.seed(12)
x1 <- runif(n)
x2 <- runif(n)
```

```
y <- sample(c(0, 1), n, replace = TRUE)

# Generate 5 new points with unknown labels
set.seed(222)
x1_new <- runif(5)
x2_new <- runif(5)

# Visualize the data
plot(x1, x2, xlab = "x1", ylab = "x2", pch = ifelse(y == 1, 17, 16),
     col = "black", cex = 1.2, lwd = 1)
points(x1_new, x2_new, col = "gray40", pch = 4, lwd = 2)
```

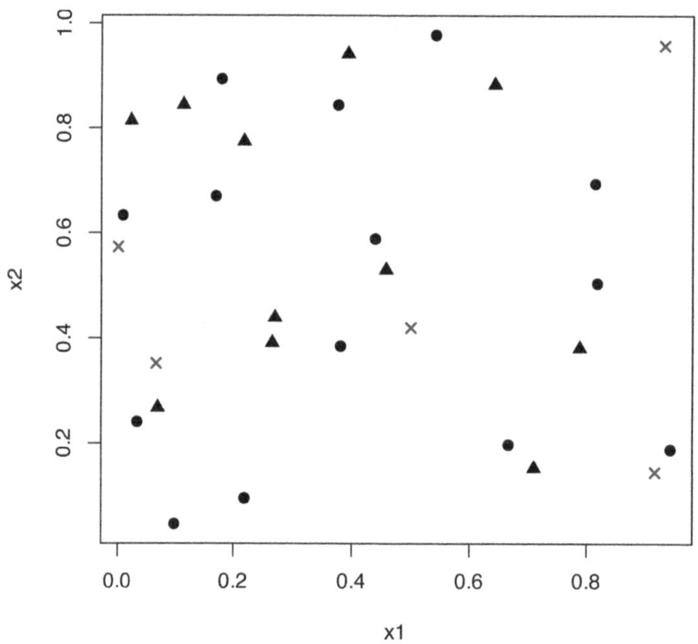

Now we apply the kNN rule with $k = 5$. For each new point, we compute its distance to every training point, find the 5 nearest neighbors, and assign the most common label among them as the predicted class:

```
k <- 5
new_labels <- c()

for (i in 1:length(x1_new)) {
  # Compute distances from new point to all training points
  distances <- sqrt((x1_new[i] - x1)^2 + (x2_new[i] - x2)^2)

  # Identify the k closest training points
  nearest <- order(distances)[1:k]

  # Extract their labels
  nearest_labels <- y[nearest]
```

```
  # Assign the most frequent label among neighbors
  new_labels[i] <- names(which.max(table(nearest_labels)))
}

print(new_labels)

## [1] "0" "1" "1" "1" "0"

# Plot training and newly classified points
plot(x1, x2, xlab = "x1", ylab = "x2",
     pch = ifelse(y == 1, 17, 16), col = "black", cex = 1.2, lwd = 1)

points(x1_new, x2_new, pch = ifelse(new_labels == 1, 17, 16),
       col = "gray40", cex = 1.5, lwd = 2)
```

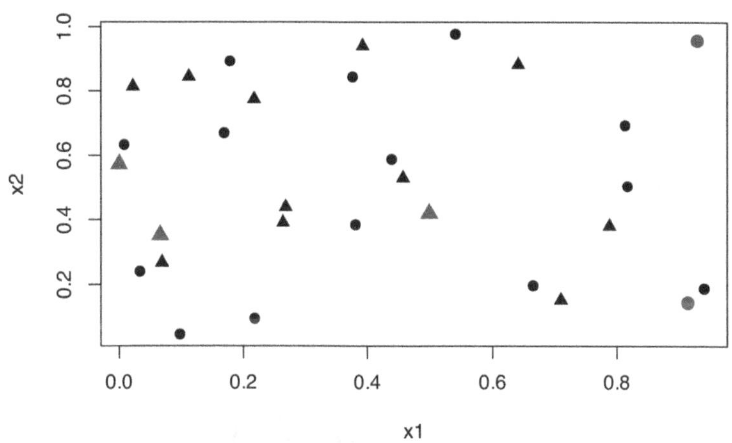

FIGURE 10.9 Classifying new points using kNN with k = 5.

As shown, each new point is assigned a label based on the majority vote among its 5 nearest neighbors. This simple implementation highlights the core idea behind kNN classification: local proximity drives prediction, with no training or model fitting required. In the next step, we apply this to our real-world handwritten digit dataset, `mnist_27`.

10.3.3 `mnist_27` Dataset

We now apply kNN to the `mnist_27` dataset. We will use the training data to predict the labels of the validation data. We will use the Euclidean distance to find the nearest neighbors. We begin by loading and scaling the data.

```
#loading the data
library(tidyverse)
library(dslabs)
library(parallel)

data("mnist_27")
```

```
#str(mnist_27)

#Scale the data.
mdata <- mnist_27$train
mdata$x_1 <- scale(mdata$x_1)
mdata$x_2 <- scale(mdata$x_2)
```

This step ensures that both features are on the same scale, which is important for distance-based algorithms like kNN.

Next, we split the data into training and validation sets, using 75% of the observations for training and the rest for validation. We then plot the two sets to visualize the separation of classes and see which points we are trying to classify.

```
set.seed(1)
ind <- sample(nrow(mdata), nrow(mdata) * 0.75, replace = FALSE)
train <- mdata[ind, ]
val <- mdata[-ind, ]

# Plot the training data and validation data together
plot(train$x_1, train$x_2,
     pch = ifelse(train$y == 7, 17, 16), col = "black",
     xlab = "x1", ylab = "x2", lwd = 0.8)

points(val$x_1, val$x_2,
       pch = 4, col = "gray40", lwd = 1.5)
```

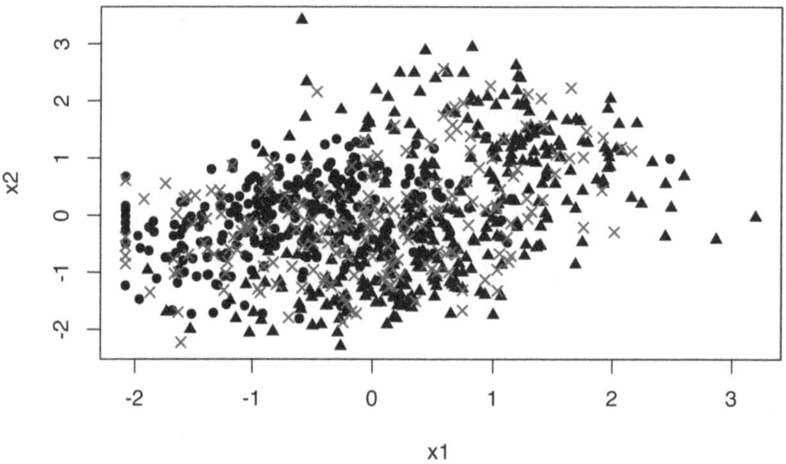

FIGURE 10.10 Training and validation sets in the mnist_27 dataset.

To perform kNN, we define a function that takes a new point and returns a predicted label based on the majority class among its k nearest neighbors in the training data.

```
# Function to determine the label for a validation set
find_label <- function(new_point, k){
```

```
  distances <- sqrt((new_point[1] - train[, 2])^2 + (new_point[2] - train[, 3])^2)
  nearest <- order(distances)[1:k]
  nearest_labels <- train$y[nearest]
  return(names(which.max(table(nearest_labels))))
}
```

To classify the validation points, we define a function that, for a given new point, computes its Euclidean distance to all training points, identifies the k nearest neighbors, and assigns the most frequent label among them. Once this function is in place, we apply it to the entire validation set and compute the prediction accuracy—the proportion of correctly classified points. Since the classes are balanced in this dataset, accuracy is an appropriate evaluation metric.

```
# Pick your k
new_data <- val[, 2:3]
new_labels <- apply(new_data, 1, function(x) find_label(x, k = 30))
acc <- mean(new_labels == val$y)
acc
```

```
## [1] 0.835
```

10.3.4 Single-run Grid Search for k

To investigate the sensitivity of model performance to the choice of k, we perform a single-run grid search. We evaluate accuracy over a range of k values (from 5 to 100) and visualize how accuracy changes with k.

```
# Same function as before, but now we pass k as an argument
find_label <- function(new_point, k){
  distances <- sqrt((new_point[1] - train[, 2])^2 + (new_point[2] - train[, 3])^2)
  nearest <- order(distances)[1:k]
  nearest_labels <- train$y[nearest]
  return(names(which.max(table(nearest_labels))))
}

# to compute accuracy for a grid of k's
acc <- c()
grid <- seq(5, 100, by = 2)

for (i in 1:length(grid)) {
  k <- grid[i]
  set.seed(1)
  ind <- unique(sample(nrow(mdata), nrow(mdata), replace = TRUE))
  train <- mdata[ind, ]
  val <- mdata[-ind, ]
  new_data <- val[, 2:3]
  new_labels <- apply(new_data, 1, function(x) find_label(x, k = k))
  acc[i] <- mean(new_labels == val$y)
```

```
}
plot(grid, acc, type = "l", xlab = "k", ylab = "Accuracy")
```

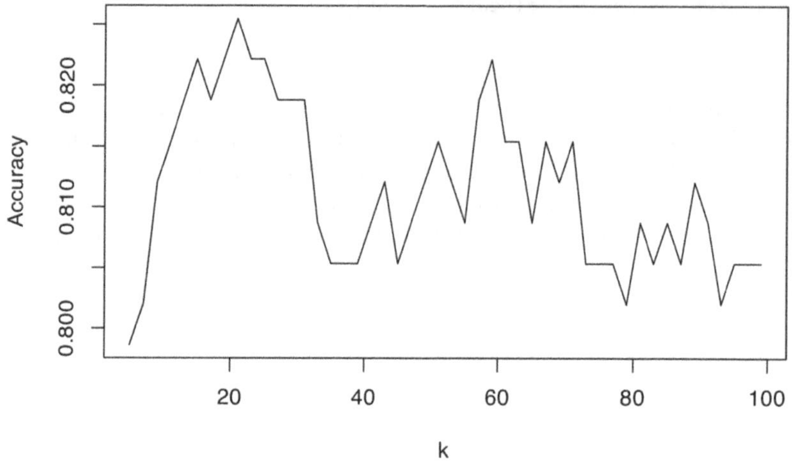

FIGURE 10.11 Validation accuracy across different k values in kNN.

The results, shown in the figure above, reveal how model performance changes with different levels of smoothing. Too small k can overfit the training data; too large k may ignore local structure. We then demonstrate an equivalent approach below using `sapply` for a more concise implementation. This achieves the same result—evaluating accuracy for different values of k— but with cleaner code.

```
# Same function as before
find_label <- function(new_point, k){
  distances <- sqrt((new_point[1] - train[, 2])^2 + (new_point[2] - train[, 3])^2)
  nearest <- order(distances)[1:k]
  nearest_labels <- train$y[nearest]
  return(names(which.max(table(nearest_labels))))
}

# Function to compute accuracy with splits
compute_accuracy <- function(k) {
  set.seed(1)
  ind <- unique(sample(nrow(mdata), nrow(mdata), replace = TRUE))
  train <- mdata[ind, ]
  val <- mdata[-ind, ]
  new_data <- val[, 2:3]
  new_labels <- apply(new_data, 1, function(x) find_label(x, k = k))
  return(mean(new_labels == val$y))
}

# Compute accuracy for different values of k using `sapply`
grid <- seq(5, 100, by = 2)
acc <- sapply(grid, compute_accuracy)
plot(grid, acc, type = "l", xlab = "k", ylab = "Accuracy")
```

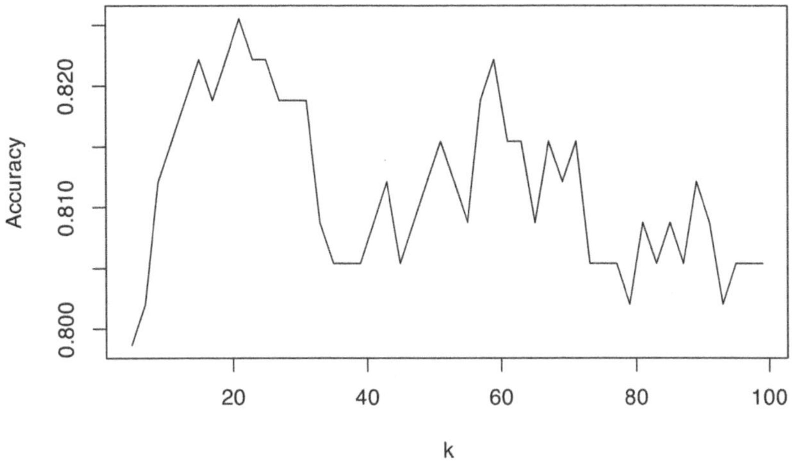

FIGURE 10.12 Validation accuracy for different k values using sapply.

10.3.5 Multi-run Grid Search (Training the Model) for k

Finally, to account for the variability that comes from random sampling, we implement a multi-run grid search. Here, for each k, we repeat the training-validation split 30 times, each with a different random seed. We then compute the average accuracy across runs. This gives a more stable estimate of how well each k performs on average. The resulting plot below shows the smoothed average validation accuracy as a function of k, helping us identify the most robust value for prediction.

```
find_label <- function(new_point, k){
  distances <- sqrt((new_point[1] - train[, 2])^2 + (new_point[2] - train[, 3])^2)
  nearest <- order(distances)[1:k]
  nearest_labels <- train$y[nearest]
  return(names(which.max(table(nearest_labels))))
}

macc <- c()
u95 <- c()
l95 <- c()
grid <- seq(5, 100, by = 2)

for (k in 1:length(grid)) {

  acc <- c()

  for (i in 1:30) {
    set.seed(i)
    ind <- unique(sample(nrow(mdata), nrow(mdata), replace = TRUE))
    train <- mdata[ind, ]
    val <- mdata[-ind, ]
    new_data <- val[, 2:3]
    new_labels <- apply(new_data, 1, function(x) find_label(x, k = grid[k]))
    acc[i] <- mean(new_labels == val$y)
```

```
  }
  macc[k] <- mean(acc)
}

plot(grid, macc, type = "l", xlab = "k", ylab = "Accuracy")
```

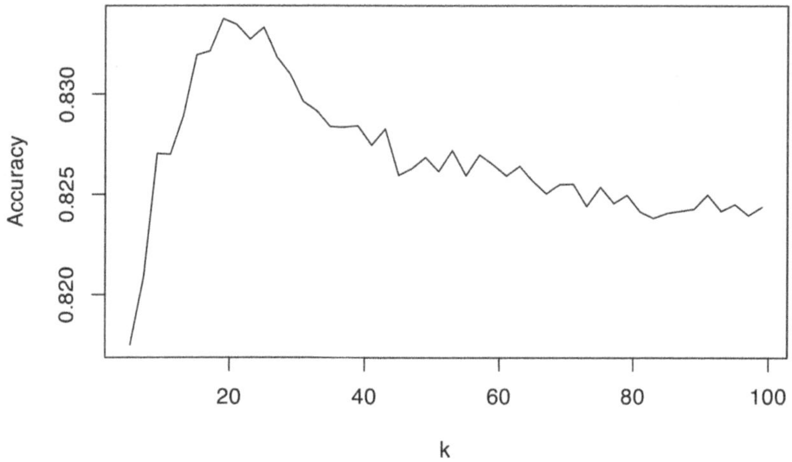

FIGURE 10.13 Average validation accuracy over multiple runs for different k values.

In the next section, we'll further improve computational efficiency by introducing parallel computing.

10.3.6 Grid search with parallel computing (`mclapply`)

To speed up our multi-run grid search, we can use parallel computing. In this implementation, we evaluate the accuracy of kNN across a range of k values by averaging performance over multiple bootstrapped training-validation splits. Each k is evaluated across 30 runs using different seeds to reduce variance. To do this efficiently, we use `mclapply` to distribute computation across available CPU cores. The function `compute_accuracy()` performs the distance calculation and majority vote prediction for a given k and seed.

```
library(doParallel)
```

```
## Loading required package: foreach

##
## Attaching package: 'foreach'

## The following objects are masked from 'package:purrr':
##
##     accumulate, when

## Loading required package: iterators
```

```
compute_accuracy <- function(params) {
  k <- params$k
```

```
  seed <- params$seed

  set.seed(seed)  # Set seed for reproducibility
  ind <- unique(sample(nrow(mdata), nrow(mdata), replace = TRUE))
  train <- mdata[ind, ]
  val <- mdata[-ind, ]
  new_data <- val[, 2:3]

  find_label <- function(new_point, k, train_data) {
    distances <- sqrt((new_point[1] - train_data[, 2])^2
                      + (new_point[2] - train_data[, 3])^2)
    nearest <- order(distances)[1:k]
    nearest_labels <- train_data$y[nearest]
    return(names(which.max(table(nearest_labels))))
  }

  new_labels <- apply(new_data, 1, function(x) find_label(x, k, train))
  return(mean(new_labels == val$y))
}

# Parallel implementation
macc_parallel <- c()
grid <- seq(5, 120, by = 1)

no_cores <- detectCores() - 1
cl <- makeCluster(no_cores)
registerDoParallel(cl)

for (k in 1:length(grid)) {
  # Create parameter list with consistent seeds
  params_list <- lapply(1:30, function(i) list(k = grid[k], seed = i))

  # Run parallel computation
  acc <- unlist(mclapply(params_list, compute_accuracy, mc.cores = no_cores))
  macc_parallel[k] <- mean(acc)
}

stopCluster(cl)

plot(grid, macc_parallel, type = "o", xlab = "k", ylab = "Accuracy",
     col = "black", lwd = 1.5)
abline(v = grid[which.max(macc_parallel)], lty = 2, col = "gray40")
```

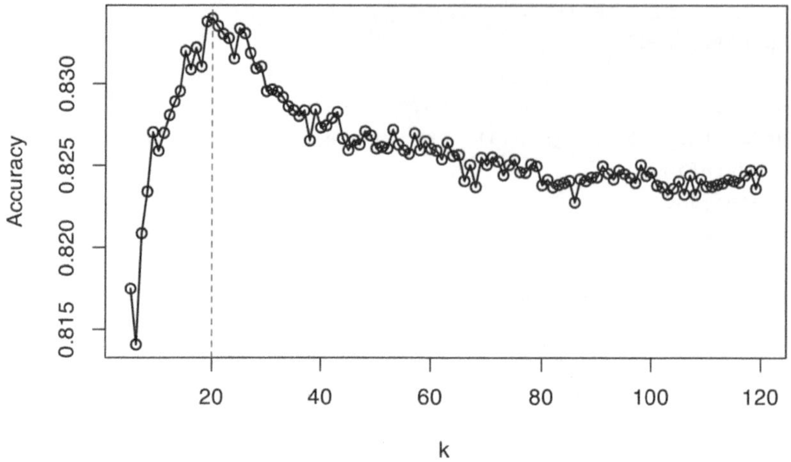

The resulting plot shows how validation accuracy varies with k, helping us identify the optimal value while leveraging faster execution through parallelism.

There are multiple packages in R for applying kNN. The `knn()` function from the `class` package is simple but requires all features to be numeric and does not handle scaling or factor variables automatically. On the other hand, the `knn3()` function from the `caret` package is more robust: it scales features internally and handles categorical variables without manual intervention. These features make it preferable for most applications.

10.3.7 `knn3` from `caret`: Single vs Parallel Core Application

To demonstrate the computational benefits of parallelism further, we compare sequential and parallel implementations of kNN grid search using `knn3()` from the `caret` package. In both versions, we train the model across a range of k values (5 to 120), repeating each experiment 30 times with different seeds.

In the **sequential version**, the code loops through each k, then iterates over 30 runs to average the validation accuracy.

In the **parallel version**, we achieve the same result but distribute the 30 runs for each k across multiple CPU cores using `mclapply`. This reduces total runtime significantly, especially when the dataset or number of hyperparameter combinations is large.

```
library(caret)
library(foreach)
library(doParallel)

# Define consistent parameters
grid <- seq(5, 120, by = 1)
n_iterations <- 30

# Function to compute accuracy that will be used in both implementations
compute_knn_accuracy <- function(k, seed) {
  set.seed(seed)
  ind <- unique(sample(nrow(mdata), nrow(mdata), replace = TRUE))
  train <- mdata[ind, ]
  val <- mdata[-ind, ]
```

```
  model <- knn3(y ~ ., data = train, k = k)
  yhat <- predict(model, val, type = "class")
  return(mean(yhat == val$y))
}

# Sequential Implementation
macc_sequential <- numeric(length(grid))
for (i in seq_along(grid)) {
  acc <- numeric(n_iterations)
  for (j in 1:n_iterations) {
    acc[j] <- compute_knn_accuracy(grid[i], j)
  }
  macc_sequential[i] <- mean(acc)
}

# Parallel Implementation
no_cores <- detectCores() - 1
cl <- makeCluster(no_cores)
registerDoParallel(cl)

# Create parameter combinations
param_grid <- expand.grid(k = grid, seed = 1:n_iterations)
macc_parallel <- numeric(length(grid))

for (i in seq_along(grid)) {
  k_params <- param_grid[param_grid$k == grid[i], ]
  acc <- unlist(mclapply(1:n_iterations,
                function(x) compute_knn_accuracy(k_params$k[x],
                k_params$seed[x]), mc.cores = no_cores))
  macc_parallel[i] <- mean(acc)
}

stopCluster(cl)

# Compare results visually
par(mfrow = c(1, 2))
plot(grid, macc_sequential, type = "o",
     xlab = "k", ylab = "Accuracy",
     main = "Sequential Implementation",
     col = "black", lwd = 1.5)
abline(v = grid[which.max(macc_sequential)], lty = 2, col = "gray40")
text(x = 70, y = 0.82,
     labels = paste("k =", grid[which.max(macc_sequential)],
                    "\nAccuracy =", round(max(macc_sequential), 5)),
     pos = 4, col = "gray40", cex = 0.50)

plot(grid, macc_parallel, type = "o",
     xlab = "k", ylab = "Accuracy",
     main = "Parallel Implementation",
```

```
    col = "black", lwd = 1.5)
abline(v = grid[which.max(macc_parallel)], lty = 2, col = "gray40")
text(x = 70, y = 0.82,
    labels = paste("k =", grid[which.max(macc_parallel)],
                   "\nAccuracy =", round(max(macc_parallel), 5)),
    pos = 4, col = "gray40", cex = 0.50)
```

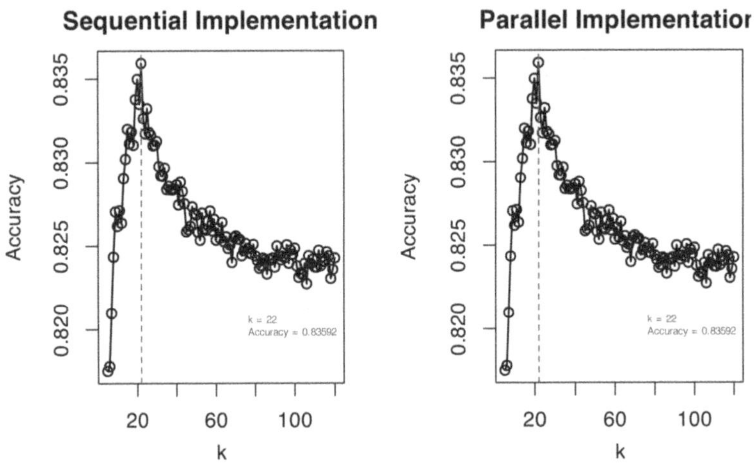

FIGURE 10.14 Comparison of sequential and parallel kNN grid search: validation accuracy across k.

```
# Print summary statistics
cat("\nSequential Implementation - Optimal k:", grid[which.max(macc_sequential)],
   "\nParallel Implementation - Optimal k:", grid[which.max(macc_parallel)])
```

```
##
## Sequential Implementation - Optimal k: 22
## Parallel Implementation - Optimal k: 22
```

The two side-by-side plots show validation accuracy across different k values under both implementations. Red vertical lines mark the value of k that gives the highest accuracy in each case.

In summary, kNN is a flexible and intuitive algorithm for classification and regression. It is entirely data-driven and requires no assumptions about the underlying distribution. The model's simplicity, however, comes with trade-offs: it can be computationally intensive, especially with large datasets, and its performance depends heavily on the choice of k and the distance metric. Nonetheless, with careful tuning and tools like parallel computing, kNN remains a practical and effective method—particularly when model interpretability or minimal assumptions are important.

10.4 Tuning in Classification

We now turn to an essential topic in classification: how to tune models and evaluate their performance. The choice of hyperparameters (like k in kNN), distance metrics, and feature scaling all affect the model's ability to generalize. But equally important is the choice of performance metrics used during training and validation.

Take kNN as an example: in our earlier applications to the `mnist_27` and `Adult` datasets, we selected the best k by maximizing accuracy. But what exactly is accuracy? And is it always the best choice? Can other metrics lead to better predictions—especially in imbalanced datasets or when different types of misclassifications carry different consequences?

These questions are central to model evaluation. To answer them, we begin with a foundational concept: the confusion matrix.

10.4.1 Confusion Matrix

Evaluating the performance of a classification model starts with understanding how predicted outcomes match the actual values. One of the most useful tools for this is the confusion matrix.

For binary classification, we often predict class labels by thresholding predicted probabilities. For example, using:

$$\hat{Y} = \begin{cases} 1 & \text{if } \hat{p}(x_1, \ldots, x_k) > 0.5 \\ 0 & \text{otherwise} \end{cases}$$

From this rule, we build a confusion matrix summarizing how many predictions are correct or incorrect:

Predicted vs. Reality	$Y = 1$	$Y = 0$
$\hat{Y} = 1$	True Positive (TP)	False Positive (FP)
$\hat{Y} = 0$	False Negative (FN)	True Negative (TN)

This table—also called the confusion table—makes it easy to visualize where the model is "confusing" one class with another.

To illustrate, consider this example.[2]

Predicted vs. Reality	$Y =$ Cat	$Y =$ Dog
$\hat{Y} =$ Cat	5	2
$\hat{Y} =$ Dog	3	3

There are 8 actual cats and 5 dogs. The model correctly predicts 5 cats and 3 dogs, but misclassifies 3 cats as dogs and 2 dogs as cats. The diagonal shows correct predictions; off-diagonal entries show errors.

10.4.2 Why Accuracy Can Be Misleading

The most commonly reported metric is accuracy, defined as:

$$\text{Accuracy} = \frac{TP + TN}{TP + FP + FN + TN}$$

But this measure can be misleading—especially when classes are imbalanced. Suppose 95 of 100 observations are cats and only 5 are dogs. A classifier that labels everything as "cat" would achieve 95% accuracy, even though it fails to detect any dogs.

Thus, when evaluating classification performance, especially during training and tuning, alternative metrics are often more appropriate.

10.4.3 Performance Measures Beyond Accuracy

Which metric should we use to tune classification models? This question becomes critical when class distributions are imbalanced, or when the cost of different types of misclassification varies. For example, in cancer detection, failing to identify a cancer patient (a false negative) can be far more serious than mistakenly flagging a healthy individual (a false positive). In contrast, in spam detection, false positives—wrongly classifying real emails as spam—may be more disruptive.

The confusion matrix allows us to define a wide range of metrics beyond simple accuracy. Here's a visual summary[3]:

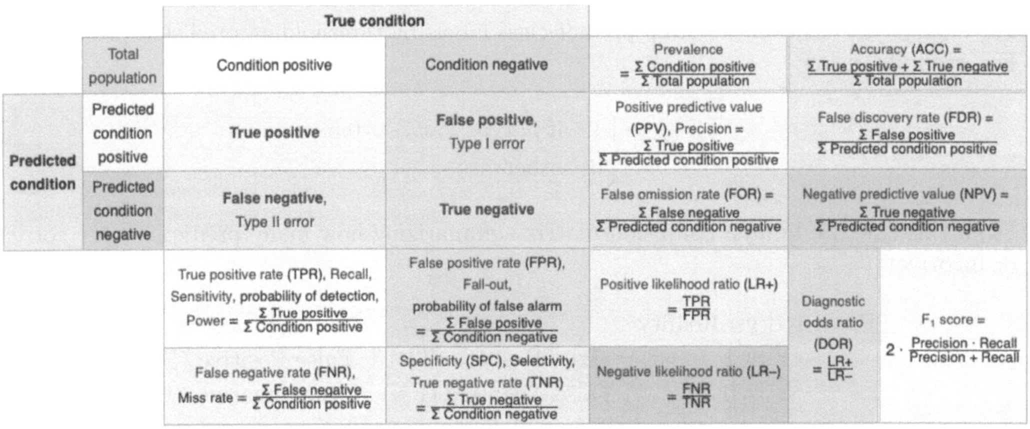

FIGURE 10.15 Confusion matrix and related classification metrics.

Let's walk through some of the most commonly used metrics, using a medical diagnosis example (e.g., cancer detection):

- *Accuracy*: Proportion of correctly classified patients (both cancerous and non-cancerous). Appropriate when the classes are roughly balanced (e.g., not worse than a 60–40 split). $(TP + TN)/n$
- *Balanced accuracy*: Average of sensitivity and specificity. Useful when classes are imbalanced. $(TP/P + TN/N)/2$
- *Precision*: Of those predicted as cancer patients, how many actually have cancer? $TP/(TP + FP)$
- *Sensitivity* (*Recall* or *True Positive Rate*): Of all patients who have cancer, how many were correctly identified? $TP/(TP + FN)$

- *Specificity* (*True Negative Rate*): Of all patients without cancer, how many were correctly identified as such? $TN/(TN + FP)$

Here's a schematic interpretation using the confusion matrix structure:

Predicted vs. Reality	$Y = \text{Cat}$	$Y = \text{Dog}$
$\hat{Y} = \text{Cat}$	TPR (Sensitivity)	FPR (Fall-out)
$\hat{Y} = \text{Dog}$	FNR (Miss Rate)	TNR (Specificity)

10.4.4 Kappa Statistic

Kappa measures how much better a classifier performs compared to random guessing. It adjusts for the expected accuracy that would occur by chance, based on marginal class frequencies.

Consider this confusion matrix:

Predicted vs. Reality	$Y = \text{Cat}$	$Y = \text{Dog}$
$\hat{Y} = \text{Cat}$	22	9
$\hat{Y} = \text{Dog}$	7	13

Total observations: $n = 51$ Observed accuracy (OA): $(22 + 13)/51 = 0.69$

To compute Kappa, we first calculate expected accuracy (EA) based on marginal totals. The contribution of true positives, for example, is:

$$\Pr(\hat{Y} = \text{Cat}) \left[\Pr(Y = \text{Cat} \mid \hat{Y} = \text{Cat}) - \Pr(Y = \text{Cat})\right]$$

This compares the conditional probability of correct classification (reflecting model performance) to the marginal probability (reflecting random chance). Similar logic applies for TN:

$$(\text{OA} - \text{EA})_{\text{TN}} = \Pr(\hat{Y} = \text{Dog}) \left[\Pr(Y = \text{Dog} \mid \hat{Y} = \text{Dog}) - \Pr(Y = \text{Dog})\right]$$

Using the joint and marginal probabilities, the general form becomes:

$$OA - EA = \frac{m_{ij}}{n} - \frac{m_i m_j}{n^2}$$

For this example:

- $OA - EA$ for TP $= 22/51 - \frac{31 \times 29}{51^2} = 0.0857$
- $OA - EA$ for TN $= 13/51 - \frac{20 \times 21}{51^2} = 0.0934$
- $1 - EA = 1 - \left(\frac{31 \times 29 + 20 \times 21}{51^2}\right) = 0.51$

So, Kappa is:

$$\text{Kappa} = \frac{0.0857 + 0.0934}{0.51} \approx 0.3655$$

This shows that the classifier performs better than chance, but there is still room for improvement.

10.4.5 Youden's J Statistic

A final useful metric is Youden's J (also known as *Informedness*), which summarizes a model's ability to avoid both false positives and false negatives:

$$J = \text{TPR} + \text{TNR} - 1$$

It ranges from 0 (no predictive ability) to 1 (perfect classification). This index is closely linked to the ROC curve, which we will examine next.

10.4.6 ROC Curve

In classification tasks, the outcome variable Y is categorical—typically coded as 0 or 1. Most classification algorithms output the predicted probability of success ($Y = 1$), not the class label directly. To obtain a class prediction, we must compare this probability to a discriminating threshold (often set at 0.5): if the predicted probability exceeds the threshold, the model predicts $Y = 1$; otherwise, it predicts $Y = 0$.

Because the threshold determines the final classification, performance measures like accuracy, sensitivity, and specificity will vary with different thresholds. To summarize model performance across all thresholds, we use the receiver operating characteristic (ROC) curve, which plots the true positive rate (TPR) against the false positive rate (FPR). A good model yields high TPR with low FPR, and the better this trade-off, the closer the curve hugs the top-left corner.

The area under the curve (AUC) provides a single number summarizing the overall performance: a value of 1 indicates perfect classification, and 0.5 indicates no better than random guessing.

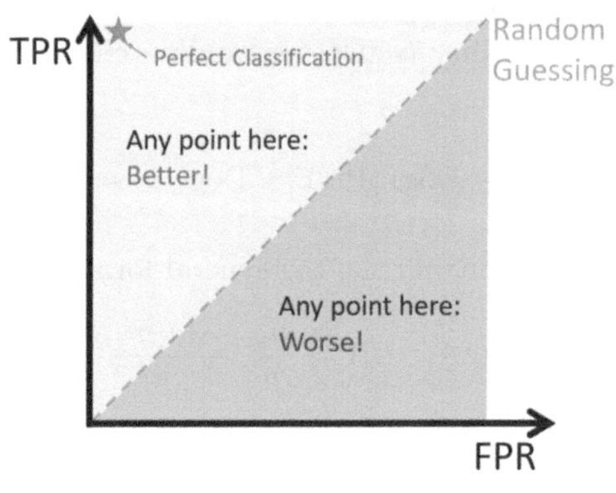

FIGURE 10.16 Visualization of the ROC curve and key diagnostic regions.

Let's begin with a simple example. Suppose we have 100 individuals—50 with $Y = 1$ and 50 with $Y = 0$. Now imagine two extreme threshold choices:

- Threshold = 0%: the model predicts everyone as $Y = 1$

$$\hat{Y} = \begin{cases} 1 & \hat{p}(x) > 0 \\ 0 & \text{otherwise} \end{cases} \Rightarrow$$

Predicted vs. Reality	$Y = 1$	$Y = 0$
$\hat{Y} = 1$	50	50
$\hat{Y} = 0$	0	0

TPR = 1, FPR = 1

- Threshold = 100%: the model predicts everyone as $Y = 0$

$$\hat{Y} = \begin{cases} 1 & \hat{p}(x) > 1 \\ 0 & \text{otherwise} \end{cases} \Rightarrow$$

Predicted vs. Reality	$Y = 1$	$Y = 0$
$\hat{Y} = 1$	0	0
$\hat{Y} = 0$	50	50

TPR = 0, FPR = 0

As we adjust the threshold from 0% to 100%, we generate a series of (TPR, FPR) pairs. Plotting these points gives us the ROC curve, which visualizes the model's ability to distinguish between classes at various threshold settings. A model with strong discriminative power will maintain a high TPR while keeping FPR low, resulting in a curve that approaches the top-left corner of the plot.

To illustrate this using real data, we now turn to the `mnist_27` dataset and apply the `knn3` model from the `caret` package. While standard kNN only produces class labels based on majority voting, this is not directly compatible with generating an ROC curve, which requires probabilistic outputs. But there is a workaround:

- **Standard kNN** returns only class labels based on the majority vote among the k nearest neighbors. It does not produce probabilities, so it cannot be used directly to generate ROC curves.
- **Modified kNN**, however, allows us to estimate probabilities by computing the proportion of neighbors that belong to each class. This proportion serves as an estimated probability for class membership and enables threshold-based classification.

```
data(mnist_27)
#str(mnist_27)

table(mnist_27$train$y) # data is well-balanced
```

```
##
##   2   7
## 379 421
```

```
#scale the data

train <- mnist_27$train
test <- mnist_27$test

train$x_1 <- scale(mnist_27$train$x_1)
train$x_2 <- scale(mnist_27$train$x_2)

test$x_1 <- scale(mnist_27$test$x_1)
test$x_2 <- scale(mnist_27$test$x_2)
```

We can modify kNN to provide probability estimates. For instance, instead of just using the majority class among the k nearest neighbors, we can calculate the proportion of neighbors belonging to each class. This proportion can act as a probability estimate for class membership.

Here, we redefine our earlier function `find_probabilities()` that, for a given test point, finds the $k = 21$ nearest neighbors and returns the proportion of each class label among them. This gives us the probability that the new observation belongs to class 2 (our positive class).

```
find_probabilities <- function(new_point, k) {
  distances <- sqrt((new_point[1] - train[, 2])^2 + (new_point[2] - train[, 3])^2)
  nearest <- order(distances)[1:k]
  nearest_labels <- train$y[nearest]
  label_proportions <- table(nearest_labels) / k
  return(label_proportions)
}

new_data <- test[, 2:3]
new_prob <- apply(new_data, 1, function(x) find_probabilities(x, k = 21))
new_prob <- t(new_prob)
```

We then apply this function to each test point to generate a set of estimated probabilities. These will be used to simulate different threshold cutoffs.

We need to decide the positive class. Snippet below prepares these probabilities for analysis by converting them into a named data frame and ensuring the correct interpretation of class labels—assigning class 2 as positive and class 7 as negative.

```
# Convert the labels in new_prob to 0 and 1
new_prob <- data.frame(new_prob)
names(new_prob) <- c("prob_2 as 1", "prob_7 as 0")
```

Here we define a function that, given a threshold, calculates the TPR and FPR by comparing the predicted labels (based on that threshold) to the actual test set labels. It returns not just TPR and FPR, but also confusion matrix components like true/false positives and negatives, allowing for richer diagnostics.

```
# Function to calculate TPR and FPR at each threshold
calculate_tpr_fpr <- function(prob, true_labels, threshold) {
  if (threshold == 0) {
    predicted_labels <- rep(1, length(prob))
  } else {
    predicted_labels <- ifelse(prob > threshold, 1, 0)
  }

  true_positive <- sum(predicted_labels == 1 & true_labels == 2)
  false_positive <- sum(predicted_labels == 1 & true_labels == 7)
  true_negative <- sum(predicted_labels == 0 & true_labels == 7)
  false_negative <- sum(predicted_labels == 0 & true_labels == 2)

  total_2 = sum(true_labels == 2)
  total_7 = sum(true_labels == 7)

  tpr <- true_positive / total_2
```

```
  fpr <- false_positive / total_7
  specificity <- true_negative / total_7

  return(c(tpr, fpr, specificity, true_positive,
           false_positive, true_negative, false_negative,
           threshold, total_2, total_7))
}
```

By applying this function over a grid of thresholds (e.g., 0.01 to 0.99), we can trace the ROC curve and calculate the AUC. This will be the focus of the next section.

10.4.7 ROC Curve and AUC

Let's now visualize the performance of our classifier using the ROC curve and compute AUC, focusing on predicting digit 2.

```
# Define a sequence of thresholds
thresholds <- seq(0, 1, by = 0.01)

# Calculate TPR and FPR for each threshold
tpr_fpr_list <- lapply(thresholds, function(threshold)
  calculate_tpr_fpr(new_prob[, 1], test$y, threshold))

# Convert the list to a matrix for easier handling
tpr_fpr <- do.call(rbind, tpr_fpr_list)
colnames(tpr_fpr) <- c("tpr", "fpr", "true_positive", "specificity",
                       "false_positive", "true_negative", "false_negative",
                       "threshold", "total_2", "total_7")

ggplot(data.frame(tpr = tpr_fpr[, 1], fpr = tpr_fpr[, 2]), aes(x = fpr, y = tpr)) +
  geom_line(color = "black") +
  geom_abline(intercept = 0, slope = 1, linetype = "dashed", color = "gray40") +
  labs(x = "False Positive Rate", y = "True Positive Rate") +
  theme_minimal()
```

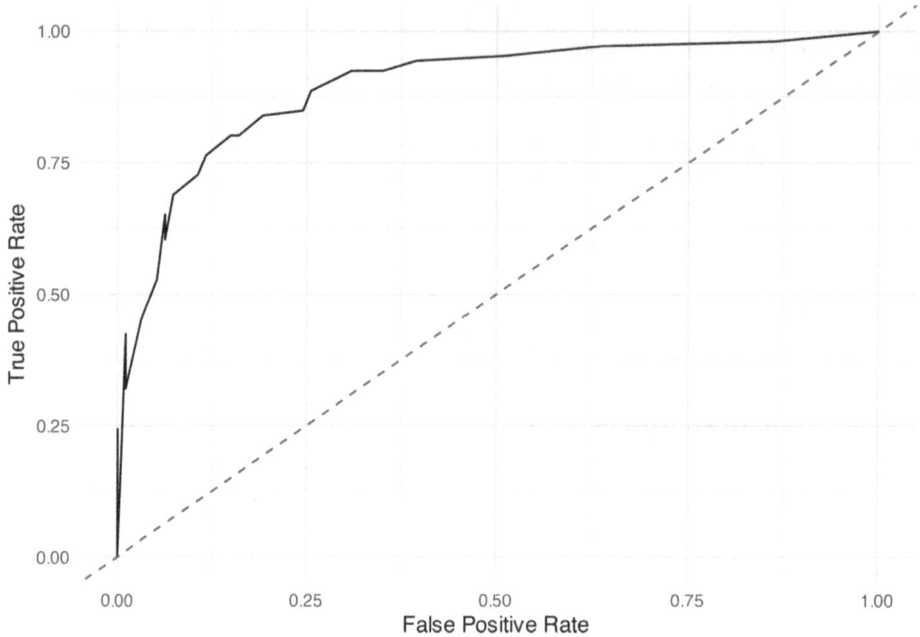

The function `do.call()` is useful when applying a function like `rbind()` to a list of outputs, as it collapses the list into a single matrix or data frame, making it easier to manipulate or plot. Combined with `lapply()`, it simplifies batch computation across a sequence of inputs—in this case, thresholds.

10.4.8 Using ROCR and pROC Packages

ROC with the ROCR Package

```
# now let's use ROCR package to plot the ROC curve
library(ROCR)
pred <- prediction(new_prob[, 1], test$y == 2)
perf <- performance(pred, "tpr", "fpr")
plot(perf, lwd = 2,
     main = "ROC curve for digit 2",
     col = "black")
abline(a = 0, b = 1, lty = 2, col = "gray40")
```

This shows the ROC curve for digit 2. Let's now calculate the AUC.

```
# And let's calculate the AUC
auc <- performance(pred, "auc")
auc@y.values[[1]]
```

```
## [1] 0.8964773
```

The AUC is approximately 0.90, indicating excellent classification performance. The ROC curve plots sensitivity (TPR) against 1 - specificity (FPR) over all possible thresholds. The AUC summarizes this curve into a single number: the probability that the classifier will rank a randomly chosen positive case higher than a negative one.

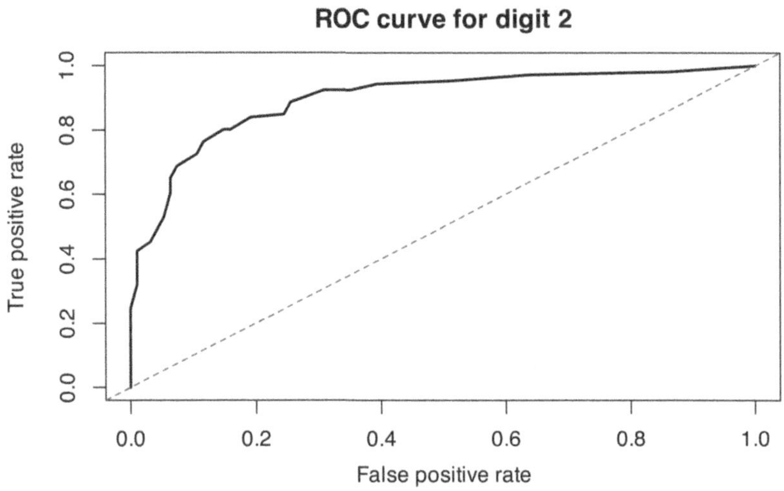

FIGURE 10.17 ROC curve for digit 2 using the ROCR package.

AUC values closer to 1 indicate strong separation between classes. A value of 0.5 means the classifier performs no better than random guessing. This measure is particularly useful because it is independent of the threshold chosen to determine the positive class and provides a single measure of performance across all classification thresholds. It is also invariant to changes in the class distribution, making it a good metric for evaluating the intrinsic quality of the predictions of the model.

ROC with the `pROC` package

```
roc_obj <- roc(test$y, new_prob[, 1], levels = c("7", "2"),
               direction = "<", percent = TRUE,
               xlim = c(100, 0), plot = TRUE,
               xlab = "False Positive Percentage", ylab = "True Postive Percentage",
               print.auc = TRUE)
```

```
ggroc(roc_obj, legacy.axes = T) +
  geom_abline(slope = 1 ,intercept = 0) + # add identity line
  xlab('100% - Specificity') +
  ylab('100% - Sensitivity') +
  theme_minimal()
```

These plots provide additional ways to visualize classifier performance and verify AUC, using slightly different conventions. Both confirm that our classifier distinguishes well between the digits 2 and 7.

While this chapter focuses primarily on binary classification—such as distinguishing digit 2 from digit 7—the methods discussed can be extended to multiclass problems. Common approaches include one-vs-rest (OvR), where separate binary classifiers are trained for each class against all others, and one-vs-one (OvO), which constructs a classifier for every pair of classes. For methods like kNN, multiclass classification is handled naturally by majority vote among neighbors. However, performance metrics such as AUC or Youden's index must be adapted or replaced with multiclass alternatives, like macro-averaged precision or Cohen's kappa, as binary-specific metrics may not fully capture performance across multiple classes.

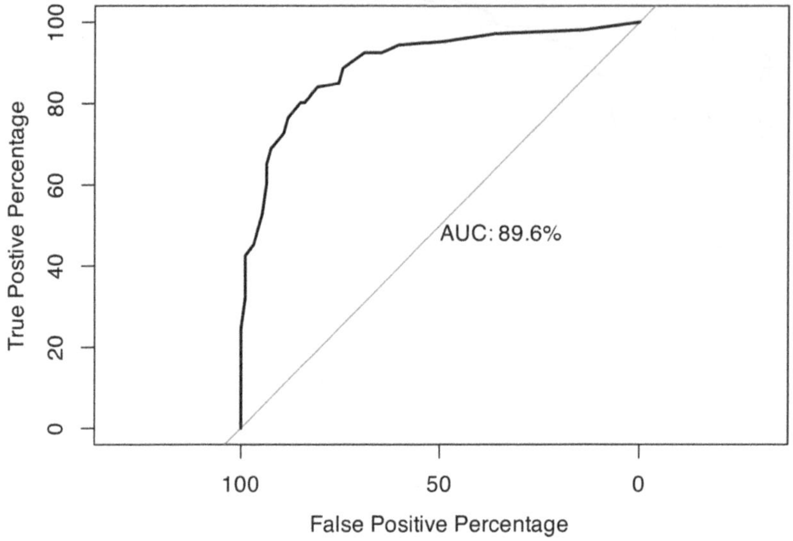

FIGURE 10.18 ROC curve for digit 2 using the pROC package with AUC displayed.

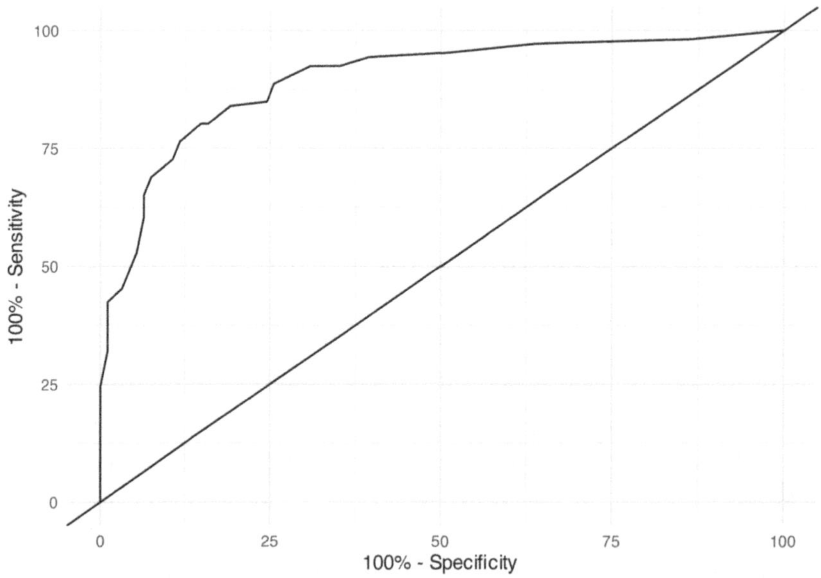

FIGURE 10.19 Specificity vs Sensitivity plot for digit 2 using the pROC package with AUC displayed.

10.4.9 Tuning kNN with AUC

So far, we have used accuracy to tune the value of k, but AUC often provides a more robust metric, especially when class balance or threshold sensitivity matters. Instead of checking how many predictions are correct, AUC evaluates how well the model ranks positive cases above negative ones across all thresholds.

In the next code block, we calculate the AUC for a range of k values using the `knn3()` function from the `caret` package. For each k, we perform 50 bootstrap iterations. In each iteration, we draw a sample with replacement from the training set, fit a kNN model, predict class probabilities on the out-of-bag sample, and calculate the AUC using the `ROCR` package.

The AUC values are then averaged over the 50 runs to get a stable performance estimate for each k. This process is parallelized across CPU cores using the `foreach` and `doParallel` packages to reduce computation time.

```
# Let's use the `knn3` model in the `caret` package to classify the data
library(caret)
library(foreach)
library(doParallel)
library(ROCR)

grid <- seq(5, 120, by = 1)

# Register the parallel back-end to use
no_cores <- detectCores() - 1  # reserve one core for the system
cl <- makeCluster(no_cores)
registerDoParallel(cl)

# Using foreach for parallel processing
mauc <- foreach(i = grid, .combine = 'c', .packages = c('caret', "ROCR")) %dopar% {

  aucc <- c()

  for (j in 1:50) {
    set.seed(j)
    ind <- unique(sample(nrow(train), nrow(train), replace = TRUE))
    traint <- train[ind, ]
    val <- train[-ind, ]

    model <- knn3(y ~ ., data = traint, k = i)
    phat <- predict(model, val, type = "prob")

    pred_rocr <- prediction(phat[, 1], val$y == 2)
    auc_rocr <- performance(pred_rocr, "auc")
    aucc[j] <- auc_rocr@y.values[[1]]
  }

  mean(aucc)

}
```

```
# Stop the cluster
stopCluster(cl)

# Plot the training AUC vs k
plot(grid, mauc, type = "o", xlab = "k", ylab = "AUC", lwd = 1.5, col = "black")
optimal_k <- grid[which.max(mauc)]
abline(v = optimal_k, lty = 2, col = "gray40", lwd = 2)
text(x = 70, y = 0.89, labels = paste("k =", optimal_k, "\nAUC =",
                round(max(mauc),5)), pos = 4, col = "gray40")
```

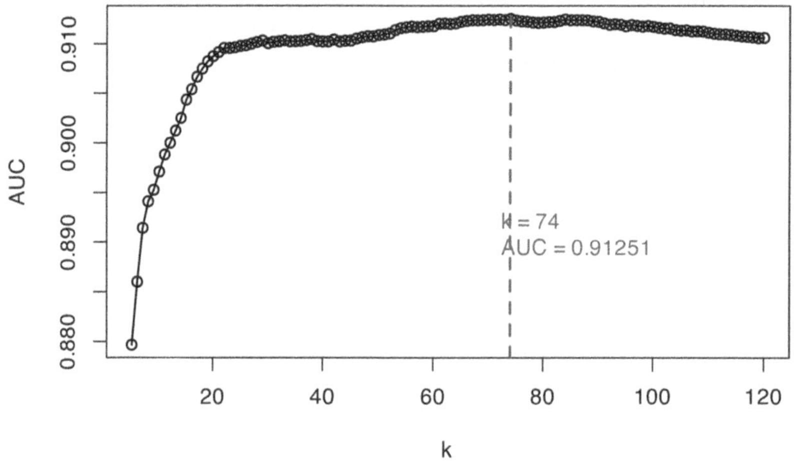

The plot shows how AUC varies with different k values. The vertical red line identifies the k that yields the highest AUC. Selecting k based on this metric rather than accuracy can lead to more consistent performance, especially when probability ranking or threshold tuning is central to the application—such as in medical diagnosis or fraud detection.

While we use AUC-based tuning of k with bootstrapped samples in this chapter, more generally, cross-validation remains a robust and widely used framework for evaluating classification models. In particular, stratified k-fold cross-validation—where each fold preserves the proportion of class labels—is especially valuable in classification settings to ensure fair representation of minority classes across folds. This approach is particularly important when classes are imbalanced or when comparing multiple models to avoid overfitting on random splits.

10.4.10 Test Score and Threshold

After selecting the optimal k value using AUC, we now assess the test performance and determine the best classification threshold. In ROC analysis, the default threshold of 0.5 may not be ideal. Instead, a more effective approach is to use **Youden's J statistic**, which balances sensitivity and specificity by maximizing the distance from the diagonal line of random guessing in the ROC space.

Youden's index is calculated as: $J = \text{Sensitivity} + \text{Specificity} - 1$ or equivalently: $J = \text{TPR} + \text{TNR} - 1$

This value ranges between 0 (no better than random) and 1 (perfect classification). The threshold corresponding to the highest Youden's J is considered optimal, as it maximizes true positives while minimizing false positives.

To implement this, we fit a kNN model using the previously selected $k = 76$, generate predicted probabilities on the test set, and calculate the AUC. Then, using the `ROCR` package, we extract sensitivity and specificity across thresholds, compute the Youden index for each, and identify the threshold that gives the maximum value.

```
model <- knn3(y ~ ., data = train, k = 76)
phat <- predict(model, test, type = "prob")

# Calculate the AUC
pred_rocr <- prediction(phat[, 1], test$y == 2)
auc_rocr <- performance(pred_rocr, "auc")
auc_rocr@y.values[[1]]
```

```
## [1] 0.9096748
```

```
# Extract the sensitivity and specificity
perf <- performance(pred_rocr, "sens", "spec")
sensitivity <- perf@y.values[[1]]
specificity <- perf@x.values[[1]]


# Calculate Youden's Index
youden_index <- sensitivity + specificity - 1

# Find the maximum Youden's index and corresponding cutoff
max_index <- which.max(youden_index)
max_youden_index <- youden_index[max_index]

youden_index[max_index] # maximum Youden's index
```

```
## [1] 0.6870735
```

```
perf@alpha.values[[1]][max_index] # optimal discrimination threshold
```

```
## [1] 0.4210526
```

After identifying the best threshold, we generate the confusion matrix using that cutoff to evaluate the final classification performance on the test set.

```
# Calculate the confusion matrix at the optimal threshold


optimal_threshold <- perf@alpha.values[[1]][max_index]
predicted_labels <- ifelse(phat[, 1] > optimal_threshold, 1, 0)

confusion_matrix <- table(predicted_labels, test$y == 2)
confusion_matrix <- confusion_matrix[c(2,1), c(2,1)]
```

```
rownames(confusion_matrix) <- c("Predicted 2", "Predicted 7")
colnames(confusion_matrix) <- c("Actual 2", "Actual 7")
confusion_matrix
```

```
##
## predicted_labels Actual 2 Actual 7
##       Predicted 2       89       17
##       Predicted 7       17       77
```

This approach highlights the advantage of using probabilistic outputs rather than relying only on hard classification. It provides flexibility in decision-making depending on the application's tolerance for false positives or false negatives. Although Youden's J offers a convenient method for identifying a single optimal threshold, threshold selection is inherently context-dependent. In applications where misclassification costs differ—for example, false negatives may be costlier than false positives in medical diagnosis—it may be preferable to select a threshold that reflects this asymmetry. This connects to broader decision-theoretic approaches and cost-sensitive classification strategies, which we revisit in later chapters.

10.5 Conclusion

Classification models are essential when outcomes fall into binary or multi-class categories. This chapter offered a comprehensive treatment of classification, using kNN as a guiding example to demonstrate key ideas in performance evaluation, threshold selection, and model tuning. We showed that while kNN is conceptually simple, it becomes powerful when paired with thoughtful implementation—using parallel computing for tuning, and evaluating results with metrics like sensitivity, specificity, AUC, and Youden's index, which are more appropriate than MSPE for binary outcomes.

We also emphasized that accuracy alone is often misleading, especially in the presence of class imbalance. More informative metrics and visualization tools—such as confusion matrices, ROC curves, and precision-recall trade-offs—are critical for assessing model performance. While the chapter focused on binary classification (e.g., digit 2 vs. 7), we noted that many of these tools can be extended to multiclass settings via strategies like one-vs-rest or majority voting.

Although we used kNN for illustration, the same framework can be applied to other widely used classifiers, such as logistic regression, decision trees, random forests, support vector machines, and neural networks. Swapping the model while keeping the outcome setup binary allows for broad experimentation and comparison. These models will be introduced throughout the book, often alongside real-world applications in economics, health, and policy—making this chapter a resource you can return to for essential building blocks.

After preparing and splitting the data—ideally using stratified sampling to maintain class balance—the workflow involves selecting a model, generating predictions, evaluating with robust metrics, tuning hyperparameters, and setting decision thresholds based on context. These steps form a general-purpose, model-agnostic roadmap for classification tasks and offer a practical foundation for the model selection strategies discussed in the next chapter.

Notes

1. http://yann.lecun.com/exdb/mnist/
2. From: https://en.wikipedia.org/wiki/Confusion_matrix
3. From: https://en.wikipedia.org/wiki/Evaluation_of_binary_classifiers

11

Model Selection and Sparsity

Model selection and variable selection are often used interchangeably, but they have distinct meanings depending on the context. In econometrics, models are typically assumed to be linear, and the primary focus is on selecting variables for interpretation. In this sense, **variable selection** is a more precise term, as it refers to choosing relevant covariates to estimate causal effects. However, when the goal shifts to prediction, **model selection** extends beyond variable choice within a linear framework—it involves comparing different modeling approaches altogether.

This chapter explores these distinctions and provides a structured approach to model selection by addressing both traditional econometric methods and machine learning techniques. We begin with a discussion of model selection in econometrics, emphasizing the importance of defining clear objectives. Are we selecting variables to improve causal interpretation, or choosing a model for better predictive accuracy? These goals require different criteria: for causal inference, unbiased estimation and interpretability take precedence, whereas for prediction, minimizing out-of-sample error is the key concern.

Next, we discuss key model assessment criteria, including adjusted R^2, Mallows' C_p, Akaike Information Criterion (AIC), Bayesian Information Criterion (BIC), cross-validation, and bootstrap methods, all aimed at balancing complexity and fit while preventing overfitting. We examine traditional variable selection techniques such as best subset selection and stepwise methods, highlighting their limitations and the motivation for penalized regression techniques. Functional form selection—through transformations like logarithmic, polynomial, and piecewise linear models—also plays a crucial role in capturing underlying relationships.

Alternative models such as penalized regressions, random forests, and other machine learning approaches offer improved predictive performance in some cases. While these methods will be covered in later chapters, this chapter focuses on foundational concepts in variable selection and model evaluation within simpler frameworks. We covered best practices for integrating machine learning (ML) into model selection while balancing accuracy, interpretability, and efficiency. We also introduce the fundamentals of sparsity and the oracle property, which provide theoretical foundations for selecting an optimal set of variables in high-dimensional data.

Although many of these methods apply to both causal inference and prediction, their interpretation depends on the specific goal. For causal inference, variable selection aims to mitigate bias and ensure valid estimation, while for prediction, the focus is on optimizing performance and generalizability. A central theme of this chapter is the importance of precise terminology—distinguishing between variable and model selection, as well as different evaluation criteria—to ensure clarity in empirical research.

DOI: 10.1201/9781003504795-11

11.1 Model Selection in Econometrics

In econometrics, **model selection** primarily refers to **variable selection**—determining which covariates to include in a regression to ensure valid causal inference and reliable predictions. Unlike in ML, where model selection often involves comparing different algorithms, econometricians typically assume a model structure and refine it by selecting appropriate explanatory variables and functional forms.

Selecting the right variables is crucial for both **causal inference** and **prediction**. In causal analysis, the goal is to avoid omitted variable bias while maintaining interpretability. Including too many variables, especially collinear ones, can obscure relationships and reduce efficiency, while excluding key covariates can lead to biased estimates. For instance, estimating the returns to education without controlling for ability or family background may overstate the effect, whereas excessive controls could dilute the true impact. For prediction, variable selection balances complexity and generalizability: too many variables risk overfitting, while an overly simple model may miss important relationships. Selection criteria such as Mallows' C_p, AIC, BIC, and adjusted R^2, cross-validation, and bootstrap resampling help identify models that generalize well.

A key distinction in econometrics is that a 'linear model' refers to linearity in parameters (coefficients), not necessarily to a linear relationship between the dependent and independent variables. While interactions, polynomial terms, or categorical variables introduce nonlinearity, the parameters enter the model linearly. Functional form selection—deciding on transformations, interactions, or structural breaks—affects both causal inference and predictive accuracy.

This section covers variable selection methods, model evaluation criteria, and functional form selection within standard econometric frameworks. While these methods apply to both causal inference and prediction, their interpretation depends on context. Later chapters will introduce advanced techniques that extend beyond linear models, but for now, we focus on the fundamentals of variable selection and precise terminology to avoid confusion.

11.2 Model Assessment Criteria

To compare models systematically, we use model assessment criteria, which balance goodness-of-fit and model complexity. These criteria are essential because they help avoid overfitting by penalizing excessive parameters, ensuring that selected models generalize well beyond the training (or in-sample) data. In economics, health, and social sciences, we typically work with a single dataset. In classical econometrics, we rely on this sample to estimate multiple linear regression models, aiming to find the "best" one for either causal inference or prediction. Since we often lack a separate test dataset, model selection is performed entirely in-sample, meaning all models are compared within the same dataset used for estimation. The challenge, however, is that assessing models based solely on their in-sample fit can lead to severe overfitting.

A natural but flawed approach would be to minimize the in-sample mean squared error (MSE) or mean squared prediction error (MSPE). However, as discussed in Chapter 6, this is problematic because in-sample MSPE consistently underestimates true out-of-sample error. The relationship between out-of-sample and in-sample MSPE is given by:

$$\mathbf{MSPE}_{out} = \mathbf{MSPE}_{in} + \frac{2}{n}\sigma^2(p+1) \tag{11.1}$$

where the second term represents overfitting bias—the extent to which in-sample MSPE fails to capture the true error (under the standard assumptions of the linear regression model with normally distributed errors).[1] The magnitude of this bias depends on several factors:

1. **Higher noise** (σ^2) increases overfitting.
2. **Larger sample size** (n) reduces overfitting.
3. **More predictors** (p) increase overfitting.

As the ratio p/n grows, overfitting becomes more severe. This means that simply minimizing in-sample MSPE results in models that are too complex and perform poorly on new data. When we fit a model using least squares, the coefficients are estimated to minimize training residual sum of squares (RSS), which means that training error systematically decreases as more variables are added, regardless of whether those variables improve prediction or causal inference. This is why metrics like training RSS and training R^2 are not reliable for comparing models with different numbers of variables—they fail to adjust for model complexity.

To address this issue, several model selection criteria introduce penalties that account for the number of predictors. These criteria include Mallows' C_p, AIC, BIC, and adjusted R^2. These approaches adjust the training error to better approximate the model's true out-of-sample performance. Among them, Adjusted R^2 is one of the simplest and most commonly used criteria in classical econometrics, particularly in settings where linear regression is the dominant framework.

11.2.1 Adjusted R^2 as a Model Selection Criterion

Adjusted R^2 modifies the standard R^2 measure to account for the number of predictors in a model. The standard R^2, or coefficient of determination, is defined as:

$$R^2 = 1 - \frac{\sum_{i=1}^{n}(y_i - \hat{y}_i)^2}{\sum_{i=1}^{n}(y_i - \bar{y})^2} \tag{11.2}$$

where y_i are observed values, $\hat{y}_i$ are predicted values, and $\bar{y}$ is the mean of y. This statistic represents the proportion of variance in y explained by the model. However, because R^2 never decreases when adding predictors, it can overstate a model's explanatory power, making it unreliable for comparing models with different numbers of variables.

Adjusted R^2 corrects for this by introducing a penalty for additional predictors, ensuring that only meaningful improvements in model fit lead to an increase. It is defined as:

$$\text{Adjusted } R^2 = 1 - \frac{(1-R^2)(n-1)}{n-p-1} \tag{11.3}$$

where n is the sample size, p is the number of predictors (excluding the intercept), and R^2 is the standard coefficient of determination. Unlike raw R^2, Adjusted R^2 increases only if a new predictor significantly reduces unexplained variation; otherwise, it decreases, discouraging the inclusion of unnecessary variables.

Adjusted R^2 is a practical tool for selecting variables, especially in causal models where researchers start with theoretically relevant predictors and refine the model by removing insignificant ones. It is also useful in prediction models, particularly with small datasets where overfitting is a concern. When the p/n ratio is high, Adjusted R^2 helps control complexity by discouraging excessive parameter inclusion.

Compared to model selection criteria such as Mallows' C_p, AIC, and BIC, Adjusted R^2 does not explicitly estimate out-of-sample prediction error but instead adjusts the explained variance to prevent overfitting. While useful for comparing models estimated on the same dataset, it is less effective for purely predictive settings, where cross-validation or penalized regression methods may be preferable. Nevertheless, in traditional econometrics, where the focus is on selecting variables in a linear regression framework, Adjusted R^2 remains a simple and effective measure for balancing model fit and complexity.

11.2.2 Mallows' C_p Statistic

Mallows' C_p statistic incorporates an estimate of the error variance into model selection is Mallows' C_p, which provides a way to evaluate the trade-off between model fit and complexity by accounting for prediction error bias. Since the true σ^2 is unknown, we estimate it from the largest model under consideration. The C_p statistic is defined as:

$$C_p = \frac{1}{n}(RSS + 2p\hat{\sigma}^2) = \frac{1}{n}\sum_{i=1}^{n}(Y_i - \hat{Y}_i)^2 + \frac{2}{n}p\hat{\sigma}^2 \tag{11.4}$$

where $\hat{\sigma}^2$ is an unbiased estimate of the error variance. A model with C_p close to p is preferred. The rule for model selection is to choose the model that minimizes C_p. The intuition behind this criterion is straightforward: it balances the in-sample MSE with a penalty for complexity. The first term in C_p measures how well the model fits the training data, while the second term penalizes excessive parameters.

This trade-off highlights the fundamental challenge of model selection: more predictors improve fit within the sample but may degrade performance on new data. We can decompose C_p into two key components:

$$C_p = \text{MSE} + \text{Penalty} \tag{11.5}$$

From one perspective, the penalty term is an adjustment for bias introduced by overfitting. From another perspective, it represents a cost imposed on models for adding parameters. Each additional predictor must justify its inclusion by reducing the MSE enough to offset this cost. If it does not, the additional complexity is unwarranted.
The penalty term $2p\hat{\sigma}^2$ corrects for the tendency of training error to underestimate test error. If $\hat{\sigma}^2$ is an unbiased estimate of σ^2, then C_p itself becomes an unbiased estimate of the true test MSE.

The practical implication is that models with smaller C_p values tend to have lower true test error, making C_p a reliable selection criterion. For comparing two models, the difference in C_p values can be expressed as:

$$\Delta C_p = \text{MSE}_1 - \text{MSE}_2 + \frac{2}{n}\hat{\sigma}^2(p_1 - p_2) \tag{11.6}$$

where the intercept term cancels out, leaving a direct comparison based on the number of predictors. This reinforces the idea that increasing p beyond a certain point introduces unnecessary complexity without improving predictive accuracy.

Mallows' C_p is particularly useful in situations where we suspect some variables in the full model may be redundant. It allows us to identify a parsimonious model—one that explains the data well without unnecessary complexity.

However, it is important to note that while C_p provides a balance between fit and complexity, it is not the only model selection criterion. Other methods, such as AIC and BIC, introduce different penalty structures and are often preferred in certain applications. By incorporating complexity adjustments into model selection, we avoid the pitfalls of naive in-sample MSE minimization and ensure that chosen models generalize well to new data.

11.2.3 Akaike Information Criterion

The AIC is a widely used model selection criterion applicable to a broad class of models estimated via maximum likelihood. It balances model fit and complexity by penalizing excessive parameters to prevent overfitting. AIC is particularly useful for comparing models that are not necessarily nested, such as in predictive modeling and model selection in statistical learning.

AIC is defined as:

$$AIC = -2 \log L + 2p \tag{11.7}$$

where L is the maximized likelihood and p is the number of estimated parameters.[2] Since σ^2 is typically unknown, it is estimated using the maximum likelihood estimate, the difference in AIC between models simplifies to:

$$\Delta AIC = n \log(\hat{\sigma}^2) + 2p \tag{11.8}$$

which highlights that AIC penalizes models based on both goodness of fit and the number of parameters (check foornote for derivation).

If all models under consideration have the same number of parameters (p), selecting the model with the lowest AIC is equivalent to selecting the model with the lowest RSS and thus the minimum MSE.[3]

AIC provides a principled way to balance model fit and complexity, ensuring that more complex models are preferred only if they yield significantly better fit. A lower AIC value indicates a better model in terms of explanatory power and generalizability. Unlike adjusted R^2, which is tied to linear regression, AIC extends to a broad range of statistical models, including nonlinear and probabilistic models.

11.2.4 Bayesian Information Criterion

Similarly, the BIC introduces a complexity penalty but does so more aggressively than AIC. It is given by:

$$BIC = -2 \log L + p \log n \tag{11.9}$$

where n is the sample size. The key difference between AIC and BIC lies in the penalty term: while AIC applies a constant penalty of $2p$, BIC increases the penalty as the sample size grows, imposing a stricter preference for simpler models when n is large. This means that BIC tends to favor more parsimonious models compared to AIC, making it particularly useful when the true model is believed to be among the candidate models.

Both AIC and BIC provide useful criteria for selecting among competing models, but they serve slightly different purposes. AIC is often preferred when the goal is prediction, as it selects models that best approximate the underlying data-generating process. BIC, by favoring simpler models, is more commonly used in settings where the true model is assumed to exist within the set of candidates. While both methods help address the trade-off between model fit and complexity, their differences highlight the importance of selecting an appropriate criterion based on the specific objectives of the analysis.

11.2.5 Cross-Validation and Bootstrap Methods

Prediction error reflects how well a model generalizes to new data. In Chapter 4, we discussed MSPE for regression, while this chapter covers misclassification rate and other predictive measures for classification. Chapter 9 introduced cross-validation and bootstrap, key resampling techniques that improve error estimation by reducing variance and improving the stability of the model. Here, we focus on cross-validation, a widely used technique that refines prediction error estimates by systematically splitting the data multiple times. We also introduce leave-one-out cross-validation (LOOCV), a special case that provides a near-unbiased error estimate but at a higher computational cost.

Cross-Validation

Cross-validation improves upon a simple validation set approach by reducing variance in error estimation. Instead of relying on a single train-test split, it repeatedly trains and evaluates models on different data subsets for a more stable generalization error estimate.

K-Fold Cross-Validation

In **K-Fold Cross-Validation**, we:
1. Divide the data into K equal-sized subsets (folds).
2. Train the model on $K-1$ folds and validate it on the remaining fold.
3. Repeat this process K times, using each fold as the validation set exactly once.
4. Compute the average error across all folds to estimate generalization error.

This method balances bias and variance by using more training data in each fold while allowing multiple validation sets. Common choices include $K = 5$ **or** $K = 10$.

Leave-One-Out Cross-Validation

LOOCV is an extreme case of **K-Fold Cross-Validation**, where $K = n$, meaning each observation is treated as its own validation set. That is:
- We fit n **models**, each leaving out exactly **one** observation.
- The model is trained on the remaining $n-1$ observations, and the left-out observation is used for validation.
- The final prediction error is the average squared error across all models:

$$CV(n) = \frac{1}{n}\sum_{i=1}^{n}(y_i - \hat{y}_{-i})^2 \tag{11.10}$$

where $\hat{y}_{-i}$ is the predicted value for i using a model trained on all data except observation i.

LOOCV has low bias since almost all data is used for training each time. However, it can have high variance because training on nearly the entire dataset may lead to unstable model estimates. Computationally, LOOCV can be costly, but for certain models—like linear regression—it has an efficient formulation that avoids fitting n separate models. We discussed this in Chapter 7.

Bootstrap

The **bootstrap** is another resampling method, but instead of splitting data into training and validation sets, it generates multiple **random samples with replacement** from the original dataset to estimate variability in model selection. The general steps are:
1. Draw B bootstrap samples from the dataset, each of the same size as the original but sampled with replacement.
2. Fit the model on each bootstrap sample.
3. Evaluate variability and stability of model estimates across different resampled datasets.

Bootstrap is particularly useful for estimating confidence intervals and assessing model uncertainty, which we covered in detail in Chapter 9. While cross-validation primarily focuses on prediction error estimation, bootstrap is more often used for uncertainty quantification in parameter estimates.

Together, these model assessment criteria provide robust approaches for evaluating model performance and ensuring reliability in causal and predictive modeling. In classic econometrics, especially in introductory courses, these criteria are commonly taught, particularly for prediction. Additionally, the integration of prediction models into causal models is becoming more common, with cross-validation and bootstrapping playing an increasingly important role. These are the main criteria we use in the further chapters of our book. After discussing various assessment methods, let's now briefly examine some simple variable selection methods in linear models.

11.3 Subset Variable Selection Methods

Subset selection methods aim to identify the most relevant predictors while keeping the model **parsimonious**, meaning as simple as possible without sacrificing explanatory power. Too many predictors make interpretation difficult, inflate coefficient variability, and increase prediction variance, reducing model stability. In some cases, the goal is not just prediction accuracy but understanding which variables matter most. Practical constraints like data availability, cost, and computational efficiency also favor smaller models. Additionally, excessive predictors can cause estimation issues such as multicollinearity. The objective is to strike a balance between fit and complexity, improving interpretability and minimizing overfitting.

11.3.1 Best Subset Selection

Best subset selection aims to identify the optimal set of predictors by evaluating all possible combinations of variables and selecting the model that best fits the data. This approach systematically considers every subset of available predictors, choosing the one that minimizes or maximizes a given evaluation criterion, such as adjusted R^2, AIC, or BIC—criteria we discussed in the previous section. Mathematically, for a regression model:

$$y = \beta_0 + \sum_{j \in S} \beta_j x_j + \varepsilon, \tag{11.11}$$

where S is a subset of p available predictors, the goal is to determine the optimal subset S^* that minimizes prediction error while maintaining a balance between model complexity and goodness-of-fit.

- Algorithm:
    1. Fit all 2^p possible models.
    2. Compute an evaluation metric such as adjusted R^2, AIC, or BIC for each model.
    3. Select the model that optimizes the chosen criterion.

Best subset selection is particularly useful when there are relatively few predictors, allowing for a comprehensive assessment of all possible models.

A common application is wage determination, where predictors such as years of experience, education level, and job training influence earnings. If we have three predictors, the algorithm evaluates all eight (2^3) possible models and selects the best-performing one.

However, the main drawback of best subset selection is its computational intensity. Since the number of models grows exponentially with p, it quickly becomes infeasible for large datasets. Additionally, without validation techniques such as cross-validation or bootstrapping, this method risks overfitting by tailoring the model too closely to the training data.

Given the computational challenges of best subset selection, stepwise methods provide a more practical alternative by adding or removing predictors sequentially based on statistical significance or model fit criteria. These methods are widely used in econometrics due to their efficiency and flexibility.

11.3.2 Forward Selection

Forward selection starts with an empty model, including only the intercept, and iteratively adds the most significant predictor at each step. The process stops when adding additional variables no longer improves the model based on a predefined criterion, such as AIC, BIC, or adjusted R^2.

- Algorithm:
    1. Begin with the intercept-only model: $y = \beta_0 + \varepsilon$.
    2. Identify the predictor that results in the largest increase in adjusted R^2 (or the smallest AIC/BIC).
    3. Continue adding variables until no remaining predictor improves the criterion.

For instance, when modeling student test scores based on socioeconomic factors, the algorithm may first include family income if it explains the most variance, then add parental education if it further improves model performance.

However, a key limitation is that forward selection may overlook interactions between variables. If two predictors are only jointly significant, the method might exclude them both, leading to a suboptimal model.

11.3.3 Backward Selection

Backward selection follows the opposite approach, starting with a full model containing all predictors and sequentially removing the least significant one until only relevant variables remain. This method assumes that the initial full model is reasonable, making it less effective if irrelevant predictors are included from the start.

- Algorithm:
  1. Start with the full model: $y = \beta_0 + \sum_{j=1}^{p} \beta_j x_j + \varepsilon$.
  2. Remove the variable that contributes least to model improvement.
  3. Repeat until all remaining predictors are statistically significant.

For example, when predicting house prices using square footage, number of rooms, location, and crime rate, backward elimination might remove crime rate first if it has little impact on property values.

The limitation of this method is that it assumes the full model is appropriate initially. If the dataset includes many irrelevant predictors, the elimination process may still retain unnecessary variables due to chance correlations.

11.3.4 Hybrid (Stepwise) Selection

Hybrid stepwise selection combines forward selection and backward elimination, dynamically adjusting the model by both adding significant variables and removing those that become insignificant.

- Algorithm:
  1. Begin with an empty model, as in forward selection.
  2. Add the most significant predictor at each step.
  3. After each addition, check if any previously included variable has become insignificant and remove it if necessary.
  4. Continue until no more improvements can be made.

This method is particularly useful in scenarios where predictors exhibit interdependencies. For instance, in credit risk modeling, the algorithm might first add borrower income as a strong predictor of loan default risk but later remove credit card debt if its effect diminishes after including another financial variable.

Despite its flexibility, hybrid selection is still subject to arbitrary significance thresholds (e.g., $p < 0.05$) and the multiple testing problem, which can lead to the inclusion of spurious relationships.

Ultimately, while best subset selection provides a theoretically optimal approach, its computational demands make it impractical for large models. Stepwise methods offer a more efficient alternative but come with trade-offs, including potential model instability and reliance on arbitrary selection thresholds. Moreover, they are prone to overfitting, as the repeated hypothesis testing at each step increases the risk of selecting variables based on noise rather than true relationships. Stepwise procedures also ignore the joint effect of excluded variables, potentially leading to suboptimal models. As econometrics continues to integrate ML techniques, more sophisticated variable selection methods, such as LASSO and ridge regression, which will be covered next chapter, are increasingly used to address these limitations.

11.4 Selecting Functional Form in Econometric Models

Model selection in econometrics involves choosing both the appropriate functional form and the relevant variables to best capture the relationship between Y and X. While parameter estimation often assumes linearity, the functional form of X and its transformation significantly impact model performance. The choice of form depends on theoretical considerations, empirical patterns, and statistical validation. In many economic applications, such as labor supply models, consumption functions, and production functions, nonlinear relationships arise naturally, requiring careful selection of transformations or breakpoints.

11.4.1 Functional Forms and Transformations

- *Linear models*: Assume a straight-line relationship, $y = \beta_0 + \beta_1 x + \epsilon$, often used as a baseline.

- *Logarithmic models*: Capture diminishing returns or exponential growth, e.g., $y = \beta_0 + \beta_1 \ln(x) + \epsilon$, common in elasticity estimation.

- *Polynomial models*: Introduce curvature, such as $y = \beta_0 + \beta_1 x + \beta_2 x^2 + \epsilon$, useful for non-monotonic relationships, such as the Laffer curve in taxation.

- *Interaction terms*: Account for combined effects of multiple variables, e.g., $y = \beta_0 + \beta_1 x_1 + \beta_2 x_2 + \beta_3 x_1 x_2 + \epsilon$.

- *Dummy variables*: Represent categorical factors, e.g., college graduate as 1 , and non-college graduate as 0.

Functional form selection is guided by theoretical expectations, exploratory data analysis, and diagnostic measures. Economists, social and health researchers, often work with models that capture complex relationships between variables, where structural changes or nonlinear patterns arise. While we have previously covered linear and probit models, here we introduce another linear framework—piecewise linear models—which provide a foundation for understanding more advanced modeling techniques. These models help address cases where relationships are not uniform across the entire range of data, making them particularly valuable in policy analysis and applied research.

Economists, social scientists, and health researchers often work with models that capture complex relationships between variables, where structural changes or nonlinear patterns arise. While we have previously covered linear and probit models, here we introduce another linear framework—piecewise linear models—which provide a foundation for understanding more advanced modeling techniques. These models help address cases where relationships are not uniform across the entire range of data, making them particularly valuable in policy analysis and applied research.

Functional form selection is guided by theoretical expectations, exploratory data analysis, and diagnostic tests. Adjusted R^2, AIC, BIC, and cross-validation provide comparative model evaluation.

11.4.2 Models with Structural Changes: Piecewise Linear Models

In economics, social sciences, and health research, relationships between variables often exhibit structural shifts rather than remaining constant across all values. Policies, institutional changes, or behavioral responses can cause abrupt or gradual changes in how an independent variable influences an outcome. **Piecewise linear models** offer a flexible way to capture such shifts by allowing different slopes before and after a critical threshold, known as a **knot** or **breakpoint**. These models are widely applied in studying taxation effects on labor supply, the impact of minimum wage policies on employment, and changes in healthcare utilization following policy reforms.

A well-known application of piecewise regression is in understanding how progressive tax systems influence labor supply. Suppose an individual's work hours y depend on their wage x, but a higher tax rate applies beyond an income threshold k. Economic theory suggests that at higher tax rates, workers may reduce their hours due to lower after-tax earnings. This relationship can be captured by a piecewise linear regression model:

$$y = \beta_0 + \beta_1 x + \beta_2 (x - k)d + \epsilon \tag{11.12}$$

where $d = 1$ if $x > k$ and 0 otherwise. The coefficient β_1 represents the effect of wages on labor supply before the tax threshold, while $\beta_1 + \beta_2$ represents the effect after the threshold, allowing for a change in slope.

Empirical studies using this framework have found that higher tax rates can reduce labor supply elasticity, particularly for secondary earners in households. Similar models have been used to examine how unemployment benefits create disincentives for job-seeking by introducing a threshold at which benefits phase out, leading to shifts in employment behavior.

Another application appears in healthcare policy, where subsidies or cost-sharing reductions change how individuals seek medical care. For instance, when out-of-pocket expenses drop below a certain income level, patients may increase their use of preventive services, creating a structural shift in healthcare consumption patterns.

Determining Breakpoints

Accurately identifying breakpoints is crucial for making valid inferences. Several statistical methods are available to determine these critical thresholds:

- *Chow test*: This test evaluates whether coefficients in a regression model differ significantly before and after a suspected breakpoint. It estimates separate regressions for each segment and tests whether the differences in coefficients are statistically significant. If the null hypothesis of no break is rejected, a structural change is confirmed.
- *Information criteria (AIC/BIC)*: When the exact location of a breakpoint is unknown, different models with varying breakpoints can be estimated and compared using AIC or BIC. The model with the lowest AIC or BIC provides the best trade-off between fit and complexity.
- *Supremum Wald test*: Unlike the Chow test, which requires pre-specifying a breakpoint, this method scans across all possible breakpoints to identify the most statistically significant

change in coefficients. It is commonly used in structural break analysis of financial and macroeconomic time series.

- *Residual sum of squares (RSS) minimization*: Another approach is to estimate multiple piecewise models across different potential breakpoints and select the one that minimizes RSS:

$$RSS(k) = \sum_{i=1}^{N}(y_i - \hat{y}_i)^2 \tag{11.13}$$

where k represents the candidate breakpoints. The value of k that yields the lowest RSS indicates the most likely location of the structural change.

- *Bayesian methods*: By incorporating prior distributions on possible breakpoints, Bayesian approaches allow estimation of both the number and location of breaks. These methods are particularly useful in settings with multiple structural changes, such as shifts in macroeconomic regimes over time.

11.4.3 Extensions and Alternative Methods

While piecewise regression provides a simple and intuitive way to model structural changes, other techniques offer greater flexibility in handling multiple breakpoints or ensuring smoother transitions between segments.

- *Spline regression*: Unlike piecewise regression, which assumes abrupt changes at breakpoints, spline models introduce smooth transitions by fitting polynomials to different segments of the data. Splines are widely used in wage distribution studies, where income returns to education or experience change gradually rather than suddenly.
- *Generalized additive models (GAMs)*: These models extend linear regression by allowing relationships to be estimated nonparametrically, making them useful when functional forms are unknown. GAMs are commonly applied in public health research to study nonlinear effects of pollution on mortality rates.
- *Regime-switching models*: Used primarily in time-series analysis, these models allow relationships to switch between different states based on underlying stochastic processes. For instance, in financial markets, asset returns may follow different dynamics during periods of high and low volatility.
- *Kernel regression*: A fully nonparametric approach, kernel regression estimates relationships without assuming a predefined functional form. This method is useful when structural changes are expected but their precise nature is unknown, such as in consumer demand studies where preferences shift based on economic cycles.

By carefully selecting an appropriate modeling approach—whether piecewise regression, splines, or more flexible ML methods—researchers can better capture nonlinear relationships, improving both predictive accuracy and causal inference.

11.5 Model Selection in Machine Learning

Model selection is essential in both traditional econometrics and ML, though their goals and approaches differ. Econometrics prioritizes causal inference, interpretability, and theoretical consistency, focusing on unbiased estimation and hypothesis testing. In contrast, ML aims for predictive accuracy, selecting models that generalize well to unseen data by balancing complexity and fit to avoid overfitting.

Unlike econometrics, which often assumes linear relationships, ML accommodates flexible, nonlinear patterns through models like decision trees, random forests, and K-nearest neighbors (KNN). This flexibility improves predictive accuracy, making careful model selection crucial. Economists typically rely on theoretical justification and statistical significance, while ML explicitly manages the bias–variance trade-off, using cross-validation to assess performance.

Evaluation metrics vary by task. Classification metrics include accuracy, precision, recall, and the F1-score, while regression relies on mean squared error (MSE), root mean squared error (RMSE), mean absolute error (MAE), and R-squared. Metric choice depends on context—e.g., accuracy may be misleading for imbalanced classification, requiring alternatives like precision, recall, or AUC. In regression, MAE can be more robust to outliers than MSPE.

Penalized regression methods, such as ridge and LASSO, control model complexity by adding penalty terms. LASSO encourages sparsity by shrinking coefficients to zero, improving interpretability, reducing computation, and improving generalization—especially in high-dimensional economic datasets. The implications of sparsity will be explored in the next section.

11.5.1 Comparing Prediction Models

Economists, social and health scientists apply and compare predictive models such as LASSO, random forests, embeddings, and deep learning using metrics like MSPE or AUC. Cross-validation or bootstrapping ensures reliable evaluation. Information criteria like AIC and BIC help balance model complexity and fit, though ML methods provide more direct measures of predictive performance. Performance metrics vary based on the problem type, whether it is regression (covered in Chapter 4) or classification (covered in Chapter 10).

Regression Problems

- *Mean Squared Error (MSPE)*: Penalizes larger errors more due to squaring.

$$MSE = \frac{1}{n}\sum_{i=1}^{n}(y_i - \hat{y}_i)^2 \tag{11.14}$$

  where y_i is the actual value, $\hat{y}_i$ is the predicted value, and n is the number of samples.

- *Root Mean Squared Error (RMSE)*: Provides errors in the same units as the target variable.

$$RMSE = \sqrt{MSE} \tag{11.15}$$

- *Mean Absolute Error (MAE)*: Less sensitive to outliers.

$$MAE = \frac{1}{n}\sum_{i=1}^{n}|y_i - \hat{y}_i| \tag{11.16}$$

Classification Problems

- *Accuracy*: Proportion of correct predictions, but can be misleading for imbalanced datasets.

$$Accuracy = \frac{TP + TN}{TP + TN + FP + FN} \tag{11.17}$$

- *Precision*: Proportion of predicted positives that are actual positives.

$$Precision = \frac{TP}{TP + FP} \tag{11.18}$$

  Important when false positives are costly.

- *Recall*: Proportion of actual positives correctly identified.

$$Recall = \frac{TP}{TP + FN} \tag{11.19}$$

  Crucial when false negatives are costly, such as in medical diagnostics.

- *F1-score*: Harmonic mean of precision and recall, balancing both.

$$F1 = 2 \cdot \frac{Precision \cdot Recall}{Precision + Recall} \tag{11.20}$$

- *Area Under the ROC Curve (AUC)*: Evaluates classification performance, particularly for imbalanced datasets, ranging from 0.5 (random guessing) to 1 (perfect classification).
- *Log Loss*: Measures classification performance when predictions are probabilities, penalizing confident incorrect predictions.

$$Log\ Loss = -\frac{1}{n}\sum_{i=1}^{n}[y_i \log(\hat{y}_i) + (1 - y_i)\log(1 - \hat{y}_i)] \tag{11.21}$$

Metric selection depends on the problem type and context, particularly in economics, social sciences, and health research. In classification tasks, accuracy can be misleading for imbalanced datasets, requiring alternative measures like precision, recall, or AUC. For example, in healthcare, a model predicting disease presence must prioritize recall to minimize false negatives, as missing a true case can have severe consequences. Similarly, in fraud detection, precision and recall matter more than accuracy, since incorrectly flagging a transaction as fraudulent (false positive) is less costly than missing an actual fraud case (false negative).

For regression tasks, the choice of metric depends on sensitivity to outliers and model objectives. MAE is often preferred over MSE in wage or income predictions, as extreme values should not disproportionately affect the evaluation. Conversely, in cost-effectiveness analysis in healthcare, where large deviations may indicate substantial policy inefficiencies, MSE or RMSE might be more appropriate. When comparing policy interventions, R-squared can indicate how well a model explains variation in an outcome, but it does not necessarily translate to better out-of-sample prediction, which is crucial for decision-making.

Besides these metrics, additional methods and concepts are essential for comparing and improving predictive models. While performance metrics like MSPE and AUC provide quantitative measures of accuracy, techniques such as boosting, cross-validation, bootstrapping, early stopping, and regularization play crucial roles in selecting and refining models. These approaches help balance predictive performance and interpretability, ensuring models generalize well to unseen data while maintaining theoretical and practical relevance.

11.5.1.1 Cross-Validation and Bootstrapping

Cross-validation and bootstrapping, which we covered in Chapter 9, are essential techniques for evaluating predictive models, ensuring they generalize well to unseen data. These methods provide more reliable assessments than a single train-test split, which can lead to over-optimistic or highly variable performance estimates depending on the specific data partition.

Cross-validation involves dividing the dataset into multiple subsets, training the model on some portions, and validating it on others. The most common approach, k-fold cross-validation, partitions the data into k subsets, using $k-1$ folds for training and one for validation, repeating the process k times. The final performance metric is the average across all folds. This method helps compare different models under the same data conditions, reducing bias in performance evaluation.

For example, consider a scenario where we aim to predict household income using demographic, employment, and other covariates. This is a regression problem, and we compare models such as ordinary least square (OLS), LASSO regression, decision trees, and k-NN. To prevent overfitting to a specific data split, we use 10-fold cross-validation, ensuring that each observation is included in both training and validation sets. For each model, we compute the MSPE across the 10 folds, then average these values to obtain the overall MSPE for that model. The model with the lowest average MSPE is typically preferred.

Bootstrapping, on the other hand, repeatedly resamples the dataset with replacement, training models on each resampled set and evaluating performance on the remaining data. This is particularly useful when working with smaller datasets, as it allows estimating the distribution of model performance across different samples.

11.5.1.2 Boosting

Boosting, another powerful ML approach which will discussed in chapter 15, iteratively fits simple models to data, progressively improving accuracy. Boosting methods naturally incorporate model selection through their iterative nature, continuously adjusting the complexity of the model to minimize prediction errors. This iterative refinement is particularly valuable in capturing complex socioeconomic relationships without requiring extensive manual specification. Methods like gradient boosting machines (GBM), XGBoost, and AdaBoost, which we will cover in Chapter 15, benefit from cross-validation to fine-tune hyperparameters, such as learning rates and the number of boosting rounds. For instance, when applying gradient boosting to predict healthcare expenditures, cross-validation helps determine the optimal number of boosting iterations to prevent overfitting.

Cross-validation, bootstrapping, and boosting collectively improve predictive reliability, allowing researchers to fine-tune models while ensuring they perform consistently across different data subsets. These techniques, alongside regularization and early stopping, form the foundation for selecting models that are both accurate and generalizable.

11.5.1.3 Early Stopping and Regularization

- *Early Stopping*: Early stopping is commonly used in neural networks and boosting methods to prevent overfitting by halting training when validation performance starts to degrade. In neural networks, this is implemented by monitoring the validation set during training and selecting the model at the point of optimal performance, often using metrics like log loss. Similarly, in boosted decision trees such as GBM and XGBoost, early stopping is applied by tracking validation error and stopping training when additional iterations no longer improve performance. While random forests do not traditionally use early stopping, some implementations allow limiting tree growth based on validation performance (i.e., stopping rules), indirectly controlling overfitting. These techniques serve as effective forms of model selection, ensuring better generalization to unseen data.
- *Regularization*: Techniques like L1 (LASSO) and L2 (ridge) regularization control model complexity by penalizing large coefficients. L1 regularization can perform feature selection by setting some coefficients to zero, which is a form of model selection at the feature level. For example, in a linear model, L1 regularization might select only the most relevant predictors, simplifying the model and potentially improving generalization.

These techniques, combined with performance metrics, provide a comprehensive approach to evaluating predictive models. Complex models like boosted decision trees and deep learning might yield higher predictive accuracy but make it harder to extract meaningful socio-economic and health related insights. While high accuracy is desirable, economists, social scientists, and health researchers must also weigh factors like interpretability, generalization, and practical applicability when selecting models for policy and decision-making.

11.5.2 Balancing Interpretability and Predictive Accuracy

While predictive accuracy is crucial, economists, social scientists, and health researchers must also consider interpretability. Highly flexible models such as deep learning and random forests may achieve superior predictive performance, but their complexity often makes it difficult to extract meaningful economic or social insights. This trade-off is particularly relevant when the goal is informing policy or understanding behavioral mechanisms.

For example, in predicting unemployment duration, a random forest model might outperform a logistic regression model in classification accuracy, but its black-box nature makes it challenging to identify which factors drive job search behavior. Similarly, in healthcare, a deep learning model predicting hospital readmission risk might outperform a logistic regression model, but policymakers may struggle to interpret which patient characteristics contribute most to readmission, making targeted interventions difficult.

Economists must clearly distinguish predictive models from causal models. A model predicting GDP growth may include a wide range of indicators that improve accuracy of the forecast, but if those indicators are endogenous (e.g., government spending influenced by expected growth), they do not establish a causal link. Misinterpreting such models as causal can lead to flawed policy decisions.

Thus, the choice of metric and model should align with the research objective. If the goal is purely predictive—such as forecasting inflation or demand for public services—ML methods with high predictive accuracy may be prioritized. However, when policy recommendations or causal inferences are required, simpler, interpretable models with well-defined assumptions are often preferable. Striking the right balance ensures that models are both actionable and theoretically sound.

11.5.3 Practical Guidelines

Economists, social scientists, and health researchers integrating ML into their model selection practices should follow key best practices to balance predictive accuracy, interpretability, and computational efficiency. Here are the best practices and considerations:

- *Start with simple models*: Begin with simpler models like linear regression or decision trees before increasing complexity (e.g., to random forests or neural networks). This aligns with Occam's razor, which prefers simpler models when they achieve similar performance.
- *Use appropriate metrics*: Select metrics suited to the problem. For imbalanced classification, precision and recall are more informative than accuracy, while MSPE is commonly used for regression problems.
- *Validate properly*: Use cross-validation to obtain reliable performance estimates, especially with small datasets. For classification tasks, stratified k-fold cross-validation ensures balanced class distributions in training and validation sets.
- *Tune hyperparameters*: Optimize model parameters using grid search or random search, leveraging tools like scikit-learn's model selection module for efficiency. Bayesian optimization can further refine hyperparameter tuning with fewer evaluations.
- *Evaluate on a separate test set*: After selecting a model, assess its performance on an independent test set to confirm its generalization ability and ensure no data leakage.
- *Consider computational resources*: Large datasets and complex models can make cross-validation and hyperparameter tuning computationally expensive.
- *Use penalized regression for complexity control*: Regularization techniques like LASSO (L1) and ridge (L2), covered in the following chapter, explicitly manage complexity and improve interpretability. LASSO's sparsity property is particularly useful for selecting the most relevant predictors while improving out-of-sample performance.
- *Compare multiple models thoughtfully*: When evaluating LASSO, random forests, embeddings, and deep learning, compare models using metrics like MSPE or AUC, carefully weighing predictive accuracy versus interpretability.

By following these guidelines, economists can effectively integrate ML techniques to improve predictive capabilities while complementing traditional econometric approaches. Thoughtful application of ML methods improves empirical research and policy analysis without compromising interpretability.

Ultimately, economists, social scientists, and health researchers should clearly distinguish predictive models from causal models, ensuring purely predictive models are not misinterpreted as causal findings. This structured approach ensures that ML models remain accurate, efficient, and interpretable for economic and social science applications.

11.6 Sparsity in Model Selection

Sparsity in model selection refers to constructing models where many parameters—such as regression coefficients—are set to zero, focusing only on the most relevant features. This approach improves interpretability, improves computational efficiency, and helps prevent overfitting, particularly in high-dimensional settings where datasets contain many redundant or irrelevant variables. In econometrics, ML, and applied social sciences, sparsity is particularly valuable when working with datasets where the number of predictors exceeds the number of observations.

The importance of sparsity lies in its ability to address the curse of dimensionality, which arises when an excessive number of predictors leads to models that fit the training data well but generalize poorly to unseen data. By selecting only the most relevant features, sparse models tend to perform better in predictive tasks, making them a crucial component of modern ML techniques. The ability to extract meaningful patterns while discarding unnecessary complexity makes sparsity not only a technical necessity but also a practical advantage, particularly in applications where interpretability matters, such as in policy evaluation and medical diagnostics.

A key advantage of sparsity is improved interpretability, as models with fewer nonzero coefficients are easier to analyze. In applied fields like economics and public health, researchers and policymakers need clear explanations of how different factors influence an outcome. Sparse models allow analysts to focus on the most important explanatory variables without being overwhelmed by noise. Another advantage is reduced overfitting, especially in high-dimensional settings where overparameterized models memorize the training data rather than capturing underlying patterns. Regularization techniques such as L1 regularization (LASSO) help address this issue by shrinking certain coefficients to zero, effectively eliminating irrelevant predictors. Sparse models also provide better generalization, as they avoid excessive complexity that can reduce predictive performance on new data. Furthermore, sparsity improves computational efficiency, which is particularly valuable when working with large datasets in genomics, finance, and natural language processing, where storing and processing dense models is impractical.

Despite these advantages, sparsity introduces potential challenges. Sparse solutions can be unstable, meaning small changes in the data may lead to different sets of selected variables. This instability poses difficulties in applications like neuroimaging and social network analysis, where reproducibility across different datasets is critical. Moreover, loss of small but important effects can occur when sparsity is applied too aggressively, potentially excluding variables that have weak but meaningful contributions to the outcome. Another challenge is hyperparameter sensitivity, as the choice of regularization strength (λ) significantly influences the sparsity level and overall model performance, requiring careful tuning through cross-validation.

Several techniques help introduce sparsity into models. One of the most widely used methods is L1 regularization (LASSO regression), which forces some regression coefficients to be exactly zero, effectively performing variable selection. Elastic net regularization, a combination of L1 and L2 penalties, balances sparsity with coefficient stability, making it useful when dealing with highly correlated predictors. Random forest and gradient boosting, covered in Chapters 14 and 15, provide an alternative approach by ranking variable importance scores, enabling researchers to eliminate the least informative predictors. Sparse principal component analysis (Sparse PCA) modifies traditional PCA by ensuring that only a subset of features contribute to each principal component, preserving interpretability while still reducing dimensionality. Stepwise Selection, a classic econometric method, also results in sparse models by iteratively adding or removing variables based on statistical criteria.

Sparsity can also be assessed using information criteria such as the AIC and BIC. AIC balances model fit and complexity, generally favoring models with more predictors, while BIC imposes a stronger penalty on additional parameters, leading to sparser solutions. Cross-validation provides another way to select sparse models, helping identify those that generalize well to unseen data. For example, k-fold cross-validation can be combined with regularization techniques like Lasso to ensure the selected model is not only sparse but also predictive.

The relevance of sparsity is particularly evident in high-dimensional datasets, where traditional econometric techniques struggle due to multicollinearity or redundant variables. In economics, sparsity is useful in wage determination models, where only a subset of variables significantly impact salaries, helping eliminate unnecessary controls. In health sciences, sparsity plays a key role in genetic studies, where researchers need to identify the few genes most associated with disease outcomes. In social sciences, sparsity-based methods are valuable in policy evaluation, as they help remove extraneous demographic variables while preserving the most meaningful predictors for causal inference and forecasting.

To evaluate sparsity in model selection, several metrics and techniques are used. One common approach is counting the number of nonzero coefficients, which provides a direct measure of how sparse a model is. A sparsity ratio, calculated as the proportion of zero coefficients, offers a quantitative assessment of sparsity. Regularization paths, which visualize coefficient shrinkage across different values of λ, also help assess how model complexity changes as regularization increases. Comparing models involves not only evaluating their predictive performance (e.g., accuracy for classification or MSE for regression) but also considering their sparsity level. In some cases, a model with slightly lower predictive accuracy may still be preferable if it is sparser and easier to interpret. Stability analysis, such as bootstrap resampling, can be used to test how consistently features are selected across different data samples, helping address concerns about model instability.

Sparsity plays a central role in modern model selection, offering clear benefits in terms of interpretability, overfitting prevention, and computational efficiency, particularly in high-dimensional settings. Techniques like L1 regularization, stepwise selection, and decision trees naturally lead to sparse models, while AIC, BIC, and cross-validation provide useful tools for selecting the best sparse models. However, achieving the right balance is crucial—over-penalization can lead to models that are too simplistic, while insufficient regularization can result in excessive complexity. By using open-source tools and applying stability checks, researchers can effectively incorporate sparsity into their analyses, ensuring that models remain both predictive and interpretable.

11.7 Oracle Properties in Model Selection

Oracle properties in statistics and ML refer to the ideal characteristics of an estimator that performs as well as if the true underlying model were known in advance. The term "oracle" is metaphorical, implying perfect knowledge of the correct model structure, which is rarely available in real-world applications. This concept is particularly relevant in high-dimensional model selection, where the number of features (p) can far exceed the number of observations (n), making variable selection and estimation a challenging task.

An estimator or model selection method is said to have oracle properties if it meets two fundamental criteria. First, it must exhibit variable selection consistency, meaning that as the sample size increases, it correctly identifies the subset of relevant variables with probability approaching one:

$$P(\hat{S} = S^*) \to 1 \quad \text{as} \quad n \to \infty \tag{11.22}$$

where $S^* = \{j : \beta_j \neq 0\}$ is the true support of the regression coefficients. This ensures that the method does not select irrelevant predictors, which would otherwise increase model complexity and reduce generalization ability.

Second, it must demonstrate parameter estimation consistency, implying that the estimated coefficients of the selected variables converge to their true values at the same asymptotic rate as if only the true model were used:

$$\sqrt{n}(\hat{\beta}_S - \beta_S) \xrightarrow{d} N(0, \Sigma_S) \tag{11.23}$$

where S denotes the selected subset of variables and Σ_S is the asymptotic variance-covariance matrix of the correctly specified model. This means that the estimator remains asymptotically efficient, achieving the optimal rate of convergence and minimizing bias and variance in parameter estimation.

Oracle properties are particularly important in model selection for high-dimensional datasets, where traditional methods struggle with variable selection and estimation precision. By ensuring that only the relevant variables are retained while discarding noise, methods with oracle properties improve model interpretability, avoid overfitting, and improve efficiency of the estimation. These benefits make them widely applicable in fields such as genomics, finance, and social sciences, where datasets often include many irrelevant or correlated variables. The practical utility of oracle properties extends beyond linear models to nonparametric methods, including those discussed in studies on "Surface Estimation, Variable Selection, and the Nonparametric Oracle Property," broadening their application to flexible modeling frameworks.

The ability of these methods to achieve oracle properties depends on several key conditions. Sparsity of the true model is a fundamental requirement, meaning that only a small subset of predictors should have nonzero coefficients. This assumption aligns with practical applications in genetics, economics, and finance, where most explanatory variables are either redundant or weakly related to the outcome. Regularity conditions on the design matrix are also necessary. For instance, in LASSO-like methods, the irrepresentable condition must hold, ensuring that the relevant variables can be distinguished from the irrelevant ones:

$$||X_{S^c}^\top X_S (X_S^\top X_S)^{-1} \text{sign}(\beta_S)||_\infty < 1. \tag{11.24}$$

If this condition is violated, LASSO may fail to recover the correct subset of predictors. Another critical factor is tuning parameter selection, as choosing the correct penalty level is essential for balancing bias and variance. Cross-validation, AIC, and BIC are commonly used to optimize this selection process. Sample size considerations are also crucial, as the estimator requires a sufficiently large number of observations relative to the number of relevant variables to ensure consistency. In high-dimensional settings, this is often formalized as:

$$\hat{\beta}_S = \beta_S + O_p(n^{-1/2}), \tag{11.25}$$

which ensures that the estimator achieves the same asymptotic efficiency as if the correct subset of variables were known beforehand.

The practical impact of oracle properties is evident in many fields where model selection is essential. In genomics, where p can reach millions while n remains in the thousands, methods like adaptive LASSO (will be covered in the next chapter) help identify genetic variants associated with diseases. The oracle property ensures that only the truly relevant variants are retained, improving the interpretability of gene-disease relationships while improving predictive power. In financial modeling, where many macroeconomic and firm-level predictors are available, SCAD has been employed to select key economic indicators for predicting stock returns. The ability to correctly exclude irrelevant variables ensures robust risk estimation and investment decision-making.

Oracle properties also have implications for social sciences and policy research, where model selection is crucial for evaluating interventions and estimating causal effects. Sparse models with oracle properties provide more reliable estimates by eliminating noise variables, making policy analysis more interpretable and actionable. Additionally, the ability to select relevant variables while ensuring efficient estimation is critical in healthcare analytics, where predictive modeling techniques are used for patient risk assessment and disease progression modeling.

A notable detail is that oracle properties are not merely theoretical but have been implemented in widely used software packages such as `scikit-learn` and `glmnet`, making them accessible to practitioners. Methods like adaptive LASSO and SCAD, available in these libraries, provide powerful tools for high-dimensional model selection while maintaining computational efficiency. However, despite their advantages, there is ongoing debate about the stability of methods with oracle properties, particularly in cases where data distributions shift or measurement errors exist. Studies in neuroimaging and ML have shown that slight variations in data collection can lead to different sets of selected predictors, affecting reproducibility. This highlights the need for robustness checks, stability selection techniques, and sensitivity analyses when applying oracle-based model selection methods in practice.

As ML and statistical modeling continue to evolve, the role of oracle properties in guiding efficient, interpretable, and robust model selection remains a fundamental aspect of high-dimensional data analysis. By leveraging these properties, researchers and practitioners can build models that balance predictive performance and interpretability, ensuring that they not only achieve strong empirical results but also provide meaningful insights into complex real-world phenomena.

Notes

1. The relationship between in-sample and out-of-sample MSPE always includes an additional term, leading to systematic in-sample underestimation (overfitting). For heteroskedastic errors:

$$\mathbf{MSPE}_{out} = \mathbf{MSPE}_{in} + \frac{2}{n}\sum_{i=1}^{n}\sigma_i^2(p+1).$$

For nonlinear models with heteroskedastic errors:

$$\mathbf{MSPE}_{out} = \mathbf{MSPE}_{in} + \frac{2}{n}\sum_{i=1}^{n}\sigma^2(X_i)\left(\frac{\partial f(X_i)}{\partial\theta}\right)^T \mathbb{E}[\hat{\theta}-\theta] + O(n^{-1}).$$

For regularized models:

$$\mathbf{MSPE}_{out} \approx \mathbf{MSPE}_{in} + \lambda\mathbb{E}\left[\sum_{j=1}^{p}\left(\frac{\partial f(X)}{\partial\theta_j}\right)^2\right] + O(n^{-1})$$

In all cases, in-sample MSPE is biased downward due to overfitting. The term $O(n^{-1})$ represents the asymptotic remainder in the MSPE expansion, vanishing as n grows. Its presence indicates that small-sample effects persist, causing in-sample MSPE to further underestimate out-of-sample MSPE when n is small.

2. For a standard regression model with normally distributed errors, the log-likelihood function is:

$$l(\beta, \sigma^2; y) = -\frac{n}{2}\log(2\pi\sigma^2) - \frac{1}{2\sigma^2}RSS$$

where σ^2 is the variance of the error terms, and RSS is the residual sum of squares.

Since σ^2 is typically unknown, it is estimated using the maximum likelihood estimate:

$$\hat{\sigma}^2 = \frac{RSS}{n}$$

Substituting this into the log-likelihood function, we obtain:

$$l(\hat{\beta}, \hat{\sigma}^2; y) = c - \frac{n}{2}\log(\hat{\sigma}^2)$$

where c is a constant independent of the model parameters. This leads to an alternative formulation of AIC:

$$AIC = n\log(\hat{\sigma}^2) + 2p - 2c$$

3. Since in the MLE framework: $MSE = \hat{\sigma}^2 = \frac{RSS}{n}$ this means that under equal parameter counts, AIC effectively reduces to a least squares criterion.

12

Penalized Regression Methods

Building on the foundational concepts introduced in previous chapters, it is important to recognize that while the ordinary least squares (OLS) method—discussed in Chapter 2 and detailed in Chapter 4—offers a robust framework for statistical inference by minimizing the sum of squared residuals and providing the best linear unbiased estimator (BLUE) under the Gauss–Markov assumptions, it may not always yield the best predictive performance in complex data scenarios. In the pursuit of flexible prediction or obtaining unbiased coefficients, we often transform variables in linear models into polynomials, create interaction terms, and apply various transformations to better capture complex relationships in the data. However, in the era of big data and complex models, traditional linear regression often falls short when faced with high-dimensional datasets where the number of predictors is large relative to the number of observations. In such scenarios, OLS estimates can become unstable, and exhibit high variance, leading to models that perform exceptionally well on training data but fail to generalize effectively due to overfitting , i.e., unbiased coefficients but without any external validity. Moreover, multicollinearity among predictors—where high correlations make the design matrix nearly singular—can significantly inflate variance estimates and destabilize coefficient interpretations, undermining the model's reliability. These challenges highlight the limitations of OLS in complex data settings and underscore the need for additional methodologies that can regularize the estimation process. By trading a small amount of bias for a significant reduction in variance, such methods can improve the model's predictive accuracy without compromising the integrity of its predictions.

In fields such as economics, social sciences, or health sciences, it is common practice to include a wide array of socioeconomic background variables and personal information in regression models—such as age, region, occupation, industry, and various other indicators. Researchers often improve these models by adding quadratic and cubic terms, as well as interactions among these indicators, to better capture nonlinear relationships and complex interactions within the data. This approach can result in models with hundreds or even thousands of features (variables) in a single OLS regression. When utilizing common surveys, especially panel data, incorporating these extensive sets of variables can lead to the very challenges previously discussed: the number of predictors (or control variables) may approach or exceed the number of observations, and multicollinearity becomes more prevalent than it might initially appear. Consequently, OLS estimates become unstable, exhibiting high variances and unreliable coefficient interpretations, which undermines the model's reliability and predictive performance. These issues further highlight the limitations of traditional OLS regression in complex data settings and underscore the need for alternative methodologies—such as penalized regression techniques—that can effectively address these challenges without compromising the integrity of the model's predictions.

Penalized regression methods, which often surpass OLS in predictive accuracy, have emerged as powerful tools in statistical modeling. Also known as shrinkage or regularization methods, these techniques focus primarily on predicting unseen data by emphasizing the critical balance between bias and variance. Rather than striving solely for unbiased estimators—the BLUE

DOI: 10.1201/9781003504795-12

in classical statistical terms—our goal shifts toward developing models that manage this balance effectively. In this chapter, we focus on how penalized regression methods is used for prediction, and walk through each step in detail—from raw data to final results. We explain how to prepare and standardize data, fit the lasso model across a grid of penalty values, use cross-validation to select the optimal penalty, and interpret the selected coefficients. In Chapter 19 and 22, we will later return to these methods in the context of causal inference, where it plays a key role in the double machine learning framework.

These methods are particularly adept at handling overfitting, a key concern when modeling complex datasets, as previously discussed. Penalized regression techniques intentionally adjust the regression coefficients to create slightly biased estimators if it results in a lower overall prediction error on new, unseen data. We explored the possibility of biased estimators as better predictors in the final section of Chapter 5, highlighting how accepting some bias can reduce variance and improve predictive performance.

The general principle of penalization in regression models involves adding a penalty term to the model's objective function—usually the sum of squared errors—to constrain or shrink the estimated regression coefficients. This penalty discourages the model from fitting too closely to the training data, helping to avoid overfitting and manage multicollinearity. By shrinking the coefficients, the model effectively reduces their variance at the cost of introducing some bias, thereby striking a balance that improves prediction accuracy on unseen data.

In practice, penalization aims to control the complexity of the model. Regularization involves adding a penalty term to the loss function used in parameter estimation, which discourages complex models by shrinking coefficient estimates and effectively preventing overfitting. The general principle is to modify the traditional loss function by introducing a penalty that constrains the magnitude of the coefficients. This penalty term includes a regularization parameter that controls the strength of the penalty, allowing us to adjust how much we want to shrink the coefficients.

While the loss function can take various forms—such as means, quantiles, expectiles, or others depending on the modeling objective—and penalization techniques can be applied across different types of models including logistic regression, splines, tree-based methods, and neural networks, in this chapter we specifically focus on linear regression models with the mean square prediction error (MSPE) as the loss function.

The general form of a regularized loss function in linear regression is:

$$\min_{\beta}\left\{L(\beta)=\sum_{i=1}^{n}\left(y_i-\beta_0-\sum_{j=1}^{p}\beta_j x_{ij}\right)^2+\lambda P(\beta)\right\} \tag{12.1}$$

where:

- y_i is the response variable.
- x_{ij} are the predictor variables.
- β are the coefficients to be estimated.
- λ is the regularization parameter controlling the strength of the penalty.
- $P(\beta)$ is the penalty function applied to the coefficients.

The choice of the penalty function distinguishes different regularization methods. L1 and L2 refer to two types of penalty functions used in regularization methods for regression models. Ridge regression applies an L2 penalty, penalizing the sum of the squares of the coefficients. This approach shrinks the coefficients toward zero without fully eliminating any variables, leading to smaller and more stable estimates—particularly useful when dealing with multicollinearity. LASSO regression employs an L1 penalty, penalizing the sum of the absolute values of the coefficients. This method not only shrinks coefficients but can also set some exactly to zero, effectively performing variable selection and simplifying the model. Elastic net regression combines both L1 and L2 penalties, blending the strengths of Ridge and LASSO using the mixing parameter, α. This technique is especially beneficial when dealing with highly correlated predictors, as it can handle multicollinearity while still enabling variable selection.

By selecting the appropriate penalized regression method and tuning its hyperparameters (e.g., λ and α), we can control the trade-off between bias and variance, increase model interpretability, and improve predictive performance. Cross-validation is commonly used to determine the optimum values for these hyperparameters. In the subsequent sections, we will examine each penalized regression method in detail.

12.1 Ridge Regression

Ridge regression is one of the earliest and most fundamental techniques in the family of penalized regression methods. It was introduced by Hoerl and Kennard in 1970 to address the problems of multicollinearity in linear regression models. When predictor variables are highly correlated, the design matrix becomes close to singular, and the OLS estimates exhibit high variance and instability. Ridge regression addresses this challenge by incorporating a penalty term into the loss function, which effectively reduces the magnitude of the coefficient estimates toward zero without completely zeroing any of them. In ridge regression, this penalty term helps in minimizing the weight of each predictor, thereby allocating greater significance to the most crucial predictors. This approach helps decrease the variance and consequently, the risk of overfitting. When the sample size is large, the ridge estimators are nearly identical to the unbiased OLS estimators, which we will demonstrate further ion text.

Consider a linear regression model with n observations and p predictors $\mathbf{y}_i = \mathbf{X}_i\boldsymbol{\beta} + \boldsymbol{\varepsilon}_i$, where $\mathbf{y} \in \mathbb{R}^n$ is the response vector, $\mathbf{X} \in \mathbb{R}^{n \times p}$ is the design matrix of predictor variables, $\boldsymbol{\beta} \in \mathbb{R}^p$ is the vector of coefficients to be estimated, $\boldsymbol{\varepsilon} \in \mathbb{R}^n$ is the error term, assumed to be independently and identically distributed with mean zero and variance σ^2.

In OLS regression, the estimates of $\boldsymbol{\beta}$ are obtained by minimizing the residual sum of squares (RSS) as a loss function, $\min_{\boldsymbol{\beta}} \sum_{i=1}^{n} \left(y_i - \beta_0 - \sum_{j=1}^{p} \beta_j x_{ij}\right)^2$. The OLS solution in matrix form is $\hat{\beta}_{\text{OLS}} = (X^\top X)^{-1} X^\top y$. In component form, $\hat{\boldsymbol{\beta}}_{\text{OLS}}$ can be expressed as $\hat{\beta}_{k,\text{OLS}} = \frac{\sum_{i=1}^{n} y_i x_{ik}}{\sum_{i=1}^{n} x_{ik}^2}$, as shown at the end of Chapter 4. In OLS, the variance of $\hat{\boldsymbol{\beta}}$ is given by $\text{Var}(\hat{\boldsymbol{\beta}}_{\text{OLS}}) = \sigma^2(\mathbf{X}^\top\mathbf{X})^{-1}$, assuming homoscedastic errors. However, OLS coefficient and variance estimates can become unstable when dealing with multicollinearity or high-dimensional data (where $p \approx n$ or $p > n$).

When multicollinearity is present, i.e., some of the predictors are highly correlated, meaning they carry nearly the same information. In this case, the columns of $\mathbf{X}$ are almost linearly dependent, making it difficult to distinguish the effects of different predictors. This increases the variance of $\hat{\boldsymbol{\beta}}$, meaning small changes in the data can lead to large changes in the estimates and predictions. Similarly, in high-dimensional data, predictions become unreliable because the matrix $\mathbf{X}^\top\mathbf{X}$ becomes close to singular (non-invertible), which occurs when there isn't enough data to provide a unique estimate for each predictor, as $p > n$. Ridge regression addresses these issues by adding a penalty term, referred as $L2$, to the OLS loss function, which stabilizes the coefficient estimates and reduces their variance.

The objective function for ridge regression is:

$$\min_{\boldsymbol{\beta}} \left\{ L(\boldsymbol{\beta}) = \sum_{i=1}^{n} \left(y_i - \beta_0 - \sum_{j=1}^{p} \beta_j x_{ij} \right)^2 + \lambda \sum_{j=1}^{p} \beta_j^2 \right\}, \tag{12.2}$$

where $\mathbf{y}_i$ represents the response variable, x_{ij} are the predictors, and $\boldsymbol{\beta} = [\beta_0, \beta_1, \ldots, \beta_p]$ is the vector of coefficients. The regularization or tuning parameter $\lambda \geq 0$ controls the strength of the penalty applied to the coefficients. The term $\|\boldsymbol{\beta}\|_2^2 = \sum_{j=1}^{p} \beta_j^2$ represents the squared $L2$ norm of the coefficient vector, which is the key component of ridge regression's penalty term.

The ridge regression objective function can also be written in matrix form as:

$$L(\boldsymbol{\beta}) = (\mathbf{y} - \mathbf{X}\boldsymbol{\beta})^\top (\mathbf{y} - \mathbf{X}\boldsymbol{\beta}) + \lambda \boldsymbol{\beta}^\top \boldsymbol{\beta} \tag{12.3}$$

where $\mathbf{X}$ is the matrix of predictor variables, $\mathbf{y}$ is the response vector, and $\boldsymbol{\beta}$ is the coefficient vector. This matrix form compactly expresses the optimization problem that ridge regression solves, while the summation form provides insight into the individual components of the solution.

The ridge regression estimator for the coefficient β_k in summation notation is given by:

$$\hat{\beta}_{k,\text{Ridge}} = \frac{\sum_{i=1}^{n} y_i x_{ik}}{\sum_{i=1}^{n} x_{ik}^2 + \lambda} \tag{12.4}$$

For all coefficients $k = 1, \ldots, p$, the solution in matrix form is $\hat{\boldsymbol{\beta}}_{\text{Ridge}} = (\mathbf{X}^\top\mathbf{X} + \lambda\mathbf{I})^{-1}\mathbf{X}^\top\mathbf{y}$.

The matrix formulation provides a general solution for all coefficients simultaneously, while the summation form gives more insight into how ridge regression operates on individual predictors.

The ridge regression coefficient for the k-th predictor can also be expressed in terms of the OLS coefficient as:

$$\hat{\beta}_{k,\text{Ridge}} = \frac{\hat{\beta}_{k,\text{OLS}}}{1 + \frac{\lambda}{\sum_{i=1}^{n} x_{ik}^2}} \tag{12.5}$$

(Most texts provide the following relationship between OLS and ridge coefficients as given in Equation 12.5. Orthogonality is beneficial but not necessary for either OLS or ridge regression. In the context of ridge regression, its primary advantage is that it simplifies the formula and interpretation of the results.)

For orthogonal covariates, we assume each predictor x_i is independent and has an equal sum of squares across observations, $\sum_{j=1}^{n} x_{ji}^2 = n$ for each predictor i. Also, the cross-product of different predictors equals zero, $\sum_{j=1}^{n} x_{ji} x_{jk} = 0$ for all $i \neq k$.

We can express the ridge estimator in terms of the OLS estimator:

$$\hat{\beta}_{k,\text{Ridge}} = \frac{n}{n+\lambda}\hat{\beta}_{k,\text{OLS}} \tag{12.6}$$

In matrix form, for orthogonal covariates, the matrix $X^\top X$ is a diagonal matrix where each diagonal element is equal to n (the number of observations), assuming the predictors are standardized (which we discuss next). Thus $X^\top X = nI$. Substituting in the formula for ridge regression, we get $\hat{\beta}_{\text{Ridge}} = \frac{n}{n+\lambda}\hat{\beta}_{\text{OLS}}$.

The simplification $\frac{n}{n+\lambda}$ serves as a good approximation and conceptual aid to understand how regularization impacts coefficient estimates, particularly emphasizing uniform shrinkage in a broad sense. In practice, ridge is particularly useful in scenarios where predictors are highly correlated or when the number of predictors is large relative to the number of observations, conditions under which OLS might perform poorly due to high variance or non-invertibility of $X^\top X = nI$. (All derivations are at the end of the chapter.)

In the OLS solution, the denominator consists only of the sum of squares of the predictor values for x_{ik}. In ridge regression, however, the regularization term λ is added to this denominator, which effectively shrinks the coefficient $\hat{\beta}_{k,\text{Ridge}}$ compared to $\hat{\beta}_{k,\text{OLS}}$. This expression shows that ridge regression shrinks the OLS coefficient by a factor that depends on both the regularization parameter λ and the variability of the predictor x_{ik}. When $\lambda = 0$, the ridge regression solution reduces to the OLS solution, meaning $\hat{\beta}_{k,\text{Ridge}} = \hat{\beta}_{k,\text{OLS}}$. As λ increases, the ridge coefficient $\hat{\beta}_{k,\text{Ridge}}$ becomes smaller, or more shrunk, compared to $\hat{\beta}_{k,\text{OLS}}$. Thus, these estimates called shrinkage estimators, which are biased but have lower variance than OLS estimates.

The purpose of this shrinkage is to stabilize the coefficient estimates by reducing their variance. This makes the model more robust to multicollinearity and high-dimensional data, where OLS can become unstable. By controlling the impact of large or highly correlated predictors, ridge regression helps to reduce overfitting and improve the model's performance on unseen data.

Similar to all MSPEs, the MSPE of the ridge estimator can also be decomposed into the sum of the squared bias and the variance as $\text{MSE}(\hat{\boldsymbol{\beta}}_{\text{Ridge}}) = \text{Bias}(\hat{\boldsymbol{\beta}}_{\text{Ridge}})^2 + \text{Var}(\hat{\boldsymbol{\beta}}_{\text{Ridge}})$.

The bias and variance of the ridge estimator can be derived both in summation form and matrix form, and these expressions help us understand how regularization affects the estimates. To compare the MSE of ridge regression and OLS, recall that the bias and variance for OLS were discussed at the end of Chapter 5.

The **variance** of the ridge estimator in summation notation for the k-th coefficient is:

$$\text{Var}(\hat{\beta}_{k,\text{Ridge}}) = \frac{\sigma^2 \sum_{i=1}^n x_{ik}^2}{\left(\sum_{i=1}^n x_{ik}^2 + \lambda\right)^2} \tag{12.7}$$

This shows that ridge regression reduces or at least maintains the variance compared to OLS, thanks to the regularization term λ, which controls the shrinkage of the coefficients. The variance of the Ridge estimator is **always** less than or equal to the variance of the OLS estimator, which can be formally written as $\text{Var}(\hat{\beta}_{k,\text{Ridge}}) \leq \text{Var}(\hat{\beta}_{k,\text{OLS}})$.

On the other hand, the **bias** introduced by ridge regression in summation form is:

$$\text{Bias}(\hat{\beta}_{k,\text{Ridge}}) = -\beta_k \frac{\lambda}{\sum_{i=1}^n x_{ik}^2 + \lambda} \tag{12.8}$$

This bias arises because ridge regression shrinks the coefficients toward zero, deviating from the true coefficient values. As λ increases, the bias increases because the model penalizes the size of the coefficients more aggressively, i.e., the coefficients are shrunk more toward zero.

In matrix form, the **variance** of the ridge estimator is $\text{Var}(\hat{\boldsymbol{\beta}}_{\text{Ridge}}) = \sigma^2(\mathbf{X}^\top\mathbf{X} + \lambda\mathbf{I})^{-1}\mathbf{X}^\top\mathbf{X}(\mathbf{X}^\top\mathbf{X} + \lambda\mathbf{I})^{-1}]$. Also, in matrix form, the **bias** introduced by ridge regression is expressed as $\text{Bias}(\hat{\boldsymbol{\beta}}_{\text{Ridge}}) = -\lambda(\mathbf{X}^\top\mathbf{X} + \lambda\mathbf{I})^{-1}\boldsymbol{\beta}$.

Both forms show that ridge regression trades off a small amount of bias for a significant reduction in variance. The key to minimizing the MSE lies in finding the appropriate λ, as ridge regression reduces variance at the cost of introducing bias, but this can lead to improved prediction performance compared to OLS, particularly when dealing with multicollinearity or high-dimensional data.

When $\lambda = 0$, ridge regression reduces to OLS, and both MSEs are the same because the bias is zero, and the variance matches that of OLS. However, when $\lambda > 0$, ridge regression introduces bias but compensates by reducing the variance, thanks to the shrinkage of the coefficients.

For small values of λ, the bias is minimal, and ridge regression behaves similar to OLS. In this scenario, ridge regression manages to reduce the variance slightly without introducing much bias. As a result, the MSE of ridge can be lower than that of OLS, meaning ridge regression is able to outperform OLS by stabilizing the coefficient estimates while maintaining low bias. On the other hand, when λ is large, the bias introduced by ridge regression increases substantially as the model shrinks the coefficients more aggressively. Although the variance decreases with larger λ, the overall MSE can increase due to the growing bias. In such cases, ridge regression may underfit the data, and its MSE can become larger than that of OLS because the model is too biased to capture the true relationships in the data.

In ridge regression, for it to have a lower MSE than OLS, the regularization parameter λ must satisfy the inequality:

$$\lambda < \frac{2\sigma^2 \sum_{i=1}^{n} x_{ik}^2}{\beta_k^2 \sum_{i=1}^{n} x_{ik}^2 - \sigma^2} \tag{12.9}$$

This inequality provides a threshold for λ. If λ is below this threshold, ridge regression will outperform OLS in terms of MSE. However, if λ exceeds this value, the bias introduced by ridge regression becomes too large, and the OLS estimator will have a lower MSE.

While this formula offers valuable theoretical insight, calculating the exact value of λ in practice is difficult. This is primarily because the formula depends on the true coefficient values β_k and the variance of the error term σ^2, both of which are unknown in real-world scenarios. The whole purpose of regression is to estimate the coefficients β_k, and we don't have access to their true values in practice. Similarly, the true variance of the error term σ^2 is also unknown and, while it can be estimated, this estimation introduces additional uncertainty. Therefore, because β_k and σ^2 are unknown, the formula for the optimal λ cannot be computed exactly, making the direct calculation of λ impractical in real-world settings.

Even if we were able to approximate β_k and σ^2, the theoretical formula for λ is based on several simplifying assumptions, such as independent and identically distributed (i.i.d.) errors, linearity, and no model misspecification. In real-world datasets, the relationships between variables are often more complex than the assumptions of the model. There may be nonlinearities, interactions, multicollinearity, or other complexities that affect model performance. Furthermore, the noise distribution may not follow the standard assumptions of homoscedasticity or normality. These complexities mean that the theoretically derived λ might not generalize well to the specific characteristics of the dataset being analyzed.

Cross-validation is a practical, data-driven method for finding λ because it directly tests how different values of λ perform on unseen data, without relying on theoretical assumptions about the true model or unknown parameters like β_k and σ^2. While the theoretical formula for λ is valuable for understanding the trade-off between bias and variance, cross-validation is the preferred method in real-world applications. It avoids the need for unknown parameters, accounts for complexities in the data, and directly optimizes the model's performance on unseen data, making it the go-to approach for tuning the λ hyperparameter in ridge regression.

In addition to cross-validation, there are several other methods that can be used to select the regularization parameter λ in ridge regression. While cross-validation remains the most commonly used method due to its flexibility and robustness, alternatives like bootstrap or analytical approaches such as generalized cross-validation, Akaike Information Criterion/Bayesian Information Criterion (AIC/BIC), SURE, empirical bayes, and the L-curve method also offer ways to estimate the performance of different λ values in real-world scenarios. These methods may be preferable depending on the context, computational considerations, or available prior information (such as knowledge of noise variance). Each method has its advantages and trade-offs, and the choice depends on the specific characteristics of your data and modeling needs.

In summary, ridge regression's performance is determined by balancing bias and variance. When λ is small, ridge regression performs similarly to OLS but with lower variance, making it advantageous in the presence of multicollinearity. However, when λ becomes too large, the bias increases, and the MSE may eventually surpass that of OLS. Therefore, selecting the optimal λ, often through cross-validation, is essential for minimizing the MSE by finding the right balance between bias and variance.

The L2 penalty in ridge regression shrinks the coefficients toward zero without setting any of them exactly to zero. This shrinkage stabilizes the coefficient estimates, reducing variance and improving predictive performance on new, unseen data. Ridge regression does not perform variable selection, unlike LASSO, but its ability to handle multicollinearity makes it a powerful tool for dealing with high-dimensional data.

12.1.1 Standardization of Predictors

It is important to consider the role of **standardization** when applying regularization. In practice, predictor variables $\mathbf{X}$ are often on different scales, which can cause larger-scaled variables to dominate the regularization penalty, leading to inconsistent shrinkage across coefficients. Regularization penalizes models for large coefficients, and the magnitude of these coefficients depends on two key factors: the strength of the relationship between the input variable x and the output variable y, and the units of measurement of x. For instance, if a predictor represents weight and is measured in grams, its coefficient will differ significantly compared to when the same variable is measured in kilograms. This is because the coefficient must adjust to reflect the unit of measurement, inflating or deflating accordingly.

To ensure that the regularization focuses on the strength of the relationship between predictors and the response variable, rather than the units of measurement, it is essential to **standardize the features**. Standardization involves transforming the variables to have a mean of 0 and a standard deviation of 1. This adjustment ensures that the regularization penalty applies equally across all predictors, allowing the coefficients to reflect only the importance of each predictor in predicting y.

Without standardization, larger-scaled variables could receive a disproportionately large penalty, distorting the model's performance. Standardizing the data allows ridge regression to apply the same regularization parameter λ across different datasets, irrespective of their scales or sample sizes, leading to more stable and interpretable results. The test and validation sets should also be standardized using the mean and standard deviation derived from the training set to avoid data leakage and ensure consistency during evaluation. Standardizing the predictors does not alter the fundamental relationship between ridge and OLS estimates in terms of their mathematical form. However, it simplifies the calculation and interpretation by ensuring that each predictor contributes equally to the penalty term, thereby standardizing the effect of regularization across all predictors.

In ridge regression, it is customary to standardize predictor variables before fitting the model. This is done by transforming each predictor as follows:

$$x_{ij}^{\text{std}} = \frac{x_{ij} - \bar{x}_j}{s_j} \tag{12.10}$$

where $\bar{x}_j$ is the mean of the j-th predictor across all n observations, and $s_j = \sqrt{\frac{1}{n}\sum_{i=1}^{n}(x_{ij} - \bar{x}_j)^2}$ is the standard deviation of the j-th predictor. This process ensures that the regularization penalty is applied consistently across all predictors, making the model's coefficients comparable and independent of the variables' units. Standardization thus plays a crucial role in ensuring the fairness and effectiveness of the regularization process, allowing ridge regression to focus on the predictive strength of each variable rather than its scale.

12.1.2 Implementation and Simulation of Ridge Regression

To begin the ridge regression simulation, the first step is **Data Preparation**. This involves centering and standardizing the predictor variables to ensure that all features are on the same scale, allowing the ridge penalty to be applied consistently across them. Additionally, if the intercept term is omitted from the model, the response variable should also be centered to maintain consistency with the standardized predictors.

Next is the **Parameter Selection** process. Here, we select a grid of λ values to test. To find the optimal λ, k-fold cross-validation is employed. This method evaluates the model's performance for each value of λ by dividing the data into training and validation sets multiple times. The value of λ that minimizes the cross-validation error is chosen as the optimal regularization parameter.

Once the optimal λ has been determined, the **Model Fitting** step begins. Using the selected λ, the ridge regression coefficients are computed, applying the regularization penalty that shrinks the coefficients toward zero without setting any of them exactly to zero.

Finally, in the **Model Evaluation** phase, the predictive performance of the ridge regression model is assessed. This can be done by evaluating its performance on a separate test set or by using additional cross-validation. The goal is to compare the ridge regression model's ability to generalize to new data, particularly in relation to an OLS model.

In the following ridge regression simulation in R, we generate synthetic data, fit a ridge regression model using the `glmnet` package, and then evaluate its performance by comparing it to OLS results. This step-by-step approach demonstrates how ridge regression can be applied in practice to address multicollinearity and improve predictive accuracy.

```
# Load necessary libraries
library(glmnet)

# 1. Generate synthetic data
set.seed(123)  # For reproducibility
n <- 100  # Number of observations
p <- 10   # Number of predictors

# Create a matrix of independent variables (X) and response variable (y)
X <- matrix(rnorm(n * p), nrow = n, ncol = p)  # Random normal data for predictors
beta <- rnorm(p)  # Random coefficients
y <- X %*% beta + rnorm(n)  # Linear relationship with noise

# 2. Split data into training and testing sets
set.seed(456)
train_idx <- sample(1:n, size = 0.8 * n, replace = FALSE)
X_train <- X[train_idx, ]
X_test <- X[-train_idx, ]
y_train <- y[train_idx]
y_test <- y[-train_idx]

# 3. Fit a Ridge regression model
# Ridge regression with alpha = 0 (L2 regularization)
ridge_model <- glmnet(X_train, y_train, alpha = 0)

# 4. Cross-validation to find the best lambda
cv_ridge <- cv.glmnet(X_train, y_train, alpha = 0)

# 5. Get the best lambda value
best_lambda <- cv_ridge$lambda.min
cat("Best lambda for Ridge regression:", best_lambda, "\n")
```

```
## Best lambda for Ridge regression: 0.2529632
```

```
# 6. Fit the final Ridge model with the best lambda
ridge_final <- glmnet(X_train, y_train, alpha = 0, lambda = best_lambda)

# 7. Make predictions on the test set
y_pred_ridge <- predict(ridge_final, s = best_lambda, newx = X_test)

# 8. Evaluate the model performance
mse_ridge <- mean((y_test - y_pred_ridge)^2)
cat("Mean Squared Error for Ridge regression on test set:", mse_ridge, "\n")
```

```
## Mean Squared Error for Ridge regression on test set: 2.315453
```

```
# 9. Compare with Ordinary Least Squares (OLS)
ols_model <- lm(y_train ~ X_train)
y_pred_ols <- predict(ols_model, newdata = data.frame(X_train = X_test))

# Mean Squared Error for OLS
mse_ols <- mean((y_test - y_pred_ols)^2)
cat("Mean Squared Error for OLS on test set:", mse_ols, "\n")
```

```
## Mean Squared Error for OLS on test set: 34.04085
```

```
# 10. Plot the cross-validation results
  plot(cv_ridge)
```

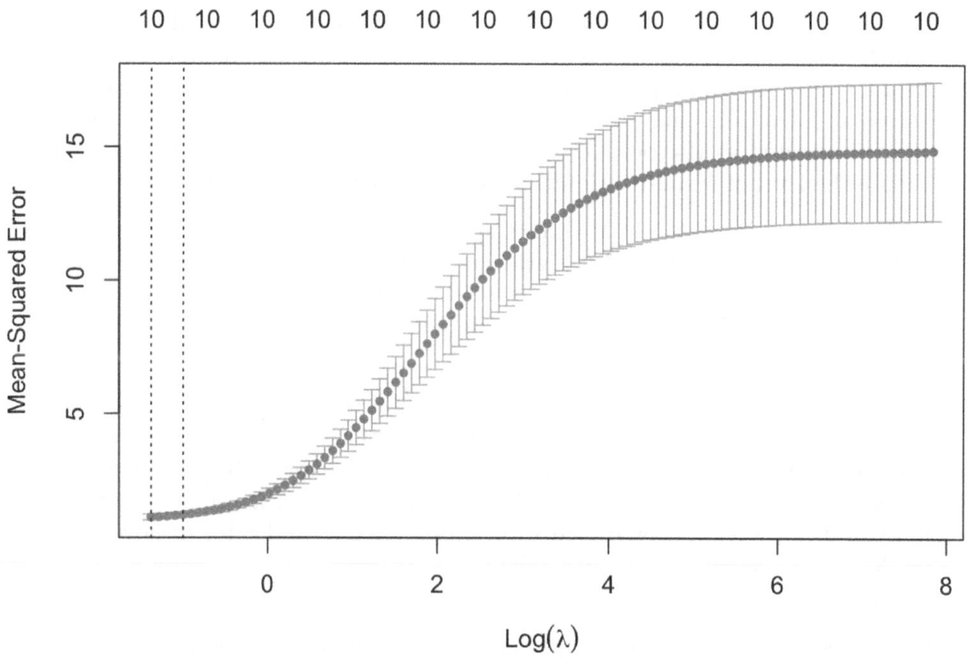

FIGURE 12.1 Cross-validation results for Ridge regression: optimal lambda selection.

In this simulation, we demonstrate how ridge regression works using synthetic data. The steps involved include generating data, fitting the ridge model, selecting the optimal regularization parameter λ via cross-validation, and evaluating the model's performance. The code uses the `glmnet` package in R, which is commonly employed for fitting generalized linear models, including ridge regression (L2 regularization) and LASSO regression (L1 regularization). Let's break down each step of the simulation:

1. *Data generation*: We begin by generating synthetic data. In this example, we create a dataset with 100 observations (rows) and 10 predictors (columns). The predictors are generated randomly from a normal distribution, and the response variable `y` is created as a linear combination of these predictors, with some additional random noise added to simulate real-world conditions where data is rarely perfectly linear. This ensures that the dataset mimics a scenario with some inherent variability and noise.

2. *Train-test split*: After generating the data, the dataset is split into two parts: 80% of the data is used for training the model and 20% is held out as a test set. This is done to assess how well the model generalizes to unseen data. The training set is used to fit the ridge regression model, and the test set is reserved for evaluating the model's performance later.

3. *Fitting ridge regression*: Ridge regression is fitted using the `glmnet()` function from the `glmnet` package. The `glmnet()` function is highly flexible and allows the user to fit both ridge and Lasso models by adjusting the `alpha` parameter. For Ridge regression, we set `alpha = 0`, which indicates that L2 regularization (Ridge) is being applied. The Ridge regression model introduces a penalty for large coefficients, shrinking them toward zero to prevent overfitting, especially in cases where multicollinearity is present among predictors.

4. *Cross-validation for λ selection*: To determine the optimal value of λ, the regularization parameter, we use the `cv.glmnet()` function, which performs ten-fold cross-validation. Cross-validation is a technique where the training data is divided into 10 parts, and the model is trained on 9 parts while the remaining 1 part is used for validation. This process is repeated 10 times, and the average performance across all folds is calculated. The value of λ that minimizes the cross-validation error is selected as the best λ. This step is crucial because it ensures that we choose the level of regularization that best balances bias and variance.

5. *Final ridge model with best λ*: Once the best λ value is identified through cross-validation, we fit the final ridge regression model using this optimal λ value. The `glmnet()` function is called again, this time with the best λ, to compute the ridge coefficients that minimize the loss function (sum of squared errors) while applying the chosen regularization penalty.

6. *Model evaluation*: To evaluate the performance of the ridge regression model, we use the test set that was set aside earlier. Predictions are made on the test data using the fitted ridge model, and the mean squared error (MSE) is calculated to assess the model's accuracy. The MSE measures the average squared difference between the actual and predicted values, providing an indicator of how well the model generalizes to unseen data. The lower the MSE, the better the model is at predicting new data.

7. *Comparison with OLS*: For comparison, an OLS regression model is also fitted using the `lm()` function. The OLS model does not apply any regularization, so it may overfit the data, especially when multicollinearity is present. The MSE of the OLS model is computed on the same test set, allowing us to directly compare the predictive performance of Ridge regression with OLS. Since ridge regression shrinks the coefficients, it typically yields a lower MSE on the test set compared to OLS, especially when multicollinearity or high-dimensional data is present.

8. *Visualization of cross-validation results*: Finally, the results of the cross-validation process are visualized using a plot. The plot displays the cross-validation errors for different values of λ, highlighting the best λ that minimizes the error. This graphical representation helps illustrate how different levels of regularization affect the model's performance. In addition, numbers shown above the vertical bars indicate the number of nonzero coefficients in the model for each λ value. For ridge regression, these numbers are typically constant (e.g., "10") because ridge does not set coefficients exactly to zero—it shrinks them continuously.

At the end of the simulation, it's important to note that the grid of λ values used in ridge regression plays a crucial role in determining the model's performance. By default, the `glmnet` package selects around 100 values of λ, typically ranging from a very large value to a very small one. The largest λ value is chosen so that all coefficients are shrunk very close to zero, meaning the model is almost entirely regularized. The smallest value of λ approaches zero, meaning the regularization effect is minimal, and the model behaves similarly to OLS. This ensures that the model is evaluated across a broad spectrum of regularization strengths. Because of this wide range, the figure above presents $\log(\lambda)$ on the x-axis to make the range of λ values easier to visualize.

While the default grid works well in most cases, you may want to customize the grid of λ values using the `lambda` argument in the `glmnet()` or `cv.glmnet()` function for finer control. Customizing the grid can be useful if you have prior knowledge about the scale of regularization that is appropriate for your data, if you want to explore a finer or coarser set of λ values, or if the default grid does not cover the range of λ values that interests you. By exploring different values of λ, Ridge regression is able to balance bias and variance, optimizing the model's predictive accuracy. Here's how you can customize the λ grid at step 3 above:

To further analyze the results of the simulation, ridge regression coefficients can be extracted and compared to OLS coefficients. This can be added in step 6 of the simulation for ridge regression and in step 9 for OLS. By using the `coef()` function in R, we can present the ridge coefficients for the selected λ, and then compare them to the OLS coefficients.

```
# Step 6: Extract Ridge regression coefficients for the best lambda
ridge_coefs <- coef(ridge_final)
print(ridge_coefs)
# Step 9: Extract OLS coefficients
ols_coefs <- coef(ols_model)
print(ols_coefs)
```

Ridge regression coefficients can be interpreted similarly to OLS coefficients, in that they represent the impact of each predictor on the response variable. However, the ridge regression coefficients are typically smaller than the OLS coefficients due to the regularization penalty, which shrinks the coefficients to reduce overfitting and stabilize the model in the presence of multicollinearity. Unlike OLS, where large coefficients can arise when predictors are highly correlated, ridge introduces a bias but reduces variance, resulting in more stable and reliable estimates. The ridge model retains all predictors, but their influence is diminished based on the strength of the regularization parameter λ. This provides a better balance between bias and variance, leading to improved predictive performance on new, unseen data compared to OLS.

In summary, this simulation demonstrates the use of ridge regression for handling multicollinearity and improving model generalization. The `glmnet` package provides an efficient and user-friendly way to apply ridge regression, select the optimal regularization parameter via cross-validation, and evaluate model performance. Ridge regression is particularly useful when predictors are highly correlated or when dealing with high-dimensional datasets, as it shrinks the coefficients to prevent overfitting while maintaining good predictive accuracy.

Ridge regression is widely implemented in various statistical software packages, including R, Python, and Stata. In R, the `glmnet` package is commonly used for ridge regression, providing a flexible and efficient framework for applying ridge and LASSO regression. In Python, the scikit-learn library includes ridge regression as part of its suite of linear models, making it easy to apply and experiment with regularization techniques. Additionally, Stata offers ridge regression functionality through specifying $\alpha = 0$ while using the elasticnet command, allowing users to fit models with L2 regularization within the familiar Stata environment. These platforms provide robust tools for applying ridge regression, making it accessible and practical for a wide range of data analysis tasks.

Ridge regression can also be extended to other types of models, such as generalized linear models (GLMs). By modifying the loss function to account for different response variables, ridge regression can be applied to logistic regression for binary outcomes, among other models. Additionally, ridge regression has a Bayesian interpretation. In this view, the coefficients have a Gaussian prior distribution with mean zero and variance proportional to $1/\lambda$, expressed as $\beta_j \sim \mathcal{N}\left(0, \frac{\sigma^2}{\lambda}\right)$. This Bayesian perspective adds another layer of understanding to how the regularization in ridge regression works, as it can be seen as imposing a prior belief that the coefficients should be small unless there is strong evidence from the data to support larger values.

Ridge regression is a powerful tool for addressing multicollinearity and improving the predictive performance of linear models by introducing a penalty term that balances the bias–variance trade-off. Its ability to provide more stable coefficient estimates makes it particularly useful for handling datasets with many correlated predictors. Although ridge regression does not perform variable selection, it effectively reduces variance in exchange for a slight bias, which often leads to better generalization to new data. Its computational efficiency, even with large datasets, is another key advantage. However, the lack of variable selection can make interpretation more challenging, as all predictors remain in the model, albeit with shrunk coefficients. Despite this limitation, ridge regression remains an essential technique for statisticians and data scientists working with complex, high-dimensional datasets.

12.2 LASSO Regression

Lasso regression, short for least absolute shrinkage and selection operator, improves traditional linear regression by incorporating a penalty term proportional to the absolute value of the coefficients. The primary purpose of LASSO is to encourage **sparsity** in the model, i.e., to shrink some coefficients to exactly zero, effectively performing **variable selection**. Sparsity, in this context, refers to the situation where a significant portion of the coefficients in a model are zero or nearzero, effectively reducing the complexity of the model and focusing on fewer, more significant variables. This concept is particularly useful in high-dimensional datasets—where the number of predictors p exceeds the number of observations n—, which often contain many predictors that are not essential for accurate predictions. LASSO's ability to induce sparsity helps in variable (feature) selection by retaining only the most relevant predictors, thus simplifying the model and improving interpretability.

We want to emphasize some basic facts about LASSO solutions. There is either a unique Lasso solution or an (uncountably) infinite number of solutions, highlighting the potential for multiple equivalent models especially in scenarios with high-dimensional data. Solutions is not unique unless $\text{rank}(X) = p$. This condition, $\text{rank}(X) = p$, means that all the columns (predictors) in the matrix X are linearly independent, with no column being a linear combination of any others, which is crucial for ensuring a unique solution in traditional regression analysis. When this condition is met, the matrix X has full column rank, and the matrix $X^\top X$ is invertible, a necessary condition for the uniqueness.

However, regardless of the number of solutions, each one will predict the same outcomes, ensuring that predictions are consistent across different models. If the regularization parameter, often symbolized by lambda, is greater than zero, all solutions will have the same total amount of shrinkage applied across their coefficients. This ensures that the predictions remain consistent across different models, even if the specific coefficients differ. These properties highlight the complexities inherent in interpreting LASSO results, particularly in cases with many predictors.[1]

One important consideration is how LASSO selects the best-fitting linear combinations of predictors, which may not necessarily correspond to the best individual predictors. This distinction is particularly important when variables are correlated, as is often the case in observational studies in economics, health, and social sciences, where potential predictors/regressors include powers and interactions of underlying variables. In such scenarios, LASSO prioritizes selecting the optimal combination of predictors rather than isolating the best individual predictors. This nuance is essential when interpreting LASSO results, as the exact variables chosen can vary across repeated samples or different data partitions in k-fold cross-validation.

It's crucial to understand that variable selection in LASSO regression can be approached from two perspectives: identifying the set of true relevant covariates for statistical inference, or using dimension reduction to improve predictive accuracy without necessarily capturing the true model. The former is known as the **oracle property**, where the LASSO estimator consistently selects the true model as the sample size increases. The latter focuses on the model's predictive performance, where the goal is to minimize the prediction error without necessarily identifying the true underlying model. The LASSO does not typically satisfy the oracle property on its own. To address the limitations of LASSO and achieve the oracle property, modifications to the basic LASSO, such as the adaptive LASSO, which is discussed next section, have been proposed. Both perspectives are essential in understanding the role of LASSO regression in practice.[2]

Major limitation of the LASSO's penalization process is its bias relative to OLS, which we will discuss in further text. To address this, a two-step procedure known as "Double LASSO" is often employed. Initially, LASSO regression selects variables by shrinking some coefficients to zero, focusing on the most relevant predictors and simplifying the model. In the second step, an OLS regression is performed on the selected predictors, providing unbiased estimates since it only involves variables deemed significant in the first step, thereby avoiding the multicollinearity typically seen with larger sets. This method, which balances variable selection with precise estimation, will be discussed further in the next chapter. Overall, LASSO is widely used for variable selection in statistical inference within economics, social sciences, and health sciences, though it is less commonly used for pure prediction due to the availability of more effective predictive models.

Let's now examine the foundational aspects of LASSO itself. The LASSO objective function is similar to ridge regression but uses the $L1$ norm of the coefficients instead of the $L2$ norm. The LASSO loss function can be written as:

$$\min_{\boldsymbol{\beta}} \left\{ L(\boldsymbol{\beta}) = \sum_{i=1}^{n} \left(y_i - \beta_0 - \sum_{j=1}^{p} \beta_j x_{ij} \right)^2 + \lambda \sum_{j=1}^{p} |\beta_j| \right\} \tag{12.11}$$

where y_i are the response variable, x_{ij} are the predictor variables, β_j are the coefficients to be estimated. λ is the regularization/shrinkage parameter that controls the strength of the

LASSO penalty. The term $\sum_{j=1}^{p} |\beta_j|$ is the $L1$ norm, which shrinks some coefficients to be exactly zero.

This optimization problem is equivalent to minimizing the sum of squares subject to the constraint $\sum_{j=1}^{p} |\beta_j| \leq s$, where higher values of s correspond to lower values of λ. When $p = 2$, this constraint becomes $|\beta_1| + |\beta_2| \leq s$, which defines the diamond-shaped region commonly illustrated in LASSO figures, such as those in Tibshirani's book or notes widely used to explain LASSO.

The LASSO optimization function is also presented with different scaling factors like $1/2$, $1/n$, or $1/2n$ in front of the loss term. These factors simply adjust the relative scaling of the residual sum of squares term, but the solutions to the optimization problem remain equivalent because the penalty term is unaffected. After standardizing predictors (i.e., centering each predictor to have mean 0 and scaling it to have unit variance), the version of the LASSO optimization problem scaled by $1/2n$ is generally preferred for its mathematical consistency and practical interpretability. All these variations are equivalent, and the function is convex, which guarantees that it always has at least one solution. However, if $p > n$, there may be multiple minimums (see Tibshirani, 2013, or Schneider & Tardivel, 2020, for more details).

For a tuning parameter $\lambda \geq 0$, solving the LASSO optimization problem produces sparse solutions, meaning that some coefficients are exactly zero. Larger values of λ result in sparser solutions. Sparsity is desirable for two key reasons: first, it performs variable selection in the fitted linear model, offering interpretability by identifying which features are most important; second, it often improves prediction when the underlying regression function can be well-approximated by a sparse linear model.

It is common to standardize the predictors before fitting a LASSO regression model to ensure that the regularization penalty is applied uniformly across all variables. Standardizing predictors is a crucial step when applying LASSO because the penalty term $\lambda \sum_{j=1}^{p} |\beta_j|$ treats all predictors equally, regardless of their scale. If the predictors are not standardized, variables with larger scales, such as income measured in thousands, may dominate the penalty term, causing their coefficients to be penalized more heavily and shrunk disproportionately toward zero. Conversely, predictors with smaller scales, such as percentages, might be penalized less and retained in the model even if they are less important.

Standardization ensures that all predictors are on the same scale by centering each variable to have a mean of zero and scaling it to have a standard deviation of one. This equalizes the influence of the predictors on the penalty term, making the LASSO regression fair in its treatment of variables. Without standardization, the model's variable selection process may be biased, leading to suboptimal or misleading results.

Moreover, standardization aligns the predictors with the normalization typically applied to the residual sum of squares in LASSO, particularly in versions of the optimization problem that use scaling by $1/n$ or $1/2n$. This consistency makes the results of LASSO more interpretable and ensures that comparisons across predictors and models are meaningful.

After standardizing the predictors and using the scaled version of the LASSO optimization problem, we encounter a key characteristic of LASSO regression: it does not have a closed-form solution. Unlike ridge or OLS regression, where the estimated coefficients $\hat{\beta}$ can be directly computed using an explicit formula, LASSO introduces an additional layer of complexity.

For ridge and OLS regression, the closed-form solution allows us to plug the data X, γ and the regularization parameter λ into a formula to compute the coefficients in a single step. However, LASSO regression applies an $L1$ penalty, which is the sum of the absolute values of the coefficients. This absolute value term disrupts the straightforward derivation of an explicit formula for the LASSO solution $\hat{\beta}_{Lasso}$, making it impossible to solve directly in one step.

To estimate the LASSO coefficients, we rely on iterative optimization algorithms. These algorithms progressively refine the solution until convergence. Two commonly used algorithms for solving the LASSO optimization problem are **Coordinate Descent** and **Least Angle Regression (LARS)**. While both methods can solve the LASSO problem, they differ in their approach. Coordinate descent focuses on iterative, localized updates to each coefficient, while LARS takes a more global approach by navigating along a solution path that optimally adjusts multiple coefficients in response to residual correlations. Despite its utility in certain contexts, LARS is less commonly used in practice compared to coordinate descent, which has become the standard for most LASSO applications due to its versatility and practical performance. Among these, coordinate descent stands out as the preferred algorithm in modern applications.

12.3 Coordinate Descent for LASSO Regression

The concept of using coordinate descent for LASSO regression was initially introduced by Fu (1998) and later explored by Daubechies et al. (2004). However, it gained significant popularity following its implementation in the glmnet package by Friedman et al. (2007, 2010). Coordinate descent can be applied as a convex optimization method when the convex function can be written as a separable function with two components: one that is convex and smooth, and another that is convex but non-smooth. Specifically, the LASSO loss function comprises two additive components. The first component, the residual sum of squares, is convex, differentiable, and smooth. The second component of the loss term ,$\sum_{j=1}^{p} |\beta_j|$, is convex but non-differentiable.

The LASSO solution is not closed-form (a closed-form solution is an explicit mathematical expression that can be computed directly) like ridge and OLS, but it can be written as:

$$\hat{\beta}_k^{LASSO} = \frac{1}{n}\left[\sum_{i=1}^{n} x_{ik}\left(y_i - \sum_{j\neq k} \tilde{\beta}_j x_{ij}\right) - \lambda\right]_+ \tag{12.12}$$

where $[z]_+$ denotes the **soft-thresholding operator**, which sets $\hat{\beta}_k$ to zero if the penalization term λ is too large compared to the residual correlation. This is why LASSO can shrink some coefficients exactly to zero.

The **coordinate descent algorithm** iteratively updates each coefficient one at a time while keeping the others fixed, repeating the process until the solution converges (i.e., the coefficients change minimally between iterations). The initialization process begins by standardizing the predictors x_j and the response y to have a mean of zero and a variance of one, ensuring all predictors are on the same scale. The initial regularization parameter, λ_{max}, is the smallest value of λ at which all coefficients are zero. As λ decreases geometrically from λ_{max} to λ_{min}, predictors sequentially enter the model based on their correlation strength, allowing the algorithm to explore the regularization path.

Now, let us discuss each step of the algorithm in detail to calculate coefficients β_j in the first iteration. Initially, all coefficients are set to zero, and the first step begins with the residual calculation.

In the first iteration, the residual is simply the response variable, $r = y$, because no predictor contributions are included (all coefficients are zero). For each coefficient β_j, the algorithm computes the correlation between the predictor x_j and $r = y$. Using this correlation, the coefficient is updated using the soft-thresholding operator:

$$\beta_j \leftarrow S\left(\frac{1}{n}\sum_{i=1}^{n} x_{ij} r_i, \lambda_{\max}\right), \tag{12.13}$$

where the soft-thresholding operator is defined as:

$$S(z, \lambda) = \text{sign}(z) \cdot \max(|z| - \lambda, 0). \tag{12.14}$$

Here, $z = \frac{1}{n}\sum_{i=1}^{n} x_{ij} r_i$ represents the scaled correlation between the predictor x_j and the response y, while $\lambda_{\max}$ acts as a threshold. If this correlation is smaller than $\lambda_{\max}$, the coefficient is set to zero, indicating that the predictor x_j does not enter the model in this iteration. Otherwise, β_j takes a nonzero value, reflecting its contribution to the model. (If $|\frac{1}{n}\sum_{i=1}^{n} x_{i1} r_i| \leq \lambda$: The coefficient β_1 is set to zero, meaning x_1 is excluded from the model. If $|\frac{1}{n}\sum_{i=1}^{n} x_{i1} r_i| > \lambda$: The coefficient β_1 is updated to a nonzero value which is the scaled correlation between the predictor and the response.

The initial regularization parameter $\lambda_{\max}$ is calculated as the largest value of λ at which all coefficients are zero. It is defined as:

$\lambda_{\max} = \max_j \left|\frac{1}{n}\sum_{i=1}^{n} x_{ij} y_i\right|.$

This value ensures that the penalty is large enough to suppress all predictors, serving as the starting point for the algorithm. After calculating $\lambda_{\max}$, the algorithm proceeds to compute coefficients for all predictors $\beta_1, \beta_2, \ldots, \beta_n$ using the same residual $r = y$ and the soft-thresholding operator.

At the end of the first iteration, some coefficients may remain zero if their correlation with y is insufficient to overcome the threshold set by $\lambda_{\max}$. The result is an initial sparse solution, where only strongly correlated predictors have nonzero coefficients.

In subsequent iterations, the regularization parameter λ is decreased according to the sequence:

$\lambda_k = \lambda_{\max} \times \left(\frac{\lambda_{\min}}{\lambda_{\max}}\right)^{\frac{k-1}{\text{nlambda}-1}}$, where $\lambda_{\min}$ is a fraction of $\lambda_{\max}$, controlled by lambda.min.ratio (default: 0.0001 for $p \leq n$ or 0.01 for $p > n$), and nlambda is the number of values in the sequence (default: 100).

As λ decreases, the residual is recalculated to incorporate contributions from nonzero coefficients from the previous step as a "warm start":

$$r = y - \sum_{j=1}^{p} x_j \beta_j^{(\text{previous})} \tag{12.15}$$

This updated residual reflects the remaining variation in y after accounting for predictors already included in the model. For each predictor, the coefficient is updated iteratively using the soft-thresholding operator with the new λ_k:

$$\beta_j \leftarrow S\left(\frac{1}{n}\sum_{i=1}^{n} x_{ij} r_i, \lambda_k\right) \tag{12.16}$$

The process continues iteratively for all coefficients and all values of λ_k, progressively building the regularization path.

The algorithm stops when the coefficients converge, as determined by a stopping criterion. Convergence is achieved when the relative change in the coefficients between iterations falls below a predefined tolerance level: $\frac{\|\beta^{(t)}-\beta^{(t-1)}\|}{\|\beta^{(t-1)}\|} < tolerance$. The default tolerance in glmnet is 10^{-4}, but it can be adjusted using the thresh parameter. This ensures that the algorithm produces a stable and reliable solution for each value of λ_k.

The coordinate descent method is particularly effective for LASSO regression due to its efficiency and suitability for high-dimensional data. By updating coefficients one at a time through an iterative process, it remains computationally efficient even with large numbers of predictors. The method leverages the soft-thresholding rule to identify the most relevant predictors while shrinking others to zero, producing a sparse solution that balances model complexity and predictive accuracy. Although coordinate descent delivers solutions for specific λ values rather than the entire solution path, its simplicity, efficiency, and adaptability make it a practical tool for LASSO and other regularization methods like elastic net, especially in data analysis where direct solutions are unavailable.

12.4 Least Angle Regression for LASSO

LARS incrementally builds the solution path by sequentially adding predictors and adjusting coefficients based on their correlation with the residuals. It is particularly efficient when the number of predictors (ρ) is large, making it a valuable tool for understanding the full solution path of LASSO regression. While LARS was originally designed for linear regression, it can be modified to solve the LASSO problem by introducing a LASSO constraint.

The LARS algorithm begins with all coefficients initialized to zero, ensuring that no predictors are included in the model initially. The predictors x_j and the response y are first standardized to have mean zero and variance one, ensuring all predictors are on the same scale. Once standardized, the coefficients are set to zero, and the residuals are initialized as $r = y$. At this stage, the algorithm proceeds to identify the most correlated predictor by computing the correlation between each predictor x_j and the residuals r as $c_j = x_j^T r$. The predictor with the highest absolute correlation, x_k, is selected since it has the strongest relationship with the unexplained variance in y.

After selecting the most correlated predictor, the algorithm moves along the least angle direction, which ensures that the residuals are minimized in the direction of x_k. The term "least angle regression" refers to the property of moving in the direction that forms the smallest angle with the residual vector. To achieve this, the active set of predictors is defined, initially containing only x_k. The active predictors, represented as X_A, and their corresponding coefficients, β_A, are used to calculate the direction vector $u_A = X_A(X_A^T X_A)^{-1} s_A$, where $s_A = \text{sign}(X_A^T r)$ captures the signs of the correlations of the active predictors with the residuals. The algorithm then computes the step size γ to determine how far to move along this direction before another predictor becomes equally correlated with the residuals:

$$\gamma = \min_{j \notin A} \left\{ \frac{C - c_j}{1 - a_j}, \frac{C + c_j}{1 + a_j} \right\},$$

where $C = \max_j |c_j|$ is the maximum absolute correlation, $c_j = x_j^T r$ is the correlation of predictor x_j with the residuals, and $a_j = x_j^T u_A$ is the projection of x_j onto the direction vector u_A. The coefficients of the active predictors are updated as $\beta_A \leftarrow \beta_A + \gamma u_A$, and the residuals are recalculated as $r \leftarrow r - \gamma u_A$. If another predictor becomes equally correlated with the residuals, it is added to the active set, and the process is repeated.

To calculate the initial value of λ, which determines the LASSO constraint, the algorithm computes the maximum absolute correlation of the predictors with the response variable. This value, referred to as $\lambda_{\max}$, is given by: $\lambda_{\max} = \max_j |x_j^T y|$, where $x_j^T y$ represents the correlation of predictor x_j with the response y. This ensures that at $\lambda = \lambda_{\max}$, all coefficients are exactly zero, providing a starting point for the solution path. The corresponding threshold t for the sum of absolute coefficients is initialized as $t = 0$ since no predictors are active at this stage.

To enforce the LASSO constraint during the algorithm, the sum of the absolute values of the coefficients, denoted as $S = \sum_{j=1}^{p} |\beta_j|$, is monitored. When this sum exceeds the threshold determined by the regularization parameter λ, it triggers a re-scaling of the coefficients to ensure compliance with the LASSO constraint. The value of δ, representing the amount of shrinkage required, is calculated as: $\delta = \frac{S-t}{|A|}$, where $|A|$ is the number of active predictors, i.e., predictors with nonzero coefficients. The coefficients are then updated as:

$$\beta_j \leftarrow \text{sign}(\beta_j) \cdot \max(0, |\beta_j| - \delta),$$

which proportionally reduces the magnitude of the coefficients and sets some coefficients to zero if their absolute value is less than δ. This ensures that the total sum of the coefficients satisfies the constraint $S \leq t$. In practical terms, λ determines the degree of regularization: higher values of λ correspond to more stringent constraints (smaller t), which shrink coefficients more aggressively and encourage sparsity. Conversely, as λ decreases, t increases, allowing more coefficients to take nonzero values and thus introducing additional predictors into the model.

The LARS algorithm updates the solution path by iteratively adding new predictors to the active set and moving their coefficients in the least angle direction while maintaining the LASSO constraint. This process continues until all predictors are included or the regularization parameter λ reaches zero. The algorithm produces the entire solution path, which tracks how coefficients evolve as λ decreases. Convergence is achieved when the coefficients stabilize, ensuring a reliable solution for each value of λ.

A key feature of LARS is its efficiency in computing the solution path for LASSO, providing valuable insights into the effects of regularization on model selection and sparsity. The threshold imposed on the sum of the absolute values of the coefficients is directly linked to λ, and as λ decreases, the threshold increases, allowing more predictors to enter the model and their coefficients to grow. By adapting the standard LARS algorithm with the LASSO constraint, the method ensures a sparse and interpretable solution, making it a powerful tool for addressing high-dimensional regression problems.

12.5 Differences Between Coordinate Descent and LARS

Here are the main differences between coordinate descent and LARS in their approach to solving LASSO regression problems. First, coordinate descent uses an iterative optimization approach where each coefficient is updated one at a time while keeping the others fixed.

This is done using the soft-thresholding operator applied to each predictor individually. In contrast, LARS builds the entire solution path in a geometric manner, incrementally adding predictors to the active set and updating coefficients of all active predictors simultaneously.

Second, in enforcing the LASSO constraint, coordinate descent uses implicit soft-thresholding to shrink coefficients. This approach ensures sparsity by setting coefficients to zero when their absolute value falls below the penalization threshold. The update for each coefficient in coordinate descent is performed as:

$$\beta_j \leftarrow \text{sign}\left(\frac{1}{n}\sum_{i=1}^{n} x_{ij} r_i\right) \cdot \max\left(0, \left|\frac{1}{n}\sum_{i=1}^{n} x_{ij} r_i\right| - \lambda\right)$$

On the other hand, LARS explicitly monitors the sum of absolute coefficients, denoted as $S = \sum_{j=1}^{p} |\beta_j|$, and ensures that this sum does not exceed a threshold t determined by the regularization parameter λ. When S exceeds t, LARS applies proportional shrinkage across all active coefficients to restore compliance with the constraint. The amount of shrinkage, δ, is calculated as: $\delta = \frac{S-t}{|A|}$, where $|A|$ is the number of active predictors. The coefficients are then updated as:

$$\beta_j \leftarrow \text{sign}(\beta_j) \cdot \max(0, |\beta_j| - \delta).$$

This explicit enforcement ensures that the total sum of the absolute coefficients satisfies the constraint $S \leq t$, while selectively shrinking or zeroing out coefficients that fail to meet the penalization criteria.

Third, the selection of predictors differs significantly. Coordinate descent cycles through all predictors sequentially and updates each one iteratively until convergence. In contrast, LARS selects the predictor most correlated with the residuals at each step and dynamically adds it to the active set when its correlation equals that of the current active predictors.

Fourth, coordinate descent does not build the entire solution path but instead computes solutions for specific values of λ. On the other hand, LARS explicitly constructs the entire solution path, showing how coefficients evolve as λ decreases.

Fifth, in terms of computational efficiency, coordinate descent is more scalable for high-dimensional datasets ($p \gg n$), as it updates one coefficient at a time without requiring matrix operations. LARS, while efficient for small to moderate datasets ($p \leq n$), involves matrix inversions for updating active predictors, which can be computationally intensive for large p.

Sixth, LARS can suffer from numerical instability in the presence of highly collinear predictors due to its reliance on matrix operations. Coordinate descent, by avoiding matrix inversion, is more stable in such scenarios.

Finally, the key use cases differ. Coordinate descent is better suited for scenarios where the focus is on high-dimensional data or when solutions for specific λ values are needed, such as in cross-validation. LARS is more appropriate for moderate-sized datasets where understanding the full regularization path and predictor selection process is important.

12.6 Implementation and Simulation of LASSO Regression

To begin the LASSO regression simulation, the first step is **Data Preparation**. This involves centering and standardizing the predictor variables to ensure that all features are on the same scale, allowing the LASSO penalty to be applied consistently across them. Additionally, if the intercept term is omitted from the model, the response variable should also be centered to maintain consistency with the standardized predictors.

Next is the **Parameter Selection** process. Here, we define a grid of λ values to explore. To determine the optimal λ, k-fold cross-validation is employed. This method evaluates the model's performance for each value of λ by dividing the data into training and validation sets multiple times. The λ that minimizes the cross-validation error is chosen as the optimal regularization parameter.

Once the optimal λ has been determined, the **Model Fitting** step begins. Using the selected λ, the LASSO regression coefficients are computed by applying the regularization penalty. Unlike ridge regression, LASSO can shrink some coefficients exactly to zero, making it an effective method for feature selection.

Finally, in the **Model Evaluation** phase, the predictive performance of the LASSO regression model is assessed. This can be done by evaluating its performance on a separate test set or through additional cross-validation. The goal is to compare the LASSO regression model's ability to generalize to new data and analyze which variables were selected by the model.

In the following LASSO regression simulation in R, we generate synthetic data, fit a LASSO regression model using the `glmnet` package, and evaluate its performance by comparing it to OLS results. This step-by-step approach demonstrates how LASSO regression can be applied in practice to perform variable selection and improve predictive accuracy.

```r
# Load the required package
library(glmnet)

# Step 1: Data Preparation
set.seed(123) # For reproducibility
n <- 100       # Number of observations
p <- 10        # Number of predictors
X <- matrix(rnorm(n * p), nrow = n, ncol = p)  # Generate random predictors
beta <- c(3, -2, 0, 0, 1.5, 0, 0, 0, 0, 0)    # True coefficients
y <- X %*% beta + rnorm(n)                     # Generate response with some noise

# Split data into training and testing sets
set.seed(456)
train_idx <- sample(1:n, size = 0.8 * n, replace = FALSE) # 80% training data
X_train <- X[train_idx, ]
X_test <- X[-train_idx, ]
y_train <- y[train_idx]
y_test <- y[-train_idx]

# Step 3: Fit a Lasso regression model
# Lasso regression with alpha = 1 (L1 regularization)
```

```
lasso_model <- glmnet(X_train, y_train, alpha = 1)

# Step 4: Cross-validation to find the best lambda
cv_lasso <- cv.glmnet(X_train, y_train, alpha = 1)

# Step 5: Get the best lambda value
best_lambda <- cv_lasso$lambda.min
cat("Best lambda for Lasso regression:", best_lambda, "\n")
```

```
## Best lambda for Lasso regression: 0.1020998
```

```
# Step 6: Fit the final Lasso model with the best lambda
lasso_final <- glmnet(X_train, y_train, alpha = 1, lambda = best_lambda)

# Step 7: Make predictions on the test set
y_pred_lasso <- predict(lasso_final, s = best_lambda, newx = X_test)

# Step 8: Evaluate the model performance
mse_lasso <- mean((y_test - y_pred_lasso)^2)
cat("Mean Squared Error for Lasso regression on test set:", mse_lasso, "\n")
```

```
## Mean Squared Error for Lasso regression on test set: 1.19065
```

```
# Step 9: Compare with Ordinary Least Squares (OLS)
ols_model <- lm(y_train ~ X_train)
y_pred_ols <- predict(ols_model, newdata = data.frame(X_train = X_test))
mse_ols <- mean((y_test - y_pred_ols)^2)
cat("Mean Squared Error for OLS on test set:", mse_ols, "\n")
```

```
## Mean Squared Error for OLS on test set: 34.29426
```

```
# Step 10: Plot the cross-validation results
# plot(cv_lasso)
```

1. *Data generation*: We begin by generating a synthetic dataset to simulate the application of LASSO regression. The dataset consists of 100 observations and 10 predictors (X), generated as random normal variables. The response variable (y) is defined as a linear combination of a subset of predictors with predefined coefficients. Additional random noise is added to y to reflect real-world variability. This setup allows us to test LASSO regression's ability to identify relevant predictors and reduce the impact of irrelevant ones.

2. *Train-Test Split*: The dataset is split into training and testing subsets to evaluate the model's performance on unseen data. Specifically, 80% of the data (80 observations) is used for training, and the remaining 20% is held out for testing. The split ensures that the model is trained on one subset and validated on another, mimicking real-world scenarios where future data is not available during model fitting.

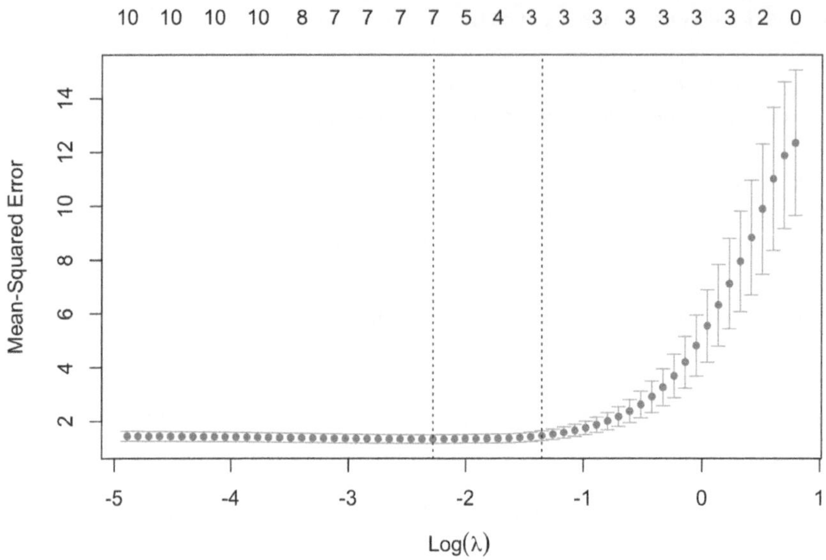

FIGURE 12.2 Cross-validation results for LASSO regression: optimal lambda selection.

3. *Fitting LASSO Regression*: The `glmnet()` function is used to fit the Lasso regression model on the training data. By setting `alpha = 1`, we specify Lasso regression, which applies an $L1$ penalty to the coefficients. This penalty encourages sparsity by shrinking irrelevant coefficients to zero. The fitting process employs **coordinate descent**, an iterative optimization algorithm. At each iteration, one coefficient is updated while holding others fixed, using the soft-thresholding operator as explained above.

4. *Cross-validation*: To determine the optimal value of λ, we use the `cv.glmnet()` function to perform 10-fold cross-validation. The training dataset is divided into 10 equally sized folds. In each fold, 90% of the data is used for training, and the remaining 10% serves as the validation set. For each fold, the model is trained on a predefined sequence of λ values, ranging from λ_{max} (the largest λ at which all coefficients are zero) to a minimum value λ_{min}, with a default of 100 λ values in total (as discussed above, we explained how λ_{max} and the other λ values are calculated). The cross-validation error (e.g., Mean Squared Error) is computed for each λ as the average validation error across all folds:

$$\text{CV Error}(\lambda) = \frac{1}{k}\sum_{i=1}^{k} \text{MSE}_i(\lambda)$$

where $k = 10$ is the number of folds, and $MSE_i(\lambda)$ is the validation error for fold i.

The optimal λ (i.e., the minimum λ) is the value of λ that minimizes the average cross-validation error. Additionally, a less regularized λ, λ_{1SE}, may also be selected. This is the largest λ within one standard error of the minimum error, favoring simpler models.

5. *Final LASSO Model*: Using the optimal λ obtained from cross-validation, we fit the final Lasso model on the training data. This step ensures that the model

incorporates the most relevant predictors while penalizing irrelevant ones. Predictions are then made on the test set using the final model.

6. *Model Evaluation*: The MSE is calculated on the test set to evaluate the predictive performance of the Lasso regression model:

$$\text{MSE} = \frac{1}{n_{\text{test}}} \sum_{i=1}^{n_{\text{test}}} (y_i - \hat{y}_i)^2$$

where n_{test} is the number of test observations, y_i is the actual response, and $\hat{y}_i$ is the predicted response. To provide a baseline, we also fit an OLS model and calculate its test set MSE. This comparison highlights Lasso's ability to improve predictive accuracy by addressing overfitting and reducing variance.

7. *Cross-validation Plot*: Finally, we visualize the cross-validation results using the `plot()` function for the `cv.glmnet()` object. This plot shows the cross-validated error for each λ value, with the optimal λ values (λ_{min} and λ_{1SE}) marked. The visualization aids in understanding how the choice of λ impacts model performance. Above each vertical bar, the numbers indicate the number of nonzero coefficients in the model at that specific λ. For LASSO regression, this number typically decreases as λ increases, reflecting how LASSO performs variable selection by shrinking some coefficients exactly to zero.

12.7 The Relationship Between the LASSO and OLS Coefficient Estimates

The relationship between the LASSO coefficient estimates, denoted as $\hat{\beta}^{\text{Lasso}}$, and the OLS coefficient estimates, $\hat{\beta}^{\text{OLS}}$, is determined by the shrinkage and sparsity induced by the LASSO penalty. While OLS estimates are obtained by minimizing the residual sum of squares without any penalty, LASSO incorporates an $L1$-norm penalty, which modifies the OLS estimates by shrinking them toward zero and, in some cases, setting them exactly to zero.

For each coefficient β_j, the LASSO estimate can be expressed as:

$$\hat{\beta}_j^{\text{Lasso}} = \text{sign}(\hat{\beta}_j^{\text{OLS}}) \cdot \max\left(0, |\hat{\beta}_j^{\text{OLS}}| - \lambda\right) \tag{12.17}$$

where $\text{sign}(\hat{\beta}_j^{\text{OLS}})$ denotes the sign of the OLS estimate, $|\hat{\beta}_j^{\text{OLS}}|$ is its absolute value, and λ is the regularization parameter controlling the amount of shrinkage applied. This equation highlights two important behaviors of LASSO regression: shrinkage and sparsity.

Shrinkage refers to the reduction in the magnitude of the coefficients compared to their OLS counterparts. When $|\hat{\beta}_j^{\text{OLS}}| > \lambda$, LASSO shrinks the coefficient by a fixed amount equal to λ, while preserving its sign. This behavior reduces the variance of the estimates and improves model generalizability, particularly in high-dimensional settings or when multicollinearity is present.

Sparsity is introduced by LASSO when $|\hat{\beta}_j^{\text{OLS}}| \leq \lambda$. In such cases, the penalty is large enough to set the corresponding LASSO coefficient exactly to zero, effectively excluding the predictor from the model. This property makes LASSO a powerful tool for variable selection, as it identifies and retains only the most relevant predictors, leading to simpler and more interpretable models.

The parameter λ plays a central role in controlling the balance between model complexity and predictive performance. When $\lambda = 0$, the LASSO estimates reduce to the OLS estimates, as no penalty is applied. As λ increases, the penalty grows, leading to greater shrinkage and more coefficients being set to zero. For very large values of λ, all coefficients may be shrunk to zero, leaving only the intercept term in the model.

The relationship between LASSO and OLS also highlights the trade-off between bias and variance. OLS estimates are unbiased but may have high variance, particularly when the predictors are highly correlated or the number of predictors (p) exceeds the number of observations (n). In contrast, LASSO introduces bias by shrinking coefficients, but this bias often leads to a substantial reduction in variance, resulting in a lower overall mean squared error (MSE) in prediction.

The intuition behind this relationship lies in LASSO's ability to impose a preference for simpler models with fewer predictors. By shrinking small coefficients and removing irrelevant predictors, LASSO achieves a balance between capturing the underlying signal in the data and avoiding overfitting. This makes it particularly effective in high-dimensional datasets, where traditional OLS may fail due to overfitting or multicollinearity.

In order to correct the bias introduced by LASSO relative to OLS, a common approach is to employ a two-step LASSO-OLS procedure. In this method, LASSO regression is first used to select a subset of relevant variables by leveraging its sparsity-inducing properties. Once the variable selection step is completed, a least squares estimator is applied to the selected variables to obtain unbiased coefficient estimates. This approach effectively separates the tasks of variable selection and coefficient estimation, mitigating the bias induced by the LASSO penalty in the final estimates. The theoretical properties and advantages of this two-step procedure have been extensively studied, particularly in Belloni and Chernozhukov (2013), who demonstrate its consistency and efficiency in high-dimensional settings.

Another widely used alternative to address the bias is the application of weighted versions of LASSO methods based on iterative schemes. A prominent example is the adaptive LASSO, introduced by Zou (2006) and further studied by Huang et al. (2008) and Van de Geer et al. (2011). The adaptive LASSO modifies the standard LASSO penalty by assigning different weights to each predictor in the penalization term. Specifically, predictors with larger estimated coefficients in an initial step are penalized less, while those with smaller coefficients are penalized more heavily. These weights are updated iteratively, recalibrating the penalization in every step until convergence. This iterative process ensures that the final solution incorporates both the variable selection and estimation accuracy, addressing some of the limitations of standard LASSO.

In summary, LASSO modifies OLS estimates by applying shrinkage and sparsity through the $L1$-penalty. The resulting coefficients balance predictive accuracy and model simplicity, making LASSO a robust and interpretable alternative to OLS, especially in complex or high-dimensional data settings.

12.8 LASSO: Key Takeaways

LASSO regression is widely implemented in various statistical software packages, making it accessible for practitioners and researchers. In R, the `glmnet` package provides a robust implementation of LASSO regression using coordinate descent. This package allows users to specify a grid of λ values and select the optimal value using cross-validation, offering flexibility and efficiency. In Python, the `scikit-learn` library includes LASSO regression as part of its suite of linear models, also employing coordinate descent for optimization. This makes it straightforward to apply LASSO regression within Python's versatile machine learning framework. Similarly, in Stata, the `LASSO` command implements LASSO regression using coordinate descent, delivering an efficient solution for high-dimensional datasets within the familiar Stata environment. These software tools provide powerful and user-friendly methods for applying LASSO regression, ensuring its practicality for a wide range of data analysis tasks.

LASSO regression can be extended and adapted to address various challenges and improve its applicability. One such extension is elastic net regression, which combines the $L1$ and $L2$ penalties, effectively mitigating limitations of LASSO when dealing with highly correlated predictors. Another variation is the adaptive LASSO, which assigns adaptive weights to the penalty terms, potentially improving the consistency of variable selection. Additionally, the group LASSO extends the standard LASSO penalty to select groups of variables together, making it particularly useful in settings where predictors naturally form groups. These extensions expand the utility of LASSO regression in diverse analytical contexts. The adaptive LASSO and elastic net will be discussed in more detail in the sections below.

In conclusion, LASSO regression is a versatile and powerful method for addressing overfitting and performing variable selection in high-dimensional datasets. By penalizing the absolute size of the coefficients, LASSO encourages sparsity, resulting in simpler and more interpretable models. Whether implemented through coordinate descent or LARS, it has become a standard tool for statisticians and data scientists. LASSO excels in automatic variable selection by shrinking irrelevant coefficients to zero, improving prediction accuracy by reducing variance through penalization, and creating sparse models that are easier to interpret. Despite its strengths, Lasso has some limitations. The penalization introduces bias in the coefficient estimates, and its selection consistency can be affected when predictors are highly correlated. Furthermore, while efficient algorithms like coordinate descent and LARS have made it computationally feasible, Lasso can face challenges with very large datasets. Nevertheless, its flexibility and utility in creating interpretable and effective models make it a cornerstone of modern statistical and machine learning practices.

The non-uniqueness of the LASSO solution underscores the importance of understanding the underlying assumptions and conditions that influence the selection of variables and the stability of the model across different datasets. In the following section, we will introduce the adaptive LASSO estimator. This variant of the traditional LASSO uses weights to adjust the penalty applied to each coefficient based on an initial estimate, typically from an ordinary least squares regression. This weighting approach improves the estimator's ability to distinguish between significant and non-significant predictors, thereby improving the adaptive LASSO's oracle properties—the estimator's ability to consistently select the true model as the sample size increases. That means it can perform as well as if the true underlying model were known in advance. In such high-dimensional settings, where LASSO might choose different sets of variables in different samples, considering the adaptive LASSO can be particularly crucial for robust statistical inference and practical applications.

12.9 Adaptive LASSO

The adaptive Lasso, introduced by Zou (2006), builds on the standard LASSO regression, which was covered in detail in the previous section. It modifies the L_1-penalty by incorporating adaptive weights for each predictor, allowing different coefficients to be penalized with varying strengths. This modification enables the adaptive LASSO to address limitations of the standard LASSO, particularly in high-dimensional data or when predictors are correlated.

Standard LASSO applies a uniform penalty to all coefficients, which can lead to suboptimal variable selection by over-penalizing important predictors and under-penalizing less significant ones. The adaptive LASSO overcomes these issues by introducing weights that adjust the penalty applied to each coefficient, improving variable selection consistency and reducing bias.

The adaptive LASSO modifies the standard LASSO loss function by introducing adaptive weights to the penalty term. The loss function is given by:

$$\min_{\boldsymbol{\beta}} \left\{ L(\boldsymbol{\beta}) = \sum_{i=1}^{n} \left(y_i - \beta_0 - \sum_{j=1}^{p} \beta_j x_{ij} \right)^2 + \lambda \sum_{j=1}^{p} w_j |\beta_j| \right\} \tag{12.18}$$

where w_j are the adaptive weights, typically defined as inversely proportional to the initial coefficient estimates. These weights allow for variable-specific penalization, improving variable selection and reducing bias while retaining the sparsity-inducing property of the $L1$-norm.

A key property of the adaptive LASSO is its ability to achieve consistency in variable selection, meaning that as the sample size increases, it correctly identifies the true set of nonzero coefficients with high probability. This consistency is a result of the adaptive LASSO's weighted penalty, which allows it to more effectively distinguish between relevant and irrelevant predictors. Additionally, the adaptive LASSO satisfies the oracle property under certain regularity conditions. This property implies that it performs as well as if the true underlying model were known in advance, achieving asymptotic unbiasedness (the estimated nonzero coefficients converge to their true values as the sample size grows) and efficient estimation for the nonzero coefficients (the estimates for the nonzero coefficients achieve the minimum possible variance, as given by the Cramér-Rao lower bound).[3]

Another significant advantage of the adaptive LASSO is its ability to reduce bias. By assigning smaller penalties to larger coefficients, it mitigates the bias inherent in standard LASSO estimates, leading to more accurate coefficient estimation. This property is particularly valuable in high-dimensional settings where over-penalization can compromise the model's performance.

In addition to reducing bias, the adaptive LASSO ensures reliable inference by achieving asymptotic normality for the estimated nonzero coefficients.[4] This property ensures that confidence intervals and hypothesis tests for the nonzero coefficients can be constructed using standard methods, making the adaptive LASSO a reliable tool for high-dimensional data analysis. The asymptotic normality further highlights the Adaptive LASSO's robustness and practical utility in applications where accurate estimation and inference are critical.

While the adaptive LASSO offers significant advantages, including improved variable selection and reduced bias, its performance depends heavily on the quality of the initial coefficient estimates. Poor initial estimates can compromise the model's effectiveness. Additionally, the computational effort required to compute weights and solve the optimization problem is greater than that for standard LASSO. The selection of the parameter γ, which determines the adaptiveness of the weights, adds another layer of complexity, requiring careful tuning for optimal performance.

The core innovation of the adaptive LASSO lies in assigning different penalty weights to each predictor. The weighted penalty term can be written as:

$$\lambda \sum_{j=1}^{p} w_j |\beta_j|$$

where w_j are the adaptive weights, typically chosen as the reciprocal of the absolute value of the initial estimates of the coefficients:

$$w_j = \frac{1}{|\hat{\beta}_j|^\gamma}$$

with $\gamma > 0$ controlling the degree of adaptiveness. This weighting ensures that predictors with larger initial coefficients are penalized less, increasing the likelihood of retaining important variables in the model.

12.10 Steps in Adaptive LASSO

The adaptive LASSO begins by obtaining initial coefficient estimates, typically using methods such as OLS, ridge regression, or standard LASSO. These initial estimates serve as the basis for calculating the penalty weights. Each weight is computed as the reciprocal of the absolute value of the corresponding initial estimate, scaled by a parameter γ to control the level of adaptiveness.

The choice of weights w_j is a critical component of the adaptive LASSO. Typically, the weights are inversely proportional to the magnitude of the initial coefficient estimates, defined as $w_j = \frac{1}{|\hat{\beta}_j^{\text{initial}}|^\gamma}$, where γ determines the degree of adaptiveness. Initial estimates can be obtained through various methods, such as OLS for smaller-dimensional data, ridge regression when multicollinearity is present, or standard LASSO in cases where sparsity is expected. The parameter γ controls the influence of the weights; smaller values of γ reduce the adaptiveness, making the adaptive LASSO more similar to standard LASSO, while larger values emphasize the differences among predictors based on their initial estimates.

The choice of weights w_j is a critical component of the adaptive LASSO. Typically, the weights are inversely proportional to the magnitude of the initial coefficient estimates, defined as $w_j = \frac{1}{|\hat{\beta}_j^{\text{initial}}|^\gamma}$, where γ determines the degree of adaptiveness. Initial estimates can be obtained through various methods, such as OLS for smaller-dimensional data, Ridge regression when multicollinearity is present, or standard LASSO in cases where sparsity is expected. An alternative approach to constructing adaptive weights is stability selection, which improves robustness in high-dimensional settings by leveraging repeated sampling to identify stable predictors.[5] The parameter γ controls the influence of the weights; smaller values of γ reduce the adaptiveness, making the adaptive LASSO more similar to standard LASSO, while larger values emphasize the differences among predictors based on their initial estimates.

The process of choosing γ involves both theoretical and data-driven considerations. Zou (2006) recommends values of $\gamma > 0$, with $\gamma = 1$ being a common choice as it balances adaptiveness without overly amplifying or diminishing the influence of the initial estimates. Alternatively, cross-validation can be employed to select the optimal γ.[6] By evaluating the model's performance across a range of γ values, the γ that minimizes the validation error can be selected. Domain-specific knowledge may also guide the choice of γ; for example, smaller values might be preferred when predictors are less correlated, while larger values can help in situations where multicollinearity is significant. In practice, γ is often fixed at 1 unless specific considerations suggest otherwise. This parameter plays a crucial role in determining the adaptiveness of the penalty and, consequently, the overall performance of the adaptive LASSO.

Once the weights are determined, the adaptive LASSO optimization problem is solved. This involves minimizing the residual sum of squares while applying the weighted penalty term. The optimization problem is expressed as:

$$\hat{\beta}^{\text{Adaptive}} = \arg\min_{\beta} \left\{ \frac{1}{2n} \sum_{i=1}^{n} \left(y_i - \sum_{j=1}^{p} x_{ij}\beta_j \right)^2 + \lambda \sum_{j=1}^{p} w_j |\beta_j| \right\} \tag{12.19}$$

Numerical techniques, such as coordinate descent, are commonly used to solve this weighted optimization problem iteratively. The solution must satisfy the Karush–Kuhn–Tucker (KKT) conditions, which characterize the optimality of the adaptive LASSO solution.[7]

The adaptive LASSO solution, like standard LASSO, does not have a closed-form expression. However, it can be written as an iterative update rule that incorporates adaptive weights into the L_1-penalty. For a given predictor k, the adaptive LASSO coefficient update can be expressed as:

$$\hat{\beta}_k^{\text{Adaptive}} = \frac{1}{n} \left[\sum_{i=1}^{n} x_{ik} \left(y_i - \sum_{j \neq k} \tilde{\beta}_j x_{ij} \right) - \lambda w_k \right]_+$$

where $w_k = \frac{1}{|\hat{\beta}_k^{\text{initial}}|^{\gamma}}$ are the adaptive weights, typically derived from initial coefficient estimates such as OLS, ridge, or LASSO, $\gamma > 0$ controls the degree of adaptiveness, with larger values emphasizing the differences among predictors, λ is the regularization parameter controlling the overall strength of the penalization, $[z]_+ = \max(0, z)$ denotes the soft-thresholding operator that ensures sparsity by shrinking some coefficients to zero.

This formulation emphasizes how the adaptive weights adjust the penalty applied to each coefficient. Predictors with larger initial coefficients (and smaller weights) are penalized less, improving variable selection consistency and reducing bias compared to standard LASSO. The iterative nature of the solution allows the adaptive LASSO to fine-tune the penalization for each predictor based on its importance.

12.11 Implementation and Simulation of Adaptive LASSO

The adaptive LASSO simulation begins with **Data Preparation**, where synthetic data is generated and divided into training and testing sets. Predictor variables are standardized to ensure consistent application of penalties across coefficients, and any necessary centering is performed.

Next is **Initial Weight Estimation**, achieved through ridge regression. The coefficients from the Ridge model are used to compute adaptive weights, defined as the inverse of the absolute Ridge coefficients. These weights guide the penalization in the Adaptive LASSO model, ensuring differential regularization.

In the **Model Fitting** step, the adaptive LASSO is applied using the computed weights. Cross-validation determines the optimal regularization parameter (λ) by minimizing prediction error over multiple folds. This parameter is then used to fit the final model.

Finally, the **Model Evaluation** phase compares the adaptive LASSO's performance, assessed via MSE, against LASSO and OLS models. This evaluation highlights the advantages of adaptive LASSO in balancing feature selection and prediction accuracy.

```
# Load necessary libraries
library(glmnet)

# 1. Generate synthetic data (Updated to match the second snippet)
set.seed(123)  # For reproducibility
n <- 100       # Number of observations
p <- 10        # Number of predictors

# Updated coefficients and response generation
X <- matrix(rnorm(n * p), nrow = n, ncol = p)  # Generate random predictors
beta <- c(3, -2, 0, 0, 1.5, 0, 0, 0, 0, 0)    # True coefficients
y <- X %*% beta + rnorm(n)                     # Generate response with noise

# 2. Split data into training and testing sets
set.seed(456)
train_idx <- sample(1:n, size = 0.8 * n, replace = FALSE)  # 80% training data
X_train <- X[train_idx, ]
X_test <- X[-train_idx, ]
y_train <- y[train_idx]
y_test <- y[-train_idx]

# 3. Estimate initial weights using Ridge regression (L2 regularization)
ridge_model <- glmnet(X_train, y_train, alpha = 0)  # Ridge with alpha = 0
ridge_coefs <- coef(ridge_model, s = 0.1)[-1]       # Get Ridge coefficients

# 4. Calculate the weights for Adaptive Lasso
weights <- 1 / abs(ridge_coefs)              # Inverse of Ridge coefficients
weights[is.infinite(weights)] <- 1e10 # Assign large weights if division by zero

# 5. Fit Adaptive Lasso model using glmnet with custom penalty factors
adaptive_lasso_model <- glmnet(X_train, y_train, alpha = 1, penalty.factor = weights)

# 6. Cross-validation to find the best lambda
cv_adaptive_lasso <- cv.glmnet(X_train, y_train, alpha = 1, penalty.factor = weights)
```

```
# 7. Get the best lambda value
best_lambda <- cv_adaptive_lasso$lambda.min
cat("Best lambda for Adaptive Lasso:", best_lambda, "\n")
```

```
## Best lambda for Adaptive Lasso: 0.5505076
```

```
# 8. Fit the final Adaptive Lasso model with the best lambda
adaptive_lasso_final <- glmnet(X_train, y_train, alpha = 1,
                       lambda = best_lambda, penalty.factor = weights)

# 9. Make predictions on the test set
y_pred_adaptive_lasso <- predict(adaptive_lasso_final,s=best_lambda,newx=X_test)

# 10. Evaluate the model performance
mse_adaptive_lasso <- mean((y_test - y_pred_adaptive_lasso)^2)
cat("Mean Squared Error for Adaptive Lasso on test set:", mse_adaptive_lasso, "\n")
```

```
## Mean Squared Error for Adaptive Lasso on test set: 1.129094
```

```
# 11. Compare with Lasso and OLS
# Lasso Model (without adaptive weights)
cv_lasso <- cv.glmnet(X_train, y_train, alpha = 1)
best_lambda_lasso <- cv_lasso$lambda.min
lasso_final <- glmnet(X_train, y_train, alpha = 1, lambda = best_lambda_lasso)
y_pred_lasso <- predict(lasso_final, s = best_lambda_lasso, newx = X_test)
mse_lasso <- mean((y_test - y_pred_lasso)^2)
cat("Mean Squared Error for Lasso on test set:", mse_lasso, "\n")
```

```
## Mean Squared Error for Lasso on test set: 1.156458
```

```
# OLS Model
ols_model <- lm(y_train ~ X_train)
y_pred_ols <- predict(ols_model, newdata = data.frame(X_train = X_test))
mse_ols <- mean((y_test - y_pred_ols)^2)
cat("Mean Squared Error for OLS on test set:", mse_ols, "\n")
```

```
## Mean Squared Error for OLS on test set: 34.29426
```

```
# 12. Plot the cross-validation results for Adaptive Lasso
#plot(cv_adaptive_lasso)
```

1. *Data Generation*: Synthetic data with 100 observations and 10 predictors (X) is generated, where each predictor follows a standard normal distribution. The response variable (y) is defined as a linear combination of predictors with random coefficients (β), plus random noise to mimic real-world variability. This setup tests the adaptive Lasso's ability to select relevant predictors.

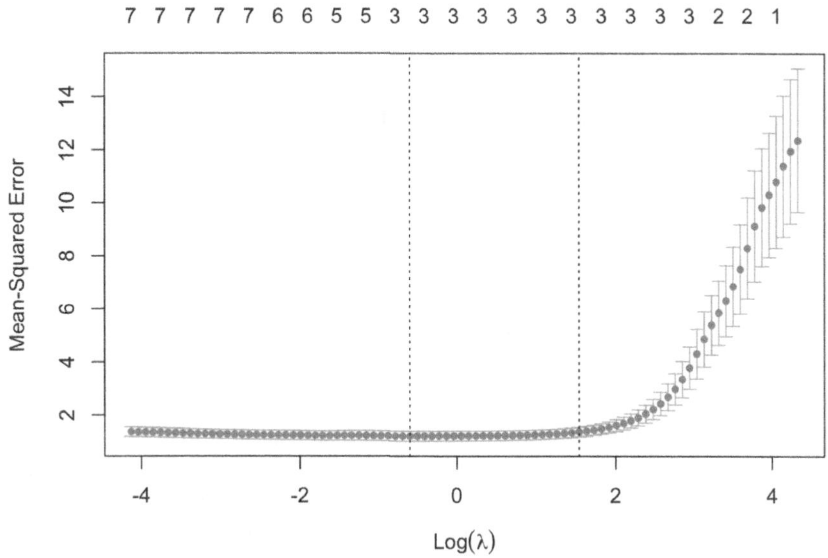

FIGURE 12.3 Cross-validation results for Adaptive Lasso: optimal lambda selection.

2. *Train-Test Split*: The dataset is split into training (80%) and testing (20%) subsets. Training data is used to fit models, while testing data evaluates performance on unseen data, simulating real-world prediction scenarios.

3. *Initial Ridge Regression*: Ridge regression ($L2$ penalty) estimates initial coefficients ($\hat{\beta}_{\text{ridge}}$) using `glmnet` with $\alpha = 0$. The ridge coefficients inform the adaptive LASSO weights:

$$w_j = \frac{1}{|\hat{\beta}_{\text{ridge},j}|}$$

Infinite weights (due to zero coefficients) are replaced with large values to avoid computational issues.

4. *Adaptive Lasso Fitting*: The adaptive LASSO model is fit using `glmnet` with $\alpha = 1$ (LASSO penalty), incorporating weights via the `penalty.factor` argument. To determine the optimal regularization parameter (λ), k-fold cross-validation is performed. The data is divided into k subsets (folds). For each fold i, the model is trained on the remaining $k - 1$ folds and validated on the i-th fold. The MSE is calculated for the predictions on the validation fold. This process is repeated for all k folds, and the average MSE across folds is computed as the cross-validation error for that λ:

$$\text{CV Error}(\lambda) = \frac{1}{k}\sum_{i=1}^{k} \text{MSE}_i(\lambda)$$

where $\text{MSE}_i(\lambda)$ is the MSE for fold i. The λ that minimizes the cross-validation error is selected as the optimal λ.

5. *Model Comparison*: Test-set MSE is computed for adaptive LASSO, LASSO (unweighted), and OLS:

$$\text{MSE} = \frac{1}{n_{\text{test}}} \sum_{i=1}^{n_{\text{test}}} (y_i - \hat{y}_i)^2$$

These metrics quantify each model's predictive accuracy and highlight adaptive LASSO's advantage in feature selection and error reduction.

6. *Cross-validation Plot*: A plot of cross-validated error for the adaptive LASSO model visualizes the effect of λ on model performance, marking the optimal λ values. This aids in understanding the regularization process and its impact on sparsity and prediction. As with standard LASSO, the numbers above each bar indicate the number of nonzero coefficients at each λ; these typically decrease as λ increases, highlighting how adaptive LASSO increasingly shrinks coefficients to zero, resulting in sparser models.
7. *Final Adaptive LASSO Model*: Using the optimal λ_{min} determined through cross-validation, the final adaptive LASSO model is refit on the training data. This model leverages the adaptive weights to selectively penalize coefficients, balancing sparsity and prediction accuracy. Predictions on the test set are made using the final model, ensuring a robust evaluation of its performance.
8. *Comparison Across Models*: Test-set MSE values from the adaptive LASSO, standard Lasso, and OLS models are compared. This comparison underscores the strengths of the adaptive Lasso in addressing multicollinearity, improving prediction accuracy, and effectively selecting relevant features, unlike the OLS model, which lacks regularization, or the standard Lasso, which applies uniform penalization.
9. *Practical Insights*: The simulation demonstrates how ridge-based weights improve Lasso's ability to prioritize predictors. Adaptive Lasso's iterative weight adjustment allows it to fine-tune variable selection, making it a powerful tool for high-dimensional regression problems where feature relevance varies significantly. The cross-validation plot further illustrates the trade-offs involved in selecting λ, offering a clear visual of the model's behavior under different regularization strengths.

12.12 Adaptive LASSO: Practical Notes

Adaptive LASSO is supported in many statistical software packages, making it accessible for advanced regression analysis. In R, the `glmnet` package enables adaptive LASSO implementation by allowing users to specify custom penalty factors, along with a grid of λ values for optimization. Cross-validation is seamlessly integrated, ensuring efficient model tuning. In Python, the `scikit-learn` library provides the flexibility to implement adaptive Lasso through custom weighting of penalties, combining ease of use with the power of Python's machine learning ecosystem. In Stata, the `lasso` command can be adapted for weighted penalization, facilitating the application of adaptive Lasso within a familiar interface. These tools make adaptive Lasso practical for researchers and practitioners handling complex datasets with variable-specific penalties.

In conclusion, adaptive Lasso is a significant advancement in regularized regression methods, addressing the limitations of standard Lasso by introducing adaptive weights. This approach allows for differential penalization across predictors, improving variable selection consistency and reducing bias. Its flexibility makes it particularly powerful for high-dimensional data analysis, balancing sparsity, interpretability, and prediction accuracy.

Notable extensions, such as the adaptive elastic net, which combines adaptive weights with L_1 and L_2 penalties to handle correlated predictors,[8] and the group adaptive LASSO, which applies adaptive penalties to groups of variables, further expand its utility. These adaptations make adaptive Lasso and its variations valuable tools in diverse applications, including genomics, finance, and social sciences, where effective variable selection and model reliability are critical.Adaptive Lasso's adaptiveness, consistency, and bias reduction properties make it a valuable addition to the toolkit of researchers and practitioners seeking robust and interpretable regression models.

12.13 Elastic Net Regression

The Elastic Net, introduced by Zou and Hastie (2005), builds on the standard Lasso regression by combining the L_1-penalty of Lasso with the L_2-penalty of Ridge regression. This hybrid penalization addresses the limitations of Lasso when handling highly correlated predictors or when the number of predictors exceeds the number of observations. By introducing the L_2-penalty, the Elastic Net encourages grouping effects among correlated variables, allowing them to be selected or excluded together, unlike Lasso, which tends to select only one variable from a group.

Standard Lasso applies a uniform L_1-penalty to all coefficients, which can lead to instability in variable selection and poor performance when predictors are correlated. The Elastic Net overcomes these issues by balancing the sparsity-inducing L_1-norm and the grouping property of the L_2-norm. This is achieved through the Elastic Net loss function, given by:

$$\min_{\beta} \left\{ L(\beta) = \sum_{i=1}^{n} \left(y_i - \beta_0 - \sum_{j=1}^{p} \beta_j x_{ij} \right)^2 + \lambda \left(\alpha \sum_{j=1}^{p} |\beta_j| + \frac{1-\alpha}{2} \sum_{j=1}^{p} \beta_j^2 \right) \right\} \tag{12.20}$$

where $\alpha \in [0, 1]$ controls the balance between the L_1-penalty ($\sum_{j=1}^{p} |\beta_j|$) and the L_2-penalty ($\sum_{j=1}^{p} \beta_j^2$), and λ is the regularization parameter controlling the overall strength of the penalization.[9]

A key property of the Elastic Net is its ability to handle multicollinearity effectively. Unlike Lasso, which can arbitrarily select one variable from a group of highly correlated predictors, the Elastic Net selects or excludes the entire group, improving variable selection stability. Additionally, it performs well in high-dimensional settings where the number of predictors far exceeds the sample size, a scenario where Ridge and Lasso often struggle.

Another advantage of the Elastic Net is its flexibility in balancing sparsity and grouping effects through the parameter α. When $\alpha = 1$, the Elastic Net simplifies to standard Lasso, emphasizing sparsity. When $\alpha = 0$, it reduces to Ridge regression, prioritizing shrinkage without sparsity. This flexibility allows practitioners to tailor the Elastic Net to the specific characteristics of their data, making it a versatile tool for regularized regression.

While the Elastic Net offers significant advantages, it introduces the complexity of tuning two parameters, α and λ, requiring careful selection to achieve optimal performance. Grid search over a range of λ and α values using cross-validation is typically employed to select these parameter.[10]

The Elastic Net's ability to combine sparsity, grouping effects, and robustness to multicollinearity makes it a powerful extension of Lasso. Its hybrid penalty structure is particularly well-suited for applications in genomics, finance, and other fields where high-dimensional data with correlated predictors is common.

12.13.1 Solution to the Elastic Net Optimization Problem

As with Ridge and Lasso, standardizing the predictor variables is essential in Elastic Net Regression to ensure that the penalty terms are applied uniformly across predictors. Standardization involves centering each predictor to have a mean of zero and scaling it to have a unit variance. This step is crucial for the Elastic Net optimization algorithm to treat all predictors equally and avoid bias due to differences in scale.

The Elastic Net solution does not have a closed-form expression like Ridge and OLS. However, it can be written as an iterative update rule that combines the L_1- and L_2-penalties. For a given predictor k, the Elastic Net coefficient update can be expressed as:

$$\hat{\beta}_k^{EN} = \frac{1}{n + \lambda(1-\alpha)} \left[\sum_{i=1}^{n} x_{ik} \left(y_i - \sum_{j \neq k} \tilde{\beta}_j x_{ij} \right) - \lambda\alpha \right]_+ \tag{12.21}$$

where $\alpha \in [0, 1]$ controls the balance between the L_1-penalty ($\lambda\alpha$) and the L_2-penalty ($\lambda(1-\alpha)$), λ is the overall regularization parameter, $[z]_+ = \max(0, z)$ denotes the soft-thresholding operator that enforces sparsity by shrinking some coefficients to zero. The solution must satisfy the Karush-Kuhn-Tucker (KKT) conditions, which characterize the optimality of the elastic net solution.[11]

This formulation highlights the interplay between the penalties. The L_1-penalty promotes sparsity by shrinking coefficients toward zero, while the L_2-penalty provides stability, especially in the presence of multicollinearity. The iterative nature of the solution allows the Elastic Net to adapt to different data structures effectively.

Elastic Net's constraint region combines the shapes of the Lasso's diamond (for $L1$ penalty) and Ridge's circle (for $L2$ penalty) into an elastic net-like shape. This allows for solutions that can both zero out coefficients (due to the $L1$ penalty) and shrink coefficients towards each other (due to the $L2$ penalty).

The Elastic Net offers several advantages, making it a powerful tool for regularized regression. It effectively handles multicollinearity by encouraging grouping, allowing correlated predictors to be selected or excluded together. Additionally, it performs variable selection, producing sparse models that are easier to interpret and computationally efficient. Its flexibility is another key strength, as the balance between Ridge and Lasso penalties can be adjusted via the parameter α, enabling users to tailor the model to their specific data characteristics.

However, the Elastic Net also has limitations. It requires the tuning of two parameters (λ and α), which can be computationally intensive and challenging in practice. The simultaneous use of L_1- and L_2-penalties can complicate the interpretation of the coefficients, particularly when both penalties contribute significantly. Additionally, improper tuning of the parameters may lead to over-penalization, potentially discarding relevant variables and reducing model performance. Despite these challenges, the Elastic Net remains a versatile and effective method for addressing high-dimensional regression problems.

12.14 Implementation and Simulation of Elastic Net

The Elastic Net simulation begins with **Data Preparation**, where synthetic data is generated and split into training and testing sets. Predictor variables are standardized to ensure the penalties are applied consistently across coefficients.

Next is the **Parameter Tuning** phase, where double cross-validation is used to select the optimal combination of the regularization parameter (λ) and the mixing parameter (α). The λ controls the overall penalty strength, while α balances the L_1- and L_2-penalties.

In the **Model Fitting** step, the Elastic Net model is trained on the training set using the optimal parameters obtained from cross-validation. This ensures that the model balances variable selection and stability effectively.

Finally, the **Model Evaluation** phase assesses the Elastic Net's predictive performance on the test set using MSE. The results are compared against Lasso, Ridge, and OLS models to highlight the Elastic Net's advantages in handling multicollinearity and selecting relevant predictors.

```
# Load necessary libraries
library(glmnet)

# 1. Generate synthetic data
set.seed(123)  # For reproducibility
n <- 100       # Number of observations
p <- 10        # Number of predictors

# Updated coefficients and response generation
X <- matrix(rnorm(n * p), nrow = n, ncol = p)  # Generate random predictors
beta <- c(3, -2, 0, 0, 1.5, 0, 0, 0, 0, 0)     # True coefficients
y <- X %*% beta + rnorm(n)                     # Generate response with noise

# 2. Split data into training and testing sets
set.seed(456)
train_idx <- sample(1:n, size = 0.8 * n, replace = FALSE)  # 80% training data
X_train <- X[train_idx, ]
X_test <- X[-train_idx, ]
y_train <- y[train_idx]
y_test <- y[-train_idx]

# 3. Fit an Elastic Net model
# Define a range of alpha values to explore
alpha_values <- seq(0, 1, by = 0.1)  # Values from 0 (Ridge) to 1 (Lasso)

# Initialize variables to store the best parameters and error
best_alpha <- NULL
best_lambda <- NULL
lowest_cv_error <- Inf
cv_elastic_net <- NULL  # Placeholder for the final cv.glmnet object
```

```
# Perform cross-validation for each alpha value
for (alpha in alpha_values) {
  cv_model <- cv.glmnet(X_train, y_train, alpha = alpha)
  current_cv_error <- min(cv_model$cvm) # Minimum cross-validation error

  # Update best parameters if current error is lower
  if (current_cv_error < lowest_cv_error) {
    lowest_cv_error <- current_cv_error
    best_alpha <- alpha
    best_lambda <- cv_model$lambda.min
    cv_elastic_net <- cv_model  # Store the best cv.glmnet object
  }
}

cat("Best alpha for Elastic Net regression:", best_alpha, "\n")

## Best alpha for Elastic Net regression: 0.6

cat("Best lambda for Elastic Net regression:", best_lambda, "\n")

## Best lambda for Elastic Net regression: 0.06711713

# 4. Fit the final Elastic Net model using the best alpha and lambda
elastic_net_final <- glmnet(X_train,y_train,alpha=best_alpha,lambda=best_lambda)

# 5. Make predictions on the test set
y_pred_elastic_net <- predict(elastic_net_final, s = best_lambda, newx = X_test)

# Evaluate the model performance
mse_elastic_net <- mean((y_test - y_pred_elastic_net)^2)
cat("Mean Squared Error for Elastic Net on test set:", mse_elastic_net, "\n")

## Mean Squared Error for Elastic Net on test set: 1.168959

# 6. Compare with Lasso, Ridge, and OLS
# Lasso Model (alpha = 1)
cv_lasso <- cv.glmnet(X_train, y_train, alpha = 1)
best_lambda_lasso <- cv_lasso$lambda.min
lasso_final <- glmnet(X_train, y_train, alpha = 1, lambda = best_lambda_lasso)
y_pred_lasso <- predict(lasso_final, s = best_lambda_lasso, newx = X_test)
mse_lasso <- mean((y_test - y_pred_lasso)^2)
cat("Mean Squared Error for Lasso on test set:", mse_lasso, "\n")

## Mean Squared Error for Lasso on test set: 1.161081

# Ridge Model (alpha = 0)
cv_ridge <- cv.glmnet(X_train, y_train, alpha = 0)
best_lambda_ridge <- cv_ridge$lambda.min
```

```
ridge_final <- glmnet(X_train, y_train, alpha = 0, lambda = best_lambda_ridge)
y_pred_ridge <- predict(ridge_final, s = best_lambda_ridge, newx = X_test)
mse_ridge <- mean((y_test - y_pred_ridge)^2)
cat("Mean Squared Error for Ridge on test set:", mse_ridge, "\n")
```

```
## Mean Squared Error for Ridge on test set: 1.341623
```

```
# OLS Model
ols_model <- lm(y_train ~ X_train)
y_pred_ols <- predict(ols_model, newdata = data.frame(X_train = X_test))
mse_ols <- mean((y_test - y_pred_ols)^2)
cat("Mean Squared Error for OLS on test set:", mse_ols, "\n")
```

```
## Mean Squared Error for OLS on test set: 34.29426
```

```
# 7. Plot the cross-validation results for Elastic Net
# plot(cv_elastic_net)
```

1. *Data Generation*: Synthetic data with 100 observations and 10 predictors (X) is generated, where each predictor follows a standard normal distribution. The response variable (y) is a linear combination of the predictors with specified coefficients ($\beta = [3, -2, 0, 0, 1.5, 0, 0, 0, 0, 0]$) and added random noise. This setup tests the Elastic Net's ability to handle correlated predictors and balance variable selection with shrinkage.

2. *Train-Test Split*: The dataset is split into training (80%) and testing (20%) subsets. Training data is used to fit models and optimize parameters, while testing data evaluates performance on unseen data, simulating real-world prediction scenarios.

3. *Cross-validation for Tuning Parameters*: To determine the optimal regularization parameters (λ and α), double cross-validation is performed:

   - *Outer Cross-validation*: A range of α values (e.g., from 0 to 1 in steps of 0.1) is explored. For each α, an inner cross-validation process selects the best λ value.
   - *Inner Cross-validation*: For each combination of training and validation sets, the Elastic Net model is trained with the specified α and a sequence of λ values. The MSE is computed for each validation fold, and the λ that minimizes the cross-validation error is chosen. The cross-validation error for each combination of λ and α is calculated as:

   $$\text{CV}(\lambda, \alpha) = \frac{1}{K} \sum_{k=1}^{K} \|\mathbf{y}^{(k)} - \mathbf{X}^{(k)} \hat{\boldsymbol{\beta}}^{(-k)}(\lambda, \alpha)\|_2^2,$$

   where $\hat{\boldsymbol{\beta}}^{(-k)}(\lambda, \alpha)$ is the Elastic Net estimate computed without the k-th fold. The combination of α and λ that minimizes the outer cross-validation error is selected.

4. *Final Elastic Net Model*: Using the best α and λ values determined through double cross-validation, the final Elastic Net model is refit on the full training data. This model balances the sparsity-inducing L_1 penalty and the stability-improving L_2 penalty, effectively handling multicollinearity and selecting relevant predictors.

5. *Test Set Predictions*: Predictions are made on the test set using the final Elastic Net model. The test-set MSE is calculated as:

$$\text{MSE} = \frac{1}{n_{\text{test}}} \sum_{i=1}^{n_{\text{test}}} (y_i - \hat{y}_i)^2,$$

where n_{test} is the number of test observations, y_i is the actual response, and $\hat{y}_i$ is the predicted response.

6. *Model Comparison*: Test-set MSE values are computed for Elastic Net, Lasso ($\alpha = 1$), Ridge ($\alpha = 0$), and OLS models. These metrics quantify the predictive accuracy of each model, highlighting Elastic Net's ability to balance variable selection and shrinkage effectively.

7. *Cross-validation Plot*: A plot of the inner cross-validation results for the optimal α value visualizes the effect of λ on model performance. This plot helps in understanding the trade-offs between sparsity and stability introduced by Elastic Net's penalties.

8. *Practical Insights*: The simulation demonstrates the flexibility and power of Elastic Net in handling correlated predictors and balancing variable selection with regularization. The use of double cross-validation ensures robust parameter selection, making Elastic Net a reliable tool for high-dimensional regression problems where multicollinearity and sparsity are key concerns. The cross-validation plot further highlights the importance of tuning parameters to achieve optimal performance.

12.15 Elastic Net: Final Notes

Elastic net is widely implemented across statistical software packages, making it a versatile tool for regression analysis. In R, the `glmnet` package provides robust support for elastic net, allowing users to specify the mixing parameter (α) and optimize the regularization parameter (λ) through cross-validation. In Python, the `scikit-learn` library includes elastic net as part of its linear model suite, offering seamless integration with Python's broader machine learning framework. In Stata, the `elasticnet` command enables elastic net regression, providing an intuitive interface for balancing L_1- and L_2-penalties. These tools ensure elastic net is accessible to researchers and practitioners working with high-dimensional and multicollinear datasets.

Elastic net effectively addresses the limitations of both ridge and LASSO regression by combining their strengths. By balancing L_1- and L_2-penalties, it offers a flexible framework for variable selection and multicollinearity handling. This makes it particularly suitable for high-dimensional datasets. Its ability to adapt to the structure of data ensures robust performance across various applications, maintaining a balance between sparsity, stability, and prediction accuracy.

Notable adaptations further extend the utility of elastic net. The adaptive elastic net refines variable selection and improves consistency by incorporating adaptive weights into the elastic net framework:

$$P_{\text{Adaptive EN}}(\boldsymbol{\beta}) = \lambda \left(\alpha \sum_{j=1}^{p} w_j |\beta_j| + \frac{1-\alpha}{2} \|\boldsymbol{\beta}\|_2^2 \right), \text{ where } 0 \leq \alpha \leq 1 \tag{12.22}$$

Additionally, the group elastic net applies penalties to predefined groups of predictors, enabling effective selection or shrinkage at the group level:

$$P_{\text{Group EN}}(\boldsymbol{\beta}) = \lambda\left(\alpha\sum_{g=1}^{G}\|\boldsymbol{\beta}_g\|_2 + \frac{1-\alpha}{2}\|\boldsymbol{\beta}\|_2^2\right), \text{where } \boldsymbol{\beta}_g \text{ is the coefficient vector for group } g.$$

This approach is particularly useful in applications where variables naturally form groups, such as gene sets in genomics or categorical predictors with multiple levels.

To conclude this chapter, penalized regression methods provide a powerful and flexible framework for addressing key challenges in modern data analysis. By introducing regularization penalties, these methods offer effective solutions for variable selection, multicollinearity, and model complexity management. Their adaptability makes them particularly useful when dealing with high-dimensional datasets, where the number of predictors may exceed the number of observations.

LASSO and adaptive LASSO are valuable tools for variable selection, promoting sparsity and improving model interpretability. When predictor variables are naturally organized into groups, methods like group LASSO and sparse group LASSO extend the benefits of regularization to structured data, facilitating both group-level and within-group selection. The choice between these techniques depends on the research objectives, data characteristics, and computational resources available.

When applying penalized regression, practitioners must carefully assess factors such as the ratio of predictors to observations, the presence of multicollinearity, and the importance of group-level inference. Additionally, the intended goal—whether it be maximizing predictive accuracy or obtaining a parsimonious model—should guide the selection of an appropriate penalty. Computational considerations are also crucial, as some penalties demand more intensive optimization routines.

Hyperparameter tuning, particularly the selection of λ, α, and γ, remains essential for achieving optimal performance. Cross-validation and other model selection techniques can help determine the most suitable settings. Ultimately, the thoughtful application of penalized regression improves both the robustness and interpretability of models, making these methods indispensable for statisticians and data scientists tackling complex analytical problems.

12.16 Technical: Derivation of Ridge Regression Estimator

Let's derive the expressions for the OLS and Ridge regression estimators using summation notation, and then demonstrate their relationship under the assumption of orthogonal covariates. In the OLS model, we seek to minimize the sum of squared residuals. The derivative of the RSS with respect to β and then solving gives us $\hat{\beta}_{\text{ls}} = \frac{\sum_{i=1}^{n} x_i y_i}{\sum_{i=1}^{n} x_i^2}$ (for details check Chapter 4)

For Ridge regression, the objective function includes a penalty term $\lambda\beta^2$, where λ is a positive constant. We minimize:

$$J(\beta) = \sum_{i=1}^{n}(y_i - x_i\beta)^2 + \lambda\beta^2$$

Taking the derivative with respect to β and setting it to zero:

$$\frac{\partial J}{\partial \beta} = -2\sum_{i=1}^{n} x_i(y_i - x_i\beta) + 2\lambda\beta = 0$$

Solving for β:

$$\sum_{i=1}^{n} x_i y_i = \sum_{i=1}^{n} x_i^2\beta + \lambda\beta$$

$$\hat{\beta}_{\text{ridge}} = \frac{\sum_{i=1}^{n} x_i y_i}{\sum_{i=1}^{n} x_i^2 + \lambda}$$

When the covariates are orthogonal, x_i are such that $\sum_{i=1}^{n} x_i x_j = 0$ for all $i \neq j$ and $\sum_{i=1}^{n} x_i^2 = n$ for each predictor (assuming each x_i has been normalized or is inherently orthogonal). The OLS and Ridge estimators simplify as:

$$\hat{\beta}_{\text{ls,i}} = \frac{\sum_{i=1}^{n} x_i y_i}{n}$$

$$\hat{\beta}_{\text{ridge,i}} = \frac{\sum_{i=1}^{n} x_i y_i}{n + \lambda}$$

So the relationship between the ridge estimator and the OLS estimator under these conditions is:

$$\hat{\beta}_{\text{ridge,i}} = \frac{n}{n + \lambda}\hat{\beta}_{\text{ls,i}}$$

This formula shows that the ridge estimator shrinks the OLS estimator toward zero by the factor $\frac{n}{n+\lambda}$. This shrinkage increases as λ increases, thus demonstrating how Ridge regression can be biased but potentially more stable or with lower variance, especially in cases where multicollinearity or limited data points pose a problem for the reliability of the OLS estimates.

Matrix Form: To derive the relationship between the Ridge regression estimator and the OLS estimator for the coefficients in matrix form, we can start by considering the following regression model: $Y = X\beta + \epsilon$, where Y is the response vector, X is the design matrix, β is the coefficient vector, and ϵ is the error term. The OLS estimator of β is given by minimizing the residual sum of squares (RSS):

$$\text{RSS} = (Y - X\beta)^\top(Y - X\beta)$$

Assuming $X^\top X$ is invertible and taking the derivative with respect to β and setting it to zero yields the normal equations $\hat{\beta}_{\text{ls}} = (X^\top X)^{-1}X^\top Y$

For Ridge regression, the cost function includes a penalty term, and is given by:

$$J(\beta) = (Y - X\beta)^\top(Y - X\beta) + \lambda\beta^\top\beta$$

Where λ is the regularization parameter. Again, taking the derivative with respect to β and setting it to zero:

$$-2X^\top Y + 2X^\top X\beta + 2\lambda\beta = 0$$

Rearranging terms:

$$(X^\top X + \lambda I)\beta = X^\top Y$$

Solving for β:

$$\hat{\beta}_{\text{ridge}} = (X^\top X + \lambda I)^{-1} X^\top Y$$

For orthogonal covariates, the matrix $X^\top X$ is a diagonal matrix where each diagonal element is equal to n (the number of observations), assuming the predictors are standardized:

$$X^\top X = nI$$

Substituting in the formula for Ridge regression:

$$\hat{\beta}_{\text{ridge}} = (nI + \lambda I)^{-1} X^\top Y$$

$$\hat{\beta}_{\text{ridge}} = \left(\frac{1}{n+\lambda} I\right) X^\top Y$$

$$\hat{\beta}_{\text{ridge}} = \frac{n}{n+\lambda}\hat{\beta}_{\text{ls}}$$

Notes

1. Tibshirani, R. J. "The lasso problem and uniqueness" (2013). Available at https://doi.org/10.48550/arXiv.1206.0313

2. For a comprehensive critique of LASSO and its adaptations in contexts with dependent covariates, consider exploring the open-source article by Freijeiro-González, Febrero-Bande, and González-Manteiga, titled "A Critical Review of LASSO and Its Derivatives for Variable Selection Under Dependence Among Covariates" (2022), Available at https://doi.org/10.1111/insr.12469

3. The regularity conditions required for these properties include: Irrepresentable Condition: $\|\mathbf{X}_{S^c}^\top \mathbf{X}_S(\mathbf{X}_S^\top \mathbf{X}_S)^{-1}\text{sign}(\boldsymbol{\beta}_S)\|_\infty < 1$, where S denotes the set of true nonzero coefficients. This condition ensures that the irrelevant predictors (S^c) are not overly correlated with the relevant predictors (S), allowing the model to isolate the true signals. Consistent Weight Selection: $w_j \to 0 \quad \text{for} \quad \beta_j \neq 0, \quad \text{and} \quad w_j \to \infty \quad \text{for} \quad \beta_j = 0$. This ensures that the weights properly penalize zero coefficients while allowing nonzero coefficients to be accurately estimated.

4. Specifically, as the sample size n increases, the estimated coefficients for the true predictors ($\boldsymbol{\beta}_S$) follow a normal distribution, which is $\sqrt{n}(\hat{\boldsymbol{\beta}}_S - \boldsymbol{\beta}_S) \xrightarrow{d} \mathcal{N}(0, \sigma^2(\mathbf{X}_S^\top \mathbf{X}_S)^{-1})$,, where $\boldsymbol{\beta}_S$ represents the true coefficients for the selected variables, σ^2 is the error variance, and $(\mathbf{X}_S^\top \mathbf{X}_S)^{-1}$ is the covariance matrix for the selected predictors.

5. In this method, weights are derived from the selection probabilities of each variable across multiple subsamples: $w_j = \frac{1}{\hat{\pi}_j^\gamma}$, where $\hat{\pi}_j$ is the estimated probability that variable j is selected across subsamples, and γ controls the adaptiveness. Stability selection provides a probabilistic perspective on variable importance, making it particularly useful in scenarios where the stability of variable selection is critical.

6. $\text{CV}(\lambda, \gamma) = \frac{1}{K}\sum_{k=1}^K \|\mathbf{y}^{(k)} - \mathbf{X}^{(k)}\hat{\beta}^{(-k)}(\lambda, \gamma)\|_2^2$, where $\hat{\beta}^{(-k)}(\lambda, \gamma)$ is the adaptive LASSO estimate computed without the k-th fold.

7. These conditions are given by: $0 = -\frac{1}{n}\mathbf{X}^\top(\mathbf{y} - \mathbf{X}\hat{\beta}) + \lambda\mathbf{w} \odot \boldsymbol{\theta}$, where $\mathbf{w} = [w_1, w_2, \ldots, w_p]^\top$, $\odot$ denotes element-wise multiplication, and $\boldsymbol{\theta}$ is the subgradient vector, accounting for the non-differentiability of the L_1-penalty. The KKT conditions ensure that the optimal solution respects the weighted penalization structure while maintaining consistency with the data.

Finally, the optimal tuning parameter λ, which determines the overall strength of the penalization, is selected using cross-validation or information criteria to ensure a balance between model complexity and predictive accuracy. The Karush-Kuhn-Tucker (KKT) conditions for the Adaptive LASSO play a critical role in defining the solution. They ensure that the optimal coefficients are consistent with the penalty weights and the data constraints. The subgradient vector ($\boldsymbol{\theta}$) adjusts for the non-smooth nature of the L_1-norm, allowing the solution to handle sparsity effectively. For further details, see Zou (2006).

8. $P_{\text{Adaptive EN}}(\beta) = \lambda\left(\alpha\sum_{j=1}^{p} w_j|\beta_j| + \frac{1-\alpha}{2}\|\beta\|_2^2\right)$, where $0 \leq \alpha \leq 1$.

9. The constraint formulation of Elastic Net is mathematically equivalent to the Elastic Net loss function. This formulation minimizes the residual sum of squares with the regularization expressed as a constraint, and written as $\min_\beta\left\{\frac{1}{2n}\|\mathbf{y} - \mathbf{X}\beta\|_2^2\right\}$ subject to $\alpha\|\beta\|_1 + \frac{1-\alpha}{2}\|\beta\|_2^2 \leq t$. In the penalized form, λ controls the strength of the penalty. In the constraint form, t determines the maximum allowable value of the combined penalties.

10. Cross-Validation Error for Tuning Parameters is calculated as: $\text{CV}(\lambda, \alpha) = \frac{1}{K}\sum_{k=1}^{K}\|\mathbf{y}^{(k)} - \mathbf{X}^{(k)}\hat{\beta}^{(-k)}(\lambda, \alpha)\|_2^2$, where $\hat{\beta}^{(-k)}(\lambda, \alpha)$ is the Elastic Net estimate computed without the k-th fold.

11. Karush-Kuhn-Tucker (KKT) conditions are expressed as the following: first, the stationarity condition requires that for each predictor k, the gradient of the loss function with respect to β_k must balance the residuals and penalties. For nonzero coefficients ($\hat{\beta}_k \neq 0$), this is given by $\frac{1}{n}\mathbf{x}_k^\top(\mathbf{y} - \mathbf{X}\hat{\beta}) + \lambda\alpha \cdot \text{sign}(\hat{\beta}_k) + \lambda(1-\alpha)\hat{\beta}_k = 0$ For zero coefficients ($\hat{\beta}_k = 0$), the residual term must lie within the regularization threshold: $\left|\frac{1}{n}\mathbf{x}_k^\top(\mathbf{y} - \mathbf{X}\hat{\beta})\right| \leq \lambda\alpha$. Second, the primal feasibility condition ensures that the Elastic Net penalty constraint is satisfied (in the constraint formulation): $\alpha\|\hat{\beta}\|_1 + \frac{1-\alpha}{2}\|\hat{\beta}\|_2^2 \leq t$, where t controls the level of shrinkage. Third, the dual feasibility condition guarantees that the dual variables align with the regularization term: $\left|\frac{1}{n}\mathbf{x}_k^\top(\mathbf{y} - \mathbf{X}\hat{\beta})\right| \leq \lambda\alpha$. Finally, the complementary slackness condition links zero coefficients with the soft-thresholding induced by the L_1-penalty. Specifically, for $\hat{\beta}_k = 0$: $\left|\frac{1}{n}\mathbf{x}_k^\top(\mathbf{y} - \mathbf{X}\hat{\beta})\right| = \lambda\alpha$. These conditions ensure that the Elastic Net solution $\hat{\beta}^{EN}$ optimally balances residual minimization with the combined effects of the L_1- and L_2-penalties, providing a robust framework for regularized regression.

13

Classification and Regression Trees (CART)

Classification and regression trees (CART) offer a versatile framework for addressing both categorical and continuous target variables within the same methodological approach. By systematically dividing the feature space into increasingly homogeneous subsets, CART models can predict outcomes by learning simple decision rules inferred from the data features. This approach is particularly useful for understanding the relationships between the features and the target, as well as for identifying the most important features for prediction.

For example, an economist studying poverty could use CART to predict household income based on demographic and labor market characteristics, identifying the most critical factors influencing economic mobility. Similarly, a health researcher analyzing disease risk could use CART to determine which patient characteristics (e.g., age, BMI, medical history) best predict the likelihood of developing a condition like diabetes. In social sciences, CART can help classify voting behavior of individuals based on socioeconomic and ideological variables.

Single decision trees serve as the foundation for more advanced tree-based models, including random forests and gradient boosting machines, but they are also effective as standalone models for classification and regression tasks. In this chapter, we will cover the key concepts of classification trees, including impurity measures, recursive partitioning, and variable importance. We will also implement a simple recursive partitioning algorithm to construct a decision tree from scratch and demonstrate how to train a tree model using the rpart package in R.

This discussion naturally follows the previous chapter on shrinkage estimators such as Lasso and Ridge, which are designed for variable selection and regularization. While those methods impose constraints to shrink coefficient estimates toward zero, decision trees take a different approach by iteratively splitting the data to form distinct prediction rules. Unlike linear models, tree-based methods handle complex, nonlinear relationships without requiring transformations, making them particularly useful when the underlying relationship between variables and the outcome is unknown or highly interactive.

Decision trees form the foundation of tree-based models. A decision tree is essentially a flowchart where:

- Each **internal node** represents a decision point based on variable values.
- Each **branch** represents the possible outcomes of that decision.
- Each **leaf node** at the end of a branch represents the final predicted outcome.

Here is an example of a simple decision tree illustrating a gambling situation:

This decision tree consists of internal nodes, branches, and leaf nodes to represent a gambling decision. The root internal node (labeled "Gamble") is the starting point, where the decision to "play" or "do not play" creates two branches. The left branch leads directly to a leaf node ("\$100k, 100%"), representing a guaranteed outcome. The right branch ("Play") leads to another internal node, where the outcome is uncertain. This second decision splits into two

DOI: 10.1201/9781003504795-13

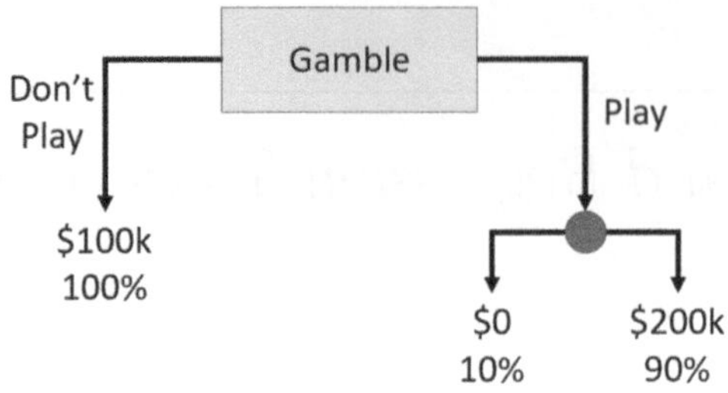

FIGURE 13.1 Example of a simple decision tree.

leaf nodes, representing a 10% chance of $0 and a 90% chance of $200k. The structure visually maps the decision process, showing how different choices lead to distinct final outcomes.

Let's illustrate how a decision tree classifies observations by recursively splitting the data to maximize homogeneity and minimize misclassification with a simple classification problem.

Intuition: Classification Trees

We begin with a dataset containing two variables, x_1 and x_2, used to classify observations into two categories: circle (0) and triangle (1).

```
y <- c(1,1,1,0,0,0,1,1,0,1)
x1 <- c(0.09, 0.11, 0.17, 0.23, 0.33, 0.5, 0.54, 0.62, 0.83, 0.88)
x2 <- c(0.5, 0.82, 0.2, 0.09, 0.58, 0.5, 0.93, 0.8, 0.3, 0.83)

data <- data.frame(y = y, x1 = x1, x2 = x2)
pch_vals <- ifelse(data$y == 1, 16, 17)
plot(data$x1, data$x2, pch = pch_vals, col = "black", lwd = 2,
     ylab = "x2", xlab = "x1")
```

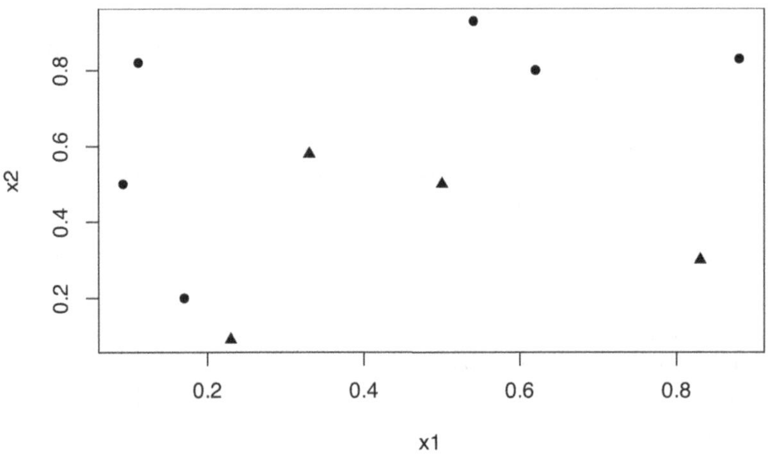

Our goal is to find the best decision rule on x_2 to correctly classify circle (0) and triangle (1). Specifically, we want to find a cutoff point on x_2 such that the maximum number of observations is correctly classified.

To minimize misclassification, we determine that the optimal cutoff point/split should be between 0.6 and 0.79. Thus, our rule becomes $x_2 < k$, where $k \in [0.6, 0.79]$.

```
pch_vals <- ifelse(data$y == 1, 16, 17)
plot(data$x1, data$x2, pch = pch_vals, col = "black", lwd = 2)
abline(h = 0.62, col = "black", lty = 5, lwd = 2)
```

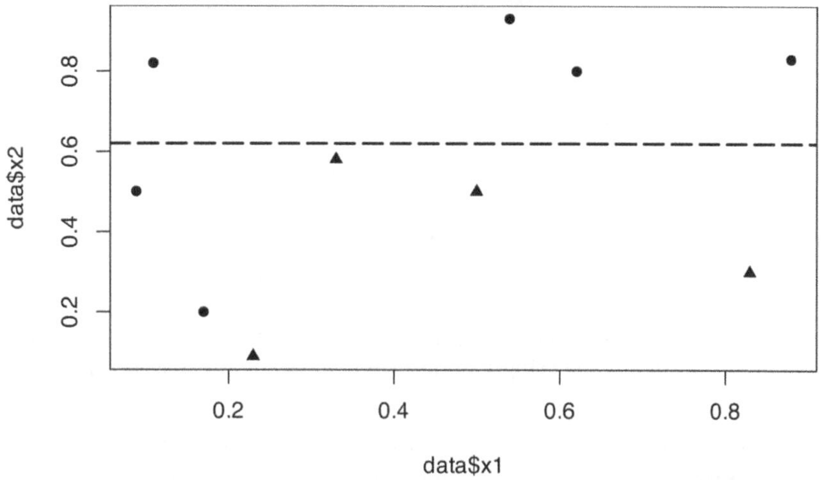

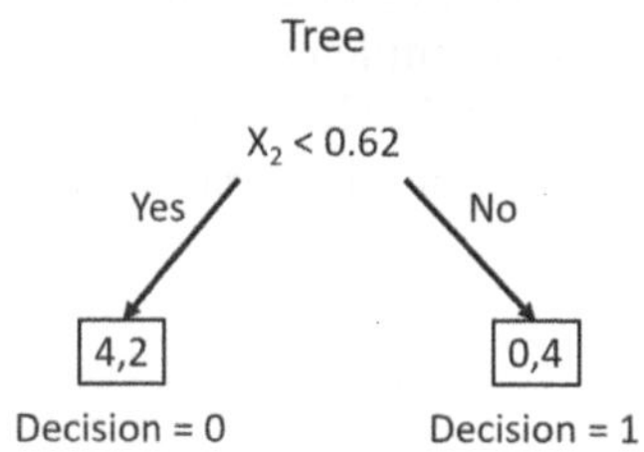

This simple rule results in two misclassified balls. We can improve our classification by adding a new rule in the area below the horizontal blue line:

```
pch_vals <- ifelse(data$y == 1, 16, 17)
plot(data$x1, data$x2, pch = pch_vals, col = "black", lwd = 2)
abline(h = 0.62, col = "black", lty = 5, lwd = 2)
abline(v = 0.2, col = "gray40", lty = 5, lwd = 2)
```

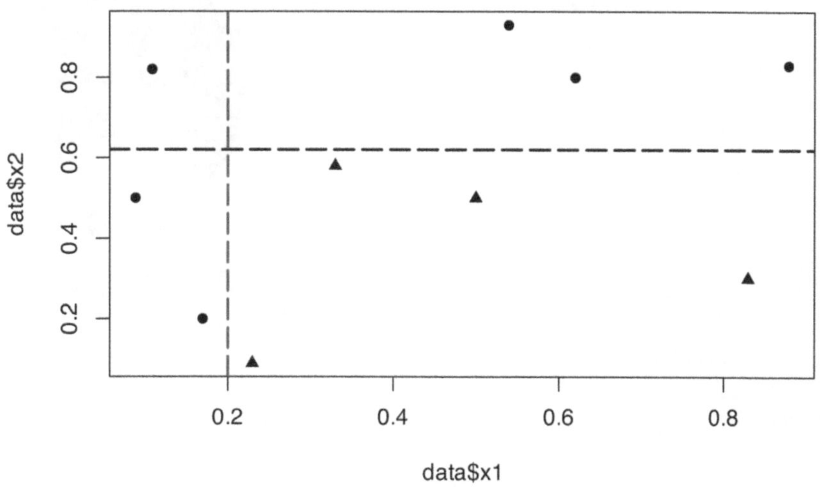

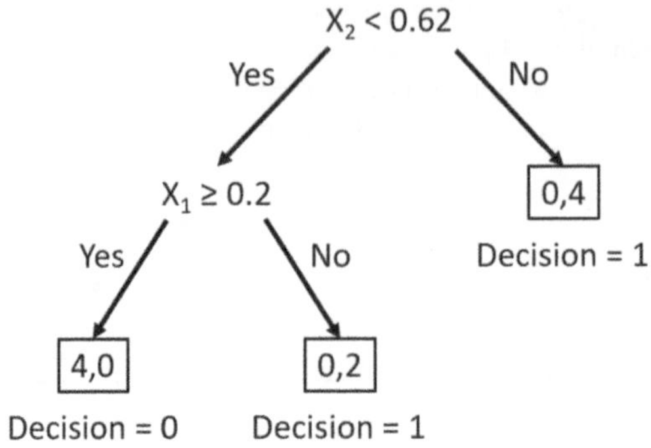

This manual classification process follows the same logic as a decision tree algorithm. The first split at $x_2 = 0.62$ represents an internal node, dividing the data into two branches. The additional split on x_1 creates a further partition, leading to the final leaf nodes that assign class labels. Instead of manually choosing splits, a decision tree learning algorithm automates this process, systematically selecting the best split at each step based on measures such as impurity reduction.

Now, let's explore how to formalize this procedure into an algorithmic approach. We will cover the key concepts of decision trees, including impurity measures, recursive partitioning, and variable importance. We will also implement a simple recursive partitioning algorithm to construct a decision tree from scratch, then discuss formal step by step algorithm, and then demonstrate how to train a classification tree model using the rpart package in R.

13.1 Basic Concepts

Let's use a simple example given by freakonometrics,[1] which uses a dataset (from Gilbert Saporta) that reports about heart attacks and decease (our binary outcome thus classification) in an ER.

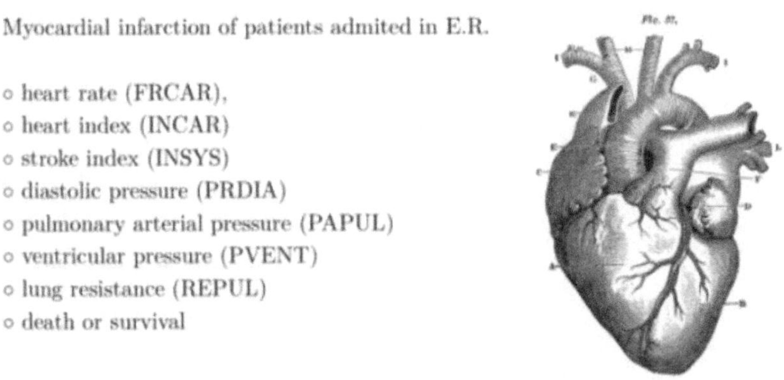


```
# Data
myocarde = read.table("http://freakonometrics.free.fr/myocarde.csv",
                      head = TRUE, sep = ";")
str(myocarde)
```

```
## 'data.frame':    71 obs. of  8 variables:
##  $ FRCAR: int  90 90 120 82 80 80 94 80 78 100 ...
##  $ INCAR: num  1.71 1.68 1.4 1.79 1.58 1.13 2.04 1.19 2.16 2.28 ...
##  $ INSYS: num  19 18.7 11.7 21.8 19.7 14.1 21.7 14.9 27.7 22.8 ...
##  $ PRDIA: int  16 24 23 14 21 18 23 16 15 16 ...
##  $ PAPUL: num  19.5 31 29 17.5 28 23.5 27 21 20.5 23 ...
##  $ PVENT: num  16 14 8 10 18.5 9 10 16.5 11.5 4 ...
##  $ REPUL: int  912 1476 1657 782 1418 1664 1059 1412 759 807 ...
##  $ PRONO: chr  "SURVIE" "DECES" "DECES" "SURVIE" ...
```

Recode the binary variable `DECEASE` to 1/0 a

```r
df <- myocarde[ ,-ncol(myocarde)]
df$y <- ifelse(myocarde$PRONO == "SURVIE", 1, 0)
```

13.1.1 Gini Index - Impurity Measure

Calculate the Gini index for y:

```r
G <- 2 * mean(df$y) * (1 - mean(df$y))
G
```

```
## [1] 0.4832375
```

There are several impurity measures, such as Gini impurity, entropy, and classification error. The Gini impurity is used in the CART algorithm to evaluate the quality of a split. It measures the probability of misclassifying a randomly chosen element if it were randomly labeled according to the distribution of labels in the node. The Gini impurity is defined as:

$$\operatorname{Gini}(D)=1-\sum_{i=1}^{C} p_i^2 \tag{13.1}$$

where D is the dataset, C is the number of classes, and p_i is the class i probability, calculated as $p_i=\frac{n_i}{N}$, where n_i is the class count and N is the total observations. The Gini impurity is minimized when the classes are perfectly mixed, and it is maximized when the classes are perfectly separated.

In a binary classification problem, where there are only two classes (let's call them class 1 and class 2), the formula simplifies to:

$$\operatorname{Gini}(D)=1-\left(p_1^2+p_2^2\right)=2p(1-p) \tag{13.2}$$

where p is the probability of class 1 in the dataset. The Gini impurity ranges from 0 (perfectly pure) to 0.5 (maximally impure).

- If all examples are from one class (say all are class 1), then $p_1=1$ and $p_2=0$, leading to $\operatorname{Gini}(D)=1-\left(1^2+0^2\right)=0$, indicating perfect purity.
- If the examples are evenly split between the two classes, then $p_1=0.5$ and $p_2=0.5$, leading to $\operatorname{Gini}(D)=1-\left(0.5^2+0.5^2\right)=0.5$, indicating maximum impurity.

As a side note, for categorical classification with C classes, Gini impurity is $1-\sum p_i^2$, where p_i is the proportion of each class. It reaches 0 when all observations belong to one class and is highest when classes are evenly distributed, e.g., $\frac{2}{3}$ for three equal classes.

13.1.2 Finding the Split

We need to find the variable that gives the highest Gini gain, which measures the improvement in purity after a split. The Gini gain is calculated as:

$$\text{Gini Gain} = G - (p_L G_L + p_R G_R) \tag{13.3}$$

where G is the Gini impurity before the split, G_L and G_R are the Gini impurities of the left and right splits, p_L and p_R are the proportions of observations in the left and right splits.

Now, let's consider the variable FRCAR and split the data into two groups: $FRCAR \leq 90$ and $FRCAR > 90$. We can calculate the Gini index for each group.

```
y = df$y
x = df$FRCAR
GL <- 2 * mean(y[x <= 90]) * (1 - mean(y[x <= 90]))
GR <- 2 * mean(y[x > 90]) * (1 - mean(y[x > 90]))
GL
```

```
## [1] 0.4444444
```

```
GR
```

```
## [1] 0.4994055
```

The overall Gini index for the split is the weighted average of the Gini index for each group. let's calculate the gain (reduction in impurity) after the split.

```
pL <- sum(x <= 90) / length(x)
pR <- sum(x > 90) / length(x)
delta <- G - (pL * GL + pR * GR)
delta
```

```
## [1] 0.01634414
```

13.1.3 Can We Find a Better Split?

Instead of choosing a fixed threshold, we check all possible splits for the variable FRCAR and calculate the Gini gain, delta, for each split.

```
# let's have a function for different splits
GI <- function(y, x, sp){
  G <- 2 * mean(y) * (1 - mean(y))
  GL <- 2 * mean(y[x <= sp]) * (1 - mean(y[x <= sp]))
  GR <- 2 * mean(y[x > sp]) * (1 - mean(y[x > sp]))
  pL <- length(x[x <= sp]) / length(x)
  pR <- length(x[x > sp]) / length(x)
  delta <- G - (pL * GL + pR * GR)
 return(delta)
}
```

Now, we apply this function to all unique values of FRCAR and calculate the Gini gain for each split, and then to find the best split:

```
# let's consider all possible splits
splits <- sort(unique(df$FRCAR))

# let's calculate the gain for each split
delta <- c()
for (i in 1:(length(splits) - 1)) {
  delta[i] <- GI(df$y, df$FRCAR, splits[i])
}

delta
```

```
##  [1] 9.998016e-03 4.978782e-04 1.082036e-05 1.041714e-03 8.855953e-05
##  [6] 7.363859e-04 2.295303e-03 2.546756e-04 1.142757e-03 2.551599e-03
## [11] 9.862318e-03 1.329134e-02 8.257492e-03 2.402430e-02 1.160767e-02
## [16] 1.634414e-02 1.352527e-02 1.229951e-02 3.109723e-03 5.692941e-03
## [21] 9.212475e-03 1.919591e-02 1.244092e-02 6.882353e-03 2.747959e-03
## [26] 6.282533e-03 1.547312e-03 1.082036e-05 4.978782e-04 9.671419e-03
## [31] 4.766628e-03
```

```
splits[which.max(delta)]
```

```
## [1] 86
```

```
plot(splits[-length(splits)], delta, type = "l",
     xlab = "Split", ylab = "Gain",
     main = "Gain for different splits",
     col = "black", lwd = 2)
abline(v = splits[which.max(delta)], col = "gray40")
text(x = 70, y = 0.015, labels = paste("Gain =", round(max(delta), 4),
                 "\nSplit =", splits[which.max(delta)]), pos = 4, col = "black")
```

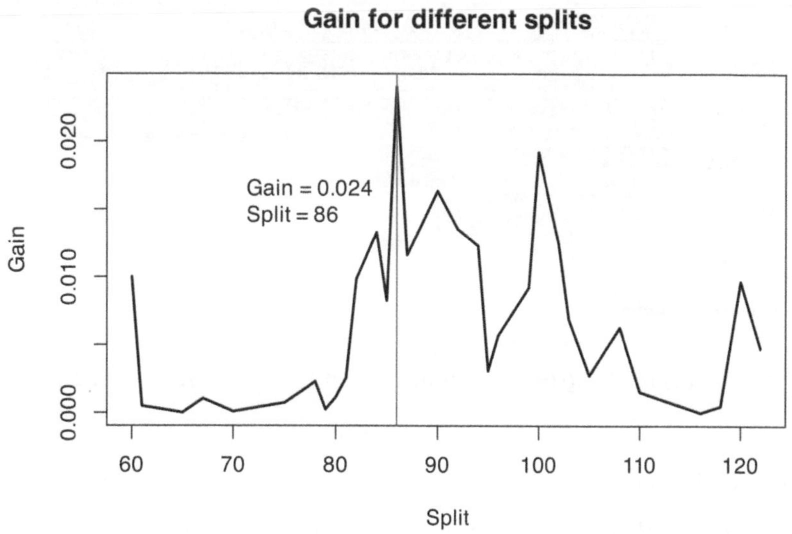

We visualize the gain for different split points.

13.1.4 Function to Find the Important Variables and Gains

We repeat this process for all variables and determine the one with the highest gain:

```
# let's have a function to find the variable that the largest delta (gain)
best.split <- function(x, y){
  splits <- sort(unique(x))
  delta <- c()
  for (i in 1:(length(splits) - 1)) {
    delta[i] <- GI(y, x, splits[i])
  }
  return(c(max(delta), splits[which.max(delta)]))
}

# apply the function to each variable
gains <- apply(df[, -ncol(df)], 2, function(x) best.split(x, df$y))
gains
```

```
##           FRCAR     INCAR      INSYS      PRDIA       PAPUL      PVENT
## [1,]  0.0240243 0.2621902  0.2832801  0.1318471  0.09890283 0.04612125
## [2,] 86.0000000 1.6800000 18.7000000 16.0000000 23.00000000 7.00000000
##            REPUL
## [1,]    0.267907
## [2,] 1093.000000
```

```
# horizontal barplot
barplot(sort(gains[1, ]), horiz = TRUE,
        col = "gray40", las = 1,
        main = "Gains", xlab = "Gains")
```

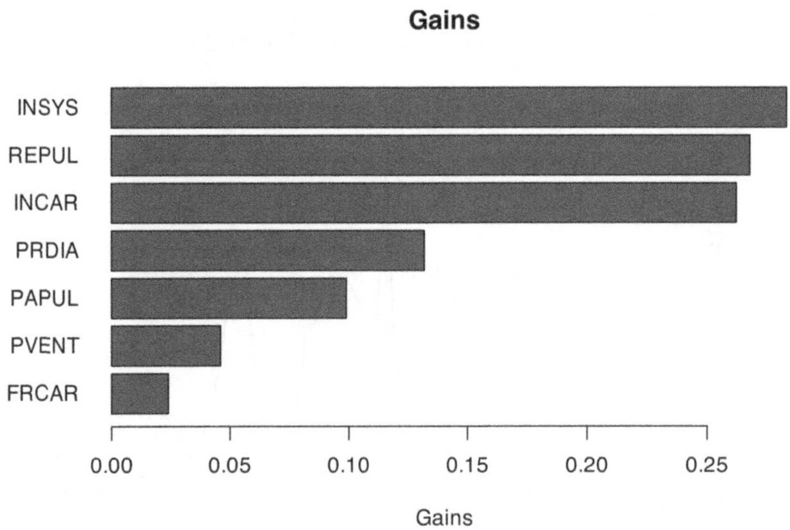

The Second Split

Our first split tells us that the best feature to split on is `INSYS` and the best split value is 86. We can now split the dataset based on this feature and value.

```
# Identify the feature for the first best split
best_feature <- names(which.max(gains[1, ]))
best_split_value <- gains[2, which.max(gains[1, ])]

# Now, split the dataset based on the best feature and best split value
left_subset <- df[df[, best_feature] <= best_split_value, ]
right_subset <- df[df[, best_feature] > best_split_value, ]
head(left_subset)
```

```
##    FRCAR INCAR INSYS PRDIA PAPUL PVENT REPUL y
## 2     90  1.68  18.7    24  31.0  14.0  1476 0
## 3    120  1.40  11.7    23  29.0   8.0  1657 0
## 6     80  1.13  14.1    18  23.5   9.0  1664 0
## 8     80  1.19  14.9    16  21.0  16.5  1412 1
## 22   110  1.66  15.1    23  31.0   6.5  1494 0
## 23    80  1.50  18.7    13  17.0  12.0   907 0
```

```
head(right_subset)
```

```
##    FRCAR INCAR INSYS PRDIA PAPUL PVENT REPUL y
## 1     90  1.71  19.0    16  19.5  16.0   912 1
## 4     82  1.79  21.8    14  17.5  10.0   782 1
## 5     80  1.58  19.7    21  28.0  18.5  1418 0
## 7     94  2.04  21.7    23  27.0  10.0  1059 1
## 9     78  2.16  27.7    15  20.5  11.5   759 1
## 10   100  2.28  22.8    16  23.0   4.0   807 1
```

Now we have two datasets, `left_subset` and `right_subset`. We can apply the `best.split` function to each subset to find the next best split.

```
# Y's
table(left_subset$y)
```

```
##
##  0  1
## 24  3
```

```
table(right_subset$y)
```

```
##
##  0  1
##  5 39
```

```
# Apply to each subset
gains_left <- apply(left_subset[, -ncol(left_subset)], 2,
                    function(x) best.split(x, left_subset$y))
gains_right <- apply(right_subset[, -ncol(right_subset)], 2,
                     function(x) best.split(x, right_subset$y))
gains_left
```

```
##              FRCAR       INCAR       INSYS       PRDIA       PAPUL       PVENT
## [1,]  0.006874299 0.01234568 0.01234568  0.0379349  0.0379349 0.06077873
## [2,] 95.000000000 1.68000000 8.80000000 16.0000000 21.0000000 2.00000000
##             REPUL
## [1,] 3.59147e-02
## [2,] 1.58300e+03
```

```
gains_right
```

```
##             FRCAR       INCAR       INSYS       PRDIA       PAPUL       PVENT
## [1,]  0.00573921 0.04475744  0.04823772  0.02152204  0.03730487  0.1149718
## [2,] 95.00000000 1.58000000 21.40000000 15.00000000 23.00000000 16.0000000
##              REPUL
## [1,] 7.929816e-02
## [2,] 1.093000e+03
```

Are we going to split on the left or right subset? We can use the same approach to decide which subset to split next.

```
max(gains_left[1, ])
```

```
## [1] 0.06077873
```

```
max(gains_right[1, ])
```

```
## [1] 0.1149718
```

Seems like we need to split on the right subset. Let's find the best feature and split value.

```
# Identify the feature for the first best split
best_feature_right <- names(which.max(gains_right[1, ]))
best_split_value_right <- gains_right[2, which.max(gains_right[1, ])]
best_feature_right
```

```
## [1] "PVENT"
```

```
best_split_value_right
```

```
## PVENT
##    16
```

Now, we can again split the dataset based on the best feature and best split value, but ...

```
left_subset_right <- right_subset[
  right_subset[, best_feature_right] <= best_split_value_right,
]
right_subset_right <- right_subset[
```

```
  right_subset[, best_feature_right] > best_split_value_right,
]
head(left_subset_right)
```

```
##    FRCAR INCAR INSYS PRDIA PAPUL PVENT REPUL y
## 1     90  1.71  19.0    16  19.5  16.0   912 1
## 4     82  1.79  21.8    14  17.5  10.0   782 1
## 7     94  2.04  21.7    23  27.0  10.0  1059 1
## 9     78  2.16  27.7    15  20.5  11.5   759 1
## 10   100  2.28  22.8    16  23.0   4.0   807 1
## 11    90  2.79  31.0    16  25.0   8.0   717 1
```

```
head(right_subset_right)
```

```
##    FRCAR INCAR INSYS PRDIA PAPUL PVENT REPUL y
## 5     80  1.58  19.7    21    28  18.5  1418 0
## 25    95  1.89  19.9    25    27  20.0  1143 0
## 70    87  2.51  28.8    16    24  20.0   765 0
```

```
table(left_subset_right$y)
```

```
##
##  0  1
##  2 39
```

```
table(right_subset_right$y)
```

```
##
## 0
## 3
```

We can continue this process until we reach a stopping criterion, such as a minimum number of samples in each node, a maximum depth, or a minimum gain. But, up to this point, we have a good understanding of the recursive partitioning—the core idea behind decision trees. Nevertheless, we did not train a tree model. We will use the `rpart` package to train the tree, later.

13.1.5 Variable Importance: Easy and Heuristic Way

We can use the Gini index to calculate the importance of each variable.

```
# let's calculate the importance of each variable

importance <- c()
for (i in 1:(ncol(df) - 1)) {
  importance[i] <- GI(df$y, df[, i], median(df[, i]))
}
```

```
sdimp <- importance/sum(importance)
names(sdimp) <- names(df)[1:ncol(df) - 1]
barplot(sort(sdimp), horiz = TRUE,
        col = "gray40", las = 1,
        main = "Variable Importance", xlab = "Importance")
```

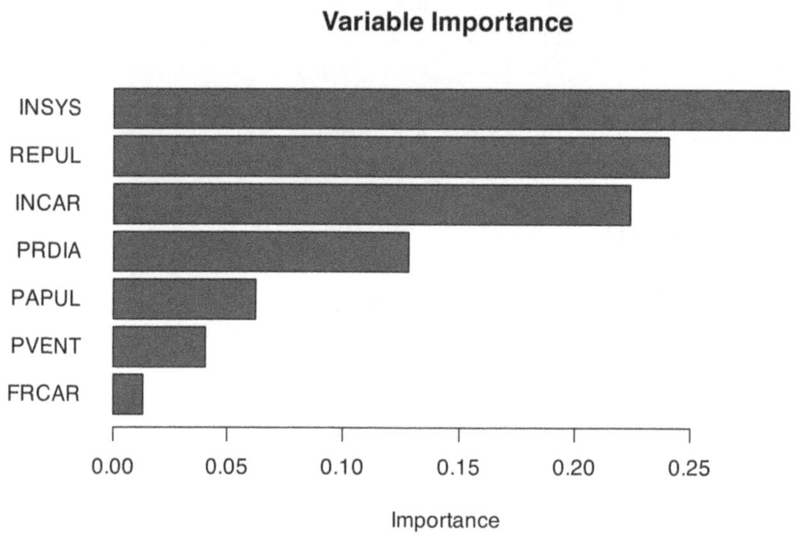

It is important to note that while using the median is a practical and efficient heuristic, it might not always result in the most optimal split, especially for variables with unique distributions or in cases where the relationship between the variable and the target is complex. In more sophisticated decision tree algorithms (like those implemented in machine learning packages), multiple potential split points are evaluated to find the one that maximizes the gain (e.g., information gain, Gini impurity decrease) rather than relying solely on the median. can we write this

13.1.6 Recursive Partitioning: The Tree Model

Recursive functions are a fundamental concept in programming, where a function calls itself to solve a problem by breaking it down into smaller, more manageable sub-problems. This technique is widely used in various applications, including tree structures, sorting algorithms, and solving mathematical problems.

Here is a very simple example to illustrate recursion: calculating the factorial of a number. The factorial of a number `n` (denoted as `n!`) is the product of all positive integers less than or equal to `n`. By definition, `0!` is `1`.

The factorial can be expressed recursively as:

`n! = n * (n-1)!` for `n > 0`. `1` if `n = 0`.

Here is how you can implement this in R:

```
factorial <- function(n) {
  if (n == 0) {
    return(1)
```

```
  } else {
    return(n * factorial(n - 1))
  }
}
print(factorial(5))
```

```
## [1] 120
```

The factorial example illustrates recursion, a key concept in building decision trees. Decision trees use recursion by repeatedly splitting the data into smaller subsets until a stopping criterion is met. Just as the factorial function calls itself to compute smaller subproblems, a decision tree algorithm recursively partitions the data, refining the model step by step. Now, we apply this idea to construct a decision tree. Using the same dataset and Gini impurity measure, we will build a tree by recursively finding the best split at each node. We will also introduce a complexity parameter to control tree depth.

```
X <- df[, -ncol(df)]
y <- df$y

GI <- function(y, x, sp){
  G <- 2 * mean(y) * (1 - mean(y))
  GL <- 2 * mean(y[x <= sp]) * (1 - mean(y[x <= sp]))
  GR <- 2 * mean(y[x > sp]) * (1 - mean(y[x > sp]))
  pL <- length(x[x <= sp]) / length(x)
  pR <- length(x[x > sp]) / length(x)
  delta <- G - (pL * GL + pR * GR)
  return(delta)
}

best.split <- function(x, y, cp) {
  splits <- sort(unique(x))
  delta <- numeric(length(splits) - 1)
  for (i in 1:(length(splits) - 1)) {
    delta[i] <- GI(y, x, splits[i])
  }
  if (length(delta) == 0 || max(delta) < cp) {
    return(list(gain = 0, split_point = NA)) # No valid split
  }
  max_gain <- max(delta)
  split_point <- splits[which.max(delta)]
  return(list(gain = max_gain, split_point = split_point))
}

recursive_split <- function(X, y, cp, depth = 0, maxDepth = 10) {
  if (ncol(X) == 0 || length(unique(y)) <= 1 || depth >= maxDepth) {
    most_common_class <- names(sort(table(y), decreasing = TRUE))[1]
    return(list("Leaf", most_common_class))
  }

  gains <- sapply(1:ncol(X), function(i) best.split(X[, i], y, cp)$gain)
```

```
  if (all(is.na(gains))) {
    most_common_class <- names(sort(table(y), decreasing = TRUE))[1]
    return(list("Leaf", most_common_class))
  }
  best_feature_idx <- which.max(gains)
  best_feature <- names(X)[best_feature_idx]
  best_split_value <- best.split(X[, best_feature_idx], y, cp)$split_point

  print(paste("Depth:", depth, "Best Feature:", best_feature, "Gain:",
              gains[best_feature_idx]))

  if (is.na(best_split_value) || gains[best_feature_idx] < cp) {
    most_common_class <- names(sort(table(y), decreasing = TRUE))[1]
    return(list("Leaf", most_common_class))
  }

  left_idx <- X[, best_feature_idx] <= best_split_value
  right_idx <- X[, best_feature_idx] > best_split_value
  left_X <- X[left_idx, ]
  right_X <- X[right_idx, ]
  left_y <- y[left_idx]
  right_y <- y[right_idx]

left_tree <- if (sum(left_idx) > 0)
  recursive_split(left_X, left_y, cp, depth + 1, maxDepth)
else
  list("Leaf", names(sort(table(left_y), decreasing = TRUE))[1])

right_tree <- if (sum(right_idx) > 0)
  recursive_split(right_X, right_y, cp, depth + 1, maxDepth)
else
  list("Leaf", names(sort(table(right_y), decreasing = TRUE))[1])

return(list(
  "Node" = list("Split" = best_split_value, "Feature" = best_feature),
  "Left" = left_tree,
  "Right" = right_tree
))
}

cp <- 0.01  # Example complexity parameter
tree <- recursive_split(X, y, cp)
```

```
## [1] "Depth: 0 Best Feature: INSYS Gain: 0.283280133210703"
## [1] "Depth: 1 Best Feature: PVENT Gain: 0.0607787274453941"
## [1] "Depth: 2 Best Feature: PVENT Gain: 0.0650887573964497"
## [1] "Depth: 3 Best Feature: INCAR Gain: 0.5"
## [1] "Depth: 1 Best Feature: PVENT Gain: 0.114971779883088"
## [1] "Depth: 2 Best Feature: INCAR Gain: 0.0452409280190363"
## [1] "Depth: 3 Best Feature: REPUL Gain: 0.01125"
```

```
## [1] "Depth: 4 Best Feature: FRCAR Gain: 0.375"
```

And, let's print the tree structure.

```
# Function to print the tree structure
print_tree <- function(node, depth=0) {
  if (is.null(node)) {
    return()
  }

  # Indent to represent the depth
  indent <- paste(rep("  ", depth), collapse = "")

  if (!is.null(node$Leaf)) {
    cat(indent, "Leaf:", node$Leaf, "\n")
  } else {
    cat(indent, "Node: If", node$Node$Feature, "<=", node$Node$Split, "\n")
    cat(indent, "Left:\n")
    print_tree(node$Left, depth + 1)
    cat(indent, "Right:\n")
    print_tree(node$Right, depth + 1)
  }
}

print_tree(tree)
```

```
##  Node: If INSYS <= 18.7
##  Left:
##    Node: If PVENT <= 2
##    Left:
##      Node: If <=
##      Left:
##      Right:
##    Right:
##      Node: If PVENT <= 15.5
##      Left:
##        Node: If <=
##        Left:
##        Right:
##      Right:
##        Node: If INCAR <= 1.19
##        Left:
##          Node: If <=
##          Left:
##          Right:
##        Right:
##          Node: If <=
##          Left:
##          Right:
##  Right:
##    Node: If PVENT <= 16
```

```
##     Left:
##       Node: If INCAR <= 1.44
##       Left:
##         Node: If <=
##         Left:
##         Right:
##       Right:
##         Node: If REPUL <= 1093
##         Left:
##           Node: If <=
##           Left:
##           Right:
##         Right:
##           Node: If FRCAR <= 86
##           Left:
##             Node: If <=
##             Left:
##             Right:
##           Right:
##             Node: If <=
##             Left:
##             Right:
##     Right:
##       Node: If <=
##       Left:
##       Right:

## NULL
```

Obviously this is not a "trained" tree! We have manually built a decision tree by recursively splitting the dataset based on the best variable at each step. However, this approach is inefficient for large datasets, as it requires manually selecting splits and handling stopping criteria.

Decision trees rely on recursive partitioning to create splits that improve classification accuracy. However, uncontrolled tree growth can lead to overfitting, where the model captures noise rather than patterns. To address this, we use pruning, which removes unnecessary splits and simplifies the tree while maintaining predictive performance. Instead of continuing with a custom implementation, we will now use the `rpart` package, which automates tree construction efficiently.

13.2 Step-by-Step Tree Construction for Binary Classification

In this section, we will outline the key steps involved in constructing a decision tree for binary classification using the Gini index as a measure of impurity. Starting from defining the problem and calculating impurity, we will explain how to choose optimal split points, perform recursive partitioning, and prune the tree to prevent overfitting. Finally, we will discuss how to make predictions for new observations based on the majority class rule within each terminal node.

1. Setup: Define the Problem and Data

For binary classification, the outcome Y has two possible classes, which we can denote as Class 0 and Class 1. We have a total of N observations with predictors $X_1, X_2, \ldots, X_p$. The objective is to construct a decision tree that maximizes class purity within each leaf, minimizing misclassifications between the two classes.

2. Define Gini Index for a Node t

The Gini Index is used to measure the impurity of a node based on the distribution of the two classes within that node. For binary classification, the Gini Index is simplified to:

$$\text{Gini}(t) = 1 - (p_0^2 + p_1^2) \tag{13.4}$$

where p_0 is the proportion of observations in Class 0 and p_1 is the proportion in Class 1 within node t. A Gini index of 0 indicates perfect purity (all observations belong to one class), while higher values indicate greater impurity. The Gini index is preferred due to its computational efficiency and its ability to handle imbalanced class distributions.[2]

3. Define Gini Gain for Splitting

To decide on the best splits, the Gini gain is calculated, which measures the reduction in Gini index achieved by making a split. It is defined as:

$$\Delta\text{Gini} = \text{Gini (parent)} - \frac{N_L}{N}\text{Gini (left)} - \frac{N_R}{N}\text{Gini (right)} \tag{13.5}$$

In this equation, N_L and N_R represent the number of observations in the left and right child nodes, respectively, and N is the number of observations in the parent node. A split with a higher Gini gain indicates a more effective separation between the two classes, and thus is preferred.[3]

4. Choose Optimal Split Point for Each Variable

For each predictor X_j, the algorithm identifies potential split points by sorting the data and evaluating midpoints between consecutive unique values. If X_1 is a continuous variable, such as income, the algorithm considers multiple split points across the range to find the one that maximizes the Gini gain.[4] For an integer variable like X_2 representing age, which ranges from 16 to 65, the algorithm tests each integer as a split point and selects the one that leads to the maximum Gini gain in the resulting nodes, ensuring the best class separation. For binary predictors like X_3, which might represent home ownership status (Yes/No), the only split point is between the two categories. In all cases, the algorithm calculates the Gini gain for each potential split and chooses the point that maximizes this gain.

Mathematically, the same Gini gain criterion is calculated at each node one by one for every variable:

$$\Delta\text{Gini} = \text{Gini (parent)} - \frac{N_L}{N}\text{Gini (left)} - \frac{N_R}{N}\text{Gini (right)} \tag{13.6}$$

Then, the optimal split point s_j^* for each variable X_j is chosen as:

$$(X_j^*, s_j^*) = \arg\max_{(X_j, s_j)} \Delta\text{Gini}(X_j, s_j) \tag{13.7}$$

This approach ensures that the chosen splits create the most distinct separation between

Class 0 and Class 1, making the resulting nodes as pure as possible. By tailoring the split criteria based on the type of predictor—continuous, integer, or binary—the algorithm effectively captures different patterns in the data to improve classification accuracy.

5. Recursive Partitioning: Split Left and Right Nodes

Once the optimal split point is determined for a variable which gives the highest gini gain, the data is divided into left and right nodes based on that split. Each of these nodes is then treated as a new parent node for subsequent splits until it reacheds to stopping criteria. Let's provide a detailed explanation of the recursive partitioning process with an example:

Step 1. **Initial Split:** X_3 is chosen as the first split because it gave the max Gini gain.
- This creates a left node (e.g., $X_3 \leq s_3$) and a right node (e.g., $X_3 > s_3$).

Step 2. **Recursive Splitting for Left and Right Nodes:** For the left node:
- Evaluate X_1, X_2, and other variables/features by trying all possible split points.
- Choose the variable and split point that gives the max Gini gain within the left node.

- For the **right node**:
  - Similarly, evaluate X_1, X_2, and other variables with all split points.
  - Choose the variable and split point that maximizes gini gain for the right node.

The left and right nodes can split on different variables (e.g., X_1 for left and X_2 for right). - Use different split points for the same variable (evenif used before) if they choose the same one.

Step 3. **Stopping Criteria:** Splitting continues recursively for left and right nodes until: - **Minimum leaf size** is reached. Splitting stops if a node contains fewer than a specified minimum number of observations (e.g., 5 or 10). This prevents overfitting by avoiding splits that result in very small, unreliable leaves.

- **Max tree depth** is reached. A pre-defined maximum depth can be set (e.g., 10 levels). Once this depth is reached, no further splits are allowed. This ensures that the tree remains interpretable and avoids becoming too complex.
- **Insufficient Gini Gain:** If the maximum possible Gini Gain (ΔGini) for any potential split falls below a certain threshold (e.g., 0.01), no further splits are made. This criterion prevents the algorithm from making splits that provide negligible improvements in class purity.

6. Define Class Predictions for Leaf Nodes

Once the tree is fully grown, each terminal node, also known as a leaf, represents a subset of the data with a predominant class. To make predictions for new observations, the decision tree uses the most frequent class within the leaf that the observation falls into. This strategy, known as the majority class rule, helps minimize the misclassification rate within each leaf by aligning predictions with the most common outcome among the observations assigned to that leaf.

Mathematically, the predicted class for a leaf t is determined by selecting the class that has the highest proportion of observations within that leaf. This can be expressed using the following formula:

$$\hat{Y}_t = \arg \max_{j \in \{0,1\}} p_j \tag{13.8}$$

In this formula, j represents the possible classes (0 or 1 in the case of binary classification), and p_j is the proportion of observations in class j within the leaf t. The notation arg max is used to indicate that the predicted class $\hat{Y}_t$ will be the one that maximizes p_j, effectively choosing the most frequent class in that leaf.

To illustrate this with a simple example, imagine a leaf with five observations classified into two possible classes, 0 and 1. Suppose the class distribution in this leaf is as follows: three observations belong to class 1, and two belong to class 0. The proportions for each class would then be $p_0 = \frac{2}{5} = 0.4$ and $p_1 = \frac{3}{5} = 0.6$. According to the majority class rule, the predicted class for this leaf would be 1, as it has a higher proportion than class 0.

The rationale behind this approach is straightforward. By assigning the most frequent class within each leaf, the tree minimizes the chances of making incorrect predictions for new observations that fall into that leaf. For instance, if the tree were to predict class 0 instead of class 1 in the previous example, it would result in a higher misclassification rate since three out of five observations actually belong to class 1. Therefore, predicting the majority class not only simplifies the decision-making process but also ensures that the predictions are as accurate as possible given the information available within each leaf.

This strategy is particularly effective in classification tasks because it leverages the natural partitioning of data created by the tree. As each split is made to maximize class purity by reducing impurity (measured by criteria like Gini Index), the resulting leaves tend to have a dominant class. Thus, predicting the most frequent class aligns with the underlying goal of decision trees: to create clear and interpretable rules that separate different classes based on the available predictors.

7. Pruning the Tree Using Cost-Complexity

In decision tree algorithms, the initial tree that is grown is often referred to as a "fully grown" or "maximal" tree. This tree is constructed by recursively partitioning the data until no further splits can significantly reduce impurity or until certain stopping criteria, such as minimum leaf size or maximum depth, are met. While this fully grown tree might fit the training data very well, it is also prone to overfitting. Overfitting occurs when the tree captures noise and details that are specific to the training data but do not generalize well to unseen data. To address this issue, a process called pruning is applied to simplify the tree and improve its ability to generalize to new data.

Pruning involves systematically reducing the size of the fully grown tree by removing branches that have little predictive power. The decision of which branches to remove is guided by a cost-complexity criterion, which balances the trade-off between the fit of the tree to the training data and its complexity. The cost-complexity function is defined as follows:

$$C_\alpha(T) = \sum_{t \in T} N_t \cdot \text{Gini}(t) + \alpha \cdot |T| \tag{13.9}$$

In this function, T represents the set of terminal nodes (or leaves), N_t is the number of observations in leaf t, and $|T|$ is the number of leaves in the tree. The parameter α is a complexity parameter that controls the penalty for the number of leaves. When α is set to zero, the penalty term disappears, and the cost-complexity function reduces to a measure of impurity alone, favoring larger trees. As α increases, the penalty for having more leaves also increases, encouraging a simpler tree with fewer branches.

There are several alternative complexity functions for binary classification in decision trees, each offering a different balance between model complexity and accuracy. One option is the entropy-based cost-complexity function[5], which is more sensitive to class distributions than Gini. Another is the misclassification error-based function[6], which directly minimizes the error rate but is less sensitive to class proportions. For imbalanced datasets, the weighted Gini function[7] helps emphasize the minority class. Similarly, a cost-sensitive complexity function[8] accounts for different misclassification costs. Lastly, the penalized likelihood function[9] balances fit and complexity based on the likelihood of the model. These alternatives make decision trees more adaptable to various classification tasks.

The pruning process involves incrementally increasing α and evaluating the cost-complexity function $C_\alpha(T)$ for the resulting trees. For each value of α, branches are removed if doing so results in a lower value of $C_\alpha(T)$. This process produces a sequence of nested trees, each simpler than the previous one. The goal is to identify the optimal tree that minimizes the cost-complexity function, striking a balance between complexity and fit.

Pruning inevitably involves a trade-off. By reducing the number of branches, we simplify the tree, making it more interpretable and less likely to overfit. However, this simplification comes at the cost of losing some fit to the training data. In essence, pruning sacrifices some accuracy on the training set in exchange for better generalization to unseen data. The key is to find the right level of complexity that maximizes performance on new data, which is typically achieved using cross-validation.

To select the optimal value of α, cross-validation is employed first. The training data is divided into several folds, and for each fold, a tree is pruned at different levels of α. The misclassification rate or another performance metric is recorded for each level of α. The optimal α^* is chosen as the one that minimizes the cross-validation error:

$$\alpha^* = \arg\min_{\alpha} \text{CV Error}(\alpha) \tag{13.10}$$

The tree corresponding to α^* is then selected as the final pruned tree. This tree represents a compromise between complexity and fit, offering better generalization performance while retaining interpretability. In this way, pruning not only simplifies the model but also improves its predictive accuracy on new data by preventing overfitting.

8. Cross-Validation to Select Optimal Tree Size

Cross-validation is used to select the optimal complexity parameter α^* that minimizes the classification error on unseen data (overfitting) by pruning the tree. The training data is divided into K folds. For each fold, a tree is trained on $K-1$ folds and validated on the remaining fold. The cross-validation error is calculated as:

$$\text{CV Error}(\alpha) = \frac{1}{K}\sum_{k=1}^{K} \text{Misclassification Rate (Validation Set } k \text{ for Tree with } \alpha) \tag{13.11}$$

The optimal α^* is chosen as the value that minimizes the cross-validation error:

$$\alpha^* = \arg\min_{\alpha} \text{CV Error}(\alpha) \tag{13.12}$$

Selecting α^* based on cross-validation ensures that the tree has the right balance between complexity and predictive accuracy.

9. Variable Importance in Decision Trees

After constructing and pruning the tree, it is often useful to assess the importance of each predictor variable in making accurate predictions. Variable importance measures how much each predictor contributes to reducing impurity across all splits in the tree, helping to identify which variables have the most influence on the classification outcome.

The importance of a variable X_j is determined by the total reduction in Gini index resulting from splits using X_j. Specifically, at each split where X_j is used, the reduction in Gini is calculated based on the difference between the Gini index of the parent node and the weighted Gini indices of the child nodes. The importance score for X_j is then obtained by summing these Gini reductions across all nodes where X_j was used for splitting.

In practice, the Gini index is calculated for each node using the observed distribution of classes within that node. For instance, if a node contains 60 observations from Class 0 and 40 from Class 1, the Gini index for that node would be computed based on these proportions. As the tree grows, each split ideally leads to child nodes with lower Gini indices, indicating greater class purity. The reductions in Gini index caused by splits using a particular variable accumulate to form the importance score for that variable.

Additionally, permutation importance can be used as an alternative method. This involves randomly shuffling the values of X_j across observations while keeping other variables unchanged, and then measuring the drop in predictive accuracy. A substantial decrease in accuracy suggests that the variable is important for making predictions.

By identifying the most influential variables, this step improves the interpretability of the tree and helps in understanding which predictors have the most significant impact on the model's decisions.

10. Final Prediction for New Observations

To predict the class of a new observation, the pruned tree is traversed starting from the root node. The observation is assigned to a leaf node based on its predictor values, following the sequence of splits defined by the tree. The predicted class for the observation is the majority class within the assigned leaf. This method ensures that the prediction reflects the most common outcome among similar past observations, providing a clear and interpretable decision rule.

To illustrate this, consider a decision tree used to predict whether a customer will default on a loan based on their income, age, and home ownership status. After pruning, the data of a new applicant is input into the tree. Starting at the root, income of the applicant, treated as a continuous variable, is compared against a series of split points to determine which path to follow. Next, the age of applicant, an integer variable ranging from 16 to 65, is evaluated to decide the subsequent branch. Finally, the binary predictor of home ownership status (Yes/No) is used to direct the applicant further down the tree. Eventually, the applicant reaches a leaf node that contains previous applicants with similar characteristics. Suppose this leaf has 70% of observations in the "No Default" class and 30% in the "Default" class. According to the majority class rule, the predicted outcome for this applicant would be "No Default." This approach ensures that the prediction reflects the most common outcome among past similar cases, making the decision both interpretable and based on historical patterns.

13.3 Classification Decision Trees with rpart

The `rpart` package automates the tree-building process by determining the optimal depth, minimum observations required for splits, and minimum size for leaf nodes. There are two ways to control this:
1. Setting control parameters in `rpart` to constrain tree growth.
2. Growing a deep tree and pruning it using cross-validation to find the best tree size.

Pruning simplifies a tree by removing splits that add little to predictive accuracy, preventing overfitting. Now, we train a decision tree using `rpart` and apply pruning to improve its performance. Here are the control parameters for the `rpart` package:

- `cp`: Complexity parameter. A split will only be made if the improvement in the model (i.e. increase in the overall goodness of split) is larger than `cp`.
- `minsplit`: the minimum number of observations required in a node to allow splitting.
- `minbucket`: the minimum number of observations in any terminal node. If only one of `min-bucket` or `minsplit` is specified, the code either sets `minsplit` to `minbucket*3` or `minbucket` to `minsplit/3`, as appropriate.
- `maxdepth`: Maximum tree depth. If `maxdepth` is not `NULL`, the tree is limited to `maxdepth` splits from the root.

The `rpart` function has many other parameters that control the tree-building process. You can find more details in the documentation by running `?rpart.control` in the R console.

Here are the default values for the control parameters:

```
rpart.control(minsplit = 20, minbucket = round(minsplit/3), cp = 0.01,  maxcompete
= 4, maxsurrogate = 5, usesurrogate = 2, xval = 10,  surrogatestyle = 0, maxdepth =
30, ...)
```

Let's train a decision tree using the `rpart` package and visualize the tree structure.

```
library(rpart)

# Train the tree

tree_model <- rpart(PRONO ~ ., data = myocarde)
print(tree_model)
```

```
## n= 71
##
## node), split, n, loss, yval, (yprob)
##       * denotes terminal node
##
## 1) root 71 29 SURVIE (0.40845070 0.59154930)
##   2) INSYS< 18.85 27  3 DECES (0.88888889 0.11111111) *
##   3) INSYS>=18.85 44  5 SURVIE (0.11363636 0.88636364)
##     6) REPUL>=1094.5 7  3 DECES (0.57142857 0.42857143) *
##     7) REPUL< 1094.5 37  1 SURVIE (0.02702703 0.97297297) *
```

```
printcp(tree_model)
```

```
##
## Classification tree:
## rpart(formula = PRONO ~ ., data = myocarde)
##
## Variables actually used in tree construction:
## [1] INSYS REPUL
##
## Root node error: 29/71 = 0.40845
##
## n= 71
##
##         CP nsplit rel error  xerror    xstd
## 1 0.724138      0   1.00000 1.00000 0.14282
## 2 0.034483      1   0.27586 0.55172 0.12140
## 3 0.010000      2   0.24138 0.51724 0.11861
```

```
# Plot the tree
library(rpart.plot) # You can use plot() but prp() is much better
prp(
  tree_model,
  type = 2,
  extra = 1,
  split.col = "black",
  split.border.col = "black",
  box.col = "gray90"
)
```

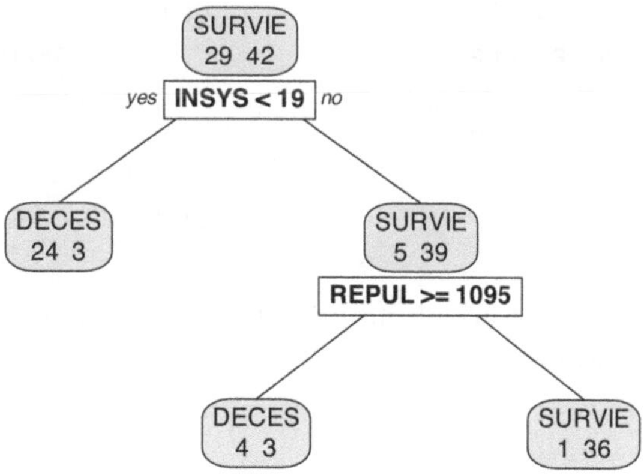

Why `rpart` chose `REPUL` in the second split, while our algorithm picked `PVENT`?

13.3.1 Pruning the Tree

To prevent overfitting, we first grow the tree fully by setting `cp = 0` and then prune it using cross-validation to find the optimal tree size. The `printcp` function displays the complexity parameter table, which shows the cross-validated error rate for different tree sizes. We can then prune the tree using the optimal complexity parameter.

```r
# Grow the tree to the maximum depth
tree_model <- rpart(PRONO ~ ., data = myocarde,
                    control = rpart.control(minsplit = 2,
                                            minbucket = 1,
                                            cp = 0
                                            )
                    )

# Let's see the tree
library(rpart.plot) # You can use plot() but prp() is much better
prp(
  tree_model,
  type = 2,
  extra = 1,
split.col = "black",
  split.border.col = "black",
  box.col = "gray90"
)
```

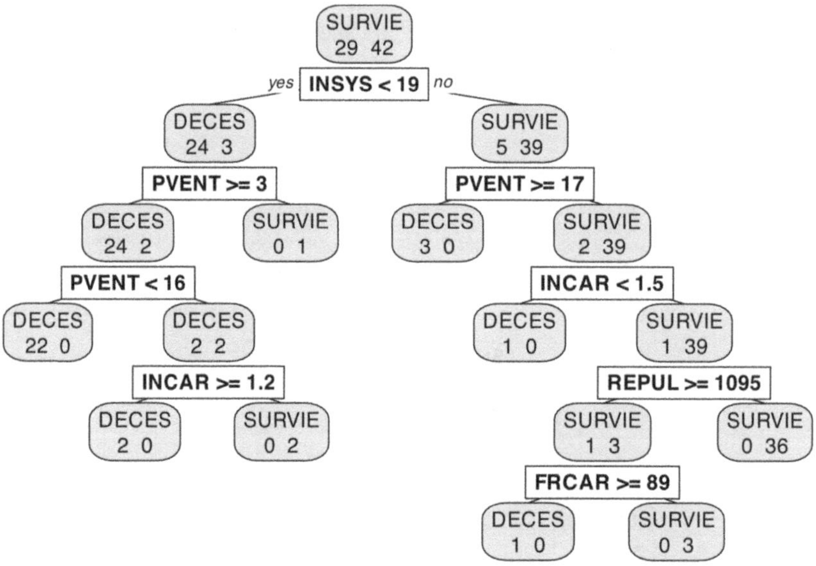

Now we are going to prune it. But now, instead of manually selecting the optimal depth, we use rpart's built-in cross-validation:

```r
printcp(tree_model)
```

```
##
## Classification tree:
## rpart(formula = PRONO ~ ., data = myocarde, control = rpart.control(minsplit = 2,
##     minbucket = 1, cp = 0))
##
## Variables actually used in tree construction:
## [1] FRCAR INCAR INSYS PVENT REPUL
```

```
##
## Root node error: 29/71 = 0.40845
##
## n= 71
##
##          CP nsplit rel error  xerror    xstd
## 1 0.724138      0   1.000000 1.00000 0.14282
## 2 0.103448      1   0.275862 0.55172 0.12140
## 3 0.034483      2   0.172414 0.44828 0.11237
## 4 0.017241      6   0.034483 0.34483 0.10107
## 5 0.000000      8   0.000000 0.34483 0.10107
```

```
min_cp = tree_model$cptable[which.min(tree_model$cptable[,"xerror"]),"CP"]
min_cp
```

```
## [1] 0.01724138
```

Now, we prune the tree using the best cp value:

```
pruned_tree_model <- prune(tree_model, cp = min_cp)
printcp(pruned_tree_model)
```

```
##
## Classification tree:
## rpart(formula = PRONO ~ ., data = myocarde, control = rpart.control(minsplit = 2,
##     minbucket = 1, cp = 0))
##
## Variables actually used in tree construction:
## [1] INCAR INSYS PVENT
##
## Root node error: 29/71 = 0.40845
##
## n= 71
##
##          CP nsplit rel error  xerror    xstd
## 1 0.724138      0   1.000000 1.00000 0.14282
## 2 0.103448      1   0.275862 0.55172 0.12140
## 3 0.034483      2   0.172414 0.44828 0.11237
## 4 0.017241      6   0.034483 0.34483 0.10107
```

```
prp(
  pruned_tree_model,
  type = 2,
  extra = 1,
 split.col = "black",
  split.border.col = "black",
  box.col = "gray90"
)
```

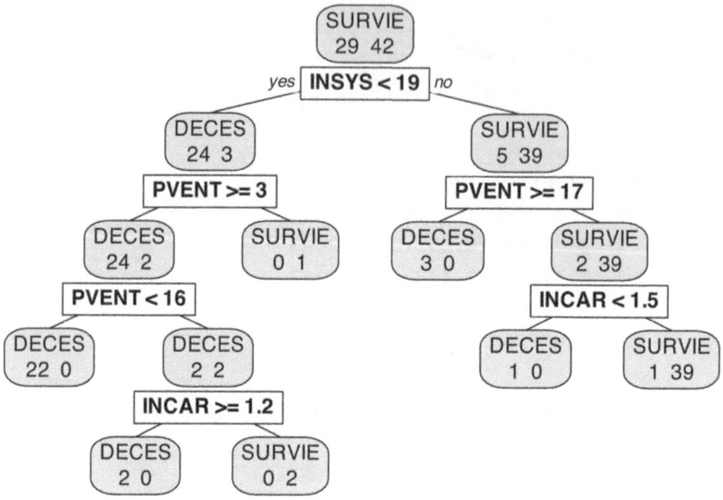

Next, we extend this to regression trees to predict continuous outcomes.

13.4 Regression Trees

In this section, we will outline the key steps involved in constructing a decision tree for regression using mean-squared error (MSE) as a measure of impurity. The process is similar to that of classification trees, with MSE replacing the Gini index to guide splits and pruning. We will explain how the tree is built, how splits are chosen to minimize MSE, and how pruning is performed to prevent overfitting. Additionally, we will discuss variable importance and the prediction process for new observations. In the section on the `rpart` package, we will cover the default values for these parameters and how to modify them.

1. Setup: Define the Problem and Data

In regression trees, the outcome Y is continuous rather than binary, with the objective being to minimize the mean-squared error within each node. We have N observations with predictors $X_1, X_2, \ldots, X_p$. The tree is constructed by iteratively splitting the data to minimize MSE, aiming to create terminal nodes that are as pure as possible in terms of outcome values.

2. Define MSE for a Node t

The impurity of a node in regression trees is measured using MSE, which is equivalent to the sum of squared errors divided by the number of observations in the node. For a given node t, MSE is defined as:

$$\text{MSE}(t) = \frac{1}{N_t} \sum_{i \in t} (Y_i - \bar{Y}_t)^2 \tag{13.13}$$

where N_t is the number of observations in node t and $\bar{Y}_t$ is the mean outcome for those observations. Lower MSE indicates greater purity in terms of outcome values, making it an effective criterion for guiding splits.

3. Define MSE Reduction for Splitting

To choose the best split at each node, the tree evaluates the reduction in MSE that would result from potential splits. For a parent node t split into left and right child nodes, MSE reduction is calculated as:

$$\Delta\text{MSE} = \text{MSE (parent)} - \frac{N_L}{N_t}\text{MSE (left child)} - \frac{N_R}{N_t}\text{MSE (right child)} \tag{13.14}$$

where N_L and N_R are the numbers of observations in the left and right child nodes, respectively. The split that maximizes ΔMSE is chosen, as this indicates the greatest improvement in node purity.

4. Choose Optimal Split Point for Each Variable

The procedure for selecting split points in regression trees is similar to that used in classification trees, but instead of Gini index, the algorithm uses MSE as the criterion for measuring impurity. For continuous variables, the algorithm tests multiple potential split points across the range of values, typically using midpoints between consecutive unique values, and evaluates each based on MSE reduction. For integer variables, each integer value is considered as a potential split point, and for binary variables, the split point is simply between the two categories. The split that results in the greatest reduction in MSE is chosen as the optimal split point. This process is repeated for all variables one by one to determine the best split at for each one.

5. Recursive Partitioning: Split Left and Right Nodes

In Step 4, the algorithm evaluates potential split points for each variable and chooses the one that provides the greatest MSE reduction. Once an optimal split point for the specific variable is determined, the data is divided into left and right child nodes. The algorithm then recursively repeats the splitting process for each child node (left and right separately), evaluating potential splits for all variables to maximize MSE reduction at each level of the tree. It then selects the best variable and split point for the left node, performs a similar process for the right node, and chooses the best option there as well. These "best" variables and split points can be different for the left and right nodes. This recursive partitioning continues until one of the stopping criteria is met: a minimum leaf size, maximum tree depth, or insufficient MSE reduction. These stopping criteria can be adjusted based on the desired complexity and interpretability of the tree. In the **rpart** section, we will discuss the default settings for these parameters and how they can be modified.

6. Define Leaf Predictions for Regression

In regression trees, the predicted outcome for each terminal node is simply the mean of the observed outcomes in that node:

$$\hat{Y}_t = \bar{Y}_t = \frac{1}{N_t}\sum_{i\in t} Y_i \tag{13.15}$$

This approach minimizes the MSE within each node and ensures that predictions are representative of the observed data.

7. Pruning the Tree Using Cost-Complexity

To prevent overfitting, the fully grown tree is pruned based on a cost-complexity function similar to the one used in classification trees, but with MSE replacing the Gini index:

$$C_\alpha(T) = \sum_{t \in T} N_t \cdot \text{MSE}(t) + \alpha \cdot |T| \tag{13.16}$$

Here, α is a complexity parameter that penalizes larger trees. As α increases, the penalty for having more leaves becomes greater, encouraging simpler trees. This method balances model fit and complexity to improve the generalization to new data.

8. Cross-Validation to Select Optimal Tree Size

Cross-validation is used to select the optimal value of α that minimizes prediction error on unseen data. The process involves splitting the data into K folds, training the tree on $K-1$ folds, and validating it on the remaining fold. For each value of α, the cross-validation error is calculated as the average MSE across all folds:

$$\alpha^* = \arg\min_{\alpha} \text{CV Error}(\alpha) \tag{13.17}$$

This ensures that the tree is neither too complex nor too simple, achieving a balance between fit and interpretability.

9. Variable Importance in Regression Trees

Variable importance is assessed based on the total MSE reduction attributable to each predictor across all splits. For each variable X_j, the importance score is computed by summing the MSE reductions at every node where X_j was used for splitting. Higher scores indicate greater importance in predicting outcomes. Specifically, for a variable X_j, the importance score is computed as:

$$\begin{aligned} \text{Importance}(X_j) &= \sum_{\text{all splits on } X_j} \Delta\text{MSE} \\ &= \sum_{\text{all splits on } X_j} \left[\text{MSE(parent)} - \frac{N_L}{N}\text{MSE(left)} - \frac{N_R}{N}\text{MSE(right)}\right] \end{aligned} \tag{13.18}$$

where N_L and N_R are the number of observations in the left and right child nodes, respectively. This approach helps identify which predictors have the most significant impact on the model's predictions.

Alternatively, permutation importance can be used, which measures the drop in predictive accuracy when a variable's values are randomly shuffled. Permutation importance is calculated by randomly shuffling the values of a variable X_j across observations while keeping all other variables unchanged. The model's predictive accuracy is then measured on a validation set. The importance score is the difference between the original accuracy and the accuracy after shuffling $\text{Importance}(X_j) = \text{MSE(shuffled)} - \text{MSE(original)}$. A larger drop in accuracy indicates a more influential variable. Both methods help identify the most influential predictors in the model.

10. Final Prediction for New Observations

To predict outcomes for new observations, the pruned tree is traversed based on the values of the predictors. Each observation is assigned to a terminal node, and the predicted outcome is the mean of observed outcomes in that node. This approach ensures that predictions are interpretable and grounded in the observed data, similar to the majority class rule used in classification trees but adapted for continuous outcomes.

13.5 Building and Interpreting Regression Trees with `rpart`: A Step-by-Step Guide

In this section, we build a regression tree using the `rpart` package in R, focusing on the `Boston` dataset from the `MASS` package. The process is divided into six steps: setting up and loading data, building the initial tree, selecting optimal complexity through cross-validation, pruning and visualizing the tree, making predictions and evaluating performance, and interpreting the results. This approach balances detail and simplicity, ensuring clarity while covering all key aspects.

Step 1: Setup and Load Data

To begin, we install and load the necessary packages—`MASS` for the dataset, `rpart` for building the tree, and `rpart.plot` for visualization. After loading the `Boston` dataset, which includes 506 observations and 14 variables, we examine its structure using basic summary functions. The outcome variable is `medv` (median house value), with other predictors such as `crim` (crime rate), `rm` (average rooms per dwelling), and `nox` (nitric oxides concentration). This setup helps us understand the characteristics of the data before constructing the tree.

```
library(MASS)
library(rpart)
library(rpart.plot)

data("Boston")
#head(Boston)   #summary(Boston)
```

Step 2: Build the Initial Tree

When building a regression tree, we need to decide on certain control parameters that influence the growth and complexity of the tree. These parameters determine when to stop splitting nodes, how deep the tree can grow, and how complex the final model will be. By specifying `method = "anova"` in the `rpart` function, we use mean-squared error (MSE) as the splitting criterion to build the tree. In this step, we discuss the key parameters, their default values, and why we might choose to modify them in this simulation. Choosing appropriate values for these parameters is crucial because setting them too high can lead to underfitting, where the tree is too simple to capture the patterns of the data. On the other hand, setting them too low can cause overfitting, making the tree excessively complex and less able to generalize to new data.

Parameters to Decide On:
- **`minsplit`:** Controls the minimum number of observations required to attempt a split. The default value is **20**, but we reduce it to **2**. - **`minbucket`:** Sets the minimum number of observations that must exist in each terminal node (leaf). The default value is **7**.
- **`maxdepth`:** Limits how deep the tree can grow. The default value is **30**.
- **`cp` (Complexity Parameter):** Prevents the tree from growing too large by requiring a minimum reduction in MSE for additional splits. The default value is **0.01**. However, we set `cp` to **0** to disable pre-pruning entirely, allowing a fully grown tree before pruning based on cross-validation.

We change the default values of `minsplit` and `cp` to build a fully expanded tree initially, capturing as much detail as possible, which we will later simplify through pruning based on cross-validation results.

```
set.seed(42)
initial_tree <- rpart(medv ~ ., data = Boston, method = "anova",
                      control = rpart.control(cp = 0, minsplit = 2))
#summary(initial_tree)
```

Step 3: Perform Cross-Validation to Select Optimal `cp`

To prevent overfitting, we perform 10-fold cross-validation using the `printcp` function, which generates a table with several important columns: `cp` (complexity parameter), `nsplit` (number of splits), `rel error` (relative error), `xerror` (cross-validated error), and `xstd` (standard deviation of the cross-validated error). This table helps us identify the optimal `cp` that minimizes cross-validation error. By selecting the `cp` with the lowest cross-validated error, we ensure the final tree is not only accurate but also generalizes well to new data. Keep in mind lower `cp` values lead to larger trees and higher `cp` leads to simpler trees.

```
#printcp(initial_tree)
best_cp <- initial_tree$cptable[which.min(initial_tree$cptable[, "xerror"]), "CP"]
cat("Optimal cp:", best_cp, "\n")
```

```
## Optimal cp: 0.00743042
```

Step 4: Prune and Visualize the Tree

After determining the optimal `cp`, we prune the tree to balance complexity and interpretability. The `prune` function removes branches that provide minimal reduction in MSE, using the selected `cp` value to simplify the tree without significantly affecting accuracy. We visualize the pruned tree with `rpart.plot`, which clearly displays split conditions and terminal nodes, making it easier to interpret the influence of different predictors on the median house value (`medv`).

Using `type = 2` ensures that labels for splits are placed inside the nodes, while `fallen.leaves = TRUE` arranges terminal nodes at the same level for better readability. This setup helps trace decision paths more clearly, enhancing the interpretability of the tree while maintaining predictive power.

```
pruned_tree <- prune(initial_tree, cp = best_cp)
rpart.plot(pruned_tree, type = 2, fallen.leaves = TRUE, cex = 0.8,
           box.col = "gray90", branch.col = "black", split.col = "black",
           main = "Pruned Regression Tree for Boston Data")
```

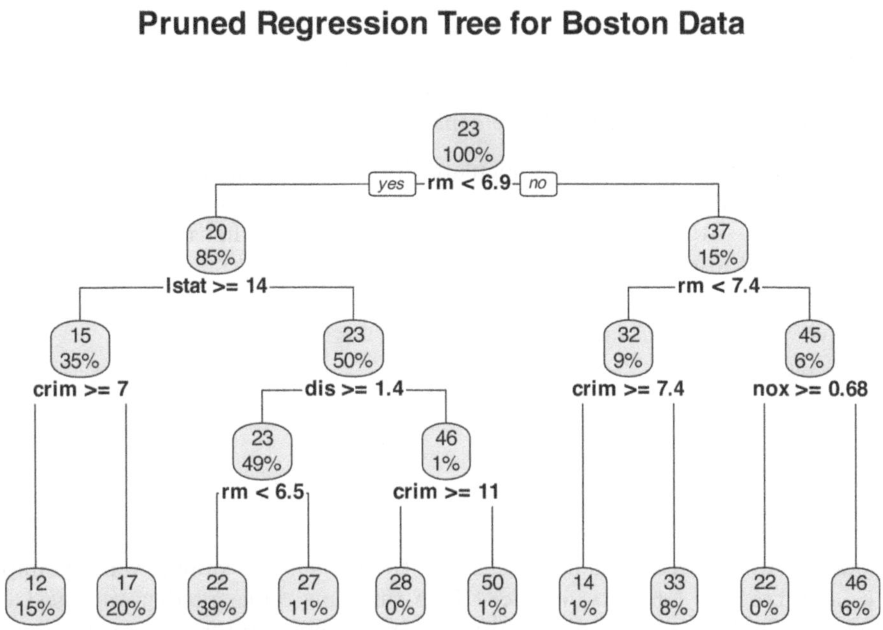

–> The pruned regression tree for the Boston data predicts the median house value (`medv`) based on factors like the average number of rooms (`rm`), percentage of lower status population (`lstat`), per capita crime rate (`crim`), and nitric oxide concentration (`nox`). Internal nodes represent decision points with split conditions, directing observations left if the condition is met ("Yes") and right if not ("No"). Each node shows the predicted `medv` at the top and the percentage of the dataset it represents below. The terminal nodes at the bottom display the average predicted `medv` for each subgroup of houses, with the percentages indicating how much of the overall data falls into that category. By focusing on a few key splits, the pruned tree maintains interpretability while effectively predicting house prices.

For example, following the path to the leftmost terminal node, we start at the root with the condition `rm < 6.9`. Observations satisfying this condition move left, leading to a predicted median house value of \$20,000, covering 85% of the data. The next split is `lstat >= 14`, where houses meeting this condition move left again, reducing the predicted `medv` to \$15,000 for 35% of the data. Further splitting on `crim >= 7` leads to the leftmost terminal node with an average `medv` of \$12,000, representing 15% of the data. This path suggests that fewer rooms, a higher percentage of lower-status population, and a high crime rate are associated with lower house prices. The pruned tree's structure, with simplified but meaningful splits, helps in interpreting how different factors influence housing values while avoiding overfitting.

Step 5: Make Predictions and Evaluate Performance

After pruning, we use the `predict` function to generate predictions for `medv` based on the pruned tree. Each prediction represents the mean `medv` of the terminal node in which an observation falls, providing a clear and interpretable estimate of house prices. For instance, consider a house with 4 rooms (`rm = 4`), 15% lower status population (`lstat = 15`), and a crime rate of 8 (`crim = 8`). Following the pruned tree, this house meets the conditions `rm < 6.9`, `lstat \ge 14`, and `crim \ge 7`, leading it to the leftmost terminal node. The predicted median house value for this path is \$12,000, reflecting the average price for similar houses in the data.

In contrast, suppose we have a house with 7 rooms (`rm = 7`), a crime rate below 7.4 (`crim = 5`), and a nitric oxide concentration below 0.68 (`nox = 0.5`). Following the pruned tree, this house takes the path `rm \ge 6.9` → `rm < 7.4` → `crim < 7.4` → `nox < 0.68`, ending at a terminal node with a predicted median house value of $46,000, the highest in the tree. This path suggests that larger houses in cleaner, safer neighborhoods command the highest prices, highlighting how the pruned tree captures meaningful patterns in the data.

To assess the tree's performance, we calculate the MSE by comparing actual and predicted `medv` values. A lower MSE indicates better predictive accuracy, providing a straightforward measure of how well the tree fits the data. This step ensures that the pruned tree not only interprets well but also predicts accurately.

```
predictions <- predict(pruned_tree, newdata = Boston)
mse <- mean((Boston$medv - predictions)^2)
cat("Mean Squared Error:", mse, "\n")
```

```
## Mean Squared Error: 11.76003
```

Step 6: Interpret Results

To understand which predictors most influence the model, we examine variable importance using `pruned_tree$variable.importance`. This function ranks predictors based on their contribution to reducing MSE, highlighting the most impactful factors for predicting `medv`. Additionally, the `pruned_tree$frame` function provides details about each node, such as the number of observations, mean `medv` per node, and the variables used for splitting. For a more in-depth examination, including split conditions and node statistics, the `summary` function offers a comprehensive view of the structure of the tree. This step helps connect the model's splits to real-world insights, enhancing the interpretability and practical relevance of the results.

```
importance <- pruned_tree$variable.importance
print(importance)
```

```
##          rm      lstat        dis        nox      indus        age    ptratio
## 23842.4392 15046.6276  6696.6716  5439.5008  5217.2458  4304.9329  4202.2984
##         tax       crim         zn        rad      black
##   4011.7488  3199.8925  1605.4575   814.2009   408.1277
```

```
# Create a bar plot for variable importance
barplot(sort(importance, decreasing = TRUE), horiz = TRUE, col = "gray60",
        main = "Variable Importance for Pruned Regression Tree",
        xlab = "Importance Score")
```

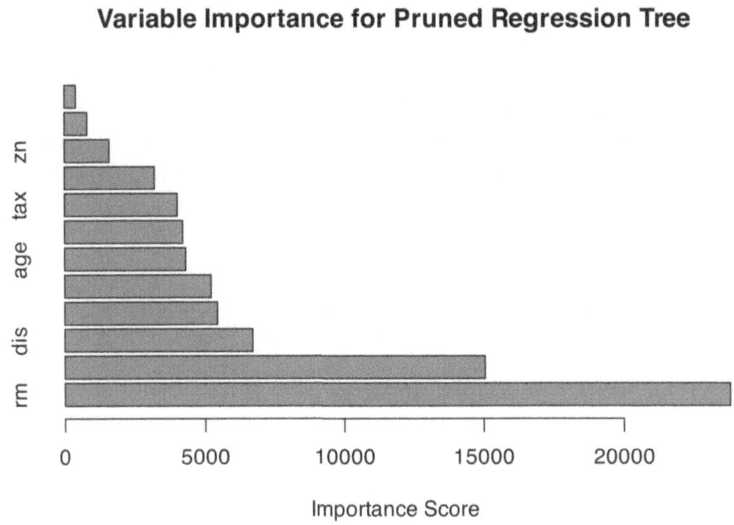


```
#print(pruned_tree$frame)
#summary(pruned_tree)
```

Here is the variable importance plot for the pruned regression tree based on the `Boston` dataset. The horizontal bars represent the importance scores of each predictor, indicating their contribution to reducing the MSE during tree construction.

From the plot, it is evident that the number of rooms (`rm`) and the percentage of lower status population (`lstat`) are the most influential variables for predicting the median house value (`medv`). Variables like crime rate (`crim`), nitric oxide concentration (`nox`), and distance to employment centers (`dis`) also play significant roles but to a lesser extent. The age of buildings (`age`) appears to have the least impact among the displayed predictors. This analysis highlights which factors most affect housing prices, providing valuable insights for further interpretation. By the way, the `rpart` package does not directly provide permutation importance scores. However, you can use the `vip` package, which is designed for computing permutation-based variable importance.

In summary, building and interpreting regression trees with `rpart` involves setting up the data, constructing the initial tree, pruning it based on cross-validation, visualizing the pruned tree, making predictions, evaluating performance, and interpreting the results. By following these steps, we can create a robust and interpretable regression model that captures the relationships between predictors and outcomes, providing valuable insights for decision-making and analysis.

13.6 Additional Considerations and Extensions for Regression Trees

In this section, we discussed common issues that may arise in real-world data when building regression trees and their solutions. These include handling missing data with surrogate splits, addressing variability through weighted observations, and using alternative splitting criteria. We also covered techniques like post-pruning, evaluating variable importance, and stabilizing models (which will be explored in the next chapter).

13.6.1 Handling Missing Data in Regression Trees

Handling missing data is an essential consideration when building regression trees, as real-world datasets often contain incomplete information. By default, the `rpart` package addresses this issue using surrogate splits. Surrogates are alternative splits that mimic the behavior of the primary split when the data for the primary splitting variable is missing. For a split based on X_j with missing values, `rpart` finds a surrogate X_k that maximizes agreement with X_j. The agreement is calculated using the following expression:

$$\text{Agreement}(X_k, X_j) = \frac{\text{Number of matches}}{\text{Total cases with } X_j \text{ missing}} \tag{13.19}$$

In practice, we can specify the use of surrogates with the following command:

```
initial_tree <- rpart(medv ~ ., data = Boston, method = "anova",
control = rpart.control(usesurrogate = 2, surrogatestyle = 1))
```

Here, `usesurrogate = 2` allows up to two surrogates per split, and `surrogatestyle = 1` prioritizes surrogates that best mimic the primary split.

13.6.2 Weighted Regression Trees

In some cases, certain observations might be more influential than others, necessitating a **weighted regression tree** approach. The **weighted MSE** is defined as:

$$\text{Weighted MSE}(t) = \frac{\sum_{i \in t} w_i (Y_i - \bar{Y}_t)^2}{\sum_{i \in t} w_i} \tag{13.20}$$

where w_i is the weight for observation i and $\bar{Y}_t$ is the weighted mean outcome in node t. By adjusting weights, we can emphasize observations based on their importance or reliability:

```
weights <- rep(1, nrow(Boston))
initial_tree <- rpart(medv ~ ., data = Boston, method = "anova", weights = weights)
```

13.6.3 Alternative Splitting Criteria

While MSE is the default splitting criterion, it can be sensitive to outliers. An alternative is the mean absolute error (MAE):

$$\text{MAE} = \frac{1}{N_t} \sum_{i \in t} |Y_i - \bar{Y}_t| \tag{13.21}$$

Though `rpart` does not directly support MAE, the `quantregForest` package offers quantile-based splitting, which is more robust to outliers:

```
install.packages("quantregForest")
library(quantregForest)
model <- quantregForest(x = Boston[, -14], y = Boston$medv)
```

Post-Pruning Analysis: Complexity Parameter Plot

Understanding the effect of the complexity parameter (`cp`) on tree size and performance is also crucial. The `plotcp` function helps visualize the cross-validation error against different `cp` values:

```
plotcp(initial_tree)
```

This plot aids in identifying a point where increasing `cp` simplifies the tree without significantly increasing error.

Importance of Interactions in Trees

CART trees naturally capture interactions between variables, but these interactions are not explicitly reported. Examining the sequence of splits can reveal such interactions. For instance, if splits involve combinations of variables like `rm` and `nox`, it suggests an interaction effect:

```
print(pruned_tree)
```

Handling Continuous vs. Categorical Variables

Proper handling of categorical variables is essential for meaningful splits. Converting binary variables to factors ensures they are treated as categorical:

```
Boston$chas <- as.factor(Boston$chas)
```

Properly specifying variable types allows the tree to produce more interpretable splits.

Interpreting Predictions in Terminal Nodes

Each terminal node in a regression tree represents a subgroup of houses with similar characteristics, and the predicted value is the average median house value (`medv`) for that subgroup. For example, if a terminal node has `medv = 30`, it implies an average predicted value of $30,000 for houses in that group. This information is accessible through:

```
pruned_tree$frame
```

Model Stability and Bootstrap Aggregation (Bagging)

CART trees can be unstable—small changes in the data might lead to different trees. To address this, bootstrap aggregating (bagging) and random forest can improve stability by averaging predictions across multiple trees built on bootstrapped samples, which we will cover next chapter.

Additional Information

There is a concept known as tree growing, where a tree is fully expanded by recursively splitting the data into smaller subsets. However, this approach is rarely used in practice because it often leads to overly complex models that overfit the data. Instead, trees are typically pruned or regularized to improve generalization.

In this chapter, we covered classification trees, including impurity measures, recursive partitioning, and variable importance. We implemented a decision tree from scratch and trained a model using `rpart`. Finally, we introduced regression trees and demonstrated how to train one using the `rpart` package.

In Stata, regression trees can be implemented using the `crtrees` package, which supports both classification and regression tasks based on the CART methodology. Another option is `treeplot`, which integrates with Python's `scikit-learn` to visualize decision and regression trees within Stata.

In Python, the most widely used package for regression trees is `scikit-learn`, which includes `DecisionTreeRegressor` for constructing and tuning tree-based models. Another option is `linear-tree`, which extends decision trees by incorporating linear models at the leaf nodes, allowing for a hybrid approach between decision trees and linear regression.

Another refinement to prevent overfitting is the honest tree, where one subset of the data determines the splits, and another estimates the outcomes at the leaf nodes. This prevents overfitting by ensuring that the same data is not used both to decide the structure of the tree and to make predictions. This method, commonly used in causal inference (e.g., generalized random forests and causal trees), ensures unbiased treatment effect estimation and will be covered in the Heterogeneous Treatment Effects chapter.

Now, we move on to ensemble learning, starting with random forests.

Notes

1. https://freakonometrics.hypotheses.org/52776

2. Other impurity measures include entropy, defined as $\text{Entropy}(t) = -p_0 \log_2(p_0) - p_1 \log_2(p_1)$, which is more sensitive to class distributions, and misclassification error, defined as $\text{Error}(t) = 1 - \max(p_0, p_1)$, which directly measures the proportion of misclassified observations.

3. Alternative gain measures include information gain based on entropy, defined as $\Delta\text{Entropy} = \text{Entropy (parent)} - \frac{N_L}{N}\text{Entropy (left)} - \frac{N_R}{N}\text{Entropy (right)}$, and misclassification error reduction, which uses the decrease in misclassification error instead of Gini or entropy.

4. For instance, if income ranges between 5,000 and 200,000, the algorithm creates a series of potential split points within this range. The number of intervals is calculated based on unique values in the dataset or by using a predefined maximum number of splits to manage computational cost. Each of these split points is evaluated by calculating the Gini Gain to determine which point provides the most effective separation between classes.

5. $C_\alpha(T) = \sum_{t\in T} N_t \cdot \text{Entropy}(t) + \alpha \cdot |T|$

6. $C_\alpha(T) = \sum_{t\in T} N_t \cdot (1 - \max(p_0, p_1)) + \alpha \cdot |T|$

7. $C_\alpha(T) = \sum_{t\in T} N_t \cdot (1 - w_0 p_0^2 - w_1 p_1^2) + \alpha \cdot |T|$

8. $C_\alpha(T) = \sum_{t\in T} N_t \cdot \sum_{i,j} C(i,j) \cdot p_i \cdot p_j + \alpha \cdot |T|$

9. $C_\alpha(T) = -\sum_{t\in T} \log(\text{Likelihood}(t)) + \alpha \cdot |T|$

14

Ensemble Learning and Random Forest

Ensemble learning is a powerful technique in machine learning that combines multiple models to improve prediction accuracy and reduce overfitting. Rather than using all predictors in one complex model, ensemble learning combines multiple simpler models, each trained on a subset of features and/or data. This approach results in a more robust learning system overall. The three main methods of ensemble learning are bagging, random forests, and boosting.

The concept behind ensemble learning is based on the idea of the "wisdom of crowds." This theory suggests that the collective decision-making of a diverse, large group is often better than that of a single expert. When using a single robust model, poor predictors tend to be eliminated during training; however, each "weak" predictor may still offer a small but meaningful contribution when combined with others. Ensemble methods keep these weaker predictors in the system—rather than discarding them—allowing them to "vote" in the final decision. This is the main reason why ensemble methods are considered robust learning algorithms in machine learning.

14.1 Bagging (Bootstrap Aggregating)

CART models are known for their sensitivity to small changes in the training data, leading to different splits and consequently, a **high variability in the model predictions**. This variance makes them unstable, especially when the data set is not large enough or when it contains many features that may not all be relevant to the prediction task. To address this issue, ensemble methods such as bagging and random forests are used to reduce the variance of the model predictions.

Bagging gets its name from **B**ootstrap **agg**regat**ing** of trees. Bagging is a technique that improves model stability by reducing variance. The key idea behind bagging is to create multiple versions of a base model, typically decision trees, trained on different random subsets of the original dataset, and then aggregate their predictions for the final decision. These models then vote (for classification) or average their predictions (for regression) to generate the final output. The steps are as follows:

1. Select number of trees (B), and the tree depth (D),
2. Create a loop (B) times to create B trees with the following step, 2a. In each iteration:
   - Generate a bootstrap sample from the original data.
   - Fit (train) a tree of depth D on that bootstrap sample.

DOI: 10.1201/9781003504795-14

3. For regression, average the predictions from all trees. For classification, take the majority vote.

Below is a simple example with the Titanic dataset, where we train B = 9 unpruned classification trees on nine bootstrap samples:

```
library(PASWR)
library(rpart)
library(rpart.plot)

# Load data
data(titanic3)

# Set up colors for plotting
box_col <- rep("gray90", 9)
n <- nrow(titanic3)
par(mfrow = c(3, 3))  # 3x3 plotting area

# Train 9 trees (B = 9)
for(i in 1:9) {
  set.seed(i * 2)
  idx <- sample(n, n, replace = TRUE)  # Bootstrap sampling
  tr <- titanic3[idx, ]

  cart <- rpart(
    survived ~ sex + age + pclass + sibsp + parch,
    data = tr,
    method = "class",
    cp = 0  # unpruned tree
  )

  # Plot each tree
  prp(cart, type = 1, extra = 1, box.col = box_col[i], branch.col = "black",
  split.col = "black")
}
```

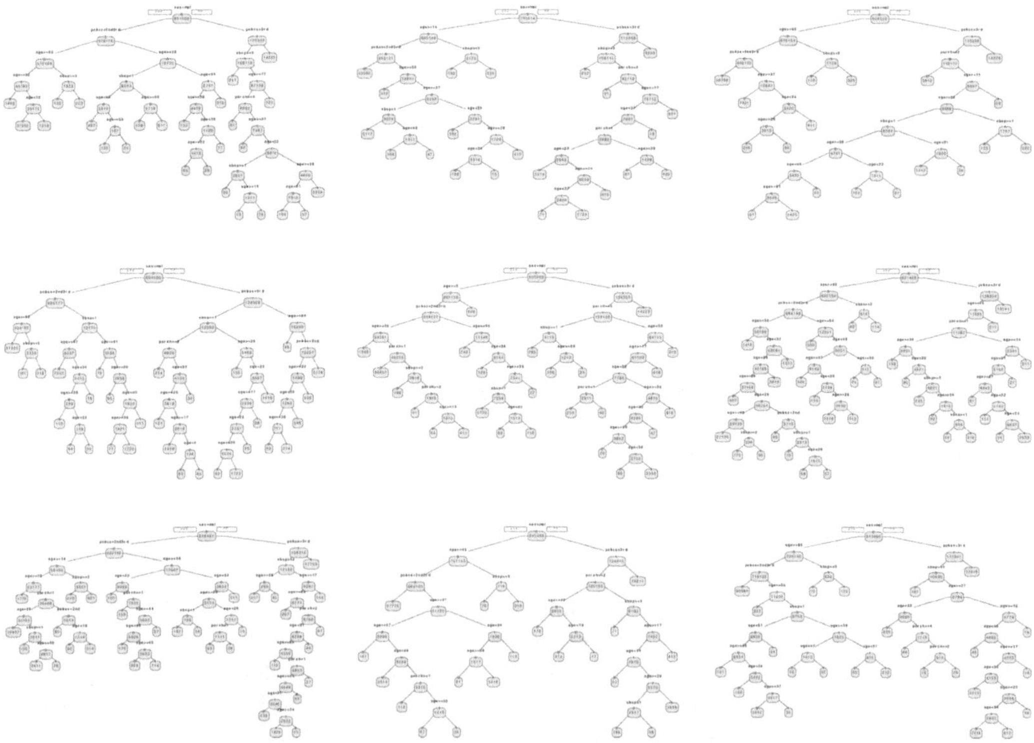

What do we do with these 9 trees?

1. In regression trees (continuous outcome), the final prediction will be the **average** of all tree predictions.
2. In classification trees (binary outcome), the final prediction is based on the **majority** vote.

Because averaging multiple predictions from bootstrap samples reduces variance, bagging often increases prediction accuracy. More importantly, compared to a single CART model, bagging results are much less sensitive to the specific training sample, resulting in impressive improvements in accuracy.

Below is an example that illustrates how to apply bagging in a classification context. We first fit a single tree model to the Titanic data, then compare its performance to a bagged model.

We first fit a single classification tree to the Titanic dataset and evaluate its performance.

```
library(ROCR)
library(rpart)

# Split into training and testing sets
set.seed(1)
ind <- sample(nrow(titanic3), nrow(titanic3) * 0.8)
train <- titanic3[ind, ]
test  <- titanic3[-ind, ]
```

```
# Fit a single decision tree (pruned by default rpart complexity parameters)
cart <- rpart(
  survived ~ sex + age + pclass + sibsp + parch,
  data = train,
  method = "class"
)

# Predicted probabilities on the test set
phat1 <- predict(cart, test, type = "prob")

# Compute AUC for the single tree
pred_rocr <- prediction(phat1[, 2], test$survived)
auc_ROCR  <- performance(pred_rocr, measure = "auc")
single_tree_auc <- auc_ROCR@y.values[[1]]
single_tree_auc
```

```
## [1] 0.8352739
```

The dataset is split into an 80% training set and a 20% test set. A single decision tree is trained using `rpart()`. We compute the area under the curve (AUC) to evaluate its classification performance. Bagging many trees can improve the performance of a single tree. Next, we train 100 trees in this example. We will use the `rpart` package to fit the trees and the `ROCR` package to calculate the AUC.

```
B <- 100  # Number of trees

# Matrix to store predicted probabilities for each tree
phat2 <- matrix(0, nrow = B, ncol = nrow(test))

for(i in 1:B) {
  set.seed(i) # For reproducibility
  idx <- sample(nrow(train), nrow(train), replace = TRUE)
  dt  <- train[idx, ]

  # Fit an unpruned classification tree
  cart_B <- rpart(
    survived ~ sex + age + pclass + sibsp + parch,
    data = dt,
    method = "class",
    cp = 0 # No pruning
  )
# Store predicted probabilities for the "survived = 1" class
  phat2[i, ] <- predict(cart_B, test, type = "prob")[, 2]
}
dim(phat2) # Output: 100 x number of test observations
```

```
## [1] 100 262
```

We create $B = 100$ bootstrap samples from the training data. Each tree is trained independently on its bootstrap sample. The result stored to `phat2` which is a 100×262 matrix. Each row corresponds to predictions from one of the 100 trees, and each column corresponds to one observation in the test set.

Now, we average the predictions across all trees to obtain the final predicted probability for each test observation. We then calculate the AUC for the bagged model to compare it with the single tree model.

```
# Average the predicted probabilities across the 100 trees
phat_f <- colMeans(phat2)

# Compute the AUC for the bagged model
pred_rocr <- prediction(phat_f, test$survived)
auc_ROCR  <- performance(pred_rocr, measure = "auc")
bagged_auc <- auc_ROCR@y.values[[1]]
bagged_auc
```

```
## [1] 0.8765668
```

Expected outcome from bagging should be the bagged model should have a higher AUC than the single tree, demonstrating improved classification performance. We can also examine how the number of trees (B) affects the AUC (or the mean squared prediction error, MSPE, in regression tasks):

```
B <- 300
phat3 <- matrix(0, nrow = B, ncol = nrow(test))
AUC   <- numeric(B)

for(i in 1:B) {
  set.seed(i)
  idx <- sample(nrow(train), nrow(train), replace = TRUE)
  dt  <- train[idx, ]

  fit <- rpart(
    survived ~ sex + age + pclass + sibsp + parch,
    data = dt,
    method = "class",
    cp = 0  # unpruned
  )

  phat3[i, ] <- predict(fit, test, type = "prob")[, 2]

  # Compute cumulative average up to the i-th tree
  phat_f <- if (i == 1) phat3[i, ] else colMeans(phat3[1:i, , drop = FALSE])

  # Calculate AUC
  pred_rocr <- prediction(phat_f, test$survived)
  auc_ROCR  <- performance(pred_rocr, measure = "auc")
  AUC[i]    <- auc_ROCR@y.values[[1]]
}

# Plot how AUC stabilizes as B grows
```

```
plot(
  AUC,
  type = "l",
  col = "black",
  lwd = 2,
  xlab = "B - Number of Trees",
  ylab = "AUC"
)
```

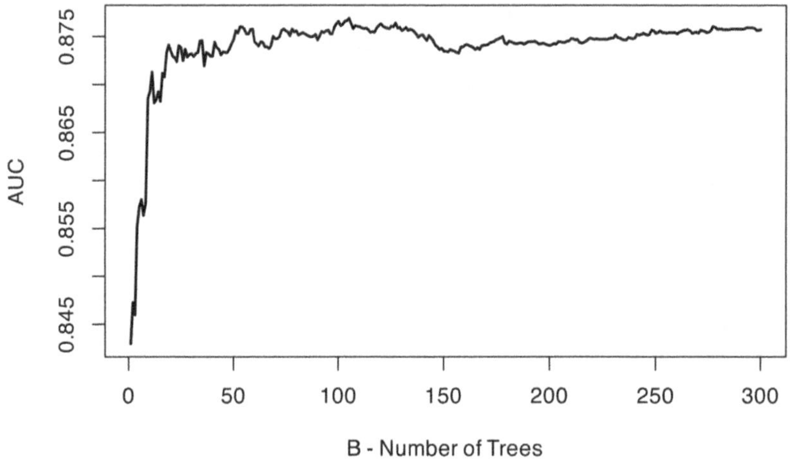

As shown by the plot, when B is large, the AUC (or error in regression) stabilizes, indicating that we do not need to perform extensive hyperparameter tuning for the number of trees in bagging. Simply, bagging requires choosing a sufficiently large B, but no fine-tuning beyond that.

14.1.1 Application with Multiple Splits

We can also use `foreach` and `doParallel` to parallelize and speed the process instead of doing the previous steps. Now, we will have multiple train-test splits and average the AUCs. We will use the `rpart` package for the trees and the `ROCR` package to calculate the AUC.

```
library(foreach)
library(doParallel)

no_cores <- detectCores() - 1
registerDoParallel(cores = no_cores)

B <- 100
AUC <- numeric(100)  # Preallocate memory for speed

for (i in 1:100) {

  set.seed(i)
  ind <- sample(nrow(titanic3), nrow(titanic3) * 0.8)
  train <- titanic3[ind, ]
```

```
  test <- titanic3[-ind, ]

  # Use foreach for the inner loop
  phat2 <- foreach(b = 1:B, .combine = 'cbind') %dopar% {
    set.seed(b) # Ensure reproducibility
    idx <- sample(nrow(train), nrow(train), replace = TRUE)
    dt <- train[idx, ]

    cart_B <- rpart(survived ~ sex + age + pclass + sibsp + parch,
                    cp = 0, data = dt, method = "class") # unpruned
    predict(cart_B, test, type = "prob")[, 2]
  }

  # Take the average
  phat_f <- rowMeans(phat2)

  # AUC
  pred_rocr <- prediction(phat_f, test$survived)
  auc_ROCR <- performance(pred_rocr, measure = "auc")
  AUC[i] <- auc_ROCR@y.values[[1]]
}

stopImplicitCluster()
```

Let's plot AUC with its mean and 95% confidence interval.

```
# Plot AUC with its mean and 95% confidence interval
plot(AUC, col = "black",
     xlab = "B - Number of trees",
     lwd = 2)
abline(h = mean(AUC), col = "gray40", lwd = 2)
abline(h = quantile(AUC, 0.025), col = "gray60", lwd = 2, lty = 2)
abline(h = quantile(AUC, 0.975), col = "gray60", lwd = 2, lty = 2)
```

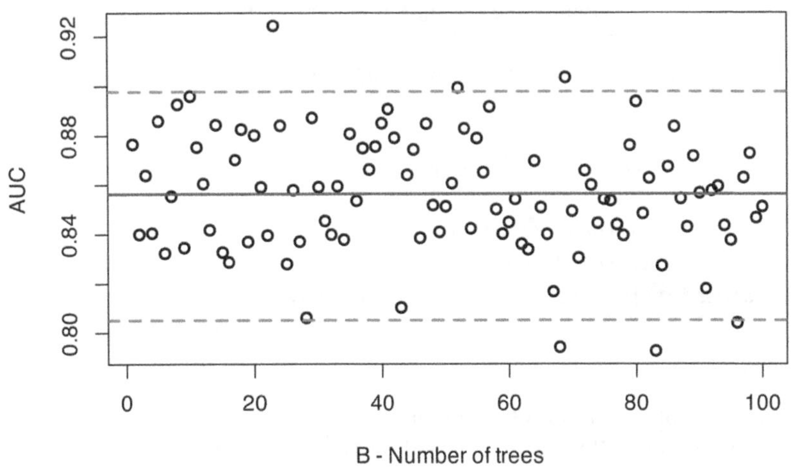

14.1.2 Tuning?

Bagging (Bootstrap AGGregatING) does not require the same level of hyperparameter tuning as some other machine learning models, such as selecting the number of layers in a neural network or the depth of individual trees in a random forest. The primary parameter in bagging is the number of bootstrap samples (or trees in the forest, B), which often does not need fine-tuning beyond ensuring it is sufficiently large. This characteristic makes bagging a very robust method with minimal setup requirements.

Regarding cross-validation, bagging inherently includes a form of it through the use of bootstrap samples. Each tree is trained on a random subset of the data (with replacement), making it unnecessary to use cross-validation to prevent overfitting or to estimate model accuracy reliably. The out-of-bag (OOB) error provides a handy estimate of prediction error (generalization error) without the need for a separate validation set or cross-validation, as the OOB samples act as a built-in test set for each tree.

Hence, instead of running two loops as we seen in the previous example, we can have one split and use the OOB error to estimate the generalization error.

14.1.3 Using Out-of-Bag Error

Bagging provides a built-in error estimate without needing cross-validation. We use OOB samples to estimate the AUC.

```
df <- titanic3

B <- 100
pred <- matrix(NA, nrow(df), B)

for (i in 1:B) {
    set.seed(i)  # Ensure reproducibility
    train_idx <- sample(nrow(df), nrow(df), replace = TRUE)
    oob_idx <- setdiff(1:nrow(df), train_idx)
    train <- df[train_idx, ]
    oob <- df[oob_idx, ]

    cart_B <- rpart(survived ~ sex + age + pclass + sibsp + parch,
                      cp = 0, data = train, method = "class")  # unpruned
    pred[oob_idx, i] <- predict(cart_B, oob, type = "prob")[, 2]
}

phat_f <- rowMeans(pred, na.rm = TRUE)

pred_rocr <- prediction(phat_f, df$survived)
auc_ROCR <- performance(pred_rocr, measure = "auc")
auc_ROCR@y.values[[1]]
```

```
## [1] 0.8515995
```

It is important to note that the OOB error is a good estimate of the generalization error, but it is not perfect. The OOB error is based on the samples that were not used to train the model, but it is still an estimate and may not be as accurate as a separate validation set or cross-validation. However, it is a useful metric for quickly estimating the generalization error without the need for additional data or computational resources. If B is large enough, the OOB error will be a good estimate of the generalization error. Pay attention to the `pred` matrix. It is a matrix of predicted probabilities for each tree. We take the average of these probabilities to get the final prediction.

14.1.4 OOB Error Rate

Here, we will use the OOB samples to estimate the error rate, which is the proportion of misclassified observations. Although we show the classification error as calculated by 1 - accuracy, it can also be shown separately for each classes. Here, instead of AUC, we can compute the classification error based on OOB predictions.

```
df <- titanic3
B <- 100
n <- nrow(df)
pred <- matrix(NA, n, B)

for (i in 1:B) {
  set.seed(i)
  train_idx <- sample(n, replace = TRUE)
  oob_idx <- setdiff(1:n, train_idx)
  train <- df[train_idx, ]
  oob <- df[oob_idx, ]

  # Train model on bootstrap sample
  cart_B <- rpart(survived ~ sex + age + pclass + sibsp + parch,
                  cp = 0, data = train, method = "class")

  # Predict for OOB observations
  oob_pred <- predict(cart_B, oob, type = "class")
  oob_pred01 <- ifelse(oob_pred == 1, 1, 0)

  # Store OOB predictions
  pred[oob_idx, i] <-  oob_pred01
}

# Calculate OOB error for each tree
oob_err <- numeric(B)
for (j in 1:ncol(pred)) {
  ind <- which(!is.na(pred[ ,j]))
  oob_err[j] <- 1 - mean(pred[ind, j] == df$survived[ind])
}

oob_error <- mean(oob_err, na.rm = TRUE)
oob_error  # This is the overall OOB error rate
```

```
## [1] 0.2151985
```

```
plot(1:B, oob_err, col = "black")
abline(h = oob_error, col = "gray40", lwd = 2)
```

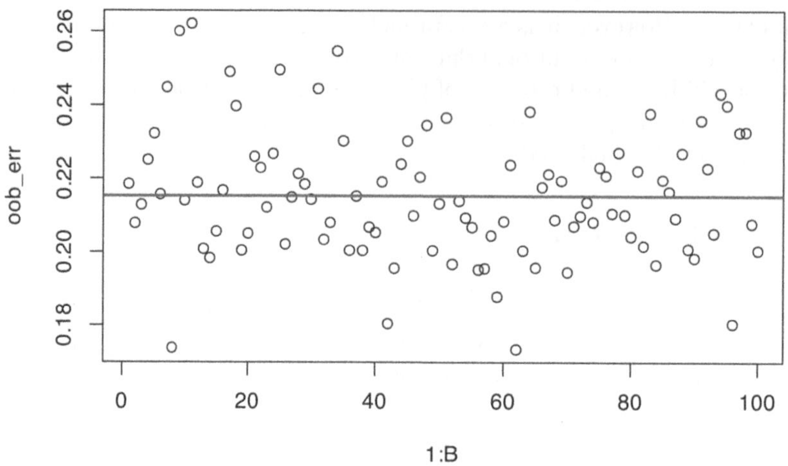

Bagging is a powerful ensemble method that significantly improves the performance of decision trees by addressing their high variance. It builds multiple models from different bootstrap samples of the training data and then combines their predictions by averaging (for regression) or voting (for classification). Bagging is a robust method with minimal setup requirements and does not require the same level of hyperparameter tuning as some other machine learning models. The primary parameter in bagging is the number of bootstrap samples (or trees in the forest, B), which often does not need fine-tuning beyond ensuring it is sufficiently large. Bagging inherently includes a form of cross-validation through the use of bootstrap samples, and the OOB error provides a handy estimate of prediction error (generalization error) without the need for a separate validation set or cross-validation.

Bagging reduces the variance of high-variance base learners (e.g., decision trees or kNN) by averaging (for regression) or voting (for classification). This significantly stabilizes predictions compared to using a single model. However, if the individual trees are highly correlated, the reduction in variance achieved by bagging is diminished, limiting its effectiveness in improving predictive performance. Moreover, bagging can be computationally expensive as it requires training multiple models, increasing processing time and memory usage. Next, we explore an even more refined ensemble method called random forests.

14.2 Random Forest

We have seen that bagging is a method of combining multiple models to improve generalization. It works by training the same model type on multiple bootstrap samples—each of the same size as the original dataset and drawn with replacement. This means some observations appear multiple times while others may be excluded. By averaging the predictions of these independent models trained on slightly different data distributions, bagging reduces variance.

The random forest algorithm extends bagging by introducing **random feature selection**. In addition to training each model on a different bootstrap sample, Random Forest randomly selects a subset of features at each split when constructing decision trees. This additional randomness helps decorrelate the trees, further improving predictive performance compared to standard bagging."

Random forest is an ensemble learning algorithm designed to improve predictive performance by combining many decision trees. For regression, it works with a dataset $\{(X_i, Y_i)\}_{i=1}^N$, where $X_i \in \mathbb{R}^p$ are predictor variables and $Y_i \in \mathbb{R}$ is a continuous response. Each tree contributes a prediction, and the final result is obtained by averaging across all trees, reducing variance and improving generalization.

The construction of a random forest begins with bootstrap sampling. For each of the B trees, a new training set S_b is generated by sampling with replacement from the original dataset:

$$S_b = \{(X_i, Y_i)\}, \quad i \sim \text{Uniform}(1, N) \tag{14.1}$$

This means each tree is trained on roughly 63% of the data, while the remaining 37%—the OOB observations—are held out and used later to estimate prediction error.

Each tree is built by recursively splitting the data. Unlike traditional decision trees, random forests introduce randomness by considering only a random subset of m predictors (where $m = \lfloor\sqrt{p}\rfloor$, rounded down, used by default for regression) at each split, instead of all p predictors. At a given node t, the algorithm selects the split that minimizes mean squared error (MSE):

$$\Delta\text{MSE} = \text{MSE}(t) - \left(\frac{N_L}{N_t}\text{MSE}(L) + \frac{N_R}{N_t}\text{MSE}(R)\right) \tag{14.2}$$

where L and R are the left and right child nodes, and N_t, N_L, N_R are their respective sample sizes.

Unlike single decision trees, which often require **pruning** to avoid overfitting, random forests intentionally grow each tree fully. The rationale is that overfitting in individual trees is addressed not by simplification, but by **averaging** across many high-variance trees. In single trees, pruning is used to control the bias–variance trade-off. In contrast, random forests manage variance through aggregation, making pruning unnecessary. Each tree may overfit its bootstrap sample, but the ensemble average tends to generalize well.

The final prediction for a new observation x is simply the average of predictions from all B trees:

$$\hat{Y}_{\text{RF}}(x) = \frac{1}{B}\sum_{b=1}^{B} T_b(x) \tag{14.3}$$

where $T_b(x)$ is the prediction from the b-th tree. This averaging ensures a stable prediction, typically with lower variance than any single tree.

Another important feature of random forests is the use of **Out-of-Bag (OOB) error** as an internal validation method. Since about 37% of the training data is excluded from each tree's bootstrap sample, those OOB observations can serve as a test set for that tree. For each data point i, its OOB prediction $\hat{Y}_i^{\text{OOB}}$ is the average of predictions from trees where i was not included in S_b. The overall OOB mean squared error is computed as:

$$\text{OOB MSE} = \frac{1}{N}\sum_{i=1}^{N}(Y_i - \hat{Y}_i^{\text{OOB}})^2 \tag{14.4}$$

This built-in estimate acts like cross-validation and is a key reason why random forests do not require separate cross-validation. The OOB error serves as a reliable measure of test error, making model evaluation more efficient.

For classification, the structure is similar, but the prediction target Y_i is categorical. At each node, the split is chosen to maximize class purity—typically using the Gini index. Each tree produces a class prediction, and the final forest prediction is the majority vote:

$$\hat{C}_{\text{RF}}(x) = \text{mode}\{T_1(x), T_2(x), \ldots, T_B(x)\} \tag{14.5}$$

The OOB classification error is calculated by comparing the true class label of each observation to its OOB-predicted class:

$$\text{OOB Error Rate} = \frac{1}{N}\sum_{i=1}^{N} \mathbb{I}\left(Y_i \neq \hat{C}_i^{\text{OOB}}\right) \tag{14.6}$$

where $\hat{C}_i^{\text{OOB}}$ is the predicted class for observation i using only trees where i was OOB.

Random forests offer several built-in advantages that make them powerful tools for predictive modeling. One of the most practical features is the use of out-of-bag (OOB) error to estimate generalization error. Since each tree in the forest is trained on a bootstrap sample that leaves out some part of data (roughly 37% of the data in `randomforest` package) those unused observations can serve as a test set for that tree. By aggregating these out-of-sample predictions across all trees, the model provides an internal and unbiased estimate of its prediction error—eliminating the need for a separate validation set or cross-validation.

Another key advantage of random forests is their ability to assess variable importance. During training, the algorithm tracks how much each predictor contributes to reducing impurity at decision nodes across the forest. By averaging these reductions, random forests highlight which features are most influential in driving predictions. This built-in diagnostic helps researchers and practitioners interpret models, detect relevant variables, and guide feature selection or further analysis.

Finally, predictions from a random forest reflect its ensemble nature. In regression tasks, each tree outputs a numeric prediction, and the forest returns the average of these values, yielding a stable and smooth estimate. In classification, each tree casts a vote for a class label, and the majority vote determines the final prediction. This aggregation makes random forests robust against noise and overfitting, while still capturing complex patterns in the data.

14.2.1 Random Forest-like Variable Selection

To incorporate a Random Forest twist into our bagging code, which originally applies a bagging approach using CART, we need to introduce variable selection into the training of each tree. Random Forest improves upon bagging by not only using bootstrap samples but also by randomly selecting a subset of the variables (features) for splitting at each node within a tree. This approach increases the diversity among the trees in the ensemble, which can lead to better performance.

Here is how we can adjust our code to include a Random Forest-like variable selection:

- **Variable Subset Selection**: At each split in the construction of a tree, randomly select a subset of the variables to consider. The number of variables, often denoted as m, is a hyperparameter. A common choice for classification problems is $m = \sqrt{p}$, where p is the total number of variables. For regression problems, a common choice is $m = p/3$. As you can see if $m = p$, then we are back to the original bagging algorithm.

- **Implementation Adjustment:** Since the `rpart` function in R does not directly support random selection of variables for each split, we typically use a Random Forest package like `randomForest` or `ranger`. However, to keep the spirit of our original implementation and manually introduce randomness in variable selection for now, we can simulate this aspect by randomly selecting a subset of variables to include in the model for each tree.

Let's see how we can adjust our code to include this Random Forest-like variable selection instead of using package.

To begin, we load the necessary libraries, clean up the environment, and handle missing values in the Titanic dataset. We then subset the data to include only relevant predictors.

```
# clean up
rm(list = ls())
# Load libraries
library(randomForest)
library(ROCR)
library(PASWR)
library(rpart)
library(rpart.plot)
library(caret)
library(ggplot2)
# NA handling
data(titanic3)
colSums(is.na(titanic3))  # Check for missing values
```

```
##    pclass  survived      name       sex       age     sibsp     parch    ticket
##         0         0         0         0       263         0         0         0
##      fare     cabin  embarked      boat      body home.dest
##         1         0         0         0      1188         0
```

```
# Subset the data
df <- titanic3[, c("survived", "sex", "age", "pclass", "sibsp", "parch")]
```

To manually implement variable selection similar to Random Forest, we build multiple decision trees, randomly selecting a subset of features at each iteration. The number of selected features follows the rule of $\sqrt{p}$ where $p$is the number of predictors. Each tree is trained on a bootstrap sample, and predictions are made on out-of-bag (OOB) observations.

```
B <- 1000
n <- nrow(df)
p <- sqrt(ncol(df) - 1)  # One column for the response variable
pred <- matrix(NA, nrow = n, ncol = B)

for (i in 1:B) {
    set.seed(i)  # Ensure reproducibility
    train_idx <- sample(n, n, replace = TRUE)
    oob_idx <- setdiff(1:n, train_idx)
```

```
  # Randomly select a subset of variables
  vars <- sample(colnames(df[, -which(colnames(df) == "survived")]),
                 size = ceiling(p), replace = FALSE)
  formula <- as.formula(paste("survived ~", paste(vars, collapse = "+")))

  train <- df[train_idx, ]
  oob <- df[oob_idx, ]

  cart_B <- rpart(formula, data = train, method = "class", cp = 0)  # unpruned
  pred[oob_idx, i] <- predict(cart_B, oob, type = "prob")[, 2]
}

# Calculate OOB AUC
phat_f <- rowMeans(pred, na.rm = TRUE, dims = 1)
pred_rocr <- prediction(phat_f, df$survived)
auc_ROCR <- performance(pred_rocr, measure = "auc")
auc_ROCR@y.values[[1]]
```

```
## [1] 0.8473894
```

This approach manually introduces feature randomness, which is the key difference between Random Forest and standard bagging. While not the most efficient way to implement Random Forest, it helps in understanding the underlying mechanism.

14.3 Implementing Random Forest in R

Instead of manually implementing the process, we can use the `randomForest` package, which automates feature selection, tree construction, and aggregation.

In our earlier examples, we manually applied bagging to single trees by `rpart`. We used the `rpart`'s internal CV to find the "optimal depth" of the tree. In other words, we did not grow the tree to the maximum possible terminal nodes or pruned it later. Now we will use `randomForest` for Titanic first, then for Hitters. Let's summarize what we know about `randomForest` to start with:

1. Unlike `rpart`, `randomForest` does not have an internal Cross Validation process.
2. The number of trees, which we denoted as B in `rpart`, is now `ntree` and has to be set by the user. Here is the definition: "*Number of trees to grow. This should not be set to too small a number, to ensure that every input row gets predicted at least a few times*".
3. Maximum number of terminal nodes trees in the forest can have, `maxnodes`. The maximum can be set to the number observations in our train set, in which case, each tree grows to its full extend. If not given, trees are grown to the maximum possible (subject to limits by `nodesize`, which is Minimum size of terminal nodes. The default values are different for classification (1) and regression (5)).
4. Number of variables randomly sampled as candidates at each split, `mtry`. There is no consensus whether it is or should be a hyperparameter in practice. If it is not

set, the default values are for classification, sqrt(p) where p is number of variables in x and for regression p/3.

14.3.1 Building a Random Forest Model on Titanic Data

Let's see how we can use `randomForest` to build a Random Forest model for the Titanic dataset.

```
df$survived <- as.factor(titanic3$survived)
df <- df[complete.cases(df), ]  #We can also handle it within RF

set.seed(42)  # for reproducibility
#rf_model <- randomForest(survived ~ ., data = df, na.action = na.omit, ntree = 1000)
rf_model <- randomForest(survived ~ ., data = df, ntree = 1000)

# Print the model
rf_model
```

```
##
## Call:
##  randomForest(formula = survived ~ ., data = df, ntree = 1000)
##                Type of random forest: classification
##                      Number of trees: 1000
## No. of variables tried at each split: 2
##
##         OOB estimate of  error rate: 20.46%
## Confusion matrix:
##     0   1 class.error
## 0 563  56   0.0904685
## 1 158 269   0.3700234
```

```
names(rf_model)
```

```
##  [1] "call"            "type"            "predicted"       "err.rate"
##  [5] "confusion"       "votes"           "oob.times"       "classes"
##  [9] "importance"      "importanceSD"    "localImportance" "proximity"
## [13] "ntree"           "mtry"            "forest"          "y"
## [17] "test"            "inbag"           "terms"
```

14.3.1.1 Out-of-Bag (OOB) Error and Performance Evaluation

We can see the development of the model, the number of trees, the number of variables randomly sampled as candidates at each split, and the OOB error. The OOB error is a built-in measure of model performance. It is computed by predicting OOB samples for each tree and averaging the errors across all trees.

```
# OOB error
head(rf_model$err.rate)
```

```
##             OOB         0         1
```

```
## [1,] 0.2423398 0.1690141 0.3493151
## [2,] 0.2416667 0.1637931 0.3492063
## [3,] 0.2513369 0.1378505 0.4031250
## [4,] 0.2416185 0.1428571 0.3795014
## [5,] 0.2353579 0.1322160 0.3792208
## [6,] 0.2317328 0.1196429 0.3894472
```

```
tail(rf_model$err.rate)
```

```
##              OOB         0         1
##  [995,] 0.2055449 0.0904685 0.3723653
##  [996,] 0.2045889 0.0904685 0.3700234
##  [997,] 0.2045889 0.0904685 0.3700234
##  [998,] 0.2045889 0.0904685 0.3700234
##  [999,] 0.2045889 0.0904685 0.3700234
## [1000,] 0.2045889 0.0904685 0.3700234
```

We can plot the OOB error rate to see how it changes as the number of trees increases.

```
plot(rf_model, col = "black", lty = 1)
```

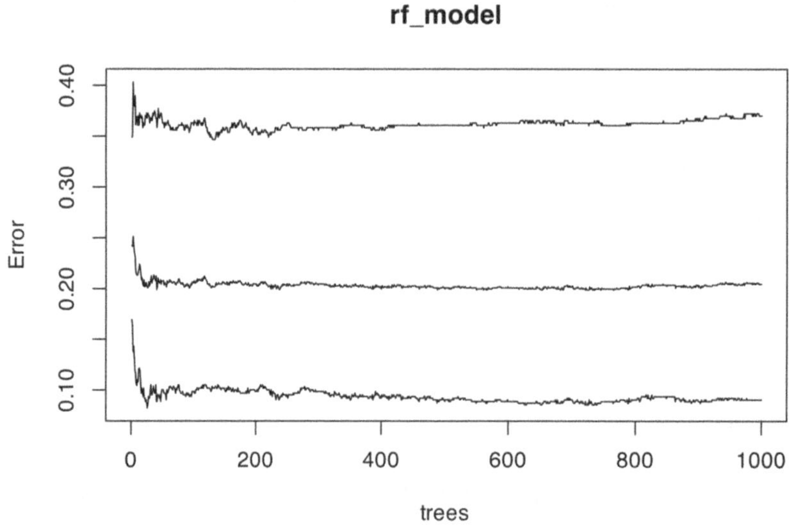

Green line is the OOB error rate for the first class, and the red line is the OOB error rate for the second class. The blue line is the overall OOB error rate.

14.3.1.2 OOB Predictions

We can use the `predict` function to get the OOB predictions from the model.

```
library(caret)
```

```
length(rf_model$predicted)
```

```
## [1] 1046
```

```
table(rf_model$predicted)
```

```
##
##   0   1
## 721 325
```

```
table(df$survived)
```

```
##
##   0   1
## 619 427
```

```
table(rf_model$predicted, df$survived)
```

```
##
##       0   1
##   0 563 158
##   1  56 269
```

```
# let's change the order of the table
cm <- table(rf_model$predicted, df$survived)
cm <- cm[c(2, 1), c(2, 1)]
cm
```

```
##
##       1   0
##   1 269  56
##   0 158 563
```

```
#lets make it more readable
confusionMatrix(rf_model$predicted, df$survived, positive = "1")
```

```
## Confusion Matrix and Statistics
##
##           Reference
## Prediction   0   1
##          0 563 158
##          1  56 269
##
##                Accuracy : 0.7954
##                  95% CI : (0.7697, 0.8195)
##     No Information Rate : 0.5918
##     P-Value [Acc > NIR] : < 2.2e-16
##
##                   Kappa : 0.5603
##
##  Mcnemar's Test P-Value : 5.048e-12
##
```

```
##            Sensitivity : 0.6300
##            Specificity : 0.9095
##         Pos Pred Value : 0.8277
##         Neg Pred Value : 0.7809
##             Prevalence : 0.4082
##         Detection Rate : 0.2572
##   Detection Prevalence : 0.3107
##      Balanced Accuracy : 0.7698
##
##       'Positive' Class : 1
##
```

`oob.times`: This records for each data point in the dataset the number of times it was not included in the bootstrap sample for a tree (i.e., the number of times it was in the OOB sample). This information can be useful for understanding how evenly the OOB samples are distributed across the dataset and can also play a role in the calculation of the OOB error estimate.

```
head(rf_model$oob.times)
```

```
## [1] 378 360 364 369 343 399
```

14.3.2 AUC

In the context of Random Forest and other ensemble learning methods, `votes` refers to the predictions made by the individual trees in the forest for each observation in the dataset. When you train a Random Forest model for classification tasks, it builds multiple decision trees. Each of these trees makes its own prediction for each observation. The `votes` are essentially these predictions; for a given observation, each tree `votes` for a particular class. You can think of the `votes` as a matrix where each row corresponds to an observation and each column corresponds to a tree in the forest. The value in each cell is the class that the corresponding tree `voted` for.

The final prediction for each observation is determined by a majority vote across all trees in the forest. This means the class that receives the most votes from the trees is chosen as the final prediction for each observation. This process is a key aspect of how Random Forest improves upon the performance of individual decision trees. By aggregating the predictions (votes) of numerous trees, Random Forest can reduce the variance of the predictions, leading to better generalization on unseen data.

For example, in a binary classification problem, if you have a Random Forest consisting of 100 trees and you are making a prediction for a single observation, it is possible that:

- 60 trees predict the observation belongs to Class 0,
- 40 trees predict the observation belongs to Class 1.

In this case, since Class 0 receives more votes (60 vs. 40), `votes` will 0.6 for Class 0 and 0.4 for Class 1. The final prediction for this observation will be Class 0.

```
dim(rf_model$votes)
```

```
## [1] 1046    2
```

```
head(rf_model$votes)
```

```
##            0          1
## 1 0.02116402 0.97883598
## 2 0.21388889 0.78611111
## 3 0.06868132 0.93131868
## 4 0.52574526 0.47425474
## 5 0.01457726 0.98542274
## 6 0.92230576 0.07769424
```

Since these are phat values, we can calculate AUC as follows:

```
pred_rocr <- prediction(rf_model$votes[, 2], df$survived)
auc_ROCR <- performance(pred_rocr, measure = "auc")
auc_ROCR@y.values[[1]]
```

```
## [1] 0.8466288
```

```
# And ROC curve
plot(performance(pred_rocr, measure = "tpr", x.measure = "fpr"),
     col = "black", lwd = 2, main = "ROC Curve")
abline(a = 0, b = 1, lwd = 2, lty = 2, col = "gray40")
```

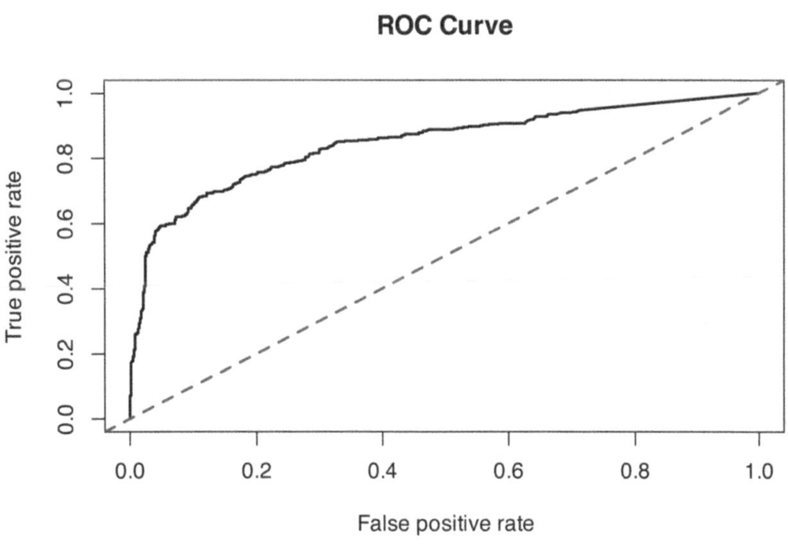

OOB AUC or Test AUC?

In the context of using Random Forest for classification tasks, Out-of-Bag (OOB) error estimation is a convenient feature. The OOB error is the average error for each training observation calculated using only the trees that did not have this observation in their bootstrap sample. This method is an unbiased estimate of the model error as if it were computed on a test set.

The OOB error estimation can be a good indicator of the performance of the model when the number of trees is sufficiently large. Therefore, we can use the OOB error as an estimate of the performance of the model without having to split the data into a separate training and test set. This is particularly useful when you have a limited amount of data, or when computational resources are constrained.

However, it is important to note that while the OOB error is a good estimate of performance, it might not always perfectly match the error you would observe on a completely separate test set. The true test of the performance of the model is how well it generalizes to unseen data. Therefore, if you have enough data, it is still considered a best practice to set aside a test set and evaluate the performance of your model on it.

For AUC, which is a measure of the ability of the model to discriminate between the positive and negative classes, the randomForest package does not directly provide an OOB estimate of AUC. However, you can calculate it manually by using the OOB predictions to compute the AUC with a package like pROC or ROCR.

Here is another way to compute OOB AUC with the `randomForest` package:

```
library(pROC)

set.seed(42)  # for reproducibility
rf_model <- randomForest(survived ~ ., data = df, ntree = 1000,
                         keep.inbag = TRUE, oob.probs = TRUE)

# Extracting OOB votes (Assuming the 2nd column corresponds to the positive class)
oob_probs <- rf_model$votes[, 2]

# Compute OOB AUC
roc_curve <- roc(df$survived, oob_probs)
auc(roc_curve)
```

```
## Area under the curve: 0.8466
```

Now we will try AUC on the test set.

```
n = nrow(df)
AUC <- rep(NA, 50)

for (i in 1:50) {
    set.seed(i)
    train_idx <- sample(n, n, replace = TRUE)
    test_idx <- setdiff(1:n, train_idx)

    train <- df[train_idx, ]
    test <- df[test_idx, ]

    rf_model <- randomForest(survived ~ ., data = train, ntree = 1000)
    phat <- predict(rf_model, test, type = "prob")[, 2]

    roc_curve <- roc(test$survived, phat)
    AUC[i] <- auc(roc_curve)
}
```

```
plot(AUC, pch = 19, col = "black", xlab = "Iteration",
     ylab = "AUC", main = "AUC on Test Set")
abline(h = mean(AUC), lwd = 2, lty = 2, col = "gray40")
```

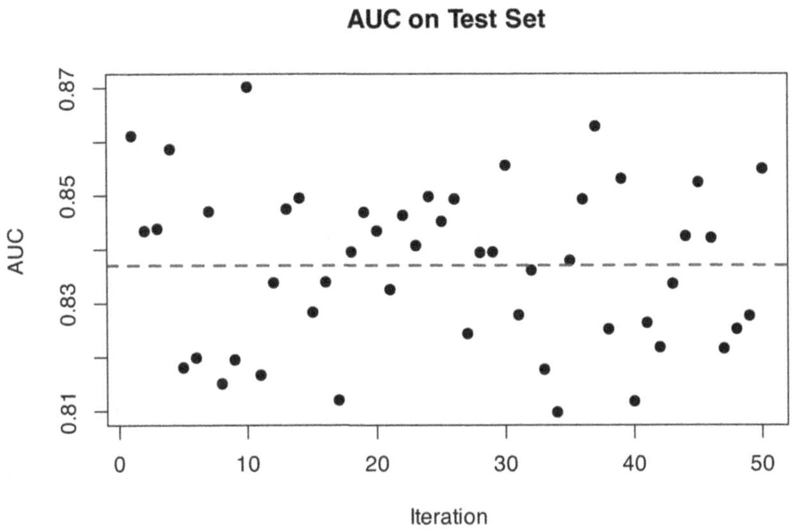

Multicore version of the same code:

```
library(doParallel)
num_cores <- detectCores() - 1

cl <- makeCluster(num_cores)
registerDoParallel(cl)

n = nrow(df)
AUC <- foreach(i = 1:100, .combine = c,
               .packages = c('randomForest', 'pROC')) %dopar% {
    set.seed(i)
    train_idx <- sample(n, n, replace = TRUE)
    test_idx <- setdiff(1:n, train_idx)

    train <- df[train_idx, ]
    test <- df[test_idx, ]

    rf_model <- randomForest(survived ~ ., data = train, ntree = 1000)
    phat <- predict(rf_model, test, type = "prob")[, 2]

    roc_curve <- roc(test$survived, phat)
    auc(roc_curve)
}

stopCluster(cl)
plot(AUC, pch = 19, col = "black", xlab = "Iteration",
     ylab = "AUC", main = "AUC on Test Set")
```

```
abline(h = mean(AUC), lwd = 2, lty = 2, col = "gray60")
abline(h = quantile(AUC, 0.025), lwd = 2, lty = 2, col = "gray40")
abline(h = quantile(AUC, 0.975), lwd = 2, lty = 2, col = "gray40")
```

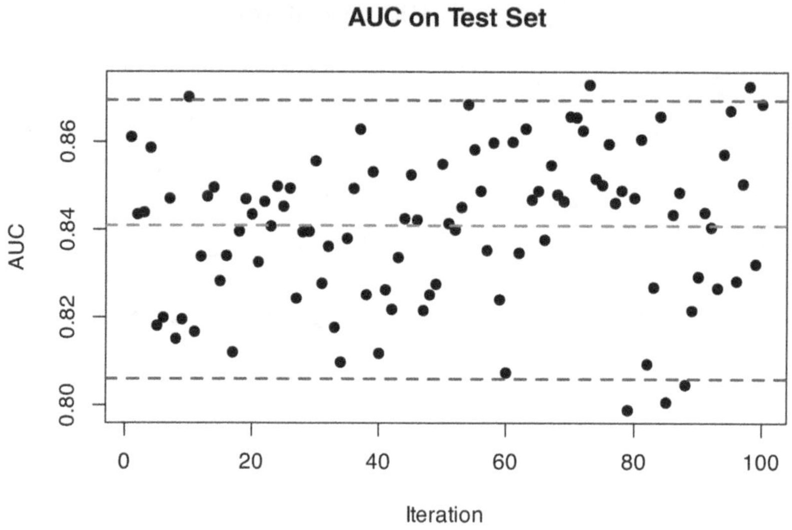

Or

```
library(randomForest)
library(pROC)
library(parallel)

n <- nrow(df)
AUC <- rep(NA, 100)

# Define the function to be applied in each iteration
iteration_function <- function(i, df) {
  set.seed(i)
  n <- nrow(df)
  train_idx <- sample(n, n, replace = TRUE)
  test_idx <- setdiff(1:n, train_idx)

  train <- df[train_idx, ]
  test <- df[test_idx, ]

  rf_model <- randomForest(survived ~ ., data = train, ntree = 1000)
  phat <- predict(rf_model, test, type = "prob")[, 2]

  roc_curve <- roc(test$survived, phat)
  auc_value <- auc(roc_curve)

  return(auc_value)
}
```

```
no_cores <- detectCores() - 1

# Use mclapply to parallelize
AUC <- mclapply(1:100, iteration_function, df = df, mc.cores = no_cores)
AUC <- unlist(AUC)

# Plot
plot(AUC, pch = 19, col = "black", xlab = "Iteration",
     ylab = "AUC", main = "AUC on Test Set")
abline(h = mean(AUC), lwd = 2, lty = 2, col = "gray60")
abline(h = quantile(AUC, 0.025), lwd = 2, lty = 2, col = "gray40")
abline(h = quantile(AUC, 0.975), lwd = 2, lty = 2, col = "gray40")
```

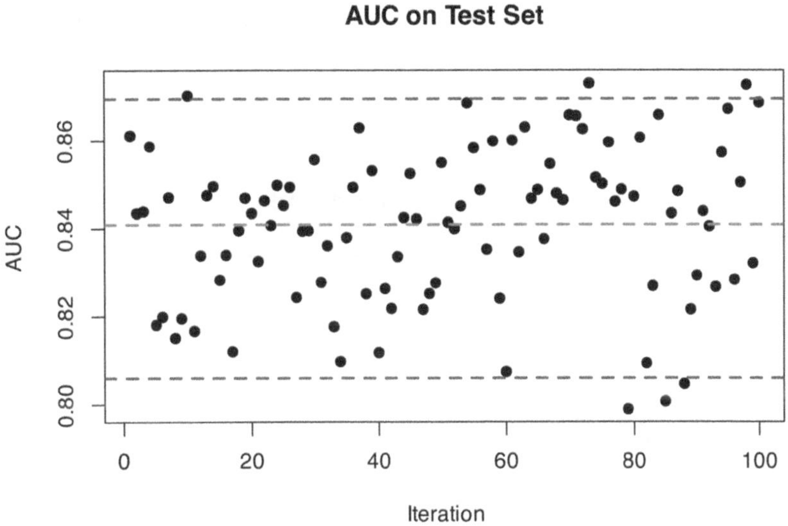

14.3.3 Regression: Hitters

Now let's see how we can use `randomForest` to build a Random Forest model for the Hitters dataset.

```
library(ISLR)
dfr <- data(Hitters)
dfr <- na.omit(Hitters)
dfr$Salary <- log(dfr$Salary)

set.seed(42)  # for reproducibility
rf_model_reg <- randomForest(Salary ~ ., data = dfr, ntree = 1000)

# Print the model
rf_model_reg
```

```
##
## Call:
##  randomForest(formula = Salary ~ ., data = dfr, ntree = 1000)
##                Type of random forest: regression
```

```
##                      Number of trees: 1000
## No. of variables tried at each split: 6
##
##           Mean of squared residuals: 0.1782892
##                     % Var explained: 77.36
```

```
# OOB error
head(rf_model_reg$mse)
```

```
## [1] 0.2860233 0.2522324 0.2576919 0.2671088 0.2551585 0.2548879
```

```
tail(rf_model_reg$mse)
```

```
## [1] 0.1784544 0.1784147 0.1783455 0.1783773 0.1782916 0.1782892
```

And the plot

```
plot(rf_model_reg, col = "black")
```

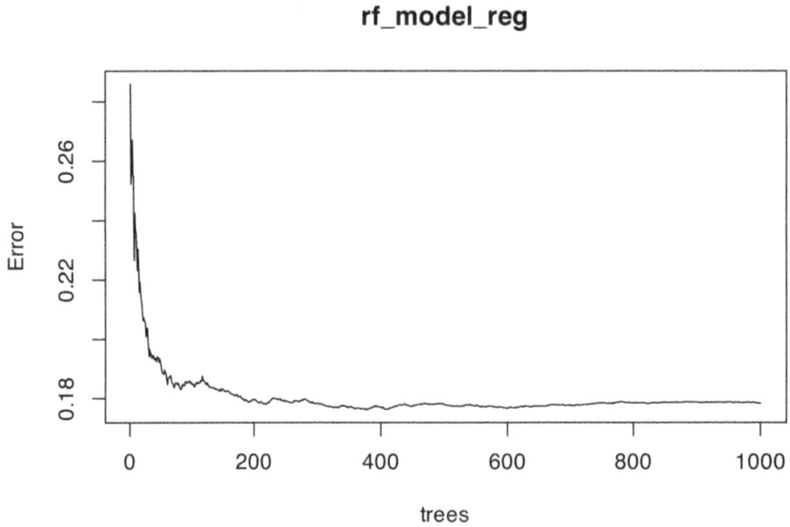

Benchmarking

We can use linear regression, decision trees, and Random Forest to predict the Salary of baseball players in the Hitters dataset. We can then compare the performance of these models using the RMSE metric.

```
library(caret)

# Linear regression

lm_model <- lm(Salary ~ ., data = dfr)
lm_pred <- predict(lm_model, newdata = dfr)
lm_rmse <- sqrt(mean((dfr$Salary - lm_pred)^2))
```

```
# Decision trees

library(rpart)

tree_model <- rpart(Salary ~ ., data = dfr)
tree_pred <- predict(tree_model, newdata = dfr)
tree_rmse <- sqrt(mean((dfr$Salary - tree_pred)^2))

# Bagging

set.seed(42)  # for reproducibility
bag_model <- randomForest(Salary ~ ., data = dfr, ntree = 1000, mtry = ncol(dfr) - 1)
bag_pred <- predict(bag_model, newdata = dfr)
bag_rmse <- sqrt(mean((dfr$Salary - bag_pred)^2))

# Put the RMSE values in a data frame
rmse_df <- data.frame(Model = c("Linear Regression", "Decision Trees", "bagging"),
                      RMSE = c(lm_rmse, tree_rmse, bag_rmse))
rmse_df
```

```
##               Model      RMSE
## 1 Linear Regression 0.5896648
## 2    Decision Trees 0.4198185
## 3           bagging 0.1787337
```

14.3.4 Importance of Variables

When working with Random Forest models, particularly in R using the `randomForest` package, you encounter options to evaluate the importance of variables (features) in making predictions. The concepts of `importance` and `localImportance` are central to understanding how your model evaluates the significance of different predictors, but they serve different purposes and provide insights at different levels.

`importance`

- Global Measure: This option, when enabled (by setting `importance=TRUE`), calculates global importance scores for each variable in the dataset. These scores are meant to reflect the overall contribution of each variable to the prediction of the model accuracy across the entire dataset.
- How It Works: The importance of a variable is usually assessed in terms of the decrease in model accuracy or increase in node impurity when the values of the variable are permuted across out-of-bag (OOB) samples. A larger decrease in accuracy or increase in impurity indicates a more important variable.
- Purpose: The global importance scores are useful for feature selection, understanding which variables contribute most to the predictions of the model, and guiding model refinement and interpretation.

`localImportance`

- Individual Measure: The `localImportance` option calculates importance scores for each variable for each individual observation. This means you get a matrix where each row represents an observation, and each column represents a variable, with the values indicating the importance of that variable for predicting that particular observation.

- How It Works: This is calculated similarly to global importance but on a per-observation basis. It can show how the importance of variables varies across individual predictions, which can be particularly useful for detailed analyses of model behavior and understanding variable contributions in specific cases.
- Purpose: Local importance scores can be invaluable for detailed diagnostics, such as explaining model predictions for individual instances or identifying why certain observations are outliers or misclassified. They provide a granular view of the decision-making process of the model.

14.3.4.1 Global Importance

Here are the key points we cover:

Mean Decrease in Impurity (MDI)

This method computes importance scores based on how much each feature contributes to homogeneity in nodes across all trees. For classification, this is often measured by the Gini impurity, and for regression, it can be measured by the variance reduction.

Imagine we have a RandomForest model trained on a dataset with 1000 samples, and we are looking at how the feature ' **X1** ' contributes to node purity in one of the trees.

1. Splitting on 'X1' in a Node: Suppose ' **X1** ' is used to split a node that contains 600 of the 1000 samples. The split divides the node into two child nodes:

- Child Node A with 400 samples
- Child Node B with 200 samples

2. Calculating Impurity Decrease: Assume the impurity (e.g., Gini impurity for classification) of the parent node before the split is 0.5. After the split:

- Child Node A has an impurity of 0.2
- Child Node B has an impurity of 0.1 The weighted impurity decrease due to the split on ' **X1**' is calculated by considering the impurit) decrease in each child node, weighted by the proportion of samples in each node relative to the parent node or the entire dataset.

3. Weight Calculation: The weight for the decrease in impurity contributed by the split on ' **X1**' would consider the proportion of total samples that went through the split. In this case, since the node being split had 600 samples out of 1000 in the dataset, the weight of this split in the overall MDI calculation for ' $X1$ ' would be 0.6(600/1000).
4. Weighted Impurity Decrease:

- The decrease in impurity for each child node is calculated. Let's say the decrease for Node A is 0.3 and for Node B is 0.4.
- The total decrease in impurity from the split is the sum of the decreases for each child, weighted by their proportion of samples: $\left(0.3 \times \frac{400}{600}\right) + \left(0.4 \times \frac{200}{600}\right)$.
- This weighted decrease is then further weighted by the proportion of samples in the parent node relative to the entire dataset: W eightedDecrease $\times \frac{600}{1000}$.

5. Aggregation: The weighted decrease in impurity from all splits on ' **x1**' across all trees in the forest is aggregated to compute the MDI for ' **X1** '.

Mean Decrease in Accuracy (MDA)

The Mean Decrease Accuracy (MDA), also known as permutation importance, evaluates the importance of a feature based on how the permutation (random shuffling) of the feature values across the dataset affects the accuracy of the model. Unlike MDI, MDA does not rely on the internal structure of the model but rather on the performance of the model with altered data. Let's walk through an example of how MDA is calculated.

1. First, we assess the accuracy of the model on the validation dataset. Assume the model achieves an accuracy of 80%.
2. We then permute the values of the `Fare` feature in the validation dataset. This means the `Fare` value for each passenger is randomly shuffled, breaking the original relationship between `Fare` and survival outcomes.
3. With the `Fare` feature permuted, we reassess the accuracy of the model on this altered dataset. Let's say the accuracy drops to 70%.
4. The decrease in accuracy, in this case, is the difference between the original accuracy (80%) and the accuracy after permuting `Fare` (70%), which is 10 percentage points.
5. To get a reliable estimate of `Fares` importance, we repeat the permutation process multiple times, each time recording the decrease in accuracy. Assume we do this five times, and we observe the following decreases in accuracy: $10\%, 9\%, 11\%, 8\%,$ and 12%.
6. Calculating MDA: The Mean Decrease in Accuracy (MDA) for 'Fare' is the average of these decreases, which would be $\frac{10\%+9\%+11\%+8\%+12\%}{5} = 10\%$. This means that, on average, permuting the 'Fare' values results in a 10% decrease in the accuracy of the model.

Points to consider

- Variable Importance Plots: These plots visually show the importance of each feature. Higher values indicate more important predictors. This can help in understanding which features are driving the predictions and possibly in feature selection for simplifying the model.
- Built-in Feature: Most Random Forest implementations, including the R randomForest package, have built-in functions to calculate and report feature importance. This is accessible via the `importance()` function
- OOB Samples for Importance: Random Forests can use out-of-bag (OOB) samples to estimate feature importance. The accuracy is measured on the OOB samples for each tree, then the values for a particular feature are permuted in the OOB samples and the accuracy is measured again. The drop in accuracy is a measure of importance of the features.
- Bias in Importance Measures: It is important to note that some methods of calculating importance can be biased, particularly in the presence of categorical variables with different levels or features that vary in scale or units. Permutation importance can sometimes be less biased in this regard.
- Model Interpretation: Even though Random Forests are considered "black box" models due to their complexity, importance scores provide a means to interpret the model by identifying which features are most influential in making predictions.

Examples

```
# Train the Random Forest model
rf_model <- randomForest(survived ~ ., data = df, ntree = 500, importance = TRUE)
importance(rf_model)
```

```
##               0          1 MeanDecreaseAccuracy MeanDecreaseGini
## sex    67.04451 101.167267             99.01523        133.87192
## age    23.08417  17.321865             32.21119         63.12850
## pclass 25.92289  39.133413             41.68083         52.38648
## sibsp  27.27390  -8.412601             18.82066         19.50711
## parch  15.68115   1.387492             16.32029         16.95344
```

`MeanDecreaseAccuracy`: This column shows the decrease in model accuracy caused by excluding a variable from the model. A larger value indicates a more important variable because its absence significantly reduces model accuracy.

`MeanDecreaseGini`: This column is related to classification problems and indicates the decrease in node impurity from splits on a variable, averaged over all trees. The Gini impurity is a measure of how often a randomly chosen element from the set would be incorrectly labeled if it was randomly labeled according to the distribution of labels in the subset. The greater the decrease in the Gini impurity, the greater the importance of the variable.

The initial columns `0` and `1` in the output from the `importance()` function in the randomForest package are class-specific measures of variable importance for classification problems. In the context of a binary classification problem, like predicting survival on the Titanic, '0' corresponds to the importance of the variable for predicting the '0' class (not survived), and '1' corresponds to the importance of the variable for predicting the '1' class (survived). These measures are calculated by permuting the values of the variable in the out-of-bag (OOB) samples and measuring the resulting increase in prediction error rate. The higher the value, the more important the variable is for predicting that particular class.

```
varImpPlot(rf_model, type = 1, main = "Variable Importance Plot")
```

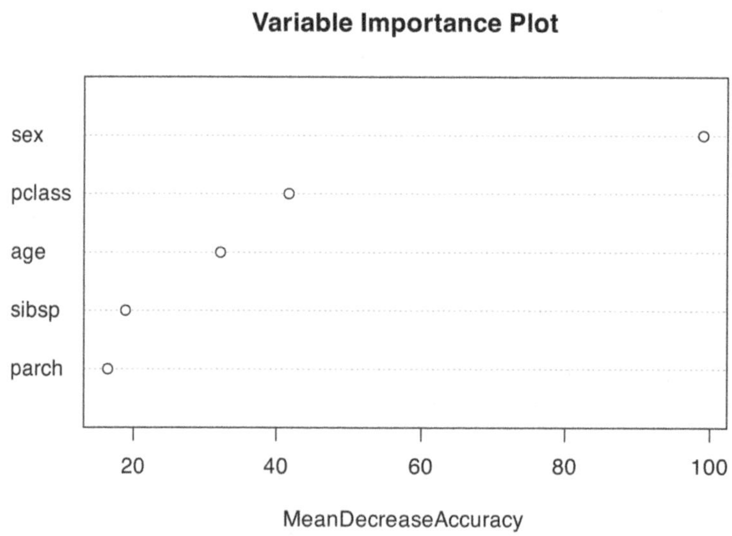


```
# Better
importance_data <- importance(rf_model)
importance_df <- as.data.frame(importance_data)
importance_df$Variable <- rownames(importance_df)

ggplot(importance_df,
```

```
    aes(x = reorder(Variable, MeanDecreaseGini),
        y = MeanDecreaseGini)) +
  geom_bar(stat = "identity") +
  coord_flip() +  # Flip coordinates for horizontal layout
  theme_minimal() +
  xlab("Variables") +
  ylab("Mean Decrease in Gini")
```

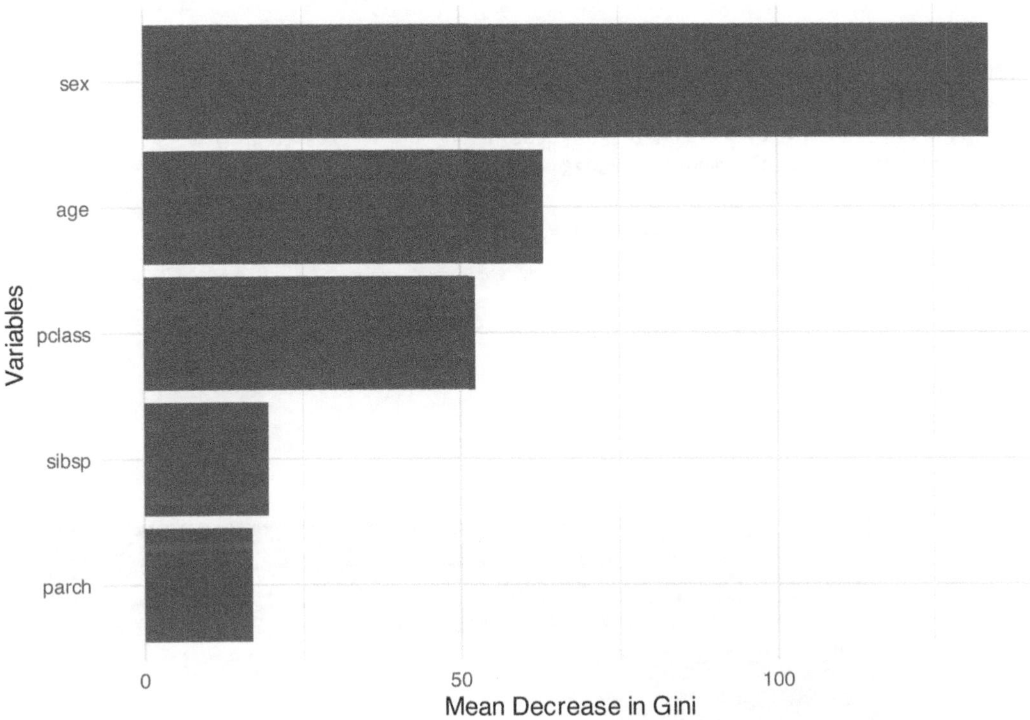

14.3.5 Local Importance

The `localImportance = TRUE` option in the `randomForest` package returns a matrix of variable importance scores at the individual level. Each row corresponds to an observation, and each column corresponds to a predictor, with cell values indicating how much that variable contributed to predicting that observation. Unlike global importance measures (e.g., Mean Decrease in Gini), local importance reveals how variable relevance can differ across individuals.

```
# Train the Random Forest model
rf_model <- randomForest(survived ~ ., data = df, ntree = 500, localImp = TRUE)

# Extract local importance scores
local_importance <- rf_model$localImportance
local_importance[, 1:5]
```

```
##                1          2           3          4           5
## sex  0.514851485 0.09195402 0.010810811 0.20114943 -0.37297297
```

```
## age     0.014851485  0.34482759  0.005405405  0.05747126  0.00000000
## pclass  0.321782178 -0.04022989 -0.172972973 -0.29885057 -0.24864865
## sibsp  -0.004950495 -0.02298851  0.037837838 -0.08620690  0.00000000
## parch   0.004950495  0.19540230  0.043243243 -0.23563218  0.01081081
```

```
# Plot the average local importance
local_importance_avg <- rowMeans(local_importance, na.rm = TRUE)
local_importance_avg <- local_importance_avg[
  order(local_importance_avg, decreasing = TRUE)
]

ggplot(
  data = data.frame(
    Variable = names(local_importance_avg),
    Importance = local_importance_avg
  ),
  aes(x = reorder(Variable, Importance), y = Importance)
) +
  geom_bar(stat = "identity") +
  coord_flip() +  # Flip coordinates for horizontal layout
  theme_minimal() +
  xlab("Variables") +
  ylab("Average Local Importance")
```

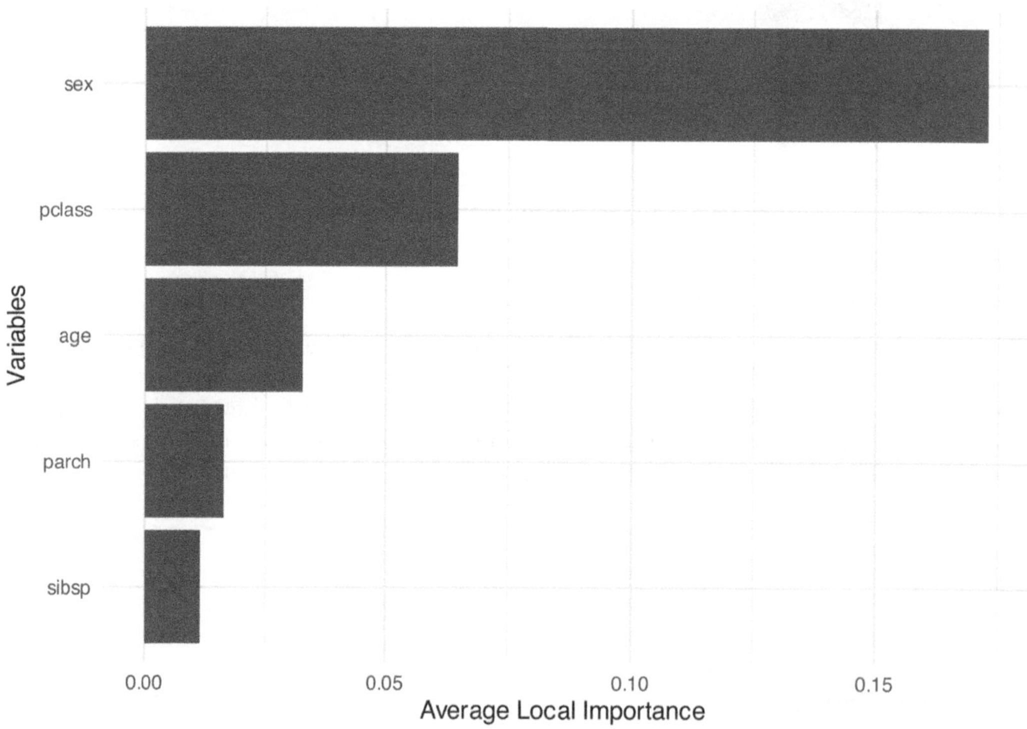

To summarize this high-dimensional output, we can take the average importance score for each variable across all observations. This gives us a sense of which variables are generally most useful across the sample, while still acknowledging that importance is computed locally.

```
local_importance_with_age <- rbind(local_importance, Age = df$age)
# Extract the local importance for 'Fare' and the values of 'Age'

age_importance <- local_importance_with_age["age", ]
ages <- local_importance_with_age["Age", ]

# Create a data frame for plotting
importance_age_df <- data.frame(Age = ages, AgeImportance = age_importance)

# Plot with ggplot2
library(ggplot2)
ggplot(importance_age_df, aes(x = Age, y = AgeImportance)) +
  geom_point(shape = 16, color = "black", alpha = 0.5) +
  geom_smooth(color = "black", fill = "gray80") +
  theme_minimal() +
 labs(title = "Local Importance of 'Age'",
      x = "Age", y = "Local Importance of Age")
```

```
## `geom_smooth()` using method = 'gam' and formula = 'y ~ s(x, bs = "cs")'
```

We can also explore how the local importance of a specific variable (e.g., Age) varies with its actual value. Plotting observation-level importance against the covariate itself can uncover nonlinear patterns or heterogeneous effects. For instance, we might observe that Age is more influential in certain age ranges. This type of plot provides more nuance than global summaries and helps us interpret variable relevance in a personalized way.

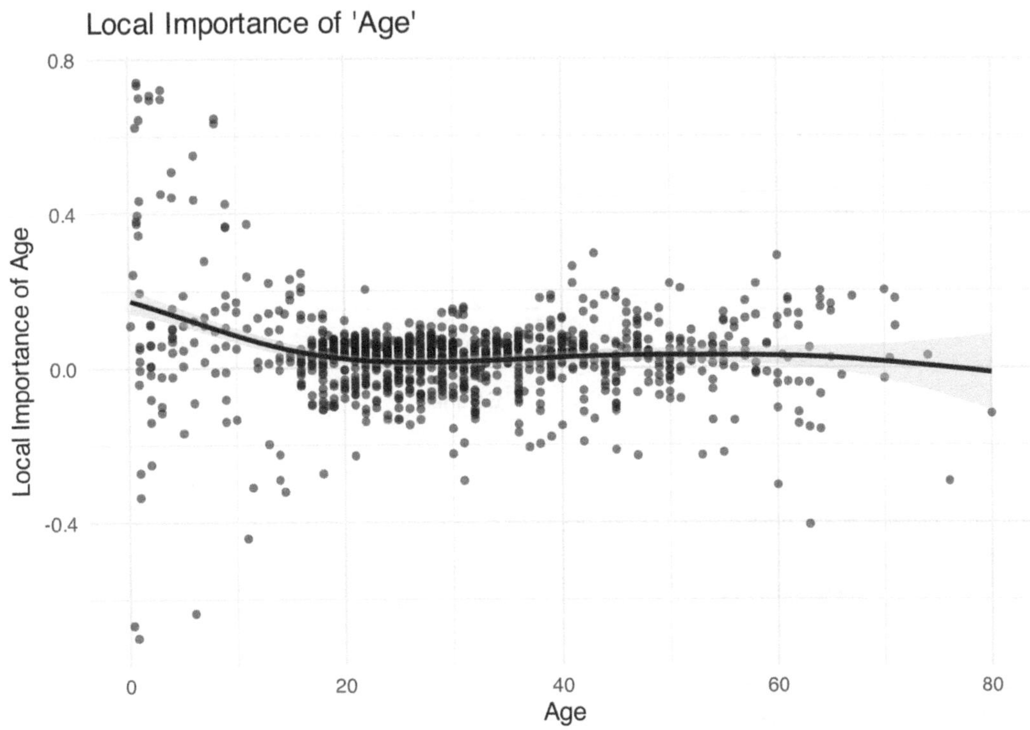

Partial Dependence Plots

Partial dependence plots (PDPs) are a powerful tool for interpreting the effects of features in a machine learning model, including random forests. They show the dependence of the target variable on one or two features, averaged over the values of the other features. PDPs can help in understanding the relationship between a feature (or features) and the predicted outcome, regardless of the model complexity.

```
library(ISLR)
dfr <- data(Hitters)
dfr <- na.omit(Hitters)
dfr$Salary <- log(dfr$Salary)

set.seed(42)  # for reproducibility
rf_model_reg <- randomForest(Salary ~ ., data = dfr, ntree = 1000)

# Extract the partial dependence

library(pdp)
```

```
## Warning: package 'pdp' was built under R version 4.3.3
```

```
pdp_cruns <- partial(rf_model_reg, pred.var = "CRuns")
plot(pdp_cruns, type = "l")
```

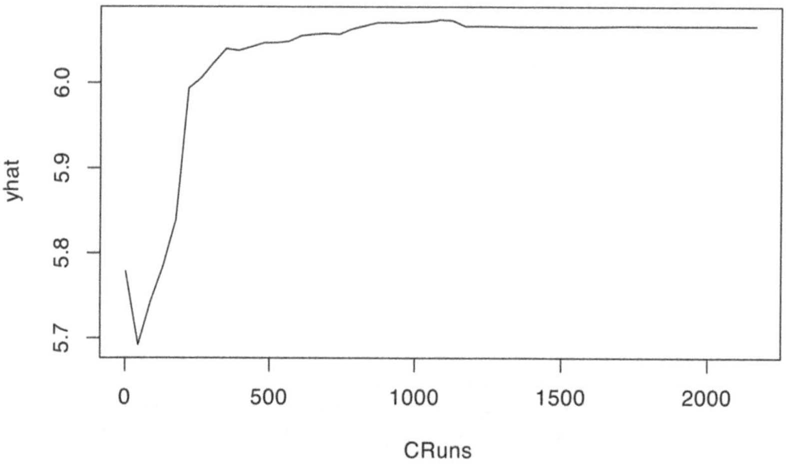

And, variable importance

```
rf_model_reg$importance
```

```
##          IncNodePurity
## AtBat        7.3455433
## Hits         8.7474467
## HmRun        2.6924395
## Runs         4.8806266
## RBI          6.1682936
```

```
## Walks          6.1472616
## Years          6.3439290
## CAtBat        38.4968438
## CHits         34.7239086
## CHmRun         7.5009988
## CRuns         34.2645475
## CRBI          21.5538919
## CWalks        16.6938438
## League         0.2817703
## Division       0.2952981
## PutOuts        3.3469015
## Assists        1.8196078
## Errors         1.7097708
## NewLeague      0.3593825
```

For regression, purity is typically measured by the variance or mean squared error (MSE) within each node. A perfectly pure node in regression would have no variance in the response variable (i.e., all instances in the node have the same value for the target variable). Therefore, the decrease in variance or MSE due to splits on a variable is a measure of the importance of the variable. In the regression context, `IncNodePurity` represents the total decrease in variance or MSE achieved by splits on each feature across all trees in the forest. A high decrease in variance due to a feature indicates that the feature is important for predicting the target variable.

14.3.6 `randomForestExplainer`

The `randomForestExplainer` package is a powerful tool for understanding and interpreting Random Forest models. It provides a suite of functions for visualizing and interpreting the results of Random Forest models. The package is built on top of the `DALEX` package, which is a popular package for model-agnostic explanations.

```
library(randomForestExplainer)
# randomForestExplainer()
min_depth_frame <- min_depth_distribution(rf_model)
importance_frame <- measure_importance(rf_model)
importance_frame
```

```
##   variable mean_min_depth no_of_nodes accuracy_decrease gini_decrease
## 1      age          1.566       14709        0.03281681      63.53371
## 2    parch          1.676        5467        0.01636135      16.52178
## 3   pclass          1.058        3813        0.06442979      53.23042
## 4      sex          0.882        1741        0.17236343     133.34472
## 5    sibsp          2.010        6934        0.01137672      18.97330
##   no_of_trees times_a_root      p_value
## 1         500           44 0.000000e+00
## 2         500           82 1.000000e+00
## 3         500          157 1.000000e+00
## 4         500          208 1.000000e+00
## 5         500            9 1.868887e-08
```

- mean_min_depth: The average minimum depth at which a variable is used to split a node. A lower value indicates that the variable tends to appear near the top of the trees and is considered more important for splitting the data.
- no_of_nodes: The total number of nodes in the forest where the variable is used for splitting. This gives an idea of how frequently the variable is used in making decisions across all trees. accuracy_decrease: The decrease in model accuracy when the variable is permuted. This value represents the importance of the variable, with larger decreases indicating more important variables. It reflects how the prediction error increases when the information provided by the variable is removed, hence showing its contribution to model accuracy.
- gini_decrease: The total decrease in Gini impurity brought by splits over this variable, summed over all trees. A higher value suggests that the variable is very effective in creating homogeneous nodes and thus is important for the model.
- no_of_trees: The number of trees in which the variable appears. This can be seen as a measure of consistency of the importance of a variable across the forest.
- times_a_root: The number of times a variable is used as the root of a tree. Being used as the root (the first split in a tree) suggests that the variable is very discriminative.
- p_value: The p-value testing the hypothesis that the permutation-based importance measure (such as accuracy decrease) is higher than expected by chance. A low p-value indicates that the importance of the variable is statistically significant, while a p-value of 1 suggests no evidence of importance beyond random chance. The p-value is calculated using a permutation test, which is a non-parametric test that assesses the statistical significance of the importance measure. The null hypothesis is that the variable is not important, and the p-value indicates the probability of observing the importance measure by random chance. A low p-value suggests that the variable is important beyond random chance, while a high p-value suggests that the variable is not important.

In the context of our results:

- Variables like sex and pclass have lower mean_min_depth values, indicating they are used in top splits and are therefore likely more important.
- The accuracy_decrease and gini_decrease for sex are relatively high, underscoring its importance in predicting the outcome.
- sex also appears as the root node in many trees (times_a_root), further highlighting its significance.
- The p_value for parch, pclass, and sex is 1, which at first glance might suggest their importance is not statistically significant beyond random chance. However, the interpretation of this p-value might require careful consideration of the context, model configuration, and how the permutation importance is calculated.

```
library(ggplot2)
theme_set(theme_minimal())

plot_multi_way_importance(importance_frame,
                          x_measure = "mean_min_depth",
                          y_measure = "gini_decrease",
                          size_measure = "p_value",
                          no_of_labels = 6)+
  scale_color_manual(values = rep("black", 10)) +
  scale_fill_manual(values = rep("gray70", 10))
```

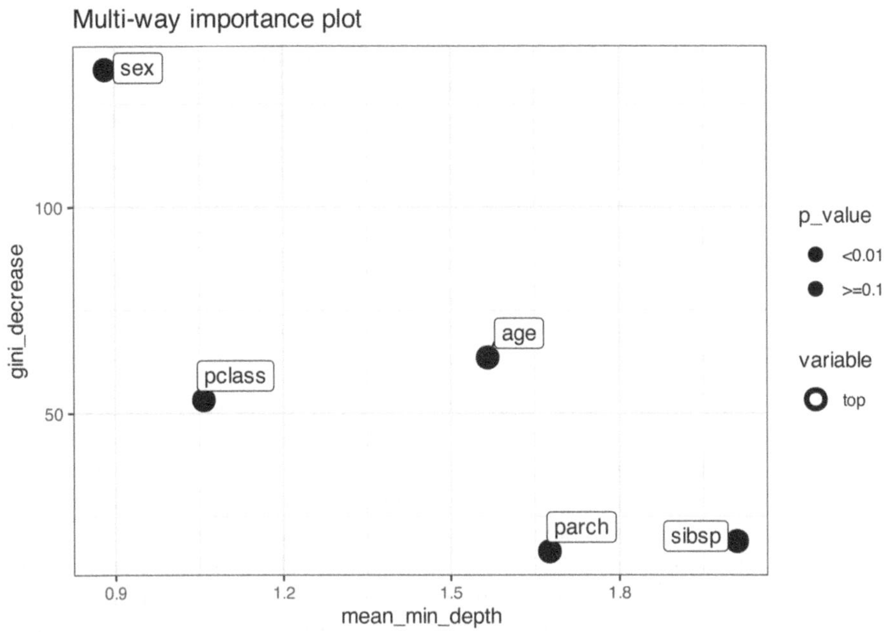


```
plot_min_depth_distribution(min_depth_frame,
                            mean_sample = "all_trees",
                            k = 20,
                            main = "Distribution of minimal depth and its mean")+
  scale_fill_manual(values = rep("gray70", 10)) +
  scale_color_manual(values = rep("black", 10))
```

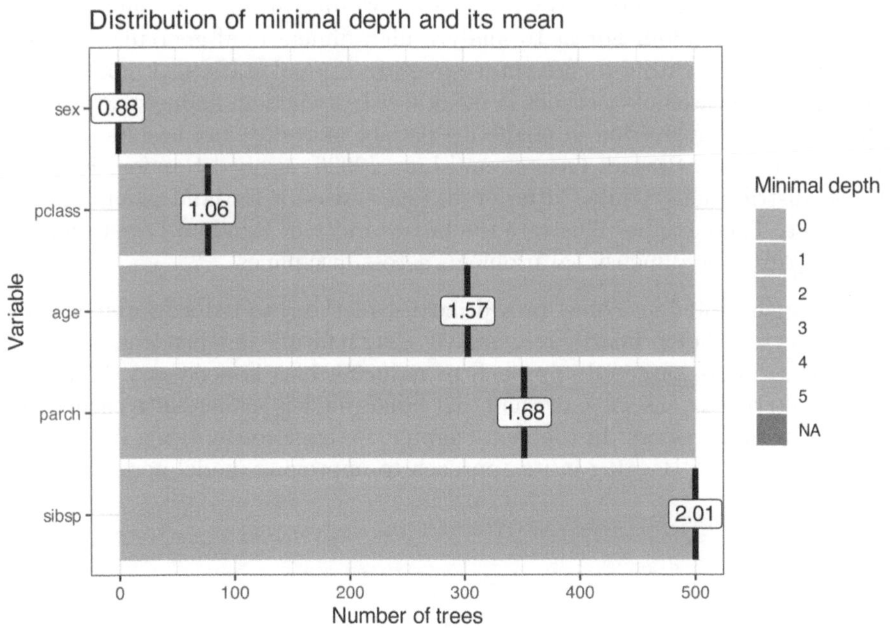

In this chapter, we learned about Bagging and Random Forest, a powerful ensemble learning method that can be used for both classification and regression tasks. We saw how to build Random Forest models in R using the `randomForest` package and how to evaluate the performance of these models using metrics like OOB error and AUC. We also learned about the importance of variables in Random Forest models and how to calculate and visualize these importance scores. Finally, we saw how to use the `randomForestExplainer` package to interpret the results of Random Forest models and understand the importance of variables in making predictions.

Beyond R, you can implement Random Forest models in both Stata and Python. In Stata, the `rforest` command provides a way to fit Random Forest models and assess variable importance, allowing for robust modeling within an econometric framework. In Python, the `scikit-learn` library offers the `RandomForestClassifier` and `RandomForestRegressor` functions, which facilitate model training and evaluation. Additionally, Python's `SHAP` and `eli5` libraries help interpret variable importance, improving transparency in predictions.

Random Forest and Bagging have been widely applied across economics, social sciences, and health sciences to improve predictive accuracy and interpretability in complex datasets. Breiman (2001) introduced these ensemble methods, showing their effectiveness in reducing variance and improving model performance. In economics, Thoplan (2014) demonstrated how Random Forests improved poverty classification based on socioeconomic indicators, while Sadorsky (2021) used ensemble methods to predict clean energy stock prices, outperforming traditional models. Mullainathan and Spiess (2017) highlighted the potential of machine learning, including Random Forest, for economic policy evaluation, emphasizing its predictive power. Athey and Wager (2019) extended Random Forest to causal inference, applying Generalized Random Forests (GRF) to estimate heterogeneous treatment effects in employment training programs.

In social sciences and health research, Random Forest has been instrumental in identifying key predictors of health outcomes. Puterman et al. (2020) employed Random Forest survival analysis to predict mortality based on economic, behavioral, and psychological factors. Feng et al. (2022) applied a geographical Random Forest approach to rank demographic and socioeconomic contributors to COVID-19 mortality in the U.S. Similarly, Boulesteix et al. (2012) used Random Forest to analyze high-dimensional genomic data for disease classification, demonstrating its advantage over traditional statistical models. Caruana et al. (2006) compared ensemble methods in medical risk prediction, finding that Random Forest outperformed logistic regression in predicting patient mortality and hospital readmission. In environmental science, Prasad, Iverson, and Liaw (2006) applied Random Forest to model tree species distributions, while Cutler et al. (2007) used it for land cover classification in remote sensing. These studies illustrate the broad utility of Random Forest and Bagging in tackling complex, high-dimensional problems across disciplines.

While Bagging and Random Forest provide robust methods to reduce variance and improve predictive accuracy, they primarily rely on averaging multiple independent models. However, in some cases, a more targeted approach to reducing bias is necessary. This leads us to Boosting, an alternative ensemble method that builds models sequentially, each learning from the errors of its predecessors. In the next chapter, we explore Boosting, its key algorithms, and how it differs from Bagging in its approach to improving model performance.

15

Boosting

Boosting, bagging, and random forests (RF) are all ensemble learning techniques that combine multiple models to improve prediction accuracy. **Bagging** (Bootstrap Aggregating) reduces variance by training multiple models on bootstrap samples and averaging their predictions. **Random forests** extend bagging by introducing feature randomness at each tree split, further decorrelating models to reduce both bias and variance.

Boosting, unlike bagging and RF, is a sequential process where each model corrects the errors of its predecessor. Instead of training models independently, boosting adjusts weights to focus on difficult-to-predict instances, reducing bias while potentially affecting variance. The key idea is to train weak learners—typically decision trees—incrementally, improving performance at each step. The process follows three main steps: (1) start with a simple model, (2) focus subsequent models on previous errors, and (3) combine predictions through weighted averaging.

This chapter covers leading boosting algorithms, including AdaBoost, gradient boosting, and XGBoost, discussing their mechanics, regularization techniques, and applications in economics and social sciences. We also explore how to implement boosting algorithms in R.

15.1 Types of Boosting Models

Boosting methods refine predictions by building models sequentially, with each step correcting the errors of the previous one. While all boosting algorithms follow this principle, they differ in how they adjust weights and minimize errors.

- *Gradient boosting machines (GBM)*: A flexible and widely used approach that minimizes a specified loss function by sequentially fitting new models to the gradient (direction of steepest descent) of the error. GBM can handle both regression and classification problems, making it a foundational boosting technique.
- *AdaBoost (adaptive boosting)*: One of the earliest boosting algorithms, AdaBoost assigns higher weights to misclassified instances, forcing subsequent models to focus on harder cases. While primarily used for classification, it laid the groundwork for modern boosting techniques.
- *XGBoost (extreme gradient boosting)*: An optimized, high-performance extension of GBM designed for speed and efficiency. It includes advanced regularization, parallelization, and handling of missing values, making it a top choice for large datasets and competitive modeling. XGBoost has become the most widely used boosting algorithm due to its efficiency, scalability, and strong predictive performance. Its ability to distribute computations across multiple machines further improves its scalability, making it ideal for handling large datasets. These advantages have cemented XGBoost as a go-to choice in

DOI: 10.1201/9781003504795-15

both machine learning competitions and real-world applications, particularly in fields like finance, healthcare, and economics. We will also show how to use this method in selection on observables.

- *LightGBM*: A highly efficient gradient boosting framework that grows trees leaf-wise instead of level-wise, leading to faster training and lower memory usage. It is particularly suited for large-scale datasets.

- *CatBoost*: Developed by Yandex, CatBoost is optimized for categorical features, reducing the need for extensive preprocessing. It is designed to handle categorical data efficiently while preventing overfitting.

- *LPBoost (linear programming boosting)*: Uses linear programming to optimize the weights assigned to weak learners, focusing on maximizing the margin between classes.

- *BrownBoost*: A variation of AdaBoost designed to handle noisy data by adjusting the weight update process to be more robust to outliers.

- *TotalBoost*: A modification of AdaBoost that ensures the total weight assigned to misclassified instances remains bounded, preventing excessive focus on hard-to-classify cases.

- *LogitBoost*: Based on logistic regression, LogitBoost modifies AdaBoost by using a weighted log-likelihood loss, making it more stable for classification tasks.

Each of these methods refines the boosting framework by addressing different challenges, such as computational efficiency, handling categorical variables, or reducing overfitting. Among them, GBM serve as the foundation for many advanced variants, including XGBoost, LightGBM, and CatBoost. Understanding GBM is essential for grasping how these enhancements improve performance. The next section explores GBM in detail.

15.2 Gradient Boosting Machines

GBM sequentially fit new models to refine predictions, with each step correcting errors from the previous ones. The "gradient" in GBM refers to the direction and magnitude of model updates, determined through gradient descent on the loss function. This allows GBM to be applied to both regression and classification problems, making it a versatile tool in ensemble learning.

While GBM is highly effective, it requires careful hyperparameter tuning to achieve optimal performance. Its flexibility comes at a computational cost, as training can be slow, especially for large datasets. However, when properly tuned, GBM often delivers strong predictive accuracy, making it a preferred choice in many real-world applications.

GBM stands out with key characteristics that set it apart from earlier boosting methods. It minimizes gradients of a specified loss function, providing flexibility in handling different problem types. Unlike AdaBoost, which primarily focuses on classification, GBM supports various loss functions, allowing for both regression and classification tasks. Additionally, it offers extensive customization through hyperparameters, enabling adaptation to different datasets and modeling needs.

In summary, GBM is a powerful and adaptable boosting algorithm known for its accuracy across diverse applications. While other boosting methods have their own strengths, GBM serves as a foundational approach, with advanced versions like XGBoost and LightGBM improving upon its efficiency and scalability. The choice of algorithm depends on the trade-offs between computational cost, accuracy, and ease of use.

15.2.1 Gradient Boosting Algorithm Steps

1. *Initialize the model*: Start with a simple model, such as a single leaf node predicting the average target value.

2. *Calculate residuals*: Compute the difference between predicted and actual values.

3. *Fit a new model to residuals*: Train a model to predict the residuals from the previous step.

4. *Update the model*: Combine the new model with previous predictions to improve accuracy.

5. *Repeat*: Iterate steps 2–4 until a stopping criterion is met (e.g., max iterations or minimal improvement).

By iteratively fitting new models to the residuals, GBM gradually improves predictions. The final output is the sum of predictions from all models, weighted by learning rates and other tuning parameters.

15.2.1.1 Example: Gradient Boosting for Regression

In this section, we demonstrate how GBM work through a step-by-step example. We will generate a synthetic dataset, build an initial regression tree, iteratively improve it by fitting new trees to residuals, and visualize the impact of boosting.

Step 1: Simulating Data and Initial Model

First, we generate a dataset where the response variable y follows a noisy sine function. We then fit a simple regression tree as the initial model.

```
library(rpart)
library(ggplot2)
library(dplyr)
library(gbm)

# First we will simulate our data
n <- 300
set.seed(1)
x <- sort(runif(n) * 2 * pi)
y <- sin(x) + rnorm(n) / 4
df <- data.frame("x" = x, "y" = y)
plot(df$x, df$y, ylab = "y", xlab = "x", col = "grey")
```

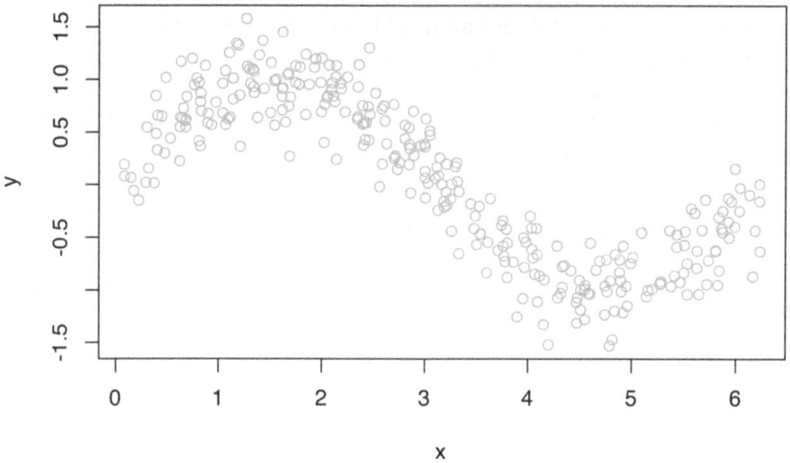

We now initialize the boosting process by fitting a single regression tree. This tree provides an initial prediction, which will be refined in subsequent steps.

```
library(rpart)

# Initiate the matrix to store the predictions
B = 100
h = 1.8 # shrinkage parameter
boost <- matrix(0, nrow = n, ncol = B) # 100 trees

# Regression tree
in_model <- rpart(y ~ x, data = df)
yp <- predict(in_model) # using the entire data

# Store the first predictions
boost[ ,1] <- yp * h

# Plot for single regression tree
plot(df$x, df$y, ylab = "y", xlab = "x", col = "grey")
lines(df$x, yp, type = "s", col = "black", lwd = 3)
```

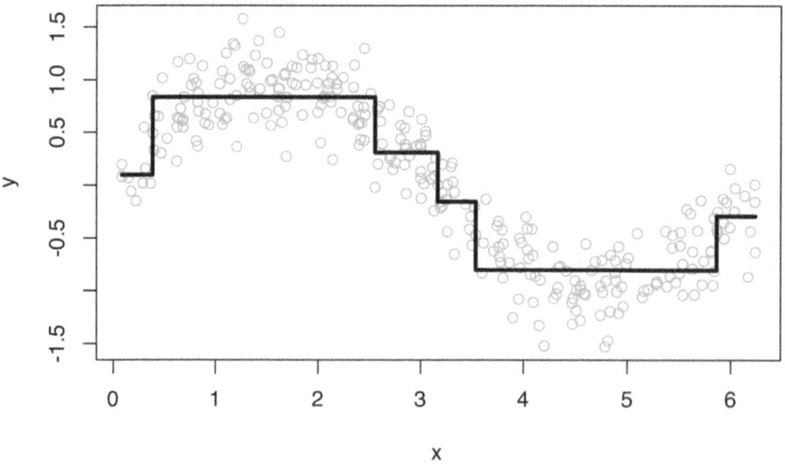

At this point, we have a single regression tree that provides a rough estimate of y. However, this model is limited and will be improved using boosting. Now, we will have a loop that will "boost" the model. What we mean by boosting is that we seek to improve yhat, i.e. $\hat{f}(x_i)$, in areas where it does not perform well by fitting trees to the residuals.

Step 2: Iterative Boosting Process The key idea in boosting is to sequentially train models that correct the residuals of previous predictions. At each iteration, we compute the residuals, fit a new tree to them, and update the overall prediction.

```
df$yr <- df$y - h * yp #First iteration

# Boosting loop for t times (trees)
for (t in 2:B) {

  # Fit a regression tree to the residuals
  fit <- rpart(yr ~ x, data = df)
  yp <- predict(fit, newdata = df)

  # Store the predictions
  boost[, t] <- h * yp

  # Update the residuals
  df$yr <- df$yr - h * yp
}

str(boost)
```

```
##  num [1:300, 1:100] 0.174 0.174 0.174 0.174 0.174 ...
```

Here, each iteration refines the previous predictions, reducing the overall error. Now, we will plot the predictions from the first 100 trees.

Step 3: Combining Predictions and Visualization

To obtain the final prediction, we sum the contributions from all boosting iterations. We then visualize how predictions evolve as more trees are added.

```
# Function to plot a single tree and boosted trees for different t
viz <- function(M) {
  # Boosting
  yhat <- apply(boost[, 1:M], 1, sum) # This is predicted y for depth M
  # Plotting
  plot(df$x, df$y, ylab = "", xlab = "", col = "grey") # Data points
  lines(df$x, yhat, type = "s", col = "gray20", lwd = 3) # line for boosting
}

# Run
viz(100)
```

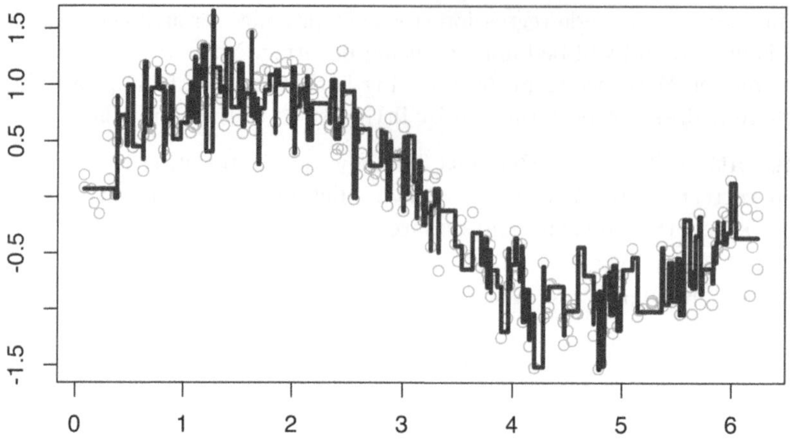

We see that as boosting progresses, the model captures the underlying pattern in the data more effectively. To make the boosting process more flexible, let's define a function that allows for tuning key parameters (B, M, and h as inputs and returns the predictions).

```
library(rpart)

# Function to predict
boosting <- function(B, M, h, df) {
  if (M > B) stop("M should be less than or equal to B")

  n <- nrow(df)
  boost <- matrix(0, nrow = n, ncol = B) # B trees
  model <- rpart(y ~ x, data = df)
  yp <- predict(model, df) # using the entire data

  boost[, 1] <- yp * h # Store the first predictions
  df$yr <- df$y - h * yp #First iteration error

  for (t in 2:B) {
    fit <- rpart(yr ~ x, data = df)
    yp <- predict(fit, newdata = df)
    boost[ ,t] <- yp * h
    df$yr <- df$yr - h * yp
  }

  yhat <- apply(boost[, 1:M], 1, sum) # This is predicted y for depth M
  return(yhat)
}
```

This function allows us to specify the number of trees (`B`), the number of iterations used for predictions (`M`), and the learning rate (`h`).

15.2.2 Hyperparameter Tuning in GBM

The performance of GBM depends heavily on the choice of hyperparameters. Below, we examine how the learning rate (`h`) and the number of iterations (`M`) affect predictions. Here are a few examples of how the predictions change as we change `h`.

```
# Plot
for (i in c(0.2, 0.5, 1, 1.4)) {
  yhat <- boosting(100, 100, i, df)
  plot(df$x, df$y, ylab = "y", xlab = "x", col = "grey")
  lines(df$x, yhat, type = "s", col = "gray20", lwd = 3)
}
```

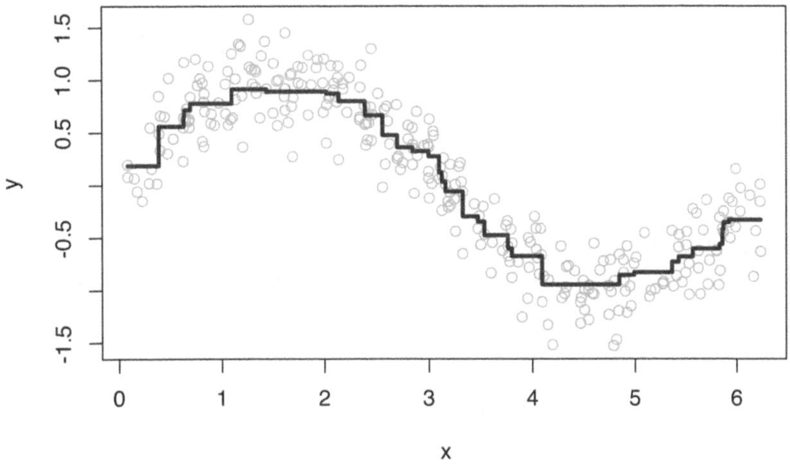

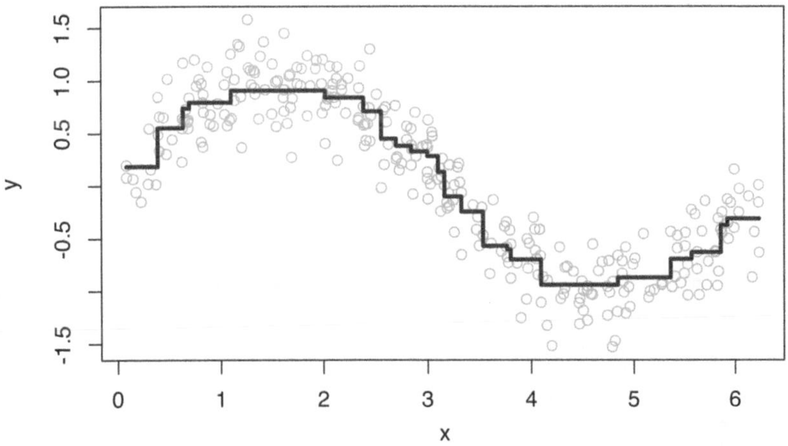

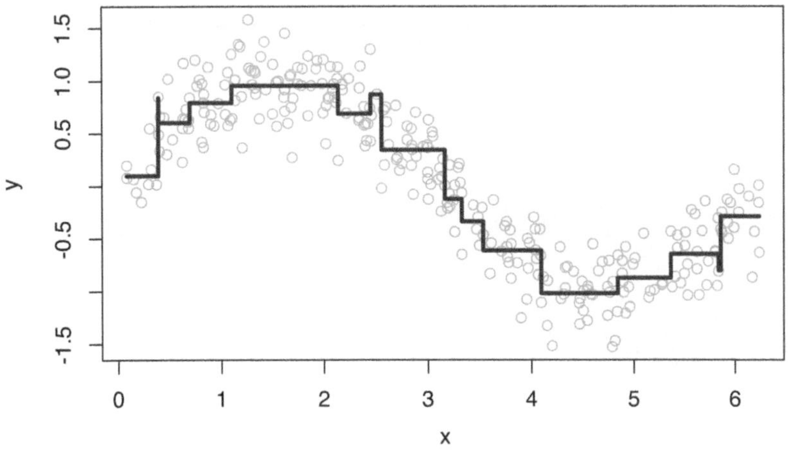

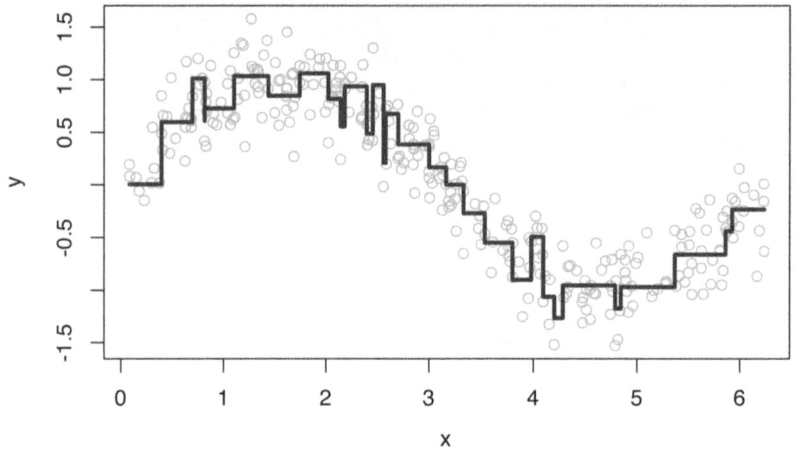

Smaller h values lead to more conservative updates, while larger values make faster updates but risk overfitting. Here are a few examples of how the predictions change as we change M.

```
# Plot
for (i in c(2, 15, 50, 200)) {
  yhat <- boosting(200, i, 0.03, df)
  plot(df$x, df$y, ylab = "y", xlab = "x", col = "grey")
  lines(df$x, yhat, type = "s", col = "gray60", lwd = 3)
}
```

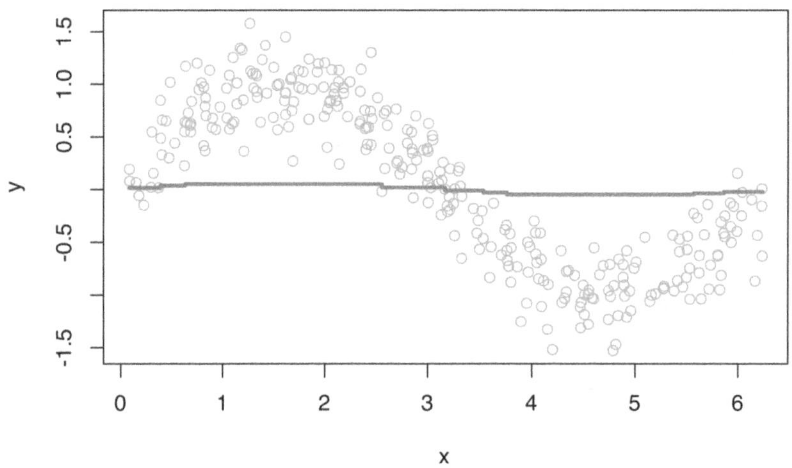

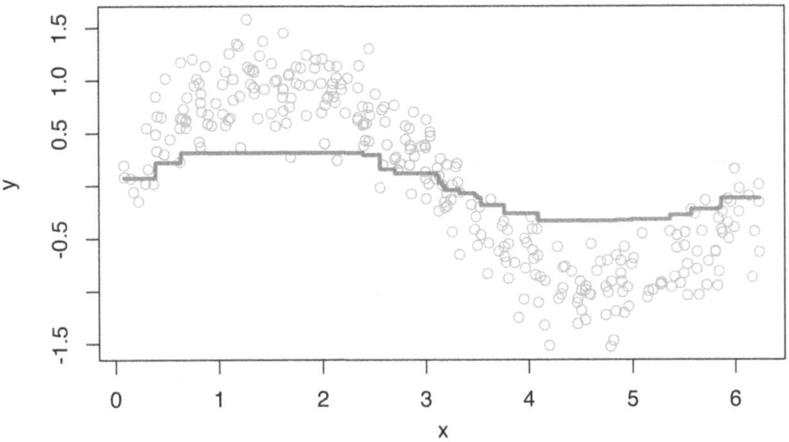

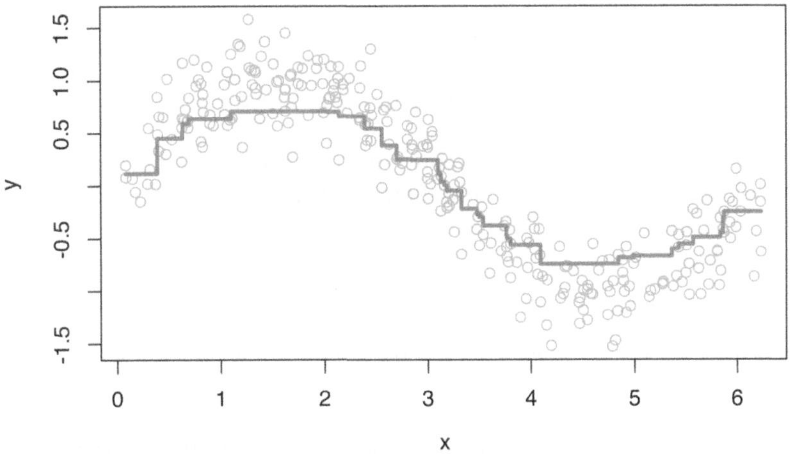

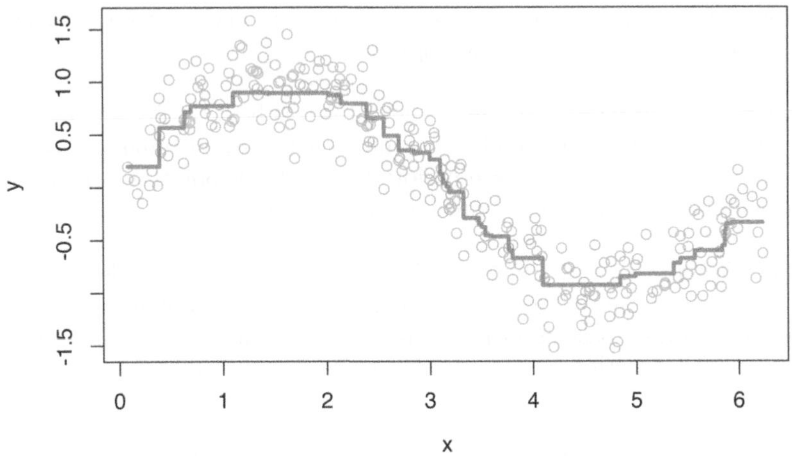

More boosting iterations refine predictions, but too many can lead to overfitting. Selecting `M` requires a balance between bias and variance.

This example illustrates how GBM iteratively refines predictions by fitting new models to residuals. We demonstrated the core boosting steps, visualized the impact of different hyperparameters, and provided a flexible function for tuning. These concepts extend directly to more advanced implementations like XGBoost and LightGBM, which build on GBM's foundation with additional optimizations for speed and efficiency.

15.3 GBM in R

Instead of manually implementing gradient boosting, we can use the `gbm` package in R, which provides a flexible and efficient way to fit gradient boosting models. The `gbm()` function allows for extensive customization through various hyperparameters, making it adaptable to a wide range of problems. It supports both regression and classification and includes built-in mechanisms for model selection and overfitting control.

The key tuning parameters in `gbm` fall into two categories:

- *Boosting parameters*: Control the number of boosting iterations (`n.trees = 100`) and the learning rate (`shrinkage = 0.1`). The learning rate determines how much each new tree contributes to the model, affecting the trade-off between speed and accuracy.

- *Tree parameters*: Define the structure of individual trees. The maximum depth (`interaction.depth = 1`) controls how complex each tree can be, while `n.minobsinnode = 10` sets the minimum number of observations required at a terminal node, affecting the model's granularity.

Internal Training and Model Selection

The `gbm` algorithm offers three internal methods for selecting the optimal number of boosting iterations:

1. *Out-of-Bag (OOB) Estimation*: Uses bootstrap sampling to estimate prediction error without requiring a separate validation set. The error improvement at each iteration is tracked using `oobag.improve`.

2. *Test Set Validation*: A fraction of the dataset is held out as a test set, regulated by `train.fraction`. The model is trained on `train.fraction * nrow(data)`, and the test set is used to evaluate performance. This is not cross-validation but can be combined with external loops for repeated testing.

3. *k-Fold Cross-validation (`cv.fold`)*: Splits the dataset into `k` folds and fits `k` models to compute cross-validation error. If `cv.folds = 5`, `gbm` fits five models and selects the best iteration based on the average error. A final model is then trained on the full dataset using the optimal number of iterations. The best iteration is reported in `cv.error`.

To introduce randomness and improve generalization, `gbm` includes the **bagging fraction (`bag.fraction`)**, which applies stochastic gradient descent by training each tree on a random subset of the data. This helps reduce overfitting while maintaining efficiency.

For more details, refer to the official documentation: GBM Vignette[1].

Example: GBM Without Hyperparameter Tuning

In this example, we fit a basic gradient boosting model using the `gbm` package without extensive hyperparameter tuning. We use the `Hitters` dataset from the `ISLR` package, which contains baseball player statistics, including salary. The goal is to predict `Salary` using all available features. Since salaries are highly skewed, we apply a logarithmic transformation for better model performance.

```
library(ISLR)
library(gbm)

# Load and preprocess data
data(Hitters)
df <- Hitters[complete.cases(Hitters$Salary), ]  # Remove missing values
df$Salary <- log(df$Salary)  # Log-transform salary

# Fit GBM model
boos <- gbm(Salary ~ ., distribution = "gaussian",
            n.trees = 5000, interaction.depth = 3,
            shrinkage = 0.01, data = df,
            bag.fraction = 0.5, cv.folds = 10)

boos  # Print model summary
```

```
## gbm(formula = Salary ~ ., distribution = "gaussian", data = df,
##     n.trees = 5000, interaction.depth = 3, shrinkage = 0.01,
##     bag.fraction = 0.5, cv.folds = 10)
## A gradient boosted model with gaussian loss function.
## 5000 iterations were performed.
## The best cross-validation iteration was 758.
## There were 19 predictors of which 19 had non-zero influence.
```

Here, we specify the **Gaussian distribution** (`distribution = "gaussian"`) since salary is a continuous variable. The model runs **5000 boosting iterations** (`n.trees = 5000`), with an **interaction depth of 3**, meaning each tree can have up to three splits. The **learning rate** (`shrinkage = 0.01`) controls how much each tree contributes to the final prediction, and **bagging fraction** (`bag.fraction = 0.5`) ensures that each tree is trained on a random 50% subset of the data. Additionally, we use **10-fold cross-validation** (`cv.folds = 10`) to evaluate model performance.

Selecting the Optimal Number of Trees

The function `gbm.perf()` helps identify the best number of boosting iterations based on cross-validation performance.

```
perf <- gbm.perf(boos, method = "cv", plot.it = FALSE)
plot(boos$cv.error, type = "l", col = "black", lwd = 2,
     ylab = "CV Error", xlab = "Number of Trees",
     main = "Cross-Validation Error vs Trees")
abline(v = perf, col = "gray40", lty = 2)
```

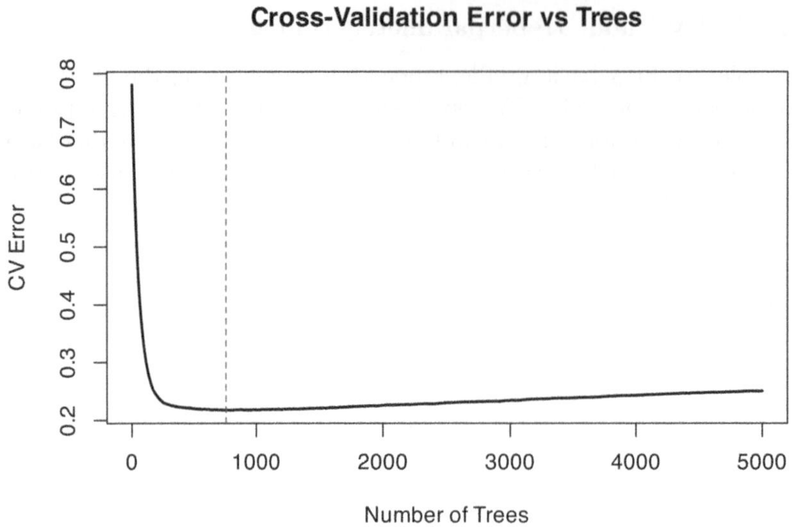

This function plots the cross-validation error and suggests the number of trees that minimize the error, helping avoid overfitting.

Visualizing Cross-Validation Error

To manually inspect the cross-validation error and determine the best iteration:

```
plot(boos$cv.error, type = "l")  # Plot cross-validation error over iterations
```

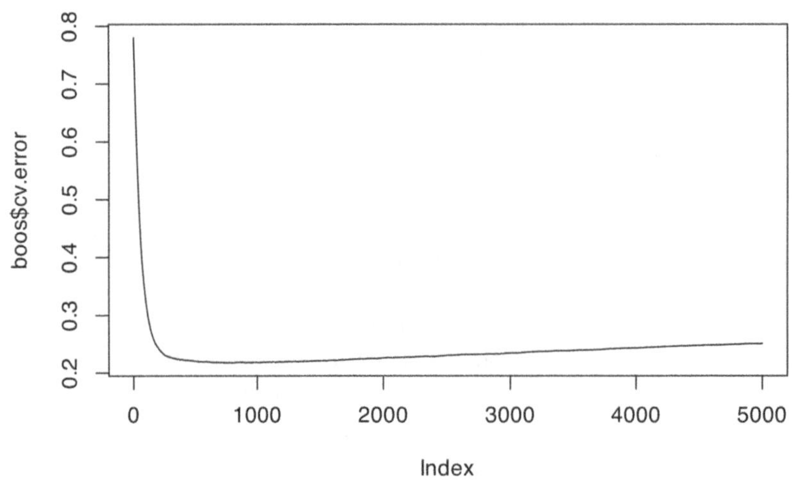


```
best <- which.min(boos$cv.error)  # Find the iteration with the lowest error
best  # Optimal number of trees
```

```
## [1] 758
```

```
boos$cv.error[best]  # Minimum cross-validation error
```

```
## [1] 0.2172724
```

The plot shows how the error decreases as more trees are added and helps identify the point where adding more trees no longer improves performance. The `best` iteration provides the number of trees that achieve the lowest error.

This example demonstrates the core workflow for fitting a GBM model in R, using cross-validation to select the best number of trees without extensive tuning.

Example: GBM with Hyperparameter Tuning

Instead of using fixed hyperparameters, we now perform hyperparameter tuning to find the best combination of parameters for the gradient boosting model. We define a grid of possible values for key parameters and evaluate model performance using cross-validation.

Step 1: Creating a Hyperparameter Grid

We set up a grid of hyperparameters to explore different values for the learning rate (`shrinkage`), tree depth (`interaction.depth`), minimum number of observations per terminal node (`n.minobsinnode`), and bagging fraction (`bag.fraction`).

```
# Load required libraries
library(dplyr)
library(gbm)

# Create hyperparameter grid
grid <- expand.grid(
  shrinkage = c(0.01, 0.1, 0.3),
  interaction.depth = c(1, 3, 5),
  n.minobsinnode = c(5, 10, 15),
  bag.fraction = c(.65, .8, 1),
  optimal_trees = 0,    # Placeholder for results
  min_RMSE = 0          # Placeholder for results
)

# Total number of hyperparameter combinations
nrow(grid)
```

```
## [1] 81
```

This grid search will evaluate different combinations of boosting and tree parameters, helping us understand their impact on model performance.

Step 2: Evaluating Hyperparameter Combinations

We iterate over the grid and fit a `gbm` model for each set of parameters, using 10-fold cross-validation to measure the root mean squared error (RMSE).

```
for (i in 1:nrow(grid)) {
  gbm.tune <- gbm(
    formula = Salary ~ .,
    distribution = "gaussian",
    data = df,
    n.trees = 2000,
    interaction.depth = grid$interaction.depth[i],
    shrinkage = grid$shrinkage[i],
    n.minobsinnode = grid$n.minobsinnode[i],
    bag.fraction = grid$bag.fraction[i],
    cv.folds = 10,
```

```
    n.cores = NULL, # Uses all available cores
    verbose = FALSE
  )

  # Store results
  grid$optimal_trees[i] <- which.min(gbm.tune$cv.error)
  grid$min_RMSE[i] <- sqrt(min(gbm.tune$cv.error))
}

# Display the top 10 best-performing models
grid %>%
  arrange(min_RMSE) %>%
  head(10)
```

```
##    shrinkage interaction.depth n.minobsinnode bag.fraction optimal_trees
## 1       0.10                 3              5         0.65            88
## 2       0.10                 1             15         1.00           500
## 3       0.10                 3              5         0.80           156
## 4       0.10                 5              5         1.00            67
## 5       0.10                 1             10         1.00           377
## 6       0.30                 1             15         1.00           201
## 7       0.10                 5              5         0.65            70
## 8       0.01                 5              5         0.65           589
## 9       0.10                 1              5         0.65           243
## 10      0.01                 5              5         0.80           647
##     min_RMSE
## 1  0.4499207
## 2  0.4505496
## 3  0.4516853
## 4  0.4522516
## 5  0.4539349
## 6  0.4541096
## 7  0.4543849
## 8  0.4543888
## 9  0.4544533
## 10 0.4544835
```

The results show the trade-off between the learning rate (`shrinkage`) and the number of boosting iterations. A smaller learning rate requires more trees for convergence, while a larger learning rate allows for fewer trees but increases the risk of overfitting. Additionally, deeper trees tend to perform better with smaller shrinkage values since they capture more complex interactions.

Step 3: Training the Final Model with Best Parameters

Once we identify the best hyperparameters, we fit a final model using them.

```
# Select best hyperparameter combination
best_params <- grid %>%
  arrange(min_RMSE) %>%
  head(1)
```

```
# Train final GBM model
gbm.final <- gbm(
  formula = Salary ~ .,
  distribution = "gaussian",
  data = df,
  n.trees = best_params$optimal_trees,
  interaction.depth = best_params$interaction.depth,
  shrinkage = best_params$shrinkage,
  n.minobsinnode = best_params$n.minobsinnode,
  bag.fraction = best_params$bag.fraction,
  train.fraction = 1, # Use full dataset
  cv.folds = 0, # No cross-validation (already tuned)
  n.cores = NULL,
  verbose = FALSE
)

gbm.final  # Display final model details
```

```
## gbm(formula = Salary ~ ., distribution = "gaussian", data = df,
##     n.trees = best_params$optimal_trees, interaction.depth =
best_params$interaction.depth,
##     n.minobsinnode = best_params$n.minobsinnode, shrinkage =
best_params$shrinkage,
##     bag.fraction = best_params$bag.fraction, train.fraction = 1,
##     cv.folds = 0, verbose = FALSE, n.cores = NULL)
## A gradient boosted model with gaussian loss function.
## 88 iterations were performed.
## There were 19 predictors of which 18 had non-zero influence.
```

This tuned model is expected to perform better than an untuned version because it optimally balances bias and variance.

Step 4: Evaluating the Tuned Model on a Test Set

Since we didn't initially create a test set, we now split the data and assess the model's performance using Root Mean Squared Prediction Error (RMSPE). We perform bootstrapped sampling 100 times to get a distribution of RMSPE values.

```
rmspe <- c()

for (i in 1:100) {
  ind <- unique(sample(1:nrow(df), nrow(df), replace = TRUE))
  train <- df[ind, ]
  test <- df[-ind, ]

  gbm.final <- gbm(
    formula = Salary ~ .,
    distribution = "gaussian",
    data = train,
    n.trees = 1000,
```

```
    interaction.depth = best_params$interaction.depth,
    shrinkage = best_params$shrinkage,
    n.minobsinnode = best_params$n.minobsinnode,
    bag.fraction = best_params$bag.fraction,
    train.fraction = 1,
    cv.folds = 0,
    n.cores = NULL,
    verbose = FALSE
  )

  yhat <- predict(gbm.final, test, n.trees = best_params$optimal_trees)
  rmspe[i] <- sqrt(mean((test$Salary - yhat)^2))
}
```

Step 5: Visualizing Model Performance

We compute and visualize the mean RMSPE along with its variability across iterations.

```
mean(rmspe)  # Mean RMSPE
```

```
## [1] 0.4933289
```

```
# Plot RMSPE distribution
plot(rmspe, col = "black")
abline(h = mean(rmspe), col = "gray20")  # Mean RMSPE
abline(h = mean(rmspe) + 2*sd(rmspe), col = "gray40", lty = 2)  # Upper bound
```

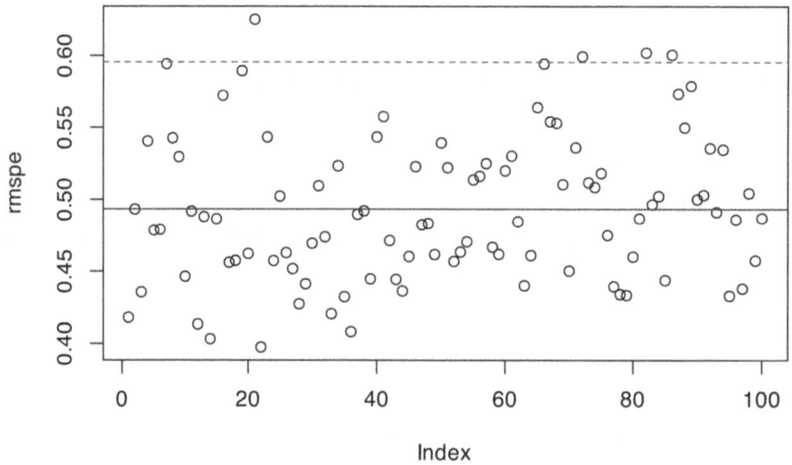

This plot gives insight into model stability. A lower RMSPE indicates better predictive accuracy, while a small standard deviation suggests consistent performance across different test sets.

15.3.1 Variable Importance in GBM

Understanding which variables influence predictions the most is crucial for interpreting gradient boosting models. The `gbm` package in R provides multiple ways to assess variable importance, including built-in measures and permutation-based methods.

1. Built-in `summary.gbm()` function.

- Most convenient: Call `summary.gbm(model)`, where 'model' is your fitted gbm model.
- Output: This provides a table of variable importance measures calculated based on the sum of improvements that are due to splits on each variable across all trees in the ensemble.
- relative_influence column: The values in this column are normalized so that they add up to 100, indicating the percentage contribution of each variable.

```
summary.gbm(gbm.final)
```

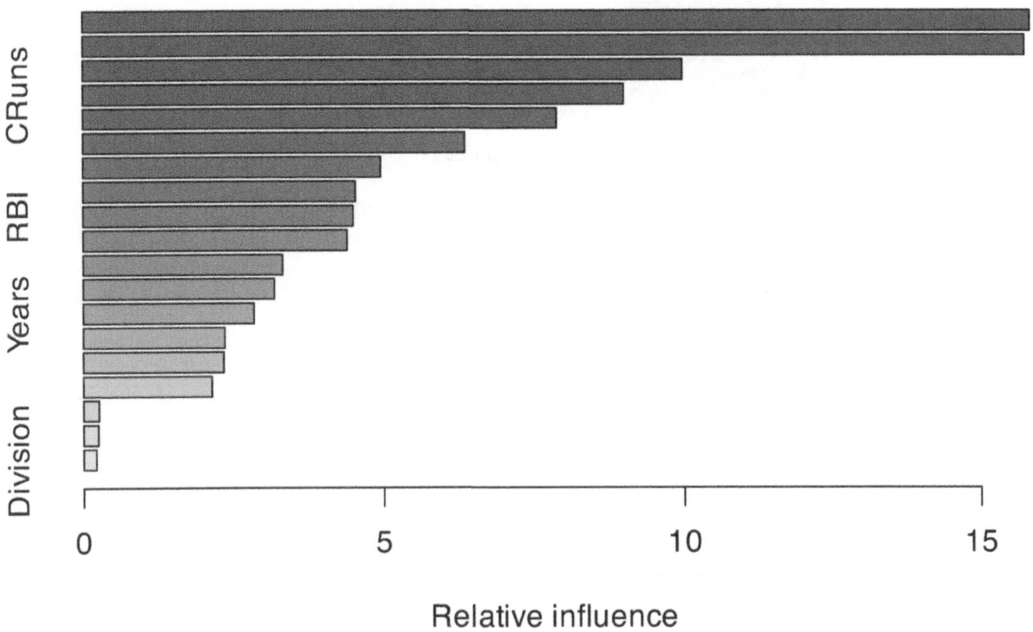

2. `cprod()` function and permutation-based importance: Permutation-based importance measures the change in model performance when a specific variable is randomly shuffled. This breaks the relationship between the variable and the outcome, helping to assess its true contribution. While computationally more intensive, this method is often considered more robust than built-in importance scores.

- Use `vip::vi()` or `vip::cprod()` :
- `vi(model, method="permutation")` calculates permutation-based importance by randomly shuffling each variable and measuring the decrease in model performance.
- `cprod(model, num_trees = "all")` is generally faster, offering a model based importance calculation as an approximation of the permutation approach

```
library(gbm)
library(vip)

vip::vip(gbm.final)
```

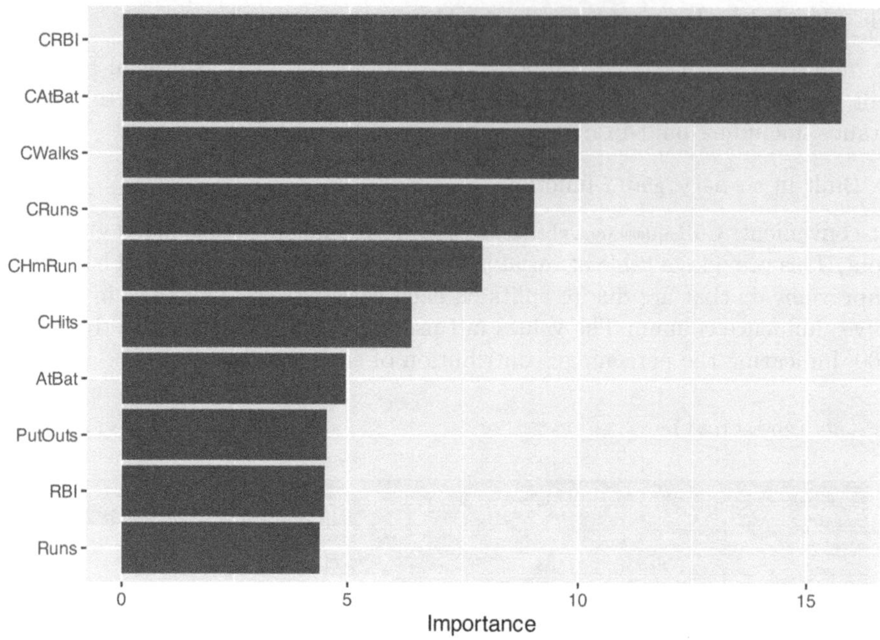

To further visualize and understand how an individual variable influences predictions, we can use partial dependence plots. For example, to examine the impact of CRuns on the predicted salary:

```
plot(gbm.final, i = "CRuns", lwd = 2, col = "black", cex.lab = 1, cex.axis = 1, cex.main = 1)
```

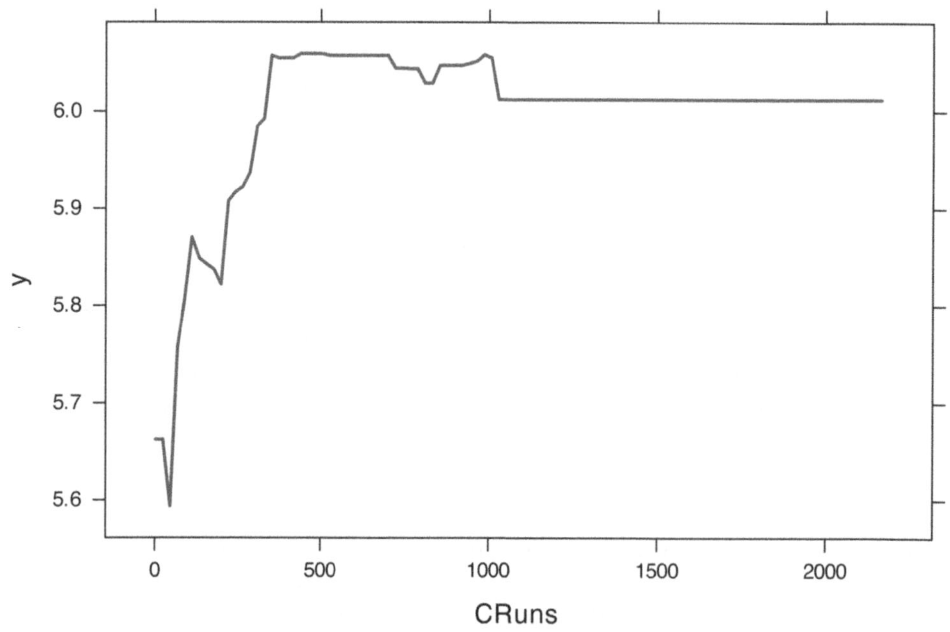


```
# plot(gbm.final, i = "CRuns")
```

These plots help interpret the functional relationship between predictors and the target variable, revealing whether the effect is linear, nonlinear, or exhibits threshold effects. They complement variable importance measures by showing how changes in a feature affect predictions, not just whether the feature is important.

By combining these methods—built-in importance scores for quick insights, permutation-based importance for robustness, and partial dependence plots for visualization—we can thoroughly assess and interpret the influence of different variables in a gradient boosting model.

15.3.2 External Training

Instead of relying on gbm's built-in cross-validation, we can perform external training by manually splitting the data, training multiple models, and evaluating their performance across different hyperparameter combinations. This approach allows for more flexibility but is computationally intensive, making it less practical than internal cross-validation for large datasets. We define a range of values for key hyperparameters. We then create a grid of all possible combinations and initialize a column to store the Root Mean Squared Prediction Error (RMSPE).

```
library(gbm)
h <- seq(0.01, 0.1, 0.01)
B <- c(100, 300, 500, 750, 900)
D <- 1:2

grid <- as.data.frame(expand.grid(D, B, h))
colnames(grid) <- c("D", "B", "h")
grid$rmspe <- rep(0, nrow(grid))
```

We perform bootstrap sampling 20 times for each parameter combination. The dataset is randomly split into training and validation sets, where we fit the gbm model on the training data and evaluate predictions on the validation set.

```
mse <- c()
sdmse <- c()

for (i in 1:nrow(grid)) {

  val_mse <- c()

  for (j in 1:20) { try({
    set.seed(j)
    ind <- sample(nrow(df), nrow(df), replace = TRUE)
    train <- df[ind, ]
    val <- df[-ind, ]

     boos <- gbm(
       Salary ~ .,
       distribution = "gaussian",
       n.trees = 1000,
```

```
      interaction.depth = grid[i, 1],
      shrinkage = grid[i, 3],
      bag.fraction = 0.8,
      data = train
    )

    yhat<- predict(boos, val, n.trees = grid[i, 2])
    val_mse[j] <- sqrt(mean((val$Salary - yhat)^2))

  }, silent = TRUE)

  }

grid$rmspe[i] <- mean(val_mse)
}
```

Once all combinations are evaluated, we identify the best-performing set of hyperparameters based on the lowest RMSPE to select the best model.

```
res <- arrange(grid, rmspe)
head(res)
```

```
##   D   B    h     rmspe
## 1 2 100 0.07 0.4586446
## 2 2 900 0.01 0.4590531
## 3 2 750 0.01 0.4592292
## 4 2 300 0.03 0.4593989
## 5 2 100 0.08 0.4600807
## 6 2 500 0.02 0.4601794
```

When it is done with a proper grid, this application requires a lot of computational power and it is also lacking test set. So, using the internal training is more practical. We have shown a regression example. The same approach can be used for classification problems. While external validation provides a more flexible approach, it is computationally expensive, especially when using a large grid. Additionally, this method does not include a final holdout test set, meaning it does not provide a true out-of-sample error estimate.

For practical applications, internal training (cross-validation within `gbm`) is often more efficient and reliable, as it is optimized for selecting the best number of trees and prevents excessive computational overhead. However, external training remains a valuable technique for testing specific hyperparameter settings or when additional control over validation splits is needed.

This example focused on regression, but the same approach applies to classification problems by adjusting the `distribution` parameter (e.g., `"bernoulli"` for binary classification).

Next, we explore AdaBoost, another boosting method with a different weighting mechanism.

15.4 AdaBoost

One of the most popular boosting algorithm is **AdaBost.M1** due to Freund and Schpire (1997). We consider two-class problem where $y \in \{-1, 1\}$, which is a categorical outcome. With a set of predictor variables X, a classifier $\hat{m}_b(x)$ at tree b among B trees, produces a prediction taking the two values $\{-1, 1\}$.

To understand how AdaBoost works, let's look at the algorithm step by step:

1. Select the number of trees B, and the tree depth D;
2. Set initial weights, $w_i = 1/n$, for each observation.
3. Fit a classification tree $\hat{m}_b(x)$ at $b = 1$, the first tree.
4. Calculate the following misclassification error for $b = 1$:

$$\mathbf{err}_{b=1} = \frac{\sum_{i=1}^{n} \mathbf{I}\left(y_i \neq \hat{m}_b\left(x_i\right)\right)}{n}$$

5. By using this error, calculate

$$\alpha_b = 0.5 \log\left(\frac{1 - err_b}{err_b}\right)$$

For example, suppose $err_b = 0.3$, then $\alpha_b = \log(0.7/0.3)$, which is a log odds or log(success/failure).

6. If the observation i is misclassified, update its weights, if not, use w_i which is $1/n$:

$$w_i \leftarrow w_i e^{\alpha_b}$$

Let's try some numbers:

```
#Suppose err = 0.2, n=100
n = 100
err = 0.2
alpha <- 0.5 * log((1 - err) / err)
alpha
```

```
## [1] 0.6931472
```

```
exp(alpha)
```

```
## [1] 2
```

So, the new weight for the misclassified i in the second tree (i.e., $b = 2$ stump) will be

```
# For misclassified obervations
weight_miss <- (1 / n) * (exp(alpha))
weight_miss
```

```
## [1] 0.02
```

```r
# For correctly classified observation
weight_corr <- (1 / n) * (exp(-alpha))
weight_corr
```

```
## [1] 0.005
```

This shows that as the misclassification error goes up, it increases the weights for each misclassified observation and reduces the weights for correctly classified observations.

7. With this procedure, in each loop from b to B, it applies $\hat{m}_b(x)$ to the data using updated weights w_i in each b:

$$\mathbf{err}_b = \frac{\sum_{i=1}^{n} w_i \mathbf{I}\left(y_i \neq \hat{m}_b\left(x_i\right)\right)}{\sum_{i=1}^{n} w_i}$$

We normalize all weights between 0 and 1 so that sum of the weights would be one in each iteration. These individual weights give higher probabilities for misclassified observations in the next tree. In the `rpart` package, for example, the weights can be assigned to each observation in the data set and are used to adjust the impurity measure when splitting the nodes of the tree. By setting the "weights" argument in the rpart function, we can assign different weights to different observations. Observations with higher weights will have a greater impact on the impurity measure and therefore, on the splitting decisions made by the algorithm.

Here is an example with the `myocarde` data that we only use the first 6 observation:

```r
library(readr)

myocarde <- read_delim("myocarde.csv", delim = ";" ,
                       escape_double = FALSE, trim_ws = TRUE,
                       show_col_types = FALSE)
myocarde <- data.frame(myocarde)
df <- head(myocarde)
df$Weights = 1 / nrow(df)
df
```

```
##    FRCAR INCAR INSYS PRDIA PAPUL PVENT REPUL  PRONO   Weights
## 1     90  1.71  19.0    16  19.5  16.0   912 SURVIE 0.1666667
## 2     90  1.68  18.7    24  31.0  14.0  1476  DECES 0.1666667
## 3    120  1.40  11.7    23  29.0   8.0  1657  DECES 0.1666667
## 4     82  1.79  21.8    14  17.5  10.0   782 SURVIE 0.1666667
## 5     80  1.58  19.7    21  28.0  18.5  1418  DECES 0.1666667
## 6     80  1.13  14.1    18  23.5   9.0  1664  DECES 0.1666667
```

Suppose that our first stump misclassifies the first observation. So the error rate

```r
# Alpha
n = nrow(df)
```

```
err = 1 / n
alpha <- 0.5 * log((1 - err) / err)
alpha
```

```
## [1] 0.804719
```

```
exp(alpha)
```

```
## [1] 2.236068
```

```
# Weights for misclassified observations
weight_miss <- (1 / n) * (exp(alpha))
weight_miss
```

```
## [1] 0.372678
```

```
# Weights for correctly classified observations
weight_corr <- (1 / n) * (exp(-alpha))
weight_corr
```

```
## [1] 0.0745356
```

Hence, our new sample weights

```
df$New_weights <- c(weight_miss, rep(weight_corr, 5))
df$Norm_weights <- df$New_weight / sum(df$New_weight) # normalizing
# Not reporting X's for now
df[, 8:11]
```

```
##    PRONO   Weights New_weights Norm_weights
## 1 SURVIE 0.1666667   0.3726780          0.5
## 2  DECES 0.1666667   0.0745356          0.1
## 3  DECES 0.1666667   0.0745356          0.1
## 4 SURVIE 0.1666667   0.0745356          0.1
## 5  DECES 0.1666667   0.0745356          0.1
## 6  DECES 0.1666667   0.0745356          0.1
```

We can see that the misclassified observation (the first one) has 5 times more likelihood than the other correctly classified observations. Let's see a simple example how weights affect Gini and a simple tree:

```
# Define the class labels and weights
class_labels <- c(1, 1, 1, 0, 0, 0)
weights <- rep(1/length(class_labels), length(class_labels))

# Calculate the proportion of each class
class_prop <- table(class_labels) / length(class_labels)
```

```
# Calculate the Gini index
gini_index <- 1 - sum(class_prop^2)
print(gini_index)
```

```
## [1] 0.5
```

```
# Change the weight of the first observation to 0.5
weights[1] <- 0.5
weights[2:6] <- 0.1

# Recalculate the class proportions and Gini index
weighted_table <- tapply(weights, class_labels, sum)
class_prop <- weighted_table / sum(weights)
gini_index <- 1 - sum(class_prop^2)
print(gini_index)
```

```
## [1] 0.42
```

Notice how the change in the weight of the first observation affects the Gini index. The weights assigned to the observations in a decision tree can affect the splitting decisions made by the algorithm, and therefore can change the split point of the tree.

In general, observations with higher weights will have a greater influence on splitting decisions of the decision tree. This is because the splitting criterion used by the decision tree algorithm, such as the Gini index or information gain, takes the weights of the observations into account when determining the best split. If there are observations with very high weights, they may dominate the splitting decisions and cause the decision tree to favor a split that is optimal for those observations, but not necessarily for the entire data set.

```
library(rpart.plot)
# Define the data
x <- c(2, 4, 6, 8, 10, 12)
y <- c(0, 1, 1, 0, 0, 0)
weights1 <- rep(1/length(y), length(y))
weights2 <- c(0.95, 0.01, 0.01, 0.01, 0.01, 0.01)

# Build the decision tree with equal weights
library(rpart)
fit1 <- rpart(y ~ x, weights = weights1,
              control =rpart.control(minsplit =1,minbucket=1, cp=0))

# Build the decision tree with different weights
fit2 <- rpart(y ~ x, weights = weights2,
              control =rpart.control(minsplit =1,minbucket=1, cp=0))
```

```
prp(fit1)
```

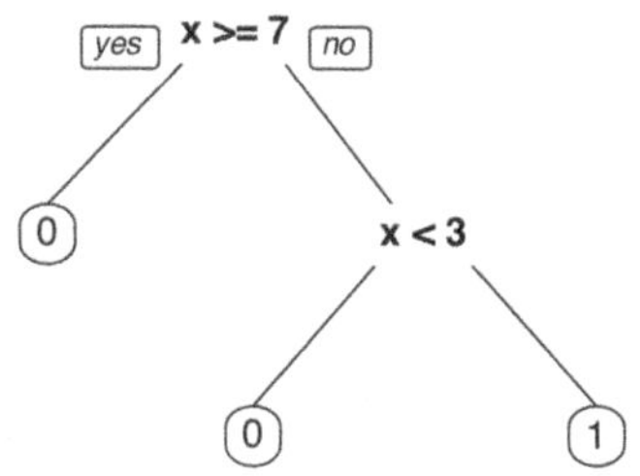


```
prp(fit2)
```

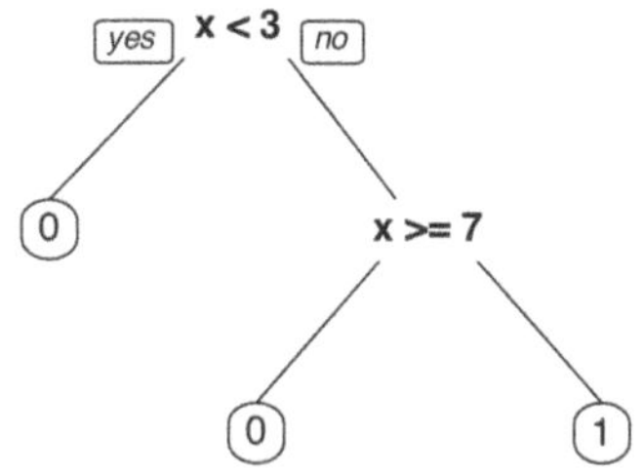

By comparing the two decision trees, we can see that the weights have a significant impact on the splitting decisions made by the algorithm. In the decision tree built with equal weights, the algorithm splits on $x <= 7$ and then on $x < 3$ to separate the two classes. In contrast, in the decision tree built with different weights, the algorithm first splits on $x < 3$ to separate the first observation with $y = 0$ from the rest of the data. This split favors the first observation, which has a higher weight, and results in a simpler decision tree with fewer splits.

Hence, observations that are misclassified will have more influence in the next classifier. **This is an incredible boost that forces the classification tree to adjust its prediction to do better job for misclassified observations.**

8. Finally, in the output, the contributions from classifiers that fit the data better are given more weight (a larger α_b means a better fit). Unlike a random forest algorithm where each tree gets an equal weight in final decision, here some stumps get more say in final classification. Moreover, "forest of stumps" the order of trees is important.

Hence, the final prediction on y_i will be combined from all trees, b to B, through a weighted majority vote:

$$\hat{y}_i = \operatorname{sign}\left(\sum_{b=1}^{B} \alpha_b \hat{m}_b(x)\right)$$

which is a signum function defined as follows:

$$\operatorname{sign}(x) := \begin{cases} -1 & \text{if } x < 0 \\ 0 & \text{if } x = 0 \\ 1 & \text{if } x > 0 \end{cases}$$

Here is a simple simulation to show how α_b will make the importance of each tree ($\hat{m}_b(x)$) different:

```
n = 1000
set.seed(1)
err <- sample(seq(0, 1, 0.01), n, replace = TRUE)
alpha = 0.5 * log((1 - err) / err)
ind = order(err)
plot(
  err[ind],
  alpha[ind],
  xlab = "error (err)",
  ylab = "alpha",
  type = "o",
  col = "black",
  lwd = 2
)
```

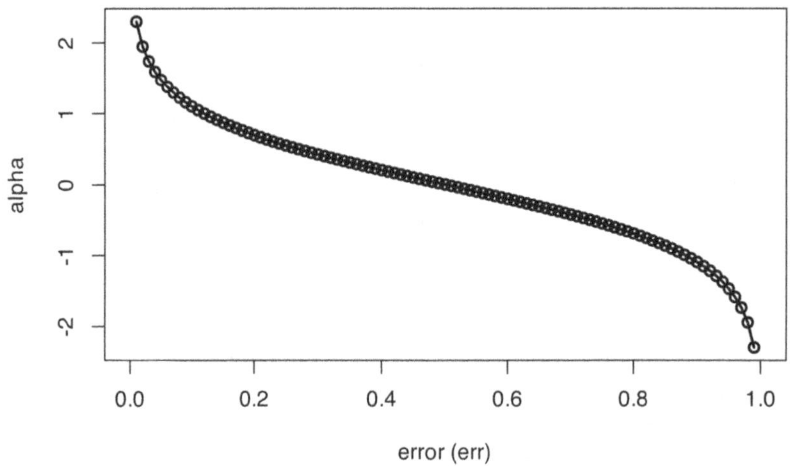

We can see that when there is no misclassification error (`err` $= 0$), "alpha" will be a large positive number. When the classifier very weak and predicts as good as a random guess (`err` $= 0.5$), the importance of the classifier will be 0. If all the observations are incorrectly classified (`err` $= 1$), our alpha value will be a negative integer.

The AdaBoost.M1 is known as a "discrete classifier" because it directly calculates discrete class labels $\hat{y}_i$, rather than predicted probabilities, $\hat{p}_i$.

What type of classifier, $\hat{m}_b(x)$, would we choose? Usually a "weak classifier" like a "stump" (a two terminal-node classification tree, i.e one split) would be enough. The $\hat{m}_b(x)$ choose one variable to form a stump that gives the lowest Gini index.

Here is our simple example with the `myocarde` data to show how we can boost a simple weak learner (stump) by using AdaBoost algorithm:

```
library(rpart)

# Data
myocarde <- read_delim("myocarde.csv", delim = ";" ,
                       escape_double = FALSE, trim_ws = TRUE,
                       show_col_types = FALSE)
```

```
myocarde <- data.frame(myocarde)
y <- (myocarde[ , "PRONO"] == "SURVIE") * 2 - 1
x <- myocarde[ , 1:7]
df <- data.frame(x, y)

# Setting
rnd = 100 # number of rounds
m = nrow(x)
whts <- rep(1 / m, m) # initial weights
st <- list() # container to save all stumps
alpha <- vector(mode = "numeric", rnd) # container for alpha
y_hat <- vector(mode = "numeric", m) # container for final predictions

set.seed(123)
for(i in 1:rnd) {
  st[[i]] <- rpart(y ~., data = df, weights = whts, maxdepth = 1, method = "class")
  yhat <- predict(st[[i]], x, type = "class")
  yhat <- as.numeric(as.character(yhat))
  e <- sum((yhat != y) * whts)
  # alpha
  alpha[i] <- 0.5 * log((1 - e) / e)
  # Updating weights
  whts <- whts * exp(-alpha[i] * y * yhat)
  # Normalizing weights
  whts <- whts / sum(whts)
}

# Using each stump for final predictions
for (i in 1:rnd) {
  pred = predict(st[[i]], df, type = "class")
  pred = as.numeric(as.character(pred))
  y_hat = y_hat + (alpha[i] * pred)
}

# Let's see what y_hat is
y_hat
```

```
##  [1]   3.132649  -4.135656  -4.290437   7.547707  -3.118702  -6.946686
##  [7]   2.551433   1.960603   9.363346   6.221990   3.012195   6.982287
## [13]   9.765139   8.053999   8.494254   7.454104   4.112493   5.838279
## [19]   4.918513   9.514860   9.765139  -3.519537  -3.172093  -7.134057
## [25]  -3.066699  -4.539863  -2.532759  -2.490742   5.412605   2.903552
## [31]   2.263095  -6.718090  -2.790474   6.813963  -5.131830   3.680202
## [37]   3.495350   3.014052  -7.435835   6.594157  -7.435835  -6.838387
## [43]   3.951168   5.091548  -3.594420   8.237515  -6.718090  -9.582674
## [49]   2.658501 -10.282682   4.490239   9.765139  -5.891116  -5.593352
## [55]   6.802687  -2.059754   2.832103   7.655197  10.635851   9.312842
## [61]  -5.804151   2.464149  -5.634676   1.938855   9.765139   7.023157
## [67]  -6.078756  -7.031840   5.651634  -1.867942   9.472835
```

```
# sign() function
pred <- sign(y_hat)

# Confusion matrix
table(pred, y)
```

```
##      y
## pred -1  1
##   -1 29  0
##   1   0 42
```

This is our in-sample confusion table. We can also see several stumps:

```
library(rpart.plot)

plt <- c(1,5,10,30, 60, 90)

p = par(mfrow=c(2,3))
for(i in 1:length(plt)){
prp(st[[i]], type = 2, extra = 1, split.col = "black",
    split.border.col = "black", box.col = "gray90")
}
```

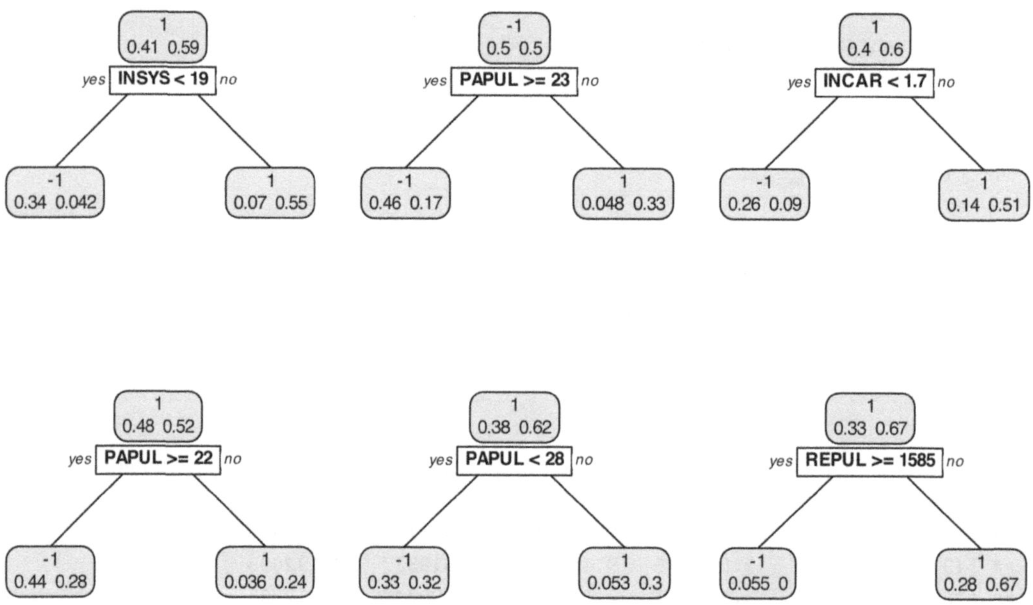


```
par(p)
```

Let's see it with the `JOUSBoost` package[2]:

```
library(JOUSBoost)
ada <- adaboost(as.matrix(x), y, tree_depth = 1, n_rounds = rnd)
summary(ada)
```

```
##                  Length Class  Mode
## alphas           100    -none- numeric
## trees            100    -none- list
## tree_depth         1    -none- numeric
## terms              3    terms  call
## confusion_matrix   4    table  numeric
```

```
pred <- predict(ada, x)
table(pred, y)
```

```
##     y
## pred -1  1
##   -1 29  0
##   1   0 42
```

These results provide in-sample predictions. When we use it in a real example, we can train AdaBoost.M1 by the tree depths (1 in our example) and the number of iterations (100 trees in our example). Note that we can also use `gbm` for adaboost type classifications.

15.5 XGBoost

XGBoost, or extreme gradient boosting, is the most efficient implementation of the gradient boosting framework, optimized for speed and scalability. A key reason for its efficiency is its ability to perform parallel computation on a single machine, significantly reducing training time compared to traditional gradient boosting methods. It supports both regression and classification problems, offering two distinct modes: a linear model and a tree-based learning algorithm. While decision trees generally excel at capturing nonlinear relationships between predictors and outcomes, XGBoost's linear model mode allows for regularization, making it useful for cases where a more traditional parametric structure may be preferable. Comparing the performance of the linear and tree-based modes can provide valuable insights into model selection, particularly in causal analyses.

One of XGBoost's standout advantages is its speed, reported to be "10 times faster than `gbm`" according to its official documentation[3]. This efficiency stems not only from its optimized tree-growing strategy but also from its ability to handle sparse data efficiently. XGBoost can work with compressed input formats, such as sparse matrices, where only nonzero values are stored in memory. This makes it particularly useful for large datasets with many zero or missing values, common in high-dimensional econometric applications. However, using this structure requires specific data preparation steps: the input features must be formatted as a matrix, and categorical variables need to be properly encoded. XGBoost supports multiple matrix types, including regular R matrices, sparse matrices from the `Matrix` package, and its own optimized `xgb.Matrix` class. These features, combined with its speed, regularization options, and adaptability to both linear and nonlinear settings, have made XGBoost the dominant boosting method in both academic and applied settings.

We start with a regression example here and leave the classification example to the next chapter in boosting applications. We will use the Ames housing data from the `AmesHousing` package to see the best "predictors" of the sale price.

```
library(xgboost)
library(mltools)
library(data.table)
library(modeldata)  # Also available via tidymodels

# Load the Ames housing dataset
data(ames)
dim(ames)
```

```
## [1] 2930   74
```

Since, the `xgboost` algorithm accepts its input data as a matrix, all categorical variables have be one-hot coded, which creates a large matrix even with a small size data. That is why using more memory efficient matrix types (sparse matrix etc.) speeds up the process. We ignore it here and use a regular R matrix, for now.

```
ames_new <- one_hot(as.data.table(ames))
df <- as.data.frame(ames_new)

# Split data into training and testing sets
set.seed(123)
ind <- sample(nrow(df), nrow(df) * 0.8)  # 80% training, 20% test
train <- df[ind,]
test <- df[-ind,]

# Prepare matrices for XGBoost
X <- as.matrix(train[, -which(names(train) == "Sale_Price")])
Y <- train$Sale_Price
```

Now we are ready for finding the optimal tuning parameters. One strategy in tuning is to see if there is a substantial difference between train and CV errors. We first start with the number of trees and the learning rate. If the difference still persists, we introduce regularization parameters. There are three regularization parameters: `gamma`, `lambda`, and `alpha`.

- `gamma`: Controls tree pruning by setting a minimum loss reduction threshold.
- `lambda`: L2 regularization term (similar to Ridge Regression).
- `alpha`: L1 regularization term (similar to Lasso Regression).

15.5.1 Regression Example

Here is our first run without a grid search. We will have a regression tree. The default booster is `gbtree` for tree-based models. For linear models it should be set to `gblinear`. The number of parameters and their combinations are very extensive in XGBoost. Please see them here: https://xgboost.readthedocs.io/en/latest/parameter.html#global-configuration. The combination of parameters we picked below is just an example.

```
#
params = list(
  eta = 0.1, # Step size in boosting (default is 0.3)
  max_depth = 3, # maximum depth of the tree (default is 6)
  min_child_weight = 3, # minimum number of instances in each node
  subsample = 0.8, # Subsample ratio of the training instances
  colsample_bytree = 1.0 # the fraction of columns to be subsampled
)

set.seed(123)
boost <- xgb.cv(
  data = X,
  label = Y,
  nrounds = 3000, # the max number of iterations
  nthread = 4, # the number of CPU cores
  objective = "reg:squarederror", # regression tree
  early_stopping_rounds = 50, # Stop if doesn't improve after 50 rounds
  nfold = 10, # 10-fold-CV
  params = params,
  verbose = 0 #silent
)
```

Let's see the RMSE and the best iteration:

```
best_it <- boost$best_iteration
best_it
```

```
## [1] 561
```

```
min(boost$evaluation_log$test_rmse_mean)
```

```
## [1] 24562.31
```

```
# One possible grid would be:
# param_grid <- expand.grid(
#   eta = 0.01,
#   max_depth = 3,
#   min_child_weight = 3,
#   subsample = 0.5,
#   colsample_bytree = 0.5,
#   gamma = c(0, 1, 10, 100, 1000),
#   lambda = seq(0, 0.01, 0.1, 1, 100, 1000),
#   alpha = c(0, 0.01, 0.1, 1, 100, 1000)
#   )

# After going through the grid in a loop with `xgb.cv`
# we save multiple `test_rmse_mean` and `best_iteration`
# and find the parameters that gives the minimum rmse
```

Now after identifying the tuning parameters, we build the best model:

```
tr_model <- xgboost(
  params = params,
  data = X,
  label = Y,
  nrounds = best_it,
  objective = "reg:squarederror",
  verbose = 0
)
```

We can obtain the top 10 influential features in our final model using the impurity (gain) metric:

```
library(vip)
vip(tr_model,
    aesthetics = list(color = "gray80", fill = "gray20"))
```

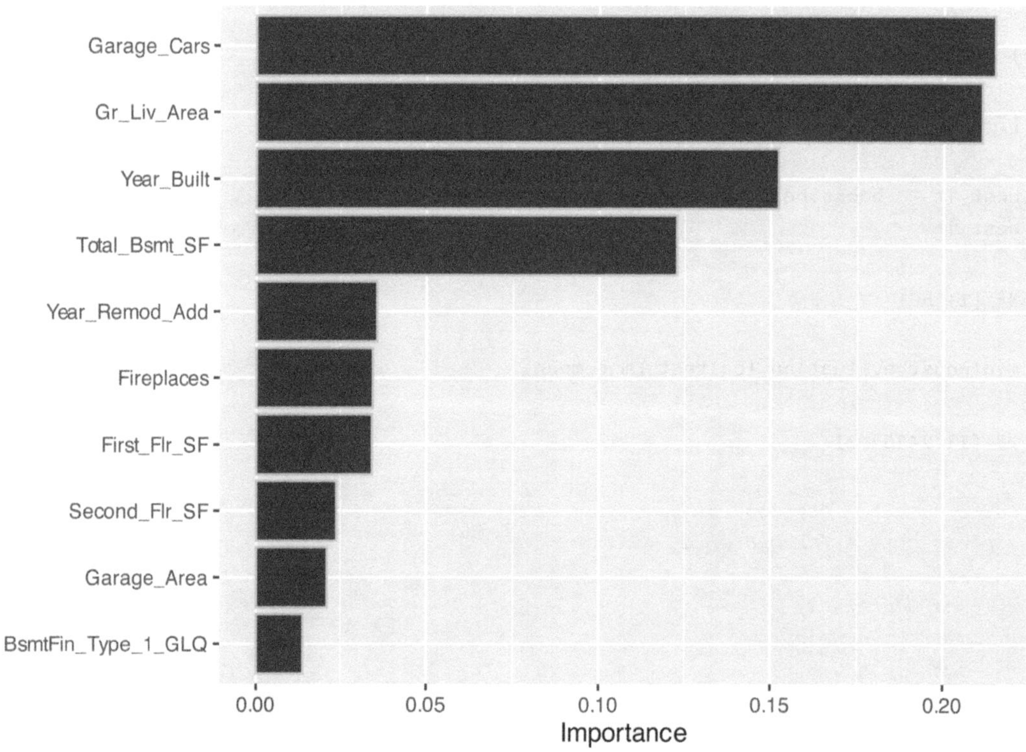

Now, we can use our trained model for predictions using our test set. Note that, again, `xgboost` would only accept matrix inputs.

```
yhat <- predict(tr_model,
               as.matrix(test[, -which(names(test) == "Sale_Price")]))
rmse_test <-
```

```
  sqrt(mean((test[, which(names(train) == "Sale_Price")] - yhat) ^ 2))
rmse_test
```

```
## [1] 22529.54
```

Note the big difference between training and test RMSPE's. This is an indication that our "example grid" is not doing a good job. We should include regularization tuning parameters and run a full scale grid search.

15.5.2 Classification Example

It uses spacial data preparation and it is not as straightforward as GBM.

Input Type: it takes several types of input data:

- Dense Matrix: R's dense matrix, i.e. `matrix`;
- Sparse Matrix: R's sparse matrix, i.e. `Matrix::dgCMatrix`;
- Data File: local data files;
- `xgb.DMatrix`: its own class (recommended).

Example from the documentation

In this example, we are aiming to predict whether a mushroom can be eaten or not (from Vignette). The data is from UCI Machine Learning Repository. The data has 22 features and 8124 observations. The target variable is `class` and it has two levels: `e` (edible) and `p` (poisonous). Mushroom data is cited from UCI Machine Learning Repository.

```
library(xgboost)

data(agaricus.train, package='xgboost')
data(agaricus.test, package='xgboost')
train <- agaricus.train
test <- agaricus.test
str(train)
```

```
## List of 2
##  $ data :Formal class 'dgCMatrix' [package "Matrix"] with 6 slots
##   .. ..@ i       : int [1:143286] 2 6 8 11 18 20 21 24 28 32 ...
##   .. ..@ p       : int [1:127] 0 369 372 3306 5845 6489 6513 8380 8384 10991 ...
##   .. ..@ Dim     : int [1:2] 6513 126
##   .. ..@ Dimnames:List of 2
##   .. .. ..$ : NULL
##   .. .. ..$ : chr [1:126] "cap-shape=bell" "cap-shape=conical" "cap-
shape=convex" "cap-shape=flat" ...
##   .. ..@ x       : num [1:143286] 1 1 1 1 1 1 1 1 1 1 ...
##   .. ..@ factors : list()
##  $ label: num [1:6513] 1 0 0 1 0 0 0 1 0 0 ...
```

Each variable is a `list` containing two things, `label` and `data`. The `label` is the target variable and `data` is the feature matrix.

```
dim(train$data)
```

```
## [1] 6513  126
```

```
dim(test$data)
```

```
## [1] 1611  126
```

In a sparse matrix, cells containing 0 are not stored in memory. Therefore, in a dataset mainly made of 0, memory size is reduced. It is very common to have such a dataset.

```
head(train$data)
```

```
## 6 x 126 sparse Matrix of class "dgCMatrix"
```

```
##    [[ suppressing 126 column names 'cap-shape=bell', 'cap-shape=conical', 'cap-shape=convex' ... ]]
```

```
##
## [1,] . . 1 . . . . . . 1 1 . . . . . . . . . 1 . . . . . . . . 1 . . . 1 . 1 .
## [2,] . . 1 . . . . . . 1 . . . . . . . . . 1 1 . 1 . . . . . . . . . . 1 . 1 .
## [3,] 1 . . . . . . . . 1 . . . . . . . . 1 . 1 . . 1 . . . . . . . . . 1 . 1 .
## [4,] . . 1 . . . . . 1 . . . . . . . . . 1 . 1 . . . . . . . . 1 . . . 1 . 1 .
## [5,] . . 1 . . . . . . 1 . . . 1 . . . . . . . 1 . . . . . . 1 . . . . 1 . . 1
## [6,] . . 1 . . . . . 1 . . . . . . . . . . 1 1 . 1 . . . . . . . . . . 1 . 1 .
##
## [1,] . . 1 1 . . . . . . . . . . . 1 . . . . 1 . . . . . . 1 . . . 1 . . . . .
## [2,] . 1 . 1 . . . . . . . . . . . 1 . . 1 . . . . . . . . 1 . . . 1 . . . . .
## [3,] . 1 . . 1 . . . . . . . . . . 1 . . 1 . . . . . . . . 1 . . . 1 . . . . .
## [4,] . . 1 . 1 . . . . . . . . . . 1 . . . . 1 . . . . . . 1 . . . 1 . . . . .
## [5,] . 1 . 1 . . . . . . . . . . . . 1 . . . 1 . . . . . . 1 . . . 1 . . . . .
## [6,] . 1 . . 1 . . . . . . . . . . 1 . . 1 . . . . . . . . 1 . . . 1 . . . . .
##
## [1,] . . 1 . . . . . . . . 1 . 1 . . . 1 . . 1 . . . . . . 1 . . 1 . . . . . .
## [2,] . . 1 . . . . . . . . 1 . 1 . . . 1 . . 1 . . . . . . 1 . . . 1 . . . . .
## [3,] . . 1 . . . . . . . . 1 . 1 . . . 1 . . 1 . . . . . . 1 . . . 1 . . . . .
## [4,] . . 1 . . . . . . . . 1 . 1 . . . 1 . . 1 . . . . . . 1 . . 1 . . . . . .
## [5,] . . 1 . . . . . . . . 1 . 1 . . . 1 . . 1 . . 1 . . . . . . . 1 . . . . .
## [6,] . . 1 . . . . . . . . 1 . 1 . . . 1 . . 1 . . . . . . 1 . . 1 . . . . . .
##
## [1,] . . . . . 1 . . . . . . 1 . .
## [2,] . . . . 1 . . . 1 . . . . . .
## [3,] . . . . 1 . . . . . 1 . . . .
## [4,] . . . . . 1 . . . . . . 1 . .
## [5,] . . 1 . . . . . 1 . . . . . .
## [6,] . . . . 1 . . . 1 . . . . . .
```

Here is our model:

Sparse Matrix

```
stSparse <- xgboost(data = train$data,
                    label = train$label,
                    max.depth = 2, eta = 1,
                    nthread = 2,
                    nrounds = 2,
                    objective = "binary:logistic")
```

```
## [1]  train-logloss:0.233376
## [2]  train-logloss:0.136658
```

Dense Matrix

```
stDense <- xgboost(data = as.matrix(train$data),
                   label = train$label,
                   max.depth = 2, eta = 1,
                   nthread = 2,
                   nrounds = 2,
                   objective = "binary:logistic")
```

```
## [1]  train-logloss:0.233376
## [2]  train-logloss:0.136658
```

xgb.DMatrix

```
dtrain <- xgb.DMatrix(data = train$data, label = train$label)
dtest <- xgb.DMatrix(data = test$data, label = test$label)
stDMatrix <- xgboost(data = dtrain,
                     max.depth = 2, eta = 1,
                     nthread = 2,
                     nrounds = 2,
                     objective = "binary:logistic")
```

```
## [1]  train-logloss:0.233376
## [2]  train-logloss:0.136658
```

Predictions

Here are your phat values:

```
predSparse <- predict(stSparse, test$data)
predDense <- predict(stDense, as.matrix(test$data))
predDMatrix <- predict(stDMatrix, dtest)

head(predSparse)
```

```
## [1] 0.28583017 0.92392391 0.28583017 0.28583017 0.05169873 0.92392391
```

```
head(predDense)
```

```
## [1] 0.28583017 0.92392391 0.28583017 0.28583017 0.05169873 0.92392391
```

```r
head(predDMatrix)
```

```
## [1] 0.28583017 0.92392391 0.28583017 0.28583017 0.05169873 0.92392391
```

You can go with AUC and or confusion matrix to evaluate the model.

```r
# AUC
library(ROCR)

predSparse <- prediction(predSparse, test$label)
aucSparse <- performance(predSparse, "auc")
aucSparse@y.values[[1]]
```

```
## [1] 0.9799301
```

```r
# Confusion Matrix
library(caret)
```

```
## Loading required package: lattice
```

```r
prediction <- unlist(predSparse@predictions)
yhat <- as.numeric(prediction > 0.5)

yhat_factor <- factor(yhat, levels = unique(test$label))
test_label_factor <- factor(test$label)

confusionMatrix(yhat_factor, test_label_factor, positive = "1")
```

```
## Confusion Matrix and Statistics
##
##           Reference
## Prediction   0   1
##          0 813  13
##          1  22 763
##
##                Accuracy : 0.9783
##                  95% CI : (0.9699, 0.9848)
##     No Information Rate : 0.5183
##     P-Value [Acc > NIR] : <2e-16
##
##                   Kappa : 0.9565
##
##  Mcnemar's Test P-Value : 0.1763
##
##             Sensitivity : 0.9832
##             Specificity : 0.9737
##          Pos Pred Value : 0.9720
##          Neg Pred Value : 0.9843
##              Prevalence : 0.4817
```

```
##          Detection Rate : 0.4736
##    Detection Prevalence : 0.4873
##       Balanced Accuracy : 0.9785
##
##        'Positive' Class : 1
##
```

15.5.3 Tuning in XGBoost

Tuning is essential for maximizing the performance of XGBoost models, striking a balance between accuracy, computational efficiency, and generalization. XGBoost offers two primary functions for training: `xgboost()`, which provides a straightforward approach, and `xgb.train()`, which allows for more control over training, including monitoring model performance at each iteration and implementing early stopping. Early stopping is particularly useful as it prevents unnecessary computation by halting training when improvements become negligible.

XGBoost parameters can be categorized into three broad groups: those that govern how individual boosting steps are performed (booster parameters), those that guide the optimization process (tuning parameters), and those that control how model performance is evaluated (evaluation parameters). Understanding these categories provides a structured approach to tuning.

Booster Parameters: Controlling the Learning Process

XGBoost supports different types of base models for boosting, most commonly tree-based models (`gbtree`) and linear models (`gblinear`). The choice of booster influences how predictions are made: trees excel at capturing complex interactions, while linear models provide a regularized alternative for structured data. A variant called `"dart"` adds dropout to trees to reduce overfitting.

The learning rate (`eta`) is one of the most important parameters, as it controls the contribution of each tree to the final prediction. A small learning rate (e.g., `0.01`) leads to a more stable model but requires more boosting iterations to converge, while larger values speed up training at the cost of potential overfitting. Tree depth (`max_depth`) also plays a crucial role—deeper trees can capture more intricate patterns but risk overfitting, whereas shallower trees promote simplicity.

Other parameters refine how trees are constructed. `min_child_weight` determines the minimum sum of weights required for a node split, helping to control tree size by discouraging splits on small sample sizes. `gamma` sets a threshold for the minimum loss reduction needed to make a further split; higher values result in more conservative trees. To introduce randomness and prevent trees from being too correlated, subsampling (`subsample`) controls the fraction of data used for each tree, while feature sampling (`colsample_bytree`) determines the fraction of features considered at each split.

Tuning Parameters: Optimizing for Performance

Beyond individual trees, XGBoost provides several ways to improve generalization and prevent overfitting. Regularization is an integral part of this process, with L2 regularization (`lambda`) applying a penalty on large weights to prevent overfitting, while L1 regularization (`alpha`) encourages sparsity by forcing some feature coefficients to zero, effectively performing variable selection. These regularization parameters are particularly useful when dealing with high-dimensional data or highly correlated features.

Another critical tuning parameter is early stopping, controlled via `early_stopping_rounds`, which allows training to halt if performance on a validation set does not improve after a specified number of rounds. This helps prevent overfitting while saving computation time. The number of boosting rounds (`n_estimators`) also needs careful tuning—higher values allow more complex patterns to be captured but should be balanced against the learning rate.

When working with imbalanced datasets, XGBoost provides `scale_pos_weight`, which adjusts the balance between positive and negative classes. This is particularly useful in classification problems where one class is underrepresented, such as fraud detection or rare event prediction.

Evaluation Parameters: Assessing Model Performance

Performance evaluation in XGBoost is highly customizable. The `eval_metric` parameter specifies the metric used to assess model quality during training. For regression tasks, root mean squared error (`rmse`) is commonly used, while classification problems often rely on log loss (`logloss`) for binary classification and area under the curve (`auc`) for imbalanced datasets. For multiclass classification, misclassification error (`merror`) or probability-based metrics like softmax loss (`mlogloss`) can be selected.

Additionally, the `objective` parameter determines the loss function optimized during training. Regression tasks typically use squared error loss (`reg:squarederror`), while classification tasks can use logistic loss (`binary:logistic`) for binary classification or softmax (`multi:softprob`) for multiclass problems. The ability to customize loss functions provides flexibility for handling different types of predictive modeling tasks.

Extending Tuning Beyond the Basics

While we have covered the core tuning concepts, XGBoost's tuning capabilities are extensive, and optimal parameter selection often depends on the specific dataset and problem type. The interplay between different parameters—such as the trade-off between `eta` and `n_estimators`, or how `max_depth` interacts with `gamma`—requires iterative testing and validation. Grid search and Bayesian optimization techniques can further refine hyperparameter selection beyond manual tuning.

For more in-depth guides on tuning XGBoost, practical examples, and real-world applications, the following resources provide excellent insights:

- **XGBoost Official Documentation** (xgboost.readthedocs.io[4])
- **"XGBoost: A Scalable Tree Boosting System"** (Chen & Guestrin, 2016) – Original research paper describing the methodology behind XGBoost.
- **Hyperparameter tuning guides** from platforms like Kaggle, which demonstrate practical parameter selection strategies.
- **The `mlr` and `caret` R packages**, which provide tools for systematic hyperparameter tuning, including grid search and Bayesian optimization.

Tuning XGBoost effectively requires a mix of theoretical understanding and practical experimentation. While this chapter provides a structured approach to tuning, further refinements often emerge from hands-on experience with different datasets and validation techniques.

15.6 The Implementation of Boosting Methods

Boosting methods have been increasingly applied across multiple disciplines, particularly in economics, social sciences, and health sciences, as a powerful tool for predictive modeling and decision-making. This section explores the implementation of boosting methods in different practical settings and with different programming languages.

Boosting algorithms have been successfully used in economic forecasting, financial modeling, and labor market analysis. Jiang, Xu, and Danowski (2025) analyzed social media texts using network analysis and boosting techniques to predict economic trends. Their study showed that boosting-based sentiment analysis improves accuracy in market behavior prediction, offering valuable insights for policymakers. Kueck et al. (2023) applied the L2-Boosting algorithm to estimate treatment effects in scenarios with numerous control variables or instruments. This approach allows for effective variable selection and shrinkage, facilitating more accurate inference in complex economic models. Their empirical studies demonstrated that L2-Boosting can effectively handle datasets where the number of parameters is large relative to the number of observations, providing robust estimates in such challenging contexts. Lindenmeyer and Torrent(2024) applied boosting techniques to monthly macroeconomic data, resulting in more accurate predictions compared to traditional forecasting models.

Piot-Lepetit (2025) applied boosting methods in sustainable food systems, showing that machine learning can optimize agrifood value chains by predicting market demand and improving supply chain efficiency. Similarly, Rahayu et al. (2025) examined the role of boosting algorithms in optimizing domestic product use within government institutions, highlighting its potential to improve resource allocation and economic efficiency.

Handayani (2025) explored the application of boosting techniques in governance and corruption analysis, finding that these methods effectively identify fraudulent financial transactions, thereby aiding anti-corruption measures. Such applications demonstrate the capacity of boosting algorithms to improve the economic decision-making and transparency.

The use of boosting methods in social sciences has primarily focused on understanding human behavior, social interactions, and policy impacts. Sharma and Bangur (2025) investigated circular economy policies and found that boosting models effectively identify factors influencing policy adoption and success. Their findings suggest that machine learning can guide policymakers in designing effective sustainability strategies.

McLean (2025) applied boosting techniques in public engagement research, exploring their effectiveness in predicting civic participation levels based on demographic and behavioral data. This approach provided more precise insights into how different social groups interact with policy initiatives, facilitating targeted interventions. In the realm of digital literacy, Boshnjaku, Krasniqi, and Kamberi (2025) used boosting models to assess the effectiveness of digital health literacy integration in education curricula. Their study found that boosting methods significantly improve predictive accuracy in understanding student engagement with digital health tools.

In healthcare, boosting algorithms have been widely used for disease prediction, medical diagnostics, and healthcare resource optimization. Rajagopal and Subramanian (2025) utilized boosting techniques to improve AI-based edge and fog computing applications in healthcare, demonstrating improved predictive capabilities for patient monitoring systems.

Aliyeva and Aliyev (2025) applied boosting models to assess the environmental impact of healthcare facilities, finding that these techniques provide robust insights into sustainability challenges and resource management in medical institutions. Yildiz and Kalayci (2025) have shown that GBDT algorithms can effectively classify medical conditions, leading to more accurate diagnoses and personalized treatment plans. Tamba, Yanti, and Tamba (2025) analyzed healthcare accessibility in Jakarta, using boosting algorithms to predict the impact of waste management policies on public health. Their findings emphasize the role of machine learning in optimizing environmental health policies.

As machine learning continues to evolve, the integration of boosting techniques in interdisciplinary research is expected to expand and can be implemented in various languages across fields.

While R provides powerful tools like `gbm` and `xgboost`, boosting can also be implemented efficiently in Stata and Python, each offering its own advantages in flexibility, scalability, and integration with econometric workflows.

In Stata, boosting is supported through built-in and user-written commands. The most direct way to implement gradient boosting is via the `gbm` command from Stata's Machine Learning Suite (`stata ml`), or through community-contributed packages like `boost` and `xgboost`.

The basic approach in Stata involves specifying the response variable and predictors, setting key hyperparameters like the number of trees (`ntrees`), learning rate (`shrinkage`), and tree depth (`maxdepth`), and evaluating performance using cross-validation. For example, a simple GBM model in Stata could be trained using:

```
ml gbm salary age experience education, ntrees(500) shrinkage(0.05) maxdepth(3) cv(10)
```

This command fits a gradient boosting model to predict `salary` using `age`, `experience`, and `education`, with 500 boosting rounds, a learning rate of 0.05, and trees with a maximum depth of 3. The inclusion of `cv(10)` ensures 10-fold cross-validation for tuning purposes.

Alternatively, XGBoost can be used in Stata through integration with Python. The `python` command in Stata allows users to call `xgboost` directly, enabling greater flexibility in parameter tuning and performance monitoring.

Python offers extensive support for boosting through libraries like XGBoost (`xgboost`), LightGBM (`lightgbm`), and CatBoost (`catboost`), making it a preferred choice for large-scale applications. The `xgboost` library, in particular, provides seamless integration with `pandas` and `scikit-learn`, allowing users to preprocess data, implement cross-validation, and fine-tune hyperparameters in a streamlined workflow.

Until now, we have covered the fundamental concepts and methodologies used in machine learning, focusing on both predictive accuracy and model generalization. We explored key ideas such as the bias–variance trade-off, overfitting, and hyperparameter tuning, followed by an in-depth discussion of classification methods and model selection techniques. We then examined regularization methods such as Lasso, which help improve model stability, and ensemble learning approaches, including bagging, random forests, and boosting, which improve predictive power by combining multiple models. Having established a strong foundation in machine learning methods, we now shift our focus to causal inference, beginning with the Rubin causal model and the framework of randomized controlled trials (RCTs). These methods provide the tools needed to move beyond correlation-based predictions and estimate causal effects in observational and experimental settings.

Notes

1. https://cran.r-project.org/web/packages/gbm/vignettes/gbm.pdf
2. https://cran.r-project.org/web/packages/JOUSBoost/vignettes/JOUS.pdf
3. see its vignette at https://xgboost.readthedocs.io/en/stable/R-package/xgboostPresentation.html
4. https://xgboost.readthedocs.io/en/stable/

16

Counterfactual Framework

Causal inference lies at the heart of understanding the effects of interventions, policies, or treatments. The term "causal inference" is sometimes misused in informal contexts to mean identifying variables/features associated with an outcome rather than establishing "true" cause-and-effect relationships. This confusion appears in some blogs, online tutorials, and discussions in data science communities, where they refer feature importance or explanatory variables as evidence of causality. As discussed in detail in Chapter 2, one of the key motivations for causal inference is the fundamental distinction between correlation and causation. While correlation can identify patterns and associations, causation establishes whether one variable influences another. This distinction is essential for guiding decisions in fields like economics, health, and social sciences, where policymakers and practitioners require actionable insights.

It is important to clarify key conceptual issues before we start the causal inference chapters. While we may observe differences in outcomes between two groups, we cannot directly attribute these differences to group-specific characteristics alone, as confounding factors, which lead to omitted variable bias, selection bias, or unobserved variables, may be driving the observed differences. Also, we cannot say for sure that a specific characteristic caused those outcome differences directly. Simply observing differences or identifying an association does not establish causality; for a characteristic, event, or variable to be considered causal, it must be possible to manipulate, treat, or intervene in it. For example, we cannot conclude that being older directly causes higher income because age itself cannot be manipulated. The same applies to characteristics such as gender or race—observed differences in outcomes may result from structural factors, confounding variables, or other unobservable variables rather than the characteristic itself. However, if we introduce an intervention—such as providing older workers with special security training or equipment—and observe that they can work night shifts more often, leading to higher earnings, then we can say that the special training or equipment caused the increase in income among older workers. The key distinction is that the treatment (training or equipment) is manipulable, allowing us to assess its causal effect on the outcome. This is the essence of causal inference: understanding the effect of an intervention, policy or treatment on an outcome by comparing the observed outcome with the outcome that would have been observed under a different treatment. Causal inference relies on several assumptions, and in these chapters, we will cover both the methods and the assumptions required for valid causal analysis.

In policymaking and business, understanding causal relationships is critical. Decisions, such as whether increasing education funding improves student outcomes or raising the minimum wage reduces poverty, hinge on knowing the direction and magnitude of these effects. In healthcare, policymakers must determine whether expanding access to preventive care reduces long-term hospitalizations or whether a new drug truly improves patient recovery rates rather than just correlating with better outcomes. Similarly, firms must evaluate whether advertising campaigns boost sales or if seasonal trends are the real cause. Causal

DOI: 10.1201/9781003504795-16

inference provides the rigorous framework needed to evaluate such policies, going beyond mere observation of statistical associations to uncover real cause-and-effect relationships.

Traditional machine learning (ML) methods, which are typically optimized for prediction, are not inherently designed to answer causal questions. Predictive models may identify spurious correlations or overlook confounding variables, leading to misleading inferences about cause and effect. To overcome these limitations, causal ML techniques—such as double/debiased ML, causal forests, and methods that integrate causal graphs—have been developed. These approaches bridge the gap between prediction and causation by enabling the estimation of causal effects while leveraging ML's strengths in handling high-dimensional, complex data.

For readers unfamiliar with the foundational concepts of causal inference, we recommend reviewing Chapter 2, where these topics, including the distinction between correlation, regression, and the causal effect, are explored in depth. This chapter will build on those concepts, focusing on methods and their practical applications. It will cover the counterfactual framework, including the definition of counterfactuals, the challenges of missing data and state, and how ML methods can estimate counterfactuals under identification assumptions. The chapter will also explore the fundamental problem of causal inference by examining the role of randomization and the interplay between experimental and observational data. Randomized controlled trials, cluster randomization, and adaptive experimental designs optimized through machine learning will also be discussed, as will partial effects, focusing on marginal effects in linear and nonlinear models and their estimation through ML-based predictions.

Selection on observables will then be examined in detail, covering matching methods, weighting strategies, and other approaches such as doubly robust. These methods address challenges posed by confounding factors when causal inference relies on observational data. Finally, the next chapter will focus on selection on unobservables, explore techniques like instrumental variables, difference-in-differences, regression discontinuity designs, and synthetic control methods. One chapter later, we will discuss heterogeneous causal methods such as double machine learning, causal forests, and related approaches. Together, these chapters provide a comprehensive understanding of causal inference techniques based on potential outcome framework and their integration with ML.

For readers seeking a deeper understanding of causal inference methods in economics, we recommend widely regarded texts such as *Causal Inference: The Mixtape* by Scott Cunningham, *The Effect* by Nick Huntington-Klein, and *Mostly Harmless Econometrics* by Angrist and Pischke, *A First Course in Causal Inference* by Peng Ding. These books provide excellent coverage of foundational concepts and techniques, from randomized controlled trials to instrumental variables, regression discontinuity, and difference-in-differences. While the purpose of these chapters is to offer a comprehensive overview of these methods, the primary focus is to demonstrate how machine learning can complement and improve traditional approaches. By providing intuitive explanations and practical insights, we aim to show how ML methods can improve causal inference in areas like high-dimensional data, flexible modeling, and estimating heterogeneous treatment effects. Let's start our discussion with the counterfactual framework.

16.1 Counterfactual Framework

Imagine studying the "causal" effect of obtaining a college degree on earnings. For an individual with a degree, we observe their income $Y_i(1)$ but not the counterfactual income $Y_i(0)$ they would have earned without the college degree; similarly, for someone without a degree, we observe their income $Y_i(0)$ but not the income $Y_i(1)$ they might have earned with a degree. In theory, the "causal" effect for each person is $Y_i(1) - Y_i(0)$, but since both states cannot be observed simultaneously, individual causal effects remain unmeasurable. Now, consider a population of N individuals—say, a quarter have a degree—and imagine a parallel universe where each person's potential outcomes (income with a degree in one universe/state; income without a degree in another universe/state) are observable. In that case, the causal effect of a college degree for each person could be determined by calculating the difference between their incomes in the actual and counterfactual states, $Y_i(1) - Y_i(0)$. By aggregating these individual causal effects across the entire population, we could compute the Average Treatment/Causal Effect of treatment (college degree, here) on outcome (income, here) as the mean of these differences, $\mathbb{E}[Y_i(1) - Y_i(0)]$. In this scenario, the average treatment effect (ATE) can also be computed as $\mathbb{E}[Y_i(1)] - \mathbb{E}[Y_i(0)]$, which is equivalent to the difference between the average income if everyone had a degree and if no one did. This framework precisely defines the population-level average treatment/causal effect of education on earnings. However, there are other causal effect measures, which we will cover in later sections.

Because of the inability to observe the outcome in both the treated state and the control state for the same individual, as well as the confounding factors that complicate causal interpretation, Donald Rubin and others developed the Rubin causal model (RCM) (Rubin, 1974; Holland, 1986). The RCM provides a formal framework for defining and estimating causal effects in the presence of the fundamental problem of causal inference, which is defined as "for each unit, only one of the two potential outcomes—$Y_i(1)$ (if treated) or $Y_i(0)$ (if untreated)—can be observed, depending on the treatment assignment. The other outcome remains unobservable, referred to as the counterfactual, creating a missing data problem. The potential outcomes framework is also known as the counterfactual framework" [1] As a result, direct calculation of the causal effects at the individual or population level is impossible. This challenge forces researchers to rely on assumptions and robust designs to approximate the unobservable counterfactuals and isolate causal effects.

A key concept in causal inference is that the unit of analysis may differ from the physical unit being studied. Consider the effect of a college degree on income: if I initially earned a low salary without a degree and later earned more after obtaining one, it might seem that we observe both potential outcomes—$Y_i(0)$ (without a degree) and $Y_i(1)$ (with a degree)—allowing us to estimate the causal effect. However, this reasoning is flawed because my "before" and "after" measurements represent two distinct units, meaning we have four potential outcomes—$Y_{i,\text{before}}(0)$, $Y_{i,\text{before}}(1)$, $Y_{i,\text{after}}(0)$, and $Y_{i,\text{after}}(1)$—with only two observed, and two missing. For instance, if my income increased over time without obtaining a degree, one might erroneously conclude that a college degree has no effect; similarly, if my income decreased after earning a degree, the effect might be falsely deemed negative. This example highlights the need to clearly define the unit in causal analysis. Causal inference requires multiple units exposed to different treatments, either by observing the same unit over time or different units simultaneously. However, having multiple units alone does not resolve the fundamental problem of causal inference.

The RCM assumes that each unit has a well-defined treatment status (D_i) and potential outcomes ($Y_i(1)$ and $Y_i(0)$). The observed outcome (Y_i) depends on the treatment received, given by:

$$Y_i = D_i \cdot Y_i(1) + (1 - D_i) \cdot Y_i(0) \tag{16.1}$$

This equation states that if $D_i = 1$, the observed outcome is the potential outcome in the treatment state ($Y_i = Y_i(1)$), and if $D_i = 0$, it is the potential outcome in the control state ($Y_i = Y_i(0)$). If both potential outcomes were observable, identifying individual treatment effects ($Y_i(1) - Y_i(0)$) and computing the population-level average treatment effect ($\mathbb{E}[Y_i(1) - Y_i(0)]$) would be straightforward.

TABLE 16.1 Potential Outcomes and Individual Treatment Effects

D	$Y(1)$ Potential Outcome Treatment state	$Y(0)$ Potential Outcome NO Treatment(Control) state	Individual Treatment Effect	Y Observed Outcome
1	$Y_T(1)$ observed	$Y_T(0)$ what-if	$Y_T(1) - Y_T(0)$	Y_T
1	$Y_T(1)$ observed	$Y_T(0)$ what-if	$Y_T(1) - Y_T(0)$	Y_T
1	$Y_T(1)$ observed	$Y_T(0)$ what-if	$Y_T(1) - Y_T(0)$	Y_T
0	$Y_C(1)$ what-if	$Y_C(0)$ observed	$Y_C(1) - Y_C(0)$	Y_C
0	$Y_C(1)$ what-if	$Y_C(0)$ observed	$Y_C(1) - Y_C(0)$	Y_C
0	$Y_C(1)$ what-if	$Y_C(0)$ observed	$Y_C(1) - Y_C(0)$	Y_C

This table illustrates the notations we discussed above, helping us understand the estimation of treatment effects when only one outcome per individual is observed. Each unit (student) has two potential outcomes: one if they receive the scholarship ($Y(1)$) and one if they do not ($Y(0)$). The table separates individuals into two groups: those who receive the scholarship ($D = 1$) and those who do not ($D = 0$). For treated students, the observed outcome is $Y_T(1)$ (earnings after receiving the scholarship, while the counterfactual outcome $Y_T(0)$ (what their earnings would have been without the scholarship; labeled "what-if")) is unobserved. In contrast, for the control group, we observe $Y_C(0)$ (earnings without the scholarship) and do not see $Y_C(1)$ (what their earnings would have been if they had received the scholarship; labeled "what-if")). The individual treatment effect is defined as the difference $Y(1) - Y(0)$. If both outcomes were observable for every student, averaging the differences would yield the ATE. However, because one outcome is always missing, we must estimate the ATE using observed data and statistical methods. This table therefore highlights the distinction between observed outcomes and unobserved counterfactuals, which is key for causal inference.

TABLE 16.2 Expected Potential Outcomes and Treatment Effects

	$E[Y(1)]$ Expected Potential Outcome at Treatment state	$E[Y(0)]$ Expected Potential Outcome at Control state	Difference	$E[Y]$
D = 1	$E[Y(1)\|D = 1]$ (Observable for Treated)	$E[Y(0)\|D = 1]$ (Unobservable for Treated)	ATT	$E[Y\|D = 1]$
D = 0	$E[Y(1)\|D = 0]$ (Unobservable for Control)	$E[Y(0)\|D = 0]$ (Observable for Control)	ATU	$E[Y\|D = 0]$
Sample	$E[Y(1)]$	$E[Y(0)]$	ATE	
		Selection Bias: $E[Y(0)\|D = 1] - E[Y(0)\|D = 0]$		Simple Diff.

The table above links our data to the Rubin causal model by showing how we compute expected potential outcomes. For example, to calculate $E[Y(1)|D=1]$, we take the observed outcomes for the treated group—the column showing $Y(1)$ for students who received the scholarship ($D = 1$s)—and compute their average. This average represents the expected outcome under treatment for the treated units. Similarly, for the control group, we average the observed outcomes in the $Y(0)$ column to obtain $E[Y(0)|D=0]$, which is the expected outcome under no treatment for these units.

The table also shows how these group-specific averages contribute to overall expectations. For instance, the overall expected potential outcome under treatment, $E[Y(1)]$, is a weighted combination of $E[Y(1)|D=1]$ and $E[Y(1)|D=0]$, even though for the control group the outcome $Y(1)$ is unobserved (thus, needs to be estimated). The same idea applies to $E[Y(0)]$ using $E[Y(0)|D=1]$ (unobserved for the treated) and $E[Y(0)|D=0]$ (observed for the control).

This process of averaging helps us understand how we move from individual data points to group-level expectations and, ultimately, to estimates of causal effects. The table also highlights the difference in expectations, such as the simple difference between $E[Y|D=1]$ and $E[Y|D=0]$, which provides a preliminary estimate of the treatment effect. However, since we only observe one potential outcome per student, unobserved counterfactuals can lead to selection bias. For instance, the difference $E[Y(0)|D=1] - E[Y(0)|D=0]$ represents how the unobserved outcomes among the treated differ from those in the control group.

This visual framework is essential as we move forward. Next, we will discuss the assumptions that let us approximate these missing counterfactuals, formally define treatment effects such as ATE, ATT, and ATU, and address selection bias. Later in the chapter, we will also cover methods like randomization, regression, matching, and weighting, all of which rely on this framework to estimate causal effects accurately.

16.1.1 Assumptions of the Rubin Causal Model

The Stable Unit Treatment Value Assumption (SUTVA) is essential for defining potential outcomes (Rubin, 1980). SUTVA consists of two key components: no interference and consistency. The no-interference assumption states that the potential outcomes of observation i are independent of the treatment assignments of all other units. In other words, a unit's potential outcomes are not affected by spillover or network effects from the treatment of others. This assumption rules out general equilibrium effects, which can be problematic in economic settings where policy interventions influence neighboring households or firms. When a treated group represents a large share of the population, assuming no spillovers may not be plausible. Consequently, policy recommendations based on randomized or natural experiments should be reassessed for broader populations, as small-sample studies may satisfy SUTVA while large-scale implementations violate it. Violations, such as spillovers or treatment diffusion, require alternative approaches, including network-based models and machine learning techniques for spillover detection.

The consistency assumption states that the observed outcome for a unit under its actual treatment status equals the corresponding potential outcome: $Y_i = D_i \cdot Y_i(1) + (1 - D_i) \cdot Y_i(0)$. This ensures that the treatment received aligns exactly with the treatment assigned, ruling out hidden variations in treatment (e.g., different treatment intensities, misreporting, or noncompliance). If multiple versions of a treatment exist but are not accounted for, causal estimates may be biased. In practice, inconsistencies arise due to implementation errors, partial treatment uptake, or measurement issues. Machine learning methods, such as anomaly detection and data imputation, can help identify and mitigate such inconsistencies, improving the robustness of causal analysis.

In the potential outcomes framework, causal identification relies on the ignorability (or unconfoundedness/conditional independence) assumption, i.e., $(Y_i(1), Y_i(0)) \perp D_i \mid X_i$, which asserts that treatment assignment is as good as random once conditioned on covariates X_i. In another words, an assignment is unconfounded if the assignment mechanism does not depend on the potential outcomes and depend only on the pre-treatment covariates. A perfect randomized controlled trial would eliminate the need for this assumption. However, in observational studies, this assumption is required to employ statistical adjustments—such as matching, weighting, and regression—to mimic randomization. Machine learning further improves these methods by improving propensity score and outcome model estimation. In high-dimensional settings, ML-based variable selection (e.g., LASSO, random forests) effectively isolates relevant confounders, mitigating biases from model misspecification or omitted variables and bolstering the credibility of causal estimates when true randomization is not feasible.

With these assumptions in place, we now turn to the most common linear causal estimand (target quantity, which is function of potential outcomes) in the RCM framework.

16.2 Average Treatment Effects

Even though we cannot observe the treatment effect for each individual, the collection of individual causal effects within a population forms a distribution. This distribution allows us to estimate key parameters such as the mean (ATE), variance, and other descriptive statistics.[2]

The ATE is the most common parameter of interest (i.e., estimand) in causal inference within the RCM framework. It represents the average of individual treatment effects, $Y_i(1) - Y_i(0)$, serving as a crucial measure in policy evaluation, program assessment, and medical research. Formally, it is defined as:

$$\delta^{ATE} = \mathbb{E}[\delta_i] = \mathbb{E}[Y_i(1) - Y_i(0)] = \mathbb{E}[Y_i(1)] - \mathbb{E}[Y_i(0)] \tag{16.2}$$

To understand this intuitively, consider a government implementing a policy, an agency introducing a new regulation, or a doctor prescribing a medication. In each scenario, we can conceptualize two states: a treatment state where all units in the population are exposed to the intervention, and a control state where none are exposed. The equation $\mathbb{E}[Y_i(1) - Y_i(0)]$ represents the population average of the individual treatment effects, while $\mathbb{E}[Y_i(1)] - \mathbb{E}[Y_i(0)]$ captures the difference in the expected outcomes for the entire population if all units were treated versus untreated. Under the SUTVA, which ensures no interference between units and consistent treatment effects, these two expressions are equivalent.

However, like the individual-level treatment effect, the ATE is inherently unknowable because it requires knowledge of both potential outcomes, $Y_i(1)$ and $Y_i(0)$, for each unit. Due to the fundamental problem of causal inference, only one of these outcomes is observable for any given unit. As a result, the ATE is not a quantity that can be directly calculated but rather one that can only be estimated with different methods (i.e., estimators) under certain statistical assumptions.[3] Each method in the RCM framework aims to estimate the ATE by leveraging different assumptions and estimators to address missing counterfactuals and isolate causal effects. In the following sections, we will explore these estimation techniques in detail.

Depending on the research question, we may also be interested in estimating other treatment effects beyond the ATE. The average treatment effect on the treated (ATT) measures the average treatment effect specifically for the subgroup of individuals who actually received the treatment. It is formally defined as:

$$ATT = \mathbb{E}[Y_i(1) \mid D = 1] - \mathbb{E}[Y_i(0) \mid D = 1] \tag{16.3}$$

In observational data, the ATT almost always differs from the ATE because individuals tend to sort into treatment based on expected gains. For instance, individuals with higher anticipated benefits from a policy or intervention may be more likely to participate, introducing endogenous selection into the treatment group. Like the ATE, the ATT is unknowable because it requires knowledge of both potential outcomes for each treated individual, which is impossible to observe directly; thus, it must be estimated using causal inference methods.

The average treatment effect on the untreated (ATU) measures the average treatment effect for those who did not receive the treatment (the control or untreated group). It is defined as:

$$ATU = \mathbb{E}[Y_i(1) \mid D = 0] - \mathbb{E}[Y_i(0) \mid D = 0] \tag{16.4}$$

As with the ATT, the ATU is also unknowable in observational data, as it requires observing both potential outcomes for individuals in the control group. Furthermore, if treatment effects are heterogeneous, the ATU will differ from the ATT, reflecting the varying impacts of treatment across different subpopulations.

Beyond the ATE, ATT, and ATU, additional parameters are used to estimate causal effects in specific contexts or with particular methods. These parameters such as conditional average treatment effect (CATE), local average treatment effect (LATE), and heteregeneous treatment effects will be explored in greater detail in the next chapter as we discuss approaches to causal inference tailored to different assumptions and data structures.

Binary outcomes, common in medical and health studies, require specialized causal estimands due to their nature. In this context, we define $\mu_1 = \Pr(Y_i(1) = 1)$ and $\mu_0 = \Pr(Y_i(0) = 1)$, representing the probabilities of an outcome under treatment and control, respectively, often referred to as causal risks in medical research. The three primary causal estimands for binary outcomes are the causal risk difference ($\delta_{RD} = \mu_1 - \mu_0$, equivalent to the ATE), the causal risk ratio ($\delta_{RR} = \mu_1/\mu_0$), and the causal odds ratio ($\delta_{OR} = \frac{\mu_1/(1-\mu_1)}{\mu_0/(1-\mu_0)}$). It is important to note that the ratio estimands (e.g., risk ratio and odds ratio) are not averages of individual causal effects unless the effects are homogeneous across individuals. These estimands provide nuanced insights into treatment effects for binary outcomes, extending beyond the simple ATE to address specific contexts, especially in medical and epidemiological research.

Before we continue, we want to highlight an important causal estimands more common in medical and health studies. Binary outcomes require specialized causal estimands due to their nature. In this context, we define $\mu_1 = \Pr(Y_i(1) = 1)$ and $\mu_0 = \Pr(Y_i(0) = 1)$, representing the probabilities of an outcome under treatment and control, respectively, often referred to as causal risks in medical research. The three primary causal estimands for binary outcomes are the causal risk difference ($\delta_{RD} = \mu_1 - \mu_0$, equivalent to the ATE), the causal risk ratio ($\delta_{RR} = \mu_1/\mu_0$), and the causal odds ratio ($\delta_{OR} = \frac{\mu_1/(1-\mu_1)}{\mu_0/(1-\mu_0)}$). It is important to note that the ratio estimands (e.g., risk ratio and odds ratio) are not averages of individual causal effects unless the effects are homogeneous across individuals. These estimands provide nuanced insights into treatment effects for binary outcomes, extending beyond the simple ATE to address specific contexts, especially in medical and epidemiological research.

16.3 Selection Bias and Heterogeneous Treatment Effect Bias

The fundamental problem of causal inference prevents the direct calculation of causal effects, so researchers often rely on survey or administrative data to estimate them in economics, health, and social sciences. A common but flawed approach is interpreting raw differences in average outcomes between groups (e.g., treated versus untreated) as the ATE of the treatment, intervention, or policy. This approach fails to account for confounding factors—factors that influence both treatment and outcome—and overlooks selection bias, arising from non-random assignment to the treatment group, as well as heterogeneous effects, caused by the differential impact of treatment on treated and control groups.

Selection bias arises when the process determining treatment assignment is correlated with potential outcomes, leading to systematic differences between groups. For example, in observational studies, treatment is often not randomly assigned, and pre-existing differences between treated and untreated groups influence both the probability of receiving treatment and outcomes. As a result, the simple difference in average outcomes captures not just the treatment effect but also these underlying differences, conflating the two and biasing estimates.

Table 16.2 breaks down the observed and unobserved components of potential outcomes, helping us see how we calculate group averages and, ultimately, causal effects. For example, the column labeled $E[Y(1)|D=1]$ represents the average outcome for students who received the scholarship. We compute this by taking each treated student's observed earnings and then averaging these values. Similarly, $E[Y(0)|D=0]$ is the average outcome for students who did not receive the scholarship, calculated by averaging their observed earnings.

When we compute the simple difference $E[Y|D=1]-E[Y|D=0]$ (as shown in the "Simple Diff." column), we are subtracting the average earnings of the control group from the treated group. However, this difference does not purely reflect the causal effect of the scholarship. Instead, it contains two components: the ATT and a term capturing selection bias.

In Table 16.2, the ATT focuses solely on the treated group. For students who receive the scholarship, the table shows their observed outcome under treatment—labeled as $E[Y(1)|D=1]$—and it also reminds us that their potential outcome if they had not received the scholarship, $E[Y(0)|D=1]$, is unobserved. The ATT captures the difference between these two values: it tells us how much, on average, the scholarship changes the earnings for the treated students compared to what their earnings would have been without the scholarship.

We can express the simple difference by adding and subtracting the unobserved counterfactual average $E[Y(0)|D=1]$ from the treated group:

$$\underbrace{\mathbb{E}[Y(1)\mid D=1]-\mathbb{E}[Y(0)\mid D=0]}_{\text{Simple Difference in average Outcomes (SDO)}} = \underbrace{(\mathbb{E}[Y(1)\mid D=1]-\mathbb{E}[Y(0)\mid D=1])}_{\text{Average Treatment Effect on the Treated (ATT)}} + \underbrace{(\mathbb{E}[Y(0)\mid D=1]-\mathbb{E}[Y(0)\mid D=0])}_{\text{Selection Bias}} \tag{16.5}$$

Here, the first term $\{E[Y(1)|D=1]-E[Y(0)|D=1]\}$ is the ATT—the true causal effect of receiving the scholarship for those who did. The second term $\{E[Y(0)|D=1]-E[Y(0)|D=0]\}$ is the selection bias, indicating that if treated students would have had different earnings than control students even without the scholarship, then part of the observed difference is due to these pre-existing differences rather than the treatment itself.

For instance, if students who receive scholarships tend to come from backgrounds that predispose them to higher earnings (reflected in a higher $E[Y(0)|D=1]$ compared to $E[Y(0)|D=0]$), then the selection bias term becomes positive. As a result, simply comparing average earnings would overstate the true effect of the scholarship. Similarly, in a health intervention study, individuals who choose to participate in a wellness program may already be more health-conscious and predisposed to better outcomes than those who opt out, regardless of the treatment itself. In both cases, observed differences in outcomes conflate the true causal effect with the effects of selection bias, making it impossible to interpret the results as causal without addressing this issue.

In some cases, we can also relate the simple difference to the ATE by a similar reasoning, leading to a decomposition where the naive difference equals the ATE plus a selection bias term that accounts for differences in potential outcomes between treated and control units.

The goal of causal inference methods is to minimize or eliminate this bias by imposing identifying assumptions and using statistical techniques to adjust for these systematic differences. In the following chapters, we will explore these methods in detail, focusing on their ability to approximate the counterfactual outcomes for valid causal interpretation.

Heterogeneous treatment effect bias arises because treatments often affect individuals or groups differently. For example, a policy intervention may benefit some subgroups more than others, resulting in varying treatment effects across the population. The difference between the ATT and the ATU reflects this heterogeneity. When multiplied by the proportion of the control group, $(1-\pi)(ATT-ATU)$, it quantifies the bias introduced by assuming homogeneous treatment effects.

To simplify analysis, researchers often assume constant treatment effects (i.e., no heterogeneity), which eliminates this bias. Even though this is a strong assumption, this is very common and plausible in social sciences and economics as we want to analyze average effects, not individual effect. That average treatment/causal effect is presented either treatment effect for average person or "homogeneous" average treatment effect for everyone. However, recent advances in machine learning allow researchers to better capture and model heterogeneous treatment effects, enabling more granular and targeted policy insights.

As we discussed above, the simple difference between the average outcomes of treated and untreated groups is often misinterpreted as the causal effect of treatment. This interpretation holds true only if selection bias and heterogeneous treatment effect bias are absent, which is rarely the case in practice. Addressing selection and heterogeneous treatment effect biases requires robust causal inference methods that impose appropriate identifying assumptions, helping to isolate the true treatment effect and ensure accurate estimates.

This simple difference in outcomes between treated $(D=1)$ and untreated groups $(D=0)$ can be decomposed ATE, selection bias, and heterogeneous bias as:

$$\underbrace{\mathbb{E}[Y_i(1) \mid D=1] - \mathbb{E}[Y_i(0) \mid D=0]}_{\text{Simple Difference in Outcomes (SDO)}} = \underbrace{\mathbb{E}[Y_i(1)] - \mathbb{E}[Y_i(0)]}_{\text{Average Treatment Effect (ATE)}} + \underbrace{\mathbb{E}[Y_i(0) \mid D=1] - \mathbb{E}[Y_i(0) \mid D=0]}_{\text{Selection Bias}} + \underbrace{(1-\pi)(ATT - ATU)}_{\text{Heterogeneous Treatment Effect Bias}} \quad (16.6)$$

where $(1-\pi)$ is the proportion of the population in the control group.[4] This decomposition highlights three key components: the true ATE, selection bias, and heterogeneous treatment effect bias. ATE represents the true causal effect averaged over the entire population. It is the effect we aim to estimate but cannot observe directly due to the fundamental problem of causal inference—we only observe one potential outcome for each unit.The second and third term captures selection and heterogeneous treatment effect bias, which will be discussed below. These biases must be accounted for to estimate the true treatment effect accurately. All the methods in RCM framework we will discuss in the following sections and next chapter aim to address these biases and isolate the causal effect of treatment on the outcome. Let's continue our discussion first discussing assumptions, then these biases and then with the parameters of interest in causal inference.

16.4 Limitations of RCM

The fundamental problem of causal inference makes ATE, ATT, ATU, LATE, and CATE unknowable since we can observe individuals in only one state (treated or untreated). To estimate these parameters, researchers rely on statistical methods under identifying assumptions by comparing treated and untreated groups. This estimation process raises several critical questions: Were units randomly assigned to treatment? Were the groups sufficiently large and similar? Was the treatment uniformly administered across all individuals? The answers to these questions determine the identification strategy and shape the choice of causal inference methods, which will be covered throughout this book. Each method seeks to overcome the limitations imposed by missing counterfactuals and to estimate treatment effects as accurately as possible given the data and assumptions at hand.

Randomized controlled trials (RCTs) are the gold standard for causal inference, ensuring that treatment assignment is independent of potential outcomes. This eliminates selection bias, allowing direct estimation of causal effects with minimal assumptions. However, RCTs are often impractical due to ethical, logistical, or financial constraints. In such cases, researchers turn to quasi-experimental methods like matching, weighting, regression adjustment, instrumental variables, difference-in-differences, and regression discontinuity to approximate randomization and address selection bias.

While the RCM is widely used in modern causal inference, it has limitations. One key criticism, highlighted by Heckman and Vytlacil (2007), its lack of external validity (cannot generalized to new settings or untested programs) and failure to provide insights into underlying mechanisms (treating interventions as "black boxes") limit its generalizability. Angrist and Pischke (2010) argue that causal estimates under RCM are context-specific, requiring synthesis across studies for broader conclusions. Additionally, RCM estimands like

ATE, ATT, and ATU may fail to guide marginal policy decisions or account for externalities, such as changes in scale or assignment mechanisms (Heckman et al., 1999).

Alternative frameworks address some of these limitations. The Roy model incorporates self-selection by modeling individual decisions based on expected returns, enabling analysis of treatment effect heterogeneity. Directed acyclic graphs (DAGs), introduced by Judea Pearl, use graphical tools and "do-calculus" to represent causal relationships and identify effects. However, both frameworks have trade-offs. Roy models rely on strong parametric assumptions. DAGs are less common in economics due to their complexity, lack of focus on estimation, and challenges in modeling interference and simultaneity (Imbens, 2020).

Finally, while machine learning cannot replace sound experimental design or credible assumptions, it improves causal analysis by improving flexibility, precision, and scalability, particularly in high-dimensional settings. As we transition to the next section on randomization, we will explore how RCTs address selection bias and why they remain the most robust method for identifying causal effects. For settings where randomization is not feasible, subsequent chapters will examine quasi-experimental methods to address the challenges of observational data and provide actionable insights for policy and research.

Notes

1. Counterfactual scenarios are a popular theme in literature, film, and TV, offering thought-provoking explorations of alternate realities. The movie "Sliding Doors" examines how a single missed train transforms a woman's life, while Robert Frost's poem "The Road Not Taken" poetically reflects on the choices that shape our paths. Similarly, "The Man in the High Castle" envisions a world where the Axis powers won World War II, and Philip Roth's "The Plot Against America" reimagines U.S. history under a Lindbergh presidency. Documentaries like "The Fog of War" analyze pivotal decisions during conflicts, illustrating how alternative choices might have changed history. These examples mirror the counterfactual frameworks in causal inference, emphasizing how different "what if" scenarios reveal the impact of specific interventions or decisions.

2. A distribution is simply a collection of data or scores on a variable, typically arranged from smallest to largest and often represented graphically. — Page 6, "Statistics in Plain English," Third Edition, 2010.

3. Keep in my mind estimator is functions (statistics) of observed data, usually denote by a hat. For the same estimand, there can be multiple estimators. Desirable properties of estimators are Unbiased/Consistency and Efficiency (low variance).

4. Detailed derivation of this equation is provided in Section 4.1.3 of Mixtape. For more details, visit https://mixtape.scunning.com/04-potential_outcomes

17

Randomized Controlled Trials

Randomization is fundamental in causal inference, ensuring that treatment and control groups are comparable. By assigning units to treatment states (D) independently of their characteristics, randomization eliminates systematic differences between groups, allowing observed outcome differences to be attributed solely to the treatment. In randomized controlled trials (RCTs), the assignment mechanism is fully known and controlled by investigators, satisfying the random assignment assumption.

Randomization, if implemented perfectly, balances observed and unobserved covariates , and potential outcomes $((Y(1), Y(0)))$, guaranteeing unconfoundedness (Rubin, 1978) without conditioning on covariates. This assignment mechanism ensure that differences between groups are not due to preexisting characteristics. By balancing covariates and potential outcomes, RCTs are considered the gold standard in causal inference, enabling researchers to attribute outcome differences directly to the treatment.

Although RCTs and A/B tests share foundational principles and are often used interchangeably, they differ in scope and application. A/B tests, widely used in business and digital settings, compare versions of an intervention—such as webpage designs, emails, or marketing campaigns—to optimize outcomes like click-through or conversion rates. In contrast, RCTs are used in fields like medicine, public health, and economics to evaluate policies, treatments, or interventions, often addressing broader, long-term effects.

Both rely on randomization to eliminate systematic differences between groups,i.e. selection bias, and estimate causal effects. A/B tests typically label groups as "variant A" (control) and "variant B" (new version), while RCTs use "treatment" and "control." While A/B tests focus on short-term business decisions, RCTs aim to generate generalizable scientific evidence. Despite these differences, A/B tests are a practical application of RCT principles, offering valuable insights for real-world optimization.

Although randomization is crucial, particularly in medical studies, it can be complex—especially when calculating confidence intervals and statistical power for studies with small samples. Additionally, there are many types of randomization, each suited to different study designs. Here, we focus on fundamental concepts relevant to causal inference and those that form the basis for understanding the other RCM methods. Despite their strengths, RCTs are not without limitations. They can be resource-intensive, ethically complex, or logistically challenging to conduct, but their ability to provide robust causal evidence makes them indispensable in many fields of study. Next, we explore the key assumptions underlying RCTs.

DOI: 10.1201/9781003504795-17

17.1 Key Assumptions

The core assumptions for valid causal inference in RCTs include random assignment, SUTVA, and perfect compliance.

Random assignment is central to the validity of an RCT and ensures that treatment status (D) is statistically independent of the potential outcomes ($Y_i(1)$ and $Y_i(0)$). Formally, under the assumption of exogeneity, $(Y_i(1), Y_i(0)) \perp D_i$. This guarantees that the distribution of both observed and unobserved characteristics is balanced between the treatment ($D = 1$) and control ($D = 0$) groups. By breaking the link between treatment assignment and characteristics that could influence the outcome, random assignment eliminates selection bias, enabling causal interpretation of the observed differences in outcomes.

Complementing this, the stable unit treatment value assumption (SUTVA) ensures two critical conditions: no interference between units, meaning one individual's treatment status does not influence another's outcome, and consistency of treatment, requiring uniform application of the treatment and accurate correspondence between observed outcomes and potential outcomes under the assigned treatment status.

Perfect compliance is another key assumption, where all units assigned to treatment receive it, and no units in the control group access the treatment. This ensures that the observed treatment status matches the assigned treatment, preserving the validity of causal estimates. Additionally, RCTs assume no attrition, requiring that participants remain in the study and provide complete data. Systematic dropout can bias results, particularly if dropout rates are related to the treatment or outcomes. Proper implementation of randomization is equally important; deviations, such as participant selection based on convenience or unblinded treatment assignment, can undermine validity of the study. Finally, sufficient sample size and statistical power are necessary to ensure that the randomization process effectively balances covariates and that the study can detect meaningful treatment effects.

Under these assumptions, RCTs are widely considered the gold standard in causal inference. They eliminate selection bias by construction, ensuring that the observed differences in outcomes between treated and control groups are attributable solely to the treatment. As a result, RCTs are routinely employed in economics, social sciences, and health sciences to evaluate the efficacy of interventions such as medical treatments, public health programs, educational policies, job training initiatives, and social welfare strategies, among others.

17.2 Randomization Inference

In randomized causal inference, two primary approaches emerge: Fisher's randomization inference and Neyman's sampling inference. Fisher (1935)'s method tests the sharp null hypothesis by computing p-values from the randomization distribution, making it ideal for experimental designs and small-sample biomedical studies where exact inference is essential. In contrast, Neyman (1923)'s framework, which introduced potential outcomes, focuses on estimating the average causal effect under a completely randomized experiment. This approach emphasizes estimation through an unbiased point estimator and conservative confidence intervals based on its sampling distribution, and is widely used in economics, social sciences, and large-scale policy evaluations. Given that Neyman's framework aligns closely with the Rubin causal model and regression-based methods—the predominant techniques in empirical research—we focus on it here, while Fisher's method, along with its simulations, is discussed later in the last section of this chapter. Because of Neyman's foundational contributions, the Rubin causal model is sometimes also called the Neyman-Rubin Model.

Consider the example of studying the causal effect of a college degree on earnings. Without random assignment, individuals who pursue higher education may differ systematically in motivation, family support, or access to resources, leading to potential bias in direct income comparisons between graduates and non-graduates. Complete random assignment, achieved through methods like lottery-based assignment or random number generators, ensures that these confounding factors are distributed equally across groups. Perfect randomization eliminates selection bias, making observed differences in income attributable solely to the causal effect of obtaining a college degree. Next, we will explore Neyman's inference, first verbally and then formally.

Neyman's repeated sampling approach provides a systematic framework for estimating the average treatment effect (ATE) in randomized experiments. It seeks to answer two fundamental questions: What would the average outcome be if all units were exposed to the treatment? How would this compare to the average outcome if all units were instead exposed to the control? These questions reflect the central focus of Neyman's causal inference approach—understanding the difference in potential outcomes caused by the treatment.

To address these questions, Neyman proposed a classic frequentist approach. The first step involves identifying an unbiased estimator of the ATE. An unbiased estimator ensures that its expected value, averaged over all possible randomizations of treatment assignment, equals the true treatment effect. The next step is to calculate the sampling variance of this estimator and derive an unbiased estimate of that variance. Using this information, confidence intervals can be constructed. These intervals, consistent with Neyman's framework, have the property that over repeated randomizations, the true treatment effect is contained within the interval at least as frequently as the stated confidence level. For example, a 95% confidence interval would include the true value at least 95% of the time. In randomized experiments, the ATE equialent to ATT and ATU as treatment assignment ensures comparability across groups. Neyman's method thus provides a statistical basis for estimating causal effects, ensuring both unbiasedness and valid inference under randomization.

Formally, under a completely randomized experiment (CRE) with N units, where N_T receive treatment ($D = 1$) and N_C receive control ($D = 0$), the potential outcomes are $Y_i(1)$ and $Y_i(0)$ for unit i, representing income under treatment and control, respectively. The ATE is defined as the difference in potential outcomes:

$$\delta^{ATE} = \mathbb{E}[Y(1)] - \mathbb{E}[Y(0)] \tag{17.1}$$

Under randomization, the observed average income in the treated group ($\bar{Y}(1) = \sum_{i=1}^{N} Y_i/N$) estimates $\mathbb{E}[Y(1)]$, and the observed average income in the control group ($\bar{Y}(0) = \sum_{i=1}^{N} Y_i/N$) estimates $\mathbb{E}[Y(0)]$. Therefore, the ATE can be identified as:

$$\hat{\delta}^{ATE} = \bar{Y}(1) - \bar{Y}(0) \tag{17.2}$$

However, due to the fundamental problem of causal inference, we cannot observe $Y_i(1)$ and $Y_i(0)$ simultaneously for any individual, making direct calculation of δ^{ATE} impossible. Neyman's (1923) theorem provides a framework for causal inference under a CRE. The ATE can be estimated using the following sample means for the treatment and control groups:

$$\hat{\bar{Y}}(1) = \frac{1}{N_T}\sum_{i=1}^{N} D_i Y_i, \quad \hat{\bar{Y}}(0) = \frac{1}{N_C}\sum_{i=1}^{N}(1-D_i)Y_i \tag{17.3}$$

and taking their difference, we obtain unbiased estimation of ATE:

$$\hat{\delta}^{ATE}_{unadjusted} = \hat{\bar{Y}}(1) - \hat{\bar{Y}}(0) = \frac{1}{N_T}\sum_{i=1}^{N} D_i Y_i - \frac{1}{N_C}\sum_{i=1}^{N}(1-D_i)Y_i \tag{17.4}$$

This unadjusted estimator relies solely on observed outcomes (Y), while adjustments incorporating covariates will be discussed in the next section.

The sample variances for the treated and control groups are given by[1]:

$$\hat{S}^2(1) = \frac{1}{N_T - 1}\sum_{i=1}^{N} D_i \left(Y_i - \hat{\bar{Y}}(1)\right)^2, \quad \hat{S}^2(0) = \frac{1}{N_C - 1}\sum_{i=1}^{N}(1-D_i)\left(Y_i - \hat{\bar{Y}}(0)\right)^2 \tag{17.5}$$

To account for the unobservable variance of treatment effects ($S^2(\delta)$), Neyman proposed a conservative variance estimator for $\hat{\delta}^{ATE}$[2]:

$$\hat{\text{Var}}_{Neyman}(\hat{\delta}^{ATE}) = \frac{\hat{S}^2(1)}{N_T} + \frac{\hat{S}^2(0)}{N_C} \tag{17.6}$$

This approach ensures unbiased estimation of δ^{ATE} and its variance and valid inference using conservative confidence intervals, even when $S^2(\delta)$ is unobservable.

Returning to the college degree example, imagine randomly assigning individuals to either pursue higher education ($D = 1$) or not ($D = 0$), irrespective of their background characteristics. In this setup, $\hat{\bar{Y}}(1)$ represents the average income for those assigned to treatment, approximating the average income the entire sample, $\bar{Y}(1)$, would achieve if everyone attended college. Similarly, $\hat{\bar{Y}}(0)$ represents the average income for those assigned to control, approximating the average income, $\bar{Y}(1)$, of the entire sample if no one attended college. Randomization ensures that any differences in outcomes between the treated and control groups can be attributed solely to the causal effect of a college degree. This framework provides unbiased estimation of δ^{ATE} and valid inference using conservative confidence intervals, ensuring robustness even under the absence of observed covariances.

The sampling distribution of $\hat{\delta}^{ATE}$ can be approximated by a normal distribution under large sample sizes and regularity conditions. Neyman's variance estimator ensures that the sample variances are consistent with population variances, which justifies a conservative large-sample confidence interval for δ^{ATE}:

$$\hat{\delta}^{ATE}_{unadjusted} \pm z_{1-\alpha/2}\sqrt{\hat{\text{Var}}_{Neyman}(\hat{\delta}^{ATE})} = \hat{\delta}^{ATE}_{unadjusted} \pm z_{1-\alpha/2} \cdot \text{SE}_{Neyman}(\hat{\delta}^{ATE}) \tag{17.7}$$

where $z_{1-\alpha/2}$ is the $1-\alpha/2$ upper quantile of the standard normal random variable.

For example, if $1 - \alpha = 0.95$ (a 95% confidence level), then $\alpha = 0.05$ and $\alpha/2 = 0.025$. The critical value $z_{1-\alpha/2} = z_{0.975}$ is approximately 1.96. Thus, the 95% confidence interval for δ^{ATE} is $\hat{\delta}^{ATE}_{unadjusted} \pm 1.96 \cdot \text{SE}_{Neyman}(\hat{\delta}^{ATE})$. This confidence interval covers δ^{ATE} with a probability of at least $1 - \alpha = 0.95$ when the sample size is large enough. By duality, it implies a test for the null hypothesis $H_0 : \delta^{ATE} = 0$, which is known as the weak null hypothesis.

Let's assume we conducted a randomized experiment where one-quarter of the population was assigned to attend college while the rest did not. After college, we observed their earnings and aimed to estimate the ATE using the equations above. The simplest way to estimate ATE is by computing the difference in mean earnings between the treated and control groups as this is complete randomization.

```
set.seed(42)  # For reproducibility
N <- 20000
NT <- 5000  # Number in treatment (college graduates)
NC <- N - NT  # Number in control (non-graduates)
# Generate earnings separately for treated and control groups
earnings_non_graduates <- rnorm(NC, mean = 50, sd = 10)
earnings_graduates <- rnorm(NT, mean = 53, sd = 10)  # Mean increased by 3
# Merge into a single dataset
data <- data.frame(
  ID = 1:N,
  Treatment = c(rep(1, NT), rep(0, NC)),
  Earnings = c(earnings_graduates, earnings_non_graduates)
)
# Neyman's inference: Point estimator and standard error
Y <- data$Earnings  # Outcome variable
D <- data$Treatment  # Treatment indicator
# Point estimator for ATE
deltahat <- mean(Y[D == 1]) - mean(Y[D == 0])
deltahat
```

```
## [1] 3.209321
```

```
# Variance and standard error using Neyman's method
vhat_Neyman <- var(Y[D == 1]) / NT + var(Y[D == 0]) / NC
sqrt(vhat_Neyman)  # Standard error
```

```
## [1] 0.1651089
```

```
# OLS estimation, coefficient and standard error
olsfit <- lm(Y ~ D, data = data)
summary(olsfit)$coef[2,1:2]
```

```
##    Estimate Std. Error
##  3.2093210  0.1644279
```

```
# Eicker-Huber-White robust SE using lm_robust (HC2) to match Neyman SE
library(estimatr)
ols_robust_hc2 <- lm_robust(Y ~ D, data = data, se_type = "HC2")
coef(summary(ols_robust_hc2))[2, "Std. Error"]
```

```
## [1] 0.1651089
```

Another approach we will cover below is regression, where we estimate ATE using ordinary least square (OLS). The estimated coefficient on treatment from OLS should match the ATE from our earlier equation, but the standard errors differ. Ordinary least squares assumes homoskedasticity, which can lead to incorrect standard errors in the presence of heteroskedasticity. To address this, we use heteroskedasticity-robust standard errors, specifically the Eicker–Huber–White (EHW) estimator, which corrects the standard errors to align with Neyman's inference. Now, we move to the next section, where we explain OLS in RCTs in detail, discuss why we need to use robust standard errors, and then introduce regression models that include additional covariates (X) to improve efficiency.

17.3 Linear Regression in RCTs

Linear regression is widely used in RCTs to estimate treatment effects because it flexibly incorporates covariates and adjusts for residual imbalances. The connection between the potential outcomes framework and linear regression is critical for understanding how treatment effects under randomization can be estimated using the OLS estimator and its inference. As explained in detail in Section 17.2, in the potential outcomes framework, the observed outcome of each unit is defined as $Y_i = D_i \cdot Y_i(1) + (1 - D_i) \cdot Y_i(0)$; that is, if $D_i = 1$, then $Y_i = Y_i(1)$, and if $D_i = 0$, then $Y_i = Y_i(0)$. This can also be written as $Y_i = Y_i(0) + (Y_i(1) - Y_i(0)) \cdot D_i$, where $Y_i(0)$ is the baseline outcome for the untreated state, and $Y_i(1) - Y_i(0)$ is the treatment effect for unit i, applied only to the treated. Under random assignment, although both potential outcomes cannot be observed for any unit, the expected treatment effect (ATE) can still be estimated.

In the previous section, we discussed Neyman's causal inference, where the ATE is estimated as the difference in average outcomes between treated and control groups, $\bar{Y}(1) - \bar{Y}(0)$, which is equivalent to the OLS estimator β_1 in a more flexible regression framework. Below, we explain how OLS can be used to estimate treatment effects in RCTs, considering both homogeneous and heterogeneous effects, as well as binary and continuous treatments. Even after including covariates and their interactions, β_1 still captures the ATE. For inference, robust standard errors should be used, as demonstrated in the simulation above. We now continue the discussion with binary and continuous treatments, excluding covariates.

When treatment effects are homogeneous, meaning $Y_i(1) - Y_i(0) = \beta_1$ for all units, the model simplifies to $Y_i = Y_i(0) + \beta_1 \cdot D_i$. This aligns with the regression equation $Y_i = \beta_0 + \beta_1 D_i + \varepsilon_i$, where $\beta_0 = \mathbb{E}[Y_i(0)]$ and $\varepsilon_i = Y_i(0) - \mathbb{E}[Y_i(0)]$ captures deviations from this baseline. Here, $\beta_1 = \beta^{\text{OLS}}_{\text{Unadjasted}}$ is an unbiased estimator of the constant treatment effect and equals the ATE.

For heterogeneous treatment effects, where $Y_i(1) - Y_i(0) = \beta_{1i}$ varies across units, the OLS estimator still captures the average treatment effect by averaging individual effects: $\beta_1^{OLS} = \beta^{\text{OLS}}_{\text{Unadjasted}} = \mathbb{E}[\beta_{1i}] = \delta^{ATE}$, as long as treatment assignment is independent of the potential outcomes.[3] Even when D_i is ordered or continuous, the OLS framework adapts to estimate the treatment effect, with the coefficient representing a convexly weighted average of the treatment effect gradient, with further details provided in the footnote.[4] This derivation shows that the regression model effectively estimates causal effects, accommodating both uniform and variable treatment responses.

Below, we present the (unadjusted) linear regression for RCTs using a more commonly used notation in recent books and lectures. This notation simplifies the presentation and is frequently applied to extend the model by including covariates for adjustment in RCTs. While this approach may appear cleaner and more intuitive, it can create conceptual confusion for some readers, particularly when distinguishing between (pre-treatment) covariate adjustment in RCTs and conditioning on covariates (including confounding variables) in observational studies. In RCTs, covariates are used to improve precision (regression adjustment), whereas in observational studies, the goal is to approximate randomization to estimate causal effects (i.e., obtaining ATEs with estimated Conditional ATEs). We will revisit these differences and explain observational studies in detail in the next chapter. Here, we present the same regression framework for clarity and alignment with standard notation.

The linear regression model for RCTs with binary treatment assignment is expressed as:

$$Y_i = \alpha + \beta D_i + \epsilon_i \iff \quad E[Y_i] = \alpha + \beta D_i \tag{17.8}$$

where D_i is a binary treatment indicator, is interpreted as follows: - For the control group $(D_i = 0)$: $E[Y_i \mid D_i = 0] = \alpha$, - For the treatment group $(D_i = 1)$: $E[Y_i \mid D_i = 1] = \alpha + \beta$

Under randomization, $E[Y_i(1)]$ and $E[Y_i(0)]$ are unbiasedly estimated by the observed means of the treatment and control groups. The ATE can then be expressed as:

$$E[Y_i(1)] - E[Y_i(0)] = E[Y_i \mid D_i = 1] - E[Y_i \mid D_i = 0] = (\alpha + \beta) - \alpha = \beta \tag{17.9}$$

This demonstrates that the regression coefficient β provides a direct and unbiased estimate of the ATE in randomized experiments. The robust standard error of this regression is also an estimate of the standard deviation of the ATE.

Robust Standard Errors in RCTs: In the regression model with binary randomization, the standard OLS variance estimator assumes homoskedasticity. The variance of $\hat{\beta}_1$[5] is:

$$\hat{\text{Var}}^{\text{OLS}}_{\text{Unadjasted}}(\hat{\beta}_1) = \frac{\hat{\sigma}^2}{\sum_{i=1}^{N}(D_i - \bar{D})^2} = \frac{\hat{\sigma}^2 N}{N_T N_C} \tag{17.10}$$

However, in RCTs, outcome variances typically differ between treatment and control groups. Since the OLS estimator assumes homoskedasticity, it underestimates the standard errors of the ATE when variances differ across groups. To correct for this, EHW robust standard errors replace the assumption of constant variance with an estimate that weights each squared residual based on its deviation from $\bar{D}$:

$$\hat{\text{Var}}^{\text{robust}}(\hat{\beta}_1) = \left(\sum_{i=1}^{N}(D_i - \bar{D})^2\right)^{-1} \sum_{i=1}^{N}(D_i - \bar{D})^2 \hat{\varepsilon}_i^2 \tag{17.11}$$

For a binary treatment, this simplifies to Neyman's variance formula, aligning with the conservative variance estimator for the ATE under randomization we discussed earlier.

$$\hat{\text{Var}}^{OLS}_{\text{robust}}(\hat{\beta}_1) = \frac{S^2(1)}{N_T} + \frac{S^2(0)}{N_C} = \hat{\text{Var}}_{\text{Neyman}}(\hat{\delta}^{ATE}) \tag{17.12}$$

Thus, robust standard errors should be used in RCTs to correctly estimate uncertainty when outcome variances differ across groups. We demonstrated this in the simulation above, where you can verify the variances and standard errors, confirming that the robust standard errors align with Neyman's variance estimator while the homoskedastic OLS standard errors underestimate variability when group variances differ.

17.4 Regression Adjustment in Randomized Experiments

Until now in this section, we have established that under a completely randomized experiment (CRE) with N units, the ATE is defined as $\delta^{ATE} = \mathbb{E}[Y(1)] - \mathbb{E}[Y(0)]$. Since we cannot observe both potential outcomes for any unit, ATE is estimated using observed sample means:

$$\hat{\delta}^{ATE}_{unadjusted} = \frac{1}{N_T}\sum_{i=1}^{N} D_i Y_i - \frac{1}{N_C}\sum_{i=1}^{N}(1 - D_i)Y_i \tag{17.13}$$

By Neyman's (1923) framework, this provides an unbiased estimate of ATE under randomization.

We showed that ATE can be estimated either by directly computing the difference between the average of outcome in treatment and control groups or by using unadjusted regression (Y on D) with robust standard errors—both yielding identical estimates and inference. However, this approach is based on unadjusted potential outcomes, and adjustments may be required for greater precision. In randomized experiments, adjusting for pre-experimental covariates improves efficiency, while in observational studies (discussed in the next chapter), adjustments also address confounding factors.

Covariate adjustment can occur at the design stage (e.g., stratified, paired randomization, or rerandomization) or at the analysis stage. In these chapters, we focus on analysis-stage adjustments—such as matching, propensity scores, weighting, and regression—that replace simple averages of potential outcomes with adjusted potential outcomes that account for pre-treatment and confounding covariates. For example, one can obtain adjusted potential outcomes by matching (using only matched pairs), by applying propensity score methods, by weighting observations, or by including covariates in a regression model. We will discuss stratification, matching, weighting, and propensity score methods (for observational studies) in the next chapter, and now continue with a detailed discussion of regression adjustment, which is widely used in practice in RCTs.

Specifically, while randomization is expected to balance covariates on average, small samples or design issues can lead to imbalances that reduce statistical power and threaten internal validity (credible treatment effect for our sample). Two main strategies can be used to construct the test statistics to test null hypothesis or to estimate ATE with confidence intervals, allowing covariate adjustments to improve precision and decrease variance (thus, the uncertainty).

17.4.1 Residual Strategy in Randomized Experiments

The residual strategy constructs the test statistic based on residuals obtained from a fitted regression model. Specifically, the outcome variable Y_i is regressed on covariates X_i to produce residuals $\hat{\varepsilon}_i$, which represent the portion of the outcome not explained by the covariates. These residuals are treated as pseudo-outcomes to test the sharp null hypothesis and calculate test statistics. All explained in detail, and demonstrated with simulation in the Fisher's inference section.

17.4.2 Regression Strategy in Randomized Experiments

Fisher (1925) introduced ANCOVA to improve estimation efficiency by regressing Y_i on $(1, D_i, X_i)$ and using the coefficient of D_i as the estimator for the ATE. Freedman (2008) reanalyzed this approach under Neyman's (1923) potential outcomes framework and found that this estimator is biased, unlike the simple difference in means, which is unbiased. Lin (2013) responded by showing that the bias of Fisher's ANCOVA (regression adjustment without interaction terms) is negligible in large samples and that including D_i, X_i, and $D_i \times X_i$ in the regression improves asymptotic efficiency.[6] Using an ANCOVA model with interactions and centered covariates $(X_i - \bar{X})$, combined with EHW robust standard errors, is recommended. The coefficient of D_i provides an unbiased estimate of the ATE, and the EHW standard error gives a valid estimate of its variability.

Now, let's discuss regression adjustment using the potential outcomes framework in detail. By presenting these approaches—obtaining potential outcomes from separate regressions, using a single regression with interactions, or applying a simpler regression without interactions—wwe aim to provide a comprehensive understanding of covariate adjustment in RCTs, even though all estimate the same ATE. While separate estimation is less commonly covered in RCT analysis, regression-based approach and notation are more standard. However, we introduce separate estimation to build intuition and establish notation within the RCT framework, which will later extend to more general methods addressing selection on observables and unobservables. Moreover, the separate regression approach forms a foundational intuition, familiarizing readers with concepts that underpin a substantial number of doubly robust and heterogeneous causal inference methods.

In separate regression approach, the potential outcomes for the treatment $(D_i = 1)$ and control $(D_i = 0)$ groups be represented as follows:

$$Y_i = \gamma_1 + \beta_1^\top X_i + \varepsilon_i \quad \text{for } D_i = 1 \tag{17.14}$$

$$Y_i = \gamma_0 + \beta_0^\top X_i + \varepsilon_i \quad \text{for } D_i = 0 \tag{17.15}$$

Here, γ_1 and γ_0 are the intercepts representing the adjusted mean outcomes for the treatment and control groups, respectively. When $\beta_1 = \beta_0 = 0$ (i.e., no covariate adjustment), these Y_i values correspond to the potential outcomes used in the unadjusted estimator , $\hat{\delta}^{ATE}_{unadjusted} = \hat{\bar{Y}}(1) - \hat{\bar{Y}}(0)$, in RCTs.

In the regression-adjusted estimator, we replace the unadjusted potential outcomes with fitted means for the covariate-adjusted potential outcomes. First, to obtain adjusted potential outcomes, we regress Y_i separately for the treatment $(D_i = 1)$ and control $(D_i = 0)$ groups:

$$\min_{\gamma_1, \beta_1} \sum_{i=1}^{n_1} \left(Y_i - \gamma_1 - \beta_1^\top X_i\right)^2, \quad \min_{\gamma_0, \beta_0} \sum_{i=1}^{n_0} \left(Y_i - \gamma_0 - \beta_0^\top X_i\right)^2 \tag{17.16}$$

From these regressions, we obtain $\hat{\gamma}_1, \hat{\beta}_1$ for the treatment group and $\hat{\gamma}_0, \hat{\beta}_0$ for the control group. Then, we obtain the fitted means for the treatment and control groups, which are:

$$\hat{Y}(1) = \hat{\gamma}_1 + \hat{\beta}_1^\top \bar{X}(1), \quad \hat{Y}(0) = \hat{\gamma}_0 + \hat{\beta}_0^\top \bar{X}(0) \tag{17.17}$$

where $\bar{X}(1)$ and $\bar{X}(0)$ are the sample means of X_i for the treated and control groups, respectively.

The adjusted ATE is then:

$$\hat{\delta}^{ATE}_{sep.adjusted} = \hat{Y}(1) - \hat{Y}(0) = \left(\hat{\gamma}_1 + \hat{\beta}_1^\top \bar{X}(1)\right) - \left(\hat{\gamma}_0 + \hat{\beta}_0^\top \bar{X}(0)\right) \tag{17.18}$$

If covariates are centered ($\bar{X}(1) = \bar{X}(0) = 0$), the adjusted ATE from separate regressions simplifies to:

$$\hat{\delta}^{ATE}_{sep.adjusted} = \hat{\gamma}_1 - \hat{\gamma}_0 \tag{17.19}$$

In this setup, $\hat{\gamma}_1$ and $\hat{\gamma}_0$ represent the adjusted mean outcomes for the treatment and control groups, respectively. The difference $\hat{\gamma}_1 - \hat{\gamma}_0$ provides an unbiased estimate of the ATE (δ^{ATE}) when covariates are centered. EHW robust standard errors are recommended for estimating the standard error of $\hat{\delta}^{ATE}$, as they account for heteroskedasticity introduced by covariate adjustment.

We want to emphasize again before implementing regression adjustment, covariates must be centered by subtracting their group-specific means. Centering ensures that $\hat{\delta}^{ATE}$ is captured entirely by the difference in intercepts ($\hat{\gamma}_1 - \hat{\gamma}_0$), simplifying interpretation. Without centering, the ATE would depend on group-specific covariate means, complicating analysis and inference.

17.4.3 Single Regression with Covariate Adjustment

Alternative approach involves a single unified regression model that incorporates interactions between treatment and covariates. This unified model offers a streamlined way to estimate the ATE and, importantly, captures heterogeneity in treatment effects.

The single regression model, often referred to as the ANCOVA model with interactions, is expressed as:

$$Y_i = \beta_0 + \beta_D D_i + \beta_X^\top X_i + \beta_{DX}^\top (D_i \cdot X_i) + \varepsilon_i \tag{17.20}$$

In this model, β_D represents the ATE when covariates X_i are centered ($\mathbb{E}[X_i] = 0$), while β_{DX} captures treatment effect heterogeneity, allowing the effect of the treatment to vary depending on the values of X_i. This approach eliminates the need for separate regressions for the treatment and control groups, combining them into a single estimation framework. To understand how the ATE is derived in this model, consider the expected outcomes for the treated group ($D_i = 1$) and the control group ($D_i = 0$).

For the treated group, the expected outcome is: $\mathbb{E}[Y_i \mid D_i = 1] = \beta_0 + \beta_D + \beta_X^\top X_i + \beta_{DX}^\top X_i$

For the control group, the expected outcome is: $\mathbb{E}[Y_i \mid D_i = 0] = \beta_0 + \beta_X^\top X_i$

The conditional treatment effect, which reflects how the treatment effect varies with X_i, is the difference between these two expected outcomes:

$$\begin{aligned}\delta(X_i) &= \mathbb{E}[Y_i \mid D_i = 1] - \mathbb{E}[Y_i \mid D_i = 0] \\ &= (\beta_0 + \beta_D + \beta_X^\top X_i + \beta_{DX}^\top X_i) - (\beta_0 + \beta_X^\top X_i) = \beta_D + \beta_{DX}^\top X_i\end{aligned} \tag{17.21}$$

To obtain the ATE, we take the expectation of $\delta(X_i)$ over X_i:

$$\hat{\delta}^{ATE}_{adj.interact} = \mathbb{E}[\delta(X_i)] = \beta_D + \beta_{DX}^\top \mathbb{E}[X_i] = \beta_D \tag{17.22}$$

As covariates are centered ($\mathbb{E}[X_i] = 0$), the interaction term drops out. This shows that centering the covariates ensures that β_D directly estimates the ATE, while β_{DX} captures any heterogeneity in treatment effects.

Alternatively, if treatment effect heterogeneity across covariates is not a concern, one could use an ANCOVA model without interactions. This simpler model assumes a constant treatment effect across all values of X_i and is expressed as:

$$Y_i = \beta_0 + \beta_D D_i + \beta_X^\top X_i + \varepsilon_i \tag{17.23}$$

In this case, the expected outcomes for the treated and control groups are:

$$\mathbb{E}[Y_i \mid D_i = 1] = \beta_0 + \beta_D + \beta_X^\top X_i, \quad \mathbb{E}[Y_i \mid D_i = 0] = \beta_0 + \beta_X^\top X_i \tag{17.24}$$

The treatment effect estimate is obtained by differencing these expected outcomes:

$$\hat{\delta}^{ATE}_{adjusted} = \mathbb{E}[Y_i \mid D_i = 1] - \mathbb{E}[Y_i \mid D_i = 0] = \beta_D \tag{17.25}$$

While this model also provides an unbiased estimate of the ATE, it does not account for treatment effect heterogeneity, making it less suitable when variation in treatment effects across covariates is expected. Consequently, the ANCOVA model with interactions is generally preferred in randomized experiments, as it accounts for treatment effect heterogeneity and improves asymptotic efficiency (as discussed at the beginning of this section) when covariates are centered and robust standard errors (i.e., sandwich or EHW estimator) are used (sec.5, Lin (2013)). This approach ensures that the coefficient of D_i provides an unbiased estimate of the ATE while capturing variations in treatment effects across subgroups.

$$\begin{aligned}
\hat{\delta}^{ATE}_{\text{sep.adjusted}} &= \hat{\gamma}_1 - \hat{\gamma}_0 = \hat{\delta}^{ATE}_{\text{int.adjusted}} \\
&= \mathbb{E}[Y_i \mid D_i = 1] - \mathbb{E}[Y_i \mid D_i = 0] \\
&= (\beta_0 + \beta_D) - \beta_0 = \beta_D = \hat{\delta}^{ATE}_{\text{adjusted}}
\end{aligned} \tag{17.26}$$

All approaches yield the same ATE because they ultimately estimate the same difference in expected outcomes between the treatment and control groups. In the separate regression approach, this difference is captured by $\hat{\gamma}_1 - \hat{\gamma}_0$, which represents the adjusted mean outcome difference after controlling for covariates. In the single regression approach, β_D directly measures the treatment effect when covariates are centered. Since all methods derive the treatment effect by comparing adjusted means, they are equal. The key reason they align is that including covariates in a regression does not alter the fundamental structure of how the treatment effect is estimated—it only adjusts for potential imbalances in covariates. When centering is applied, the contribution of the interaction term to the ATE disappears, and confirming their equivalence.

17.4.4 Simulation: Covariate Adjustment in RCTs

This simulation compares various methods for estimating the ATE in a RCT, where college graduation is the treatment and earnings is the outcome. Family income is included as a covariate because it influences both participation in randomization and other related factors, making it important for adjustment. We employ four approaches: (1) OLS without covariates, replicating robust OLS results of previous simulation; (2) OLS with covariates that assumes a common (homogeneous) treatment effect; (3) OLS with an interaction term—recommended single regression method—to allow for heterogeneous treatment effects; and (4) separate regressions for the treatment and control groups. Given that family income affects treatment assignment, the model assuming a common effect is less plausible, even though covariates should be balanced in true randomization, as imbalances can occur in finite samples. Notably, the ATE estimates from the heterogeneous model and the separate regressions are identical as expected, confirming the consistency between these methods. We use EHW (sandwich; robust in Stata) standard errors for reliable inference under heteroskedasticity and recommend bootstrapping to calculate standard errors for the separate regressions approach, especially in observational studies.

```r
# Set seed for reproducibility
set.seed(42)
# Define sample sizes
N <- 20000
NT <- 5000  # Number in treatment (college graduates)
NC <- N - NT  # Number in control (non-graduates)
# Generate earnings for treated and control groups
  # Higher mean for graduates
earnings_non_graduates <- rnorm(NC, mean = 50, sd = 10)
earnings_graduates <- rnorm(NT, mean = 53, sd = 10)
# Generate family income (X) for both groups
family_income_non_graduates <- rnorm(NC, mean = 70, sd = 20)  # Lower mean
family_income_graduates <- rnorm(NT, mean = 95, sd = 20)  # Higher mean
# Merge into a single dataset
data <- data.frame(
  ID = 1:N,
  Treatment = c(rep(1, NT), rep(0, NC)),
  X = c(family_income_graduates, family_income_non_graduates),
  Earnings = c(earnings_graduates, earnings_non_graduates)
)

# Neyman's inference: Point estimator and standard error
Y <- data$Earnings # Outcome variable
D <- data$Treatment # Treatment indicator
X <- data$X  # Covariate
# Load necessary library
library(estimatr)

# (1) OLS estimation without covariates (robust standard error)
ols_robust_hc2 <- lm_robust(Y ~ D, data = data, se_type = "HC2")
ate_ols <- coef(ols_robust_hc2)[2]
se_ols <- coef(summary(ols_robust_hc2))[2, "Std. Error"]

# Center X
data$X_centered <- data$X - mean(data$X)

# (2) OLS with covariates (no interaction)
ols_covrobust_hc2 <- lm_robust(Y ~ D + X_centered,
                               data = data, se_type = "HC2")
ate_cov <- coef(ols_covrobust_hc2)[2]
```

```
se_cov <- coef(summary(ols_covrobust_hc2))[2, "Std. Error"]

# (3) OLS with interactions
ols_introbust_hc2 <- lm_robust(Y ~ D + X_centered + D * X_centered,
                               data = data, se_type = "HC2")
ate_int <- coef(ols_introbust_hc2)[2]
se_int <- coef(summary(ols_introbust_hc2))[2, "Std. Error"]

# (4) Separate regressions for ATE estimation
ols_treated <- lm(Y ~ X_centered, data = data, subset = (D == 1))
ols_control <- lm(Y ~ X_centered, data = data, subset = (D == 0))
gamma_1 <- coef(ols_treated)[1]
gamma_0 <- coef(ols_control)[1]
ate_sep <- gamma_1 - gamma_0 # ATE from separate regressions

# Display results
results <- data.frame(
  Method = c("OLS Unadjusted", "OLS w/ Covariates (No Interaction)",
             "OLS w/ Interactions", "Separate Regressions"),
  ATE_Estimate = c(ate_ols, ate_cov, ate_int, ate_sep),
  # Separate regressions do not directly give SE
  EHW_Standard_Error = c(se_ols, se_cov, se_int, NA)
)
print(results)
```

```
##                              Method ATE_Estimate EHW_Standard_Error
## 1                    OLS Unadjusted     3.209321          0.1651089
## 2 OLS w/ Covariates (No Interaction)     3.188720          0.1872812
## 3               OLS w/ Interactions     3.175287          0.2143932
## 4              Separate Regressions     3.175287                 NA
```

In practice, covariate adjustment in practice rarely changes the conclusion of the unadjusted estimator, therefore usually only reported in secondary analysis. However, it is essential to adjust for covariates to improve precision and efficiency, particularly in small samples or when covariates are imbalanced.

17.4.5 Missing Data

Regression adjustment improves precision and efficiency in randomized experiments but has limitations. It can lead to selective model specification in pursuit of significant results, raising validity concerns. Unlike linear regression, models like Logit and Probit lack justification from randomization (Freedman, 2008b). For binary or count outcomes, linear models may not accurately capture causal effects, and some argue that regression adjustment undermines the integrity of randomization. One of the most common issues is missing data, which can arise in both covariates and outcomes. In this section, we discuss strategies for handling missing data in RCTs briefly.

When covariates have missing values in RCTs, a key question is how to adjust for them. Zhao and Ding (2022) show that as long as mean imputation is used and missingness indicators are included as predictors, the specific imputation method does not affect efficiency. This ensures that the adjusted estimator remains asymptotically more efficient than the unadjusted one.

For a single partially observed covariate X, let $R_i = 1$ if X_i is observed and $R_i = 0$ if missing. The OLS model for covariate adjustment is:

$$Y_i \sim D_i + X_i R_i + R_i + D_i(X_i R_i + R_i) \tag{17.27}$$

Since $X_i R_i$ equals X_i when observed and zero when missing, imputation choice does not affect the final estimate, holding regardless of the missing data mechanism.

In practice, missingness often occurs in both covariates and outcomes. While missing covariates can be adjusted without concern for imputation choice, missing outcomes pose a greater challenge, as imputation quality affects conclusions. Under the missing at random assumption, missingness depends only on observed covariates and treatment status, not the missing outcome itself. We estimate the probability of observation (propensity score) and apply inverse probability weights ($1/e_i$) in weighted least squares (WLS) regression to adjust for outcome missingness. IPW methods, which correct for selection bias, will be discussed in the next chapter in the context of selection on observables.

17.5 Fisher's Randomization Inference

Fisher's randomization inference is a framework for testing the sharp null hypothesis that a treatment has no effect on any individual, i.e.,

$$H_0 : Y_i(1) = Y_i(0) \quad \forall i \tag{17.28}$$

Unlike traditional statistical inference, which relies on population assumptions, Fisher's approach is based solely on the assignment mechanism of treatment. The outcomes are assumed to be fixed, and the only randomness comes from the treatment assignment. This principle, first recognized by R.A. Fisher (1925, 1935), establishes randomization as the foundation for credible causal inference. Fisher was particularly interested in testing sharp null hypotheses, where the treatment has no effect on any unit. The key idea is to compare the observed test statistic (e.g., the difference in means) to its distribution under all possible random assignments of treatment.

Fisher's exact test follows three key steps. First, researchers specify the null hypothesis, which assumes that the treatment has no effect on any individual. Second, they choose a test statistic that quantifies the treatment effect. Finally, they measure extremeness by computing the p-value under the randomization distribution, which represents the likelihood of observing a test statistic as extreme as the one obtained, assuming the null hypothesis is true.

The test statistic is a function of the data used to assess the consistency of the null hypothesis with the observed outcomes. The most common test statistic in Fisher's inference is the difference in means, given by:

$$\hat{\delta} = \frac{1}{N_T} \sum_{i:D_i=1} Y_i - \frac{1}{N_C} \sum_{i:D_i=0} Y_i \tag{17.29}$$

where N_T and N_C are the number of treated and control units, respectively. Under the sharp null $Y_i(1) = Y_i(0)$, we can reassign treatment labels and recompute $\hat{\delta}$ across all possible random assignments. (in R the following function also calculates this test `t.test var . equal = TRUE`)

Other valid test statistics include the Studentized statistic (`t.test var . equal = FALSE`), Wilcoxon rank-sum statistic (`wilcox.test`), and Kolmogorov–Smirnov statistic (`ks.test`). In the case of binary outcomes, Fisher's exact test simplifies to a contingency table comparison (`fisher.test`).

17.5.1 p-value Calculation and Simulation

After stating the sharp null hypothesis and defining the most common test statistic (the difference in means), we can generate the randomization distribution and calculate the p-value Under the sharp null, we randomly permute D_i (the treatment assignment) while keeping the observed outcomes fixed, then recalculate $\hat{\delta}$ for each permutation. This process is repeated many times (e.g., 10,000 times). The collection of all possible test statistics under randomization forms the randomization distribution, which represents the values that the test statistic could take under the null hypothesis. The p-value is then computed as:

$$p = \frac{\#(\hat{\delta}^{\text{sim}} \geq \hat{\delta}^{\text{obs}})}{\text{Total Simulations}} \tag{17.30}$$

where $\hat{\delta}^{\text{sim}}$ represents test statistics from simulated random assignments.

Beyond simple hypothesis testing, Fisher's framework allows constructing Fisher intervals, which define a range of plausible values for the causal effect. These intervals include all possible values of the treatment effect corresponding to test statistics whose p-values fall within a predefined range.

Fisher's randomization inference provides an exact test for treatment effects, making no distributional assumptions, and is particularly useful in small-sample experiments where traditional asymptotic methods may be unreliable. Since it relies solely on the randomization process, it avoids the biases and model dependence of parametric approaches. By constructing a randomization distribution, it allows researchers to rigorously test sharp null hypotheses and compute precise p-values without relying on large-sample approximations. However, this method can be computationally intensive for large datasets due to the combinatorial complexity of reassigning treatment labels across many possible permutations. Despite this, Fisher's inference remains a powerful tool in experimental settings, offering a robust alternative to parametric tests and providing credible, assumption-free causal conclusions.

```
# Set seed for reproducibility
set.seed(42)
# Define sample sizes
N <- 200
NT <- 50 # Number in treatment (college graduates)
NC <- N - NT  # Number in control (non-graduates)
# Generate earnings for treated and control groups
earnings_non_graduates <- rnorm(NC, mean = 50, sd = 10)
earnings_graduates <- rnorm(NT, mean = 53, sd = 10)  # Mean increased by 3
# Merge into a single dataset
data <- data.frame(
  ID = 1:N,
  Treatment = c(rep(1, NT), rep(0, NC)),
  Earnings = c(earnings_graduates, earnings_non_graduates)
)
```

```
# Compute observed difference in means
obs_diff <- mean(data$Earnings[data$Treatment == 1]) -
  mean(data$Earnings[data$Treatment == 0])

# Permutation test: shuffle treatment labels and compute differences
num_simulations <- 10000
simulated_diffs <- numeric(num_simulations)

for (i in 1:num_simulations) {
  shuffled_treatment <- sample(data$Treatment)  # Shuffle treatment labels
  simulated_diffs[i] <- mean(data$Earnings[shuffled_treatment == 1]) -
    mean(data$Earnings[shuffled_treatment == 0])
}

# Compute p-value
p_value <- mean(abs(simulated_diffs) >= abs(obs_diff))

# Display results
cat("Observed Difference in Means:", obs_diff, "\n")
cat("P-value:", p_value, "\n")

# Plot histogram of simulated differences
hist(simulated_diffs, breaks = 30, col = "blue",
     border = "black", probability = TRUE,
     main = "Fisher's Randomization Inference: College Graduation on Earnings",
     xlab = "Simulated Differences in Mean Earnings", ylab = "Density")
abline(v = obs_diff, col = "red", lwd = 2, lty = 2)
legend("topright", legend = c("Observed Difference"),
       col = "red", lwd = 2, lty = 2)


# Set seed for reproducibility
set.seed(42)
# Define sample sizes
N <- 200
NT <- 50 # Number in treatment (college graduates)
NC <- N - NT  # Number in control (non-graduates)
# Generate earnings for treated and control groups
earnings_non_graduates <- rnorm(NC, mean = 50, sd = 10)
earnings_graduates <- rnorm(NT, mean = 53, sd = 10)  # Mean increased by 3
# Merge into a single dataset
data <- data.frame(
  ID = 1:N,
  Treatment = c(rep(1, NT), rep(0, NC)),
  Earnings = c(earnings_graduates, earnings_non_graduates)
)
# Compute observed difference in means
obs_diff <- mean(data$Earnings[data$Treatment == 1]) -
  mean(data$Earnings[data$Treatment == 0])
```

```
# Permutation test: shuffle treatment labels and compute differences
num_simulations <- 10000
simulated_diffs <- numeric(num_simulations)

for (i in 1:num_simulations) {
  shuffled_treatment <- sample(data$Treatment) # Shuffle treatment labels
  simulated_diffs[i] <- mean(data$Earnings[shuffled_treatment == 1]) -
    mean(data$Earnings[shuffled_treatment == 0])
}

# Compute p-value
p_value <- mean(abs(simulated_diffs) >= abs(obs_diff))

# Display results
cat("Observed Difference in Means:", obs_diff, "\n")
```

```
## Observed Difference in Means: 3.050242
```

```
cat("P-value:", p_value, "\n")
```

```
## P-value: 0.0584
```

```
# Plot histogram of simulated differences
hist(simulated_diffs, breaks = 30, col = "gray",
     border = "black", probability = TRUE,
     main = "Fisher's Randomization Inference: College Graduation on Earnings",
     xlab = "Simulated Differences in Mean Earnings", ylab = "Density")
abline(v = obs_diff, col = "black", lwd = 2, lty = 2)
legend("topright", legend = c("Observed Difference"),
       col = "black", lwd = 2, lty = 2)
```

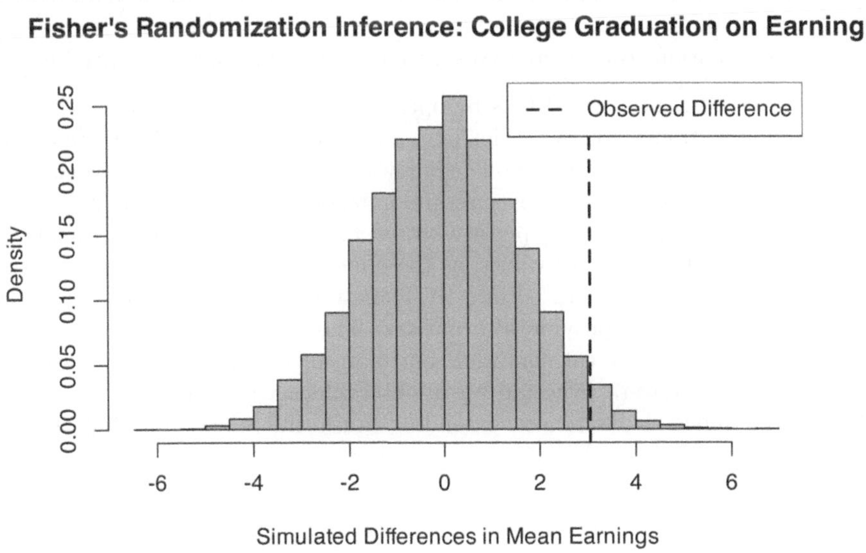

This simple simulation demonstrates Fisher's randomization inference applied to the effect of college graduation on earnings. Under the sharp null hypothesis (i.e., college graduation has no effect on earnings), we randomly assigned individuals to treatment (college graduates) and control (non-graduates) groups while keeping their earnings fixed.

We then computed the observed difference in mean earnings between the two groups and generated a randomization distribution by repeatedly shuffling treatment labels and recalculating the difference. The p-value represents the proportion of simulated differences that are as extreme as or more extreme than the observed difference. In this case, the computed p-value = 0.0584, indicating strong evidence against the null hypothesis.

The histogram of simulated differences shows the distribution under the null, with the observed difference marked in red. If the observed difference lies far in the tail of this distribution, we reject the null hypothesis, suggesting that college graduation significantly impacts earnings. The next section will discuss the challenges of RCTs, including ethical dilemmas, practical limitations.

17.6 Challenges in Randomized Controlled Trials

Ensuring the validity of RCTs depends on proper implementation, as deviations from experimental protocols can bias causal estimates. Differential dropout rates, contamination between groups, and other disruptions can distort treatment effects. For example, if attrition is higher in the treatment group, estimated effects may be misleading. Cluster randomization, which assigns groups rather than individuals (e.g., schools instead of students), simplifies logistics but introduces challenges like intracluster correlation and heterogeneous treatment effects, reducing statistical power and requiring advanced methods like mixed-effects models.

Beyond statistical challenges, RCTs face ethical, financial, and generalizability constraints. Ethical concerns arise when withholding beneficial treatments is unjustifiable, while high costs and strict inclusion criteria limit their applicability to broader populations. Additional complications include attrition, noncompliance, and spillovers. Missing data from attrition can bias results, while noncompliance and spillovers violate core assumptions, necessitating approaches like intention-to-treat analyses, instrumental variables, or spatial models.

Machine learning (ML) improves RCTs by optimizing design and addressing imperfections. Adaptive randomization improves efficiency, and ML techniques help predict attrition, estimate heterogeneous effects, and model spillovers, making RCTs more robust in complex settings. While this book will not cover rerandomization or multi-arm experiments, it will discuss stratification, clustering, propensity scores, and inverse probability weighting (IPW) in the context of selection on observables, as these methods are widely used in observational studies and increasingly incorporated into RCT analysis. When randomization is infeasible, quasi-experimental designs provide alternatives. The next section explores selection on observables, which balances treatment and control groups using measured covariates, while the following chapter covers selection on unobservables with methods like instrumental variables and difference-in-differences, extending causal inference beyond RCTs.

Notes

1. The covariance between treatment and control outcomes, $S(1,0) = \frac{1}{N-1}\sum_{i=1}^{N}(Y_i(1) - \bar{Y}(1))(Y_i(0) - \bar{Y}(0))$, and the variance of treatment effects, $S^2(\delta) = \frac{1}{N-1}\sum_{i=1}^{N}(\delta_i - \bar{\delta})^2$, are not identified since $Y_i(1)$ and $Y_i(0)$ are never jointly observed.

2. Neyman's theorem states that the true variance of the ATE estimator is:

$$\text{Var}(\hat{\delta}^{ATE}) = \frac{S^2(1)}{N_T} + \frac{S^2(0)}{N_C} - \frac{S^2(\delta)}{N}.$$

Since $S^2(\delta)$ is unobservable, the conservative variance estimator overestimates the true variance:

$$\mathbb{E}[\hat{\text{Var}}(\hat{\delta}^{ATE})] - \text{Var}(\hat{\delta}^{ATE}) = \frac{S^2(\delta)}{N} \geq 0$$

with equality holding only if δ_i is constant across all units.

3. If treatment effects heterogeneous(vary) across units, the OLS estimate of the treatment effect can be expressed as:

$$\beta^{\text{OLS}}_{\text{Unadjasted}} = \frac{\mathbb{E}[(D_i - \mu)Y_i]}{\text{Var}[D_i]} = \frac{\text{Cov}[D_i, \varepsilon_i]}{\text{Var}[D_i]} + \frac{\mathbb{E}[D_i(D_i - \mu)\beta_{1i}]}{\text{Var}[D_i]} = \mathbb{E}[\beta_{1i}] \equiv \delta^{ATE}$$

where $\mu = \mathbb{E}[D_i]$: the average treatment assignment, β_{1i}: the individual treatment effect, $\beta_{\text{OLS}} = \mathbb{E}[\beta_{1i}]$ is ATE under random assignment and independence assumptions. This derivation demonstrates that, even with heterogeneous treatment effects, the OLS estimator remains unbiased for the ATE as long as $(\varepsilon_i, \beta_{1i}) \perp D_i$.

4. In cases where treatment D is ordered or continuous (e.g., different dosages), the treatment assignment remains independent of the potential outcomes $(D \perp \{Y(d)\}_{d \in \mathcal{D}})$. This ensures that the expected outcome given treatment $D = d$ equals the potential outcome $Y(d)$, $\mathbb{E}[Y \mid D = d] = \mathbb{E}[Y(d)]$. A saturated regression of Y on dummy variables for all values of D (or a nonparametric regression for continuous D) traces the average structural function $\mathbb{E}[Y(d)]$. Alternatively, a simple regression of Y on D identifies a convexly weighted average of $\partial\mathbb{E}[Y(d)]/\partial d$ or its discrete version. The OLS coefficient β_{OLS} can be expressed as:

$$\beta_{\text{OLS}} = \int_{-\infty}^{\infty} \omega(\tilde{d})\frac{\partial\mathbb{E}[Y(d)]}{\partial d}d\tilde{d}, \quad \text{or} \quad \beta_{\text{OLS}} = \sum_{k=1}^{K} \omega_k \frac{\mathbb{E}[Y(d_k)] - \mathbb{E}[Y(d_{k-1})]}{d_k - d_{k-1}}$$

where the weights $\omega(\tilde{d}) = \frac{\text{Cov}[1\{D \geq \tilde{d}\}, D]}{\text{Var}[D]}$ for continuous D or $\omega_k = \frac{(d_k - d_{k-1}) \cdot \text{Cov}[1\{D \geq d_k\}, D]}{\text{Var}[D]}$ for discrete D. This approach allows RCTs with continuous or ordered treatments to identify the average structural function or treatment effect gradient.

5. For binary treatment, $\sum_{i=1}^{N}(D_i - \bar{D})^2 = N_T N_C/N$, which leads to the standard OLS variance formula. This assumes a common residual variance, $\hat{\sigma}^2$, across all units. Robust SEs, however, allow for different residual variances in the treated and control groups, making them consistent with Neyman's variance estimator.

6. Freedman (2008) also noted that the asymptotic variance of $\hat{\tau}_F$ may exceed that of the simple difference in means when $n_1 \neq n_0$ and that the OLS standard error is inconsistent with the true standard error of $\hat{\tau}_F$ under CRE. Lin (2013) showed that the EHW standard error is a valid, conservative estimator for the true standard error of $\hat{\tau}_F$ under CRE.

18

Selection on Observables

While randomized controlled trials (RCTs) remain the gold standard for establishing causal relationships, they are often impractical or unfeasible in real-world settings due to several challenges we discussed in the previous chapter. Moreover, noncompliance by administrators, where experimental protocols are not strictly followed, can undermine randomization. Similarly, noncompliance among treated individuals, such as those failing to adhere to the intervention, or among controls, who may seek out the treatment despite being in the control group, can distort the validity of causal estimates. In such cases, researchers must turn to alternatives that leverage observational data to approximate the conditions of a controlled experiment.

Randomization ensures that treatment assignment is independent of potential outcomes, eliminating selection bias and enabling straightforward causal inference. However, in observational data, treated and untreated groups often differ systematically in observed characteristics, introducing selection bias. To address this, researchers rely on the assumption that treatment assignment is independent of potential outcomes after adjusting for certain covariates (X), effectively making the assignment "as good as random" within groups sharing the same values of X. This approach allows researchers to infer causal effects from observational data by mimicking the conditions of randomization, provided that all relevant confounders are accounted for. Under the selection on observables framework, researchers use methods such as matching or regression adjustment for observable differences between treated and untreated groups. This framework relies on the assumption of conditional independence and overlap to estimate causal effects.

Starting with this chapter and continuing over the next three, we examine the selection on observables framework, which assumes that treatment assignment is independent of potential outcomes after conditioning on observed covariates. We begin by outlining the key identification assumptions—conditional independence and common support—required for valid causal inference. The next chapter then explores various regression estimation methods, including regression adjustment and machine learning-based approaches such as double/debiased LASSO in detail. Through theoretical discussions, empirical examples, and simulations, this chapter provides a structured guide for estimating treatment effects in observational data settings using regression. In the next chapters, we will also cover methods such stratification, matching, propensity score, inverse probability weighting, and doubly robust estimators.

We want to emphasize that selection on observables methods rely on the assumption that treated and untreated units differ only in ways we can observe, making causal effect estimates from these methods open to debate and potentially less credible. While these approaches are data-driven, widely used in econometrics and machine learning, and valuable for practical applications, it is crucial to recognize that in economics, social sciences, and health research, where human behavior is central, countless unobservable factors influence actions and decisions, limiting our ability to control for all relevant variables. When selection on observables is not sufficient because unobserved factors also influence treatment assignment

DOI: 10.1201/9781003504795-18

and outcomes, researchers may rely on methods that address selection on unobservables. Following these three chapters, we focus on models designed to handle unobservable factors.

18.1 Assumptions and Definitions Behind Selection on Observables

Our main goal is to estimate the average treatment effect (ATE), but we can also estimate measures like ATT and ATU as causal effects. Estimating causal effects under selection on observables (SOO) relies on two key assumptions. These assumptions are crucial for ensuring that treatment assignment is "as good as random", allowing researchers to isolate the causal effect of treatment from confounding factors:

18.1.1 Conditional Independence Assumption (Unconfoundedness)

"Selection on observables" assumption posits that treatment assignment is independent of the potential outcomes, $(Y_i(1), Y_i(0))$, once we condition on observed covariates, X_i. Mathematically, this is expressed as:

$$(Y_i(1), Y_i(0)) \perp D_i \mid X_i \tag{18.1}$$

where $D_i = 1$ if the individual receives treatment and $D_i = 0$ otherwise. This implies that any systematic differences in outcomes between treated and untreated/control groups can be fully explained by differences in X_i, eliminating the influence of confounding variables.

For example, in a scholarship study where X_i covers parental income, university type, and SAT scores, individuals with identical X_i have the same chance of receiving a scholarship, independent of their unobserved earning potential. Similarly, in a study of job training programs, if X_i includes age, income, and education level, SOO assumes that individuals with the same X_i are equally likely to receive treatment regardless of their potential outcomes. In health research, when X_i consists of age, pre-existing conditions, and socioeconomic status, SOO assumes that people with the same X_i are equally likely to participate in a preventive care program, regardless of factors not captured in X_i.

In essence, conditional independence assumption (CIA) tells us that:

- Potential outcomes are independent of D_i, conditional on covariates.
- This is often referred to as conditional unconfoundedness or conditional independence.
- In simpler terms, once we control for X_i (observed covariates), treatment assignment is as good as random.
- Practically, controlling for X_i eliminates selection bias, making causal interpretation plausible under SOO methods.

18.1.2 Common Support Assumption (Overlap Condition)

The overlap assumption ensures that every individual has a nonzero probability of being either treated or untreated, given their covariates. Formally, this is expressed as:

$$0 < P(D_i = 1 \mid X_i = x) < 1, \quad \forall x \tag{18.2}$$

This condition guarantees that both treated and untreated units are available for each level of X_i, allowing meaningful comparisons. Without overlap, it becomes challenging to compare units with similar covariates, and standard selection-on-observables methods (which we will cover in this chapter) may yield invalid causal estimates. However, alternative approaches—such as regression discontinuity and instrumental variables (which we will cover in the next chapter)—can sometimes identify causal effects even when the overlap assumption does not hold.

18.2 Defining the (Conditional) Average Treatment Effect

Why did CATE leave ATE at the altar?

Because ATE kept averaging out their differences, and CATE was tired of all the behind-the-scenes covariate adjustments!

In the previous section on randomization, we demonstrated that under the identification assumption of random assignment, selection bias is eliminated, providing an unbiased estimate of the ATE. Specifically:

$$\delta^{ATE} = E[Y_i(1) - Y_i(0)] = E[Y_i(1)] - E[Y_i(0)] = E[Y_i \mid D = 1] - E[Y_i \mid D = 0] \quad (18.3)$$

This simple difference in means of observable outcomes between the treated and untreated groups in observed data serves as an estimate of the ATE.

When treatment assignment is not random and depends on observable characteristics, the CIA and common support condition allow us to estimate the conditional average treatment effect (CATE) for any given covariate X_i. By averaging these CATEs over the distribution of X_i in the population, we can recover the average causal/treatment effect, ATE $= \mathbb{E}[\text{CATE}(X_i)]$. This framework ensures causal estimation is possible even in selection in observable settings, provided the key assumptions hold.

Under the SOO framework, the CIA assumption implies:

$$E[Y_i(1) - Y_i(0) \mid X_i] = E[Y_i(1) \mid X_i, D_i = 1] - E[Y_i(0) \mid X_i, D_i = 0] \quad (18.4)$$

This means that within groups with the same values of X_i, treatment assignment is "as good as random." For a specific value of $X_i = x$, we can define the CATE as:

$$\text{CATE}(x) = E[Y_i(1) - Y_i(0) \mid X_i = x] = E[Y_i \mid X_i = x, D_i = 1] - E[Y_i \mid X_i = x, D_i = 0] \quad (18.5)$$

Under the common support condition, where all values of X_i are observed in both treated and untreated groups, we can aggregate the CATE across the population to estimate the ATE:

$$\delta^{ATE} = \sum_x \text{CATE}(x) P(X_i = x) \quad (18.6)$$

Similarly, we can calculate the average treatment effect on the treated (ATT) and the average treatment effect on the controls (ATC) using the distributions of X_i in the treated and control groups:

$$\delta^{ATT} = \sum_x \text{CATE}(x)P(X_i = x \mid D_i = 1), \quad \delta^{ATC} = \sum_x \text{CATE}(x)P(X_i = x \mid D_i = 0) \tag{18.7}$$

These weighted averages allow for robust causal estimation by leveraging the observed covariates and addressing biases in observational data.[1]

These identification assumptions and result are common to various estimation methods, which we will cover below, for causal effects. Regression methods such as OLS and double-lasso adjusts for confounding by modeling the relationship between Y_i, D_i, and X_i, while subclassification groups units into strata based on covariates X_i and estimates treatment effects within each stratum. Matching pairs treated and untreated units with similar X_i values. Other methods, such as inverse probability weighting (IPW), use propensity scores to reweight observations, ensuring comparability between treated and control groups. Doubly robust estimators combine regression and weighting for added robustness, while machine learning techniques, like causal forests, estimate CATEs flexibly and handle complex covariate structures. Spline or kernel methods offer smooth adjustments for X_i without assuming a specific functional form. These approaches differ in how they condition on X_i and weight to estimate CATEs, providing versatility for a range of data and study designs.

Example: To understand the concepts of ATE and CATE, let's consider a study estimating the effect of receiving a college scholarship on annual earnings. Suppose the exact same amount of scholarships are allocated randomly among students, i.e., completely randomized experiment, ensuring that selection bias is removed. In this randomized setting, we observe that, on average, 25% of students receive the scholarship. The overall average annual earnings for students who received the scholarship (the treated group) are \$50,000, while for those who did not (the control group), the average earnings are \$40,000. The ATE is calculated as:

$$\text{ATE} = E[Y(1) - Y(0)] = E[Y \mid D = 1] - E[Y \mid D = 0] = 50,000 - 40,000 = 10,000$$

This indicates that, on average, receiving a scholarship increases annual earnings by \$10,000. Keep in mind, individual impacts can vary widely—some students may see little to no benefit, while others might experience tens of thousands of dollars in increased earnings. The ATE represents the average effect across all individuals, summarizing these various outcomes into a single measure.

Since randomization is not available, we rely on observational data, recognizing that multiple factors may influence earnings. However, for simplicity in this example, we assume university type (X_i) captures the key observable differences affecting both treatment assignment and outcomes. For example, private, public, and non-profit colleges might serve distinct populations, with varying effects of scholarships on earnings. Let's calculate the CATEs for public and non-profit colleges, given the following data:

For public colleges, the average annual earnings of scholarship recipients are \$48,000, while for non-recipients, they are \$40,000. This gives a CATE for public colleges of:

$$\text{CATE}_{\textbf{Public}} = E[Y(1) - Y(0) \mid X = \text{Public}] = 48,000 - 40,000 = 8,000$$

For non-profit colleges, scholarship recipients earn an average of \$42,000 annually, compared to \$36,000 for non-recipients. The CATE for non-profit colleges is:

$$\text{CATE}_{\text{Non-Profit}} = E[Y(1) - Y(0) \mid X = \text{Non-Profit}] = 42,000 - 36,000 = 6,000$$

For private universities, scholarship recipients earn an average of \$65,000 annually, compared to \$50,000 for non-recipients. This gives a CATE for private universities of:

$$\text{CATE}_{\text{Private}} = E[Y(1) - Y(0) \mid X = \text{Private}] = 65,000 - 50,000 = 15,000$$

Assume 50% of students attend public colleges, 30% attend private universities, and 20% attend non-profit colleges. The ATE is then calculated as a weighted average of the CATEs:

$$\text{ATE} = 0.5 \cdot 8,000 + 0.3 \cdot 12,000 + 0.2 \cdot 6,000 = 4,000 + 4,000 + 1,200 = 9,700$$

This example illustrates how the ATE aggregates heterogeneous treatment effects (conditional ATEs) across subgroups, weighted by their population shares. While the overall ATE suggests a \$10,000 increase in earnings due to the scholarship, the CATEs highlight differences in impacts (average causal/treatment effect) by university type. Students at private universities experience the largest gains, likely due to stronger networks or better resources, whereas public college students benefit moderately, and non-profit college students see smaller gains. These differences highlight the importance of considering heterogeneity in treatment effects to better inform policy decisions and target interventions more effectively.

In the following sections, we will explore various approaches under the selection on observables framework: regression methods such as ordinary least squares (OLS) and double lasso, stratification and matching, propensity score, weighting approaches,and doubly robust method to estimate causal effects. These methods provide flexible tools for addressing the challenges of causal inference in observational data.

18.3 Regression-Based Estimation Methods

In an observational setting, we adapt the same regression adjustment ideas from the randomized framework discussed in the previous chapter, but rely on the crucial assumption that treatment assignment is "as good as random" once we condition on all relevant, observed covariates. This section explains how regression adjustment helps us estimate causal effects when randomization is absent, clarifies how we move from conditional effects to an overall ATE, and discusses issues of inference and robust standard errors. Conceptually, by using covariates that influence both treatment assignment and outcomes (confounding and pretreatment variables), regression adjustment (or any predictive modeling) effectively simulates random assignment within each covariate stratum. Once we account for these confounders, the treated and untreated groups become comparable in expectation, so the observed outcome differences reflect the average causal effect, and removes the selection bias. We illustrate these points using the scholarship example, yet emphasize that the approach applies generally.

In a randomized trial, treatment assignment is physically controlled, ensuring that—on average—treated and untreated groups only differ because of the intervention. In observational research, however, we cannot manipulate treatment directly. Instead, we assume that once we condition on observed characteristics X, no unobserved confounders remain. This assumption, often termed "selection on observables," means that within each $X = x$, the treatment is "as good as" random. Under this assumption, we can treat each subgroup $X = x$ as if an internal experiment were carried out there, letting us compare treated and untreated outcomes within each subgroup.

When X is discrete (for instance, public vs. private colleges), it is straightforward to define subgroup-specific differences and then average them up. In the scholarship example, we separately compute the difference in earnings between treated and untreated students for public, private, and non-profit colleges, then weight each difference by the share of students in that subgroup to form an overall ATE. If X is continuous or contains multiple variables, the principle remains the same—though we replace that discrete summation with an integral over the distribution of X. In practice, regression-based methods carry out this averaging automatically. Lets discuss two separate regression methods to estimate the ATE.

18.3.1 Separate Regressions for Treated and Control Groups

A first approach, closely mirroring the randomized setting, is to fit two separate regressions—one for the treated group ($D_i = 1$) and one for the untreated group ($D_i = 0$). Concretely, we write:

$$Y_i = \gamma_1 + \beta_1^\top X_i + \varepsilon_i, \quad D_i = 1 \tag{18.8}$$

$$Y_i = \gamma_0 + \beta_0^\top X_i + \varepsilon_i, \quad D_i = 0 \tag{18.9}$$

Once these regressions are fitted, we obtain the fitted potential outcomes $\hat{Y}(1 \mid X_i) = \hat{\gamma}_1 + \hat{\beta}_1^\top X_i$ and $\hat{Y}(0 \mid X_i) = \hat{\gamma}_0 + \hat{\beta}_0^\top X_i$. For each individual i with covariates X_i, the conditional average treatment effect (CATE) is $\hat{Y}(1 \mid X_i) - \hat{Y}(0 \mid X_i)$.

The overall average treatment effect (ATE) follows by averaging this individual-level difference across the entire sample:

$$\widehat{\text{ATE}} = \frac{1}{n}\sum_{i=1}^{n}\left[\hat{Y}(1 \mid X_i) - \hat{Y}(0 \mid X_i)\right] \tag{18.10}$$

When X is centered in each treatment group, the ATE can be read directly off the difference in intercepts $\hat{\gamma}_1 - \hat{\gamma}_0$, just like in a RCT we discussed in the previous chapter. If X is not centered, the average difference over the sample still yields the correct ATE. Valid inference—through standard errors and confidence intervals—requires a robust variance estimate (e.g., Eicker–Huber–White) or bootstrapping—which is more common—to accommodate heteroskedasticity.

In separate regressions for regression adjustment in observational studies, we generate similar or nearly identical predicted adjusted outcomes for treated and control groups, effectively removing selection and omitted variable bias. Separate regressions naturally allow the form of Y on X to differ between treatment and control groups. This method is often used when you want maximum flexibility in modeling outcomes under each treatment status, and extends seamlessly to machine learning methods under selection on observables. If you believe all relevant confounders are observed in X, you can adopt any predictive modeling strategy—whether a simple linear model, random forests, neural networks, or other algorithms—and fit it separately to the treated and untreated samples. Having done so, you predict an individual's outcome assuming treatment,$\hat{Y}(1 \mid X_i)$, using the "treated" model and individual's outcome assuming treatment,$\hat{Y}(0 \mid X_i)$, using the untreated model.

The difference in those two predicted outcomes is that individual's estimated treatment effect, $\hat{\delta}_i = \hat{Y}(1 \mid X_i) - \hat{Y}(0 \mid X_i)$. Averaging these differences across i yields $\widehat{\text{ATE}}$. In the machine-learning literature, this two-model approach is often called a T-learner, reflecting the idea that you train two separate models (one for each treatment state), which will be discussed in meta-learners chapter in detail.

18.3.2 Single Regression with Covariate Adjustment

A common and straightforward way to estimate the ATE is through OLS regression. This method provides a direct and interpretable approach to causal effect estimation under the selection-on-observables assumption. It is important to emphasize again that this approach relies on the assumption that all relevant confounders are observed and included in X_i. If unobserved confounders influence both treatment and outcomes, bias may still persist—necessitating alternative methods, which we will cover in the next chapter. In this approach, we specify the outcome model as:

$$Y_i = \beta_D \cdot D_i + f(X_i) + \varepsilon_i \tag{18.11}$$

where Y is the outcome, D is the treatment indicator, β_D represents the treatment effect, and X includes all covariates whose influence on Y is captured by $f(X)$. The OLS estimator for β_D provides an adjusted estimate of the ATE, denoted as $\hat{\delta}^{ATE} = \hat{\beta}_D$. The derivation is identical to the case where treatment is randomized, except that here, adjustment for X accounts for selection-on-observables. Since standard OLS may overfit when X is high-dimensional, we later introduce other methods, such as debiased/double (ML) lasso. First, however, we discuss the single linear regression adjustment model, which serves as the baseline approach.

$$Y_i = \alpha + \delta\, D_i + X_i^\top \beta + \varepsilon_i \tag{18.12}$$

Here, δ is the difference between treated and untreated after partialing out the effect of X_i. If no interaction terms are added, this model imposes a constant treatment effect—i.e., it treats $\delta^{CATE}(x) = \delta$ for all x. In that simplified world, the ATE is exactly δ. Mechanically, once you fit the regression, the difference in fitted outcomes between $D = 1$ and $D = 0$ is always $\hat{\delta}$, irrespective of X_i. Averaging over i in the sample therefore yields the same $\hat{\delta}$. This approach can be viewed as "automatically" integrating (or summing) the effect across all X_i values in the data. In other words, the fitted model simultaneously estimates the conditional mean of Y for treated and untreated units across the full range of X_i, then (under the selection-on-observables assumption) "averages" those conditional means over the observed distribution of X to yield a consistent ATE. Since OLS estimates are obtained by minimizing residual variance, this adjustment effectively removes selection bias due to imbalance in observables between treated and control groups.[2]

If one wants to allow for genuine heterogeneity, an interacted model of the form

$$Y_i = \alpha + \delta\, D_i + X_i^\top \beta + (D_i \times X_i^\top)\gamma + \varepsilon_i \tag{18.13}$$

provides a richer description, where the difference between treatment and control depends on X_i. The main ideas of averaging over subgroups remain, but the summation becomes slightly more involved. The conclusion, however, is identical: to recover an overall ATE, we compute the mean difference in fitted outcomes across the distribution of X_i, which can be done analytically if the variables are centered, or by simply averaging predicted effects in the sample. More flexible or nonparametric estimators (e.g., random forests or series expansions) can approximate $\delta^{CATE}(x)$ even when the relationship between X and outcomes is complex, but the principle remains the same: once $\delta^{CATE}(x)$ is estimated for all relevant values of x, you take the expectation of those conditional estimates over your data's distribution of X. This averaging yields the ATE, which is the sample-level effect of treatment.

Turning back to the scholarship example, once we consider different university types (X_i) as key observed covariates, we treat each subgroup (e.g., public, private, or non-profit) as though the scholarship were randomly allocated within that group. Estimating these subgroup-specific, or conditional, treatment effects and then weighting them by the share of each subgroup in the population yields the overall ATE. Alternatively, a single OLS regression including a dummy for scholarship receipt and controls for university type (plus other relevant variables) would provide a direct estimate of ATE under the assumption that within each observed subgroup, scholarship assignment is as good as random. If we wished to further explore heterogeneous effects, we would extend the model to include interactions, but the basic principle remains the same: regression-based adjustment in observational settings follows the same template as in randomized experiments, with the critical difference that we must now assume unobserved confounders do not invalidate our as-if-random assumption for each subgroup $X = x$.

The link between conditional treatment effects (CATE) and the overall ATE is easiest to see when X is discrete, as in the scholarship example: we sum or weight each subgroup's treatment effect $\delta^{CATE}(x)$ by the proportion of units who fall into that subgroup. When X is continuous or high-dimensional, we instead integrate (or sum) the estimated effects over all observed X_i values.[3] In practice, you do not generally have to do an explicit integral. Regression models—particularly in their single-regression form—implement this averaging in one step, yielding a coefficient (δ or an average of $\delta + X_i^\top \gamma$) that corresponds to the population-level effect of treatment.

18.3.3 Inference and Robust Standard Errors

Regardless of which regression specification is used, valid statistical inference depends on an appropriate variance estimator. In observational studies, heteroskedasticity is common, so robust (EHW) standard errors are recommended. These account for potentially unequal error variances across units. Cluster-robust standard errors may be warranted if data exhibit intra-group correlation (e.g., if multiple observations share the same institution). Confidence intervals follow the usual construction:

$$\widehat{\delta} \pm z_{1-\alpha/2} \times \mathrm{SE}(\widehat{\delta}) \tag{18.14}$$

where $z_{1-\alpha/2}$ is from a standard normal or t-distribution.

Ultimately, the assumption that conditioning on X renders treatment assignment "as good as random" is critical. If unobserved confounders remain, none of these regression-based strategies will fully eliminate bias. Still, under selection on observables, the same frameworks that applied to RCTs can be deployed with only minor modifications—namely, stratifying or partialing out X to ensure that treated and untreated groups are comparable. Whether one uses separate regressions for treated and control groups or a single regression that pools all observations, the mechanics of going from CATE to ATE simply involve averaging predicted differences across the sample's distribution of X. If X is limited to a small set of categories, that average appears as a weighted sum of subgroup estimates; with continuous or high-dimensional covariates, the summation naturally extends to an integral or a regression-based average, providing a unified framework for causal inference in observational studies.

Beyond robust standard errors, the delta method provides an approximation for the standard error of the ATE when it is a nonlinear function of estimated regression coefficients. This nonlinearity can arise in several ways: (1) Nonlinear transformations of coefficients—for example, when ATE is expressed as a ratio of parameters, such as in elasticity estimation (β_1/β_2), requiring an approximation of its variance; (2) Treatment-covariate interactions—when interaction terms ($D_i \times X_i^\top \gamma$) are included, the marginal treatment effect varies with X_i, making ATE a nonlinear function of coefficients; and (3) Quadratic or higher-order terms of confounders—if the regression includes squared terms (X_i^2) or other nonlinear transformations, the standard error of ATE needs adjustment. While useful for analytical variance estimation, the accuracy of the delta method depends on the validity of a first-order Taylor expansion, and in some cases, bootstrapping may be more reliable.[4] Bootstrapping provides another flexible approach to inference by resampling the data to approximate the sampling distribution of δ. This is particularly common when separate regressions are estimated for treated and control groups, allowing for nonparametric variance estimation. However, it can also be applied in a single regression framework by resampling observations with replacement and re-estimating the treatment effect across bootstrap samples as we covered in bootstrapping chapter.

Before concluding this section with a simulation, we emphasize once again that the validity of causal treatment effect estimates hinges on key assumptions: conditional independence (no unobserved confounders once X is included), common support (positive probability of treatment at all X values), and correct model specification (adequately capturing nonlinearity and interactions). Researchers often test alternative specifications to ensure robustness. OLS provides consistent estimates under these assumptions, but its accuracy depends on correctly measuring and including all relevant covariates. Even small omissions can introduce omitted variable bias if excluded factors are correlated with both treatment and outcome. While regression adjustment remains a widely used and transparent approach, its reliability ultimately depends on the selection-on-observables assumption.

This simulation generates a synthetic dataset with a continuous outcome, covariates drawn from a multivariate normal distribution, and a binary treatment variable generated via a logistic model. The outcome is then constructed as a linear function of the treatment, covariates, and an error term. Various regression models are estimated to recover the ATE. First, separate regressions are run for treated and control groups, and the ATE is computed as the difference in predicted outcomes at the mean covariate values of each group. Next, a single regression with adjustment for covariates is estimated using robust standard errors, and several interaction models are fitted—including models with treatment-by-covariate interactions and centered covariates—to assess heterogeneity and compare the consistency of the estimated treatment effect.

```
library(MASS)         # For multivariate normal distribution
library(lmtest)       # For hypothesis testing
library(boot)         # For bootstrapping
library(estimatr)     # For robust standard errors with lm_robust
set.seed(123) # Set seed for reproducibility
n <- 1000  # Sample size
p <- 3     # Number of covariates
# Generate covariates X ~ N(0, I)
```

```
X <- mvrnorm(n, mu = rep(0, p), Sigma = diag(p))
# Generate treatment assignment (D) from a logistic probability model
D_prob <- plogis(X %*% runif(p, -1, 1))  # Compute propensity scores
# D_prob <- plogis(X %*% c(0.5, -0.3, 0.2))  # Probability of treatment
D <- rbinom(n, 1, D_prob)  # Treatment indicator
# Generate outcome Y with treatment effect and confounders
gamma_true <- 3  # True treatment effect
beta <- runif(p, -1, 1)  # True coefficients for X
Y <- gamma_true * D + X %*% beta + rnorm(n)  # Outcome model
# Separate Regressions for Treated and Control Groups
treated_model <- lm(Y ~ X, subset = (D == 1))  # Regression for treated group
control_model <- lm(Y ~ X, subset = (D == 0))  # Regression for control group
X_treated_mean <- colMeans(X[D == 1,])  # Mean covariates for treated group
X_control_mean <- colMeans(X[D == 0,])  # Mean covariates for control group
#fitted potential outcomes
Y1_hat <- predict(treated_model, newdata = data.frame(t(X_treated_mean)))
Y0_hat <- predict(control_model, newdata = data.frame(t(X_control_mean)))
# Compute ATE using the average of the difference of pot.outcome of units
ATE_separate <- mean(Y1_hat - Y0_hat)
ATE_separate
```

```
## [1] 2.990066
```

```
# Single OLS Regression with Adjustment for X (Robust SE HC2)
ols_robust_hc2 <- lm_robust(Y ~ D + X, se_type = "HC2")  # Robust SEs (HC2)
ATE_robust <- coef(ols_robust_hc2)["D"]
ATE_robust
```

```
##        D
## 2.989047
```

```
coef(summary(ols_robust_hc2))[2, "Std. Error"]  # Standard error for D in OLS model
```

```
## [1] 0.06491384
```

```
# Interaction Model: D * X (Robust SE HC2)
interaction_model <- lm_robust(Y ~ D + X + D:X, se_type = "HC2")  # Interaction model
ATE_interaction <- coef(interaction_model)["D"]
ATE_interaction
```

```
##        D
## 2.993935
```

```
coef(summary(interaction_model))[2, "Std. Error"]  # Standard error for D in OLS model
```

```
## [1] 0.06502589
```

```
# Interaction Model: D * (X - X_mean) (Centered Interaction Model)
X_mean <- colMeans(X)  # Compute mean of each covariate
X_centered <- sweep(X, 2, X_mean, "-")  # Center X
interaction_model_centered<-lm_robust(Y ~ D + X_centered + D:X_centered,
                                       se_type = "HC2")
ATE_interaction_centered <- coef(interaction_model_centered)["D"]
ATE_interaction_centered
```

```
##        D
## 2.990066
```

```
coef(summary(interaction_model_centered))[2, "Std. Error"]
```

```
## [1] 0.06488983
```

```
# Higher-Order Interaction Model: X^2 + X^3 + Cross-product-terms
X_sq <- X^2  # Squared terms
X_cube <- X^3  # Cubed terms
X_cross <- X[,1] * X[,2] + X[,1] * X[,3] + X[,2] * X[,3]  # Cross-product terms
higher_order_model <- lm_robust(Y ~ D + X + X_sq + X_cube +
                           D:X + D:X_sq + D:X_cube + X_cross, se_type = "HC2")
ATE_higher_order <- coef(higher_order_model)["D"]
ATE_higher_order
```

```
##        D
## 2.981863
```

```
coef(summary(higher_order_model))[2, "Std. Error"]
```

```
## [1] 0.1031995
```

Although the simulation uses generated data, the same steps can be applied to any dataset by uploading it instead. As demonstrated, the ATE from separate regressions (computed as the difference in expected outcomes) closely aligns with that from the centered interaction model, echoing results from the randomization chapter in our previous work. In practice, the treatment effect coefficient (D) from a simple regression is typically similar, both in point estimates and confidence intervals, to those obtained from more complex interacted models, which is why many studies only report the single-regression (D) model. However, when higher-order terms and numerous interactions are added—potentially increasing the number of predictors beyond the sample size—techniques like double lasso become crucial for proper variable selection and inference, a topic we will explore next.

To address cases where X is high-dimensional or difficult to specify, we next introduce double machine learning methods, which remains a linear approach but selects relevant covariates efficiently.

Notes

1. For continuous covariates, the ATE can be expressed as an integral $\delta^{ATE} = \int \text{CATE}(x) f_{X_i}(x)\,dx$, where $f_{X_i}(x)$ is the density of X_i. Similarly, δ^{ATT} and δ^{ATC} use $f_{X_i|D_i=1}(x)$ and $f_{X_i|D_i=0}(x)$, respectively.

2. Keep in mind the key identifying assumption: $E[\varepsilon_i|D_i, X_i] = E[\varepsilon_i|X_i]$ ensures that treatment assignment is independent of unobserved determinants of the outcome, given X_i. OLS estimation of δ follows the standard least squares formula $\hat{\delta} = (Z'Z)^{-1}Z'Y$ where Z includes both the treatment variable D and the covariates X. This adjustment accounts for selection bias due to observed covariates but does not address bias from unobserved confounders. The OLS estimate $\hat{\delta}$ isolates the variation in Y that is orthogonal to X, effectively controlling for selection on observables. The equivalence between OLS adjustment and residualization via the Frisch-Waugh-Lovell theorem is discussed in more detail at the end of this chapter.

3. When covariates are continuous or high-dimensional, the transition from the CATE to the ATE follows the same core logic but replaces summation over discrete subgroups with integration (or summation) over the entire distribution of X. Formally, the ATE in the presence of continuous (or multiple) covariates is:

$$\text{ATE} = E[\delta^{CATE}(X)] = \int \delta^{CATE}(x)\,dF_X(x) \tag{18.15}$$

where $\delta^{CATE}(x) = E[Y(1) - Y(0) \mid X = x]$ and $F_X(x)$ is the distribution of X. If X has multiple continuous components (for example, age and years of education), then the integral is taken over their joint distribution. The principle behind regression-based methods is that one can estimate $\delta^{CATE}(x)$ (or δ in a simpler setting without interactions) for all x at once by fitting a model that relates Y to D, X, and possibly interactions, then averaging over the empirical distribution of X.

4. For a detailed discussion, see the end of the chapter.

19

Double Machine Learning

This chapter builds directly on the previous one, which introduced causal inference under selection on observables using regression adjustment and traditional covariate control methods. While those tools work well in low-dimensional settings with a moderate number of covariates, they face limitations when dealing with high-dimensional data or when model selection is required. In this chapter, we introduce modern, high-dimensional methods—particularly double LASSO and double machine learning (DML)—that address these challenges by combining regularized estimation with orthogonalization and sample splitting. These techniques enable valid inference for treatment effects even when the number of covariates is large, and they form a core part of the modern causal ML toolkit.

We begin by reviewing the key idea behind double LASSO and partialling out, then explain the logic and mechanics of DML1 and DML2 estimators. Using simulations, we show how these methods correct for regularization bias and maintain valid inference through orthogonal score construction and cross-fitting. Readers will learn how to implement both cross-validated and plug-in penalty approaches for tuning, how to adapt the methods with different learners such as trees and random forests, and how to compute robust standard errors. By the end of the chapter, readers will be able to apply DML tools in high-dimensional causal settings and understand when and why they work.

To estimate causal/treatment effect, we often start with a model of the form:

$$Y = \delta \cdot D + f(X) + \varepsilon \tag{19.1}$$

where Y is the outcome, D is the treatment, X is a (potentially large) set of covariates that may confound the relationship between D and Y, and δ is the average treatment effect of interest. While $f(X)$ could be any function in principle, implementing LASSO typically requires us to assume linear structure so that we can apply penalized linear regression methods in high dimensions.

When $p \geq n$ (or p is large relative to n), ordinary least squares (OLS) can overfit or fail to run due to singular design matrices. LASSO (the ℓ_1-penalized least squares estimator) addresses this by shrinking coefficients toward zero, effectively selecting a smaller set of covariates. This helps manage high-dimensional confounding while reducing overfitting, as previously discussed in the penalized regression methods chapter. However, it also introduces shrinkage bias, which can distort parameter estimates (here, average treatment effect) if left uncorrected. Several techniques we cover in this section tackle the bias problem by combining LASSO estimator, partialling-out procedure, and orthogonalization to achieve debiased estimated parameters with valid inference.

The Frisch–Waugh–Lovell (FWL) theorem provides the foundation for partialling out, which is central to double LASSO methods, we will discuss in this section. In a linear model where $p \leq n$, FWL states that the coefficient on the treatment variable can be obtained by residualizing both the outcome and the treatment on the covariates and then regressing the

DOI: 10.1201/9781003504795-19

residualized outcome on the residualized treatment. This procedure effectively estimating the causal/treatment effect,δ from the model, $Y_i - E[Y \mid X] = \delta.(D - E[D \mid X]) + \epsilon$. This ensures that the estimated treatment effect retains the same standard errors and confidence intervals as those obtained from direct OLS regression, making inference straightforward and comparable to traditional methods. We recommend to review the the detailed discussion with simulation in the last section.

Double LASSO methods extend this concept to high-dimensional settings ($p \geq n$), where selecting relevant covariates is crucial to avoid overfitting thus decreasing sahrinkage bias while still permitting valid inference. These methods are also used for causal or treatment-effect estimation. They include double LASSO with partialling out, double LASSO with double selection, and a more recent approach orthogonal/DML LASSO (DML1 and DML2). While cross-validated DML2 is often recommended in high-dimensional applications (Chernozhukov et al. (2018)), it can be computationally intensive. In practice, other double LASSO approaches remain valid and efficient for estimating δ. Below, we outline the main variations of double LASSO and how we can implement to estimate average causal/treatment effect, δ.

We want to warn readers that while LASSO is useful for penalized estimation in sparse models, helping to identify variables strongly associated with an outcome and improving prediction accuracy. However, naively using LASSO for inference on model parameters can be problematic since it is designed for prediction, not hypothesis testing. To ensure valid inference, variable selection should be applied separately to the reduced-form equations (outcome and treatment models), and all selected variables should be included in the final OLS estimation to reduce omitted variable bias when estimating causal effects. Also, using LASSO for inference may require a different penalty selection (λ) than what is optimal for prediction, which we will cover the plug-in penalty section below.

Double LASSO: Partialling-Out Model

A popular approach to estimate δ in the partially linear setting relies on partialling out (Belloni, Chernozhukov, and Hansen, 2014a). Let,

$$Y_i = X_i^\top \gamma + \tilde{Y}_i, \quad D_i = X_i^\top \pi + \tilde{D}_i \tag{19.2}$$

where $\hat{\gamma}$ and $\hat{\pi}$ are coefficients estimated via LASSO regressions of Y on X and D on X, respectively. The residuals

$$\tilde{Y}_i = Y_i - X_i^\top \hat{\gamma}, \quad \tilde{D}_i = D_i - X_i^\top \hat{\pi} \tag{19.3}$$

represent the variation in Y and D that is not explained by X. The key insight is that once we remove the portion of Y and D explained by X, a simple regression of $\tilde{Y}_i$ on $\tilde{D}_i$ identifies the average treatment effect, δ:

$$\tilde{Y}_i = \delta\, \tilde{D}_i + \xi_i \tag{19.4}$$

This procedure is sometimes called "partialling out" because it resembles running a regression of Y on D and X but in two steps: first, partial out X from Y and D, then regress the residuals. The advantage is that LASSO handles the high-dimensional X in each of these auxiliary regressions. A limitation is that LASSO may exclude variables with small but nonzero coefficients or struggle in highly correlated designs. Nevertheless, the final residuals $\tilde{Y}_i$ and $\tilde{D}_i$ still isolate the variation needed to estimate δ.

Double LASSO: Double Selection Model

In the Double Selection version (Belloni, Chernozhukov, and Hansen, 2014b), the first two steps—regressing Y on X and D on X—remain the same. However, in the final step, one includes the selected covariates from both LASSO regressions alongside $\tilde{D}_i$. Formally, let X_{selected} be the union of covariates chosen by LASSO in the two separate regressions; the final model becomes

$$\tilde{Y}_i = \delta\, \tilde{D}_i + X_{\text{selected}}^{\top}\theta + \xi_i \tag{19.5}$$

Here, $\tilde{Y}_i$ is again the residual of Y on X, and $\tilde{D}_i$ is the residual of D on X. Including all selected covariates ensures that any predictor deemed important for Y or D is present in the final stage, mitigating omitted-variable bias and preserving valid standard errors.

Whereas partialling out conceptually resembles a direct regression of $\tilde{Y}$ on $\tilde{D}$, double selection is closer to an OLS of Y on D plus the combined set of selected covariates from the regression adjustment section. Both are variations on the double LASSO framework and can be used to obtain a consistent estimate of the average treatment effect, δ.

19.1 Double Machine Learning

Standard LASSO estimates are biased due to the ℓ_1-penalty shrinking coefficients toward zero. Debiased LASSO (also known as desparsified or de-sparsified LASSO) corrects this shrinkage by adding a carefully constructed adjustment term, pioneered by Zhang and Zhang (2014), van de Geer et al. (2014), and Javanmard and Montanari (2014). In essence, debiased LASSO applies a correction term to the initial LASSO estimates to remove shrinkage bias. While it can be used for inference across all coefficients, it focuses on correcting specifically only LASSO's penalty-induced bias.

A more recent and generalized approach to debiase LASSO estimates is DML (Chernozhukov et al., 2018). Although DML can incorporate various ML methods (e.g., random forests, gradient boosting), we emphasize LASSO in this section, then discuss decision trees and random forest. Debiased LASSO corrects the shrinkage bias in a single high-dimensional linear model, whereas Orthogonal/DML frameworks explicitly account for potential causal structures (e.g., treatment, outcome, high-dimensional controls) and often use sample splitting and cross-fitting to bolster inference. We highly recommend consulting the last section for a reader-friendly explanation of the orthogonalization idea, as well as more theoretical details of DML.

DML is designed for estimating causal parameters in high-dimensional settings and is built around orthogonalized estimating equations, which safeguard the final estimate against small errors in the nuisance functions (the parts of the model not containing the main parameter of interest; for example, $f(X)$). When using LASSO specifically, we fit high-dimensional regressions to learn these nuisance functions (here $E[Y \mid X]$ and $E[D \mid X]$, estimated in the second step), then orthogonalize the residuals (implementing the third step) to ensure that the final treatment effect estimate is unbiased. Meanwhile, DML applies sample splitting (cross-fitting) to ensure that the data used to learn $\hat{Y}(X)$ and $\hat{p}(X)$ are not reused for the regression of the residuals. By doing so, bias from overfitting is mitigated, and valid inference on δ is obtained.

Several software packages implement DML across Stata, R, and Python. In Stata, the most common tools include `lasso` (with `selection(plugin)`), `ds` (double selection), `pdslasso` (Post-Double Selection Lasso), and `dml` (for DML1 and DML2 estimators). In R, key packages include `hdm`, which provides `rlasso()` for plug-in penalty selection, `DoubleML`, which supports various ML models including Lasso, and `grf` for causal forests. In Python, `DoubleML` mirrors the R implementation, offering DML1, DML2, and IV-DML with Lasso, random forests, and boosting, while `econml` from Microsoft provides flexible causal ML models, including Lasso-based DML.

In this section, we will cover DML with Lasso in great detail, building our own simulation from the ground up. While we will eventually use `hdm::rlasso()`, our approach will first provide step-by-step explanations, covering default settings of different programs, theoretical foundations, and practical considerations. This ensures that readers fully grasp the mechanics of DML before applying prebuilt implementations. By deeply understanding the method, readers will find it easier to extend DML to IV estimation, random forest-based DML, and even Deep Learning-based DML. Additionally, as DML is an evolving field, this foundation will enable readers to follow the latest research and updates in the literature with confidence.

19.1.1 Brief Practical Framework (DML1 and DML2 for LASSO)

1. **Split** the sample into K folds.

For each fold k:

2a. **Use $K-1$ folds** to estimate the nuisance functions: **Fit** $\hat{Y}_{-k}(X)$ by running LASSO or another ML method to predict Y.

- **Fit** $\hat{p}_{-k}(X)$ by running LASSO or another ML method to predict D.
- **Select** the "best" tuning parameter λ.

When using LASSO, the tuning parameter λ can be selected via cross-validation (will show in first simulation below), plug-in method (i.e., direct calculation with formula based on Belloni, Chernozhukov, and Hansen (2014); will discuss and show in another simulation below).

2b. In fold k:

- **Compute** cross-fitted residuals:

$$\tilde{Y}_i = Y_i - \hat{Y}_{-k}(X_i), \quad \tilde{D}_i = D_i - \hat{p}_{-k}(X_i) \tag{19.6}$$

3. **Estimate** the treatment effect δ by regressing $\tilde{Y}_i$ on $\tilde{D}_i$:

- In **DML1**: Perform this OLS fold by fold, producing one δ estimate per fold, then average across folds.
- In **DML2**: Pool the residuals from all folds first, then run a single OLS of $\tilde{Y}$ on $\tilde{D}$.

This approach balances flexibility (any suitable ML can be used) with reliable inference on δ. The orthogonal score structure ensures that moderate errors in the estimated nuisance functions $\hat{Y}(X)$ and $\hat{p}(X)$ do not contaminate the final treatment effect estimate.

Alternative methods such as ridge regression, adaptive LASSO, and the elastic net can also be employed in nuisance function estimation (in Step 2). Adaptive LASSO modifies penalty weights to reduce bias, while elastic net combines the LASSO penalty with a ridge component, improving performance in the presence of highly correlated covariates. Other related techniques, such as the Dantzig selector and square-root LASSO, provide additional flexibility depending on the specific data structure and model requirements. Moreover, in both the double LASSO - partialling-out model and the double LASSO - Double Selection model, post-LASSO can be applied to refit OLS on the covariates selected by LASSO, reducing shrinkage bias (in nuisance function estimation) while maintaining selection advantages.[1] The next parts of this chapter—covering standard errors, step-by-step implementation for DML1 and DML2, and simulation—will help you see these methods in action.

19.1.2 Inference and Robust Standard Errors

For LASSO to perform effectively in these settings, two key conditions must be met: approximate sparsity and the irrepresentable condition. Approximate sparsity ensures that only a small subset of covariates significantly influence the outcome and treatment models, allowing LASSO to focus on the most relevant variables. The irrepresentable condition guarantees that the selected covariates are not excessively correlated with the unselected ones, preventing LASSO from mistakenly omitting important variables or selecting irrelevant ones due to high collinearity. These conditions are fundamental for LASSO to consistently identify the relevant covariates and produce accurate treatment effect estimates. In practice, researchers assess these assumptions to ensure the reliability and validity of LASSO-based estimates. We will outline specific conditions and methods for testing them after presenting the simulation results.

Across all three methods, the final step involves an OLS regression of residuals, which is equivalent to a standard regression of the treatment effect. The asymptotic distribution of this estimator follows:

$$\hat{\beta} \sim N\left(\beta, \widehat{\text{Var}}(\hat{\beta})\right) \tag{19.7}$$

where the variance is estimated using a heteroskedasticity-robust approach. In practice, this variance estimator is computed using Eicker–Huber–White (HC3/EHW) robust standard errors[2].[3] This robust variance estimator (EHW) enables the construction of confidence intervals:

$$\hat{\beta} \pm z_{\alpha/2}\sqrt{\widehat{\text{Var}}(\hat{\beta})} \tag{19.8}$$

and facilitates hypothesis testing using the test statistic:

$$t = \frac{\hat{\beta}}{\sqrt{\widehat{\text{Var}}(\hat{\beta})}} \tag{19.9}$$

which follows an approximate standard normal distribution under the null hypothesis $H_0 : \beta = 0$.

19.1.3 Step-by-step DM1-LASSO and Simulation

Now, we will discuss the step-by-step procedure for DML1-LASSO and DML2-LASSO, which we believe will provide a deeper understanding of the details before proceeding to the simulation. Lets assume we have a dataset with $n = 10,000$ observations and $p = 100$ covariates, and also choose five fold for outer loop, and ten fold for inner loop.

Step-by-step Procedure for DML1-LASSO:

Step 1: K-Fold Splitting for Cross-Fitting (Outer Loop, K=5)

- **Partition the dataset into $K = 5$ disjoint folds** (each with $n = 10,000/5 = 2,000$ observations).
- For each **fold** k:
  - **Training data**(I_k^c): The **remaining 8,000** observations.
  - **Validation data**(I_k): The **2,000** left-out observations.

Step 2: Estimating the Nuisance Functions (on Training Data, I_k^c) We estimate two separate Lasso regressions in each training set:

(A) Predict Y (Outcome Model) (Lasso with Inner Cross-Validation, $K = 10$)

- **Partition the training set (8,000 obs) into $K = 10$ inner folds.**
- **For each inner fold j (Nested CV, $K = 10$):**
  - Train Lasso regression on **7,200 observations** (training within training); Validate on the left-out **800 observations.**
  - Choose the "best" λ that minimizes prediction error (e.g., MSE).
- **Train final Lasso model with the best λ on all 8,000 training observations.**
- Compute predictions for the **2,000 validation observations** using:

$$\hat{Y}_k(X) = \hat{\beta}'X$$

- This gives an estimate of the expected outcome for each individual without treatment effects (potential outcome in control state).

(B) Predict D (Propensity Score Model) (Lasso with Inner Cross-Validation, $K = 10$)

- Follow the **same nested cross-validation procedure** as for the outcome model, but now for predicting D:

$$\hat{p}_k(X) = P(D = 1|X)$$

- This helps control for selection bias in treatment assignment.

At the end of Step 3, we obtain: $\hat{Y}_k(X)$: The predicted outcome for the validation data. and $\hat{p}_k(X)$: The estimated probability of treatment for the validation data.

Step 3: Estimating $\hat{\gamma}_{0,k}$ (On Validation Data, I_k)

For each **outer fold k (on the 2,000 validation observations)**:

- Compute the residuals:

$$\tilde{Y} = Y - \hat{Y}_k(X), \quad \tilde{D} = D - \hat{p}_k(X)$$

- Estimate $\hat{\gamma}_{0,k}$ by running the residualized regression:

$$\tilde{Y} = \hat{\gamma}_{0,k}\tilde{D} + \text{error}$$

At the end of Step 3, we obtain **5 different ATE estimates** $\hat{\gamma}_{0,k}$ (one per fold).

Step 4: Aggregating the Estimates

- Compute the final **cross-fitted ATE** by averaging over the 5 folds:

$$\tilde{\gamma}_0 = \frac{1}{5}\sum_{k=1}^{5}\hat{\gamma}_{0,k}$$

19.1.4 Simulation for DML1

In this simulation we generate a dataset with an outcome, a treatment variable, and a set of covariates. The treatment is assigned based on a logistic model using the covariates, and the outcome is determined by both the treatment effect and the covariates, plus random noise. We then split the data into five folds for cross-fitting. For each fold, we use the remaining data to fit LASSO regressions that predict the outcome and the treatment probability. These predictions are used to calculate residuals, which serve as inputs for an OLS regression to estimate the treatment effect within that fold. Finally, we average the fold-specific estimates to obtain the overall average treatment effect and compute robust standard errors for inference.

```
# Load required libraries
library(glmnet)       # For Lasso regression
library(MASS)         # For generating multivariate normal data
library(caret)        # For cross-validation and data partitioning
library(estimatr)     # For robust standard errors

# Set seed for reproducibility
set.seed(123)

# === Step 1: Generate Simulated Data (Y, D, X) ===
N <- 10000  # Number of observations
p <- 10     # Number of covariates

# Generate covariates X ~ N(0, I)
X <- mvrnorm(N, mu = rep(0, p), Sigma = diag(p))

# Generate treatment variable D using a logistic model
D_prob <- plogis(X %*% runif(p, -1, 1))  # Compute propensity scores
D <- rbinom(N, 1, D_prob)  # Assign treatment based on propensity

# Generate outcome Y with treatment effect
gamma_true <- 2  # True treatment effect
beta <- runif(p, -1, 1)  # True coefficients for X
Y <- gamma_true * D + X %*% beta + rnorm(N)  # Outcome model

# === Step 2: Split Data into K=5 Folds ===
K <- 5
folds <- createFolds(1:N, k = K, list = TRUE)

# Initialize storage for residuals and ATE estimates
```

```
Y_resid <- rep(NA, N)
D_resid <- rep(NA, N)
ate_k <- rep(NA, K)

# Step 3: Cross-Fitting to Estimate Nuisance Functions and Compute ATE Per Fold
for (k in 1:K) {

  # Training and validation split
  train_idx <- unlist(folds[-k])  # Use all but k-th fold for training
  val_idx <- folds[[k]]  # Use k-th fold for validation

  # Training Data
  X_train <- X[train_idx, ]
  Y_train <- Y[train_idx]
  D_train <- D[train_idx]

  # Validation Data
  X_val <- X[val_idx, ]

  # ---- Step 3A: Fit Lasso for Outcome Model (Y ~ X) ----
  # Cross-validation to choose lambda
  lasso_y <- cv.glmnet(X_train, Y_train, alpha = 1, nfolds = 10)
  # Predict on validation set
  Y_hat_val <- predict(lasso_y, X_val, s = "lambda.min")

  # ---- Step 3B: Fit Lasso for Propensity Model (D ~ X) ----
  # Logistic Lasso
  lasso_d <- cv.glmnet(X_train, D_train, alpha = 1, family = "binomial", nfolds=10)
  # Predict propensity
  D_hat_val <- predict(lasso_d, X_val, s = "lambda.min", type = "response")
# alpha = 0 for Ridge, alpha = 1 for Lasso, alpha = 0.5 for Elastic Net
  # ---- Compute Residuals ----
  Y_resid[val_idx] <- Y[val_idx] - Y_hat_val
  D_resid[val_idx] <- D[val_idx] - D_hat_val

  # ---- Step 3C: Estimate ATE for This Fold Using OLS ----
  ate_k[k] <- coef(lm(Y_resid[val_idx] ~ D_resid[val_idx] - 1))[1]  # No intercept
}

# === Step 4: Compute Final ATE by Averaging Over Folds ===
ate_dml1 <- mean(ate_k)

# === Step 5: Compute Standard Errors Using HC3 Robust SEs ===
dml1_model <- lm_robust(Y_resid ~ D_resid - 1, se_type = "HC3")  #Uses all residuals

# Print results
cat("Estimated ATE using DML1:", ate_dml1, "\n")

## Estimated ATE using DML1: 1.994326
```

```
cat("Standard Error (HC3):", dml1_model$std.error[1], "\n")
```

```
## Standard Error (HC3): 0.023927
```

```
cat("95% CI:", dml1_model$conf.low[1], "to", dml1_model$conf.high[1], "\n")
```

```
## 95% CI: 1.947591 to 2.041395
```

19.1.5 Step-by-step Procedure for DML2-LASSO

Step 1, **Step 2** are same.

Step 3: Compute Residuals and Estimate $\tilde{\theta}_0$ (DML1 vs. DML2 Difference)

For each fold k (on the 2,000 validation observations):

- Compute residualized Y (**regression-adjusted outcome**):

$$\tilde{Y}_i = Y_i - \hat{Y}_k(X_i), \quad \text{for } i \in I_k$$

- Compute residualized D (**regression-adjusted treatment**):

$$\tilde{D}_i = D_i - \hat{p}_k(X_i), \quad \text{for } i \in I_k$$

At the end of Step 3, we have 10,000 residuals for Y and D, covering the full dataset.

Step 4: Estimate the Final ATE

(Keep in mind **DML1:** Estimate ATE separately for each fold ($\hat{\gamma}_{0,k}$), then **average over the 5 folds.**)

- **DML2:** Estimate ATE in one step, using the **full set of 10,000 residuals**:

$$\tilde{Y} = \tilde{\theta}_0 \tilde{D} + \text{error}$$

In both methods, the final ATE estimate comes from regressing residualized Y on residualized D, but DML1 does it per fold before averaging, while DML2 does it in one step using all residuals at once.

19.1.6 Simulation for DML2

This simulation also creates a dataset with an outcome, treatment, and covariates. Like before, the treatment is generated using a logistic model and the outcome is a function of both treatment and covariates with added noise. Here, instead of tuning the Lasso model within each fold, we determine an optimal lambda with cross-validation (`cv.glmnet` and `lambda.min` which we explained in detailed in lasso chapter) from the full training data for both the outcome and treatment models. The data is then divided into five folds, and in each fold the fixed lambda is used to predict the outcome and treatment probabilities. Residuals are computed and combined to estimate the treatment effect with robust standard errors. The results from DML2 yield an ATE that is very close to the one from DML1, and DML2 is generally recommended for its simplicity and consistency.

```
# Load required libraries
library(glmnet)      # For Lasso regression
library(MASS)        # For generating multivariate normal data
library(caret)       # For cross-validation
library(estimatr)    # For HC3 robust SEs (lm_robust)
set.seed(123) # Set seed for reproducibility

# === Step 1: Generate Simulated Data (Y, D, X) ===
N <- 10000  # Number of observations
p <- 10     # Number of covariates

# Generate covariates X ~ N(0, I)
X <- mvrnorm(N, mu = rep(0, p), Sigma = diag(p))

# Generate treatment variable D using a logistic model
D_prob <- plogis(X %*% runif(p, -1, 1))  # Compute propensity scores
D <- rbinom(N, 1, D_prob)  # Assign treatment based on propensity

# Generate outcome Y with treatment effect
gamma_true <- 2  # True treatment effect
beta <- runif(p, -1, 1)  # True coefficients for X
Y <- gamma_true * D + X %*% beta + rnorm(N)  # Outcome model

# === Step 2: Define Fold Assignments (Standardized) ===
K <- 5
folds <- createFolds(1:N, k = K, list = TRUE)

# === Step 3: Compute a Fixed `lambda.min` Using Full Data ===
cv_lasso_y <- cv.glmnet(X, Y, family = "gaussian", alpha = 1, nfolds = 10)  #for Y
lambda_fixed_y <- cv_lasso_y$lambda.min  # Store optimal lambda for outcome model

cv_lasso_d <- cv.glmnet(X, D, family = "binomial", alpha = 1, nfolds = 10)  #for D
lambda_fixed_d <- cv_lasso_d$lambda.min  # Store optimal lambda for propensity model

# === Step 4: Cross-Fitting with Fixed Lambda ===
Y_resid <- rep(NA, N)
D_resid <- rep(NA, N)

for (k in 1:K) {

  # Training and validation split
  train_idx <- unlist(folds[-k])  # Use all but k-th fold for training
  val_idx <- folds[[k]]  # Use k-th fold for validation

  # Training Data
  X_train <- X[train_idx, ]
  Y_train <- Y[train_idx]
  D_train <- D[train_idx]
```

```
  # Validation Data
  X_val <- X[val_idx, ]

  # ---- Fit Lasso for Outcome Model (Y ~ X) with Fixed Lambda ----
  lasso_y <- glmnet(X_train, Y_train, family = "gaussian", alpha = 1,
                    lambda = lambda_fixed_y)
  Y_hat_val <- predict(lasso_y, X_val)

  # ---- Fit Lasso for Propensity Model (D ~ X) with Fixed Lambda ----
  lasso_d <- glmnet(X_train, D_train, family = "binomial", alpha = 1,
                    lambda = lambda_fixed_d)
  D_hat_val <- predict(lasso_d, X_val, type = "response")

  # ---- Compute Residuals ----
  Y_resid[val_idx] <- Y[val_idx] - Y_hat_val
  D_resid[val_idx] <- D[val_idx] - D_hat_val
}

# === Step 5: Estimate ATE Using HC3/EHW robust SE ===
dml2_model_robust <- lm_robust(Y_resid ~ D_resid, se_type = "HC3")

# === Step 6: Print Results ===
cat("\nEstimated ATE using DML2:", coef(dml2_model_robust)[2], "\n")
```

```
##
## Estimated ATE using DML2: 1.99448
```

```
cat("Standard Error (lm_robust):", dml2_model_robust$std.error[2], "\n")
```

```
## Standard Error (lm_robust): 0.02392815
```

```
cat("95% CI (lm_robust):", dml2_model_robust$conf.low[2], "to",
    dml2_model_robust$conf.high[2], "\n")
```

```
## 95% CI (lm_robust): 1.947576 to 2.041384
```

19.1.7 DML2-LASSO with the Plug-In Penalty

Above, we implemented DML1 and DML2 step by step using cross-validation (CV) to select the tuning parameter λ in Lasso. Cross-validation remains a widely used approach, particularly in prediction-focused machine learning applications. It is computationally feasible even with large datasets, and we fully understand its implementation, having covered all details earlier in penalized regressions-lasso chapter.

However, in causal inference, our goal is not merely to achieve high predictive accuracy but to obtain valid statistical inference (i.e., correct confidence intervals and standard errors). The cross-validation approach is not necessarily always optimal for inference because it minimizes prediction error, whereas in our setting, we need a tuning parameter that ensures valid post-selection inference in high-dimensional models.

The **plug-in penalty method** provides a theory-driven alternative to cross-validation for selecting λ. Instead of relying on data-driven minimization of prediction error, the plug-in approach selects λ based on probability bounds that control the impact of noise in high-dimensional settings. This ensures that the resulting Lasso estimator achieves the right level of sparsity while preserving valid inference properties.

The plug-in method is default in Stata's `lasso` with `selection(plugin)` and R's `hdm::rlasso()` function. Thus, readers should be familiar with how these packages select the tuning parameter when using Lasso for causal inference.

How the Plug-In Method Works

The plug-in approach follows the theoretical framework proposed by Belloni, Chernozhukov, and Hansen (2012, 2014, REStud) and ensures that λ satisfies high-dimensional probability bounds, limiting the influence of noise. Remember, in standard Lasso, we solve: $\min_\beta \frac{1}{2n}\|Y - X\beta\|_2^2 + \lambda\|\beta\|_1$. The penalty hyper-parameter λ balances fit and sparsity. The challenge is choosing λ so that you retain the true signals while shrinking or eliminating noise variables. CV chooses λ by minimizing out-of-sample prediction error, while the plug-in method selects λ based on a theoretical noise bound. In a high-dimensional model (where the number of predictors p may be large relative to n), results from high-dimensional probability tell us that the maximum of the empirical correlations between the predictors and the noise behaves roughly like $\sqrt{2\log(p)/n}$. To control for this, Belloni, Chernozhukov, and Hansen (equation 2.13, 2014, REStud) sets the penalty level λ on the order of:

$$\lambda \asymp c\,\hat{\sigma}\sqrt{n}\,\Phi^{-1}\left(1 - \frac{\gamma}{2p}\right) \tag{19.10}$$

where $\hat{\sigma}$ is an estimator of the noise standard deviation; $\Phi^{-1}(\cdot)$ is the inverse normal CDF (finds a critical value from the standard normal distribution), ensuring the probability bound holds; γ is a small significance level (often chosen based on theory); c is a constant (default 1.1) to ensure the bound holds with high probability; n is the number of observations, to control sample size effects; p is the number of covariates, to account for multiple testing in high dimensions; γ is a small significance level, controlling false positives (default as 0.1).

Unlike cross-validation, which minimizes prediction error, this theoretical approach targets the property of valid inference after variable selection. This is crucial for applications where we care not only about prediction but also about the correctness of subsequent statistical tests and confidence intervals. Plug-in method also avoids the computational cost of cross-validation, making it ideal for large datasets. Guarantees valid inference by ensuring λ controls the probability of false selection. Matches the behavior of Stata (`selection(plugin)`) and R (`hdm::rlasso()`), aligning with widely used implementations.

After the initial Lasso selection using the theory-driven λ, `rlasso` often performs a "post-Lasso" refit. This means that once the relevant variables are selected, the model is refitted using OLS on the selected variables, reducing the bias introduced by Lasso's shrinkage. In some cases, an additional adjustment is made by incorporating penalty loadings, which introduce covariate-specific penalization based on heteroskedasticity and non-Gaussian error structures.[4] The inclusion of penalty loadings refines variable selection and can be iteratively updated in an iterated Lasso procedure.[5] However, we do not discuss penalty loadings in detail here, as they go beyond the scope of this section. Readers interested in these refinements can refer to equation 2.13 and Appendix Algorithm 1 of the Belloni, Chernozhukov, and Hansen (2014, REStud) paper.

19.1.8 Simulation for DML2 with Plug-In Penalty

Here we generated the same data as above (but normally we upload our data) and applied the DML2 method using the plug-in penalty method for selecting the tuning parameter λ in Lasso. We called `hrm` package and used the `hdm::rlasso()` function, which implements the plug-in penalty method by default. We Fit Lasso for Outcome Model (Y ~ X) , and Fit Lasso for Treatment Model (D ~ X). Here instead of cross-fitting (k=10 fold); using data (all raw data but Kth fold) package calculated λ ; then used post lasso as last step, then obtain cross-fitted residuals using k-th fold. It does the same process using all 5 fold. Then combine all residuals, and implemented regular OLS with cross-fitted residuals, and EHW robust standard errors.

We generate the same data as in the previous section (though normally, we would upload our dataset) and apply the DML2 method using the plug-in penalty for selecting the tuning parameter λ in Lasso. We call the `hdm` package and use the `hdm::rlasso()` function, which implements the plug-in penalty method by default. This approach ensures valid post-selection inference without relying on CV.

Step 1: same as DML2 with CV.
- The dataset is split into $K = 5$ **folds**, each with $2,000$ **validation samples** and $8,000$ **training samples**.
- Training data: I_k^c **(all but the K-th fold)** is used for estimating nuisance functions.

Step 2: Compute the plug-in penalty λ (on training data, I_k^c)**
- The **plug-in formula** for λ is applied using the training data. This calculation **does not depend on Lasso fitting**. It is computed directly from the **training sample Fit Lasso for Outcome Model ($Y \sim X$) Using Plug-In λ**
- Use **only the training data** I_k^c. - Apply **Lasso regression with the precomputed plug-in** λ.computed using only the training data - Select relevant variables by shrinking some coefficients to zero.

Fit Lasso for Treatment Model ($D \sim X$) Using Plug-In λ
- Again, use the **same precomputed** λ (computed in Step 2). - Select relevant variables and estimate $\hat{D}$.

Post-Lasso Refitting (on Training Data, I_k^c)
- After Lasso selects variables, refit **OLS** using only the selected variables (i.e., calculate OLS coefficients for selected variables).

Step 3: Predict on Validation Data (I_k)
- Use the **post-lasso refitted model** (OLS coefficients for selected variables) to predict $\hat{Y}$ and $\hat{D}$ on the $2,000$ **validation samples**. Compute cross-fitted residuals:

$$\tilde{Y} = Y - \hat{Y}, \quad \tilde{D} = D - \hat{D}$$

At the end of Step 3, we have 10,000 residuals for Y and D, covering the full dataset.

Step 4: same as DML2 with cross-validation.

- A regular OLS regression is performed using the cross-fitted residuals.

- Standard errors are computed using EHW robust standard errors to account for heteroskedasticity.

```
# RUN until Step 2 from DML1-CV simulation above to and then run this code
library(hdm)          # For direct rlasso implementation
# === Step 2: Define Fold Assignments (Cross-Fitting) ===
K <- 5
folds <- createFolds(1:N, k = K, list = TRUE)

# === Step 3: Compute Residuals using `hdm::rlasso()` ===
Y_resid <- rep(NA, N)
D_resid <- rep(NA, N)

for (k in 1:K) {
    # Training and validation split
    train_idx <- unlist(folds[-k])  # Use all but k-th fold for training
    val_idx <- folds[[k]]  # Use k-th fold for validation

    # Training Data
    X_train <- X[train_idx, ]
    Y_train <- Y[train_idx]
    D_train <- D[train_idx]

    # Validation Data
    X_val <- X[val_idx, ]

    # ---- Fit Lasso for Outcome Model (Y ~ X) using `hdm::rlasso()` ----
    lasso_y <- rlasso(X_train, Y_train)
    Y_hat_val <- predict(lasso_y, X_val)

    # ---- Fit Lasso for Treatment Model (D ~ X) using `hdm::rlasso()` ----
    lasso_d <- rlasso(X_train, D_train)
    D_hat_val <- predict(lasso_d, X_val)

    # ---- Compute Residuals ----
    Y_resid[val_idx] <- Y[val_idx] - Y_hat_val
    D_resid[val_idx] <- D[val_idx] - D_hat_val
}

# === Step 4: Estimate ATE Using HC3/EHW robust SE ===
dml2_model_robust <- lm_robust(Y_resid ~ D_resid, se_type = "HC3")

# === Step 5: Print Final Results ===
cat("\nEstimated ATE using DML2 (`hdm::rlasso`):", coef(dml2_model_robust)[2], "\n")
```

```
##
## Estimated ATE using DML2 (`hdm::rlasso`): 1.99418
```

```
cat("Standard Error (lm_robust):", dml2_model_robust$std.error[2], "\n")
```

```
## Standard Error (lm_robust): 0.02366582
```

```
cat("95% CI (lm_robust):", dml2_model_robust$conf.low[2], "to",
    dml2_model_robust$conf.high[2], "\n")
```

```
## 95% CI (lm_robust): 1.94779 to 2.04057
```

19.1.9 Sparsity in DML-LASSO

Sparsity implies that only a few covariates in X are truly influential in predicting the outcome Y or the treatment variable D. In DML, the core assumption is one of "approximate sparsity," meaning that the true underlying functions $f(X)$ and $g(X)$ can be *well-approximated* by a sparse linear model. In Lasso $f(X)$ represents the function that links the covariates X to the outcome Y, while $g(X)$ captures the relationship between X and the treatment variable D, often reflecting the propensity score model. In practical terms, even if a dataset contains many covariates, only a limited number drive the key relationships, which is crucial for the Lasso method. Lasso effectively identifies these important predictors while discarding less relevant ones, thereby preventing the model from over-shrinking the influence of significant variables.

Text in endnote[6] explains this concept mathematically by showing that the best sparse approximation of $f(X)$ yields an error that diminishes as the number of effective covariates grows more slowly than the sample size N.

When the sparsity condition holds, Lasso serves as a valid estimator for the nuisance functions $\hat{Y}(X)$ and $\hat{p}(X)$. This results in minimal bias from omitted variables and maintains the Neyman orthogonality condition, which ensures that small inaccuracies in the estimation of these nuisance functions do not lead to substantial bias in the final estimate of $\hat{\delta}$. On the other hand, while Lasso is effective at identifying important predictors, it can sometimes shrink the coefficients of truly important variables too much, especially in the presence of highly correlated predictors. This can lead to underestimation of the true effects of these variables. Nonetheless, Lasso remains valuable in practice for its ability to perform variable selection.

This discussion underpins the use of three Lasso-based approaches: double Lasso – Partialling-Out, Double Lasso – Double Selection, and Double Machine Learning (DML1 & DML2). The sparsity assumption is vital because it allows Lasso to approximate the nuisance functions accurately while keeping selection bias under control. Following Bühlmann and Van De Geer (2011) and Belloni, Chernozhukov, and Hansen (2014), the sparsity condition for the Double Lasso-Partialling-Out and Double Selection- methods is typically framed as requiring that $\frac{s}{\sqrt{N}/\ln p}$ remains small, ensuring that the model correctly identifies important variables without overfitting.

In contrast, Chernozhukov et al. (2018) demonstrate that the use of cross-fitting in DML1 and DML2 relaxes the sparsity condition to $\frac{s}{N/\ln p}$. Because N is much larger than $\sqrt{N}$, cross-fitting permits a greater number of covariates while still yielding valid inference. For instance, with $s = 4$, $p = 10$, and $N = 10,000$, the sparsity condition is approximately 0.092 for high-dimensional Lasso but drops to 0.000921 for cross-fitting DML, meaning that cross-fitting permits a larger number of covariates while maintaining valid inference. As a result, DML2 is often recommended when there is concern about meeting the sparsity requirement.

```
N <- 10000
p <- 10
s <- 4  # Example: Lasso selects 4 variables

sparsity_standard <- s / (sqrt(N) / log(p)) # Double Lasso - Partialling-Out
sparsity_crossfit <- s / (N / log(p)) # Double Machine Learning (DML1, DML2)

cat("Sparsity Condition (Standard):", sparsity_standard, "\n")
```

```
## Sparsity Condition (Standard): 0.0921034
```

```
cat("Sparsity Condition (Cross-Fitting):", sparsity_crossfit, "\n")
```

```
## Sparsity Condition (Cross-Fitting): 0.000921034
```

19.2 DML2 -Decision Trees

In Chapter 14, we introduced decision trees as a flexible nonparametric approach for prediction. Here, we extend their use to causal inference by implementing a DML approach with cross-fitting.

This simulation estimates the ATE using decision trees. The dataset consists of an outcome variable Y, a treatment variable D, and covariates X. The treatment assignment follows a logistic model, and the outcome is a linear function of the treatment and covariates with added noise. We use decision trees to estimate two models: the outcome model $Y \sim X$ and the treatment model $D \sim X$. The treatment model is a classification tree predicting the probability of receiving treatment.

In the decision tree regression for $Y \sim X$, the `rpart` package constructs a single tree that recursively partitions the feature space into smaller regions. Unlike linear models, which assume a specific functional form between Y and X, decision trees adaptively split the data based on feature interactions and nonlinearities. The model chooses the optimal split at each node by minimizing the residual sum of squares (RSS), ensuring that the resulting partitions maximize within-node homogeneity.

By default, the `rpart` package in R uses complexity parameter (`cp = 0.01`) to prune unnecessary splits and prevent overfitting. Each split is only retained if it reduces RSS by at least 1%. Additionally, the minimum bucket size (`minbucket = 7` for regression) ensures that terminal nodes contain at least 7 observations, limiting excessive branching. The tree is grown deeply and then pruned back based on cost-complexity criteria. The final prediction for Y is obtained by assigning each observation to the mean outcome within its terminal node:

$$\hat{Y}(X) = \frac{1}{N_r} \sum_{i \in r} Y_i \tag{19.11}$$

where N_r is the number of observations in region r, and the sum is taken over all observations in that region.

For the treatment model $D \sim X$, we apply a classification decision tree using the Gini impurity criterion to determine optimal splits. Unlike regression trees, which minimize RSS, classification trees split on the feature that maximally reduces impurity across child nodes. The treatment probability (propensity score) is then estimated by calculating the proportion of treated observations within each terminal node:

$$\hat{P}(D = 1 \mid X) = \frac{\sum_{i \in r} \mathbf{I}(D_i = 1)}{N_r} \tag{19.12}$$

where $\mathbf{I}(D_i = 1)$ is an indicator function for treated individuals in region r, and N_r is the number of observations in that region.

To prevent excessive overfitting, the tree is constrained using `minbucket = 10` and `cp = 0.001`, ensuring that each leaf has at least 10 observations and that additional splits must improve model fit significantly. Unlike random forests, which smooth probability estimates by averaging over multiple trees, decision trees estimate treatment probabilities in a stepwise manner, leading to sharp discontinuities between regions.

To reduce overfitting, we apply cross-fitting by splitting the data into K-folds. In each fold, decision trees are trained on a subset of the data, and predictions are made on the held-out fold as we discussed above. The residuals from these models (i.e., differences between actual and predicted values) are then used in a final regression to estimate the treatment effect, controlling for confounding.

This method improves robustness by reducing overfitting and mitigating bias from direct regression of Y on D. The final step uses heteroskedasticity-robust standard errors to provide valid inference. By using decision trees, this approach flexibly models nonlinear relationships but may suffer from high variance and sensitivity to small changes in data, especially compared to ensemble methods like random forests.

```r
# Load required libraries
library(MASS)       # For generating multivariate normal data
library(caret)      # For cross-validation
library(estimatr)   # For HC3 robust SEs (lm_robust)
library(rpart)      # For decision trees
set.seed(123)       # For reproducibility

# === Step 1: Generate Simulated Data (Y, D, X) ===
N <- 10000   # Number of observations
p <- 10      # Number of covariates

# Generate covariates X ~ N(0, I)
X <- mvrnorm(N, mu = rep(0, p), Sigma = diag(p))

# Generate treatment variable D using a logistic model
D_prob <- plogis(X %*% runif(p, -1, 1))
D <- rbinom(N, 1, D_prob)

# Generate outcome Y with treatment effect
gamma_true <- 2
beta <- runif(p, -1, 1)
Y <- gamma_true * D + X %*% beta + rnorm(N)
```

```
# === Step 2: Define Fold Assignments (Cross-Fitting) ===
K <- 5
folds <- createFolds(1:N, k = K, list = TRUE)

# === Step 3: Compute Residuals using decision trees (rpart) ===
Y_resid <- rep(NA, N)
D_resid <- rep(NA, N)

for (k in 1:K) {
  # Define training and validation indices
  train_idx <- unlist(folds[-k])
  val_idx <- folds[[k]]

  # Prepare validation data as a data frame with proper column names
  val_data <- as.data.frame(X[val_idx, ])

  # Outcome model: Fit a regression tree for Y ~ X
  train_data_y <- cbind(data.frame(Y = Y[train_idx]), as.data.frame(X[train_idx, ]))
  outcome_tree <- rpart(as.formula("Y ~ ."), data = train_data_y,
                        minbucket = 10, cp = 0.001)
  Y_hat_val <- predict(outcome_tree, newdata = val_data)

  # Treatment model: Fit a classification tree for D ~ X
  # Convert D to factor to use classification mode
  train_data_d <- cbind(data.frame(D = factor(D[train_idx])),
                        as.data.frame(X[train_idx, ]))
  treatment_tree <- rpart(as.formula("D ~ ."), data = train_data_d,
                          method = "class", minbucket = 10, cp = 0.001)
  # Obtain predicted probabilities for class "1"
  pred_probs <- predict(treatment_tree, newdata = val_data, type = "prob")
  D_hat_val <- pred_probs[, "1"]

  # Compute residuals
  Y_resid[val_idx] <- Y[val_idx] - Y_hat_val
  D_resid[val_idx] <- D[val_idx] - D_hat_val
}

# === Step 4: Estimate ATE Using HC3/EHW robust SE ===
dml2_model_robust <- lm_robust(Y_resid ~ D_resid, se_type = "HC3")

# === Step 5: Print Final Results ===
cat("\nEstimated ATE using DML2 (decision trees):", coef(dml2_model_robust)[2], "\n")
```

```
##
## Estimated ATE using DML2 (decision trees): 1.620865
```

```
cat("Standard Error (lm_robust):", dml2_model_robust$std.error[2], "\n")
```

```
## Standard Error (lm_robust): 0.02828379
```

```
cat("95% CI (lm_robust):", dml2_model_robust$conf.low[2], "to",
    dml2_model_robust$conf.high[2], "\n")
```

```
## 95% CI (lm_robust): 1.565423 to 1.676307
```

Decision trees likely gave the worst results in the DML framework due to several well-known limitations of single decision trees, especially in high-dimensional or noisy data settings:

1. *High variance and instability*:
   - Decision trees tend to overfit the training data because they create very specific splits based on the given sample.
   - Unlike ensemble methods (e.g., random forests), a single tree lacks regularization, making it sensitive to small changes in the training data.
   - This instability can lead to large errors in the residualized treatment ($\hat{D} - D$) and outcome ($\hat{Y} - Y$) models, which propagate to the final treatment effect estimate.
2. *Limited predictive power for complex relationships*:
   - Decision trees partition the data into discrete regions, making them inefficient at capturing smooth, continuous relationships between variables.
   - If Y and D are influenced by interactions or nonlinear effects that are not well captured by splits, the residualization step will be poor, leading to biased or inconsistent treatment effect estimates.
3. *Overfitting to the training data*:
   - Trees tend to perfectly fit training data when grown deep, but this does not generalize well to new samples.
   - This means the estimated functions $\hat{Y}(X)$ and $\hat{D}(X)$ may be highly inaccurate when evaluated on held-out folds, worsening cross-fitting results.
4. *Poor handling of continuous treatment probabilities*:
   - In the treatment model $D \sim X$, a classification tree is used to predict propensity scores $P(D = 1 \mid X)$.
   - Unlike logistic regression or random forests, decision trees only create stepwise splits, which can result in poorly estimated probabilities.
   - This leads to high residual variance, reducing the effectiveness of the second-stage regression.

Random forests mitigate many of these issues by averaging multiple decision trees trained on bootstrapped samples and random feature subsets. This reduces variance, improves generalization, and provides smoother propensity score estimates for treatment assignment. Since DML relies on accurate nuisance function estimates ($\hat{Y}(X)$ and $\hat{D}(X)$), a poorly performing base model (like a single decision tree) leads to biased or highly variable ATE estimates.

In summary, decision trees struggle due to overfitting, poor smoothness, and high variance, which weakens the residualization step in DML, making them less reliable for causal inference.

19.3 DML2: Random Forest

In Chapter 14, we covered random forests for prediction. Here, we extend their use to causal inference by implementing a DML approach with cross-fitting.

This simulation estimates the ATE using random forests. The dataset consists of an outcome variable Y, a treatment variable D, and covariates X. The treatment assignment follows a logistic model, and the outcome is a linear function of the treatment and covariates with added noise. We use random forests to estimate two models: the outcome model $Y \sim X$ and the treatment model $D \sim X$. The treatment model is a regression forest predicting the probability of receiving treatment.

In the random forest regression for $Y \sim X$, the `randomForest` package constructs an ensemble of decision trees to predict Y based on the covariates X. Unlike a single decision tree, which partitions the feature space into non-overlapping regions, a random forest builds multiple trees on different bootstrap samples of the data and averages their predictions. This process helps reduce overfitting and improves generalization. Each individual tree is grown using a subset of randomly selected features at each split rather than considering all features, which introduces decorrelation between trees and further improves robustness.

By default, the `randomForest` package in R sets the number of trees to 500 (`ntree = 500`) and selects $\lfloor \sqrt{p} \rfloor$ features for each split when performing regression (`mtry = floor(sqrt(p))`). The trees are grown deep, with splits determined by minimizing the residual sum of squares (RSS) at each node until a stopping criterion is met. The default minimum node size (`nodesize = 5` for regression) controls the depth of the trees, preventing overfitting by requiring at least 5 observations in a terminal node before stopping further splits.

Increasing the number of trees (`ntree`) generally reduces variance, leading to more stable predictions. While 500 trees is often sufficient, increasing this to 1000 or more can help in high-dimensional settings or when working with noisy data. However, returns diminish beyond a certain point, as additional trees provide only marginal improvements in performance but increase computational cost. A practical approach is to start with 500 trees, monitor the out-of-bag (OOB) error, and increase `ntree` until the error stabilizes.

The minimum node size (`nodesize`) controls the granularity of splits and influences the bias–variance trade-off. The default value of 5 results in deeper trees that capture fine-grained patterns but may overfit. Setting `nodesize = 10` or higher prunes the trees earlier, reducing overfitting by forcing more generalization. If the data is noisy, increasing `nodesize` (e.g., 10-20) can improve robustness by smoothing predictions. A common rule of thumb is to set `nodesize` to approximately 1% of total observations, though the optimal choice often depends on cross-validation. In causal inference, larger node sizes are preferred to avoid overfitting the treatment and outcome models, ensuring more reliable effect estimates.

Once all trees are grown, the final prediction for Y is obtained by averaging the predictions from all trees:

$$\hat{Y}(X) = \frac{1}{B} \sum_{b=1}^{B} \hat{f}^{(b)}(X) \tag{19.13}$$

where B is the number of trees, and $\hat{f}^{(b)}(X)$ is the prediction from the b-th tree.

The same procedure is applied to estimate $D \sim X$, but since D is binary (treatment assignment), classification random forest is used instead. The splitting criterion in classification trees is different—by default, it minimizes the Gini impurity rather than RSS. Instead of averaging numerical predictions, classification forests output class probabilities, with the predicted probability for treatment ($D = 1$) given by:

$$\hat{P}(D = 1 \mid X) = \frac{1}{B} \sum_{b=1}^{B} \hat{P}^{(b)}(D = 1 \mid X) \tag{19.14}$$

where each tree $\hat{P}^{(b)}$ provides a probability estimate based on the proportion of $D = 1$ observations in each terminal node. This probability serves as the estimated propensity score, which is used in the residualization step for double machine learning.

To reduce overfitting, we apply cross-fitting by splitting the data into K-folds. In each fold, random forests are trained on a subset of the data, and predictions are made on the held-out fold as we discussed above. The residuals from these models (i.e., differences between actual and predicted values) are then used in a final regression to estimate the treatment effect, controlling for confounding.

This method improves robustness by reducing overfitting and mitigating bias from direct regression of Y on D. The final step uses heteroskedasticity-robust standard errors to provide valid inference. By using random forests, this approach flexibly captures complex nonlinear relationships while ensuring that selection bias from observed covariates is accounted for.

```
# Load required libraries
library(MASS)          # For generating multivariate normal data
library(caret)         # For cross-validation
library(estimatr)      # For HC3 robust SEs (lm_robust)
library(randomForest)  # For random forest modeling
set.seed(123)          # Set seed for reproducibility

# === Step 1: Generate Simulated Data (Y, D, X) ===
N <- 10000   # Number of observations
p <- 10      # Number of covariates

# Generate covariates X ~ N(0, I)
X <- mvrnorm(N, mu = rep(0, p), Sigma = diag(p))

# Generate treatment variable D using a logistic model
D_prob <- plogis(X %*% runif(p, -1, 1))  # Compute propensity scores
D <- rbinom(N, 1, D_prob)                # Assign treatment based on propensity

# Generate outcome Y with treatment effect
gamma_true <- 2                         # True treatment effect
beta <- runif(p, -1, 1)                  # True coefficients for X
Y <- gamma_true * D + X %*% beta + rnorm(N)  # Outcome model

# === Step 2: Define Fold Assignments (Cross-Fitting) ===
K <- 5
folds <- createFolds(1:N, k = K, list = TRUE)
```

```r
# === Step 3: Compute Residuals using random forest ===
Y_resid <- rep(NA, N)
D_resid <- rep(NA, N)

for (k in 1:K) {
  # Define training and validation indices
  train_idx <- unlist(folds[-k])
  val_idx <- folds[[k]]

  # Prepare data for outcome model
  train_data_y <- data.frame(Y = Y[train_idx], as.data.frame(X[train_idx, ]))
  val_data <- as.data.frame(X[val_idx, ])

  # Fit random forest for outcome (Y ~ X)
  rf_y <- randomForest(Y ~ ., data = train_data_y, ntree = 1000, nodesize = 10)
  Y_hat_val <- predict(rf_y, newdata = val_data)

  # Prepare data for treatment model
  train_data_d <- data.frame(D = D[train_idx], as.data.frame(X[train_idx, ]))

  # Fit random forest for treatment (D ~ X)
  rf_d <- randomForest(D ~ ., data = train_data_d, ntree = 1000, nodesize = 10)
  D_hat_val <- predict(rf_d, newdata = val_data)

  # Compute residuals
  Y_resid[val_idx] <- Y[val_idx] - Y_hat_val
  D_resid[val_idx] <- D[val_idx] - D_hat_val
}

# === Step 4: Estimate ATE Using HC3/EHW robust SE ===
dml2_model_robust <- lm_robust(Y_resid ~ D_resid, se_type = "HC3")

# === Step 5: Print Final Results ===
cat("\nEstimated ATE using DML2 (random forest):", coef(dml2_model_robust)[2], "\n")
```

```
##
## Estimated ATE using DML2 (random forest): 1.950607
```

```r
cat("Standard Error (lm_robust):", dml2_model_robust$std.error[2], "\n")
```

```
## Standard Error (lm_robust): 0.0246481
```

```r
cat("95% CI (lm_robust):", dml2_model_robust$conf.low[2], "to",
    dml2_model_robust$conf.high[2], "\n")
```

```
## 95% CI (lm_robust): 1.902292 to 1.998922
```

19.4 Technical Notes: From FWL to DML

19.4.1 Frisch–Waugh–Lovell Theorem

The Frisch–Waugh–Lovell (FWL) theorem explains why ordinary least squares (OLS) regression adjustment effectively isolates the treatment effect by controlling for confounders. It shows that in a multiple regression, you can isolate the impact of one variable by removing the effects of the other variables from both the dependent variable and the variable of interest. This process—known as partialing out—consists of regressing the variable of interest on the other predictors, then using those residuals in a second regression with the dependent variable. This yields the same coefficient as a full multiple regression. In essence, orthogonalization—the process of transforming variables so that they do not share any common variation with the other predictors—allows you to clearly see the unique contribution of each variable in the model.

For instance, in

$$y = X_1\beta_1 + X_2\beta_2 + \varepsilon \tag{19.15}$$

the FWL theorem says β_1 can be obtained by first regressing both y and X_1 on X_2 to get residuals orthogonal to X_2. If

$$M_{X_2} = I - X_2(X_2^\top X_2)^{-1}X_2^\top \tag{19.16}$$

is the residual-maker matrix, then

$$\tilde{y} = M_{X_2}y, \quad \tilde{X}_1 = M_{X_2}X_1 \tag{19.17}$$

Regressing $\tilde{y}$ on $\tilde{X}_1$ alone provides the same β_1 as the full model. In double machine learning, a similar approach is used to create an orthogonal score by residualizing both the outcome and the treatment with respect to controls, thereby removing confounder effects.

Turning to causal or treatment effects, consider the OLS model:

$$Y_i = \alpha + \delta D_i + X_i^\top \beta + \varepsilon_i \tag{19.18}$$

where Y_i is the outcome, D_i is the treatment, X_i represents observed confounders, and ε_i captures unobserved factors. Under

$$E[\varepsilon_i \mid D_i, X_i] = E[\varepsilon_i \mid X_i] \tag{19.19}$$

OLS provides a consistent estimate of the average treatment effect (ATE), δ. The FWL theorem offers an equivalent way to compute δ by residualizing both Y_i and D_i on X_i, then regressing the outcome residuals on the treatment residuals. This makes clear that once X_i is accounted for, any remaining variation in D_i should be independent of unobserved determinants of Y_i, mimicking a randomized experiment conditional on X_i.

However, if there are unobserved confounders correlated with both treatment and outcome, simple regression adjustment cannot eliminate all bias, making methods like instrumental variables or difference-in-differences necessary. Also note that both FWL and OLS rely on standard assumptions (linearity, exogeneity, etc.): violations can lead to biased or inefficient estimates.

As we discussed before, in the direct OLS approach, under the assumption $E[\varepsilon_i \mid D_i, X_i] = E[\varepsilon_i \mid X_i]$, OLS provides an unbiased estimate of δ. Standard practice is to use robust (Eicker-Huber-White) standard errors to account for potential heteroskedasticity.

In the FWL two-step approach, we first residualize Y_i and D_i with respect to X_i, then regress these residuals. The new error term reflects only variation in Y_i not explained by X_i. Because D_i is also residualized against X_i, the new error term (from residualized regression) is uncorrelated with the residualized treatment, and robust standard errors remain valid. Although the computation differs, both methods yield the same estimate of δ and identical statistical inference.

The following R simulation shows that both the full OLS model and the FWL approach yield identical treatment coefficients and robust standard errors.

```
library(MASS)       # for mvrnorm
library(sandwich)  # for robust covariance estimates
set.seed(123)
N <- 10000; p <- 10
X <- mvrnorm(N, rep(0, p), diag(p))
D_prob <- plogis(X %*% runif(p, -1, 1))
D <- rbinom(N, 1, D_prob)
gamma_true <- 2
beta <- runif(p, -1, 1)
Y <- gamma_true * D + X %*% beta + rnorm(N)

# Full OLS model
full_model <- lm(Y ~ D + X)
cat("Full OLS coefficient (D):", coef(full_model)["D"], "\n")
```

```
## Full OLS coefficient (D): 1.996462
```

```
cat("Robust SE:", sqrt(diag(vcovHC(full_model, type = "HC1")))["D"], "\n\n")
```

```
## Robust SE: 0.02366501
```

```
# FWL approach:
# 1. Residualize Y and D on X
res_Y <- resid(lm(Y ~ X))
res_D <- resid(lm(D ~ X))
# 2. Regress residualized Y on residualized D
fwl_model <- lm(res_Y ~ res_D - 1)
cat("FWL coefficient (D):", coef(fwl_model)["res_D"], "\n")
```

```
## FWL coefficient (D): 1.996462
```

```
cat("Robust SE:", sqrt(diag(vcovHC(fwl_model, type = "HC1")))["res_D"], "\n")
```

```
## Robust SE: 0.02365199
```

19.4.2 From FWL to Double Machine Learning

In this section, we bridge the classical FWL theorem with DML to illustrate how robust causal inference is achieved through orthogonalization. We first review the FWL approach—where the outcome $Y$and treatment D are residualized with respect to covariates X—and then show how these ideas extend naturally to modern DML methods using flexible machine learning tools such as LASSO. This unified framework underscores the importance of partialling out nuisance variation to isolate the causal effect of interest.

We adopt the notation from Chernozhukov et al. (2018) for our linear model:

$$Y = D\theta + X\beta + \varepsilon \tag{19.20}$$

where the goal is to estimate θ, the coefficient on D. According to the FWL theorem, we can "partial out" the effects of X by first regressing both Y and D on X and then using their residuals. Specifically:

1. **Regress Y on X** and let $\hat{g}(X)$ be the fitted values. The residual becomes

$$\tilde{Y} = Y - \hat{g}(X)$$

2. **Regress D on X** and let $\hat{m}(X)$ be the fitted values. The residual is

$$\tilde{D} = D - \hat{m}(X)$$

3. **Run OLS of $\tilde{Y}$ on $\tilde{D}$**: According to FWL, the regression of Y on D controlling for X is equivalent to regressing $\tilde{Y}$ on $\tilde{D}$. This gives

$$\tilde{Y} = \theta\tilde{D} + \tilde{\varepsilon}$$

To obtain θ by OLS, we set up the first-order condition (the moment condition) by differentiating the sum of squared residuals with respect to θ. That condition is

$$\frac{\partial}{\partial\theta}\sum_i \left(\tilde{Y}_i - \theta\tilde{D}_i\right)^2 = 0$$

Differentiating (and Dividing by -2), we have

$$\sum_i \tilde{D}_i\left(\tilde{Y}_i - \theta\tilde{D}_i\right) = 0$$

In expectation (or in sample average form), we write the moment condition as

$$E\left[\tilde{D}(\tilde{Y} - \theta\tilde{D})\right] = 0$$

Expanding this condition gives

$$E[\tilde{D}\tilde{Y}] - \theta E[\tilde{D}^2] = 0$$

Solving for θ, we have

$$\theta = \frac{E[\tilde{D}\tilde{Y}]}{E[\tilde{D}^2]}$$

In a sample, this matches the OLS estimate from regressing $\tilde{Y}$ on $\tilde{D}$. This derivation shows step by step how the coefficient on D is obtained by solving the moment condition arising from the residualized variables, and θ reflects the causal/treatment effect of D net of the controls.

This derivation also reveals the orthogonal score in the context of OLS. Defining

$$\psi(W;\theta) = \tilde{D}\,(\tilde{Y} - \theta\,\tilde{D})$$

we see that once Y and D are residualized on X, the score depends only on the variation of Y and D that is uncorrelated with X. In other words, the orthogonality condition is given by:

$$E[\psi(W;\theta)] = E\big[\tilde{D}(\tilde{Y} - \theta\,\tilde{D})\big] = 0$$

This idea underlies the orthogonal score used in double machine learning, protecting the θ estimate from small errors in the nuisance function(s) $\hat{g}(X)$ and $\hat{m}(X)$ (here $\eta = 1$ since we use a single set of nuisance functions, as opposed to using cross-fitted residuals in DML).

Notably, the coefficient θ in the final expression, and orthogonal score above coincides with the "debiased" or "partialled-out" estimator and orthogonal score function discussed in Chernozhukov et al. (2018), particularly in Sections 19.4 (and endnote 3 therein). The partialing-out procedure and the DML1 and DML2 algorithms' score function is exactly the same as equation 4.55 in Section 4-partially linear models, just framed here in a simpler linear setting.

Having established that the FWL theorem yields an orthogonal score and unbiased treatment effect estimates by residualizing Y and D, we now turn to DML. In this approach, we incorporate advanced ML techniques to estimate the nuisance functions, while preserving the orthogonality property that safeguards our causal estimates even in high-dimensional settings.

Extending to Double Machine Learning

Double or debiased ML (DML) extends the classical FWL theorem to settings where flexible machine learning methods can be used. The main idea is to estimate the outcome Y and the treatment D separately using ML techniques, then obtain residuals by partialling out the contribution of X, and finally regress the outcome residual on the treatment residual. Formally, one first estimates:

$$\hat{g}(X) \approx E[Y \mid X] \quad \text{and} \quad \hat{m}(X) \approx E[D \mid X] \tag{19.21}$$

using methods such as LASSO. One then forms the residuals

$$\tilde{Y} = Y - \hat{g}(X) \quad \text{and} \quad \tilde{D} = D - \hat{m}(X) \tag{19.22}$$

and regresses $\tilde{Y}$ on $\tilde{D}$ to obtain the treatment-effect parameter θ. This procedure generalizes the standard partialling-out approach to high-dimensional or complex data scenarios while preserving a key principle: orthogonality.

Orthogonality arises because the residuals from the outcome and treatment regressions remove most of the variation associated with X. As a result, the estimator of the ATE is robust to moderate errors in estimating these nuisance functions. In a typical treatment-effects context, the goal is to identify a parameter θ in the partially linear model

$$Y = \theta\, D + f(X) + \varepsilon$$

where $f(X)$ is a nuisance function (which is any part of the model not containing the main parameter of interest). Unbiased estimation of θ requires proper accounting for $f(X)$. In DML, one estimates the nuisance functions $\hat{g}(X)$ for the outcome and $\hat{m}(X)$ for the treatment, and then constructs an orthogonal score of the form

$$\psi(W;\theta,\eta) = \big(Y - \hat{g}(X)\big)\,\big(D - \hat{m}(X)\big) - \theta\,\big(D - \hat{m}(X)\big)^2 \tag{19.23}$$

where $W = (Y, D, X)$ and η denotes the collection of nuisance functions (this equation can be shortened as $(\tilde{Y} - \theta\,\tilde{D})\tilde{D}$, which is the same as the equation obtained above using FWL, and is exactly the same as equation 4.55 in Chernozhukov et al. (2018)). The final estimate of θ is obtained by solving the following moment condition for θ (similar to the final OLS step) of

$$E[\psi(W;\theta,\eta)] = 0 \tag{19.24}$$

Because the derivative of this moment condition with respect to the nuisance functions is zero at the true parameter, small errors in estimating $\hat{g}(X)$ or $\hat{m}(X)$ do not substantially bias θ.[7] This orthogonality property is a key feature of DML, ensuring that the treatment effect estimate is robust to moderate errors in the nuisance functions.

A key element of DML is controlling overfitting, which can bias the residuals if the ML procedure (in estimating nuisance functions) captures too much of the true relationship between X and Y or D. Overfitting in $\hat{g}(X)$ may inadvertently capture variation in Y that is due to D, leading to downward-biased $\tilde{Y}$, while overfitting in $\hat{m}(X)$ can reduce the variance in $\tilde{D}$ and increase the variance of the final estimator (θ here). To mitigate these issues, DML employs sample splitting (or cross-fitting), partitioning the data into folds. In another words, the ML models are trained on all but one fold, and the held-out fold is used to generate out-of-sample predictions (to calculate residuals) to prevent bias for the nuisance functions. Repeating and combining these predictions systematically across all folds to estimate the final model correct overfitting bias.

Naive ML approaches that fail to account for orthogonality (i.e., directly implementing LASSO or random forest in a partially linear model instead of residualizing Y and D) may yield biased estimates because moderate errors in the nuisance functions can cause large errors in the final parameter of interest. DML's orthogonalization explicitly addresses this issue. By removing the main signal of X from both Y and D, the procedure yields residuals that are, loosely speaking, "purified" of spurious variation and captures mainly the variation attributable to the relationship between Y and D.

In DML, bias from overfitting is mitigated via cross-fitting and orthogonality condition, while the use of LASSO in the training stage may also introduce shrinkage bias. LASSO (or similar regularized linear methods) is a natural choice in DML because it effectively handles high-dimensional settings by regularizing and selecting the most relevant covariates. For example, one can estimate $\hat{g}(X)$ by regressing Y on X using LASSO and $\hat{m}(X)$ by regressing D on X (employing logistic LASSO when D is binary). LASSO penalizes large coefficients, thereby controlling model complexity but also biasing estimates toward zero. This shrinkage bias is mitigated in practice by using cross-validation (or a similar tuning procedure) to select the penalty parameter λ, which balances bias and variance to minimize out-of-sample prediction error during the training of the nuisance functions. To emphasize, LASSO in training stage of DML may still exhibit finite-sample shrinkage bias for inference, the orthogonality property in DML ensures that the final estimate of θ remains unbiased even if the nuisance function estimates are moderately imprecise.[8]

DML1 (Definition 3.1)

Below is a concise restatement of **DML1** and **DML2** (Definitions 3.1 and 3.2 in Chernozhukov et al. 2018) that complements the discussion above. We focus on the key algorithmic steps and how they fit into the framework of orthogonalized estimating equations. For detailed implementation, please refer back to the practical instructions and code in the main text.

1. **Partition the Data**
   Randomly split the sample into K folds $\{I_k\}_{k=1}^K$, each containing n/K observations (assuming n is divisible by K for simplicity).
2. **Estimate Nuisance Functions** and **Form Cross-Fitted Residuals** For each fold $k \in \{1, \ldots, K\}$:
   - **Train** machine learning estimators $\hat{\eta}_{-k}$ (e.g., $\hat{g}_{-k}(X)$ and $\hat{m}_{-k}(X)$) on all folds except k.
   - **Predict** on fold k, producing out-of-sample estimates for that fold.
3. **Construct the Orthogonal Moment for Each Fold**
   Using the held-out fold k, form the score function $\psi(W_i; \theta, \hat{\eta}_{-k})$ (see the previous text for the explicit formula). The estimator $\hat{\theta}_k$ on fold k solves the empirical moment equation
   $$\frac{1}{|I_k|} \sum_{i \in I_k} \psi(W_i; \theta, \hat{\eta}_{-k}) = 0$$
   which corresponds to Equation (3.38) in Chernozhukov et al. (2018).
4. **Combine Fold-Specific Estimates**
   Average the solutions $\hat{\theta}_k$ across all folds to obtain the final DML1 estimator
   $$\hat{\theta}_{\text{DML1}} = \frac{1}{K} \sum_{k=1}^{K} \hat{\theta}_k$$

This fold-by-fold approach produces one estimate per fold and then aggregates them. The moment equation is solved separately within each hold-out set, ensuring that nuisance function estimation and treatment-effect estimation use disjoint subsets of data.

DML2 (Definition 3.2)

1. **Partition and Estimate Nuisance Functions**
   As in DML1, randomly split the data into K folds and for each fold k, train $\hat{\eta}_{-k}$ on all but the k-th fold, then predict on fold k.

2. **Estimate Nuisance Functions** and **Form Cross-Fitted Residuals**
Using $\hat{\eta}_{-k}$, construct residuals $\tilde{Y}_i$ and $\tilde{D}_i$ for all $i \in I_k$. Gather these residuals for all folds into a single sample.

3. **Solve a Single Moment Condition**
Unlike DML1, DML2 combines all folds' residuals at once and solves a single orthogonal moment equation over the entire sample (Equation (3.41) in Chernozhukov et al. (2018)):

$$\frac{1}{n}\sum_{k=1}^{K}\sum_{i\in I_k}\psi(W_i;\theta,\hat{\eta}_{-k})=0$$

yielding a single $\hat{\theta}_{\text{DML2}}$.

4. **Approximate Solutions**
If an exact solution is not possible, an approximate ϵ-solution is employed (Equation (3.42) in Chernozhukov et al.), ensuring $\hat{\theta}_{\text{DML2}}$ is still valid under mild conditions.

19.4.2.1 Connection to Orthogonality and FWL

Both DML1 and DML2 rely on the same orthogonal score construction discussed above: the partialling out of X from Y and D, followed by a moment condition that is insensitive (to first order) to errors in the nuisance functions. DML1 solves the orthogonal equation fold by fold, then averages, while DML2 pools all cross-fitted residuals and solves one global equation. In either case, the fundamental principle is that once Y and D are "purified" of the variation explained by X, the remaining variation identifies the causal parameter θ. By using cross-fitting, both procedures mitigate overfitting bias and maintain valid inference in high-dimensional or complex data settings. Because of orthogonality, small inaccuracies in the nuisance regressions have a limited effect on the final parameter estimates, offering a powerful tool for causal inference when there are many covariates or complex relationships between regressors and outcomes.

19.4.3 Debiased LASSO

Standard LASSO estimates are biased due to the ℓ_1-penalty shrinking coefficients toward zero. Debiased LASSO (also known as desparsified or de-sparsified LASSO) corrects this shrinkage by adding a carefully constructed adjustment term. Pioneered by Zhang and Zhang (2014), van de Geer et al. (2014), and Javanmard and Montanari (2014), the procedure starts with the LASSO estimate $\hat{\beta}_{\text{Lasso}}$, then updates it via

$$\hat{\beta}_{dl}=\hat{\beta}_{\text{Lasso}}+\frac{1}{n}\hat{\Theta}X^{\top}\big(Y-X\hat{\beta}_{\text{Lasso}}\big) \tag{19.25}$$

Here, $\hat{\Theta}$ approximates Σ^{-1}, with $\Sigma=\frac{1}{n}X^{\top}X$. In high-dimensional settings, $\hat{\Theta}$ is often estimated by regressing each column of X on the other columns (nodewise regression) or other high-dimensional precision-matrix estimators. This one-step correction removes much of the bias introduced by LASSO's penalty, allowing asymptotically valid confidence intervals for low-dimensional parameters (such as a specific treatment coefficient) even when p is large.

Notes

1. Post-LASSO is implemented by first running LASSO for variable selection and then refitting an unpenalized OLS regression on the selected covariates. In the partialling-out model, this applies to the regressions of Y and D on X before computing residuals. In the Double Selection Model, Post-LASSO is applied separately in the first two steps before the final OLS regression of Y on D and the selected covariates.

2. In DML (Chernozhukov et al., 2018), $\hat{\omega}_i$ is derived from the Neyman orthogonality condition, ensuring that small errors in nuisance function estimation do not affect the final treatment effect estimate. The final variance of $\hat{\beta}$ is estimated using a heteroskedasticity-robust formula that incorporates both the residual variation and the influence function correction term:

$$\widehat{\text{Var}}(\hat{\beta}) = \left(\frac{1}{n}\sum_{i=1}^{n} \hat{\omega}_i^2\right)$$

where $\hat{\omega}_i$ represents the influence function correction term that adjusts for regularization bias. The corresponding heteroskedasticity-robust standard error is given by:

$$\widehat{\text{SE}}(\hat{\beta}) = \sqrt{\frac{1}{n}\sum_{i=1}^{n} \hat{\omega}_i^2}$$

In practice, this variance estimator is computed using EHW robust standard errors, but with additional corrections to account for the impact of machine learning-based nuisance function estimation. The "additional corrections" in DML come from using cross-fitting and Neyman orthogonality, which remove bias from nuisance function estimation. This adjustment ensures that the inference for $\hat{\beta}$ remains valid despite using LASSO or other regularization techniques for covariate selection and treatment assignment modeling.

3. To estimate $\hat{\omega}_i$ in Debiased LASSO (Zhu & Hastie, 2004; Javanmard & Montanari, 2014), the correction is done using a nodewise LASSO regression to estimate the inverse covariance matrix of the selected variables. This provides an estimate of how much LASSO's shrinkage affects inference.

4. Penalty loadings modify the standard plug-in formula by assigning different penalty intensities to different covariates, ensuring that shrinkage is applied more appropriately across variables. The adjusted penalty formula includes a diagonal matrix of penalty loadings, where each loading $\hat{l}_j$ is estimated as $\hat{l}_j = \sqrt{E_n[x_{ij}^2 \epsilon_i^2]}$. See Belloni, Chernozhukov, and Hansen (2014, REStud) for details.

5. The iterated LASSO method refines the penalty loadings over multiple steps to ensure that the sparsity structure properly accounts for heteroskedasticity. This approach is implemented through Algorithm 1 in the Appendix of Belloni, Chernozhukov, and Hansen (2014, REStud). The algorithm updates the penalty loadings using the Post-Lasso estimator and stops when convergence criteria are met.

6. Mathematically, approximate sparsity implies – The best sparse approximation of $f(X)$ satisfies for simple Lasso models (in low-dimensional settings as described by Hastie, Tibshirani, and Friedman (2009)):$\min_{\|\beta\|_0 \leq s} \mathbb{E}\left[(f(X) - X'\beta)^2\right] = O(s/N)$ where s is the effective number of relevant covariates and N is the sample size. This means that the error from using only the best s variables diminishes as s grows more slowly than N. Lasso is effective if most of the explanatory power in $f(X)$ and $g(X)$ is concentrated in a small number of covariates.

7. Derivation of the Orthogonality Condition via the Gateaux Derivative: Consider the moment function

$$\psi(W;\theta,\eta) = \big(D - m(X)\big)\Big(Y - g(X) - \theta\,(D - m(X))\Big)$$

with $\eta = (g, m)$ and the true nuisance functions $\eta_0 = (g_0, m_0)$, where $g_0(X) = E[Y \mid X]$ and $m_0(X) = E[D \mid X]$. The population moment condition at the true parameter θ_0 is

$$E\big[\psi(W;\theta_0,\eta_0)\big] = 0$$

To study the sensitivity of this moment condition to small perturbations in the nuisance functions, we consider a directional perturbation $h = (h_g, h_m)$ and define the Gateaux derivative as

$$\frac{\partial}{\partial\eta} E\big[\psi(W;\theta_0,\eta)\big]\Big|_{\eta=\eta_0}[h] = \lim_{t\to 0} \frac{E\big[\psi(W;\theta_0,\eta_0 + t\,h)\big] - E\big[\psi(W;\theta_0,\eta_0)\big]}{t}$$

Since $E[\psi(W;\theta_0,\eta_0)]=0$, the orthogonality condition (or Neyman orthogonality) requires that

$$\frac{\partial}{\partial\eta}E\big[\psi(W;\theta_0,\eta)\big]\Big|_{\eta=\eta_0}[h]=0$$

for all admissible directions h. This means that first-order changes in the nuisance functions (i.e., small estimation errors) do not affect the moment condition—any impact is only of second order. This insensitivity is central to double machine learning, as it protects the estimate of θ from moderate errors in estimating the nuisance functions $g_0(X)$ and $m_0(X)$.

8. Methods like post-LASSO (refitting the model on variables selected by LASSO; default in "hrm" package "rlasso") or debiased LASSO(next section) provide additional corrections for valid hypothesis testing and confidence intervals, thereby mitigating the residual shrinkage bias in coefficient estimates directly.

20

Matching Methods

Building on the previous two chapters—where we introduced regression-based adjustment under selection on observables and extended it to high-dimensional settings using double machine learning—this chapter focuses on alternative strategies for addressing confounding without relying entirely on parametric regression models. While regression methods offer flexibility, they impose functional form assumptions that may not align with how treatment is actually assigned in observational data.

Here, we introduce stratification, matching, and the propensity score framework as complementary tools that aim to create balance between treated and control groups based on observed covariates. These approaches offer intuitive, design-based solutions to the problem of selection bias and help approximate the conditions of a randomized experiment. We show how to implement stratification and nearest-neighbor matching, then turn to estimating propensity scores and using them to evaluate balance and overlap. We also demonstrate how to integrate modern machine learning tools, such as XGBoost, to estimate propensity scores for matching. In the next chapter, we build on this foundation by introducing inverse probability weighting (IPW) and doubly robust estimators, which combine modeling and weighting strategies to further improve robustness and efficiency. Together, these methods expand the causal inference toolkit for observational studies where treatment is not randomly assigned.

20.1 Subclassification or Stratification

Subclassification, also known as stratification, is a method in causal inference that divides units into discrete groups (subclasses) based on covariates (X_i) and estimates treatment effects within each group. The treatment effect within each subclass is calculated as the difference in mean outcomes between treated and untreated units, and the overall treatment effect is obtained by taking a weighted average across subclasses. The average treatment effect (ATE) is computed as:

$$\text{ATE} = \sum_{s=1}^{S} w_s \cdot \text{CATE}_s \tag{20.1}$$

where S is the number of subclasses, w_s is the proportion of units in subclass s, calculated as:

$$w_s = \frac{n_s}{n} \tag{20.2}$$

where n_s is the number of units in subclass s, and n is the total number of units in the sample. CATE_s is the conditional average treatment effect within subclass s.

DOI: 10.1201/9781003504795-20

You determine S (the number of subclasses) by dividing the data into strata based on the values or ranges of the covariates X_i. For a single categorical variable, S is the number of unique categories. For multiple categorical variables, S is the number of unique category combinations. For continuous variables, S is chosen based on quantiles or equal-sized groups. Typically, subclasses are created to ensure covariate balance, with 5–10 subclasses commonly recommended in practice.

For example, in the college scholarship case in the previous chapter, we could subclassify by university type (public, private, or non-profit) and estimate CATE_s within each group. The overall ATE is then obtained by weighting these subclass effects by the distribution of X_i, ensuring that the effect reflects the population composition. The number of subclasses, S, can be determined based on distinct categories for categorical covariates or by dividing continuous covariates into quantiles or equal-sized groups. In the previous chapter, we calculated subclass effects and showed the weighted ATE.

Subclassification provides a clear way to satisfy the Conditional Independence Assumption by assuming treatment is "as good as random" within each subclass. This assumption enables causal inference from observational data when selection on observables holds. For example, Cochran (1968) demonstrated that subclassification into a modest number of groups (e.g., five strata) can remove over 90% of bias in observational studies. Subclassification based on propensity scores, which summarize multiple covariates into a single score ($P(D_i = 1 \mid X_i)$), offers a more flexible alternative for handling continuous or high-dimensional covariates (Rosenbaum and Rubin, 1984). We will cover this approach later.

Despite its strengths, subclassification has notable limitations. First, it requires discrete covariates or the discretization of continuous variables, which can lead to information loss and arbitrary cutoff choices. For example, discretizing parental income into income brackets may ignore nuanced differences in earnings. Second, subclassification struggles with multi-level or non-discrete outcomes, such as treatment effects that vary continuously with X_i. For these cases, methods like regression, matching, or machine learning approaches offer greater flexibility and precision.

The simulation implements subclassification by dividing the sample into five strata based on family income, a key covariate influencing both treatment assignment and earnings. Within each stratum, the treatment effect—defined as the difference in mean earnings between college graduates and non-graduates—is estimated. These within-stratum effects are then aggregated to compute the overall weighted average treatment effect on the treated (ATT). The weighting ensures that subclass effects contribute proportionally to the total estimate, aligning with the approach outlined earlier. The results illustrate how stratification accounts for covariate differences when estimating causal effects.

```r
# Load necessary libraries
library(dplyr)
library(estimatr)

# Set seed for reproducibility
set.seed(42)

# Define sample sizes
N <- 20000
NT <- 5000  # Number in treatment (college graduates)
NC <- N - NT  # Number in control (non-graduates)
```

```r
# Generate earnings for treated and control groups
earnings_non_graduates <- rnorm(NC, mean = 50, sd = 10)
earnings_graduates <- rnorm(NT, mean = 53, sd = 10) # Mean increased by 3

# Generate family income (X) for both groups
family_income_non_graduates <- rnorm(NC, mean = 70, sd = 20)  # Lower mean
family_income_graduates <- rnorm(NT, mean = 95, sd = 20)  # Higher mean

# Merge into a single dataset
data <- data.frame(
  ID = 1:N,
  Treatment = c(rep(1, NT), rep(0, NC)),
  X = c(family_income_graduates, family_income_non_graduates),  # Covariate
  Earnings = c(earnings_graduates, earnings_non_graduates) # Outcome
)

# Define quantile-based strata for family income (X)
data <- data %>%
  mutate(Stratum = ntile(X, 5))  # Divide into 5 strata

# Estimate treatment effect within each stratum
results <- data %>%
  group_by(Stratum) %>%
  summarize(
    ATT = mean(Earnings[Treatment == 1]) - mean(Earnings[Treatment == 0]),
    N_treat = sum(Treatment == 1),
    N_control = sum(Treatment == 0)
  )

# Compute overall weighted ATT
weighted_ATT <- sum(results$ATT * (results$N_treat + results$N_control) / N)

# Display results
print(results)
```

```
## # A tibble: 5 x 4
##   Stratum   ATT N_treat N_control
##     <int> <dbl>   <int>     <int>
## 1       1  3.19     140      3860
## 2       2  3.06     406      3594
## 3       3  3.34     745      3255
## 4       4  3.29    1254      2746
## 5       5  3.05    2455      1545
```

```r
cat("Overall Weighted ATT:", weighted_ATT, "\n")
```

```
## Overall Weighted ATT: 3.184567
```

Alternatives, such as matching, build upon the intuition of subclassification by directly pairing treated and untreated units, not discrete groups, based on their covariates or propensity scores. Matching avoids the need to discretize variables, making it suitable for continuous covariates and high-dimensional settings. We will explore these methods below, highlighting their ability to estimate treatment effects while addressing their challenges.

20.2 Matching Methods

Matching is another method for estimating treatment effects that addresses the fundamental problem of causal inference. Matching offers a solution by imputing the "missing" outcomes based on observed data, pairing each treated unit with the closest untreated unit (or vice versa) based on their covariates (X_i). Matching addresses the shortcomings of subclassification by directly pairing treated and untreated units based on their similarity in covariates, regardless of whether the covariates are continuous or high-dimensional.

The goal of matching is to find untreated units (j) that are most similar to treated units (i) in terms of their covariates. For each treated unit i, the outcome of the closest untreated unit j serves as an estimate for the unobserved $Y_i(0)$. The difference between the observed outcome $Y_i(1)$ and the imputed $Y_i(0)$ provides an estimate of the treatment effect for that unit. This method assumes that, based on observable characteristics, treated and untreated individuals are nearly identical, so any difference in outcomes is attributed to the causal effect of the treatment.

The ATT is then calculated as:

$$\hat{\delta}_{\text{ATT}} = \frac{1}{N_1} \sum_{D_i=1} (Y_i(1) - Y_j(0)) \tag{20.3}$$

where $Y_j(0)$ is the observed outcome of the closest untreated unit to i, and N_1 is the number of treated units.

Alternatively, matching can use the average outcomes of the M closest matches to improve robustness:

$$\hat{\delta}_{\text{ATT}} = \frac{1}{N_1} \sum_{D_i=1} \left(Y_i(1) - \frac{1}{M} \sum_{m=1}^{M} Y_{j_m}(0) \right) \tag{20.4}$$

where $Y_{j_m}(0)$ represents the outcomes of the m-th closest untreated match to i.

We could impute potential outcomes for control units similar way and define the ATE and ATU equivalently. However, ATT is the most common causal effect, matching methods estimate

Lets illustrate this method with the following example. We want to estimate the causal effect of a college scholarship on annual earnings using exact 1-to-1 matching. Matching pairs treated ($D_i = 1$) and untreated ($D_i = 0$) individuals with identical covariates (X_i). Consider the following data:

TABLE 20.1 Data for Matching Example

ID	D_i	Parental Income	GPA	SAT Score	Univ. Type	Field	Earnings
1	1	High	3.8	1400–1600	Public	STEM	60000
2	0	High	3.8	1400–1600	Public	STEM	50000
3	1	Medium	3.2	1201–1400	Private	Business	72000
4	0	Medium	3.2	1201–1400	Private	Business	62000
5	1	Low	3.0	1000–1200	Non-Profit	Arts	45000
6	0	Low	3.0	1000–1200	Non-Profit	Arts	37000

The individual CATE for each treated unit is computed as the difference between their observed outcome ($Y(1)_i$) and the outcome of the matched control unit ($Y(0)_j$): $\text{CATE}_1 = 60,000 - 50,000 = 10,000, \quad \text{CATE}_3 = 72,000 - 62,000 = 10,000, \quad \text{CATE}_5 = 45,000 - 37,000 = 8,000$.

The ATT is the average of these effects: $\hat{\delta}_{\text{ATT}} = \frac{1}{3}(10,000 + 10,000 + 8,000) = 9,333$

This example illustrates how matching imputes unobserved potential outcomes and provides an estimate of the treatment effect. Here, the scholarship increases earnings for treated students by approximately $9,333 on average.

Alternatively, matching can improve robustness by using the average outcomes of the M closest untreated matches for each treated individual with various methods. This method accounts for the possibility that multiple units in the control group are similar to a treated unit based on their covariates. In larger datasets, averaging the outcomes of these similar control units provides a better proxy for the unobserved $Y_i(0)$, reducing sensitivity to individual variability and noise. After imputing these averages, individual CATEs can be calculated and aggregated to estimate the ATT, providing a more stable and reliable estimate of treatment effects.

The goal of matching is to compare treated and untreated units that are similar in their covariates (X_i), thereby minimizing bias due to confounding. However, as the dimensionality of X_i increases, the likelihood of finding exact matches diminishes—a phenomenon known as the curse of dimensionality. Additionally, matching requires a clear definition of "closeness" between units, which depends on the chosen distance metric. Different metrics for "closeness" between units (check footnote for their equations) lead to different matching methods such as exact, nearest neighbor, bandwidth, propensity score, caliper and optimal matching. "Closeness" is defined by:

- *Exact matching*: Matches are only made when covariates are identical between treated and untreated units. Thus,"closeness" if perfect. This approach becomes impractical with high-dimensional or continuous X_i.
- *Normalized Euclidean distance*: Distances are scaled by the inverse of the sample variance, suitable for normally distributed X_i, but less effective for binary data.[1]
- *Mahalanobis distance*: Distances are scaled by the inverse of the covariance matrix, providing a robust measure for normally distributed covariates but performing poorly with binary variables.[2]
- *Genetic matching*: Uses optimization algorithms to minimize covariate imbalance across all variables, directly improving balance. (Diamond and Sekhon)[3]

Besides the exact matching method discussed above, nearest neighbor matching is one of the most widely used approaches. In this method, each treated unit is matched to the untreated unit with the closest covariates, based on a chosen distance metric such as Euclidean or Mahalanobis distance. This non parametric approach makes no assumptions about the functional form of $Y(X)$, offering a flexible way to estimate causal effects. The method can also be extended to K-nearest neighbors, where the K closest untreated units are selected for each treated unit, and their outcomes are averaged to estimate $\hat{Y}_i(0)$. This generalization reduces variance while retaining the non parametric nature of the method.

In bandwidth matching, for each treated unit $i \in T$, all untreated units $j \in U$ whose covariates fall within a defined bandwidth h are selected as matches, such that $X_j \in [X_i - h, X_i + h]$. The outcomes of these untreated units are then averaged to estimate the counterfactual for the treated unit, $\hat{Y}_i(0)$. This method ensures that only untreated units with covariate values sufficiently close to those of the treated unit are considered valid matches, reducing bias. However, if the bandwidth h is too narrow, some treated units may lack valid matches, leading to higher variance. Conversely, wider bandwidths include more observations, reducing variance but potentially introducing bias by allowing less similar matches.

Choosing the bandwidth involves a bias–variance trade-off and is critical for the reliability of the causal estimates. Methods such as cross-validation can assist in selecting the optimal bandwidth by minimizing prediction error in a validation sample. This approach balances the trade-off by systematically testing different bandwidths and choosing the one that maximizes matching accuracy. Studies like Imbens and Wooldridge (2009) and Stuart (2010) emphasize the importance of bandwidth selection and its impact on matching performance in observational studies.

Another common approach is **Propensity Score Matching (PSM)**, which matches units based on their propensity scores, $P(D_i = 1 \mid X_i)$, instead of directly on covariates. The propensity score represents the probability of receiving treatment given covariates. This technique reduces the dimensionality problem and ensures comparability, and it will be discussed in detail in the next section.

Caliper Matching imposes a constraint on the maximum allowable difference in propensity scores between matched units. Formally, a treated unit i is only matched to a control unit j if $|P(X_i) - P(X_j)| \leq \delta$ where δ is a pre-specified caliper width. This method prevents poor matches and improves balance but may reduce the number of matched pairs.

Optimal Matching selects pairs by minimizing the total distance between matched units across the entire sample rather than one-to-one nearest neighbor selection. This approach solves a global optimization problem, ensuring the best overall balance while preserving sample size.

20.2.1 Factors influence the implementation of matching methods

Several factors influence the implementation of matching methods. One key decision is the size of M, the number of control units matched to each treated unit. Smaller M minimizes bias since additional matches are less similar, but larger M reduces variance by increasing the matched sample size.

Another consideration is whether matching is performed with or without replacement. Matching with replacement means multiple treated units can match to the same control unit. This may reduces bias by allowing control matches to be reused for multiple treated units. However, this introduces dependence among matched controls, complicating inference and leading to larger standard errors. Matching without replacement avoids this issue but may result in unmatched treated units, reducing the effective sample size.Also, the ordering of the treated units matters, as the first treated unit matched to a control unit may influence subsequent matches. Randomizing the order of treated units can mitigate this issue.

Handling ties is another practical issue. When multiple controls are equally "close" to a treated unit, one approach is to select a control randomly, though this lacks consistency. Alternatively, averaging the outcomes of tied controls provides a more stable estimate. Finally, the choice of which treatment effect to estimate—ATE, ATT, or ATU—depends on the research question and the availability of suitable matches for treated or control units. Each decision involves a trade-off between bias, variance, and the practical goals of the analysis.

20.2.2 Assessing Balance

After matching, the covariate distributions for the treatment and control groups should be comparable to ensure valid causal estimates. Assessing balance is a crucial step in validating the results of matching. It ensures that the treated and control groups are comparable, making it possible to attribute differences in outcomes to the treatment rather than pre-existing differences in covariates. This involves two key diagnostics: overlap and balance checks.

Overlap, or common support, ensures that the treated and control groups have sufficiently overlapping covariate distributions. Without overlap, some treated units may lack suitable matches in the control group, or vice versa, making it impossible to estimate causal effects for those units. Overlap can be assessed visually by comparing histograms or density plots of covariates or propensity scores for treated and control groups. Regions with no overlap indicate areas where causal estimates are unreliable. To address this, researchers often restrict the sample to the overlapping region by trimming unmatched units, focusing on subsets of the data where treated and control units are comparable. (The formal condition for overlap is provided in the assumptions section.)

Balance checks evaluate whether covariate distributions are similar between treated and control groups after matching. Common methods for assessing balance include:

- **Mean Differences**: Comparing the means of covariates between groups before and after matching to ensure any differences are minimized post-matching.
- **Standardized Bias**: A more robust measure is the standardized bias, which evaluates the difference in covariate means between the treatment and control groups, scaled by the standard deviation of the covariate in the treated group. For a covariate X_i, the standardized bias is calculated as:

$$\text{Bias}_{X_i} = \frac{\bar{X}_t - \bar{X}_c}{\sigma_t} \tag{20.5}$$

where $\bar{X}_t$ and $\bar{X}_c$ are the mean values of X_i for the treated and control groups, respectively, and σ_t is the standard deviation of X_i in the full treated group. Comparing the standardized bias before and after matching provides a clearer indication of balance, with smaller post-matching bias values reflecting improved comparability between the groups. Yes, the statement is supported by commonly cited thresholds in the causal inference literature. Standardized bias values below 10% are widely regarded as acceptable for covariate balance, as suggested by Cochran and Rubin (1973) and supported by subsequent studies like Austin (2009).

- **Graphical Diagnostics**: such as **Love plots** (Love, 2002), which display standardized bias for each covariate before and after matching, and **density plots** (Rosenbaum and Rubin, 1983), which compare the distributions of covariates or propensity scores across groups, are essential tools for evaluating balance and overlap.

- **Variance Ratios**: compare the variability of covariates between treated and control groups. A variance ratio close to 1 indicates similar variability, which is essential for ensuring that treatment effects are estimated consistently across groups. Rubin (2001) suggests using variance ratios as a diagnostic for balance, with acceptable thresholds typically between 0.8 and 1.25. While variance ratios alone do not confirm balance, they complement other diagnostics, such as standardized bias and graphical tools, by highlighting differences in variability that could affect matching quality.

Balance checks demonstrate whether matching has successfully eliminated systematic differences in covariates. Many studies assess balance by presenting tables of covariate means and p-values before and after matching. However, p-values can be misleading as they are sensitive to sample size changes and may not reliably indicate balance improvement. It is best practice to report these checks before and after matching, focusing on standardized bias and graphical methods rather than relying solely on p-values, which are sensitive to sample size changes and less informative for balance assessment.

In the simple simulation, we compare different matching methods for estimating the treatment effect of college graduation on earnings while controlling for family income, and additional a continuous covariate and a binary covariate. First, we check the initial balance of these covariates between treated (college graduates) and control (non-graduates) groups using standardized mean differences. Since there are initial differences between the two groups, matching is applied to improve balance. We implement three matching methods: nearest-neighbor matching, full matching, and Mahalanobis distance matching. Nearest-neighbor matching assigns each treated unit to the closest control based on observed covariates. Full matching allows each treated unit to be matched to multiple control units and vice versa, improving the use of available data. Mahalanobis distance matching uses a distance metric to find the most similar pairs based on multiple covariates. After matching, we reassess balance and find that all methods reduce covariate imbalance, though their effectiveness varies. We then estimate the average treatment effect on the treated (ATT) by comparing mean earnings between matched treated and control groups. Finally, we visualize the family income distribution before and after matching using density plots, allowing us to assess how well each method aligns the distributions of the treated and control groups. You can apply this process to any actual data; this simulation simply demonstrates the matching procedures.

```
# Load necessary libraries
library(MatchIt)
library(dplyr)
library(tableone)
library(ggplot2)

# Set seed for reproducibility
set.seed(42)

# Define sample sizes
N <- 2000
NT <- 500
NC <- N - NT
# Generate earnings for treated and control groups
earnings_non_graduates <- rnorm(NC, mean = 50, sd = 10)
earnings_graduates <- rnorm(NT, mean = 53, sd = 10)
```

```r
# Generate covariates
family_income_non_graduates <- rnorm(NC, mean = 70, sd = 20)
family_income_graduates <- rnorm(NT, mean = 95, sd = 20)
# Generate 1 continuous variable
cont_var1_non_graduates <- rnorm(NC, mean = 5, sd = 2)
cont_var1_graduates <- rnorm(NT, mean = 6, sd = 2)
# Generate 1 binary variables
binary_var1_non_graduates <- rbinom(NC, 1, 0.4)
binary_var1_graduates <- rbinom(NT, 1, 0.6)
# Merge into a single dataset
data <- data.frame(
  ID = 1:N,
  Treatment = c(rep(1, NT), rep(0, NC)),
  FamilyIncome = c(family_income_graduates, family_income_non_graduates),
  ContVar1 = c(cont_var1_graduates, cont_var1_non_graduates),
  BinaryVar1 = c(binary_var1_graduates, binary_var1_non_graduates),
  Earnings = c(earnings_graduates, earnings_non_graduates)
)

# STEP 1: Check Initial Covariate Balance
table1 <- CreateTableOne(vars = c("FamilyIncome", "ContVar1", "BinaryVar1"),
      strata = "Treatment", data = data, test = TRUE)
print(table1)
```

```
##                           Stratified by Treatment
##                            0             1             p      test
##   n                         1500           500
##   FamilyIncome (mean (SD)) 69.77 (20.40) 94.68 (19.56) <0.001
##   ContVar1 (mean (SD))      4.98 (2.05)   6.07 (2.08)  <0.001
##   BinaryVar1 (mean (SD))    0.40 (0.49)   0.58 (0.49)  <0.001
```

```r
# STEP 2: Matching Methods Without Propensity Scores
# 1. Nearest-Neighbor Matching (based on covariates)
nn_match <- matchit(Treatment ~ FamilyIncome + ContVar1 + BinaryVar1,
                    data = data, method = "nearest")
matched_data_nn <- match.data(nn_match)
# 2. Full Matching (each treated unit matched to multiple control units)
full_match <- matchit(Treatment ~ FamilyIncome + ContVar1 + BinaryVar1,
                      data = data, method = "full")
matched_data_full <- match.data(full_match)
# 3. Mahalanobis Distance Matching
mahalanobis_match <- matchit(Treatment ~ FamilyIncome + ContVar1 + BinaryVar1,
                    data = data, method = "nearest", distance = "mahalanobis")
matched_data_mahalanobis <- match.data(mahalanobis_match)

# STEP 3: Check Covariate Balance After Matching
balance_nn <- CreateTableOne(vars = c("FamilyIncome", "ContVar1", "BinaryVar1"),
           strata = "Treatment", data = matched_data_nn, test = TRUE)
balance_full <- CreateTableOne(vars = c("FamilyIncome", "ContVar1", "BinaryVar1"),
```

```
                     strata = "Treatment", data = matched_data_full, test = TRUE)
balance_mahalanobis <- CreateTableOne(
  vars = c("FamilyIncome", "ContVar1", "BinaryVar1"),
  strata = "Treatment",
  data = matched_data_mahalanobis,
  test = TRUE
)
print(balance_nn)
```

```
##                                 Stratified by Treatment
##                                  0              1              p      test
##   n                                500            500
##   FamilyIncome (mean (SD)) 87.35 (16.02) 94.68 (19.56) <0.001
##   ContVar1 (mean (SD))      5.72 (2.05)   6.07 (2.08)   0.006
##   BinaryVar1 (mean (SD))    0.51 (0.50)   0.58 (0.49)   0.026
```

```
print(balance_full)
```

```
##                                 Stratified by Treatment
##                                  0              1              p      test
##   n                               1500            500
##   FamilyIncome (mean (SD)) 69.77 (20.40) 94.68 (19.56) <0.001
##   ContVar1 (mean (SD))      4.98 (2.05)   6.07 (2.08)  <0.001
##   BinaryVar1 (mean (SD))    0.40 (0.49)   0.58 (0.49)  <0.001
```

```
print(balance_mahalanobis)
```

```
##                                 Stratified by Treatment
##                                  0              1              p      test
##   n                                500            500
##   FamilyIncome (mean (SD)) 86.25 (16.20) 94.68 (19.56) <0.001
##   ContVar1 (mean (SD))      5.72 (1.97)   6.07 (2.08)   0.005
##   BinaryVar1 (mean (SD))    0.57 (0.49)   0.58 (0.49)   0.949
```

```
# STEP 4: Estimate Treatment Effect After Matching
estimate_ATT <- function(matched_data) {
  mean(matched_data$Earnings[matched_data$Treatment == 1]) -
  mean(matched_data$Earnings[matched_data$Treatment == 0])
}
cat("ATT (Nearest Neighbor):", estimate_ATT(matched_data_nn), "\n")
```

```
## ATT (Nearest Neighbor): 3.323404
```

```
cat("ATT (Full Matching):", estimate_ATT(matched_data_full), "\n")
```

```
## ATT (Full Matching): 3.61869
```

```
cat("ATT (Mahalanobis Matching):", estimate_ATT(matched_data_mahalanobis), "\n")
```

```
## ATT (Mahalanobis Matching): 3.72427
```

```
# STEP 5: Visualizing Matching Performance
# Sample a smaller subset for plotting (e.g., 200 observations)
plot_data <- data %>% sample_n(200)
# Visualizing Family Income Distribution Before Matching
plot_before <- ggplot(data, aes(x = FamilyIncome, fill = as.factor(Treatment))) +
  geom_density(alpha = 0.5, color = "black", adjust = 0.5) +
  scale_fill_manual(values = c("gray80", "black")) +
  labs(title = "Before Matching", fill = "Treatment") +
  theme_minimal()
# Visualizing Family Income Distribution After Matching
plot1 <- ggplot(matched_data_nn, aes(x = FamilyIncome,
                                     fill = as.factor(Treatment))) +
  geom_density(alpha = 0.5, color = "black", adjust = 0.5) +
  scale_fill_manual(values = c("gray80", "black")) +
  labs(title = "After Nn Matching", fill = "Treatment") +
  theme_minimal()
plot2 <- ggplot(matched_data_full, aes(x = FamilyIncome,
                                       fill = as.factor(Treatment))) +
  geom_density(alpha = 0.5, color = "black", adjust = 0.5) +
  scale_fill_manual(values = c("gray80", "black")) +
  labs(title = "After Full Matching", fill = "Treatment") +
  theme_minimal()
plot3 <- ggplot(matched_data_mahalanobis, aes(x = FamilyIncome,
                                fill = as.factor(Treatment))) +
  geom_density(alpha = 0.5, color = "black", adjust = 0.5) +
  scale_fill_manual(values = c("gray80", "black")) +
  labs(title = "After Mahalanobis Matching", fill = "Treatment") +
  theme_minimal()
library(gridExtra)
library(grid)
# Create the arranged plot with a title
grid.arrange(
  arrangeGrob(
    plot_before, plot1, plot2, plot3,
    ncol = 2, nrow = 2,
    top = textGrob("Family Income Distribution", gp = gpar(fontsize = 16,
                                                  fontface = "bold"))
  )
)
```

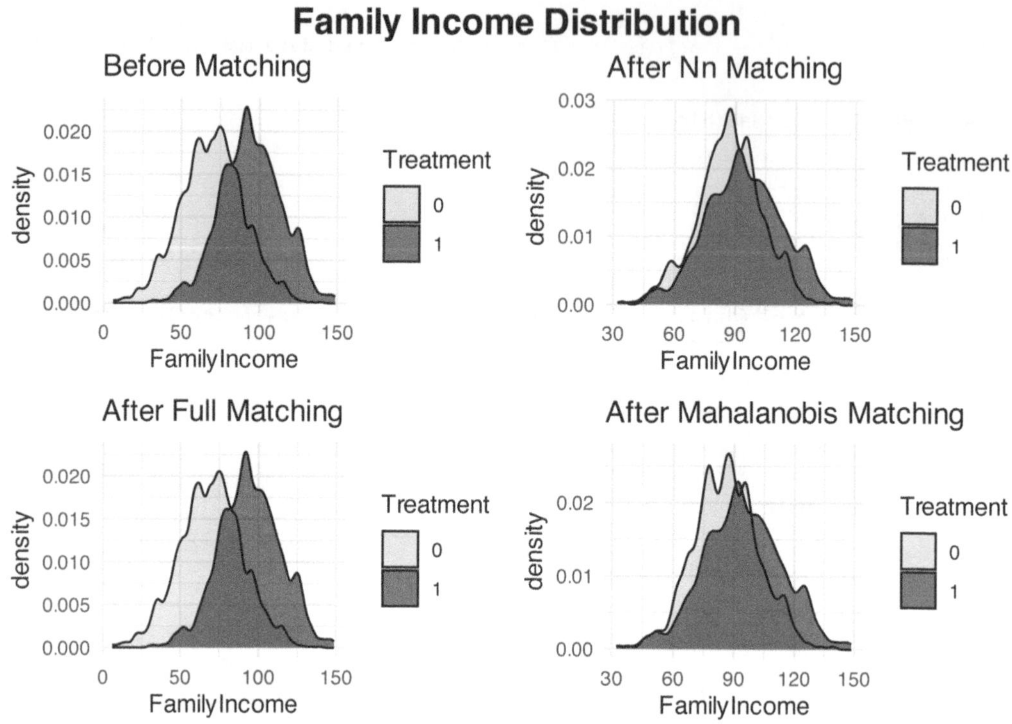

20.2.3 Matching: Benefits and Limitations

While matching is a powerful tool for causal inference, it comes with several caveats and challenges:

1. *Bias from Imperfect Matches*: Matching suffers from bias due to imperfect matches, particularly in high-dimensional settings. According to Abadie and Imbens (2002), the bias is of order $N^{-1/\dim X}$, where N is the sample size and $\dim X$ is the number of covariates. Bias remains small when the control group is much larger than the treated group ($N_0 \gg N_1$) and when $\dim X = 1$. However, as $\dim X$ increases, the bias becomes comparable to or even dominates the standard error. Bias correction methods, such as the Abadie-Imbens correction, are essential in such cases.

2. *Inference Challenges*: Conducting valid inference with matching is complex. Abadie and Imbens (2006) provide an asymptotic variance formula for matching estimators, but standard techniques like bootstrapping fail due to the discrete nature of the matching procedure (Abadie and Imbens, 2008). Advanced methods for variance estimation are required to ensure accurate inference.

3. *Efficiency Limitations*: Matching is less efficient with a fixed the ratio of control units to treated units, (R), in the matching process. However, as $R \to \infty$, efficiency improves, especially when combined with bias correction (Lin, Ding, and Han, 2023). This underscores the value of large samples and careful bias adjustment in making matching methods more effective.

4. *Computational Complexity*: Matching can be computationally intensive, particularly with large datasets or high-dimensional covariates. Algorithms must handle both the distance calculations and the optimization problem of finding the best

matches, which can be resource-intensive. One of the biggest challenges in matching is the curse of dimensionality, where increasing the number of covariates can make it difficult to find good matches. As seen in this simulation, we only used three covariates for matching, but as the number of covariates increases, computational challenges and poor overlap may arise. To address this, we recommend several adjustments: if computational limits arise, substantially decreasing N can help (like we decreased N to 2000 in this simulation); for visualization, we used `plot_data <- data %>% sample_n(200)` to reduce dataset size used for plotting; and to smooth density plots, we used `adjust = 0.5` to modify bandwidth and avoid overfitting. These adjustments improve computational feasibility and visualization while maintaining interpretability.

These caveats emphasize the importance of careful implementation and adjustment to improve the reliability of matching. Despite its limitations, matching remains a valuable tool in causal inference, offering notable advantages over standard regression methods.

Matching methods are widely used in causal inference to estimate treatment effects by creating comparable groups of treated and untreated units based on observed covariates. Traditional regression models, as discussed in the previous chapter, also rely on the assumption that selection into treatment is based only on observable characteristics, meaning that any bias can be eliminated by controlling for these variables. Given the relative simplicity of regression, why not use it instead of matching? Matching offers several advantages that standard regression does not.

First, matching does not impose a linear functional form on the relationship between treatment and outcome. If the Conditional Independence Assumption (CIA) holds but the relationship is nonlinear, matching remains a consistent estimator, whereas regression may introduce bias due to incorrect functional form specification (Imbens & Rubin, 2015). Second, matching explicitly addresses the common support problem by ensuring that treated units are only compared to similar untreated units. Unlike regression, which extrapolates across the entire sample, matching makes it clear whether sufficient overlap exists between treated and control observations. Without proper overlap, regression may estimate effects in regions where no comparable untreated units exist, leading to unreliable results (Rosenbaum & Rubin, 1983).

Another key advantage is that matching assigns different weights to observations when constructing the counterfactual. In ordinary least squares (OLS), all control units contribute to estimating the expected counterfactual for treated units. In contrast, matching ensures that only untreated units with similar covariates to each treated unit influence the counterfactual estimation, reducing the risk of extrapolation bias. However, matching methods can also face challenges, particularly when dealing with high-dimensional covariates. If multiple covariates each have several discrete levels, the number of possible combinations can grow exponentially, leading to sparse data in many strata (Abadie & Imbens, 2011). This issue, often referred to as the curse of dimensionality, makes it difficult to find exact matches for treated units, highlighting the importance of alternative approaches such as propensity score matching or weighting methods to address these limitations.

The `MatchIt` package in R provides a comprehensive framework for implementing various matching techniques, including nearest-neighbor, optimal, and full matching. The package offers flexibility in defining distance metrics, such as Euclidean or Mahalanobis distance, and supports adjustments like calipers or trimming to improve balance. MatchIt also facilitates balance diagnostics, such as love plots and standardized bias checks, ensuring robust comparisons between groups (MatchIt Documentation[4]).

In a broader context, Stuart (2010) provides a comprehensive review of matching methods, highlighting their role in reducing confounding and improving comparability in observational studies. The review covers various approaches, including nearest-neighbor matching, optimal matching, and full matching, while discussing their respective advantages and limitations. Stuart emphasizes the importance of overlap and balance diagnostics, such as standardized mean differences, to validate matching results and reduce bias. More recent reviews, such as King and Nielsen (2019), critique some traditional matching practices and advocate for refined methodologies like coarsened exact matching and genetic matching, which optimize balance and address the curse of dimensionality. For readers seeking practical implementation guides, the `MatchIt` package documentation ([Ho et al., 2007]) provides a detailed explanation of matching techniques and their application in R. Together, these resources offer a robust foundation for understanding and applying matching methods effectively in causal inference.

In addition to R, Stata and Python offer powerful tools for implementing matching methods in causal inference. Stata's `teffects` command provides built-in functions for propensity score matching, inverse probability weighting, and nearest-neighbor matching, allowing researchers to estimate treatment effects with robust standard errors (Abadie & Imbens, 2016). For more advanced options, the `psmatch2` and `cem` (Coarsened Exact Matching) packages extend Stata's capabilities with additional balance diagnostics and matching algorithms (Leuven & Sianesi, 2003). In Python, the `causalml` and `pymatch` libraries offer flexible implementations of propensity score matching, machine learning-based matching, and balance checks, enabling users to leverage modern statistical techniques for causal inference (Chen et al., 2020). These tools, along with extensive documentation and methodological advancements, provide researchers with robust alternatives to implement matching across different software environments.

Despite their flexibility, traditional matching methods face challenges in settings with numerous covariates or sparse overlap. Propensity score matching (PSM) offers an elegant solution by reducing the dimensionality of the matching problem. Instead of matching directly on covariates, PSM matches units based on their propensity scores, $P(D_i = 1 \mid X_i)$, which represent the probability of treatment given the covariates. This approach simplifies the matching process, improves computational efficiency, and ensures balance across a single propensity score rather than multiple covariates. In the next section, we explore PSM and its application in causal inference.

20.3 Propensity Score Matching

Propensity score matching (PSM) is a popular method for estimating causal effects in observational studies. It simplifies matching by summarizing all covariates (X_i) into a single propensity score, $p(X_i) = P(D_i = 1 \mid X_i)$, representing the probability of receiving treatment given the covariates (Rosenbaum and Rubin, 1983). (The propensity score is also known as the conditional treatment probability, and $e(X_i)$ is also very common notation.)

The first step in implementing PSM is to obtain propensity scores, which we will discuss below. Once the propensity scores are estimated, they can be used to calculate treatment effects through various methods. These include direct regression adjustment, stratified regressions, matching methods, inverse probability weighting (IPW), or machine learning-based approaches such as causal random forests, Bayesian additive regression trees (BART), doubly robust or double machine learning (DML). Among these, matching methods are the most commonly used. PSM addresses the dimensionality problem in traditional matching by ensuring balance across a single score rather than multiple covariates. The counterfactual outcome for treated units is approximated by the average outcome of matched untreated units with similar propensity scores, under the assumption that no unobserved factors significantly affect both treatment assignment and the outcome. The treatment effect is estimated as the average difference in outcomes between treated units and their matched controls. Meanwhile DML is increasingly gaining popularity in economics due to its ability to handle high-dimensional data and account for model selection and regularization.

Rosenbaum and Rubin's (1983) foundational work introduced two critical propositions that establish the theoretical basis for propensity scores, particularly their role in balancing covariates and enabling causal inference in observational studies:

Proposition 1: If the propensity score is defined as $p(X_i) = P(D_i = 1 \mid X_i)$, then $D_i \perp X_i \mid p(X_i)$, meaning treatment assignment D_i is independent of covariates X_i given the propensity score.[5]

Proposition 2: If the Conditional Independence Assumption (CIA) holds, such that treatment assignment is independent of potential outcomes given covariates X_i, $D_i \perp (Y_i(0), Y_i(1)) \mid X_i$, then treatment assignment is also independent of potential outcomes given the propensity score $p(X_i)$, $D_i \perp (Y_i(0), Y_i(1)) \mid p(X_i)$. This result implies that adjusting for the scalar propensity score $p(X_i) = P(D_i = 1 \mid X_i)$ is sufficient for eliminating bias due to observed confounders, simplifying the estimation of causal effects.[6]

These propositions highlight that propensity scores provide a sufficient statistic for balancing observed covariates between treated and control groups under the CIA. This simplifies matching, stratification, or weighting procedures by reducing the dimensionality of covariates.

20.3.1 Estimating the Propensity Score

The true propensity score is typically unknown and needs to be estimated in the data prior to implementing in causal effect estimation. Estimating the propensity score involves two critical decisions: selecting the estimation model and determining which variables to include in the model. The propensity score, defined as $p(X_i) = P(D_i = 1 \mid X_i)$, represents the conditional probability of receiving treatment given observed covariates.

The propensity score frequently estimated using traditional parametric methods such as logistic regression or probit models, or recently with more flexible machine learning algorithms like random forests, gradient boosting, or neural networks. Eventhough it is possible if we apply nonparametric methods like series or kernel regression the curse of dimensionality kicks back in w.r.t. the estimation of propensity score. The choice of method depends on the complexity of the dataset and the relationships between the covariates and treatment assignment. Parametric models are suitable for smaller datasets or when the relationships are relatively simple, while machine learning methods are better equipped to handle nonlinearity, interactions, and high-dimensional covariates.

The CIA requires that, conditional on the propensity score, the outcome variable(s) must be independent of treatment. Heckman, Ichimura, and Todd (1997) demonstrate that omitting important variables can severely bias the resulting treatment effect estimates. Therefore, variable selection should prioritize covariates that influence both the likelihood of treatment assignment (D_i) and the outcome variable (Y_i). Including irrelevant variables does not generally harm the validity of the estimates but can reduce precision, whereas omitting relevant variables leads to biased estimates.

For example, in estimating the effect of a college scholarship on earnings, we can use a logit model to calculate the propensity score, $p(X_i) = P(D_i = 1 \mid X_i)$. The logit model specifies:

$$p(X_i) = \frac{\exp(\beta_0 + \beta_1\text{Parental Income} + \beta_2\text{GPA} + \beta_3\text{SAT Score} + \beta_4\text{University Type} + \beta_5\text{Field of Study})}{1 + \exp(\beta_0 + \beta_1\text{Parental Income} + \beta_2\text{GPA} + \beta_3\text{SAT Score} + \beta_4\text{University Type} + \beta_5\text{Field of Study})}$$

This formulation ensures that the propensity score $p(X_i)$ is always between 0 and 1, making it a valid probability. By summarizing the relationship between these covariates and the likelihood of receiving a scholarship, the logit model simplifies the matching process. Using these scores, treated and control groups can be balanced to estimate the causal effect of scholarships on earnings.

After obtaining propensity scores, typically we need to assess overlap. Overlap, or common support, ensures that treated and control units have similar propensity scores, making meaningful comparisons possible. This step is conducted before matching by examining the distribution of propensity scores across treated and control groups. Diagnostics such as histograms, density plots, or propensity score overlap plots help identify regions where treated units lack comparable controls (or vice versa). If overlap is poor, trimming or restricting the sample to the region of common support is often necessary.

Once the propensity scores are estimated and assessed for overlap, they can be used to calculate treatment effects through various methods. Lets discuss matching methods, then IPW.

20.3.2 Matching Methods in PSM

The purpose of using the propensity score $p(X_i)$ is to identify and match treated and control observations that are "close" or nearly identical in terms of their likelihood of treatment, ensuring comparability for estimating causal effects. Several matching techniques are available after estimating propensity scores, each with its own strengths and considerations:

- *Nearest-Neighbor Matching*: This method matches each treated unit to the control unit with the closest propensity score. It can be implemented as one-to-one matching or extended to K-nearest neighbors, where the average outcomes of the K closest controls are used. This approach is straightforward but may not always ensure optimal balance.
- *Caliper Matching*: A caliper sets a maximum allowable threshold (caliper) between the propensity scores of treated and control units. For example, if the caliper is set to 0.1, treated and control units are matched only if their propensity score difference is ≤ 0.1). Matches outside this threshold are excluded, reducing the risk of poor matches. A variant, *radius matching*, adapts the caliper size as a fixed proportion of the standard deviation of the propensity score, improving flexibility and maintaining balance.
- *Optimal Matching*: Optimal matching minimizes the total distance between treated and control pairs across the sample. It considers all possible pairings and selects the combination that achieves the best overall balance. This method is computationally intensive but provides the best possible matches, reducing bias and improving precision. Optimal matching can be implemented using algorithms like the Hungarian algorithm or genetic matching, which optimize the matching process to achieve the closest possible matches.
- *Stratification and Interval Matching*: Units are grouped into strata or intervals based on their propensity scores, and treatment effects are estimated within each stratum. This

method pairs multiple treated and control units within each interval, maximizing data usage while minimizing bias.

- *Kernel Matching*: This nonparametric approach assigns weights to all control units based on their distance in propensity scores from each treated unit, with closer matches receiving higher weights. Common kernels include Gaussian and Epanechnikov. Variants such as *local linear matching (LLM)* extend this approach by fitting local regressions to further refine the counterfactual estimation.
- *Full Matching*: Full matching combines the strengths of nearest-neighbor and optimal matching by ensuring that all treated units are matched to control units and vice versa. This method minimizes bias by maximizing the number of matched pairs, improving balance and reducing variance.

The choice of matching method depends on the context of your study, the characteristics of your data, and your research objectives. Below are common choices for the selection.

TABLE 20.2 Recommended Matching Methods Based on Scenarios

Scenario	**Recommended Methods**
Few treated units, large control group	Nearest-neighbor matching, caliper or optimal matching
High-dimensional covariates	Propensity score matching, kernel matching, machine learning
Poor overlap in covariates	Caliper matching, trimming
Maximizing sample size	Full matching, inverse probability weighting
Focus on ATT	Nearest-neighbor matching, caliper matching
Focus on ATE	Kernel matching, IPW

Machine learning methods offer flexible and powerful tools for estimating propensity scores, particularly in settings with high-dimensional covariates or complex relationships between covariates and treatment assignment. Traditional parametric methods like logistic or probit regression assume linearity and additivity, which may not hold in real-world data. While nonparametric methods such as kernel or nearest-neighbor matching provide more flexibility in estimating propensity scores. However, these nonparametric approaches suffer from the curse of dimensionality, where their performance degrades as the number of covariates increases, making them less practical for high-dimensional datasets often encountered in observational studies. In the `MatchIt` package, users can also estimate propensity scores directly using classic methods like logistic or probit models, caliper or optimal matching by specifying these options in the `distance` argument, seamlessly integrating propensity score estimation with matching implementation.

In contrast, machine learning approaches such as random forests, gradient boosting machines (GBMs), and neural networks can model nonlinearities and interactions without pre-specifying the functional form and high-dimensional datasets. For instance, random forests use decision trees to capture complex patterns, while GBMs iteratively refine the propensity score model by minimizing prediction errors. GBMs offer powerful methods for estimating propensity scores, particularly in datasets with nonlinear relationships or high-dimensional covariates. Several GBM implementations are widely used. Traditional GBM, available in the `gbm` package, constructs an ensemble of decision trees iteratively, minimizing a loss function like logistic loss for binary treatment assignment. **XGBoost**, implemented in the `xgboost` package, extends GBM with additional features like L1/L2 regularization, tree pruning, and parallel processing, making it faster and less prone to overfitting. **LightGBM**, supported by the `lightgbm` package, further optimizes gradient boosting by using histogram-based learning and leaf-wise tree growth, improving speed and memory efficiency for large datasets. **CatBoost**, accessible through the `catboost` package, is designed for datasets with categorical variables, allowing direct handling of such data without extensive preprocessing. These GBM frameworks not only improve the accuracy of propensity score estimation but also provide robust handling of complex interactions and scalability for large observational studies.

Neural networks are particularly suited for very high-dimensional datasets, leveraging layers of transformations to capture intricate dependencies between covariates. Another powerful method, Super Learner, combines multiple algorithms to create an ensemble estimator, selecting the optimal combination based on cross-validation. These machine learning methods improve the precision of propensity score estimation. However, they often require larger datasets and careful parameter tuning to avoid overfitting. Each method has its strengths and limitations, and researchers should carefully consider these factors when selecting an appropriate matching strategy. The choice of method should prioritize balance, overlap, and the research question at hand, ensuring that the treatment effect estimates are reliable and interpretable.

20.3.3 Assessing Balance in PSM

After performing propensity score matching (PSM), it is crucial to evaluate whether the matched sample achieves balance between treated and control groups. The primary goal is to ensure that differences in covariates, which could bias treatment effect estimates, are minimized. While balance diagnostics in matching focus on covariates directly, in PSM, we assess balance both in terms of covariates and propensity scores to confirm that the matching procedure has successfully adjusted for pre-treatment differences.

Standardized bias, variance ratios, and t-tests are commonly used to compare covariates before and after matching. A standardized bias below 10% is often considered acceptable. While t-tests are sometimes reported, they are sensitive to sample size and may not always be informative. Graphical methods such as **love plots** (which visualize standardized mean differences) and **density plots** (which compare the distribution of propensity scores between groups) are widely used. The `MatchIt` package in R provides built-in balance diagnostics via the `summary(matchit_object)` function, which reports standardized mean differences, variance ratios, and overall balance metrics.

In empirical research, papers typically present balance tables showing covariate means and standardized differences before and after matching, often including multiple matching models for comparison. The most common approach for comparing different matching models (e.g., nearest neighbor, caliper, or kernel matching) is to assess how well each method reduces covariate imbalance while retaining sufficient sample size. Researchers may report overall balance improvement across models and select the approach that achieves the best trade-off between balance and efficiency.

20.4 Calculation of Treatment Effects

After estimating the propensity score $p(X_i)$, we first check for overlap to ensure that treated and control units have common support. Next, we perform matching to create comparable groups, followed by balance assessment to confirm that covariates are well-matched. Once we have a balanced sample, treatment effects can be estimated using various approaches, including direct regression adjustment, matching-based blocking methods, and reweighting techniques. Each method has its own strengths and assumptions, influencing the interpretation of the estimated causal effects.

20.4.1 Regression Adjustment Including Propensity Score

One approach to estimating treatment effects is to include the propensity score $p(X_i)$ as a covariate in a regression model. This method aims to control for any residual differences between treated and control units that may persist even after matching. The general specification is:

$$Y_i = \alpha + \delta D_i + f(p(X_i)) + \varepsilon_i \tag{20.6}$$

where Y_i is the outcome, D_i is the treatment indicator, and $p(X_i)$ is the estimated propensity score. The function $f(p(X_i))$ can be specified flexibly, such as linear, quadratic, or spline transformations of $p(X_i)$. The coefficient δ represents the treatment effect, assuming that $p(X_i)$ fully captures confounding effects.

A common implementation of this approach is regression adjustment, where a linear specification of $f(p(X_i))$ is used:

$$Y_i = \beta D_i + \gamma p(X_i) + \varepsilon_i \tag{20.7}$$

Here, β captures the treatment effect δ, and $p(X_i)$ is assumed to adjust for any remaining imbalances. While regression adjustment on the full sample without matching is possible, it is not recommended, as it does not ensure covariate balance. Instead, applying regression after matching provides a more robust estimate by refining balance and reducing model dependence.

By including $p(X_i)$ as a covariate, the regression effectively controls for differences in covariates between treated and control groups. Under constant treatment effects, this adjustment is sufficient, as long as the propensity score is correctly specified. If $p(X_i)$ is estimated using a linear probability model, controlling for $p(X_i)$ is numerically equivalent to controlling for X_i linearly, since $p(X_i) = P(D_i = 1|X_i) = X_i'\beta$. Thus, whether we control directly for X_i or for $p(X_i)$, the result is identical when using a linear probability model for estimating the propensity score.

When treatment effects are heterogeneous (vary across individuals), regression adjustment using $p(X_i)$ does not estimate the ATE directly but rather a variance-weighted average of effects. This happens because observations with extreme $p(X_i)$ values contribute disproportionately to the estimation due to larger residual variances in outcome prediction.

Mathematically, under heterogeneous treatment effects δ_i, the regression estimator recovers:

$$\delta_{\text{OLS}} = \frac{Var(p(X_i))}{Var(p(X_i)) + Var(\varepsilon_i)} E[\delta_i^{ATE}] \tag{20.8}$$

where the weight depends on the relative variance of $p(X_i)$ and the outcome residuals ε_i. This means that regression adjustment does not estimate a simple ATE but rather a weighted treatment effect, which is influenced by the variability in treatment assignment probabilities.

Regression adjustment and propensity score matching are closely related, as both attempt to eliminate confounding by conditioning on observed characteristics. While matching explicitly constructs comparable groups, regression adjustment models differences parametrically. When treatment effects are constant and $p(X_i)$ is estimated correctly, both approaches yield numerically similar results. However, with heterogeneous effects, weighting in regression adjustment leads to estimates that may not correspond directly to ATE, requiring additional robustness checks such as re-weighting or DML approaches.

20.4.2 Matching and Stratification on Propensity Score

Instead of matching directly on covariates, we use the estimated propensity score $p(X_i)$ to construct comparable groups. The process follows the same principles as the matching and stratification methods discussed in detail above but applied to $p(X_i)$ rather than the original covariates.

Matching assigns each treated unit to the closest control unit based on $p(X_i)$, ensuring comparability in treatment likelihood. Untreated units with $p(X_i)$ values outside the range of treated observations are discarded to maintain common support. This approach simplifies balancing by reducing the dimensionality to a single score while retaining the advantages of the matching techniques described earlier.

Stratification (or blocking) divides observations into bins based on their estimated propensity scores and estimates treatment effects within each bin. The treatment effect is then aggregated across bins, weighting by the number of observations. While this method can estimate both ATE and ATT, it is primarily used for ATT, given that matching typically focuses on treated individuals. The ATT estimate follows:

$$\delta_{\text{ATT}} = \sum_{k=1}^{K} \frac{N_{1k}}{N_1} \left(\bar{Y}_k(1) - \bar{Y}_k(0)\right) \tag{20.9}$$

where N_{1k} represents the number of treated units in stratum k, and $\bar{Y}_k(1)$, $\bar{Y}_k(0)$ denote the mean outcomes for treated and control units within the stratum. This method ensures balance within each stratum, reducing bias and improving precision.

20.4.3 Estimation of Standard Errors in PSM

Accurate estimation of standard errors is crucial in PSM to ensure valid inference. Unlike regression-based methods that rely on asymptotic approximations, matching introduces complexities due to resampling and discrete treatment assignment, necessitating specialized techniques for variance estimation. Several approaches have been proposed to improve inference in PSM, including bootstrapping, Lechner's variance approximation, and the Abadie-Imbens correction.

Bootstrapping for Standard Error Estimation
Bootstrapping is a widely used resampling technique that repeatedly draws samples with replacement from the original dataset to compute standard errors empirically. This approach accounts for sampling variability and can be applied to PSM by repeatedly matching observations within each resampled dataset and calculating treatment effects across iterations

(Austin & Small, 2014). However, bootstrapping in PSM is debated because the resampling process may not fully account for the discrete nature of matching, potentially underestimating standard errors (Abadie & Imbens, 2008). Despite this limitation, bootstrapping remains a practical tool for variance estimation, especially when alternative methods are computationally infeasible.

Variance Approximation by Lechner
Lechner (2002) introduced a variance approximation method specifically designed for matching estimators. His approach constructs standard errors by incorporating the variability introduced by the matching process itself, adjusting for the fact that each treated unit is paired with a weighted set of control units. This method provides a more computationally efficient alternative to bootstrapping while maintaining robust inference properties.

Abadie and Imbens (2006) Correction for Standard Errors
Recognizing the limitations of standard bootstrapping in matching settings, Abadie and Imbens (2006) developed a robust analytical method to estimate standard errors in PSM. Their approach explicitly accounts for the fact that nearest-neighbor matching leads to a discrete set of matched pairs, violating the assumptions underlying conventional variance estimators. The Abadie-Imbens correction derives an analytical variance formula that adjusts for matching discrepancies and improves the reliability of inference in causal effect estimation. Their framework has been widely adopted in empirical research and implemented in statistical software, including Stata's `teffects psmatch` and R's `Matching` package.

These methods highlight the challenges of inference in PSM and offer different trade-offs between computational complexity and statistical rigor. While bootstrapping provides a flexible empirical approach, Lechner's approximation offers computational efficiency, and the Abadie-Imbens correction ensures robustness by explicitly accounting for matching characteristics. Researchers must carefully choose the most appropriate method depending on sample size, matching approach, and computational constraints.

We will present the simulation and a detailed explanation of the simulation for PSM incorporating a machine learning method below. In that simulation, if you replace Step 2 with the following code, it will simulate and demonstrate including the next figure how to perform regular PSM using logistic regression instead of a machine learning approach.

```
# STEP 2: Propensity Score Estimation using Logistic Regression (PSM)
ps_model <- glm(Treatment ~ FamilyIncome + ContVar1 + BinaryVar1,
                data = data, family = binomial)

# Predict propensity scores
data$PS <- predict(ps_model, type = "response")
```

Additionally, the figure "Propensity Score Distribution using Logit" presents the output of different matching methods using logistic regression-based propensity scores to visualize propensity score overlap before and after matching. This allows for a direct visual comparison between traditional PSM and machine learning-based PSM, highlighting the potential benefits of using more flexible machine learning models for propensity score estimation.

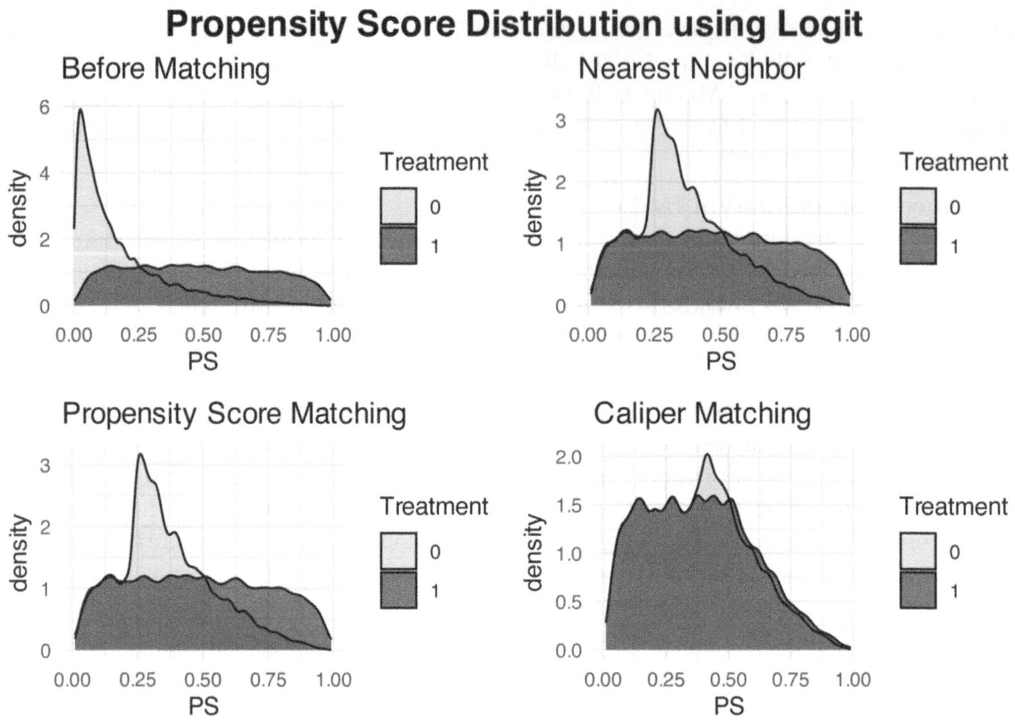

20.5 Integrating XGBoost with MatchIt for Propensity Score Matching

Combining machine learning with traditional matching techniques can improve causal inference by improving the estimation of propensity scores. The `MatchIt` package in R allows users to incorporate externally computed propensity scores, such as those generated by `XGBoost`, into its matching framework. This approach leverages the predictive power of machine learning while ensuring balance between treated and control groups through matching. Below is an example demonstrating how to estimate propensity scores using `XGBoost` and integrate them into `MatchIt` for PSM.

```
# Load required libraries
library(xgboost)
library(MatchIt)

# Example dataset (replace with your dataset)
data <- your_dataset
treatment <- as.numeric(data$treatment) # Binary treatment variable (0/1)
covariates <- as.matrix(data[, c("covariate1", "covariate2", "covariate3")])
# Covariates

# Step 1: Estimate Propensity Scores Using XGBoost
dtrain <- xgb.DMatrix(data = covariates, label = treatment)
```

```
# Train XGBoost model for propensity score estimation
xgb_model <- xgboost(
  data = dtrain,
  objective = "binary:logistic",  # Logistic regression for binary treatment
  nrounds = 100,                  # Number of boosting iterations
  eta = 0.1,                      # Learning rate
  max_depth = 3,                  # Tree depth
  verbose = 0                     # Suppress training output
)

# Predict propensity scores
data$propensity_score <- predict(xgb_model, covariates)

# Step 2: Use Pre-Computed Propensity Scores in MatchIt
match <- matchit(
  formula = treatment ~ 1, # No need to specify covariates here
  method = "nearest",        # Choose the matching method (e.g., nearest-neighbor)
  distance = data$propensity_score,  # Provide pre-computed propensity scores
  data = data
)

# Step 3: Assess Matching Quality
summary(match)                    # Summary of matching results
matched_data <- match.data(match) # Extract the matched dataset
```

After uploading data, **Estimating Propensity Scores with XGBoost**: The model predicts the probability of treatment assignment ($p(X_i)$) based on observed covariates. Unlike traditional logistic regression, gradient boosting captures complex relationships between covariates and treatment assignment, potentially improving the accuracy of propensity scores.

Integrating with MatchIt: Instead of estimating propensity scores within MatchIt, we can supply pre-computed scores using the `distance` argument. This enables flexible use of machine learning models for score estimation while still leveraging `MatchIt`'s robust matching procedures.

Choosing a Matching Method: The example above applies nearest-neighbor matching, but users can modify this to caliper matching, optimal matching, or other methods depending on their research needs. See the MatchIt Matching Methods vignette[7] for details on different techniques.

Evaluating Matching Quality: After matching, balance diagnostics should be checked using `summary(match)`. A well-matched dataset should exhibit reduced covariate imbalance between treated and control groups.

Interpreting and Analyzing Results: Once balance is achieved, treatment effects can be estimated from the matched dataset, ensuring a more valid causal inference.

Let's continue our discussion using the simulation demonstrating the use of PSM with XgBoost propensity score estimation to estimate the effect of college graduation on earnings while controlling for family income, a continuous variable, and a binary covariate. The dataset consists of 20,000 observations, where treated units (college graduates) and control units (non-graduates) differ in their observed characteristics. To ensure comparability between these groups, we implement three matching methods: nearest-neighbor matching, PSM, and caliper matching. Instead of traditional logistic regression, we use XGBoost, a powerful gradient boosting algorithm, to estimate propensity scores, allowing for more flexible modeling of treatment assignment. These scores are then used to match treated and control units, reducing bias in estimating the average treatment effect on the treated (ATT).

To assess the effectiveness of matching, we first check the initial balance of covariates, finding systematic differences between treated and control groups. After applying matching, we reassess balance and find that all methods improve comparability, though caliper matching ensures the best balance by restricting matches within a defined threshold of propensity scores. We then estimate ATT by comparing mean earnings between matched groups, with the caliper-matched sample providing the most reliable estimate due to better covariate balance. Finally, we visualize the propensity score distributions before and after matching, illustrating how each method improves overlap between treated and control units. This simulation highlights the importance of using machine learning for propensity score estimation and demonstrates the practical implementation of different matching techniques for causal inference.

```r
# Load necessary libraries
library(MatchIt)
library(dplyr)
library(tableone)
library(ggplot2)
library(xgboost)

# Set seed for reproducibility
set.seed(42)

# Define sample sizes
N <- 20000
NT <- 5000
NC <- N - NT

# Generate earnings for treated and control groups
earnings_non_graduates <- rnorm(NC, mean = 50, sd = 10)
earnings_graduates <- rnorm(NT, mean = 53, sd = 10)

# Generate covariates
family_income_non_graduates <- rnorm(NC, mean = 70, sd = 20)
family_income_graduates <- rnorm(NT, mean = 95, sd = 20)

# Generate 1 continuous variable
cont_var1_non_graduates <- rnorm(NC, mean = 5, sd = 2)
cont_var1_graduates <- rnorm(NT, mean = 6, sd = 2)

# Generate 1 binary variable
binary_var1_non_graduates <- rbinom(NC, 1, 0.4)
binary_var1_graduates <- rbinom(NT, 1, 0.6)
```

```
# Merge into a single dataset
data <- data.frame(
  ID = 1:N,
  Treatment = c(rep(1, NT), rep(0, NC)),
  FamilyIncome = c(family_income_graduates, family_income_non_graduates),
  ContVar1 = c(cont_var1_graduates, cont_var1_non_graduates),
  BinaryVar1 = c(binary_var1_graduates, binary_var1_non_graduates),
  Earnings = c(earnings_graduates, earnings_non_graduates)
)

# STEP 1: Check Initial Covariate Balance
table1 <- CreateTableOne(vars = c("FamilyIncome", "ContVar1", "BinaryVar1"),
          strata = "Treatment", data = data, test = TRUE)
print(table1)

##                              Stratified by Treatment
##                               0              1              p      test
##   n                           15000          5000
##   FamilyIncome (mean (SD)) 70.08 (20.24) 95.01 (20.18) <0.001
##   ContVar1 (mean (SD))      4.99 (1.98)   5.99 (1.99)  <0.001
##   BinaryVar1 (mean (SD))    0.40 (0.49)   0.60 (0.49)  <0.001

# STEP 2: Propensity Score Estimation using XGBoost
covariates <- as.matrix(data[, c("FamilyIncome", "ContVar1", "BinaryVar1")])
dtrain <- xgb.DMatrix(data = covariates, label = data$Treatment)

xgb_model <- xgboost(
  data = dtrain,
  objective = "binary:logistic",
  nrounds = 100,
  eta = 0.1,
  max_depth = 3,
  verbose = 0
)

# Predict propensity scores using XGBoost
data$PS <- predict(xgb_model, covariates)

# STEP 3: Nearest Neighbor Matching (1:1)
nn_match <- matchit(Treatment ~ 1, data =data,method="optimal",distance=data$PS)
matched_data_nn <- match.data(nn_match)

# STEP 4: Propensity Score Matching (PSM) using XGBoost-based scores
psm_match <- matchit(Treatment ~ 1, data =data,method ="nearest",distance=data$PS)
matched_data_psm <- match.data(psm_match)

# STEP 5: Caliper Matching
caliper_match <- matchit(Treatment ~ 1, data = data, method = "nearest",
                caliper = 0.1, distance = data$PS)
```

```
matched_data_caliper <- match.data(caliper_match)

# STEP 6: Check Covariate Balance After Matching
balance_nn <- CreateTableOne(vars = c("FamilyIncome", "ContVar1", "BinaryVar1"),
            strata = "Treatment", data = matched_data_nn, test = TRUE)
balance_psm <- CreateTableOne(vars = c("FamilyIncome", "ContVar1", "BinaryVar1"),
            strata = "Treatment", data = matched_data_psm, test = TRUE)
balance_caliper <- CreateTableOne(vars = c("FamilyIncome","ContVar1","BinaryVar1"),
              strata = "Treatment", data = matched_data_caliper, test = TRUE)

print(balance_nn)
```

```
##                              Stratified by Treatment
##                               0             1             p      test
##   n                            5000          5000
##   FamilyIncome (mean (SD)) 87.41 (15.46) 95.01 (20.18) <0.001
##   ContVar1 (mean (SD))      5.73 (1.93)   5.99 (1.99)  <0.001
##   BinaryVar1 (mean (SD))    0.54 (0.50)   0.60 (0.49)  <0.001
```

```
print(balance_psm)
```

```
##                              Stratified by Treatment
##                               0             1             p      test
##   n                            5000          5000
##   FamilyIncome (mean (SD)) 87.43 (15.45) 95.01 (20.18) <0.001
##   ContVar1 (mean (SD))      5.73 (1.92)   5.99 (1.99)  <0.001
##   BinaryVar1 (mean (SD))    0.53 (0.50)   0.60 (0.49)  <0.001
```

```
print(balance_caliper)
```

```
##                              Stratified by Treatment
##                               0             1             p      test
##   n                           3725          3725
##   FamilyIncome (mean (SD)) 88.34 (16.66) 88.16 (17.19)  0.647
##   ContVar1 (mean (SD))      5.68 (1.90)   5.69 (1.99)   0.859
##   BinaryVar1 (mean (SD))    0.55 (0.50)   0.54 (0.50)   0.926
```

```
# STEP 7: Estimate Treatment Effect After Matching
estimate_ATT <- function(matched_data) {
  mean(matched_data$Earnings[matched_data$Treatment == 1]) -
  mean(matched_data$Earnings[matched_data$Treatment == 0])
}

cat("ATT (Nearest Neighbor):", estimate_ATT(matched_data_nn), "\n")
```

```
## ATT (Nearest Neighbor): 3.143483
```

```
cat("ATT (Propensity Score Matching):", estimate_ATT(matched_data_psm), "\n")

## ATT (Propensity Score Matching): 3.136654

cat("ATT (Caliper Matching):", estimate_ATT(matched_data_caliper), "\n")

## ATT (Caliper Matching): 3.258506

# STEP 8: Visualize Propensity Score Overlap Before and After Matching
plot1 <- ggplot(data, aes(x = PS, fill = as.factor(Treatment))) +
  geom_density(alpha = 0.5, color = "black", adjust = 0.5) +
  scale_fill_manual(values = c("gray80", "black")) +
  labs(title = "Before Matching", fill = "Treatment") +
  theme_minimal()

plot2 <- ggplot(matched_data_nn, aes(x = PS, fill = as.factor(Treatment))) +
  geom_density(alpha = 0.5, color = "black", adjust = 0.5) +
  scale_fill_manual(values = c("gray80", "black")) +
  labs(title = "Nearest Neighbor", fill = "Treatment") +
  theme_minimal()

plot3 <- ggplot(matched_data_psm, aes(x = PS, fill = as.factor(Treatment))) +
  geom_density(alpha = 0.5, color = "black", adjust = 0.5) +
  scale_fill_manual(values = c("gray80", "black")) +
  labs(title = "Propensity Score Matching", fill = "Treatment") +
  theme_minimal()

plot4 <- ggplot(matched_data_caliper, aes(x = PS, fill = as.factor(Treatment))) +
  geom_density(alpha = 0.5, color = "black", adjust = 0.5) +
  scale_fill_manual(values = c("gray80", "black")) +
  labs(title = "Caliper Matching", fill = "Treatment") +
  theme_minimal()

# Combine plots into a 2x2 layout with a title
library(gridExtra)
library(grid)
grid.arrange(
  arrangeGrob(
    plot1, plot2, plot3, plot4,
    ncol = 2, nrow = 2,
    top = textGrob("Propensity Score Distribution",
        gp = gpar(fontsize = 16, fontface = "bold"))
  )
)
```

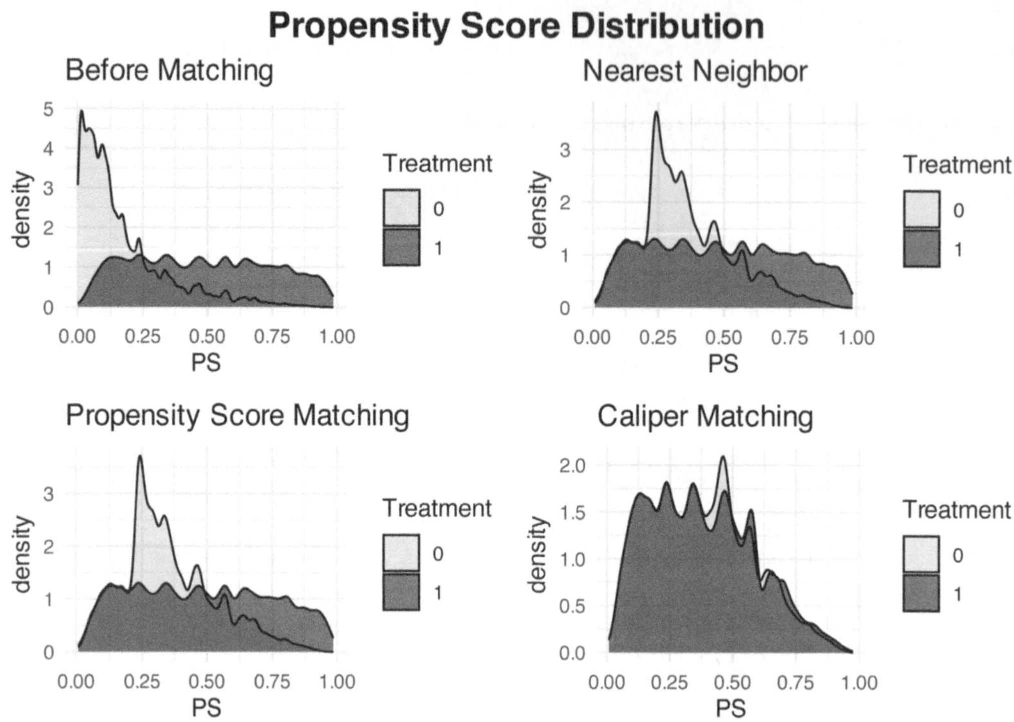

Integration of `XgBoost` combines the flexibility of machine learning for estimating propensity scores with the structured framework of `MatchIt`, allowing researchers to implement data-driven matching strategies. Other machine learning models, such as **LightGBM**, **CatBoost**, and **SuperLearner**, can also be used for propensity score estimation and incorporated into `MatchIt` in the same manner.

In Python, packages such as `causalml` and `pymatch` provide tools for propensity score estimation and matching, enabling users to apply machine learning techniques like gradient boosting and deep learning for more refined estimates.

Similarly, in Stata, the `psmatch2` and `teffects` commands allow for flexible propensity score estimation and matching, while `lassopack` can be used to incorporate machine learning-based selection of covariates in high-dimensional settings. These tools extend the application of matching beyond traditional logistic regression, making machine learning-powered causal inference more accessible across different statistical software.

Several studies have explored the use of XGBoost in propensity score estimation for matching-based causal inference. A study on cardiovascular outcomes among diabetes patients compared machine learning models, including XGBoost, to traditional logistic regression for generating propensity scores. The results showed that XGBoost provided superior covariate balance, closely approximating the performance of logistic regression while improving flexibility in capturing complex relationships (PMC, 2023). Similarly, researchers proposed an XGBoost-based approach for propensity score computation in the presence of label-corrupted data, demonstrating that clustering-based sampling improved the robustness of estimates, particularly under artificial data corruptions (Arxiv, 2018). In the domain of transportation research, XGBoost was employed to estimate propensity scores in a study analyzing the impact of interfering vehicles on traffic speed, highlighting the model's effectiveness in producing reliable causal inferences (Arxiv, 2023). These studies illustrate the growing application of XGBoost in propensity score estimation, offering improved balance diagnostics and robustness in observational studies compared to conventional methods.

In this chapter, we explored a range of matching methods for causal inference, including stratification, nearest-neighbor matching, full and optimal matching, and propensity score matching. We discussed how these methods construct comparable treatment and control groups under the selection on observables framework, with a focus on balance diagnostics and implementation across software platforms. We also demonstrated how machine learning tools like XGBoost can be integrated into matching workflows for more accurate propensity score estimation. In the next chapter, we build on this foundation by introducing IPW and doubly robust estimators—methods that combine weighting and modeling strategies to further strengthen causal identification in observational studies.

Notes

1. Normalized Euclidean Distance: $d_{\text{Normalized}}(i, j) = \sqrt{\sum_{k=1}^{K} \frac{(X_{ik} - X_{jk})^2}{s_k^2}}$, where s_k^2 is the sample variance of covariate k, ensuring equal scaling across covariates.

2. Mahalanobis Distance: $d_{\text{Mahalanobis}}(i, j) = \sqrt{(X_i - X_j)^\top \Sigma^{-1} (X_i - X_j)}$, where Σ is the covariance matrix of the covariates, accounting for correlations between them.

3. Genetic Matching Distance: $d_{\text{Genetic}}(i, j) = \sqrt{(X_i - X_j)^\top W (X_i - X_j)}$, where W is a diagonal matrix of covariate weights optimized via a genetic algorithm to minimize imbalance between treated and control groups.

4. https://cran.r-project.org/web/packages/MatchIt/vignettes/matching-methods.html

5. Proof of Proposition 1: $P(D_i = 1 \mid X_i, p(X_i)) = P(D_i = 1 \mid X_i) = p(X_i)$, since $p(X_i)$ is defined as $P(D_i = 1 \mid X_i)$. Similarly: $P(D_i = 0 \mid X_i, p(X_i)) = 1 - p(X_i)$. Thus, D_i is conditionally independent of X_i given $p(X_i)$, i.e., $D_i \perp X_i \mid p(X_i)$.

6. Proof of Proposition 2: Given $D_i \perp (Y_i(0), Y_i(1)) \mid X_i$ (CIA), we know $P(D_i = 1 \mid X_i, Y_i(0), Y_i(1)) = P(D_i = 1 \mid X_i) = p(X_i)$. Using the law of iterated expectations: $P(D_i = 1 \mid p(X_i), Y_i(0), Y_i(1)) = E[P(D_i = 1 \mid p(X_i), X_i, Y_i(0), Y_i(1)) \mid p(X_i), Y_i(0), Y_i(1)]$. Since $D_i \perp (Y_i(0), Y_i(1)) \mid X_i$, it simplifies to $P(D_i = 1 \mid X_i) = p(X_i)$. Thus $P(D_i = 1 \mid p(X_i), Y_i(0), Y_i(1)) = p(X_i)$, showing $D_i \perp (Y_i(0), Y_i(1)) \mid p(X_i)$.

7. https://cran.r-project.org/web/packages/MatchIt/vignettes/matching-methods.html

21

Inverse Weighting and Doubly Robust Estimation

This chapter continues our exploration of causal inference under the selection on observables framework, building on the matching and propensity score methods introduced in the previous chapter. While those approaches create balance by pairing or weighting units based on covariates, they often face challenges with high-dimensional data or when overlap is limited. Here, we turn to inverse probability weighting (IPW) and doubly robust (DR) estimators—two techniques that extend the logic of propensity scores to construct weighted samples that approximate randomized experiments.

These methods help reduce reliance on strong modeling assumptions by reweighting or augmenting the data in ways that address potential confounding. We begin by introducing IPW, which uses estimated treatment probabilities to balance treated and control groups across observed covariates. We then turn to doubly robust estimators, such as augmented IPW (AIPW), which combine outcome modeling and weighting to offer consistent estimates even if only one of the two models is correctly specified. Through theoretical explanations, simulations, and implementation in R, this chapter provides practical tools to estimate average treatment effects and diagnose potential pitfalls in applied settings.

21.1 Inverse Probability Weighting

IPW, or inverse probability weighting, adjusts for selection bias by weighting each observation by the inverse of its treatment probability, given covariates. This makes treated and control groups comparable, mimicking randomization. For applied economists, it is vital in policy evaluation, like assessing job training impacts on earnings, where data is not randomly assigned.

This approach, based on Horvitz and Thompson (1952), assigns weights to each observation based on the estimated propensity score $p(X_i)$ at the individual level, which is the probability of receiving treatment given covariates. For each individual in treatment group ($D_i = 1$) the weight is $\omega_i = \frac{1}{p(X_i)}$. This adjusts for how "likely" they were to receive treatment based on their covariates. For each individual in the control group ($D_i = 0$), the weight is $\omega_i = \frac{1}{1-p(X_i)}$. This adjusts for how "likely" they were to not receive treatment based on their covariates.

Each individual's outcome (Y_i) is reweighted by their assigned weight. For example, a treated individual with $p(X_i) = 0.2$ would have their outcome weighted by $\frac{1}{0.2} = 5$, reflecting their relative rarity in the treated group. A control individual with $p(X_i) = 0.8$ would have their outcome weighted by $\frac{1}{0.2} = 5$ for their likelihood of not being treated. These weights adjust for the likelihood of treatment or control assignment, ensuring that treated and control groups become comparable in the reweighted pseudo-population.

DOI: 10.1201/9781003504795-21

With unconfoundedness, positivity, and SUTVA assumptions, the average treatment effect (ATE) is defined as the difference between the expected outcomes under treatment and control and is estimated using the IPW method as:

$$\begin{aligned}\delta_{\text{ATE}} &= \mathbb{E}[Y(1) - Y(0)] = \mathbb{E}[Y(1)] - \mathbb{E}[Y(0)] \\ &= \mathbb{E}\left[\frac{D_i Y_i}{p(X_i)}\right] - \mathbb{E}\left[\frac{(1-D_i)Y_i}{1-p(X_i)}\right] = \frac{1}{N}\left(\sum_{i=1}^{N}\frac{D_i Y_i}{p(X_i)} - \sum_{i=1}^{N}\frac{(1-D_i)Y_i}{1-p(X_i)}\right)\end{aligned} \tag{21.1}$$

This method reweights each observed outcome, adjusting for selection bias. However, extreme propensity scores can lead to instability, so weights are often normalized to sum to 1 within each group:

$$\omega_{i_{Treated}}^{\text{norm}} = \frac{\omega_i}{\sum_{i=1}^{N}\omega_i D_i}, \quad \omega_{i_{Control}}^{\text{norm}} = \frac{\omega_i}{\sum_{i=1}^{N}\omega_i(1-D_i)} \tag{21.2}$$

The Hajek estimator is a stabilized alternative to the Horvitz-Thompson estimator, reducing sensitivity to extreme weights:

$$\delta_{\text{ATE}}^{\text{Hajek}} = \frac{\sum_{i=1}^{N} D_i \frac{Y_i}{p(X_i)}}{\sum_{i=1}^{N}\frac{D_i}{p(X_i)}} - \frac{\sum_{i=1}^{N}(1-D_i)\frac{Y_i}{1-p(X_i)}}{\sum_{i=1}^{N}\frac{1-D_i}{1-p(X_i)}} \tag{21.3}$$

In most practical settings, the Hajek estimator is favored for its improved stability and reduced variance.
As with unconfoundedness and SUTVA, which we discussed earlier, the positivity assumption is crucial for IPW. The positivity condition (overlap condition) ensures that every individual has a nonzero probability of receiving either treatment or control, meaning $0 < p(X_i) < 1$ for all X_i. If some individuals always receive treatment ($p(X_i) = 1$) or never do ($p(X_i) = 0$), there are no comparable observations in the other group, making it impossible to estimate counterfactual outcomes. Violations lead to unstable inverse probability weights, where $\frac{1}{p(X_i)}$ or $\frac{1}{1-p(X_i)}$ become excessively large, increasing variance and reducing reliability. To address this, researchers often trim extreme propensity scores (e.g., below 0.05 or above 0.95), restrict analysis to regions of common support, or use stabilized weights to reduce sensitivity to extreme values. When positivity violations are severe, alternative approaches like matching or entropy balancing may be preferable. Ensuring adequate overlap between treated and control groups by inspecting propensity score distributions is crucial before applying IPW.

IPW retains all observations, provided there is sufficient overlap in propensity scores between treated and control units. Common diagnostics include checking for extreme weights and assessing balance post-weighting. Popular implementations in R include the `WeightIt` and `twang` packages, which streamline weight estimation and diagnostic evaluation.

IPW is a simple method with a strong theoretical foundation, providing a smooth approach to causal estimation by explicitly defining the target population and achieving global covariate balance. It also extends naturally to more complex settings, making it a flexible tool in observational studies. However, IPW is highly sensitive to misspecification of the propensity score model, which can lead to biased estimates if key confounders are omitted or functional forms are incorrect. Additionally, when propensity scores are close to 0 or 1, extreme weights can cause instability, increasing variance and reducing reliability. Addressing these issues often requires ad hoc adjustments, such as trimming extreme weights or using stabilized weighting methods, adding a layer of complexity to its implementation.

The IPW simulation demonstrates how to estimate the causal effect of college graduation on earnings by reweighting observations based on their propensity scores. This approach addresses selection bias by adjusting the distribution of treated and control groups, ensuring that they are comparable in terms of observed characteristics. The dataset consists of 20,000 individuals, with treated units representing college graduates and control units representing non-graduates. Key covariates include family income, a continuous variable, and a binary covariate, which influence the probability of treatment assignment.

To implement IPW, we first estimate propensity scores using logistic regression, modeling the probability of college graduation based on observed covariates. Using these estimated probabilities, we compute IPW weights, where treated individuals receive a weight of 1 divided by their propensity score, and control individuals receive 1 divided by one minus their propensity score. This reweighting process adjusts the sample so that the distribution of covariates in the control group mimics that of the treated group. Additionally, we incorporate the Hajek estimator, a normalized version of IPW that ensures weights sum to one within each group, which improves efficiency in finite samples.

One of the primary challenges in IPW estimation is positivity violations, where some observations have propensity scores close to zero or one, leading to excessively large weights. To address this, we apply weight trimming, removing observations with propensity scores below 0.01 or above 0.99 to improve numerical stability and mitigate undue influence from extreme values. After computing weights, we conduct balance checks before and after weighting to assess whether the reweighted sample achieves better covariate balance between treated and control groups.

To estimate the treatment effect, we use weighted regression models, incorporating IPW and Hajek-IPW as weights in a regression of earnings on treatment status. If the weighting process is effective, the estimated treatment effect should closely reflect the true causal impact of college graduation on earnings. However, excessive weight variability can inflate standard errors and lead to unreliable inference.

To further evaluate the quality of the weighting process, we visualize weight distributions and balance diagnostics. Summary statistics and quantiles provide insight into the range and presence of extreme weights. Histograms and boxplots reveal overall weight distributions, highlighting potential outliers. Scatter plots help examine the relationship between propensity scores and weights, indicating whether extreme weights are concentrated in certain regions of the score distribution. If many extreme weights are present, it may suggest poor overlap between treated and control groups, necessitating adjustments such as weight truncation or re-specification of the propensity score model.

This simulation highlights both the strengths and limitations of IPW as a causal inference method. While weighting can effectively reduce selection bias and improve balance, it is sensitive to model specification and the presence of extreme weights. Careful diagnostics, such as trimming, balance checks, and visualization, are essential to ensure that IPW produces reliable and interpretable estimates of treatment effects.

```r
# Load necessary libraries
library(WeightIt)  # For IPW estimation
library(survey)    # For weighted regression
library(dplyr)     # For data manipulation
library(tableone)  # For balance assessment
library(ggplot2)   # For visualization
```

```r
# Set seed for reproducibility
set.seed(42)

# Define sample sizes
N <- 20000
NT <- 5000
NC <- N - NT

# Generate earnings for treated and control groups
earnings_non_graduates <- rnorm(NC, mean = 50, sd = 10)
earnings_graduates <- rnorm(NT, mean = 53, sd = 10)

# Generate covariates
family_income_non_graduates <- rnorm(NC, mean = 70, sd = 20)
family_income_graduates <- rnorm(NT, mean = 95, sd = 20)

# Generate 1 continuous variable
cont_var1_non_graduates <- rnorm(NC, mean = 5, sd = 2)
cont_var1_graduates <- rnorm(NT, mean = 6, sd = 2)

# Generate 1 binary variable
binary_var1_non_graduates <- rbinom(NC, 1, 0.4)
binary_var1_graduates <- rbinom(NT, 1, 0.6)

# Merge into a single dataset
data <- data.frame(
  ID = 1:N,
  Treatment = c(rep(1, NT), rep(0, NC)),
  FamilyIncome = c(family_income_graduates, family_income_non_graduates),
  ContVar1 = c(cont_var1_graduates, cont_var1_non_graduates),
  BinaryVar1 = c(binary_var1_graduates, binary_var1_non_graduates),
  Earnings = c(earnings_graduates, earnings_non_graduates)
)

# --- STEP 2: Estimate Propensity Scores Using Logistic Regression ---
ps_model <- glm(Treatment ~ FamilyIncome + ContVar1 + BinaryVar1,
                data = data, family = binomial)
data$PS <- predict(ps_model, type = "response") # Predict propensity scores

# --- STEP 3: Compute IPW Weights ---
# Standard weights
data$IPW <- ifelse(data$Treatment == 1, 1/data$PS, 1/(1 - data$PS))
# Normalize weights (Hajek Estimator)
data$IPW_Hajek <- ifelse(data$Treatment == 1,
                         data$IPW / sum(data$IPW[data$Treatment == 1]),
                         data$IPW / sum(data$IPW[data$Treatment == 0]))

# --- STEP 4: Trim Extreme Weights (Address Positivity Violation) ---
trim_threshold <- 0.01  # Drop observations with p(X_i) < 0.01 or > 0.99
```

```
data_trimmed <- data %>% filter(PS > trim_threshold & PS < (1 - trim_threshold))

# --- STEP 5: Balance Checks Before and After Weighting ---
table1_unweighted <- CreateTableOne(vars=c("FamilyIncome","ContVar1","BinaryVar1"),
                                   strata = "Treatment", data = data, test = TRUE)
table1_weighted <- svydesign(~1, weights = ~IPW, data = data) %>%
  svytable(~Treatment, design = .)

print(table1_unweighted)   # Balance before weighting
```

```
##                              Stratified by Treatment
##                               0              1              p      test
##   n                           15000          5000
##   FamilyIncome (mean (SD)) 70.08 (20.24) 95.01 (20.18) <0.001
##   ContVar1 (mean (SD))      4.99 (1.98)   5.99 (1.99)  <0.001
##   BinaryVar1 (mean (SD))    0.40 (0.49)   0.60 (0.49)  <0.001
```

```
print(table1_weighted)     # Balance after IPW
```

```
## Treatment
##        0        1
## 19964.70 19596.37
```

```
# --- STEP 6: Estimate Treatment Effects Using Weighted Regression ---
ipw_model <- svydesign(~1, weights = ~IPW, data = data)
ipw_estimate <- svyglm(Earnings ~ Treatment, design = ipw_model)

hajek_model <- svydesign(~1, weights = ~IPW_Hajek, data = data)
hajek_estimate <- svyglm(Earnings ~ Treatment, design = hajek_model)

cat("ATT (IPW):", coef(ipw_estimate)[2], "\n")
```

```
## ATT (IPW): 3.219614
```

```
cat("ATT (Hajek IPW):", coef(hajek_estimate)[2], "\n")
```

```
## ATT (Hajek IPW): 3.219614
```

```
# --- STEP 7: Visualization of Weight Distribution and Balance ---

# (A) Summary Statistics for IPW Weights to assess outliers
summary(data$IPW)
```

```
##    Min. 1st Qu.  Median    Mean 3rd Qu.    Max.
##   1.001   1.067   1.201   1.978   1.664  98.676
```

```
quantile(data$IPW, probs = c(0.01, 0.05, 0.95, 0.99))
```

```
##        1%        5%       95%       99%
##  1.005573  1.015336  5.077119 14.756247

# (B) Histogram of IPW Weights
plot_IPW <- ggplot(data, aes(x = IPW)) +
  geom_histogram(bins = 50, fill = "gray50", color = "black", alpha = 0.8) +
  labs(title = "Histogram", x = "IPW Weight", y = "Frequency") +
  theme_minimal()

# (C) Boxplot of IPW Weights to visually detect outliers
plot_IPWoutliers <- ggplot(data, aes(y = IPW)) +
  geom_boxplot(fill = "gray70", color = "black", outlier.colour = "gray20") +
  labs(title = "Boxplot", y = "IPW Weight") +
  theme_minimal()

# (D) Scatter Plot: IPW Weights vs. Propensity Scores by Treatment
plot_comparison <- ggplot(data, aes(x = PS, y = IPW, shape = as.factor(Treatment)))
+ geom_point(alpha = 0.5, color = "black") +
  labs(title = "Comparison",
       x = "Propensity Score", y = "IPW Weight", shape = "Treatment") +
  theme_minimal()
# Combine the three figures in a single row
library(gridExtra)
grid.arrange(plot_IPW, plot_IPWoutliers, plot_comparison, ncol = 3)
```

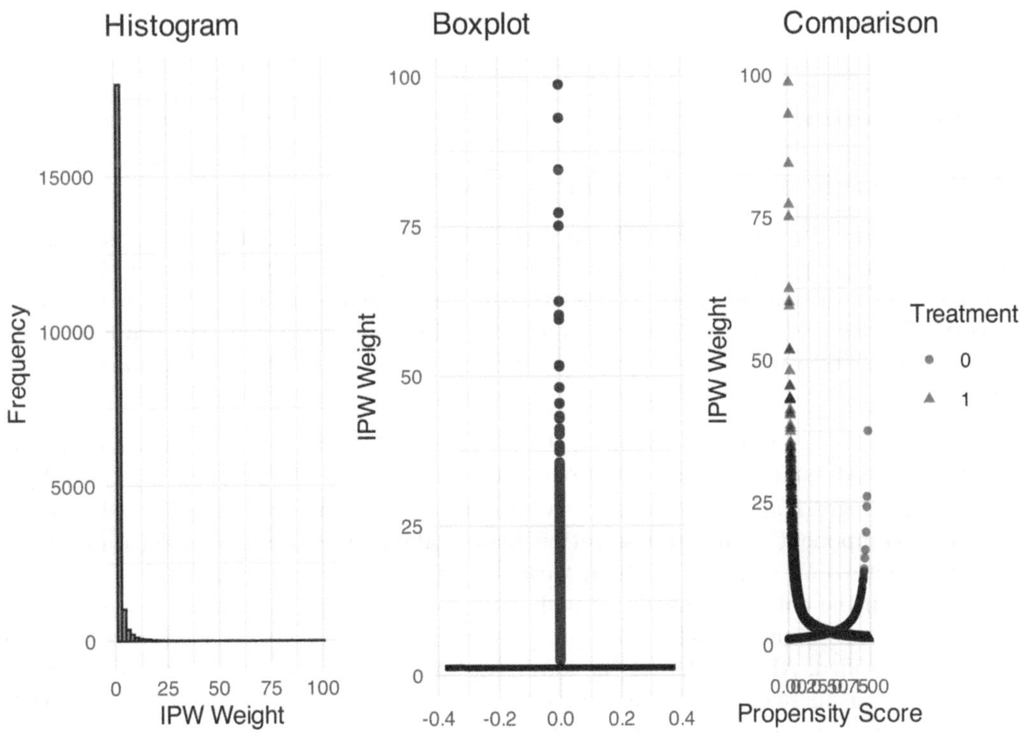

Machine learning driven approaches improve IPW by reducing reliance on strong functional form assumptions, improving covariate balance, and stabilizing weights. Methods like covariate balancing propensity score, entropy balancing, targeted maximum likelihood estimation, and DR estimators such as augmented IPW (AIPW) which we discuss below have gained traction in applied econometrics and causal inference, making IPW more robust and scalable in high-dimensional settings.

Several statistical packages facilitate IPW across different programming environments. In Stata, the primary package for IPW is `teffects ipw`, which estimates treatment effects using inverse probability weighting, along with `ipw` for more flexible weight estimation and balance diagnostics. The `psmatch2` and `causaltrt` packages also offer IPW functionality with additional propensity score-based approaches. In R, besides `MatchIt`, the `WeightIt` package provides a generalized framework for estimating various types of weights, including IPW, while `cobalt` supports comprehensive balance diagnostics and visualization. The `survey` package allows for weighted regression modeling with IPW. In Python, the `econml` package from Microsoft offers advanced causal inference methods, including IPW and doubly robust estimation, while `causalml` provides implementations for both IPW and machine learning-based weighting approaches. Additionally, `statsmodels` supports logistic regression for propensity score estimation, which can be used to compute inverse probability weights manually.

Recent research has explored ways to improve the reliability of IPW estimators in applied economics. Słoczyński, Uysal, and Wooldridge (2023) examine the equivalence of IPW, augmented inverse probability weighting (AIPW), and inverse probability weighted regression adjustment (IPWRA). They demonstrate that employing covariate balancing techniques improves the performance of IPW estimators, offering practical guidance for applied researchers. Similarly, Ma and Wang (2018) address challenges posed by small denominators in IPW, proposing a robust inference procedure that adapts to various asymptotic distributions, thereby improving estimator reliability in empirical economic analysis.

21.2 Doubly Robust Estimators: AIPW

DR estimators improve upon IPW by incorporating an additional outcome regression model alongside the propensity score model. Their key advantage is that they provide a valid estimate of the treatment effect if either the propensity score model or the outcome model is correctly specified, offering protection against misspecification. This DR property makes them particularly useful in observational studies, where modeling assumptions are often uncertain.

The augmented IPW (AIPW) estimator is a widely used DR method for estimating the ATE. It remains consistent as long as at least one of the two models—the propensity score $\hat{p}(X_i)$ or the outcome regression $\hat{f}(X_i)$—is correctly specified. If the propensity score model is correct, the IPW component adjusts for selection bias, making the outcome model unnecessary; however, pure IPW suffers from high variance, especially when $p(X_i) \approx 0$ or $p(X_i) \approx 1$. Conversely, if the outcome regression model is correct, the regression term provides a valid treatment effect estimate, making the propensity score unnecessary. When both models are correctly specified, AIPW achieves the lowest variance, improving efficiency. If both models are correct, it is more efficient than using either method alone. Read the derivation of AIPW and the proof of its doubly robust property in the last section.

AIPW builds on IPW but, instead of weighting raw outcomes Y, it weights residuals $Y - f(X)$, ensuring robustness even under model misspecification. The adjustment term $f_1(X) - f_0(X)$ further corrects any remaining bias if the regression model is accurate. This structure makes AIPW a direct extension of IPW, incorporating regression-based correction to improve robustness and efficiency. Together, these components make the AIPW estimator doubly robust, meaning it yields an unbiased estimate of the ATE if either the propensity score model or the outcome model is correct, but not necessarily both. The estimator is expressed as:

$$\hat{\delta}_{AIPW}^{ATE} = \frac{1}{n}\sum_{i=1}^{n}\left[\frac{D_i(Y_i - \hat{f}_1(X_i))}{\hat{p}(X_i)} - \frac{(1-D_i)(Y_i - \hat{f}_0(X_i))}{1-\hat{p}(X_i)} + \hat{f}_1(X_i) - \hat{f}_0(X_i)\right] \quad (21.4)$$

where $\hat{p}(X_i)$ is the estimated propensity score, $\hat{f}_1(X_i) = \hat{E}[Y|X_i, D = 1]$ is the predicted outcome for treated units, and $\hat{f}_0(X_i) = \hat{E}[Y|X_i, D = 0]$ is the predicted outcome for control units. ($\hat{e}(X_i)$ and $\hat{m}(X_i)$ are also very common notation for propensity score and outcome models, respectively.)

The AIPW estimator stands out because it combines Inverse Probability Weighting (IPW) with regression adjustment. The first two terms in the formula use IPW to account for selection bias by weighting the residuals $(Y - \hat{f}(X))$ instead of the raw outcomes. Instead of just multiplying Y by weights as in IPW, AIPW first removes the predictable part of Y using $f(X)$ .This IPW residual adjustment ensures balance between treated and untreated groups by creating a pseudo-outcomes where treatment assignment is independent of covariates. This residual is then weighted by $1/\hat{p}(X_i)$ (for treated) and $1/1 - \hat{p}(X_i)$ (for controls).

The final term, $\hat{f}_1(X) - \hat{f}_0(X)$, adds a regression-based correction by explicitly modeling the conditional outcomes for treated and control groups. This regression adjustment ensures the estimator remains consistent even if the propensity score model is misspecified, provided the outcome regression models $\hat{f}_1(X)$ and $\hat{f}_0(X)$ are correctly specified. Together, these components make the AIPW estimator doubly robust, meaning it yields an unbiased estimate of the ATE if either the propensity score model or the outcome model is correct, but not necessarily both. If both models are correct, it is more efficient than using either method alone

Best practices for implementing the AIPW estimator include ensuring covariate balance, verifying the overlap assumption, and using cross-fitting to improve efficiency and prevent overfitting. Checking covariate balance and overlap is essential, just as discussed in the matching and IPW sections, to ensure that treatment and control groups are comparable and that the propensity scores are well-distributed, avoiding extreme values near 0 or 1. Another crucial step is cross-fitting, a widely recommended technique in machine learning that mitigates overfitting when estimating the propensity score $\hat{p}(X)$ and the outcome models $\hat{f}_1(X)$ and $\hat{f}_0(X)$. The simplest form of cross-fitting involves randomly splitting the sample into two parts, using one half to estimate the propensity score and outcome models, and the other half to compute treatment effects, before swapping the roles and averaging the results. This approach ensures that the estimator does not overfit to a particular subset of the data while still using the full sample efficiently. cross-fitting significantly improves precision and derives formal error bounds, making it a key methodological improvement over traditional estimation approaches.

In R, the AIPW package is specifically designed for implementing the augmented IPW estimator. This package facilitates the estimation of average causal effects by combining propensity score modeling with outcome regression, offering a doubly robust approach.

It supports the use of machine learning algorithms for flexible covariate adjustment and includes features like cross-fitting to improve estimation precision. The package is suitable for both randomized controlled trials and observational studies. Simulation studies have shown that cross-fitting significantly reduces bias and improves confidence interval coverage for doubly robust estimators fitted with machine learning algorithms. To check the method, simulations, and comparisons with other methods, see (PMC Article[1]).

Beyond AIPW, several advanced doubly robust estimators have emerged, incorporating flexible machine learning techniques to improve efficiency and robustness. Targeted maximum likelihood estimation (TMLE) and the doubly robust learning (DRL) framework are among the most widely used approaches. TMLE, available in the tmle R package, applies iterative bias correction to improve causal effect estimation, making it particularly effective in complex observational studies. We will discuss DL-learning more in the chapter on metal-learners.

TMLE refines the initial outcome regression estimate using a clever covariate adjustment based on the estimated propensity score:

$$\hat{\tau}_{TMLE} = \frac{1}{N} \sum_{i=1}^{N} [\mu_1^*(X_i) - \mu_0^*(X_i)] \tag{21.5}$$

where $\mu_1^*(X_i)$ and $\mu_0^*(X_i)$ are updated using a logistic regression-based correction term.

For a detailed introduction to TMLE, see Gruber & van der Laan (2012). Practical implementation is available in the `tmle` R package (Lendle et al., 2017) and its extensions in the `tlverse` ecosystem (TLearn R Package).

DRL methods, on the other hand, leverages machine learning algorithms (e.g., penalized regression, neural networks, random forests) to estimate both the propensity score and outcome models instead of relying on a single model being correct like in AIPW, ensuring robustness and efficiency. This approach is particularly useful in high-dimensional settings with complex interactions and nonlinear relationships. For a detailed discussion on DLR methods for ATE, see the 2024 DRL Tutorial[2]. However, since DRL is primarily designed for CATE and heterogeneous treatment effect estimation, it will be covered in the chapter on heterogeneous causal effects.

```r
# Load necessary libraries
library(SuperLearner)  # For flexible ML-based outcome modeling
library(survey)        # For weighted estimation
library(dplyr)         # For data manipulation
library(ggplot2)       # For visualization

# Set seed for reproducibility
set.seed(42)

# Define sample sizes
N <- 20000
NT <- 5000
NC <- N - NT

# Generate earnings for treated and control groups
earnings_non_graduates <- rnorm(NC, mean = 50, sd = 10)
earnings_graduates <- rnorm(NT, mean = 53, sd = 10)
```

```
# Generate covariates
family_income_non_graduates <- rnorm(NC, mean = 70, sd = 20)
family_income_graduates <- rnorm(NT, mean = 95, sd = 20)

# Generate 1 continuous variable
cont_var1_non_graduates <- rnorm(NC, mean = 5, sd = 2)
cont_var1_graduates <- rnorm(NT, mean = 6, sd = 2)

# Generate 1 binary variable
binary_var1_non_graduates <- rbinom(NC, 1, 0.4)
binary_var1_graduates <- rbinom(NT, 1, 0.6)

# Merge into a single dataset
data <- data.frame(
  ID = 1:N,
  Treatment = c(rep(1, NT), rep(0, NC)),
  FamilyIncome = c(family_income_graduates, family_income_non_graduates),
  ContVar1 = c(cont_var1_graduates, cont_var1_non_graduates),
  BinaryVar1 = c(binary_var1_graduates, binary_var1_non_graduates),
  Earnings = c(earnings_graduates, earnings_non_graduates)
)

# --- STEP 2: Estimate Propensity Scores Using Logistic Regression ---
ps_model <- glm(Treatment ~ FamilyIncome + ContVar1 + BinaryVar1,
                data = data, family = binomial)
data$PS <- predict(ps_model, type = "response")  # Predict propensity scores

# --- STEP 3: Compute Inverse Probability Weights ---
# Standard IPW
data$IPW <- ifelse(data$Treatment == 1, 1 / data$PS, 1 / (1 - data$PS))

# --- STEP 4: Estimate Outcome Models ---
# Model outcome separately for treated and control groups
outcome_model_treated <- lm(Earnings ~ FamilyIncome + ContVar1 + BinaryVar1,
                            data = data[data$Treatment == 1, ])
outcome_model_control <- lm(Earnings ~ FamilyIncome + ContVar1 + BinaryVar1,
                            data = data[data$Treatment == 0, ])

# Predict counterfactual outcomes
data$Y1_hat <- predict(outcome_model_treated, newdata = data)  # Predicted Y(1)
data$Y0_hat <- predict(outcome_model_control, newdata = data)  # Predicted Y(0)

# --- STEP 5: Compute AIPW Estimator ---
data$AIPW <- (data$Treatment * (data$Earnings - data$Y1_hat) / data$PS) -
        ((1 - data$Treatment) * (data$Earnings - data$Y0_hat) / (1 - data$PS)) +
             (data$Y1_hat - data$Y0_hat)
AIPW_ATE <- mean(data$AIPW)

cat("ATE (AIPW):", AIPW_ATE, "\n")
```

```
## ATE (AIPW): 3.21644
```

```
# --- STEP 6: Cross-Fitting for Robustness ---
K <- 5  # Number of folds
folds <- sample(rep(1:K, length.out = N))
AIPW_estimates <- numeric(K)

for (k in 1:K) {
  train_data <- data[folds != k, ]
  test_data <- data[folds == k, ]

# Train propensity score model using covariates: FamilyIncome, ContVar1, BinaryVar1
  ps_model_k <- glm(Treatment ~ FamilyIncome + ContVar1 + BinaryVar1,
                    data = train_data, family = binomial)
  test_data$PS_k <- predict(ps_model_k, newdata = test_data, type = "response")

# Train outcome models on the training fold for treated and control groups
  outcome_model_treated_k <- lm(Earnings ~ FamilyIncome + ContVar1 + BinaryVar1,
                                data = train_data[train_data$Treatment == 1, ])
  outcome_model_control_k <- lm(Earnings ~ FamilyIncome + ContVar1 + BinaryVar1,
                                data = train_data[train_data$Treatment == 0, ])

# Predict counterfactual outcomes in the test fold
  test_data$Y1_hat_k <- predict(outcome_model_treated_k, newdata = test_data)
  test_data$Y0_hat_k <- predict(outcome_model_control_k, newdata = test_data)

# Compute AIPW for the fold
test_data$AIPW_k <- (test_data$Treatment * (test_data$Earnings - test_data$Y1_hat_k)
/ test_data$PS_k) - ((1-test_data$Treatment)*(test_data$Earnings-test_data$Y0_hat_k)
/ (1 - test_data$PS_k)) + (test_data$Y1_hat_k - test_data$Y0_hat_k)
AIPW_estimates[k] <- mean(test_data$AIPW_k)
}

AIPW_crossfit <- mean(AIPW_estimates)
cat("ATE (Cross-Fitted AIPW):", AIPW_crossfit, "\n")
```

```
## ATE (Cross-Fitted AIPW): 3.227918
```

```
# --- STEP 7: Visualization and Diagnostics ---

## (A) Propensity Score Distribution using grayscale tones
plot_PS <- ggplot(data, aes(x = PS, fill = as.factor(Treatment))) +
  geom_density(alpha = 0.5, color = "black") +
  scale_fill_manual(values = c("gray80", "gray30")) +
  labs(title = "Propensity Score Distribution", fill = "Treatment") +
  theme_minimal()

## (B) Distribution of AIPW Estimates
plot_AIPW <- ggplot(data, aes(x = AIPW)) +
  geom_histogram(bins = 50, fill = "gray50", alpha = 0.7) +
```

```
  labs(title ="Distribution of AIPW Estimates",x="AIPW Estimate",y="Frequency") +
  theme_minimal()
# Combine the two figures in a single row
library(gridExtra)
grid.arrange(plot_PS, plot_AIPW, ncol = 2)
```

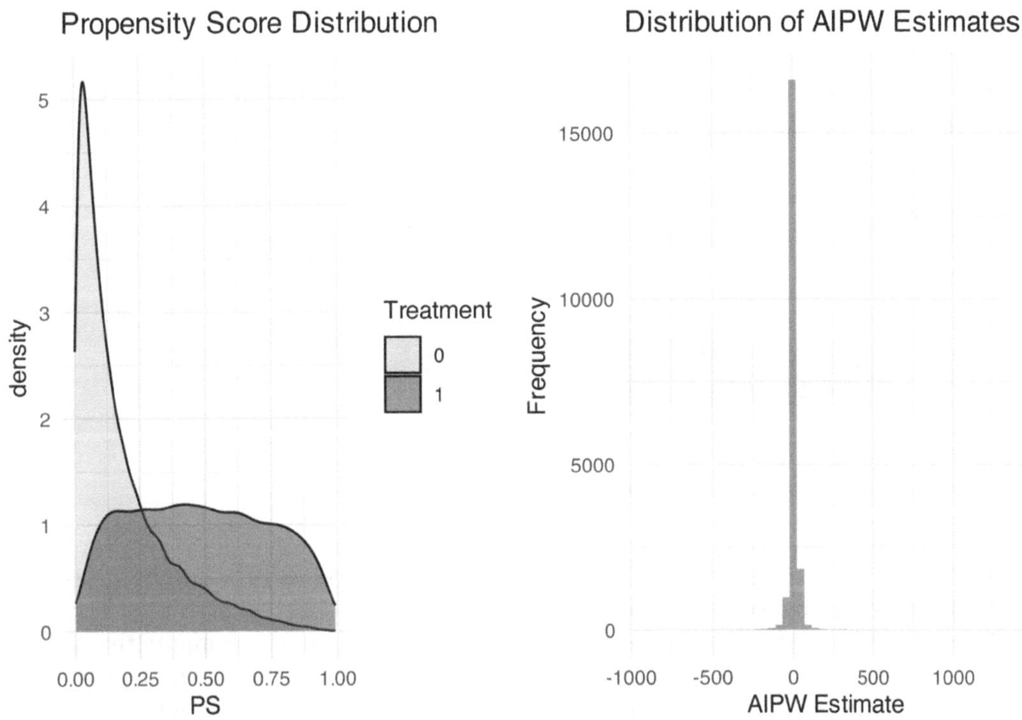


```
## (C) Estimated Counterfactual Outcomes using FamilyIncome as x-axis
ggplot(data, aes(x = FamilyIncome)) +
  geom_point(aes(y = Y1_hat, color = "Y(1) Estimate"), alpha = 0.5) +
  geom_point(aes(y = Y0_hat, color = "Y(0) Estimate"), alpha = 0.5) +
  labs(title = "Predicted Counterfactual Outcomes",
      x ="Family Income", y = "Predicted Earnings") +
  scale_color_manual(values = c("Y(1) Estimate" = "gray80",
                                "Y(0) Estimate" = "gray20")) +
  theme_minimal()
```

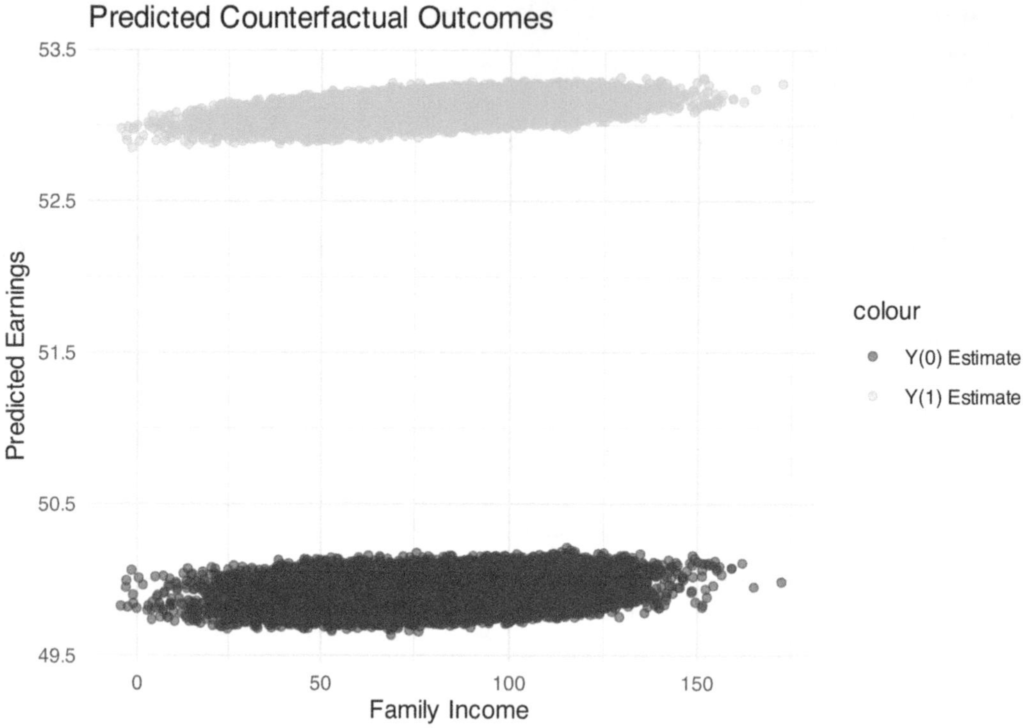

This simulation uses the same dataset as earlier to estimate propensity scores via logistic regression and compute IPW for causal inference. Outcome models are then estimated separately for the treated and control groups to predict counterfactual outcomes. The AIPW estimator combines these predictions with the observed outcomes to compute the ATE. To improve robustness, a cross-fitting procedure is implemented by partitioning the data into 5 folds and averaging the AIPW estimates across folds. Finally, the simulation visualizes the propensity score distribution and the distribution of the AIPW estimates using grayscale plots, and it displays the predicted counterfactual outcomes by Family Income, providing comprehensive diagnostics and insights into the quality of the weighting and the balance achieved.

Now, let's show and discuss the comparison and bootstrapped standard deviations of these methods in a simple simulation, demonstrating that while the naive estimator is biased, the IPW estimator is unbiased with higher variance, and the AIPW estimator is unbiased with lower variance.

```
set.seed(42)
n_sim <- 100
n <- 1000
k <- 5
results <- data.frame(
  ate_true = rep(2, n_sim),
  ate_naive = numeric(n_sim),
  ate_ipw = numeric(n_sim),
  ate_aipw = numeric(n_sim)
)
```

```
for (sim in 1:n_sim) {
  X1 <- rnorm(n, mean = 0, sd = 1)
  X2 <- rnorm(n, mean = 0, sd = 1)
  propensity <- 1 / (1 + exp(-(0.5 + 0.3 * X1 - 0.2 * X2)))
  D <- rbinom(n, 1, propensity) #Treatment assignment
  epsilon <- rnorm(n, mean = 0, sd = 1)
  Y <- 10 + 2 * D + 0.5 * X1 + 0.5 * X2 + epsilon # Outcome model
  data <- data.frame(Y, D, X1, X2)

  # Naive estimate
  mean_Y_treated <- mean(data$Y[data$D == 1])
  mean_Y_control <- mean(data$Y[data$D == 0])
  ate_naive <- mean_Y_treated - mean_Y_control

  # Estimate propensity score on entire data for IPW
  propensity_model <- glm(D ~ X1 + X2, data = data, family = binomial)
  propensity_hat <- predict(propensity_model, type = "response")

  # IPW estimates
  treated_weights <- data$D / propensity_hat
  mu1_ipw <- sum(data$Y[data$D == 1] *
        treated_weights[data$D == 1]) / sum(treated_weights[data$D == 1])

  control_weights <- (1 - data$D) / (1 - propensity_hat)
  mu0_ipw <- sum(data$Y[data$D == 0] * control_weights[data$D == 0]) /
    sum(control_weights[data$D == 0])

  ate_ipw <- mu1_ipw - mu0_ipw

  # Cross-fitting for AIPW
  folds <- cut(1:n, breaks = k, labels = FALSE)
  ate_aipw_folds <- numeric(k)
  for (fold in 1:k) {
    data_train <- data[folds != fold, ]
    data_test <- data[folds == fold, ]

    # Estimate propensity score on train data
    propensity_model_train <- glm(D ~ X1 + X2, data = data_train, family = binomial)
    propensity_hat_test <- predict(propensity_model_train, newdata = data_test,
                                   type = "response")

    # Estimate outcome model on train data
    outcome_model_train <- lm(Y ~ D + X1 + X2, data = data_train)

    # Create new data sets for potential outcomes for test data
    newdata_treated <- data_test
    newdata_treated$D <- 1
    mu1_hat_test <- predict(outcome_model_train, newdata = newdata_treated)
```

```
    newdata_control <- data_test
    newdata_control$D <- 0
    mu0_hat_test <- predict(outcome_model_train, newdata = newdata_control)

    # Compute AIPW estimates for test data
term1_aipw1 <- (data_test$D / propensity_hat_test) * (data_test$Y - mu1_hat_test) +
    mu1_hat_test
    mu1_aipw_test <- mean(term1_aipw1)

term1_aipw0 <- ((1 - data_test$D) / (1 - propensity_hat_test)) *
      (data_test$Y - mu0_hat_test) + mu0_hat_test
    mu0_aipw_test <- mean(term1_aipw0)

    ate_aipw_test <- mu1_aipw_test - mu0_aipw_test
    ate_aipw_folds[fold] <- ate_aipw_test
  }

  # Average AIPW estimate across folds
  mean_ate_aipw <- mean(ate_aipw_folds)

  # Store results
  results$ate_naive[sim] <- ate_naive
  results$ate_ipw[sim] <- ate_ipw
  results$ate_aipw[sim] <- mean_ate_aipw
}

# Summarize results
mean_naive <- mean(results$ate_naive)
sd_naive <- sd(results$ate_naive)
mean_ipw <- mean(results$ate_ipw)
sd_ipw <- sd(results$ate_ipw)
mean_aipw <- mean(results$ate_aipw)
sd_aipw <- sd(results$ate_aipw)

cat("Naive estimate: mean =", mean_naive, ", sd =", sd_naive, "\n")
```

```
## Naive estimate: mean = 2.05036 , sd = 0.07109387
```

```
cat("IPW estimate: mean =", mean_ipw, ", sd =", sd_ipw, "\n")
```

```
## IPW estimate: mean = 1.99533 , sd = 0.06077192
```

```
cat("AIPW estimate: mean =", mean_aipw, ", sd =", sd_aipw, "\n")
```

```
## AIPW estimate: mean = 1.995857 , sd = 0.05961602
```

In this simulation, we compare naive, IPW, and AIPW estimators to assess the causal effect under confounding and their standard deviations. We run 100 replications of a data-generating process with 1,000 observations each. For each simulation, we first estimate the naive difference in means between treated and control groups. Next, we compute propensity scores using logistic regression and derive IPW to adjust for confounding. The standard IPW estimate is obtained by reweighting the outcomes. We then implement a cross-fitting procedure for the AIPW estimator by partitioning the data into 5 folds; in each fold, we fit the propensity score and outcome models on the training set and predict potential outcomes for the test set. The final AIPW estimate is the average of the fold-specific estimates. This cross-fitted AIPW estimator is intended to provide a doubly robust estimate of the average treatment effect that combines the strengths of both outcome regression and IPW, thereby offering improved robustness to model misspecification.

Running the simulation, we expect the naive estimate to be biased due to confounding, the IPW estimate to be unbiased but with higher variance, and the AIPW estimate to be unbiased with lower variance—demonstrating its efficiency.The results show the mean and standard deviation of the naive, IPW, and AIPW estimates across the 100 simulations, highlighting the benefits of using doubly robust methods in causal inference.

21.3 Challenges of Selection on Observables

While the selection-on-observables framework provides a robust foundation for causal inference, it is not without significant limitations. These challenges arise from the assumptions and practical constraints inherent in this approach, which can compromise the validity of causal estimates.

Unverifiable assumptions: The selection-on-observables framework assumes that all confounding variables influencing both treatment assignment (D_i) and potential outcomes ($(Y_i(0), Y_i(1))$) are observed and included in the covariates (X_i). This assumption, known as the Conditional Independence Assumption (CIA), is critical for unbiased causal inference but inherently untestable. Two main issues threaten the validity of this framework: selection bias and post-treatment bias.

Selection bias is essentially confounding, where a variable (Z_i) influences both the treatment and the outcome. If any confounder is omitted, the identification assumption fails, leading to biased estimates. For example, in estimating the effect of college scholarships on earnings, unobserved factors such as intrinsic motivation or family support may impact both scholarship receipt and future earnings. Failing to account for these confounders violates the CIA assumption. While including covariates correlated with unobserved confounders can mitigate this bias, it does not fully eliminate the problem. Indirect evidence, such as balance tests, can help validate the assumption but cannot conclusively confirm it. We discussed selection bias in detail in the previous chapter. The concept applies similarly to both completely random and "as good as" random studies.

Post-treatment bias occurs when variables affected by the treatment are included in the analysis. Introducing post-treatment variables undermines comparability by, in effect, comparing treated and untreated groups on outcomes influenced by the treatment. For instance, in the college scholarship example, conditioning on post-treatment variables such as college GPA or internship participation—both of which are influenced by receiving the scholarship—distorts the estimated effect. These variables act as outcomes of the treatment, and conditioning on them invalidates the causal interpretation. Post-treatment bias is problematic even in randomized experiments and is only avoidable if the treatment does not influence the variable being conditioned on—a rare scenario in practice.

Post-treatment bias can arise in several ways: controlling for post-treatment variables, using proxies measured after treatment, or selecting observations based on post-treatment criteria. For example, excluding students with low GPAs after scholarship receipt, or removing people with no income from the sample creates a selection problem, as these excluded units might systematically differ in unobserved ways.

The overall lesson is clear: **do not condition on post-treatment variables**. Including such variables or removing units based on post-treatment characteristics introduces bias by design and undermines causal inference. Researchers must carefully select covariates, ensuring they address confounding while avoiding post-treatment variables, to uphold the validity of SOO methods. However, the reliance on untestable assumptions means robustness checks and alternative identification strategies remain essential for reliable causal analysis in all observational studies.

Functional Form Misspecification: The relationship between covariates (X_i), treatment (D_i), and outcomes (Y_i) may be complex or nonlinear. Simple regression models, which assume a specific functional form, may fail to capture these complexities, leading to biased estimates. For example, if the effect of a scholarship on earnings varies nonlinearly with parental income or academic preparedness, a linear regression model may oversimplify these interactions, producing inaccurate results. Flexible approaches, such as nonparametric or machine learning methods, can help address this limitation.

Lack of Overlap: SOO methods require sufficient overlap in the covariate distributions of treated and untreated groups. When treated and untreated units do not share comparable covariate profiles, causal estimates become unreliable. For example, if only students from well-funded public schools receive scholarships, it becomes challenging to estimate the effect of scholarships on students from underfunded schools due to the lack of comparable untreated individuals in that subgroup. Techniques like propensity score matching or trimming can partially address this issue by restricting analysis to regions of overlap, but these approaches may reduce sample size and statistical power.

Dimensionality Problems:In high-dimensional settings with many covariates, SOO methods like matching may struggle to balance treated and untreated groups effectively. This problem is exacerbated in small samples, where achieving covariate balance becomes even more difficult. For instance, if a scholarship program's eligibility depends on multiple criteria, such as parental income, test scores, and extracurricular activities, traditional matching methods may fail to ensure comparability across all dimensions. Machine learning tools, such as causal forests, offer promising solutions by handling high-dimensional covariates more flexibly.

Despite its strengths, we face all these significant challenges to estimate causal effect in the selection on observables methods. Careful design choices, such as selecting pre-treatment covariates, avoiding post-treatment variables, and adopting flexible modeling techniques, are crucial to mitigate these issues. However, the limitations of selection on observables methods underscore the importance of complementing them with robustness checks and alternative identification strategies to ensure reliable causal inference.

Over the past four chapters, we have developed a comprehensive set of tools for estimating causal effects under the selection on observables framework. Starting with regression adjustment, extending to modern high-dimensional methods like DML, and continuing with matching, IPW, and doubly robust estimators, these methods offer flexible strategies for controlling confounding when all relevant covariates are observed. However, they fundamentally rely on the unconfoundedness assumption—that, conditional on observed variables, treatment assignment is as good as random. When this assumption fails due to unmeasured or unobservable confounders, alternative identification strategies are needed. In the chapters that follow, we turn to methods designed to handle selection on unobservables—specifically, instrumental variables, difference-in-differences, synthetic control and regression discontinuity designs—which provide powerful tools for credible causal inference in more various observational settings.

21.4 Technical Note: Derivation and Doubly robust of the AIPW Estimator

To address the limitations of both regression-based and IPW estimators, we combine both approaches to construct the AIPW estimator. This ensures that the estimator remains unbiased as long as either the propensity score model or the outcome model is correctly specified.

Remember for ATE, the outcome regression estimator is $\hat{\delta}^{ATE}_{REG} = E[\hat{f}_1(X) - \hat{f}_0(X)]$. However, this estimator may be biased due to selection on unobservables. For ATE, we obtain the inverse probability weighting estimator is $\hat{\delta}^{ATE}_{IPW} = E\left[\frac{DY}{p(X)} - \frac{(1-D)Y}{1-p(X)}\right]$. This estimator adjusts for treatment selection bias by weighting observations based on their propensity scores. However, pure IPW is known to suffer from high variance, particularly when $p(X) \approx 0$ or $p(X) \approx 1$, leading to instability in estimation.

To combine regression adjustment with weighting, we introduce $f_1(X)$ within the IPW framework by adding and subtracting it inside the expectation. We rewrite $E[Y(1)]$ of IPW by add and subtract $f_1(X)$ inside the fraction:

$$E[Y(1)] = E\left[\frac{D(Y - f_1(X) + f_1(X))}{p(X)}\right] \tag{21.6}$$

Expanding the equation and using the linearity of expectation, $E[aX + bY] = aE[X] + bE[Y]$, we get:

$$E[Y(1)] = E\left[\frac{D(Y - f_1(X))}{p(X)} + \frac{Df_1(X)}{p(X)}\right] = E\left[\frac{D(Y - f_1(X))}{p(X)}\right] + E\left[\frac{Df_1(X)}{p(X)}\right] \tag{21.7}$$

To the second term we apply the law of iterated expectations, $E[A] = E[E[A|B]]$, where $A = \frac{Df_1(X)}{p(X)}$ and $B = X$, we can write:

$$E\left[\frac{Df_1(X)}{p(X)}\right] = E\left[E\left[\frac{Df_1(X)}{p(X)}\Big|X\right]\right] = E[f_1(X)] \tag{21.8}$$

Since both $f_1(X)$ and $p(X)$ are functions of X, they are deterministic given X (once X is given (fixed), $f_1(X)$ is a specific value and $p(X)$ is a fixed probability, and both are no longer random), allowing us to factor them out in the inside expectation:

$$E\left[\frac{Df_1(X)}{p(X)}\Big|X\right] = f_1(X) \cdot \frac{1}{p(X)} \cdot E\left[D\Big|X\right] \tag{21.9}$$

Using the definition of the propensity score which is the conditional probability of treatment given covariates, $p(X) = P(D = 1|X) = E[D|X]$, and substituting above, $p(X)$s will cancel out.

Substituting back the second term and using the linearity of expectation again, we obtain:

$$\begin{aligned}\mathbb{E}[Y(1)] &= \mathbb{E}\left[\frac{D(Y - f_1(X))}{p(X)}\right] + \mathbb{E}\left[\frac{Df_1(X)}{p(X)}\right] \\ &= \mathbb{E}\left[\frac{D(Y - f_1(X))}{p(X)}\right] + \mathbb{E}\left[f_1(X)\right] = \mathbb{E}\left[\frac{D(Y - f_1(X))}{p(X)} + f_1(X)\right] \end{aligned} \tag{21.10}$$

Similarly, we can rewrite $E[Y(0)]$ of IPW by add and subtract $f_0(X)$ inside the fraction, and follow the same steps as above to obtain:

$$E[Y(0)] = E\left[\frac{(1 - D)(Y - f_0(X))}{1 - p(X)} + f_0(X)\right] \tag{21.11}$$

Now, taking the difference and using linearity of expectations again, we obtain the AIPW estimator for the ATE as:

$$ATE = E[Y(1)] - E[Y(0)] = E\left[\frac{D(Y - f_1(X))}{p(X)} + f_1(X) - \frac{(1 - D)(Y - f_0(X))}{1 - p(X)} - f_0(X)\right] \tag{21.12}$$

$$= E\left[\frac{D(Y - f_1(X))}{p(X)} - \frac{(1 - D)(Y - f_0(X))}{1 - p(X)} + f_1(X) - f_0(X)\right] \tag{21.13}$$

Thus, the AIPW estimator is given by:

$$\hat{\delta}^{ATE}_{AIPW} = \frac{1}{n}\sum_{i=1}^{n}\left[\frac{D_i(Y_i - \hat{f}_1(X_i))}{\hat{p}(X_i)} - \frac{(1 - D_i)(Y_i - \hat{f}_0(X_i))}{1 - \hat{p}(X_i)} + \hat{f}_1(X_i) - \hat{f}_0(X_i)\right] \tag{21.14}$$

Why is AIPW doubly robust?

The estimator is unbiased for $\tau = E[Y(1) - Y(0)]$ if at least one of the two models is correctly specified:

- If the **outcome model** $\hat{f}_d(X)$ is correct, AIPW is unbiased **even if $\hat{p}(X)$ is wrong.**
- If the **propensity score model** $\hat{p}(X)$ is correct, AIPW is unbiased **even if $\hat{f}_d(X)$ is wrong.**

Case 1: Outcome Model $\hat{f}_d(X)$ is Correct

If $\hat{f}_d(X) = E[Y|X, D = d]$, we check the expectation of AIPW:

$$E[\hat{\delta}^{ATE}_{AIPW}] = E\left[\hat{f}_1(X) - \hat{f}_0(X) + \frac{D(Y - \hat{f}_1(X))}{\hat{p}(X)} - \frac{(1 - D)(Y - \hat{f}_0(X))}{1 - \hat{p}(X)}\right] \tag{21.15}$$

Since the residuals $Y - \hat{f}_d(X)$ have mean zero when $\hat{f}_d(X)$ is correctly specified, the second and third terms vanish:

$$E[\hat{\delta}^{ATE}_{AIPW}] = E[\hat{f}_1(X) - \hat{f}_0(X)] = E[Y(1) - Y(0)] = \tau \tag{21.16}$$

Thus, AIPW is unbiased even if $\hat{p}(X)$ is incorrect.

Case 2: Propensity Score Model $\hat{p}(X)$ is Correct

If $\hat{p}(X) = P(D = 1|X)$, we check the expectation of AIPW:

$$E[\hat{\delta}^{ATE}_{AIPW}] = E\left[\frac{DY}{\hat{p}(X)} - \frac{(1-D)Y}{1-\hat{p}(X)} + \hat{f}_1(X) - \hat{f}_0(X) - \frac{D\hat{f}_1(X)}{\hat{p}(X)} + \frac{(1-D)\hat{f}_0(X)}{1-\hat{p}(X)}\right] \tag{21.17}$$

Rearrange:

$$E\left[\frac{DY}{\hat{p}(X)} - \frac{(1-D)Y}{1-\hat{p}(X)}\right] - E\left[\frac{D\hat{f}_1(X)}{\hat{p}(X)} - \frac{(1-D)\hat{f}_0(X)}{1-\hat{p}(X)}\right] + E[\hat{f}_1(X) - \hat{f}_0(X)] \tag{21.18}$$

Using the law of iterated expectations, the first expectation simplifies to $E[Y(1) - Y(0)] = \tau$, while the second term cancels out. This gives:

$$E[\hat{\delta}^{ATE}_{AIPW}] = \tau \tag{21.19}$$

Thus, AIPW is **unbiased even if $\hat{f}_d(X)$ is incorrect**, as long as $\hat{p}(X)$ is correct.

The **AIPW estimator is doubly robust** because it provides an **unbiased estimate of the treatment effect** as long as either: 1. The **outcome model** $\hat{f}_d(X)$ is correct, OR 2. The **propensity score model** $\hat{p}(X)$ is correct.

Notes

1. https://pmc.ncbi.nlm.nih.gov/articles/PMC8796813/
2. https://arxiv.org/html/2406.00853v1

22

Selection on Unobservables and DML-IV

In our previous chapters on selection on observables, we reviewed methods such as regression, double machine learning, matching, propensity score weighting, and doubly robust approaches. Each of these techniques assumes that all confounders—factors that simultaneously influence both the treatment decision and the observed outcome—are measurable and observable. However, in many practical applications, even with vast amounts of data and numerous variables, unmeasured confounders may still affect both treatment assignment and outcomes. In fields like economics, social sciences, and health research, countless unobservable factors drive human behavior—common examples include motivation, ability, cultural influences, and social networks—necessitating alternative quasi-experimental approaches when observable data alone falls short. These unobservable confounding variables present a major challenge in causal inference, as they limit our ability to reliably isolate true causal effects; unobservable variables may be inherently unmeasurable or simply missing from the available data.

This chapter briefly outlines classical econometric strategies designed to account for unobservable confounders. These methods, while insightful, involve trade-offs and rely on strong assumptions such as valid instruments, parallel trends, or known thresholds. We cover the core ideas of four major approaches: instrumental variables (IV), which addresses endogeneity by introducing an external variable that affects the outcome solely through its impact on treatment; difference-in-differences (DiD), which exploits before-and-after comparisons between treated and untreated groups under the parallel trends assumption; regression discontinuity (RD), which leverages a cutoff point to compare units just above and below the treatment threshold; and synthetic control, which constructs a comparison unit from a weighted combination of controls when a traditional group is unavailable.

While these methods might not always yield the causal effects we typically discuss, such as the average treatment effect (ATE), they can offer valuable insights into different treatment effects specific to each model. It is crucial to understand the assumptions of each model and the particular causal effects being estimated. Although this may not be ideal, identifying these specific effects can inform the design of targeted policies and interventions, ultimately contributing to more nuanced decision-making.

It is important to note that these chapters do not aim to provide exhaustive derivations or proofs of these classical methods. Instead, the main focus is on how machine learning (ML) can refine, extend, and improve their application. In this chapter, we discuss how ML techniques can aid in instrument selection, strengthen parallel trends assessments, and ultimately strengthening causal inference in the presence of unobservable confounders.

DOI: 10.1201/9781003504795-22

22.1 Instrumental Variables

Observational studies in economics often face the endogeneity problem, where treatment assignment or policy adoption is correlated with unobserved factors that also affect outcomes. When unconfoundedness—relying on the assumption that all relevant confounders are observable—fails, IV methods emerge as a powerful solution. The central idea behind this method is to find an instrumental variable that influences treatment assignment while remaining independent of unobservable/unmeasured confounders and having no direct effect on the outcome except through its effect on treatment. This approach effectively extracts variation in the treatment that is free from the bias introduced by unobserved factors, allowing for more credible causal inference.

Randomization, as discussed in Chapter 17, is a fundamental tool in economics, political and social science, and health for establishing causality. Randomized controlled trials (RCTs) have been used in labor economics to assess the impact of job training programs, in political science to evaluate voter mobilization efforts, and in development economics to measure the effects of cash transfer programs on poverty alleviation. These experiments ensure that treatment assignment is exogenous, minimizing confounding issues and providing robust causal estimates. However, as we also discussed, randomization is not always feasible due to ethical, logistical, or political constraints.

In cases where actual randomization is not possible, researchers turn to exogenous randomization from natural experiments, policy changes, or other external shocks. Historically rooted in economics and widely adopted in the social sciences and genetics, natural experiments serve as instrumental variables for addressing endogeneity. Natural experiments occur when external circumstances or policy changes create conditions that mimic random assignment, offering exogenous variation in treatment assignment that researchers can exploit to estimate causal effects. For instance, researchers have used instruments like distance to college or quarter of birth to estimate the returns to education, and random supply shocks such as weather events or natural disasters to assess policy impacts, and many we will discuss later. From the Rubin causal model perspective, an instrument offers a means to indirectly assign treatment in a way that is plausibly exogenous, thus facilitating the identification of causal effects under specific conditions.

We use the term treatment broadly, originating from health sciences but applicable to economic policies, social programs, and exogenous shocks that impact outcomes. Whether referring to a subsidy, a regulatory intervention, or a natural disaster, treatment in this context encompasses any externally imposed variation that influences the variables of interest.

The validity of an instrumental variable hinges on assumptions that are inherently difficult to fully verify. Researchers assess the plausibility of an IV using a combination of theoretical justification, various empirical tests, sensitivity analyses, and robustness checks. While they rely on economic theory and contextual information to support the relevance and excludability of an an instrument, no formal test can confirm with absolute certainty that the instrument is completely independent of unmeasured confounders. Ultimately, although these methods help build confidence in the validity of a carefully chosen IV, they cannot guarantee it, so careful interpretation of the results remains essential.

The earliest concept of IV is generally attributed to Philip G. Wright and his son Sewall (1928), with IV methods traditionally formulated within structural equation models (SEMs) that rely on the key assumption that the instrument is uncorrelated with the structural error terms. These classical approaches typically assume homogeneous treatment effects, and methods such as two-stage least squares (2SLS) have proven powerful under these linear conditions. However, caution is warranted when applying 2SLS to nonlinear models—whether involving quadratic treatment effects or logistic regressions for binary treatments—as several open questions remain. For example, if the true outcome model includes interactions between covariates and the treatment, suggesting heterogeneous treatment effects, or if data are clustered, the interpretation and consistency of the standard 2SLS estimator may be compromised, necessitating alternative estimation strategies like random effects models. In landmark papers during the 1990s, Angrist, Imbens, and Rubin advanced the field by linking IV methods to the potential outcomes framework in causal inference. Motivated by the issue of noncompliance, where individuals fail to follow assigned treatments or treatments are not perfectly assigned (e.g., in randomized controlled trials), they showed that IV offers a way to still estimate causal effects, even when randomization fails to produce perfectly controlled treatment assignment. This shift not only allowed for the accommodation of heterogeneous effects but also clarified the assumptions necessary for causal interpretation, addressing some of the limitations inherent in traditional SEM approaches.

Moreover, recent advances in machine learning have improved the application of IV methods. Techniques such as LASSO and random-forest help in the data-driven selection of valid instruments from a large pool of candidates, while methods like post-LASSO and double selection refine the estimation process by mitigating omitted variable bias. Additionally, machine learning tools—including random forests, gradient boosting, and deep neural networks (as seen in Deep IV approaches)—enable flexible, non-linear modeling of the relationship between instruments and endogenous regressors, as well as the assessment of instrument validity. In this section, we will explore these classical IV strategies, and consider practical examples that illustrate various IV methods in estimating causal effects. If you are already familiar with IV from econometrics, you can jump to the double ML IV section.

22.1.1 Core IV Concepts

IV methods offer a solution to the endogeneity problem in observational studies, where treatment assignment is often correlated with unobserved factors that also affect outcomes. In the traditional linear model, the outcome Y_i is expressed as

$$Y_i = \beta_0 + \beta_1 D_i + \beta_2' X_i + \varepsilon_i \tag{22.1}$$

with D_i representing the endogenous treatment variable and X_i exogeneous covariates. Here, the target parameter is β_1 which is causal/treatment effect; however, because the error term ε_i is correlated with D_i (as we cannot assume independent/unconfoundedness), direct ordinary least squares (OLS) estimation of β_1 is biased. Keep in mind in selection in observable chapters, we implement various methods to address this bias, such as regression, matching, weighting, and DML. However, these methods rely on the assumption that all confounders are observable, which may not always be the case.

To address this, IV methods introduce a vector of instruments Z_i that must satisfy two key conditions. First, the instrument must be highly correlated, meaning it is highly related with the endogenous regressor D_i. Second, it must have exclision restriction—that is, Z_i should have no direct effect on the outcome Y_i except through its effect on D_i. In a sense, the instrument allows us to extract the variation in the treatment that is free from the bias introduced by unobserved confounders. When the dimension of Z_i is one, the model is just-identified, which is the most common case; if there are multiple instruments, the model is over-identified.

In the just-identified case without covariates, the IV estimator for β_1 is computed as the ratio of the covariance between Y_i and the instrument Z_i to the covariance between D_i and Z_i. The IV estimator of β_1 is given by the ratio of covariances[1]:

$$\hat{\beta}_{1,\text{IV}} = \frac{\text{Cov}(Y_i, Z_i)}{\text{Cov}(D_i, Z_i)} = \frac{\frac{1}{n}\sum_{i=1}^{n}(Y_i - \bar{Y})(Z_i - \bar{Z})}{\frac{1}{n}\sum_{i=1}^{n}(D_i - \bar{D})(Z_i - \bar{Z})} \tag{22.2}$$

There are two common interpretations of the IV estimator—one based on the indirect least squares (ILS) approach (please see the footnote for details)[2]Essentially, since we cannot directly regress Y on D due to unobservable confounders, we use Z—a variable highly correlated with D—to isolate the variation in D that is independent of these confounders, employing a two-stage process to estimate the effect on Y. In the first stage, we predict the treatment using OLS:

$$\hat{D}_i = \hat{c}_{20} + \hat{c}_{21}Z_i \tag{22.3}$$

and in the second stage, we replace the actual treatment with $\hat{D}_i$ in the outcome equation:

$$Y_i = \beta_0 + \beta_1\hat{D}_i + v_i \tag{22.4}$$

The OLS estimate of β_1 in this second stage, $\hat{\beta}_{1,2SLS}$, is equivalent to the IV estimator, $\hat{\beta}_{1,IV}$. Specifically, we can derive that

$$\hat{\beta}_{1,2SLS} = \frac{\text{Cov}(Y_i, Z_i)}{\text{Cov}(D_i, Z_i)} \tag{22.5}$$

which matches the ratio-of-covariances formulation of the IV estimator. Moreover, this estimator is also identical to the ILS estimator, demonstrating that under these conditions, the IV, ILS, and 2SLS estimators are equivalent.

The discussion extends naturally to the case where covariates are included.[3] In the 2SLS framework, the first stage includes the covariates:

$$\hat{D}_i = \hat{c}_{20} + \hat{c}_{21}Z_i + \hat{c}_{2X}'X_i \tag{22.6}$$

and the second stage is then

$$Y_i = \beta_0 + \beta_1\hat{D}_i + \beta_2'X_i + v_i \tag{22.7}$$

The OLS estimate of β_1 in this second stage is the 2SLS estimator, and it can be shown that the IV, ILS, and 2SLS estimators are also equivalent under these conditions.

Thus far, we have discussed instrumental variables within the context of structural equation models, which assume homogeneous treatment effects. While these traditional methods have been valuable, they have notable shortcomings—such as difficulties in accommodating non-linear models (e.g., quadratic effects or binary treatments estimated via logit), limited capacity to handle interactions between covariates and treatment, inflexible to extend to more complex settings,and challenges in managing clustered data. All these shortcomings motivate our transition to the potential outcomes framework, which accommodates heterogeneous effects and offers a clearer basis for causal interpretation.

22.1.2 IV in the Potential Outcomes Framework and RCTs

IV methods became particularly valuable with the implementation to the randomized experiments that suffer from noncompliance—a situation in which individuals fail to follow their assigned treatments, or treatments are not perfectly implemented. In such experiments, traditional methods of causal inference face significant challenges because the treatment actually received, D_i, does not perfectly align with the treatment assignment, Z_i. Motivated by this issue, Angrist, Imbens, and Rubin (1996, JASA) connected the IV approach to the potential outcomes framework, demonstrating that random assignment itself can serve as a strong instrument. By doing so, they showed that even in the presence of noncompliance, IV methods can identify local causal effects by focusing on a subgroup of individuals—known as compliers—who adhere to their assigned treatment. For more detailed discussion of connecting potential framework to IV, please see also the last section.

Consider a randomized trial for a new drug intended to improve blood pressure control. In this trial, 1,000 patients are randomly assigned to either the treatment group ($Z_i = 1$) or the control group ($Z_i = 0$), with 500 patients in each group. The treatment assignment Z_i indicates whether a patient is assigned to take the new drug. However, noncompliance is evident: in the treatment group, only 300 out of 500 patients actually take the drug ($D_i = 1$), while in the control group, 50 out of 500 patients take the drug despite being in the control group. Thus, the observed treatment uptake rates are 0.6(=300/500) for the treatment group and 0.1(=50/500) for the control group. The outcome Y_i is a binary indicator of improvement in blood pressure, with an average outcome of 0.40 for all patients in the treatment group (including both those who take and do not take the treatment while being assigned to the treatment group) and 0.25 for those in the control group (including both those who take and do not take the treatment while being assigned to the control group).

There are two common naïve approaches to estimating causal effects in the presence of noncompliance. The **first approach** compares outcomes based on the actual treatment received, thus calculates average treatment effect (the effect of treatment):

$$\delta^{\text{ATE}} = E[Y_i|D_i = 1] - E[Y_i|D_i = 0] \tag{22.8}$$

The ATE is the difference between "average outcome of treatment for all units actually are treated" and "average outcome of treatment for all units who actually are not treated". However, this approach suffers from selection bias because the decision to take the drug(treatment) is not completely randomized as there are people who take the drug even if they are in control group. When noncompliance occurs, simply comparing groups by the treatment actually received can lead to biased estimates because the decision to comply is self-selected, breaking the benefits of randomization. Patients who opt to take the drug may differ systematically in unobserved ways (e.g., in their health status or motivation) from those who do not, so the resulting difference in means does not yield a valid estimate of the causal effect.

The **second approach** uses the randomized treatment assignment to compute the ITT causal/treatment effect (the effect of treatment assignment):

$$\delta^{\text{ITT}} = E[Y_i|Z_i = 1] - E[Y_i|Z_i = 0] \tag{22.9}$$

Since Z_i is randomly assigned, this difference is unbiased for the effect of being assigned to take the drug. The ITT effect is the difference between "average outcome of treatment for all units in the treatment group" and "average outcome of treatment for all units in the control group". In our example, ITT is $\delta_{\text{ITT}} = 0.40 - 0.25 = 0.15$.

Yet, while the ITT approach preserves randomization by comparing outcomes based on assigned treatment, it only estimates the effect of the assignment rather than the treatment received. ITT procedure gives a valid estimate of the causal effect of the assignment on outcome (effectiveness), but not the effect of the treatment received on outcome (efficacy). This may not be the effect of the treatment itself, as it includes the effects of both taking the drug and being assigned to take it (i.e., there are people who does not comply the assignment rules). The ITT effect is often smaller than the true treatment effect because it includes the effects of noncompliance and other factors that may dilute the treatment effect.

To overcome these limitations, the **instrumental variable approach** is employed. Instead of comparing groups solely based on D_i or Z_i, IV offer a method to recover the causal effect of the treatment among those who comply with their assignment. In this context, the random assignment Z_i serves as an instrument for the treatment received D_i. Under the potential outcomes framework, define $Y_i(1)$ and $Y_i(0)$ as the potential outcomes under treatment and control, respectively, and $D_i(1)$ and $D_i(0)$ as the potential treatment statuses when assigned to treatment or control. Patients who take the drug if assigned ($D_i(1) = 1$); and do not take it if not assigned ($D_i(0) = 0$) are called as compliers (while those who do not follow their assignment—either always-takers ($D_i(1) = D_i(0) = 1$) or never-takers ($D_i(1) = D_i(0) = 0$)—are called noncompliers).

Before introducing the LATE estimator, it is important to outline the key assumptions required for its identification. First, independence of Z_i ensures that the instrument is as good as randomly assigned, this assumtion is plausible in the context of a randomized experiment. Second, the existence of compliers (i.e., a valid first stage) requires that the instrument meaningfully affects treatment uptake. This assumption is generally uncontroversial and can be empirically tested by regressing D_i on Z_i and checking for statistical significance. Third, the monotonicity assumption states that there are no defiers—individuals who would take the treatment when assigned to control but refuse it when assigned to treatment. While typically very reasonable, this assumption is untestable. Finally, the exclusion restriction requires that the instrument affects the outcome only through treatment. Unlike the first-stage condition, this assumption is impossible to verify empirically and is often the most controversial (particularly in observational studies which is discussed below). These assumptions collectively enable the identification of the local average treatment effect (LATE).

The IV estimator (Wald estimator) for the LATE, with the aforementioned assumptions, is given by the ratio:

$$\delta_{Wald}^{\text{LATE}} = \frac{E[Y_i|Z_i = 1] - E[Y_i|Z_i = 0]}{E[D_i|Z_i = 1] - E[D_i|Z_i = 0]} \tag{22.10}$$

represents the average change in the outcome that can be attributed to a change in treatment status among those individuals whose treatment behavior is influenced by the instrument Z (whether assigned to treatment or control group by experimenter or exogenously). In other words, the numerator captures the difference in average outcomes between those who were assigned to take the treatment ($Z = 1$) and those who were not ($Z = 0$), while the denominator measures the corresponding difference in treatment uptake. This ratio tells us the causal effect of the treatment on the outcome specifically for the subgroup of compliers—those whose decision to take the treatment is driven by the assignment.[4] While LATE itself measures the treatment effect only among compliers, the denominator provides an implicit measure of the proportion of compliers in the sample. This is why a weak first-stage (small denominator) can lead to imprecise and unstable LATE estimates, as the estimator is dividing by a small number (see the last section for a detailed proof).

For our drug trial, the difference between treatment uptake in treatment group and treatment uptake in control group (proportion of compliers). This ratio is:

$$E[D_i|Z_i = 1] - E[D_i|Z_i = 0] = 0.6 - 0.1 = 0.5$$

and, as calculated above, the ITT treatment effect is 0.15. Hence, the LATE is:

$$\delta_{\text{LATE}} = \frac{0.15}{0.5} = 0.30$$

This result implies that among the compliers—patients who take the drug only if assigned—the new drug increases the probability of improvement in blood pressure level by 30 percentage points. Keep in mind IV will not give us the ATE—it will give us an average treatment effect for a particular subgroup (LATE).

Up to this point, our approach has relied on a single binary instrument (Z_i) and a binary treatment (D_i). Under these conditions, along with the necessary assumptions, we have been able to estimate the ITT effect and the LATE. However, in many empirical applications, it is desirable to generalize this framework to accommodate more complex scenarios, such as non-binary treatments and instruments, the inclusion of covariates (particularly when Z_i is not randomly assigned), and the use of multiple instruments to strengthen identification.

Up to this point, our approach has relied on a single binary instrument (Z_i) and a binary treatment (D_i). Under these conditions, and with the necessary assumptions, we can estimate the ITT effect and the LATE. However, many empirical applications call for a more general framework that accommodates non-binary treatments and instruments, incorporates covariates (especially when Z_i is not randomly assigned), and uses multiple instruments to strengthen identification.

A widely used method for such generalizations is 2SLS. When additional covariates X_i are included, the estimation proceeds in two stages:

1. **First Stage:**
   Regress the endogenous treatment variable on the instrument(s) and covariates to obtain the predicted treatment:

$$D_i = \alpha_1 + \beta_1 Z_i + \gamma_1 X_i + \epsilon_{1i} \tag{22.11}$$

Here, $\hat{D}_i$ represents the fitted values that capture only the exogenous variation in D_i driven by Z_i after controlling for X_i.

2. **Second Stage:**
   Regress the outcome on the predicted treatment and the same covariates:

$$Y_i = \alpha_2 + \beta_2 \hat{D}_i + \gamma_2 X_i + \epsilon_{2i} \tag{22.12}$$

The coefficient $\hat{\beta}_2$ provides an estimate of the causal effect—interpreted as LATE among compliers.

This method is called Two-Stage Least Squares (2SLS) because it involves estimating two separate OLS regressions sequentially. Unlike OLS, which suffers from bias due to endogeneity when treatment is correlated with unobserved factors, 2SLS isolates variation in treatment that is induced by the instrument—while controlling for covariates—allowing for a more credible causal estimate of treatment effects.

The intuition behind 2SLS is as follows:

1. The first-stage coefficient $\hat{\beta}_1$ measures the average change in the treatment variable D_i induced by a one-unit change in the instrument Z_i, holding the covariates X_i constant. This ensures that D_i is predicted only from the exogenous variation in Z_i.
2. The variation in the fitted values $\hat{D}_i$ reflects only the component of D_i that is driven by Z_i after accounting for X_i. This removes any endogeneity present in D_i due to omitted variables or selection bias.
3. The second-stage coefficient $\hat{\beta}_2$ provides an unbiased estimate of the causal effect of D_i on the outcome Y_i for compliers—the group of individuals whose treatment status changes in response to the instrument. Since the instrument changes treatment randomly, the estimated treatment effect applies specifically to those who comply with their assigned Z_i.
4. Ultimately, $\hat{\beta}_2$ should be interpreted as the effect of a one-unit increase in the predicted treatment $\hat{D}_i$ on the outcome Y_i, but only for the subset of individuals for whom Z_i induces a change in D_i. This is why the LATE interpretation is necessary in IV settings, distinguishing it from the standard ATE.

Unlike the Wald estimator, which relies on a binary instrument and binary treatment, 2SLS can handle continuous, categorical, or multi-valued instruments and treatments. When treatment assignment is not purely randomized, controlling for covariates helps reduce residual confounding and improve precision. In such cases, 2SLS allows the inclusion of additional covariates in both the first and second stages. 2SLS can also handle settings where more than one instrument is available. This is useful when a single instrument may be weak or when multiple sources of exogenous variation exist. If multiple instruments are used, the first-stage regression includes all instruments, and the second stage continues to use the predicted values from the first stage. Overall, 2SLS is a more general framework that allows for greater flexibility in estimation, making it the preferred approach when binary instruments and treatments are not sufficient for capturing causal relationships.

We want to emphasize the importance of careful inference: while separate two-stage regressions will yield the same treatment effect estimate, the standard errors, confidence intervals, and p-values will be incorrect. In 2SLS, while the estimation conceptually proceeds in two steps, in practice, we do not simply regress the first-stage fitted values $\hat{D}_i$ in a second step as if they were observed data. Instead, the entire system of equations is estimated simultaneously to ensure that standard errors (,confidence intervals and p-values) are correctly computed, correcting for the fact that $\hat{D}_i$ is an estimated quantity. This is done using instrumental variables regression, which accounts for the first-stage estimation uncertainty in the second stage, leading to corrected standard errors. In practice, statistical software packages estimate 2SLS using a single command (e.g., `ivreg` in R, `ivregress` in Stata, or `2SLS` in Python), which automatically adjusts the standard errors appropriately. Thus, while 2SLS is conceptually a two-step procedure, it is statistically estimated as a single model to ensure proper inference.

Before we conclude our discussion of IV in randomized control studies, we want to highlight an important distinction between controlling for a variable in regression adjustment and using it as an IV. While both approaches aim to address confounding, they differ fundamentally in how they isolate variation in the treatment variable. The following section explains this key difference and its implications for causal inference.

In the previous RCT chapter, we observed that when controlling for a confounder X_i in a regression to estimate the effect of D_i on Y_i, the estimated coefficient β_1 captures the relationship between Y_i and the part of D_i that is not explained by X_i, meaning it reflects the residual variation in D_i after accounting for X_i. This is because in an OLS regression, the coefficient on D_i measures its effect on Y_i after adjusting for other included covariates.

In contrast, when we use an instrument Z_i to estimate the effect of D_i on Y_i via 2SLS, the estimated coefficient β_{12} captures the relationship between Y_i and only the part of D_i that is explained by Z_i—i.e., the predicted values from the first-stage regression. This means that rather than relying on the naturally occurring variation in D_i (which may be endogenous), we isolate only the exogenous variation in D_i that is driven by Z_i. This ensures that the estimated effect is free from confounding, allowing for a valid causal interpretation of β_{12} as the LATE among compliers.

In this section, up until now, we have focused on how IV estimators help address noncompliance which leads to bias in causal effects in randomized experiments, where some units fail to follow their assigned treatment. IV estimators provide a solution to this problem by isolating exogenous variation in treatment, and estimating local average treatment effect. In the following section, we shift our focus to a more common application of IV methods—addressing omitted variable bias in observational data.

22.1.3 IV in the Observational Data

IV serve as a critical tool for addressing selection bias when treatment assignment is not random. In previous discussions, we explored how IV methods handle non-random noncompliance in experiments, where treatment assignment was random but actual treatment uptake was not. This lecture extends the IV framework beyond experimental settings to observational studies, where treatment is not assigned randomly, and selection bias is a more severe challenge.

Until nowin various chapters, we have explored ways to address omitted variable and selection bias, including randomization, which ensures unbiased treatment assignment, conditioning on observables, which controls for measured confounders, and conditioning on unobservables, where panel data methods help mitigate bias from time-invariant factors. IV methods offer an alternative solution by leveraging an external source of variation—an instrument Z_i—that influences the outcome only through its effect on treatment, isolating exogenous variation for causal inference.

In many social science and economic settings, treatment assignment is not random. However, instruments serve as exogenous nudges/shocks, influencing treatment uptake without directly affecting the outcome. This allows us to estimate the LATE by isolating only the variation in D_i driven by the instrument Z_i. The estimated causal effect applies specifically to those individuals whose treatment status would change in response to the instrument. To credibly identify causal effects using IV, we rely on four key assumptions, which mirror those discussed in the RCT section. Here, we formally present them:

Assumption I: Independence of the Instrument: In an experimental setting, the instrument Z_i is randomly assigned. However, in observational settings, randomization may not hold. Instead, we assume conditional independence:

$$(Y_{0i}, Y_{1i}, D_{0i}, D_{1i}) \perp\!\!\!\perp Z_i | X_i \tag{22.13}$$

This means that, after conditioning on covariates X_i, the instrument is as good as randomly assigned. This assumption parallels the conditional independence of the treatment assumption in RCTs and the selection on observables framework.

Assumption II: First-Stage Relevance: The instrument must meaningfully induce variation in treatment:

$$0 < P(Z_i = 1) < 1 \quad \text{and} \quad P(D_i(1) = 1) \neq P(D_i(0) = 1) \tag{22.14}$$

This ensures that Z_i is strongly correlated with the endogenous treatment variable D_i, allowing us to identify a causal effect.

Assumption III: Monotonicity: The instrument should shift individuals only in one direction—toward treatment, but never away from it:

$$D_i(1) \geq D_i(0) \quad \forall i \tag{22.15}$$

This eliminates the possibility of defiers, individuals who would do the opposite of what the instrument assigns.

Assumption IV: Exclusion Restriction: The instrument should affect the outcome only through its impact on treatment, meaning:

$$Y(D_i = 1, Z_i = 1) = Y(D_i = 1, Z_i = 0) \tag{22.16}$$

$$Y(D_i = 0, Z_i = 1) = Y(D_i = 0, Z_i = 0) \tag{22.17}$$

This means Z_i must not have a direct effect on Y_i, aside from influencing D_i. The exclusion restriction is often the most controversial and difficult to justify in observational studies, as it requires a strong argument that the instrument does not affect the outcome through any other pathway.

In observational studies, identifying causal effects using IV relies on the same four key assumptions—independence of the instrument, first-stage relevance, monotonicity, and exclusion restriction—as in experimental settings. Since treatment assignment is not random, IV methods help isolate exogenous variation in treatment, allowing researchers to estimate the LATE. 2SLS remains the primary estimation method, enabling the correction of endogeneity while controlling additional covariates when necessary.

The estimation process follows the same two-stage procedure outlined earlier (equations 19.2.6 and 19.2.7). The first stage predicts treatment using the instrument, ensuring that only exogenous variation is used, while the second stage regresses the outcome on the predicted treatment values. This approach corrects for omitted variable bias and selection issues that would otherwise confound the treatment effect. The interpretation remains the same: the estimated effect applies only to compliers—those whose treatment status is influenced by the instrument.

One key difference in observational studies is that we often rely on natural experiments, policy changes, or historical events as instruments rather than controlled experimental designs. Additionally, observational data often require stronger justification for the exclusion restriction, as ensuring that the instrument affects the outcome only through treatment is more challenging. The concerns around weak instruments are particularly relevant in this setting, emphasizing the need for strong first-stage relationships and proper statistical inference. As discussed earlier, 2SLS must be estimated as a single system rather than as separate regressions to obtain valid standard errors, confidence intervals, and p-values. While the core methodology remains unchanged, observational studies demand careful instrument selection and extensive robustness.

Finding a valid instrument requires institutional knowledge and a deep understanding of the mechanisms driving the variable of interest. A strong instrument must be highly correlated with treatment while affecting the outcome solely through its influence on treatment. Identifying such an instrument is often challenging and requires both theoretical justification and empirical validation. One fundamental limitation of IV methods is that the exclusion restriction is untestable, meaning researchers must build a convincing argument for why their instrument satisfies this condition. Because there is no definitive test to confirm whether an instrument affects the outcome only through treatment, the credibility of an IV study hinges on the strength of its theoretical and contextual foundation.

Many influential studies in economics and social sciences have leveraged instrumental variables derived from natural experiments, policy changes, historical events, and supply-side shocks to identify causal effects. One of the most well-known examples is Card (1990), who used the Mariel Boatlift as an instrument to study the effect of immigration on wages, exploiting the sudden influx of Cuban migrants to Miami as an exogenous labor supply shock. Similarly, Angrist and Krueger (1991) used quarter of birth as an instrument for educational attainment, relying on compulsory schooling laws that dictate when students can legally drop out. Another widely cited instrument is distance to college, used by Card (1995) to estimate the returns to education, assuming that individuals living closer to a college are more likely to enroll, yet distance itself does not directly affect earnings.

Policy-driven variation has also been a rich source of instruments. Angrist (1990) used Vietnam draft lottery numbers to estimate the effects of military service on earnings, leveraging the random assignment of draft eligibility as an exogenous source of variation in veteran status. Similarly, Currie and Gruber (1996) exploited Medicaid expansions to evaluate the impact of public health insurance on infant and child health outcomes. Another example is Acemoglu, Johnson, and Robinson (2001), who used historical settler mortality rates as an instrument for institutional development, arguing that European colonization strategies shaped modern economic institutions in a way that persists today.

Supply-side shocks often serve as strong instruments in studies of credit constraints and economic development. Klemperer and Meyer (1986) used oil price shocks as an instrument for firms' investment decisions, while Autor, Dorn, and Hanson (2013) relied on China's export growth as an exogenous shock to study the impact of trade exposure on local labor markets. Natural disasters have also been utilized as instruments; for instance, Miguel, Satyanath, and Sergenti (2004) used rainfall variation as an instrument for economic shocks to examine the link between economic downturns and civil conflict in Africa.

More recent work has expanded the range of instrumental variables used in applied research. Nunn and Qian (2014) used historical potato cultivation patterns as an instrument for long-term agricultural productivity and economic development, demonstrating how historical shocks can provide exogenous variation. Dinkelman (2011) exploited the rollout of electrification in South Africa as an instrument to estimate the effect of household access to electricity on female labor force participation. Aizer and Doyle (2015) used random assignment of public defenders in the U.S. court system to study the impact of legal representation on case outcomes, while Dobkin et al. (2018) used hospital admission thresholds as an instrument to examine the causal effect of hospital care on long-term health outcomes. More recently, Goldsmith-Pinkham, Hull, and Kolesár (2020) provided new methods for assessing the strength and validity of instruments, helping researchers navigate common IV pitfalls in modern applications.

For a comprehensive review of instrumental variable strategies across disciplines, Borjas (2021) and Nakamura and Steinsson (2018) discuss recent advances and applications of IV in labor and macroeconomics. Mogstad and Torgovitsky (2018) provide an updated discussion of IV estimation in settings with treatment effect heterogeneity, extending the standard LATE framework. Meanwhile, Young (2022) critiques common IV approaches and offers guidance on addressing weak instruments and overfitting concerns. Kang et al. (2024) provide an in-depth examination of identification and inference when dealing with potentially invalid instruments, addressing challenges in satisfying standard IV assumptions. Levis, Kennedy, and Keele (2024) discuss nonparametric identification and efficient estimation of causal effects using instrumental variables, presenting advanced techniques for settings with minimal parametric assumptions. These reviews illustrate how instrumental variable methods continue to evolve, providing new insights while refining best practices in applied econometrics.

22.1.4 Weak Instrument

Even when an instrument satisfies the independence and exclusion restriction assumptions, it may still fail to be useful if it does not induce enough variation in treatment. A weak instrument provides little leverage for identifying causal effects, leading to large standard errors and potentially biased estimates. Thus, even with a well-justified instrument, researchers must ensure that it has a sufficiently strong first-stage relationship with treatment to make IV estimation reliable. Using a weak instrument can lead to biased estimates and large standard errors. Next, we discuss how to determine whether an instrument is weak.

In IV estimation, the strength of the instrument is crucial for obtaining reliable and unbiased estimates. An instrument is considered weak if it has a minimal impact on the endogenous explanatory variable D_i, leading to instability in the IV estimator. This weakness can result in estimates that are biased toward those obtained from OLS regression, with standard errors that are underestimated, thereby compromising statistical inference. Notably, this bias persists even as the sample size increases, making it a significant concern in empirical research (Stock & Yogo, 2005; Staiger & Stock, 1997).

In IV estimation, the strength of the instrument is crucial for obtaining reliable and unbiased estimates. An instrument is considered weak if it has a minimal impact on the endogenous explanatory variable D_i, leading to instability in the IV estimator. This weakness can result in estimates that are biased toward those obtained from OLS regression, with standard errors that are underestimated, thereby compromising statistical inference. Notably, this bias persists even as the sample size increases, making it a significant concern in empirical research (Stock & Yogo, 2005; Staiger & Stock, 1997).

To assess the strength of an instrument, researchers often examine the first-stage regression, where the endogenous variable D_i is regressed on the instrument Z_i and any additional covariates X_i. The key statistic from this regression is the F-statistic, which tests the null hypothesis that the instrument has no explanatory power over the endogenous variable. A commonly used rule of thumb, as suggested by Staiger and Stock (1997), is that an F-statistic less than 10 indicates a weak instrument.[5] However, this threshold is not absolute; the appropriate critical value may vary depending on the specific context and the number of instruments used. Stock and Yogo (2005) provide more refined critical values for detecting weak instruments, emphasizing that the conventional threshold may be insufficient in certain cases.

For a more comprehensive discussion on the implications of weak instruments and alternative approaches, readers are encouraged to consult Andrews, Stock, and Sun (2018), which recommends using identification-robust Anderson-Rubin confidence intervals when only a single instrument is available. Addressing weak instruments is critical, as they can lead to biased parameter estimates and invalid inference, ultimately undermining the credibility of empirical results. Other techniques to handle weak instruments include LIML estimators, which reduce bias and improve inference in weak instrument settings.[6] ML methods offer data-driven solutions for identifying and constructing stronger instruments, helping to mitigate the challenges posed by weak instruments—an approach we will explore in the next section.

IV methods are a powerful solution to selection bias, but their credibility hinges on strong assumptions. In observational settings, the challenge is not only finding valid instruments but also justifying their exogeneity and ensuring their strength. Weak instruments and violations of the exclusion restriction can lead to substantial bias, making careful instrument selection essential. 2SLS provides a flexible way to implement IV, incorporating multiple instruments, covariates, and continuous treatments.

22.2 Double Machine Learning IV

In econometrics, IVs are essential for addressing endogeneity, such as unobservable (unmeasured, hidden) confounders, when estimating causal effects. However, weak instruments—IVs with low correlation to the endogenous variable—can lead to biased estimates, inconsistent inference, and wide confidence intervals. DML offers a framework to handle high-dimensional confounders and weak instruments (Z) while maintaining valid statistical inference. This section focuses on the machine learning methods for IV. We'll explore the algorithm, practical implementation, and illustrate the process with an R simulation tailored for applied researchers.

22.2.1 Structural Model and Instrumental Variables Estimation

We consider the structural equation model (partially linear IV model):

$$Y = D\delta + f(X) + \varepsilon \tag{22.18}$$

where Y is the outcome variable, D is the endogenous treatment variable, and δ is the parameter of interest. The function $f(X)$ captures the effect of covariates X, while ε represents the error term. Since D is endogenous, we introduce an instrumental variable (external factor, exogenous shock) Z that satisfies:

$$D = Z\pi + g(X) + \eta \tag{22.19}$$

where π denotes the first-stage effect of Z on D, $g(X)$ accounts for the dependence of D on X, and η is the first-stage residual. Given that Z may also depend on X, we model:

$$Z = h(X) + \nu \tag{22.20}$$

where $h(X)$ captures the dependence of Z on X, and ν is the residual in the instrument equation.

22.2.2 Double Machine Learning (DML-IV) Estimation

To estimate δ, we use a DML approach (i.e., using cross-fitted residualized forms), requiring three nuisance functions:

1. $m_1(X) = E[Y|X] = f(X)$
2. $m_2(X) = E[D|X] = g(X)$
3. $m_3(X) = E[Z|X] = h(X)$

r_1 corresponds to $\tilde{Y}$, the residualized outcome after removing the conditional expectation $m_1(X) = E[Y|X]$. In the notation often used in double machine learning, residualizing means partialing out the effect of covariates, so we have:

$$r_1 = Y - m_1(X) = \tilde{Y} \tag{22.21}$$

Similarly:

$$r_2 = D - m_2(X) = \tilde{D}, \quad r_3 = Z - m_3(X) = \tilde{Z} \tag{22.22}$$

Thus, the estimator for δ can be written as:

$$\hat{\delta} = \frac{\mathbb{E}[r_3 r_1]}{\mathbb{E}[r_3 r_2]} = \frac{\mathbb{E}[\tilde{Z}\tilde{Y}]}{\mathbb{E}[\tilde{Z}\tilde{D}]} \tag{22.23}$$

which is equivalent to a 2SLS regression using residualized variables, ensuring orthogonality between $\tilde{Z}$ and the error term.

22.2.3 Neyman Orthogonality and Robustness

The Neyman orthogonality condition ensures that small errors in estimating $m_1(X)$, $m_2(X)$, and $m_3(X)$ do not introduce bias in δ. This is crucial in double machine learning, making the estimator robust to overfitting and regularization errors.

The condition is expressed through the score function:

$$\mathbb{E}[\psi(W; \theta, \eta)] = 0 \tag{22.24}$$

where $W = (Y, D, Z, X)$ represents the observed data, $\theta = \delta$ is the parameter of interest, and $\eta = (m_1, m_2, m_3)$ denotes the nuisance functions.

A common score function for DML-IV is:

$$\psi(W; \delta, \eta) = (Y - m_1(X) - \delta(D - m_2(X)))(Z - m_3(X)) \tag{22.25}$$

This ensures that residualized variables satisfy orthogonality, making the estimation of δ robust to errors in the machine learning-based nuisance estimates. The Neyman-orthogonality condition holds if small errors in estimating the nuisance functions do not affect the moment equation. Mathematically, this means:

$$\frac{\partial \mathbb{E}[\psi(W; \delta, \eta)]}{\partial \eta} = 0 \tag{22.26}$$

ensuring that small errors in nuisance function estimation do not impact the final estimator.

To verify that the Neyman orthogonality condition holds, we take partial derivatives of the score function with respect to the nuisance functions $m_1(X), m_2(X), m_3(X)$:

$$\frac{\partial \psi}{\partial m_1(X)} = -(Z - m_3(X)) \tag{22.27}$$

This term should have an expectation of zero to ensure robustness.

$$\frac{\partial \psi}{\partial m_2(X)} = \delta(Z - m_3(X)) \tag{22.28}$$

Since $m_2(X)$ estimates $E[D|X]$, any misestimation should not affect the expected score function.

$$\frac{\partial \psi}{\partial m_3(X)} = -(Y - m_1(X) - \delta(D - m_2(X))) \tag{22.29}$$

This ensures that any error in instrument modeling does not introduce bias into the estimation of δ. Using these derivatives, we define residualized variables for estimation:

$$r_1 = Y - \hat{m}_1(X), \quad r_2 = D - \hat{m}_2(X), \quad r_3 = Z - \hat{m}_3(X) \tag{22.30}$$

The moment condition that should hold in estimation is:

$$\mathbb{E}\left[(r_1 - \delta r_2) r_3\right] = 0 \tag{22.31}$$

This formulation ensures that even if $\hat{m}_1(X), \hat{m}_2(X), \hat{m}_3(X)$ are estimated imperfectly, the moment equation remains valid.

Below in the simulation, we explicitly compute the score function and check whether the Neyman-orthogonality condition is satisfied. Normally, when using built-in R, Stata, or Python packages for DML-IV, this step is not shown explicitly. However, we believe covering it here helps readers understand the underlying mechanics and allows them to verify orthogonality if needed.

It is recommended to use the "partial-out" score function (the one above and same as equation 4.59, Chernozhukov et al. (2018)) over an alternative formulation (the one below and same as equation 4.59, Chernozhukov et al. (2018)), as it involves only conditional mean functions, which can be efficiently estimated using machine learning. The alternative score function is: $\psi(W; \delta, \eta) = (Y - D\delta - m_1(X))(Z - m_3(X))$

We adopt a generalized DML-IV algorithm, extending the approach used in DML-Lasso, as detailed in the lasso section. This allows flexibility in implementing various machine learning methods such as lasso, random forests, and other regularized estimators for nuisance function estimation.

22.2.4 Algorithm: DML-IV

Step 1: Inputs

- **Observed Data:**

$$W_i = (Y_i, D_i, Z_i, X_i), \quad i = 1, \ldots, N$$

where Y_i is the outcome, D_i is the endogenous treatment, Z_i is the instrument, and X_i represents control variables.

- **Machine Learning Estimation of Nuisance Functions:**
  To estimate conditional expectations, define the following models:
  - $m_1(X) = E[Y|X]$ (Outcome model)
  - $m_2(X) = E[D|X]$ (Treatment model)
  - $m_3(X) = E[Z|X]$ (Instrument model)

  By defining them this way, the Neyman-orthogonal score function is implicitly implemented through residualization and 2SLS estimation. In simulations, we use Lasso as a baseline, but other ML methods such as random forest, neural networks, and gradient boosting can also be applied. The choice of estimator depends on the complexity of the data and the desired level of flexibility.

Step 2: Train ML Predictors Using Cross-Fitting

- **Split the Data** into K folds: $\{I_k\}_{k=1}^K$, ensuring equal partitioning. A typical choice is $K = 5$ cross-fitting.
- **Train the Nuisance Models** $\hat{m}_{1,k}(X)$, $\hat{m}_{2,k}(X)$, and $\hat{m}_{3,k}(X)$, excluding observations in fold I_k to prevent overfitting. Each model is trained on $N - N/K$ observations.
  - We can tune hyperparameters using cross-validation or apply plug-in penalty selection methods like rlasso.
  - The plug-in penalty method is the preferred approach, as discussed in detail in the DML-Lasso section.
- **Compute Cross-Fitted Residuals** for each $i \in I_k$:

$$r_{1i} = Y_i - \hat{m}_{1,k}(X_i) \quad \text{(Residualized Outcome)}$$

$$r_{2i} = D_i - \hat{m}_{2,k}(X_i) \quad \text{(Residualized Treatment)}$$

$$r_{3i} = Z_i - \hat{m}_{3,k}(X_i) \quad \text{(Residualized Instrument)}$$

  Cross-fitting mitigates overfitting by ensuring that each observation's nuisance function estimate is derived from a model trained on different data. This technique is crucial in DML as it reduces bias in the final estimator.

Step 3: Estimate the Causal Parameter δ_0

- Solve for δ_0 using the moment equation:

$$\frac{1}{N}\sum_{k=1}^{K}\sum_{i\in I_k}\psi(W_i;\delta_0,\hat{m}_{1,k},\hat{m}_{2,k},\hat{m}_{3,k}) = 0$$

  where:

$$\psi(W;\delta,\eta) = (r_1 - \delta r_2)r_3$$

- This step follows the DML2 approach, where residuals from all folds are merged before estimating δ_0 using instrumental variables regression (2SLS).

Step 4: Implementing in Practice (Example: R Simulation)

- The estimation is equivalent to 2SLS using residualized variables:
  - **First stage:** Residualized treatment r_2 is regressed on residualized instrument r_3.
  - **Second stage:** Residualized outcome r_1 is regressed on the predicted values from the first stage.
  - The coefficient on r_2 provides the IV estimate of δ_0.
  - Standard errors and confidence intervals from IV regression are valid if the first-stage is strong (i.e., the instrument r_3 strongly predicts r_2).
  - If the first-stage is weak, standard IV inference may be misleading. In such cases, consider methods for Inference with Weak Instruments, such as Anderson-Rubin test, Kleibergen-Paap statistic, or conditional likelihood ratio (CLR) inference.

22.2.5 DML2-IV Simulation using LASSO

In this simulation, we illustrate the complete DML-IV framework for estimating a causal effect in a high-dimensional setting. We begin by generating synthetic data that mimics a realistic empirical environment: the outcome is influenced by both the endogenous treatment and a set of high-dimensional continious and binary covariates, while the instrument is strongly related to the treatment but only weakly related to the outcome. This setup captures the common challenge in empirical work where endogeneity must be addressed through valid instruments.

Once the data are generated, we estimate the necessary nuisance functions using Lasso. To avoid overfitting, we employ cross-fitting: the data are divided into multiple folds, and for each fold, the nuisance functions are estimated on the remaining observations. After obtaining the cross-fitted residuals, we proceed to the causal parameter estimation via 2SLS regression. Here, the key insight is that by regressing the residualized outcome on the residualized treatment—using the residualized instrument as an instrument—we effectively solve the moment equation that embodies the Neyman-Orthogonality condition. This condition guarantees that small errors in the nuisance function estimates do not influence the final causal effect estimate. We further verify this by checking that the expectation of the score function's derivatives with respect to the nuisance functions is near zero.

We also assess the strength of the instrument through a weak instrument test. By running a first-stage regression of the residualized treatment on the residualized instrument and computing a robust F-statistic (using heteroskedasticity-robust standard errors), we can determine whether the instrument provides sufficient variation to yield reliable IV estimates. A robust F-statistic well above the conventional threshold suggests that the instrument is strong; otherwise, one might need to consider inference methods tailored to weak instruments.

Although our example uses simulated data to clearly demonstrate each step of the DML-IV process, you can simply replace the simulated data with your data and proceed with the cross-fitting procedure described in Step 2.

```
# Load required libraries
library(MASS)          # For generating multivariate normal data
library(estimatr)      # For HC3 robust SEs (lm_robust)
library(AER)           # For IV regression (2SLS)
library(hdm)           # For rlasso (Plug-in Penalty Selection)

set.seed(123)          # Ensure reproducibility

# === Step 1: Define Data ===
N <- 10000  # Number of observations
p <- 10     # Number of covariates (high-dimensional X)
# Split X into 1/3 continuous and 2/3 binary covariates
p_cont <- round(p / 3)  # Number of continuous covariates
p_bin <- p - p_cont     # Number of binary covariates
# Continuous covariates ~ N(0,1)
X_cont <- mvrnorm(N, mu = rep(0, p_cont), Sigma = diag(p_cont))
# Binary covariates ~ Bernoulli(0.5)
X_bin <- matrix(rbinom(N * p_bin, 1, 0.5), nrow = N, ncol = p_bin)
# Combine X into a single matrix
X <- cbind(X_cont, X_bin)
# Generate Instrument Z (Strongly Correlated with D, Nearly Uncorrelated with Y) ===
Z <- 0.7 * (X_cont[,1]) + 0.3 * rnorm(N)
# Generate Treatment D (Strongly Correlated with Z) ===
D_prob <- plogis(25 * Z + 0.8 * X %*% runif(p, -0.5, 0.5))  # First-stage
D <- rbinom(N, 1, D_prob)  # Binary treatment
# Generate Outcome Y (Affected by D, Weakly Related to Z) ===
delta_true <- 2  # True effect of D on Y
beta <- runif(p, -1, 1)  # Coefficients for X
Y <- delta_true * D + X %*% beta + 0.1 * Z + rnorm(N)  # Small correlation between
  # Z and Y
# === Check Correlations ===
cat("Correlation(Y, D):", cor(Y, D), "\n")
```

```
## Correlation(Y, D): 0.3697644
```

```
cat("Correlation(Z, D):", cor(Z, D), "\n")
```

```
## Correlation(Z, D): 0.797973
```

```
cat("Correlation(Z, Y):", cor(Z, Y), "\n")
```

```
## Correlation(Z, Y): 0.2315305
```

```
# === Define Neyman-Orthogonal Score Function ===
score_function <- function(Y, D, Z, m1, m2, m3, delta) {
  return((Y - m1 - delta * (D - m2)) * (Z - m3))
}
# === Step 2: Define Fold Assignments (Cross-Fitting) ===
```

```r
K <- 5  # Number of cross-fitting folds
folds <- split(sample(1:N), rep(1:K, length.out=N))  # Create K folds
# Initialize residual vectors
Y_resid <- rep(NA, N)  # Residuals for Y
D_resid <- rep(NA, N)  # Residuals for D
Z_resid <- rep(NA, N)  # Residuals for Z
# === Step 3: Estimate Nuisance Functions and Compute Residuals ===
for (k in 1:K) {
    # Define training (all but k-th fold) and validation (k-th fold) indices
    train_idx <- unlist(folds[-k])
    val_idx <- folds[[k]]
    # Training Data
    X_train <- X[train_idx, ]
    Y_train <- Y[train_idx]
    D_train <- D[train_idx]
    Z_train <- Z[train_idx]
    # Validation Data
    X_val <- X[val_idx, ]
    # ---- Fit Lasso for Outcome Model (Y ~ X) ----
    lasso_y <- rlasso(X_train, Y_train)
    Y_hat_val <- predict(lasso_y, X_val)
    # ---- Fit Lasso for Treatment Model (D ~ X) ----
    lasso_d <- rlasso(X_train, D_train)
    D_hat_val <- predict(lasso_d, X_val)
    # ---- Fit Lasso for Instrument Model (Z ~ X) ----
    lasso_z <- rlasso(X_train, Z_train)
    Z_hat_val <- predict(lasso_z, X_val)
    # ---- Compute Cross-Fitted Residuals ----
    Y_resid[val_idx] <- Y[val_idx] - Y_hat_val  # r1 = Y - m1(X)
    D_resid[val_idx] <- D[val_idx] - D_hat_val  # r2 = D - m2(X)
    Z_resid[val_idx] <- Z[val_idx] - Z_hat_val  # r3 = Z - m3(X)
}
# === Step 4: Solve Moment Equation Using 2SLS ===
library(sandwich) # Load necessary packages for robust standard errors
library(lmtest)
# The estimator is given by:
# delta_hat = E[r3 * r1] / E[r3 * r2], which is solved via IV regression
dml_iv_model <- ivreg(Y_resid ~ D_resid | Z_resid)  # 2SLS estimation
# Compute robust standard errors
robust_se <- coeftest(dml_iv_model, vcov = vcovHC(dml_iv_model, type = "HC3"))
# Extract and Print Results with Robust SEs
cat("Estimated IV effect using DML-IV:", robust_se[2, 1])
```

```
## Estimated IV effect using DML-IV: 2.242062
```

```r
cat("Robust Standard Error:", robust_se[2, 2], "\n")
```

```
## Robust Standard Error: 0.06641111
```

```
cat("95% CI:", confint(dml_iv_model, vcov. = vcovHC(dml_iv_model, type = "HC3"))
[2, ])
```

```
## 95% CI: 2.111222 2.372903
```

```
# === Verify the Moment Condition (Check Neyman Orthogonality) ===
delta_hat <- coef(dml_iv_model)[2]  # Estimated delta from 2SLS
# Compute derivatives of the score function
d_score_m1 <- -Z_resid
d_score_m2 <- delta_hat * Z_resid
d_score_m3 <- -(Y_resid - delta_hat * D_resid)
# Compute residualized expectations (should be close to zero)
orthogonality_m1 <- mean(d_score_m1)
orthogonality_m2 <- mean(d_score_m2)
orthogonality_m3 <- mean(d_score_m3)
# Orthogonality Condition Check (Should be close to 0)
cat("E[d_score_m1]:", orthogonality_m1, "\n")
```

```
## E[d_score_m1]: 2.453253e-05
```

```
cat("E[d_score_m2]:", orthogonality_m2, "\n")
```

```
## E[d_score_m2]: -5.500345e-05
```

```
cat("E[d_score_m3]:", orthogonality_m3, "\n")
```

```
## E[d_score_m3]: 0.0001864073
```

```
# Compute the expected value of the score function (should be close to zero)
moment_check <- mean(score_function(Y_resid, D_resid, Z_resid, orthogonality_m1,
              orthogonality_m2, orthogonality_m3, delta_hat))
cat("Moment Equation Check (Should be close to 0):", moment_check)
```

```
## Moment Equation Check (Should be close to 0): 7.050893e-08
```

```
# === Weak Instrument Test (Compute First-Stage F-statistic) ===
# First-Stage Regression using lm()
first_stage <- lm(D_resid ~ Z_resid)
# Compute heteroskedasticity-robust variance-covariance matrix
robust_vcov <- vcovHC(first_stage, type = "HC3")
# Compute robust standard errors
robust_se <- sqrt(diag(robust_vcov))
# Compute robust t-statistic for Z_resid coefficient
t_stat_robust <- coef(first_stage)[2] / robust_se[2]
# Compute robust F-statistic
f_stat_robust <- t_stat_robust^2  # Since F = t^2 in single regressor case
cat("Robust First-Stage F-statistic (Weak Instrument Test):", f_stat_robust)
```

```
## Robust First-Stage F-statistic (Weak Instrument Test): 2444.535
```

22.2.6 DML2-IV Simulation using RANDOM FOREST

Following is a simulation of a complete DML-IV simulation that uses random forest to estimate the nuisance functions instead of Lasso. You can review random forest as covered in Chapter 14 and the DML2 method with random forest in the previous chapter. We use the exact same synthetic data as above. Cross-fitting is performed by splitting the data into folds. For each fold, we estimate the outcome, treatment, and instrument models using random forest on the training set and predict on the held-out data to obtain cross-fitted residuals. These residuals are then used in a 2SLS regression, just as in the previous example, to solve the moment equation and obtain an estimate of the causal (LATE) effect, δ. Robust standard errors are computed to ensure valid inference.

```
# Load required libraries
library(MASS)          # For generating multivariate normal data
library(estimatr)      # For HC3 robust SEs (lm_robust)
library(AER)           # For IV regression (2SLS)
library(randomForest) # For Random Forest estimation of nuisance functions
library(sandwich)      # For robust standard errors
library(lmtest)        # For robust inference

set.seed(123)          # Ensure reproducibility

# === Step 1: Define Data ===
N <- 10000  # Number of observations
p <- 10     # Number of covariates (high-dimensional X)

# Split X into 1/3 continuous and 2/3 binary covariates
p_cont <- round(p / 3)  # Number of continuous covariates
p_bin <- p - p_cont     # Number of binary covariates

# Continuous covariates ~ N(0,1)
X_cont <- mvrnorm(N, mu = rep(0, p_cont), Sigma = diag(p_cont))
# Binary covariates ~ Bernoulli(0.5)
X_bin <- matrix(rbinom(N * p_bin, 1, 0.5), nrow = N, ncol = p_bin)
# Combine X into a single matrix
X <- cbind(X_cont, X_bin)

# Generate Instrument Z (Strongly Correlated with D, Nearly Uncorrelated with Y)
Z <- 0.7 * (X_cont[,1]) + 0.3 * rnorm(N)

# Generate Treatment D (Strongly Correlated with Z)
D_prob <- plogis(25 * Z + 0.8 * X %*% runif(p, -0.5, 0.5))
D <- rbinom(N, 1, D_prob)

# Generate Outcome Y (Affected by D, Weakly Related to Z)
delta_true <- 2
beta <- runif(p, -1, 1)
Y <- delta_true * D + X %*% beta + 0.1 * Z + rnorm(N)

# === Step 2: Define Fold Assignments (Cross-Fitting) ===
```

```
K <- 5  # Number of cross-fitting folds
folds <- split(sample(1:N), rep(1:K, length.out = N))

# Initialize residual vectors
Y_resid <- rep(NA, N)  # Residuals for Y
D_resid <- rep(NA, N)  # Residuals for D
Z_resid <- rep(NA, N)  # Residuals for Z

# === Step 3: Estimate Nuisance Functions and Compute Residuals using Random Forest
# ===
for (k in 1:K) {
    # Define training (all but k-th fold) and validation (k-th fold) indices
    train_idx <- unlist(folds[-k])
    val_idx <- folds[[k]]

    # Training Data
    X_train <- X[train_idx, ]
    Y_train <- Y[train_idx]
    D_train <- D[train_idx]
    Z_train <- Z[train_idx]

    # Validation Data
    X_val <- X[val_idx, ]

    # ---- Fit Random Forest for Outcome Model (Y ~ X) ----
    rf_y <- randomForest(x = X_train, y = Y_train)
    Y_hat_val <- predict(rf_y, newdata = X_val)

    # ---- Fit Random Forest for Treatment Model (D ~ X) ----
    rf_d <- randomForest(x = X_train, y = D_train)
    D_hat_val <- predict(rf_d, newdata = X_val)

    # ---- Fit Random Forest for Instrument Model (Z ~ X) ----
    rf_z <- randomForest(x = X_train, y = Z_train)
    Z_hat_val <- predict(rf_z, newdata = X_val)

    # ---- Compute Cross-Fitted Residuals ----
    Y_resid[val_idx] <- Y[val_idx] - Y_hat_val  # r1 = Y - m1(X)
    D_resid[val_idx] <- D[val_idx] - D_hat_val  # r2 = D - m2(X)
    Z_resid[val_idx] <- Z[val_idx] - Z_hat_val  # r3 = Z - m3(X)
}

# === Step 4: Solve Moment Equation Using 2SLS ===
dml_iv_model <- ivreg(Y_resid ~ D_resid | Z_resid)
# Compute robust standard errors
robust_se <- coeftest(dml_iv_model, vcov = vcovHC(dml_iv_model, type = "HC3"))
```

```
# Extract and Print Results with Robust SEs
cat("Estimated IV effect using DML-IV (Random Forest):", robust_se[2, 1], "\n")
```

```
## Estimated IV effect using DML-IV (Random Forest): 2.108413
```

```
cat("Robust Standard Error:", robust_se[2, 2], "\n")
```

```
## Robust Standard Error: 0.07169011
```

```
cat("95% CI:", confint(dml_iv_model, vcov. = vcovHC(dml_iv_model,
                                         type = "HC3"))[2, ], "\n")
```

```
## 95% CI: 1.967593 2.249233
```

22.3 Closing Remarks: ML for IV

IV methods address endogeneity in observational studies where treatment assignment is not ignorable. Traditional approaches such as 2SLS rely on linear models, but modern data environments require more flexible methods. ML methods like DML-IV improve IV estimation by modeling nonlinearities, handling high-dimensional covariates, and identifying heterogeneous treatment effects, expanding IV beyond classical assumptions. Recent nonparametric IV methods extend IV estimation to nonlinear and non-separable models. DeepIV use deep learning to approximate structural functions under endogeneity, while Kernel IV provides a nonparametric alternative when both first- and second-stage relationships are complex.

Recent work, such as Qian et al. (2024) "Learning Decision Policies with Instrumental Variables through Double Machine Learning"[7], implemented DML-IV on various real-world datasets and compared it to other IV regression methods, including Deep IV, DeepGMM, Kernel IV, and DFIV. The real-world datasets include aeroplane ticket sales and pricing data in a ticket demand scenario (Hartford et al., 2017), the IHDP dataset (747 units, 470 training samples, available at IHDP[8]), and the PM-CMR dataset (2132 counties, 1350 training samples, available at PM-CMR[9]). The findings show that DML-IV outperforms the other IV regression methods across all datasets. The authors also demonstrate that a computationally efficient version of DML-IV can estimate local treatment/causal effects as accurately as standard DML-IV on low-dimensional datasets. We recommend reading this paper if you are interested in a detailed comparison of different machine learning methods for IV regression, along with their implementation (Code available at GitHub[10], using PyTorch).

Deep IV (Hartford et al., 2017) Deep IV: A flexible approach for estimating individual causal effects[11], which also uses neural networks but may suffer from bias in high-dimensional settings, is noted for the fact that DML-IV's use of cross-fitting and orthogonal scores provides better bias reduction than DEEP IV. The Python library for Deep IV, built on Keras, is available at GitHub[12], highlighting practical implementation differences.

Chen et al. (2021) in "Debiased/Double Machine Learning for Instrumental Variable Quantile Regressions" demonstrate how DML techniques can be used for quantile treatment effects, revealing how the impact of a treatment varies across. They further demonstrate the robustness of DML-IVQR through Monte Carlo experiments and on the 401(k) dataset, it provides detailed quantile treatment effects, standardized by outcome SD, showing varying effects across wealth quantiles.

IV regression trees (IV Trees) and generalized random forests (GRF) allow localized treatment effect estimation, uncovering how the LATE varies across different subpopulations. Understanding heterogeneous LATE effects has critical policy implications, which we will cover in the next chapter. Many real-world interventions—such as education subsidies, job training programs, or medical treatments—do not affect all individuals equally. ML-based IV methods can detect and prioritize subgroups where policy interventions are most effective, ensuring efficient allocation of resources. By systematically searching across covariate spaces, these methods allow policymakers to better target interventions.

Several software tools support machine learning-based IV methods across different programming environments. In R, the `DoubleML` package (documentation[13]) implements DML-IV, while `AER::ivreg()` provides classical IV estimation, and `grf` supports random forest-based IV. In Stata, `ddml` (documentation[14]) and `crhdreg` (Yuya Sasaki's crhdreg[15]) implement DML-IV with LASSO. Standard IV methods are available via `ivregress 2sls`, while `ivlasso` supports LASSO-based IV estimation, and `cfreg` links to Chernozhukov's DML-IV framework.

In Python, the `econml` package (documentation[16]) includes DMLIV, DeepIV, and random forest-based IV estimators. The DeepIV approach (Keras-based) is available on GitHub[17], while DML-IV (PyTorch) is at GitHub[18]. For Julia, the `IVModels.jl` package provides IV estimators for scalable computation. These tools enable efficient implementation of IV methods, from classical econometric techniques to modern machine learning approaches.

ML strengthen IV estimation by capturing nonlinearities, selecting relevant instruments, and identifying local average treatment and heterogeneous effects, but valid inference requires careful adjustment. Cross-fitting, orthogonality conditions, and robust standard errors are essential to prevent overfitting and bias. As empirical economics embraces big data, combining ML with IV methods provides deeper causal insights—provided that the core IV assumptions (relevance, exogeneity, exclusion restriction) remain rigorously defended.

22.4 Technical: Connecting IV to potential outcomes to 2sls

We begin by defining the framework for using an instrument in the presence of noncompliance. For each unit i, we consider a binary instrument $Z_i = z$ defined as follows:

- $Z_i = 1$ if unit i is assigned to the treatment group,
- $Z_i = 0$ if unit i is assigned to the control group.

Now, let's define $D_i(z)$ which indicates the potential treatment statuses of each unit (either "take" or "not take" the treatment) given $Z_i = z$. This "potential treatment status" happens as not everyone will follow all the randomized control trial (A/B testing) process (if it was complete/perfect assignment then everyone who is assigned to treatment group will receive treatment,and whoever assigned control group will not). Specifically, we have four potential treatment statuses for each unit i:

- When $Z_i = 1$: $D_i(1) = 1$ is "takes treatment" if the unit is in treatment group OR $D_i(1) = 0$ is "not take the treatment" even if the unit is in treatment group

- When $Z_i = 0$: $D_i(0) = 0$ is "not take the treatment" if the unit is in control group OR $D_i(0) = 1$ is "take the treatment" although the unit is in control group

As assigning units (by experimenter) to treatment or control groups doesnot mean each unit will actually be compliant. During the experiment there may be noncompliance—a situation in which individuals fail to follow their assigned treatments, or treatments are not perfectly implemented. Thus The observed treatments D_i is related to these potential treatments $D_i(1)$ and $D_i(0)$ by:

$$D_i = Z_i \cdot D_i(1) + (1 - Z_i) \cdot D_i(0)$$

This equation simply states that if a unit is assigned to the treatment group ($Z_i = 1$), we observe $D_i = D_i(1)$. However, we cannot observe what would have happened—whether the unit would have taken the treatment or not—had it been assigned to the control group instead. Similarly, if the unit is assigned to the control group ($Z_i = 0$), we observe $D_i = D_i(0)$.

Based on the values of potential outcomes,$D_i(1)$ and $D_i(0)$, units are classified into four types:

- *Compliers:* Units for which $D_i(1) = 1$ and $D_i(0) = 0$ (i.e $D_i(1) > D_i(0)$); they take the treatment if the unit is in treatment group, and do "not take the treatment" if the unit is in control group.
- *Always-takers:* Units for which $D_i(1) = D_i(0) = 1$; they take the treatment regardless of assigned to treatment or control group.
- *Never-takers:* Units for which $D_i(1) = D_i(0) = 0$; they do not take the treatment regardless of assignment.
- *Non-compliers/Defiers:* Units for which $D_i(1) = 0$ and $D_i(0) = 1$; these units act contrary to the assignment (i.e., they take the treatment when not assigned and vice versa).

Because we only observe one of $D_i(1)$ or $D_i(0)$ for each individual (the fundamental problem of causal inference), we cannot directly determine each unit's compliance type. However, under certain assumptions we can identify the proportion of each type in the RCT sample.

Assumptions for Estimating Compliance Proportions

To recover the causal effect among compliers, we require several key assumptions:

1. *Independence of the Instrument:*

$$(Y_i(0), Y_i(1), D_i(0), D_i(1)) \perp Z_i$$

This means that the instrument is as good as randomly assigned, so that its effect on both outcomes and treatment statuses is not confounded by other variables.

2. *First-Stage (Relevance):*
   The instrument must induce variation in the treatment:

$$0 < P(Z_i = 1) < 1 \quad \text{and} \quad P(D_i(1) = 1) \neq P(D_i(0) = 1)$$

In other words, there must be some compliers for the instrument to be useful.

3. *Monotonicity:*
   We assume that

$$D_i(1) \geq D_i(0) \quad \text{for all } i$$

This eliminates defiers (thus $\pi_D = 0$, meaning no one systematically does the opposite of what the instrument assigns.

Under these assumptions, we can estimate the proportions of each type. For example: - The proportion of always-takers is given by:

$$\pi_A = E[D_i \mid Z_i = 0]$$

since among those not encouraged, any unit taking the treatment must be an always-taker (assuming monotonocity/no defiers). - The proportion of compliers is:

$$\pi_C = E[D_i \mid Z_i = 1] - E[D_i \mid Z_i = 0]$$

- And the proportion of never-takers is:

$$\pi_N = 1 - E[D_i \mid Z_i = 1]$$

With these proportions in hand—and with one further assumption (the exclusion restriction, which we describe next)—we can move from the intention-to-treat (ITT) effect to a causal effect for compliers.

From ITT to the Local Average Treatment Effect

The exclusion restriction assumes that the instrument affects the outcome only through its effect on the treatment. Formally, for each unit i:

$$Y_i(D, Z = 1) = Y_i(D, Z = 0) \quad \text{for all} \quad D = 0, 1$$

This means that, apart from its effect on the treatment received, the instrument has no direct effect on the outcome.

Under these assumptions, the (LATE) is defined as the average causal effect of the treatment among compliers:

$$\tau_{\text{LATE}} = E[Y_i(1) - Y_i(0) \mid D_i(1) > D_i(0)]$$

Because we cannot observe compliance types directly, LATE is estimated using the ratio of the difference in mean outcomes by instrument value to the difference in mean treatment uptake by instrument value. That is,

$$\tau_{\text{LATE}} = \frac{E[Y_i \mid Z_i = 1] - E[Y_i \mid Z_i = 0]}{E[D_i \mid Z_i = 1] - E[D_i \mid Z_i = 0]}$$

An alternative way to express this estimator, when the instrument Z_i is binary, is by using a ratio-of-coefficients (or covariances) representation:

$$\hat{\beta}_{1,\text{IV}} = \frac{\text{Cov}(Y_i, Z_i)}{\text{Cov}(D_i, Z_i)}$$

Since for a binary Z_i the difference in conditional means is equivalent to the covariance (up to a scaling factor), these two expressions are mathematically identical.

To see the connection verbally: the numerator, $E[Y_i \mid Z_i = 1] - E[Y_i \mid Z_i = 0]$, represents the ITT—the effect of the instrument on the outcome—effect on the outcome, while the denominator, $E[D_i \mid Z_i = 1] - E[D_i \mid Z_i = 0]$, reflects how much the instrument shifts the probability of receiving the treatment. Their ratio isolates the effect of the treatment on the outcome for the specific group which is called as compliers- whose action changed based on treatment assignment.

Expressing ITT as a Weighted Average

It is also informative to note that the overall ITT effect can be viewed as a weighted average of the causal effects for different subgroups:

$$\text{ITT} = \tau_{\text{LATE}} \cdot \pi_C + \tau_A \cdot \pi_A + \tau_N \cdot \pi_N$$

where τ_A and τ_N are the (typically zero) treatment effects for always-takers and never-takers, respectively. Under the exclusion restriction, the instrument has no effect on the outcome for always-takers and never-takers, so their contributions drop out, leaving:

$$\text{ITT} = \tau_{\text{LATE}} \cdot \pi_C$$

Thus, rearranging gives the LATE as:

$$\tau_{\text{LATE}} = \frac{\text{ITT}}{\pi_C} = \frac{E[Y_i \mid Z_i = 1] - E[Y_i \mid Z_i = 0]}{E[D_i \mid Z_i = 1] - E[D_i \mid Z_i = 0]}$$

This detailed discussion—incorporating definitions, potential outcomes, assumptions (independence, first stage, monotonicity, exclusion restriction), and the derivation of LATE both as a ratio of differences in means and via the ratio-of-covariances approach—provides a comprehensive overview of the identification of causal effects using instrumental variables. This proof shows that the denominator of LATE corresponds to the proportion of compliers in the population. This also highlights why weak instruments (small denominator) lead to imprecise LATE estimates, as dividing by a small π_C amplifies statistical noise.

Notes

1. When the instrument Z_i is binary, this simplifies to the Wald estimator:

$$\hat{\beta}_{1,\text{IV}} = \frac{\bar{Y}_1 - \bar{Y}_0}{\bar{D}_1 - \bar{D}_0}$$

where $\bar{Y}_1$ and $\bar{Y}_0$ are the sample means of the outcome for the two groups defined by Z_i, and $\bar{\pi}_1$ and $\bar{\pi}_0$ are the corresponding means of the treatment variable.

2. The first is the ILS approach, which involves estimating two reduced-form regressions:

$$Y_i = c_{10} + c_{11} Z_i + u_{1i}$$

$$D_i = c_{20} + c_{21} Z_i + u_{2i}$$

The ILS estimator is then obtained as

$$\hat{\beta}_{1,\text{ILS}} = \frac{\hat{c}_{11}}{\hat{c}_{21}}$$

This ratio of reduced-form coefficients can be interpreted as the ratio of the intent-to-treat (ITT) estimates in a randomized trial with a binary treatment.

3. In the ILS approach: with covariates, the reduced-form regressions become

$$Y_i = c_{10} + c_{11}Z_i + c_{1X}'X_i + u_{1i}$$

$$D_i = c_{20} + c_{21}Z_i + c_{2X}'X_i + u_{2i}$$

and the corresponding ILS estimator remains $\hat{\beta}_{1,\text{ILS}} = \hat{c}_{11}/\hat{c}_{21}$.

4. An alternative way to express the LATE Wald-estimator is through a ratio of coefficients approach. The first-stage equation models the relationship between the endogenous treatment D_i and the instrument Z_i:

$$D_i = \alpha_1 + \beta_1 Z_i + \epsilon_{1i}$$

The reduced-form equation captures how the instrument affects the outcome Y_i:

$$Y_i = \alpha_2 + \beta_2 Z_i + \epsilon_{2i}$$

By taking the ratio of the two coefficients, we obtain the LATE estimator:

$$\delta_{\text{LATE}} = \frac{\hat{\beta}_2}{\hat{\beta}_1} = \frac{\text{Effect of } Z_i \text{ on } Y_i}{\text{Effect of } Z_i \text{ on } D_i}$$

This ratio-based interpretation aligns with the previous derivation of LATE as the causal effect of treatment on the outcome for compliers. Thus, $\hat{\beta}_1$ captures the first-stage effect of the instrument Z_i on treatment D_i, while $\hat{\beta}_2$ represents the reduced-form effect of the instrument on the outcome Y_i. The LATE estimator, given by the ratio $\hat{\beta}_2/\hat{\beta}_1$, isolates the causal effect of treatment on the outcome for compliers—those who take the treatment only when assigned to it. As seen from the equation, if $\beta_1 = 0$, then the effect of Z in the reduced-form equation must also be zero. If the first-stage coefficient β_1 is zero (or very small) while the reduced-form effect β_2 is significant, this implies that the instrument is affecting the outcome through a channel other than the treatment. This would be a violation of the exclusion restriction.

5. To assess whether an instrument is weak, we compare two models: Full first-stage model: Includes the instrument Z_i and covariates X_i: $D_i = \alpha_1 + \beta_Z Z_i + \beta_1 X_{1i} + \beta_2 X_{2i} + \epsilon_{1i}$ and Restricted model: Excludes the instrument:$D_i = \alpha_1 + \beta_1 X_{1i} + \beta_2 X_{2i} + \epsilon_{1i}$. Using an F-test, we check if adding Z_i significantly improves the fit: $F = \frac{(R^2_{\text{full}} - R^2_{\text{restricted}})/k_{\text{full}}}{(1-R^2_{\text{full}})/(n-k_{\text{full}}-1)}$. Rule of thumb: If $F \geq 10$, the instrument is sufficiently strong.

6. LIML and Fuller estimators provide alternatives to 2SLS) in the presence of weak instruments, reducing bias and improving inference, particularly in small samples. LIML, derived from a likelihood-based framework, is less biased than 2SLS, especially when multiple instruments are used. The Fuller estimator, a modification of LIML (Fuller, 1977), applies a shrinkage factor to further correct small-sample bias, improving efficiency. These estimators are frequently used in health economics and epidemiology, particularly in Mendelian randomization and policy-based IV studies (e.g., Medicaid expansions, hospital admissions) where instrument strength may be weak. While LIML is less sensitive to over-identification bias, its computational demands make it less common than 2SLS. Readers interested in a detailed discussion of these methods are encouraged to consult Stock & Yogo (2005) for weak instrument diagnostics and the relative performance of LIML and Fuller estimators.

7. https://arxiv.org/abs/2405.08498

8. https://www.fredjo.com/

9. https://doi.org/10.23719/1506014

10. https://github.com/shaodaqian/DML-IV

11. http://proceedings.mlr.press/v70/hartford17a.html

12. https://github.com/jhartford/DeepIV

13. https://docs.doubleml.org/r/stable/

14. https://statalasso.github.io/docs/ddml/
15. https://sites.google.com/site/yuyasasaki/Home/stata/stata-command-crhdreg
16. https://econml.azurewebsites.net/
17. https://github.com/jhartford/DeepIV
18. https://github.com/shaodaqian/DML-IV

23

Heterogeneous Treatment Effects

Why didn't the Heterogeneous Treatment Effect (HTE) follow the standard treatment plan? Because it knew one size doesn't fit all!

In previous chapters, we primarily focused on the average treatment effect (ATE) as a measure of causal impact across an entire population. However, ATE alone is not sufficient when treatment effects vary across individuals or groups. A policy based solely on ATE assumes uniform effects, which can lead to inefficiencies or unintended harm—some subgroups may benefit significantly, while others may see no effect or even experience adverse consequences. To address this, we introduced the conditional average treatment effect (CATE) in Chapter 18, which accounts for treatment heterogeneity by conditioning on observable characteristics. Estimating CATE allows for more precise policy targeting and personalized treatment assignment, improving decision-making in applied settings. With increasingly rich datasets and advanced machine learning tools, the estimation of treatment heterogeneity has gained widespread interest. In this chapter, we expand our discussion beyond individual-level heterogeneity to treatment effects at the group level and other measures of subgroup effects. A key objective is to systematically identify which subgroups benefit the most (or least) from an intervention. Machine learning can uncover this heterogeneity by modeling treatment effects as $\delta^{CATE}(X) = \mathbb{E}[Y(1) - Y(0) \mid X]$, helping researchers and policymakers optimize interventions in a data-driven manner.

23.1 Parametric Approaches to Estimating Heterogeneous Treatment Effects

Parametric approaches represent a classical starting point for analyzing heterogeneous treatment effects. These methods assume a specific functional form for how the treatment effect varies with observed covariates across individuals or groups. These approaches provide interpretable and structured ways to estimate treatment heterogeneity while leveraging existing statistical techniques. A common strategy is to include interaction terms between the treatment indicator and covariates in aregression model. Other parametric methods involve stratified analysis, generalized linear models (GLMs), or random coefficients models. We discuss each of these approaches in turn.

23.1.1 Interaction Models in Linear Regression

A simple way to capture treatment effect heterogeneity is by including interaction terms between the treatment indicator and covariates in a linear regression. In a randomized trial or observational study, one can estimate models like:

DOI: 10.1201/9781003504795-23

$$Y_i = \alpha + \delta D_i + \beta X_i + \tau(D_i \times X_i) + \epsilon_i \tag{23.1}$$

where D_i is the treatment indicator, X_i is a covariate (or vector of covariates), and τ is the coefficient on the interaction term. The **CATE** in this model is given by:

$$\tau(X) = \frac{\partial \mathbb{E}[Y \mid D, X]}{\partial D} = \delta + \tau X \tag{23.2}$$

If $\tau \neq 0$, then the treatment effect depends on X, indicating heterogeneity in responses, even at the individual level.

This framework highlights that heterogeneity can be captured not only through predefined subgroups but also at the individual level when X is a continuous or categorical variable. When X is categorical, it defines subgroups, making subgroup analysis a special case of this interaction model. Instead of estimating separate regressions for different groups, we can use a single regression with group-treatment interactions, ensuring more efficient estimation while maintaining interpretability.

Variables influencing outcomes exist at multiple levels. At the individual level, demographics (age, gender, race, education, and income), behavioral factors (risk preferences, prior treatment, and medication adherence), and health conditions (chronic disease, BMI, and cognitive ability) matter. At the household level, socioeconomic status (income, parental education, and dependents) and structure (single-parent household, and siblings) play a role. The community level includes location-based factors (urban/rural, healthcare access, and infrastructure) and policy exposure (state laws, minimum wage, and economic conditions). At the institutional level, workplace characteristics (firm size, industry, and unions) and school factors (class size, teacher experience, and funding) shape outcomes. By incorporating these factors in an interaction model, we capture how treatment effects vary across these dimensions while maintaining a unified framework within a single regression.

Interaction models are easy to implement (using ordinary least squares (OLS) or GLM), and the coefficient τ directly measures how the treatment effect changes per unit change in X. However, with many potential covariates, testing all interactions can lead to multiple-comparison problems and overfitting, so usually, a few key moderators are pre-specified to avoid spurious findings.

23.1.2 Stratified or Subgroup Analysis

So far, we have discussed parametric methods for estimating heterogeneous treatment effects, primarily through interaction models. A natural extension of this approach is the group average treatment effect (GATE), which measures treatment effects for predefined or data-driven subgroups. For subgroup analysis, we can estimate treatment effects using a single regression model with group-treatment interactions or, alternatively, separate regressions for predefined groups to capture systematic heterogeneity. We will discuss parametric models in this section, while the next sections will cover nonparametric approaches for more flexible estimation of treatment heterogeneity.

23.1.2.1 Single Regression

Often, policymakers and researchers are interested in the average treatment effect for a particular subgroup rather than a fully individualized effect. This is known as the GATE. Formally, if we define a subgroup (or grouping variable) $G_i = g$ for unit i (e.g., income group, region, or risk category), the GATE for group g is:

$$\delta_g = E[Y_i(1) - Y_i(0) \mid G_i = g] \tag{23.3}$$

which represents the expected treatment effect for individuals in group g. GATE is essentially a conditional ATE for a well-defined subset of the population.

Parametric estimation of GATE is straightforward when group membership is observable. One can estimate separate treatment effects within each group or include treatment–group interaction dummies in a regression. For instance, with a binary group indicator $G_i \in \{0, 1\}$, one could extend the earlier model to:

$$Y_i = \beta_0 + \beta_1 D_i + \beta_2 G_i + \gamma D_i \cdot G_i + f(x_i)'\boldsymbol{\beta} + \varepsilon_i \tag{23.4}$$

Here, β_1 is the treatment effect for the baseline group ($G_i = 0$), and γ captures the additional effect for the group with $G_i = 1$. Thus, $\delta_{G=0} = \beta_1$ and $\delta_{G=1} = \beta_1 + \gamma$. More generally, with multiple groups, one can include a full set of interactions $D \times I\{G = g\}$ for each group g to estimate each δ_g (often with one group as a reference).

If treatment is randomly assigned (or ignorably assigned after controlling for covariates), the difference in mean outcomes between treated and control units within each group provides an unbiased estimator of δ_g. Empirically, this is equivalent to stratifying the sample by group and computing the difference in outcomes—a method widely used in social sciences for moderation analysis.

23.1.2.2 Separate Regressions

A widely used and related approach to interactions is to conduct subgroup analyses by stratifying the sample. This means estimating the treatment effect separately within levels of a categorical variable or within slices of a continuous variable (e.g., by age group, by gender, by income quantile, by health status).

Subgroup analysis involves running separate regressions for each subgroup:

$$Y_i = \beta_0^g + \beta_1^g D_i + f(x_i)'\boldsymbol{\beta} + \varepsilon_i, \quad \text{for each group } g \tag{23.5}$$

For instance, researchers might report an estimated treatment effect (and confidence interval) for men and for women separately, then perform a statistical test for interaction to see if those effects differ significantly.

These two approaches—interaction models and separate regressions—are intuitive and easy to communicate, but they are not identical and must be chosen carefully. In a single regression model with group-treatment interactions, the coefficients of all covariates that are not interacted with treatment remain constant across individuals, implying a homogeneity assumption in how these covariates affect outcomes across groups. Standard errors are also pooled accordingly.

In contrast, separate regressions allow all coefficients, including those of control variables, to vary across groups. This distinction matters because it can lead to different conclusions. If covariates influence groups differently or if omitted variables affect groups in distinct ways, a single regression model may impose misleading restrictions, while separate regressions naturally accommodate group-specific effects.

Thus, the choice between these methods should be made carefully, considering whether the homogeneity assumption of the single regression is appropriate for the given context. If the goal is to identify systematic differences in treatment effects across groups, separate regressions may be more suitable. However, if the focus is on a specific subgroup or on interactions with a particular covariate, a single regression with interactions may be more appropriate.

Here is a R simulation to illustrate the difference between a single regression with interaction terms and separate regressions for each group. The key idea is to showing a synthetic dataset where a treatment effect and covariates effect groups differently, then compare the estimated coefficients and standard errors across the two approaches.

```
# Load required packages
library(estimatr)  # For lm_robust()
library(dplyr)
# Set seed for reproducibility
set.seed(42)
# Generate simulated data
n <- 1000  # Total sample size
group <- sample(c(0,1), n, replace=TRUE)  # Binary group indicator
treatment <- sample(c(0,1), n, replace=TRUE)  # Treatment assignment
x <- rnorm(n) # Covariate
error <- rnorm(n) # Random noise

# Define true treatment effects for each group
beta1_g0 <- 2  # Treatment effect for Group 0
beta1_g1 <- 4  # Treatment effect for Group 1
beta_x_g0 <- 1  # Effect of x for Group 0
beta_x_g1 <- 2  # Effect of x for Group 1

# Generate outcome variable with different effects by group
y <- ifelse(group == 0,
            5 + beta1_g0 * treatment + beta_x_g0 * x + error,
            8 + beta1_g1 * treatment + beta_x_g1 * x + error)

# Create data frame
data <- data.frame(y, treatment, group, x)

# Single regression with interaction terms using robust standard errors
single_reg <- lm_robust(y ~ treatment * group + x, data=data)
#summary(single_reg)

# Extract coefficients
b_treat <- coef(single_reg)["treatment"]  # Treatment effect for group 0
b_interact <- coef(single_reg)["treatment:group"]  # Interaction term
b_treat_g1 <- b_treat + b_interact  # Treatment effect for group 1

# Separate regressions for each group using robust standard errors
reg_group0 <- lm_robust(y ~ treatment + x, data = data %>% filter(group == 0))
reg_group1 <- lm_robust(y ~ treatment + x, data = data %>% filter(group == 1))
#summary(reg_group0)
#summary(reg_group1)
```

```r
# Compare estimated coefficients
coef_comparison <- data.frame(
  Method = c("Single Regression (G=0)", "Single Regression (G=1)",
             "Separate Regression (G=0)", "Separate Regression (G=1)"),
  Treatment_Effect = c(b_treat, b_treat_g1, coef(reg_group0)["treatment"],
                       coef(reg_group1)["treatment"]),
  X_Effect = c(coef(single_reg)["x"], coef(single_reg)["x"],
               coef(reg_group0)["x"], coef(reg_group1)["x"])
)
print(coef_comparison)
```

```
##                      Method Treatment_Effect  X_Effect
## 1   Single Regression (G=0)         1.984070 1.4705540
## 2   Single Regression (G=1)         4.092837 1.4705540
## 3 Separate Regression (G=0)         2.020198 0.9947716
## 4 Separate Regression (G=1)         4.003187 1.9645039
```

The treatment effect for the baseline group (G=0) in the single regression is represented solely by the treatment coefficient, while the effect for group 1 is computed as the sum of the treatment and the treatment-by-group interaction coefficients. As seen, the treatment effects for both groups in the single regression and the corresponding effects from separate regressions are similar but not identical. Additionally, while the coefficient for x remains constant in the single regression, it varies between groups in the separate regressions, reflecting distinct effects of x across subgroups. This example illustrates the importance of understanding the implications of each approach and choosing the most appropriate method based on the research question and context.

Unfortunately, it is well known that forking paths (data forking) can lead to data snooping issues, where researchers test multiple analytical choices—such as variable selection, model specification, or subgroup definitions—until they find a significant result, increasing the risk of false discoveries. The same issue applies when identifying subgroups – if many groups are examined, some may show spurious differences. To address this, in clinical trials, important subgroups are often defined *a priori* (e.g., by disease severity) based on mechanism and plausibility and analysts check whether the treatment's efficacy is consistent across these groups.[1] In development economics, randomized controlled trials (RCTs) often adopt a similar approach by preregistering subgroup analyses based on theoretically motivated stratifications, such as household wealth or baseline education levels, to ensure that findings are not driven by post-hoc data mining.[2] Preregistration helps mitigate concerns of multiple hypothesis testing and publication bias by committing researchers to planned analyses before observing outcomes. In both medical and development RCTs, transparent reporting of heterogeneous effects strengthens the credibility of policy-relevant conclusions. If many subgroups are explored without pre-specification, there is a risk of finding false positives (spurious heterogeneity), so corrections or validation in hold-out samples are advised.

Researchers, especially in health research, often distinguish *quantitative* interactions (treatment effects differ in magnitude across subgroups but have the same sign) from *qualitative* or *crossover* interactions (treatment helps one group but harms another). Detecting qualitative interactions is particularly important for policy – e.g. finding that a program benefits low-income individuals but has negative effects for high-income individuals would signal a need for targeted implementation.[3]

Subgroup analyses on their own do not adjust for multiple confounders simultaneously (they typically split by one variable at a time), but when combined with regression adjustment, they can be powerful for communicating heterogeneity in an accessible way (e.g., a table of treatment effects by region, or a forest plot of effects in various patient subpopulations). Subgroup analysis is easy to interpret for policymakers (who can say "the program works for group A but not for group B"), but each subgroup estimate is effectively an average for that group and may miss continuous gradations in effect.

23.1.3 Parametric Generalized Linear Models

Beyond linear regression, parametric outcome models can incorporate interactions to allow treatment effect heterogeneity. For example, in a logistic regression model (a GLM for binary outcomes with a logit link), a treatment-by-covariate interaction ($D \times X$) enables treatment effects to vary by subgroup. Similarly, in survival analysis (Cox proportional hazards models), an interaction of treatment with a covariate allows hazard ratios to change across groups. These parametric models are particularly useful when the outcome is non-linear (binary, count, survival time), ensuring an appropriate model while capturing heterogeneity.

A key consideration is estimating the marginal treatment effect (MTE) rather than just the ATE. While ATE gives a population-wide estimate, MTE describes how the treatment effect changes with observable characteristics, making it useful for assessing effect heterogeneity.

Example: Marginal Treatment Effect in a Logistic Regression
Consider a binary outcome Y (e.g., employment) and a treatment D (e.g., receiving job training). We allow treatment effects to vary by years of education X. A logistic regression model:

$$\log\left(\frac{P(Y=1)}{1-P(Y=1)}\right) = \beta_0 + \beta_1 D + \beta_2 X + \beta_3 (D \times X) \tag{23.6}$$

The term β_3 captures how treatment effects vary by education level. The **marginal treatment effect** at a given X is computed as the partial derivative:

$$\frac{\partial P(Y=1|D,X)}{\partial D} \tag{23.7}$$

This provides insight into how treatment effects change across different values of X. If $\beta_3 > 0$, treatment is more effective for higher-educated individuals. Keep in mind, in a linear model, the marginal effect is simply the estimated coefficient of the independent variable, since the relationship is additive and constant.

Software Implementation

- **Stata**: Use `logit` or `logistic` with interaction terms, then apply `margins` for marginal effects.[4]

```
logit employment training education c.training#c.education
margins, dydx(training) at(education=(10(2)20))
```

- **R**: Use `glm()` for logistic models and `margins` package for marginal effects.[5]

```
model <- glm(Y ~ D * X, family = binomial, data = mydata)
library(margins)
marginal_effects <- margins(model, variables = "D")
summary(marginal_effects)
```

For penalized models, `glmnet` can select relevant interactions.

- **Python**: Use `statsmodels` for logistic regression and `marginaleffects` for marginal effects.

```
import statsmodels.api as sm
from statsmodels.formula.api import logit
model = logit("Y ~ D * X", data=df).fit()
print(model.summary())
```

For penalization, `sklearn.linear_model.LogisticRegressionCV` with `l1` penalty (Lasso) can be used.

Davis and Heller (2019) illustrate how standard researcher-specified interactions to search for subgroup heterogeneity methods can miss nuanced heterogeneity. They evaluated a summer jobs program for youth, implementing both the standard interaction model and causal forest, a machine learning approach which we will cover below. They demonstrate that the causal forest predictions successfully identify heterogeneity in employment treatment effects that typical interaction effects would miss. We highly recommend not only reading the paper but also reviewing their very good supplemental document, which provides a clear comparison of standard heterogeneity methods and this new approach, making it an insightful and accessible resource.

23.1.4 Random Coefficients Models

A more flexible parametric approach involves **random coefficient models**, where treatment effects vary across individuals but follow an assumed distribution:

$$Y_i = \beta_0 + (\beta_1 + u_i)D_i + \beta_2 X_i + \varepsilon_i, \quad u_i \sim \mathcal{N}(0, \sigma^2) \tag{23.8}$$

Here, u_i represents individual-specific deviations from the average treatment effect, allowing for unobserved heterogeneity. These models are useful when heterogeneity is thought to arise from latent factors not captured in the observed covariates.

Advantages and Limitations of Parametric Approaches

The theoretical underpinning in these models is that if the specified interactions capture the true effect, one can consistently estimate subgroup-specific effects by ordinary least squares or maximum likelihood. Parametric methods provide interpretable estimates and require fewer data than fully nonparametric or machine learning-based approaches. In practice, parametric models require the researcher to pre-specify which interactions (or sub-groups) to examine.

If the parametric form is correct, interaction models provide unbiased and efficient estimates of subgroup effects and are easy to interpret. Classical hypothesis tests on interaction coefficients inform whether heterogeneity along that covariate is statistically significant. However, a key limitation is potential misspecification: if the true heterogeneity is more complex (non-linear or involving higher-order interactions), a simple linear interaction can misestimate the true pattern of (x). Misspecification can lead to biased estimates of treatment heterogeneity. Additionally, these models often require careful consideration of which covariates to include and how to model interactions to avoid omitted variable bias or overfitting. Regularization techniques (like Lasso) can be used to select a sparse set of important interactions, as introduced in earlier chapters. This helps mitigate the curse of dimensionality in purely parametric models by shrinking unimportant interaction effects toward zero.

Despite these challenges, parametric HTE models remain popular for their simplicity. They are often the first approach economists or health researchers try when looking for subgroup differences. They are simple and interpretable. Their limitations (risk of model misspecification and limited ability to capture complex interactions) inspire the use of the more advanced methods discussed below, but they set the stage for understanding what HTE means substantively (e.g., identifying which observable factors might make a difference in treatment response). Let's explore semi-parametric techniques in the following section.

23.2 Double Machine Learning for Conditional Average Treatment Effects

Earlier, we discussed parametric methods for estimating heterogeneous treatment effects, including interaction models and subgroup analysis. While these approaches provide interpretability, they rely on strong functional form assumptions that may overlook complex treatment heterogeneity. We then introduced parametric GATE, which examines treatment variation at the subgroup level. However, predefined group structures may not fully exploit the richness of available data, and estimating interactions in high-dimensional settings can be challenging.

Purely parametric models risk misspecification when heterogeneity is complex. Semiparametric methods offer a middle ground by using machine learning to flexibly estimate nuisance functions—such as regression or propensity scores—while targeting a low-dimensional, structured effect. DML for CATE provides a flexible approach that integrates these techniques while preserving valid statistical inference. Built on Neyman orthogonality and cross-fitting, DML mitigates overfitting and bias, ensuring robust estimation of heterogeneous treatment effects even in high-dimensional settings. Traditional parametric methods, such as regression models with interaction terms, often fall short when dealing with high-dimensional confounders, nonlinearities, and complex interactions that influence both treatment assignment and outcomes. Model misspecification can introduce bias, limiting the accuracy of treatment heterogeneity estimates. DML provides a more robust alternative for studying treatment effect heterogeneity. We also note that related approaches like targeted maximum likelihood estimation (TMLE) and partial linear models with ML fall in this category, but DML has become especially influential in economics.

The DML Framework for CATE Estimation

DML extends the standard partially linear model for treatment effects ($Y_i = \delta D_i + f(X_i) + \varepsilon_i$) we covered in Chapter 19 by incorporating heterogeneity. The key theoretical idea behind DML is to set up an estimation problem with orthogonal moment conditions. In a simplified setting, suppose the outcome can be decomposed as:

$$Y_i = g(X_i) + \delta(X_i).D_i + \varepsilon_i \tag{23.9}$$

where Y_i is the outcome, D_i is the binary treatment indicator, X_i represents observed covariates, ε_i is the error term. Moreover, $g(X_i) = E[Y_i \mid X_i]$ captures how baseline outcomes vary with covariates and $\delta(X_i) = E[Y(1) - Y(0) \mid X_i]$ is the true heterogeneous treatment effect function we aim to learn. Both $g(\cdot)$ and $\delta(\cdot)$ are a priori unknown and potentially complex, need to be estimated. Using machine learning to estimate these components can introduce bias, but DML corrects for this using Neyman orthogonality and cross-fitting. If we also define $e(X_i) = P(D_i = 1 \mid X_i)$ as the propensity score (for treatment), the following moment condition holds at the true functions:

$$E\Big[\big(Y_i - g(X_i) - \delta(X_i)(D_i - e(X_i))\big)\big(D_i - e(X_i)\big)\Big] = 0 \tag{23.10}$$

Intuitively, this Neyman orthogonality condition says that if we remove the systematic outcome variation due to X_i (through g) and center the treatment assignment (through e), then any remaining correlation between those residuals indicates a mis-specified $\delta(\cdot)$. In other words, $\delta(X)$ can be identified as the function that makes the residual outcome uncorrelated with the residual treatment, conditional on X. This orthogonality property is powerful: it implies that small estimation errors in $g(\cdot)$ or $e(\cdot)$ (the "nuisance" functions) will not heavily bias the estimation of $\delta(\cdot)$, at least to first order (Neyman orthogonality).

In practice, DML for CATE proceeds through the following steps:

Step 1: Estimating Nuisance Functions (First Stage)
DML first estimates two nuisance functions using flexible econometrics and ML methods (e.g., logistic regressions, random forests, gradient boosting (XGBoost), and neural networks):
- **Outcome model**: $\hat{g}(X) = \mathbb{E}[Y \mid X]$, estimated via a predictive classifier or regression model.
- **Propensity score**: $\hat{e}(X) = P(D = 1 \mid X)$, also estimated using a predictive regression model.

Since ML methods tend to **overfit**, directly plugging these estimates into treatment effect models leads to biased estimates. To address this, DML applies cross-fitting.

Step 2: Cross-Fitting to Reduce Bias
Instead of using the same data for estimating nuisance functions and treatment effects, cross-fitting splits the data into K folds:
1. Train $\hat{g}(X)$ and $\hat{e}(X)$ on one subset.
2. Estimate treatment effects on a separate subset to prevent overfitting.
3. Rotate the subsets, repeating the process across all folds.
Cross-fitting prevents overfitting by separating nuisance estimation from CATE estimation.

Step 3: Orthogonalized Score Function (Second Stage Estimation)
Using the cross-fitted nuisance estimates, we construct residualized versions of the outcome and treatment:

$$\tilde{Y} = Y - \hat{g}(X)$$

$$\tilde{D} = D - \hat{e}(X)$$

Once the residuals $(\tilde{Y}, \tilde{D})$ are computed, we estimate the heterogeneous treatment effect function $\tau(X)$ by solving:

$$\tilde{Y} = \tau(X)\tilde{D} + \varepsilon$$

This regression is performed usually using OLS to capture heterogeneity in $\tau(X)$.

In this way, we obtain an orthogonalized estimator of $\tau(X)$:

$$\tau(X) = \arg\min_{\tau} \sum_{i} \left(\tilde{Y}_i - \delta(X_i)\tilde{D}_i\right)^2 \tag{23.11}$$

Since $\tilde{D}$ is **orthogonal** to errors in $\hat{g}(X)$ and $\hat{e}(X)$, this approach removes bias introduced by machine learning errors.

The DML approach has attractive theoretical guarantees. Under regularity conditions, the estimated $\hat{\delta}(x)$ is asymptotically unbiased and can achieve approximate normality, enabling confidence intervals for functionals of $\delta(x)$ or for $\delta(x)$ at a point with Neyman orthogonality. The use of cross-fitting is crucial here: it breaks the dependence between the estimated nuisances and the individual data points used to estimate δ, thereby satisfying the orthogonality condition and enabling $\sqrt{n}$-consistent estimation even when using very flexible learners for g and e (Chernozhukov et al., 2018).

Despite its complexity, DML has become feasible thanks to modern computing and has been implemented in various software (e.g., the DoubleML package in Python and R). However, there are computational challenges: one must train multiple machine learning models and iterate across folds, which can be intensive for large datasets. Tuning the learners for g and e is also critical – if either model is very poor, variance can inflate (though bias remains controlled by orthogonality). Moreover, when the covariate space is high-dimensional, one might still need to impose some structure on $\delta(x)$ (such as modeling $\delta(x)$ with a simpler ML method or basis expansion) to avoid the curse of dimensionality.

Meta-learners like the R-learner and DR-learner provide an alternative approach to estimating heterogeneous treatment effects, which we will discuss in the next chapter. Unlike DML, which emphasizes orthogonalization for valid statistical inference, meta-learners focus on flexible, nonparametric estimation of treatment heterogeneity but is not primarily designed for inference. These methods trade off interpretability for adaptability in capturing complex treatment variation.

Software implementations for these methods are readily available: the DML package in R/Python provides a high-level interface for DML, and Microsoft's EconML library in Python offers R-learner and other metalearners for CATE with built-in cross-fitting. Researchers comfortable with Stata can still incorporate these techniques by exporting prediction results from ML models (e.g., use Python/R to get $\hat{e}(X)$, then use Stata for the orthogonalized regression).

Semiparametric methods still impose some structure (e.g., a specified moment condition or basis). In contrast, fully nonparametric approaches aim to let the data speak as freely as possible about heterogeneity in δ , with minimal functional assumptions. We will explore these methods in the next section.

23.3 Fully Nonparametric Methods for HTE Estimation

Fully nonparametric approaches impose minimal structural assumptions on how treatment effects vary with covariates. Instead of specifying a functional form or a small set of interactions, these methods allow the data to reveal complex heterogeneity. Traditional techniques include nonparametric smoothing, matching, and weighting, alongside modern causal machine learning methods like recursive partitioning (causal trees) and ensemble methods (causal forests). These methods can approximate any heterogeneity pattern given sufficient data, though they may suffer from interpretability challenges and higher variance. Next, we discuss key methods and recent applications.

23.3.1 Matching and Weighting Methods for HTE Estimation

Matching and weighting are classic nonparametric causal inference techniques,we covered in Chapter 20, adapted here to estimate heterogeneous effects. The basic idea is to compare treated and control units who are similar in covariates to get an estimate of the treatment effect for those covariate profiles. Instead of fitting a global model, matching/weighting construct local approximations of the counterfactual. For heterogeneous effects, one can do matching within subgroups or even at the individual level.

Matching-Based HTE Estimation method pairs treated and control units with similar observed characteristics to estimate treatment effects as we covered in Chapter 20 . It relies on the assumption that treatment assignment is as good as random after conditioning on observables. HTE estimation via matching involves estimating treatment effects separately for subgroups or individuals. Common techniques include:

- Nearest Neighbor Matching: Matches each treated unit to the closest control unit in covariate space.
- Propensity Score Matching: Matches units with similar estimated treatment probabilities.
- Genetic Matching: Uses optimization to find optimal covariate balance.

To estimate GATE, matching can be performed separately for each subgroup, allowing for treatment effects to be estimated nonparametrically at the group level.

$$\tau_g = \frac{1}{N_g} \sum_{i \in g} \left(Y_i(1) - Y_i(0)\right), \tag{23.12}$$

where N_g is the number of matched units in group g.

Propensity Score Weighting (IPW): Propensity score methods typically aim at ATE, but we can modify them to get CATE. For instance, to estimate the treatment effect for a specific subgroup defined by $X = x$, one could use inverse probability weights that emphasize observations near x. In practice, one might stratify the sample by the propensity score (which is a function of X) – this is essentially subclassification. Within each propensity score stratum (which groups units with similar X), we can estimate an effect. If those strata are narrow, each yields an approximately homogeneous effect estimate; by looking at how the estimate changes across strata, we observe heterogeneity.

Another approach is **caliper or kernel matching on the propensity score**: for a given treated unit i, find control units with close propensity scores (or covariates) and compute the difference in outcomes; that gives an individual-level estimate $\hat{\tau}_i$. One can then aggregate those $\hat{\tau}_i$ for those with similar X to see patterns. Matching on the *covariates* themselves (rather than the propensity) is also used: e.g. exact matching on some key discrete attributes (like matching each treated woman with a similar control woman) ensures heterogeneity by that attribute is accounted for without model assumptions.

In summary, matching and weighting methods can estimate HTE by isolating comparable groups and computing treatment effects within them. They are nonparametric in spirit – no functional form, just pairwise comparisons – but they can be combined with parametric elements (like a propensity score model, which is usually logistic = parametric, although one can use boosted trees to estimate propensity = semi-parametric). These approaches are intuitive and complement machine learning: indeed, some modern ML algorithms for HTE, like *causal forests*, can be seen as an automated way of doing many localized matchings or weightings (with the forest learning the appropriate neighborhoods for averaging).

23.3.2 Kernel-Based Methods for HTE Estimation

One direct approach to estimate a CATE function is through kernel smoothing or localized regression. Kernel method is a smoothing approach where weights are assigned based on the similarity of observations, typically using kernel functions. In the context of HTE, kernel regression can be used to directly estimate treatment effects continuously across covariates, rather than forming discrete matches. Essentially, one can run two nonparametric regressions: $\hat{\mu}_1(x) = \frac{\sum_{i:D_i=1} K_h(X_i-x)Y_i}{\sum_{i:D_i=1} K_h(X_i-x)}$ and $\hat{\mu}_0(x) = \frac{\sum_{i:D_i=0} K_h(X_i-x)Y_i}{\sum_{i:D_i=0} K_h(X_i-x)}$, where K_h is a kernel weighting function around x. The CATE is then $\hat{\tau}(x) = \hat{\mu}_1(x) - \hat{\mu}_0(x)$.

This is a fully nonparametric estimator of the treatment effect function, using local averaging. For example, if X is one-dimensional (say age), this reduces to the idea of estimating the treatment effect at each age by comparing outcomes of treated and control individuals near that age (perhaps using a Gaussian kernel weight). If X is high-dimensional, kernels suffer from the curse of dimensionality, so in practice one might use dimensionality reduction or only a subset of covariates for the kernel. Nonetheless, this approach has been used in some policy evaluations; for instance, Ichimura and Heckman's work in the 1990s on active labor programs used kernel weighting to estimate distributional impacts. In the past 5–10 years, less manual kernel smoothing is done (since tree-based methods have become popular), but some recent biostatistics papers have revisited kernel methods to estimate CATE because they are straightforward to analyze theoretically.

An extension is local linear matching: fit a local linear model within a neighborhood of x, which can reduce bias at boundaries. These methods are nonparametric in that they make no global assumption about $\tau(x)$; however, they require large data to get stable estimates, and confidence intervals often rely on asymptotic approximations. Software-wise, one can implement kernel CATE in R with packages like `np` (nonparametric kernel methods) or even by writing a short routine.

Kernel methods can also estimate GATEs by smoothing treatment effects within a neighborhood of $X = x_0$", producing local GATE estimates. Instead of hard-threshold stratification, kernel smoothing provides a nuanced view of treatment effect variation. In some applications, such as epidemiology, kernel density estimation helps define groups based on continuous risk factors (e.g., blood pressure), avoiding arbitrary cutoff. However, one can also form actual discrete groups by clustering X values via kernel methods. For example, use kernel density estimation on X to find modes (clusters of covariate space), then treat those clusters as groups and estimate GATEs for each.

From a methodological standpoint, kernel-based estimation of heterogeneity is statistically well-understood (it falls under nonparametric regression), but in practice it's less used in recent empirical work compared to trees/forests, likely because trees are easier to interpret. That said, some recent papers do use BART (Bayesian Additive Regression Trees) which is akin to a kernel method in that it's a sum-of-trees model that can capture smooth nonlinear heterogeneity. BART has been used for personalized medicine studies to estimate patient-specific treatment effects and then those are averaged for groups. Yet, we will not cover BART in this book as it is mainly a Bayesian method.

In sum, kernel-based GATE is a more technical approach and not as frequently highlighted in economics literature by name. It resides in the toolbox of statisticians. It's important to remember it as a link between fully parametric subgroup analysis and black-box ML: one can always do something like "loess" (locally weighted regression) to map out treatment effect heterogeneity, then communicate it in terms of groups ("for instance, the treatment effect is about 2 units at X=10, and about 5 units at X=20"). For completeness, we include it in this section, but the star methods for fully nonparametric HTE in economics and related fields have been the tree-based algorithms, to which we turn next.

Notes

1. Read "Developing a Protocol for Observational Comparative Effectiveness Research: A User's Guide, Chapter 3" by Varathan et al. (2013) for a detailed discussion on this topic.

2. Read randomized control trials registration in economics from https://docs.socialscienceregistry.org/

3. Read "Considerations when assessing heterogeneity of treatment effect in patient-centered outcomes research" by Lesko et al. (2018) for a detailed discussion on this topic.

4. For explanation and various examples read https://www.stata.com/manuals/rmargins.pdf. For penalized models, `lasso logit` can be used.

5. For explanation and various examples read https://cran.r-project.org/web/packages/margins/vignettes/Introduction.html

24

Causal Trees and Forests

In the previous chapter, we introduced the idea that treatment effects may vary across individuals or subgroups, and discussed how to summarize this variation using quantities like the group average treatment effect (GATE). Let (X_i, D_i, Y_i) denote the observed data, where X_i are covariates, $D_i \in \{0, 1\}$ indicates treatment status, and the observed outcome is defined by the standard potential outcomes framework:

$$Y_i = D_i Y_i(1) + (1 - D_i) Y_i(0) \tag{24.1}$$

The individual treatment effect, $\tau_i = Y_i(1) - Y_i(0)$, is not directly observed, so we rely on estimation strategies to uncover average effects either for the whole population or specific subgroups. When treatment effects are heterogeneous, estimating the GATE for subgroup g, defined as

$$\tau_g = \mathbb{E}[Y_i(1) - Y_i(0) \mid i \in g] \tag{24.2}$$

provides more targeted insights. However, identifying meaningful subgroups in advance can be challenging.

This chapter turns to methods designed to estimate such heterogeneous effects in a flexible, data-driven way. One powerful non-parametric approach is the causal tree, which uses recursive partitioning to uncover subgroups—or "leaves"—where treatment effects differ. Unlike traditional HTE methods that require researchers to define subgroup structures ahead of time, causal trees learn the structure from the data, making them especially valuable in settings where heterogeneity is complex or unknown.

As these groups are determined based on covariates X, we can express the treatment effect as a conditional average treatment effect (CATE), which captures how treatment effects vary across different covariate profiles. The CATE for a covariate vector $X = x$ is defined as:

$$\tau(x) = \mathbb{E}[Y_i(1) - Y_i(0) \mid X_i = x] \tag{24.3}$$

Causal trees aim to estimate CATE by partitioning the covariate space into subgroups that are as homogeneous as possible in terms of treatment effects. Each "leaf" of the tree represents a subgroup of observations with similar covariate profiles, and the estimated treatment effect for each leaf can be interpreted as an approximation of CATE for that covariate region.

In practice, the CATE for a leaf ℓ defined by a covariate-based partition Π can be estimated as:

$$\hat{\tau}(\ell) = \bar{Y}_{\ell,D=1} - \bar{Y}_{\ell,D=0} = \frac{\sum_{i \in \ell, D_i=1} Y_i}{n_{\ell,1}} - \frac{\sum_{i \in \ell, D_i=0} Y_i}{n_{\ell,0}} \tag{24.4}$$

DOI: 10.1201/9781003504795-24

where $D_i = 1$ indicates treatment and $D_i = 0$ indicates control, and $n_{\ell,1}$ and $n_{\ell,0}$ are the numbers of treated and control observations in leaf ℓ.[1]

Our goal is to minimize the error in estimating $\tau(\ell)$, which would ideally be measured by:

$$\text{MSE} = \mathbb{E}\left[\sum_{\ell \in \pi} \frac{N_\ell}{N}(\hat{\tau}(\ell) - \tau(\ell))^2\right] \tag{24.5}$$

However, since $\tau(\ell)$ is unknown, we cannot directly compute mean squared error (MSE). There are various causal tree methods that address this issue, as well as estimating CATE, by incorporating different split functions, outcome adjustments, and other techniques to improve the reliability of (conditional) treatment effect estimates. We will discuss some of these methods in the following sections, but for now, we focus on the most reliable one for inference — the Honest Causal Tree (CT-H) method.

Causal trees function similarly to decision trees but are optimized to split the sample based on treatment effect heterogeneity rather than minimizing outcome prediction errors. However, standard regression trees (CART) do not provide accurate estimates of τ_g and CATE because their objective function — minimizing MSE of outcome,Y_i — is misaligned with the goal of unbiased treatment effect estimation. As we covered in Chapter 13, in CART, MSE is calculated using only outcomes Y, not treatment effects — the difference in the mean of potential outcomes. This discrepancy necessitates a different or adjusted risk function that prioritizes the precision of treatment effect estimates within each leaf of the tree.

By using the adjusted empirical mean squared error (EMSE) criterion, the CT-H method ensures that the estimated CATEs are both unbiased and stable, balancing the fit of the model with the reliability of the treatment effect estimates, and allowing us for inference.

24.1 Honest Causal Tree (CT-H) Method

An influential nonparametric approach for HTE is the causal tree method (Athey & Imbens, 2016). Causal trees adapt the idea of decision trees (like CART) to the goal of estimating heterogeneous treatment effects rather than predicting outcomes. The basic idea is to recursively partition the covariates/features into subgroups (leaves of a tree) that exhibit different treatment effects.

The causal tree algorithm works as follows: it searches for splits in covariates that maximize the difference in treatment effects between subgroups within the resulting leaves, while also ensuring each leaf has a mix of treated and control units for a valid comparison. At a high level, the splitting criterion is tailored to find maximal heterogeneity in τ — the treatment effect. The algorithm rewards splits that create large differences in estimated treatment effects across leaves (indicating heterogeneity) and penalizes splits that lead to high variance within leaves (Athey & Imbens, 2016). In contrast to a standard regression tree that would split to reduce the estimation error of the outcome overall, a causal tree chooses splits that most improve the precision of treatment effect estimates.

A crucial innovation in causal trees by Athey & Imbens is the concept of honest estimation. The dataset is typically split into two parts: one part is used solely to decide the tree structure (i.e., to choose splits, implement cross-validation, and perform pruning), and the other is used to estimate the treatment effects within the final leaves. By separating the "training" of the tree from the "estimation" of effects, the honest causal tree avoids the problem of overfitting the treatment effect estimates to noise in the data. This approach provides more reliable treatment effect estimates for each leaf and allows for valid inference (e.g., one can calculate standard errors for the leaf-wise $\hat{\tau}_\ell$ using only the estimation subsample).

Before discussing the algorithm, let's define the honest criterion (risk) function. The expected mean squared error (EMSE) for treatment effect criterion in the CT-H method is designed to minimize error in the estimation of treatment effects by incorporating both the variance of outcomes within leaves and an adjustment for the split proportions. As mentioned above, since $\tau(\ell)$ is unknown, we cannot directly compute MSE. Instead, Athey & Imbens propose a modified version — the EMSE — to assess the accuracy of our estimates. The EMSE criterion adjusts for both bias and variance using separate samples for training and estimation, providing a more reliable way to estimate and compare treatment effects across leaves.

The unbiased estimator of EMSE for honest causal trees is:

$$-\widehat{\text{EMSE}}_\tau(S^{tr}, N^{est}, \Pi) = \frac{1}{N^{tr}} \sum_{i \in S^{tr}} \hat{\tau}^2(X_i; S^{tr}, \Pi) - \left(\frac{1}{N^{tr}} + \frac{1}{N^{est}}\right) \sum_{\ell \in \Pi} \left(\frac{S^2_{S^{tr}_{treat}}(\ell)}{p} + \frac{S^2_{S^{tr}_{control}}(\ell)}{1-p} \right) \tag{24.6}$$

Where:

- S^{tr} = Training sample for tree construction.
- S^{est} = Estimation sample for estimating treatment effects.
- Π = Partition of the data into leaves.
- N^{tr} and N^{est} = Numbers of observations in the training and estimation samples, respectively.
- $\hat{\tau}(X_i; S^{\text{tr}}, \Pi)$ = Estimated treatment effect for observation i based on the training sample and the partition Π.
- $S^2_{S^{tr}_{control}}(\ell)$: Within-leaf variances of outcomes in the control group for leaf ℓ.
- $S^2_{S^{tr}_{treat}}(\ell)$: Within-leaf variances of outcomes in the treatment group for leaf ℓ.
- p = Proportion of treated observations.

The EMSE estimator represents a balance between two key components: the first component is the adjusted MSE, which captures the squared treatment effect estimates across different subgroups (leaves) and reflects how well the model captures treatment effect heterogeneity. This term essentially focuses on maximizing the differences in treatment effects between leaves, encouraging splits that reveal meaningful heterogeneity. The second component is a variance term that acts as a penalty for splits leading to high variance within leaves, ensuring that the treatment effect estimates remain stable and are not driven by noise. This penalty term is particularly important as it adjusts for the variance within each leaf for both treated and control groups, making the resulting estimates more reliable.

By balancing these two components, the EMSE criterion rewards splits that create large differences in $\hat{\tau}(\ell)$ (indicating heterogeneity in treatment effects) while penalizing splits that result in high variance within leaves, especially when the leaves contain only a small number of observations. This approach prevents overfitting and ensures that the identified subgroups have meaningful and stable treatment effect estimates.

The optimal splitting criterion (risk function) is to choose the split (only in the training data) that maximizes the reduction in estimated EMSE:

$$\Delta\text{EMSE} = \text{EMSE(parent node)} - \frac{N_L}{N}\text{EMSE(left child)} - \frac{N_R}{N}\text{EMSE(right child)} \quad (24.7)$$

Where: N_L, N_R: Numbers of observations in the left and right child nodes.

The splitting process stops when no further improvement in EMSE is possible. Similar to CART, stopping conditions are applied to prevent overfitting and to ensure reliable treatment effect estimates. As we covered in detail in that chapter, splitting stops when at least one of the following conditions is met: Minimum leaf size, which ensures that each leaf has enough observations; Maximum tree depth, which controls the complexity of the tree; and Insufficient EMSE reduction, which stops splitting if further splits do not significantly reduce EMSE.

We want to highlight a few key points. There is an adjusted version of this estimator that introduces a parameter α; however, in the `causaltree` package, α is set to $\frac{1}{2}$ by default, resulting in the same estimator unless you change the default setting.[2] In cross-validation, we use the same estimator but apply it to the CV fitting set instead of the entire training set. We will provide a detailed explanation of the derivation of this estimator in a separate section at the end of the chapter. For now, let's illustrate how this calculation works with a simple example.

Suppose we consider every possible integer value between 16 and 65 as potential split points for X_1 (representing age). For each possible split s_1 (e.g., 16, 17, 18, ... 65), we first divide the data into left ($X_1 \leq s_1$) and right ($X_1 > s_1$) nodes, then calculate the EMSE for the parent node, as well as for the left and right child nodes. After that, we determine the optimal split s_1, which is the one that maximizes the reduction in EMSE (ΔEMSE). This exhaustive search ensures that the selected split point provides the best treatment effect heterogeneity between the left and right nodes.

Recall that for a given leaf ℓ, the treatment effect is estimated as:

$$\hat{\tau}(\ell) = \frac{\sum_{j\in\ell} D_j Y_j}{N_{\ell,1}} - \frac{\sum_{j\in\ell}(1-D_j)Y_j}{N_{\ell,0}} \quad (24.8)$$

In this expression, $D_j = 1$ indicates that observation j is treated, while $D_j = 0$ indicates that it is a control. The terms $N_{\ell,1}$ and $N_{\ell,0}$ represent the numbers of treated and control observations in leaf ℓ, respectively. Y_j denotes the observed outcome for observation j. This formula captures the difference in average outcomes between treated and control units within the same leaf, effectively reflecting the treatment effect for that subgroup. For an observation i in leaf ℓ, the predicted treatment effect $\hat{\tau}(X_i)$ is simply the calculated treatment effect $\hat{\tau}(\ell)$ for that leaf. We will later use these individual treatment effects to calculate the treatment effect for the parent node, as it is a weighted average of the treatment effects in its child nodes.

To illustrate how the estimator works, let's go through a step-by-step example. Suppose we have a dataset with a total of 10,000 observations, which we divide into a training sample of 6,000 observations (including 3,800 treated and 2,200 control units) and an estimation sample of 4,000 observations (with 1,800 treated and 2,200 control units). The dataset contains multiple predictors, such as $X_1, X_2, \ldots$. Our goal is to recursively split the data using an optimal splitting criterion based on the EMSE to maximize treatment effect heterogeneity.

For all subsequent steps, including cross-validation and pruning, we will only use the training sample.

To begin, let's consider a potential split on X_1 (for example, age) at a threshold s_1 of 30. Each observation X_i in the training sample is assigned to a leaf $\ell(X_i)$ based on the current partitioning defined by the tree Π. This split creates two child nodes: the left node (L), which includes observations with $X_1 \leq 30$, and the right node (R), which includes observations with $X_1 > 30$.

For this example, let's assume that when we split the training sample, the left node contains 2,500 observations, with 1,500 treated and 1,000 control units, while the right node contains 3,500 observations, with 2,300 treated and 1,200 control units. The average outcomes for the left node are 5.5 for treated units and 3.0 for control units. Similarly, for the right node, the average outcomes are 6.0 for treated units and 4.5 for control units.

These splits allow us to examine the treatment effect heterogeneity between the subgroups more closely. By calculating the difference in average outcomes between treated and control units within each node, we can estimate the treatment effects for the left and right nodes. For the left node, the estimated treatment effect is:

$$\hat{\tau}(L) = 5.5 - 3.0 = 2.5$$

For the right node, it is:

$$\hat{\tau}(R) = 6.0 - 4.5 = 1.5$$

To further assess the quality of this split, we calculate the EMSE for each node. First, we determine the proportion of treated observations in the training sample, which is $p = \frac{3800}{6000} = 0.633$.

We assume the following variances for the nodes: for the left node, the variance for treated observations is $S^2_{treat}(L) = 4$ and for control observations it is $S^2_{control}(L) = 5$. For the right node, the variance for treated observations is $S^2_{treat}(R) = 6$ and for control observations it is $S^2_{control}(R) = 4$.

Using these values, we calculate the EMSE for the left node as follows:

$$-\widehat{\text{EMSE}}_\tau(L) = 2.5^2 - \left(\frac{1}{6000} + \frac{1}{4000}\right)\left(\frac{4}{0.633} + \frac{5}{0.367}\right) \approx 6.25 - 0.0083 = 6.2417$$

Similarly, for the right node, the EMSE is calculated as:

$$-\widehat{\text{EMSE}}_\tau(R) = 1.5^2 - \frac{1}{2400}\left(\frac{6}{0.633} + \frac{4}{0.367}\right) \approx 2.25 - 0.0085 = 2.2415$$

Next, we compute the EMSE for the parent node before the split. To do this, we first find the average treatment effect for the parent node by taking a weighted average of the treatment effects in the child nodes:

$$\hat{\tau} = \frac{5.5 \times 1500 + 6.0 \times 2300}{3800} - \frac{3.0 \times 1000 + 4.5 \times 1200}{2200} \approx 5.8 - 3.82 = 1.98$$

The EMSE for the parent node is then:

$$-\widehat{\text{EMSE}}_\tau(\text{parent}) = 1.98^2 = 3.92$$

Finally, we calculate the improvement in EMSE due to the split by comparing the EMSE of the parent node with the weighted EMSEs of the child nodes:

$$\Delta\text{EMSE} = 3.92 - \left(\frac{2500}{6000} \cdot 6.2417 + \frac{3500}{6000} \cdot 2.2415\right) \approx 0.02$$

This positive ΔEMSE indicates that the split increases treatment effect heterogeneity, suggesting that separating the data based on X_1 is beneficial. However, if ΔEMSE is too small or if any stopping criteria—such as minimum leaf size, maximum tree depth, or insufficient EMSE reduction—are met, the split should not be pursued.

If we proceed with the split, we apply the same process recursively to other predictors, such as X_2. For X_2, we would evaluate potential split points at different thresholds, calculate the EMSE improvement for each possible split, and select the split that maximizes ΔEMSE. This process continues recursively until no significant improvement in EMSE is observed or until one of the stopping criteria is met.

The key insight from this approach is that the EMSE criterion ensures splits create meaningful subgroups with distinct treatment effects. Additionally, the variance penalty in EMSE prevents the selection of splits that lead to high within-group variance, which helps maintain the reliability and stability of treatment effect estimates.

After constructing the initial tree using the training sample, the next step is to evaluate the performance of the tree and determine the optimal tree size. This process involves cross-validation to select the best tree size with pruning that balances model complexity and predictive accuracy. By using cross-validation, we can assess the generalization performance of the tree and avoid overfitting to the training data.

Lets outline the detailed steps involved in all the process honest causal tree algorithm:

24.1.1 CT-H Algorithm

The Honest Causal Tree aims to partition data into leaves where treatment effects $\tau(\ell)$ differ, using the following approach:
- The **training sample** (S^{tr}) is used to build the tree structure, perform cross-validation, and prune the tree.
- The **estimation sample** (S^{est}) is used only in the final step to estimate treatment effects, ensuring unbiased estimates.
- **Cross-validation** is applied to tune pruning parameters and avoid overfitting using only the training data.
- The **-EMSE criterion** is used to choose splits that maximize heterogeneity while controlling variance.

Our data consists of (X_i, D_i, Y_i) for $i = 1, \ldots, N$, where:
- X_i represents covariates.
- $D_i \in \{0, 1\}$ is the treatment indicator.
- Y_i is the observed outcome.

Step 1: Split the Data into Training and Estimation Samples
- **Input:** Full dataset $D = \{(X_i, D_i, Y_i)\}_{i=1}^N$ of total size N.
- **Process:**

- Randomly split D into two disjoint subsets:
- **Training sample** S^{tr} of size $N^{tr} \approx N/2$.
- **Estimation sample** S^{est} of size $N^{est} \approx N/2$.
- Ensure $S^{tr} \cap S^{est} = \emptyset$.
- **Output:** Training and estimation samples:
- $S^{tr} = \{(X_i, D_i, Y_i)\}_{i \in I^{tr}}$.
- $S^{est} = \{(X_i, D_i, Y_i)\}_{i \in I^{est}}$.
- **Intuition:** The training sample is used exclusively for building, validating, and pruning the tree. The estimation sample is reserved for the final unbiased estimation of treatment effects.

Step 2: Further Split Training Data for Cross-Validation
- **Input:** Training sample S^{tr} of size N^{tr}.
- **Process:**
- Divide S^{tr} into $K = 10$ folds for cross-validation (default in `causaltree`).
- For each fold k:
- Create a **cv-fit subset** $S_{tr}^{\text{cv-fit},k} = S^{tr} \setminus S_{tr}^{k}$.
- Create a **cv-val subset** $S_{tr}^{\text{cv-val},k} = S_{tr}^{k}$.
- **Output:** Pairs of cv-fit and cv-val subsets for each fold.
- **Intuition:** Using 10-fold cross-validation helps to tune pruning parameters by growing trees on cv-fit subsets and validating them on cv-val subsets to prevent overfitting.

Step 3: Grow the Tree on the cv-fit Subset
- **Input:** cv-fit subset $S_{tr}^{\text{cv-fit},k}$ of size $N_{tr}^{\text{cv-fit}}$.
- **Process (Recursive Partitioning):**
- Start with one leaf (the entire $S_{tr}^{\text{cv-fit},k}$).
- For each current leaf ℓ:
- Calculate statistics such as the number of units, estimated treatment effects $\hat{\tau}(\ell)$, and within-leaf variances for treated and control groups using only the cv-fit subset.
- For each covariate X_j and potential split t:
- Compute EMSE before and after the split based on the cv-fit subset.
- Evaluate gain Δ as the difference between EMSE before and after the split.
- Choose the split that maximizes Δ if $\Delta > 0$ and meets the stopping criteria.
- **Stopping Rules:**
- **Minimum leaf size:** Ensure each leaf has enough treated and control units.
- **Maximum tree depth:** Prevents the tree from becoming overly complex.
- **Insufficient EMSE reduction:** Stop splitting if the reduction in EMSE is below a predefined threshold.
- **Output:** A full tree T_k for each fold k.
- **Intuition:** Growing deep trees on cv-fit subsets captures all possible heterogeneity, and the -EMSE criterion ensures meaningful splits by rewarding large differences in $\hat{\tau}$ and penalizing high variance.

Step 4: Honest Cross-Validation to Determine Pruning Parameter
- **Input:** Trees T_k from each fold and cv-val subsets $S_{tr}^{\text{cv-val},k}$.
- **Process:**
- Generate a sequence of subtrees by pruning splits with the smallest gain.
- Use a cost-complexity criterion to evaluate each subtree: $C_\alpha(T) = -\text{EMSE}(T) - \alpha|T|$ where $|T|$ is the number of leaves and α is the complexity penalty.
- Select the optimal complexity parameter α^* based on the average validation EMSE across folds.

- **Output:** Optimal α^* for pruning.

- **Intuition:** Cross-validation balances model fit and complexity by selecting α^* to prevent overfitting.

Step 5: Grow and Prune the Final Tree on Full Training Sample
- **Input:** Full training sample S^{tr} and optimal α^*.
- **Process:**
- Grow a full tree using all of S^{tr}.
- Prune the tree based on α^* to get a final pruned tree T^*.
- **Output:** Pruned tree T^* with optimal partitions Π^*.
- **Intuition:** Growing the tree on the full sample maximizes power, while pruning with α^* prevents overfitting.

Step 6: Estimate Treatment Effects with Estimation Sample
- **Input:** Pruned tree T^* and estimation sample S^{est}.
- **Process:**
- For each observation in S^{est}, traverse T^* to assign it to a leaf ℓ.
- Compute the treatment effect for each leaf ℓ using only the estimation sample:

$$\hat{\tau}(\ell; S^{est}) = \bar{Y}^{\text{est}}_{\ell,D=1} - \bar{Y}^{\text{est}}_{\ell,D=0}$$

where:
- $\bar{Y}^{\text{est}}_{\ell,D=1} = \frac{1}{N^{\text{est}}_{\ell,1}} \sum_{i\in\ell,D_i=1} Y_i$ is the average outcome for treated units in leaf ℓ.
- $\bar{Y}^{\text{est}}_{\ell,D=0} = \frac{1}{N^{\text{est}}_{\ell,0}} \sum_{i\in\ell,D_i=0} Y_i$ is the average outcome for control units in leaf ℓ.
- **Optional: Compute Variance for Treatment Effect Estimates**
- For each leaf ℓ, calculate the variance for treated and control groups:

$$S^2_{\text{treat}}(\ell; S^{est}) = \frac{1}{N^{\text{est}}_{\ell,1} - 1} \sum_{i\in\ell,D_i=1} \left(Y_i - \bar{Y}^{\text{est}}_{\ell,D=1}\right)^2$$

$$S^2_{\text{control}}(\ell; S^{est}) = \frac{1}{N^{\text{est}}_{\ell,0} - 1} \sum_{i\in\ell,D_i=0} \left(Y_i - \bar{Y}^{\text{est}}_{\ell,D=0}\right)^2$$

- Estimate the variance of $\hat{\tau}(\ell; S^{est})$ as:

$$\text{Var}(\hat{\tau}(\ell)) \approx \frac{S^2_{\text{treat}}(\ell)}{N^{\text{est}}_{\ell,1}} + \frac{S^2_{\text{control}}(\ell)}{N^{\text{est}}_{\ell,0}}$$

- **Output:**
  - Estimated treatment effects $\hat{\tau}(\ell; S^{est})$ for each leaf ℓ. Each unit assigned to leaf ℓ has the same the leaf-level treatment effect $\hat{\tau}(\ell; S^{est})$.
  - **Optional:** Variance estimates $\text{Var}(\hat{\tau}(\ell))$ for each leaf ℓ.

- **Intuition:**
  - Using only the estimation sample prevents bias in the treatment effect estimates.

  - Variance estimates provides standard errors for treatment effect estimates, helping assess their reliability.

Step 7: Compute ATE for the Tree - Input: Number of units in each leaf and the total. Each unit has the same leaf-level treatment effect. $\hat{\tau}(\ell; S^{est})$

Process: Weight Each Leaf's Treatment Effect by Its Sample Proportion:
- Compute the proportion of the estimation sample in each leaf:

$$w_\ell = \frac{N_\ell^{\text{est}}}{N^{\text{est}}}$$

where N_ℓ^{est} is the number of observations in leaf ℓ, and N^{est} is the total number of observations in the estimation sample.

Aggregate Across Leaves to Compute ATE:
- Compute the tree-level ATE as the weighted average of leaf-level treatment effects:

$$\hat{\tau}^{\text{tree}} = \sum_\ell w_\ell \hat{\tau}(\ell; S^{est})$$

Output:
- Estimated ATE for the tree, summarizing the overall causal effect learned from the tree structure.

Step 8: Predict for New Data (Optional)
- **Input:** New observation X_{new} and pruned tree T^*.
- **Process:**
- Traverse T^* using X_{new} to assign it to a leaf ℓ.
- Use $\hat{\tau}(\ell; S^{est})$ to predict the treatment effect for X_{new}.
- **Output:** Predicted treatment effect for the new observation.
- **Intuition:** The pruned tree generalizes to new data using honest estimates.

24.1.2 Simulation: Honest Causal Tree

In this example, we first generate a synthetic dataset of 10,000 observations with meaningful covariates: age (X1) uniformly sampled between 16 and 65, college graduation status (X2) as a binary indicator, and parent income (X3) uniformly distributed between 40 and 100 (in thousands). The treatment indicator (D) represents whether a student received a scholarship, and the outcome (Y) is defined in thousands of dollars as income, with values constrained between 50 and 150. Heterogeneous treatment effects are introduced so that older individuals, college graduates, and those with higher parental income receive a larger boost in income if treated.

Next, we split the data into training and estimation samples, ensuring that each half contains a balanced mix of treated and control observations. This split is critical for honest estimation: the training data is used to build the tree, while the estimation data is reserved for calculating unbiased treatment effects within each subgroup.

We then fit an honest causal tree using the `honest.causalTree` function, specifying options such as the causal tree splitting rule, honest splitting and cross-validation, and restrictions on tree depth and minimum leaf size. Cross-validation (with 10 folds) is used to tune the model and determine the optimal complexity parameter. For a comprehensive guide to all options and parameters available in causalTree, you can visit the official GitHub repository (https://github.com/susanathey/causalTree). Finally, we prune the tree based on these results and visualize it using `rpart.plot`, which provides clear node labels, conditional treatment effect estimates, and counts of observations per leaf.

```
# install.packages("devtools")
# devtools::install_github('susanathey/causalTree')
library(causalTree)
library(rpart.plot)

# Set a seed for reproducibility
set.seed(42)

# Generate synthetic data
N <- 10000  # Sample size
X1 <- sample(16:65, N, replace = TRUE)  # Age between 16 and 65
X2 <- rbinom(N, 1, 0.6)  # College graduation (1 = graduated, 0 = not graduated)
X3 <- runif(N, 40, 100)  # Parent income in thousands (between 40k and 100k)
D <- rbinom(N, 1, 0.5)  # Treatment indicator (0 = no , 1 = received scholarship)

# Define treatment effect heterogeneity based on covariates
tau <- 2 * (X1 > 40) + 15 * (X2 == 1) + 2 * (X3 > 60)
# Define income (Y) in thousands with heterogeneity and add noise
Y0 <- 5 + 1 * X1 + 5 * X2 + 0.3 * X3 + rnorm(N, 0, 5)  # Control outcome
Y1 <- Y0 + tau  # Treated outcome (with scholarship)
Y <- ifelse(D == 1, Y1, Y0)  # Observed outcome

# Ensure income is within 50k to 150k
Y <- pmin(pmax(Y, 50), 150)

# Combine data into a data frame
data <- data.frame(Y, D, X1, X2, X3)

# Split data into treatment and control indices
n <- nrow(data)
trIdx <- which(data$D == 1)
conIdx <- which(data$D == 0)

# Create training and estimation samples (similar to the example)
train_idx <- c(sample(trIdx, length(trIdx) / 2),
               sample(conIdx, length(conIdx) / 2))
train_data <- data[train_idx, ]
est_data <- data[-train_idx, ]

# Fit an honest causal tree
honestTree <- honest.causalTree(
  formula = Y ~ X1 + X2 + X3,
  data = train_data,
  treatment = train_data$D,
  est_data = est_data,
  est_treatment = est_data$D,
  split.Rule = "CT",         # Causal tree splitting rule
  split.Honest = TRUE,       # Honest splitting
  HonestSampleSize = nrow(est_data),  # Honest sample size
```

```
  split.Bucket = TRUE,        # Use bucket splits for categorical variables
  xval = 10,                  # 10-fold cross-validation
  cv.option = "CT",          # Use cross-validation for pruning
  cv.Honest = TRUE,           # ENable honest cross-validation
  cv.alpha = 0.5,            # Alpha parameter for honest cross-validation
  minsize = 100,              # Minimum leaf size
  maxdepth = 5,               # Maximum tree depth
  cp = 0.0,                  # Complexity parameter for pruning
  propensity = NULL           # No propensity score provided
)
```

```
## [1] 6
## [1] "CTD"
```

```
# honestTree$cptable # Display the complexity parameter table
# Prune the tree based on cross-validation results
opcp <- honestTree$cptable[,1][which.min(honestTree$cptable[,4])]
pruned_tree <- prune(honestTree, cp = opcp)
# pruned_tree$cptable # Display the pruned tree complexity parameter table

# Plot the pruned tree using rpart.plot for better visualization
rpart.plot(pruned_tree,
           type = 3,              # Shows split labels and leaf nodes with outcomes
           extra = 101,           # Displays fitted treatment effect in each node
           under = TRUE,          # Show the number of observations under each node
           faclen = 0,            # No abbreviation for factor levels
           box.palette = 0,            # Black and white nodes
           shadow.col = 0,             # No shadow (pure B/W)
           main = "Pruned Honest Causal Tree")
```

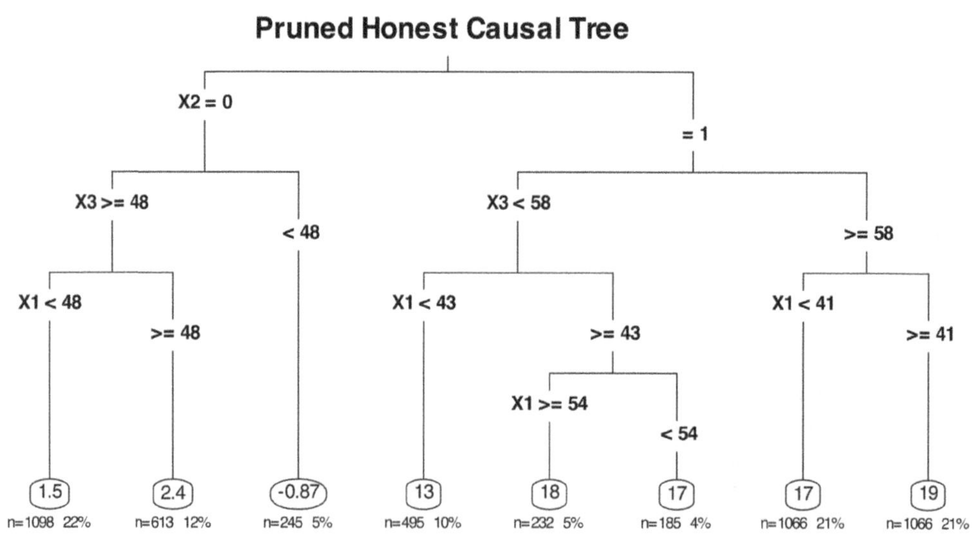

```r
# Summary of the pruned tree
# summary(pruned_tree)

# Generate predicted treatment effects for the estimation sample
predictions <- predict(pruned_tree, newdata = est_data)
# Calculate the average treatment effect (ATE)
ATE <- mean(predictions)
print(ATE)
```

```
## [1] 11.00319
```

24.1.3 "Causal" Random Forests: Ensemble Methods for HTE and Inference

Causal random forests build on causal trees by averaging many trees to create a more robust nonparametric estimator of $\tau(x)$ (Wager & Athey, 2018). This approach is analogous to how random forests extend CART: by aggregating over many randomized trees, one reduces overfitting and improves predictive stability. Yet, in causal trees and forests, the goal is not just to predict outcomes, but to estimate treatment effects, their heterogeneity, and to provide valid statistical inference.

As we discussed in Chapter 14, fitting a regression forest involves creating multiple decision trees, each trained on a different random subset of the data. For any given point Y_i, the prediction of the forest is the average of the predictions made by these individual trees. Unlike the causal tree methods that require cross-validation for model selection, random forests have a built-in mechanism for performance evaluation using out-of-bag (OOB) samples. Since each tree is trained on only part of the data, the observations excluded from the training set (the OOB samples) can be used to test that accuracy of the tree. The OOB error, calculated by averaging the prediction errors for these OOB samples across all trees, provides an internal estimate of the performance of the model without the need for separate cross-validation. So, in causal forests, the OOB error can be used to assess the predictive accuracy of the model, eliminating the need for cross-validation.

The single causal tree approach by Athey and Imbens (2016) estimates heterogeneous treatment effects for each leaf of the "best" tree. Causal forests retain the flexibility of trees—capturing complex interactions and nonlinear heterogeneity—while dramatically improving precision by averaging many "reasonable" trees, which smooths out the jagged partitioning of any single tree and produces a more stable $\hat{\tau}(x)$ function. This approach avoids the challenge of choosing a single "best" tree, as Breiman (2001) suggested, by building many reasonable trees and averaging their predictions to reduce variance and sharp decision boundaries. The causal random forest algorithm, proposed by Wager and Athey (2018), not only yields more precise treatment effect estimates but also facilitates statistical inference; under regularity conditions such as honesty and appropriate subsampling, the forest estimator for $\tau(x)$ is asymptotically Gaussian. This means that, in addition to point estimates, standard errors and confidence intervals can be derived—often automatically in implementations like the `grf` package in R—allowing researchers to assess the statistical significance of heterogeneous effects.

The "causal" random forest algorithm builds on the algorithm enforces honesty by using separate subsamples for determining splits and for estimating treatment effects within leaves, preventing overfitting to the training data. Then, the ensemble averaging step combines the estimated of (conditional) treatment effect of many trees by averaging them, leading to a more stable and precise estimate of the treatment effect function $\tau(x)$. All steps are similar to the regular forest discussed in Chapter 14, with the honesty idea introduced earlier in the causal tree section.

Formally, for a forest with B trees, the CATE at a given point x is estimated as:

$$\hat{\tau}_{\mathrm{CF}}(x) = \frac{1}{B}\sum_{b=1}^{B} \hat{\tau}^{(b)}(x) \tag{24.9}$$

where $\hat{\tau}^{(b)}(x)$ is the treatment effect predicted by the b-th tree for an observation with features x (obtained similar way we discussed in the previous section). This ensemble approach reduces variance and yields a more reliable estimate of $\tau(x)$ than any single tree could provide.

Wager and Athey (2018) introduced two procedures for directly estimating heterogeneous treatment effects using causal forests. The first, known as double-sample trees (DST), builds honest trees by using separate samples for training and estimation—similar to the causal tree method we discussed earlier, but without the need for cross-validation since out-of-bag error is available. The second procedure, propensity trees (PT), uses a propensity score calculated on the full sample to guide splits via a Gini criterion (as covered in Chapters 13 and 14) and then estimates heterogeneous treatment effects. They also provide finite-sample results using k-nearest neighbors. Although forest-based methods have proven effective for treatment effect estimation in terms of predictive error—as seen with BART methods—we focus on the clear framework and simulations provided by Wager and Athey (2018), which directly estimate treatment effects. We will not include simulations or code for these specific methods here, because a newer, high-performance implementation called generalized random forests (GRF) will be discussed in the next section. This overview highlights the evolution of methods for directly estimating treatment effects and sets the stage for the GRF approach, for which we will provide code and simulations.

24.2 Generalized Random Forests

Athey, Tibshirani, and Wager (2019) introduced the GRF algorithm, which extends the causal random forest framework to handle a broader range of causal inference models for estimating heterogeneous treatment effects. The GRF algorithm can estimate the ATE, the CATE, the conditional average treatment effect on the treated CATT, the conditional average treatment effect on the untreated (CATU), and overlap-weighted average treatment effects. Additionally, it is capable of estimating these effects in partial linear models, for binary treatments, multiple treatment categories (referred to as multi-arm causal forests), instrumental variables, quantile effects, and survival models.

The GRF algorithm is implemented in the `grf` package in R, which serves as a comprehensive and efficient tool for estimating treatment effects and conducting inference across various causal inference settings. The package offers a range of functionalities, including hypothesis testing and inference for treatment effect estimates for any specified sample, "predictions" of CATE at unit level, variable importance measures, and diagnostics for before-and-after analyses. The package is regularly updated with new features and improvements by a team of developers and contributors. They also maintain a detailed and user-friendly web page at https://grf-labs.github.io/grf/index.html, which we highly recommend for further details and practical examples.

In this section, we will focus on the basic idea of the GRF for causal forest model, its algorithm, and how to use the `grf` package in R to estimate treatment effects. We will cover the doubly robust method for estimating ATE, building on the concepts discussed earlier. Before proceeding, we recommend revisiting the concepts of doubly robust methods—such as the AIPW orthogonalization, and partial linear models—to better understand the material.

Furthermore, we suggest reading the original paper by Nie and Wager (2021), which introduces the R-learners (residual learners) approach for estimating heterogeneous treatment effects. Our explanation will primarily be based on this paper and the GRF package web page and documentation. Interestingly, the GRF web page, also explains the GRF (especially causal forest) method using the R-Learner approach, as it offers a more accessible and less notation-heavy explanation compared to the main paper. This alignment in explanation makes it easier for practitioners to understand and implement the method using the GRF package.

Before proceeding to the main idea behind the GRF method, let's briefly recap what we covered in the causal/treatment effect chapters. We observed outcomes Y_i, a binary treatment W_i, and covariates X_i (we changed the notation from D to align with the `grf` package). Our primary interest was in estimating the ATE:

$$\tau = \mathbb{E}[Y_i(1) - Y_i(0)] \tag{24.10}$$

To estimate τ without assuming a linear model, we used the partially linear model:

$$Y_i = \tau W_i + f(X_i) + \epsilon_i, \quad \mathbb{E}[\epsilon_i \mid X_i, W_i] = 0 \tag{24.11}$$

By defining the propensity score $e(x) = \mathbb{E}[W_i \mid X_i = x]$ and the conditional mean outcome $m(x) = \mathbb{E}[Y_i \mid X_i = x] = f(x) + \tau e(x)$, we rewrote the model as:

$$Y_i - m(x) = \tau(W_i - e(x)) + \epsilon_i \tag{24.12}$$

This formulation allowed us to estimate τ using residual-on-residual regression:

$$\tilde{Y}_i = Y_i - \hat{m}(X_i), \quad \tilde{W}_i = W_i - \hat{e}(X_i) \tag{24.13}$$

and then regress:

$$\tilde{Y}_i \sim \tau \tilde{W}_i \tag{24.14}$$

(this can be written as $\text{lm}(\tilde{Y} \sim \tilde{W})$) in R. Also, for continuous W, we can estimate τ with partial linear model as:
$\hat{\tau} = \frac{\sum_i \tilde{Y}_i \tilde{W}_i}{\sum_i \tilde{W}_i^2}$. We can also estimate the constant ATE using one of the nonparametric models, the AIPW method; for details, check the chapter and footnote.[3] (Keep in mind that we covered numerous nonparametric and parametric methods to estimate the constant ATE, but here we emphasize AIPW as it is used in `grf`.) However, assuming a constant τ is restrictive. To allow for non-constant treatment effects, we extended the model to:

$$Y_i = \tau(X_i) W_i + f(X_i) + \epsilon_i \tag{24.15}$$

where $\tau(X_i)$ represents the CATE. In the previous chapter, we estimated HTE using parametric models with pre-specified subgroups and explored causal trees upto now in this chapter. The key innovation of causal trees is to replace the standard mean-squared error of outcome splitting criterion with one that focuses on directly maximizing treatment effect heterogeneity and minimizing error in treatment effect estimates while using "honest" samples.

The GRF extends this idea further by providing a flexible, nonparametric model for estimating conditional average treatment effects allowing for heterogeneous treatment effects (HTE). The causal forest method can be understood as a combination of ideas from Breiman's Random Forests (2001) and Robinson's partially linear model (1988), and Nie and Wager's L-learner model (2021). The key innovation lies in modifying the tree-building phase, allowing heterogeneous treatment effects even at the unit level, and in the estimation phase, which captures and estimates various treatment effects and valid inferences nonparametrically for any specified sample by enabling the use of different random forest models.

In the building phase, GRF models create trees by selecting covariate splits that maximize the squared difference in subgroup treatment effects. The model specific criterion ensures that the splits focus on maximizing heterogeneity in treatment effects rather than just outcome variance. Each model such as causal forest, multi-outcome, survival forest implement different criterion and its approximate calculation (using gradient tree). For instance, "theoretical" splitting criterion for causal forests is $\Delta(C_1, C_2) = \frac{n_{C_1} \cdot n_{C_2}}{n_P^2} \cdot (\hat{\tau}_{C_1} - \hat{\tau}_{C_2})^2$. We will discuss its implementation and the use of an approximate criterion instead of the theoretical one in the next section, focusing primarily on causal forests for clarity.

Once the trees are built, causal forests employ an adaptive weighting scheme: weights $\alpha_i(x)$ are determined based on how often an observation falls into the same leaf as the target point x across all trees. The adaptive weights are defined as:

$$\alpha_i(x) = \frac{1}{B} \sum_{b=1}^{B} \frac{\mathbf{I}(X_i \text{ is in the same leaf as } x \text{ in tree } b)}{\text{L}} \tag{24.16}$$

where L is the number of observations in the leaf which contains x. These weights are used to run a "forest"-localized version of Robinson's residual-on-residual regression, allowing for flexible and localized estimation of treatment effects. The estimation of the CATE $\tau(x)$ at unit level is done using these weights:

$$\tau(x) = \text{lm}(Y_i - \hat{m}^{(-i)}(X_i) \sim W_i - \hat{e}^{(-i)}(X_i), \text{weights} = \alpha_i(x)) \quad \text{where} \quad X_i \in \mathcal{N}(x) \tag{24.17}$$

where $\hat{m}^{(-i)}(X_i)$ is the estimated conditional mean outcome, Y_i, without using the i-th observation; and $\hat{e}^{(-i)}(X_i)$ is the estimated propensity score without using the i-th observation. These leave-one-out estimates are part of the honesty design, ensuring unbiasedness by not reusing the same data point for both fitting and prediction. This approach essentially runs a localized OOB residual-on-residual regression weighted by the forest's structure (in other words regress the centered outcomes on the centered treatment indicator using weights which can be understood as data-adaptive kernel using only neighbour observations), allowing for a nonparametric estimation of (unit level) treatment effects, effectively "predicting" unit level CATE.

Unlike the honest tree and causal forest discussed in the previous section—where the CATE estimator at the unit level was the mean difference in average outcomes between treatment and control groups within each leaf (and thus the same for all units in that leaf)—the Causal Forest in GRF uses an R-learner estimator with adaptive weights and OOB residuals, estimating CATE using all units in the same neighborhood. The unit level estimated treatment effect at a given point x is "predicted" as:

$$\hat{\tau}(x) = \frac{\sum_{i=1}^{n} \alpha_i(x) \left(Y_i - \hat{m}^{(-i)}(X_i)\right) \left(W_i - \hat{e}^{(-i)}(X_i)\right)}{\sum_{i=1}^{n} \alpha_i(x) \left(W_i - \hat{e}^{(-i)}(X_i)\right)^2} \tag{24.18}$$

where $\alpha_i(x)$ represents the adaptive weights. The term $Y_i - \hat{m}^{(-i)}(X_i)$ is the OOB residual for the outcome, with $\hat{m}^{(-i)}(X_i)$ being the predicted outcome excluding observation i. Similarly, $W_i - \hat{e}^{(-i)}(X_i)$ is the out-of-bag residual for the treatment, using a leave-one-out approach.

In practice, estimating (C)ATE requires identifying neighborhoods of x, which are not known in advance, computing adaptive weights $\alpha_i(x)$ for each point in a sample, and then calculating a weighted average of treatment effects. However, since adaptive weights are computed using a model-specific split function, the process requires thousands of trees and involves a repetitive procedure of calculating weights and then estimating treatment effects, making it computationally expensive. To improve efficiency, the GRF algorithm employs an approximate splitting criterion and a gradient tree approach. It also precomputes certain statistics during tree building, which are then used to estimate CATE directly, rather than computing $\alpha_i(x)$ for each unit and applying the $\hat{\tau}$ equation explicitly. We will discuss this process in detail in the algorithm section and its derivation in a separate section later.

The result of causal forest is $\hat{\tau}$, which can be more accurately written as $\hat{\tau}^{(-i)}(X_i)$, representing the initial treatment effect estimate for each observation. Note that this differs from the tree and forest models discussed earlier. The `grf` package automatically computes adaptive weights, pseudo-outcomes, and other necessary statistics for various causal forest methods. The final point prediction of treatment effects is obtained using the `predict(<causal_forest>)` command.

After obtaining the point prediction of treatment effect, the final average treatment effect estimate for a given sample is calculated using the doubly robust (AIPW) method. The doubly robust method combines the predictions (unit level treatment effects) from the causal forest with the propensity score to ensure unbiasedness and efficiency in treatment effect estimation in specified sample. The sample can be new test sample, out-of-bag sample, treated sample or any other sample), the final average treatment effect estimate for that specified sample is calculated using doubly robust (AIPW) method.[4] This process enable the estimation of various treatment effects, valid inference, treatment effect prediction, and other pre- and post-estimation assessments. To get the doubly robust estimates of average treatment effects with the `grf` package, we use just `average_treatment_effect()` command, which returns the estimated ATE for specified sample and standard error.

In essence, Causal Forest extends estimating CATE allowing heterogeneous treatment effects nonparametrically by integrating machine learning (adaptive trees and random forests) with econometrics ("leave-one-out residuals"-on-"leave-one-out residuals" doubly robust regression). Beyond CATE, the GRF framework can estimate other parameters. It can handle quantile estimation by modifying the splitting criteria to focus on quantiles rather than means. It can also be adapted to estimate LATE in instrumental variables (IV) settings by incorporating specialized splitting and estimation rules. Additionally, GRF extends to survival analysis by adjusting the splitting criteria for censored data, demonstrating its flexibility across a wide range of econometric applications.

24.2.1 GRF Algorithm

Step 1. The Data Setup

You start with a dataset containing:

- **Covariates** X_i: Variables for each unit i.
- **Outcome** Y_i: Outcome for each unit i.
- **Treatment** W_i: A binary treatment indicator ($0 =$ control, $1 =$ treated); or continuous.
- **Cleaning:** The user must preprocess the data (e.g., handle missing values, encode variables) to make it usable before passing it to `causal_forest`.

```
## Load necessary libraries
library(grf)

## Load your data
# For example, reading a CSV file
data <- read.csv("your_data.csv")

## Examine the data
str(data)     # Check structure and types of variables
summary(data) # Get summary statistics to identify missing values or anomalies

# Convert categorical variables to dummy/binary indicators if needed
data$category_var <- as.factor(data$category_var)
data <- model.matrix(~ . - 1, data = data)  # One-hot encoding without intercept

## Define features, outcome, and treatment
X <- as.matrix(data[, !colnames(data) %in% c("Y", "W")])  # Features
Y <- data$Y  # Continuous outcome variable
W <- data$W  # Binary treatment variable

## Check covariate balance
# Standardize features for balance checking
X_standardized <- scale(X)
summary(X_standardized)

## Ensure data is ready for causal_forest
# Confirm binary treatment and continuous outcome
table(W)
summary(Y)
```

Optional Predictions: $\hat{m}(X_i)$ **and** $\hat{e}(X_i)$

GRF can incorporate preliminary predictions to improve accuracy (these are also used in "doubly robust" part):

- $\hat{m}(X_i)$: Predicted outcome for individual i, often estimated using a separate regression forest on (X, Y).
- $\hat{e}(X_i)$: Predicted probability of treatment (propensity score), estimated from (X, W) using another regression forest or any other predictive model.

For external computation and overlap testing, you can calculate these predictions before running GRF. If omitted, GRF will estimate regression forest as default using all covariates internally.

```
# Example of external estimation (optional)
w.forest <- regression_forest(X, W) # can use any other methods
w.hat <- predict(w.forest)$predictions

y.forest <- regression_forest(X, Y) # Outcome regression,can use any other methods
y.hat <- predict(y.forest)$predictions

# Pass externally calculated w.hat and y.hat to GRF:
cf <- causal_forest(X, Y, W, W.hat = w.hat, Y.hat = y.hat)

# If the propensity scores are well-distributed, the histogram will show most values
# away from 0 and 1. However, if the scores cluster at the extremes, the overlap
# assumption is likely violated.
hist(W.hat, main = "Propensity Score Distribution",
     xlab = "Scores")  # Check overlap assumption
hist(W.hat[W==1], col="gray", alpha=0.5,
     main="Propensity Scores: Treated vs Control", xlab="Propensity Score")
hist(W.hat[W==0], col="black", alpha=0.5, add=TRUE)
legend("topright", legend=c("Treated","Control"), fill=c("grey","black"))
```

Step 2: Initialize Forest Parameters

Before building the causal forest (or other forest models.), GRF requires tuning parameters, which control subsampling, tree size, and splitting behavior. These are set in the `grf` package:

`num.trees`: Number of trees B in the forest (default $=$ 2000 in `grf`).

`sample.fraction`: Fraction of data subsampled for each tree (default $=$ 0.5). Subsamples are drawn without replacement.

`mtry`: Number of features considered at each split (default $=$ `min(ceiling(sqrt(p) + 20), p)`, where p is the number of features).

`min.node.size`: Minimum number of observations per leaf (default $=$ 5), controlling tree depth.

`honesty`: Whether to use honest splitting (default $=$ `TRUE`), splitting the subsample into two parts for structure and estimation.

```
# Building a causal forest with default options
forest <- causal_forest(         # Fit a causal forest
  X, Y, W,                       # Features, outcome, and treatment
  num.trees = 2000,              # Default number of trees
  min.node.size = 5,             # Default minimum node size
  sample.fraction = 0.5,         # Default fraction of data per tree
  mtry = floor(sqrt(ncol(X))),         # Default number of covariates per split
  honesty = TRUE,                # Default to use honest estimation
  honesty.fraction = 0.5,        # Default fraction for honest estimation
  honesty.prune.leaves = TRUE,   # Default to prune leaves
  W.hat = NULL,                 # Default to estimate propensity scores internally
```

```
  Y.hat = NULL,                # Default to estimate outcome model internally
  alpha = 0.05,                # Default confidence interval level
  imbalance.penalty = 0,       # Default to no penalty for imbalance
  ci.group.size = 2,           # Default trees averaged for confidence intervals
  tune.parameters = "none"     # Default to no parameter tuning
)
```

Step 3: Building Each Tree ($b = 1$ to B)

A GRF consists of B trees (e.g., $B = 2000$), where each tree is built on a random subsample of the data. Then for Each tree, grf randomly divides this subsample for tree b into two evenly- sized, nonoverlapping halves:

$S_b^{\textbf{split}}$: Used to determine the tree's structure (splitting rules).

$S_b^{\textbf{est}}$: Used to estimate treatment effects within the leaves.

Step 3A: Building the Tree: Splitting Nodes

The "theoretical" splitting criterion for causal forests directly targets treatment effect heterogeneity but is computationally expensive due to repeated estimation. The gradient tree approach replaces this with an efficient approximation, using pseudo-outcomes,$\rho_i = (\hat{W}_i - \bar{W}_P)\left(\hat{Y}_i - \bar{Y}_P - (\hat{W}_i - \bar{W}_P)\hat{\beta}_P\right)/Var(\hat{W}_i)$, and the approximate criterion $\hat{\Delta}(C_1, C_2) = \sum_{j=1}^{2} \frac{1}{n_{C_j}} \left(\sum_{i \in C_j} \rho_i\right)^2$, allowing causal forests to scale to large datasets while preserving accuracy.[5]

During tree construction (see Algorithm 2, step 5; Athey, S., Tibshirani, J., & Wager, S. (2019)), these pseudo-outcomes ρ_i are used in the regression step (equation (9)) to choose splits that maximize differences in the treatment effect estimates between child nodes using standard CART regression split algorithm (which we covered in chapter 13. This approach directly targets heterogeneity by using a gradient-based approximation to the change in the target parameter when a split is made. Start with the full dataset as the root node.

- Compute pseudo-outcomes ρ_i for all observations in the parent node one time only.
- For each feature and split point, calculate $\hat{\Delta}(C_1, C_2)$ using the pseudo-outcomes. (That is performing a standard CART regression split on the pseudo-outcomes, not on the original outcomes)
- Choose the split that maximizes $\hat{\Delta}(C_1, C_2)$. (This is splitting parent node into two splits that maximizes the approximate criterion for causal forests)
- Recursively apply this process to child nodes until stopping criteria (e.g., minimum node size) are met.

Repeat step 3A for all B trees. `grf` does not store the value of the criterion it uses to determine splits.

Step 3B: Compute weights, sufficient statistics for Each Unit

For a specific tree b, once the tree structure is built using S_b^{split}, the next step is to assign points from the estimation sample S_b^{est} to specific leaves of that tree. Each point i in S_b^{est} is dropped ("pushed down") the tree according to the splitting rules until it reaches a particular leaf l. Each leaf contains training examples that fell into it during tree construction, forming the "neighborhood" of x in that tree.

This process is repeated for all B trees. Using this information, the program creates a list of neighboring training points for x, assigning adaptive weights denoted as $\alpha_i(x)$. program also calculates sufficient statistics for each leaf, which are used to estimate treatment effects efficiently. There are some other information such as variable splits, out-of-bag information is saved by the program. Yet, causal forest command in `grf` package does not print these information directly. You can check variable importance using `get_variable_importance()` in the `grf` package. If you want to estimate ATE for causal forest (or other forest models) by AIPW, you can continue next steps.

Step 4: "Predict" Point Treatment Effects for Each Unit

When using the `predict(<causal_forest>)` command, the program estimates the treatment effect at a specific point x in the sample using one of two equations. The output of `predict(<causal_forest>)` is the unit level CATE. The more efficient second method is used for causal forests. The program decides which strategy to use based on the forest type internally.

Normally, a prediction is then made using this weighted list of neighbors, following the appropriate method for the type of forest. For instance, the best linear predictor for causal forest in the neighborhood, weighted by $\alpha_i(x)$, is given by:

$$\hat{\tau}(x) = \frac{\sum_{i=1}^{n} \alpha_i(x)\tilde{Y}_i\tilde{W}_i}{\sum_{i=1}^{n} \alpha_i(x)\tilde{W}_i^2} \tag{24.19}$$

However, in most cases, lean and tree level "precomputed sufficient statistics", significantly speeding up predictions by avoiding the need to explicitly compute $\alpha_i(x)$. The point predictions $\tau(x)$ for a target sample $X = x$ are given by:

$$\hat{\tau}(x) = \frac{\frac{1}{B}\sum_{b=1}^{B}\left(Y\bar{W}_b(x)\bar{w}_b(x) - \bar{Y}_b(x)\bar{W}_b(x)\right)}{\frac{1}{B}\sum_{b=1}^{B}\left(\bar{W}_b^2(x)\bar{w}_b(x) - \bar{W}_b(x)\bar{W}_b(x)\right)} \tag{24.20}$$

Both of these equation is used to estimate the treatment effect at a specific point x in the sample and to compute unit-level CATE for each unit individually. Second one implicitly includes $\alpha_i(x)$ and estimates $\hat{\tau}(x)$ efficiently by using pre-computed averages instead of direct calculation which is computationally costly. In a later section, we show step by step how the second equation is equivalent to the first, explaining the meaning of each component.

Step 5: Estimate Average Treatment Effects for a Specified Sample

After computing initial treatment effect estimates for each observation using `predict(<causal_forest>)`, the final ATE for a specified sample is estimated using `average_treatment_effect()`. This function calculates the ATE and its standard error based on the equation 4 for any specified sample. You can access doubly robust(AIPW) scores $\hat{\Gamma}_i$ with `get_scores(<causal_forest>)` command. Under regularity conditions, the average of the doubly scores is an efficient estimate of the average treatment effect. The function `average_treatment_effect()` will give a warning if overlap seems to be an issue.

Although this command directly provides results, the following explains the calculation process. Instead of computing the ATE directly from the equation, you can estimate it approximately using the following simple process:

$$\hat{\tau}^{ATE}_{CausalForest} = E[Y|X, W = 1] - E[Y|X, W = 0] = [\hat{Y} + (1 - \hat{W})\hat{\tau}] - [\hat{Y} - \hat{W}\hat{\tau}] \tag{24.21}$$

where $\hat{Y}$ is obtained from the outcome regression forest internally, $\hat{W}$ is obtained from the propensity score forest, $\hat{\tau}$ is obtained from `predict` after running the `causal_forest` command (as discussed in the previous step; $\hat{\tau}$).

You can also retrieve $\hat{Y}$ and $\hat{W}$ using the `predict` command after optional predictions in Step 1. The `average_treatment_effect()` function returns the doubly robust estimation of the ATE for the specified sample along with its standard error, following this straightforward calculation. By default, `grf` computes ATE using OOB estimates, considering only trees where x was not included in S_b.

For further details on computing point predictions using precomputed sufficient statistics within the OptimizedPredictionStrategy, refer to the last section of this chapter and in GRF Developing Documentation.[6] For more details on CATE estimation, check this guide.[7]

24.2.2 Simulation with GRF package in R

The `grf` package in R is well-suited for estimating heterogeneous treatment effects with continuous, binary, and categorical covariates. We begin by simulating a dataset with a continuous outcome (`Y`), a binary treatment indicator (`W`), and covariates: ten continuous, ten binary, and three categorical variables converted to dummy indicators. Treatment effects are defined to vary based on these covariates, adding realistic complexity to the simulation. Before applying causal forests, we examine the data structure, summary statistics, and missingness patterns. If the missing data does not differ systematically between treated and control groups, we can keep them, as the `grf` package can handle missing values.

```r
set.seed(42)
library(grf)

# Define sample size
N <- 2000

# Generate covariates
X_cont <- matrix(rnorm(N * 10), ncol = 10)  # 10 continuous variables
X_bin <- matrix(rbinom(N * 10, 1, 0.5), ncol = 10)  # 10 binary variables

# Generate categorical variables
C1 <- sample(letters[1:3], N, replace = TRUE)  # 3 categories: a, b, c
C2 <- sample(letters[1:4], N, replace = TRUE)  # 4 categories: a, b, c, d
C3 <- sample(letters[1:2], N, replace = TRUE)  # 2 categories: a, b

# Convert categorical variables to binary (dummy coding)
X_cat <- model.matrix(~ C1 + C2 + C3)[, -1]  # Remove intercept column

# Combine all covariates
X <- cbind(X_cont, X_bin, X_cat)

# Generate binary treatment variable
W <- rbinom(N, 1, 0.5)  # 50% probability of treatment

# Define true heterogeneous treatment effects
tau <- 2 * X_cont[, 1] - X_bin[, 3] + ifelse(C1 == 'a', 1, -1)
```

```
# Generate continuous outcome variable with treatment effect
Y <- 5 + X %*% rnorm(ncol(X)) + tau * W + rnorm(N)

# Check structure of the data
str(X)
```

```
##  num [1:2000, 1:26] 1.371 -0.565 0.363 0.633 0.404 ...
##  - attr(*, "dimnames")=List of 2
##   ..$ : chr [1:2000] "1" "2" "3" "4" ...
##   ..$ : chr [1:26] "" "" "" "" ...
```

```
summary(Y)
```

```
##        V1
##  Min.   :-15.952
##  1st Qu.: -1.846
##  Median :  1.425
##  Mean   :  1.382
##  3rd Qu.:  4.635
##  Max.   : 15.404
```

```
table(W)
```

```
## W
##    0    1
##  969 1031
```

After generating simulated data (or uploading real data) and performing basic checks on the data structure and summary statistics, we can assess the overlap assumption by examining the distribution of propensity scores. Estimating propensity scores using a regression forest and plotting their distribution allows us to verify that treated and control units share common support. If the propensity scores for both groups overlap substantially, it suggests that the overlap assumption holds; otherwise, lack of overlap may indicate issues with identifying treatment effects reliably.

```
# Estimate propensity scores to check overlap
propensity_forest <- regression_forest(X, W)
W.hat <- predict(propensity_forest)$predictions
hist(W.hat, main = "Propensity Score Distribution",
     xlab = "Scores")  # Check overlap assumption
```

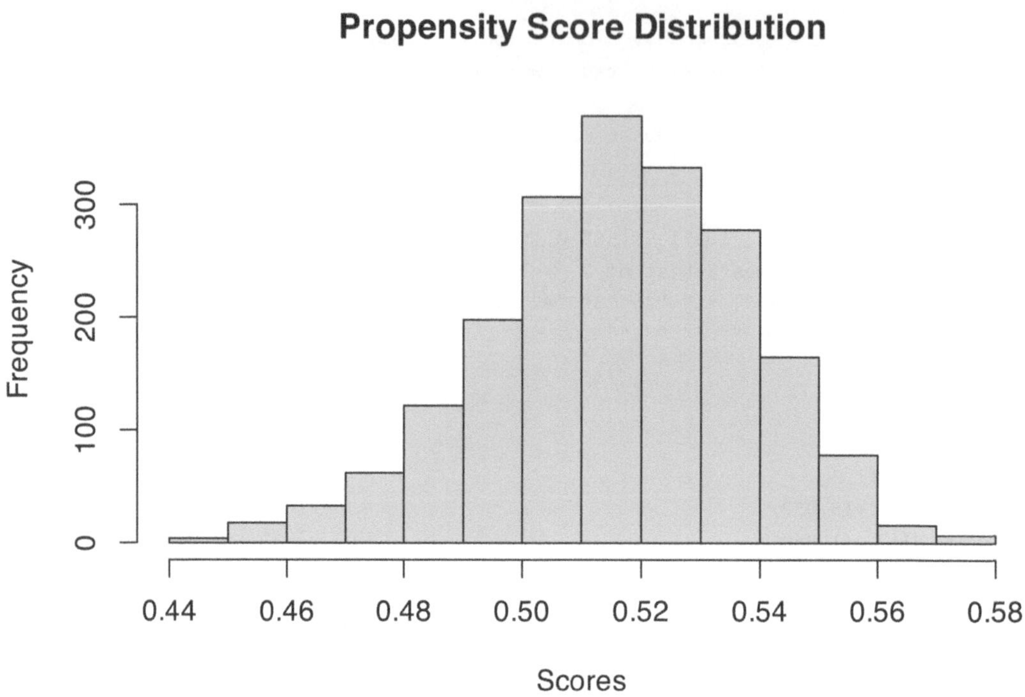


```
hist(W.hat[W==1], col="gray", alpha=0.5,
    main="Propensity Scores: Treated vs Control", xlab="Propensity Score")
hist(W.hat[W==0], col="black", alpha=0.5, add=TRUE)
legend("topright", legend=c("Treated","Control"), fill=c("grey","black"))
```

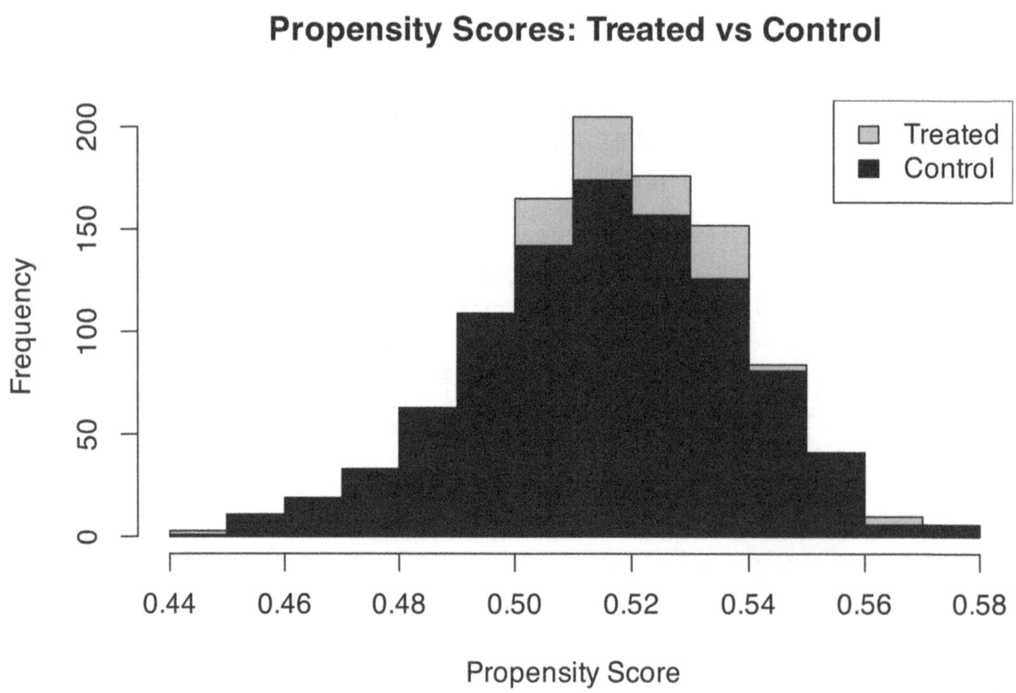

After checking overlap condition, we proceed by fitting a causal forest to estimate heterogeneous treatment effects, allowing parameter customization to improve model performance and interpretability.

```
# Causal forest estimation with common parameter adjustments
tau.forest <- causal_forest(
  X, Y, W,
  num.trees = 3000,                  # Increase number of trees for stability
  min.node.size = 10,                # Larger nodes to reduce overfitting
  sample.fraction = 0.5,             # More data per tree to lower variance
  mtry = floor(sqrt(ncol(X))),    # Number of cov. per split(default:sqrt of total)
  honesty = TRUE,                    # Use separate samples for splits and estimation
  honesty.fraction = 0.5,            # Fraction of data for honest estimation
  honesty.prune.leaves = TRUE,       # Prune leaves that do not improve estimates
  tune.parameters = "all"            # tune all parameters by cross-validation.
)
```

The causal forest identifies the most important variables contributing to treatment effect heterogeneity, which can also be visualized effectively using the bar chart below. The `variable_importance` function in the `grf` package identifies which covariates most significantly influence the heterogeneity of treatment effects rather than predicting the outcome directly. It works by measuring how often each covariate is used to create splits in the trees that improve the estimation of treatment effects. A higher importance score suggests that a covariate substantially contributes to explaining why treatment effects differ across individuals. For instance, if the first column of the output shows "Variable 1" with a score of 0.7, it means that this covariate frequently helps refine the ability of the model to estimate treatment effects accurately. In practical terms, a score of 0.7 implies that variations in Variable 1 are strongly associated with differences in treatment effects, guiding researchers to focus on such influential covariates for deeper analysis.

```
# Variable importance (top 10 most important) for causal forest
vi <- variable_importance(tau.forest)
top_10 <- order(vi, decreasing = TRUE)[1:10]
vi_top_10 <- vi[top_10]
#colnames(X)[ranked.vars[1:10]]
barplot(vi_top_10, main = "Top 10 Most Important Variables", xlab = "Variables",
        ylab = "Importance", col = "grey", las = 2, names.arg = top_10)
```

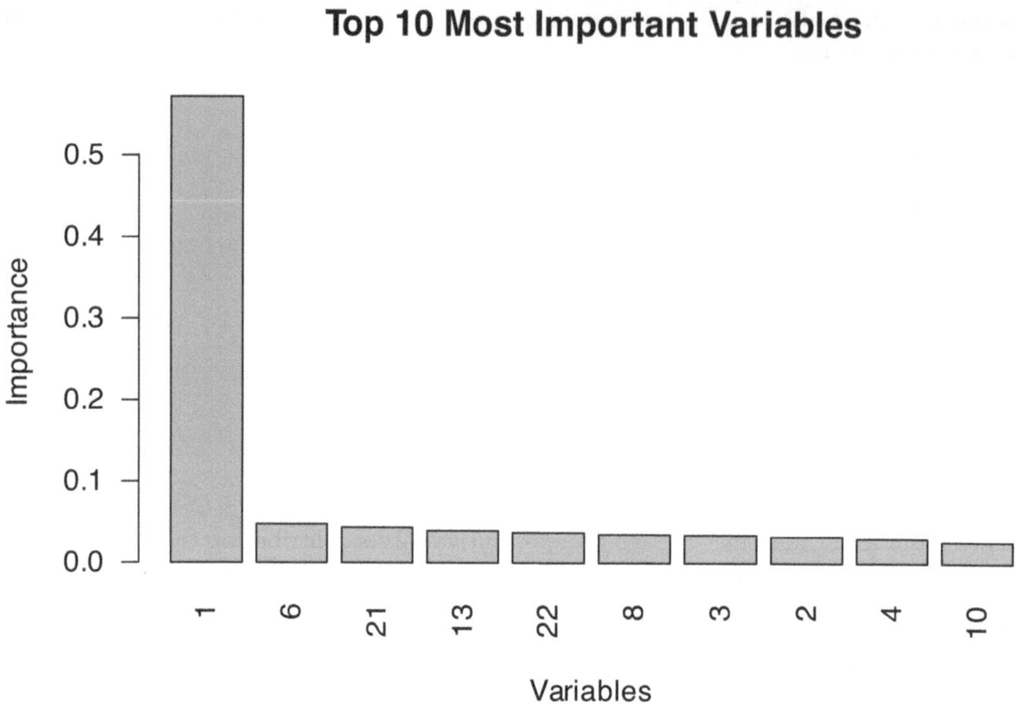

We can also predict treatment effects using OOB predictions and a test sample. OOB predictions allow us to assess the generalization ability of the causal forest without using a separate validation set. The histogram of OOB predictions displays the distribution of estimated treatment effects, helping us understand the variability in effects captured by the model.

For the test sample, the code first defines `X.test`, a matrix with 101 rows, where the first covariate is varied systematically between –2 and 2 to explore how treatment effects change along this dimension. Treatment effects are then predicted using the trained causal forest, and their distribution is visualized. The plot of predicted treatment effects against the first covariate illustrates how the effect varies with changes in that covariate. The additional line plot compares these predictions to a simple function of the covariate (using `pmax`), providing a visual check for how well the model captures the underlying treatment effect pattern. This approach helps assess if the causal forest can detect meaningful and systematic variations in treatment effects across different levels of key covariates.

```
# Estimate treatment effects for the training data using out-of-bag prediction
#GRF uses OOB predictions to assess generalization to unseen data.
tau.hat.oob <- predict(tau.forest)
# Visualize distribution of estimated treatment effects
hist(tau.hat.oob$predictions)
```

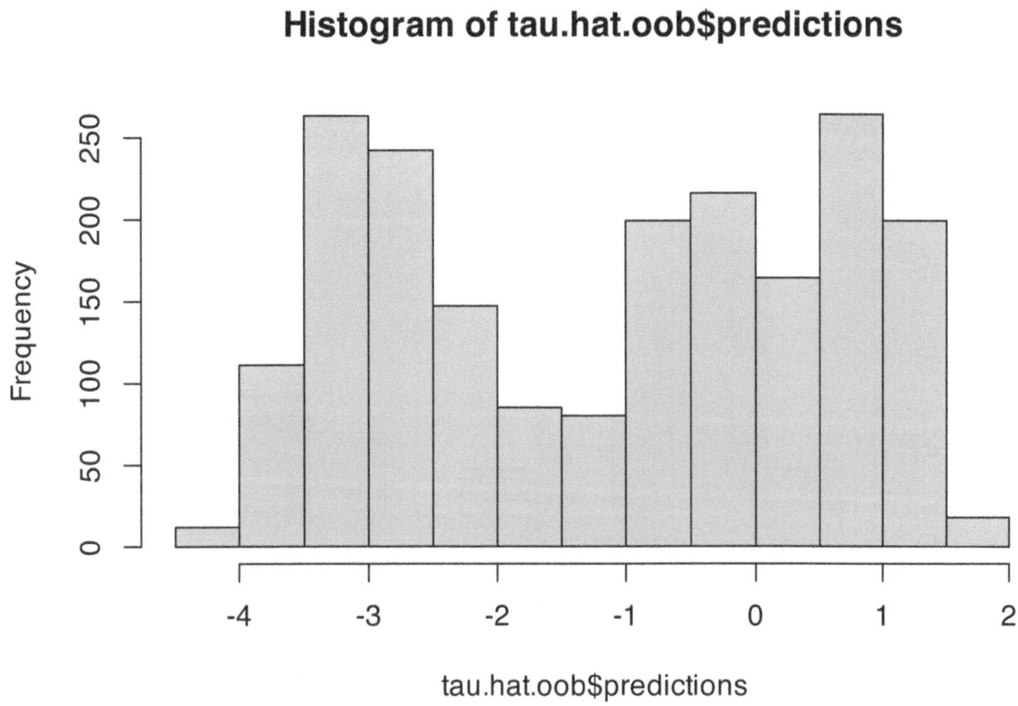


```
# Estimate treatment effects for the test sample
#X.test <- X  # Replace with separate test data if available
# Generate test data for prediction
# 10 continuous, 10 binary, 3 categorical variables
p <- 10 + 10 + (3 - 1) + (4 - 1) + (2 - 1)
# 101 test points with the same number of covariates as in X
X.test <- matrix(0, 101, p)
# Vary the first covariate between -2 and 2
X.test[, 1] <- seq(-2, 2, length.out = 101)

tau.hat.test <- predict(tau.forest, X.test)
hist(tau.hat.test$predictions, col = "gray", border = "black")
```

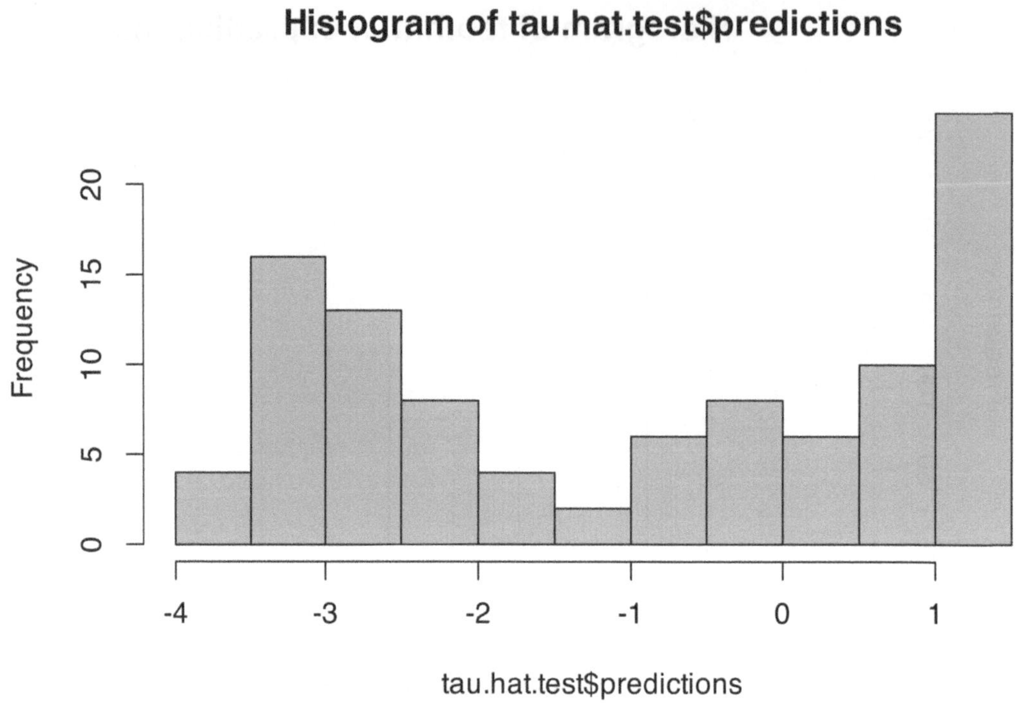


```
plot(X.test[, 1], tau.hat.test$predictions,
     ylim = range(tau.hat.test$predictions, 0, 2),
     xlab = "x", ylab = "tau", type = "l", col = "black", lty = 1)
lines(X.test[, 1], pmax(0, X.test[, 1]), col = "black", lty = 2)
```

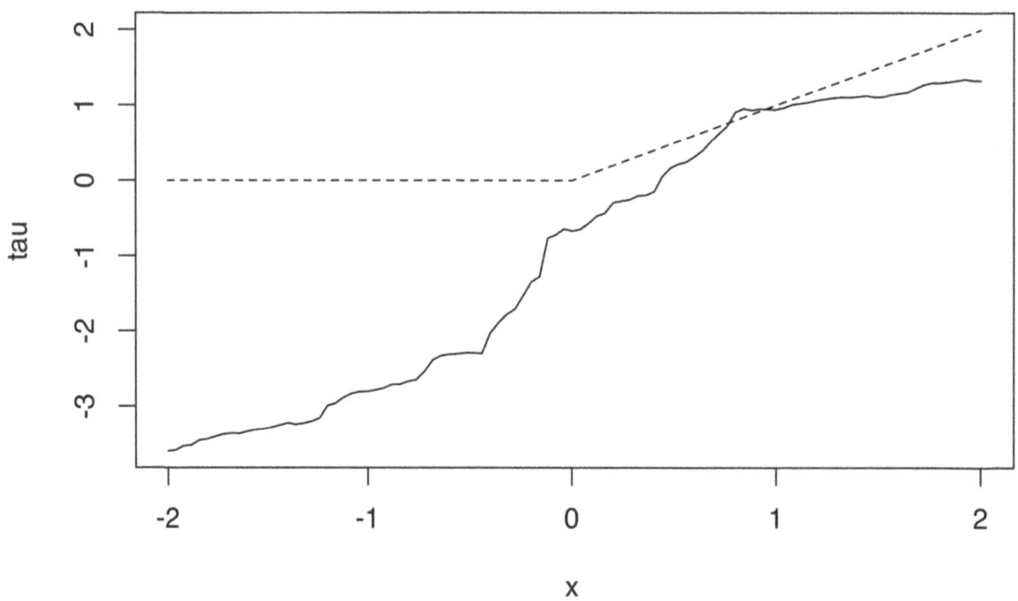

This code estimates the doubly robust CATE for the full sample and the treated sample (CATT) to assess treatment effect heterogeneity. It increases the number of trees to 4,000 for more precise estimates and predicts treatment effects for the test sample, including confidence intervals to capture uncertainty. The plot shows predicted treatment effects against the first covariate, with solid lines for point estimates and dashed lines for 95% confidence intervals, helping to visualize how treatment effects vary across different levels of the covariate. If the CATE estimates (solid line) are close to the reference line, it might suggest that treatment effects vary systematically with the first covariate.

```
# Estimate the conditional average treatment effect on the full sample (CATE)
average_treatment_effect(tau.forest, target.sample = "all")
```

```
##    estimate    std.err
## -1.1478063  0.1359256
```

```
# Estimate the conditional average treatment effect on the treated sample (CATT)
average_treatment_effect(tau.forest, target.sample = "treated")
```

```
##   estimate   std.err
## -1.150325  0.140198
```

```
# Add confidence intervals for CATE based on the first covariate
tau.forest <- causal_forest(X, Y, W, num.trees = 4000)

# Predict CATE and estimate variance for confidence intervals
tau.hat <- predict(tau.forest, X.test, estimate.variance = TRUE)
sigma.hat <- sqrt(tau.hat$variance.estimates)

# Plot CATE estimates for the first covariate with 95% confidence intervals
plot(X.test[, 1], tau.hat$predictions,
     ylim = range(tau.hat$predictions + 1.96 * sigma.hat,
                  tau.hat$predictions - 1.96 * sigma.hat, 0, 2),
     xlab = "First Covariate", ylab = "CATE", type = "l", col = "black", lty = 1,
     main = "CATE with 95% CI")

lines(X.test[, 1], tau.hat$predictions + 1.96 * sigma.hat, col = "black", lty = 2)
lines(X.test[, 1], tau.hat$predictions - 1.96 * sigma.hat, col = "black", lty = 2)

lines(X.test[, 1], pmax(0, X.test[, 1]), col = "black", lty = 3)
```

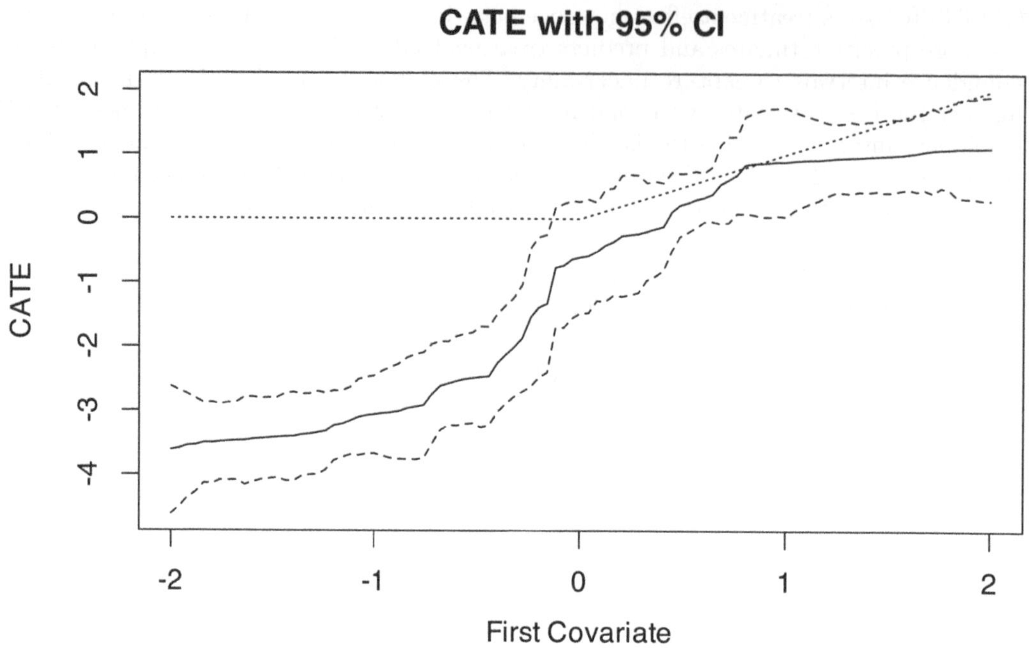

The ATE summarizes the overall impact of treatment by providing a single number that represents the average difference in outcomes between treated and control groups, without considering how treatment effects may vary across different covariates. In contrast, a best linear projection (BLP) provides a simple and interpretable summary of how CATEs vary with the most important covariates. The BLP estimates a simple and interpretable doubly robust linear model of the form:

$$\tau(X_i) = \beta_0 + A_i\beta \tag{24.22}$$

where A_i irepresents a set of covariates selected based on their variable importance. By focusing on the most important covariates, the BLP provides a clear and doubly robust summary of how treatment effects change across different subgroups. This approach helps which factors drive treatment effect heterogeneity and allows for more targeted and interpretable assesment.

```
# Estimate variable importance for covariates
vi <- variable_importance(tau.forest)

# Rank covariates based on importance and select top 5
ranked.vars <- order(vi, decreasing = TRUE)
top_5_vars <- X[, ranked.vars[1:5]]

# Compute the best linear projection of CATEs on the top 5 covariates
blp <- best_linear_projection(tau.forest, top_5_vars)

# Display the results
print(blp)
```

```
##
## Best linear projection of the conditional average treatment effect.
## Confidence intervals are cluster- and heteroskedasticity-robust (HC3):
##
##              Estimate Std. Error t value Pr(>|t|)
## (Intercept) -1.126275   0.129419 -8.7026   <2e-16 ***
## V1           1.943153   0.137771 14.1042   <2e-16 ***
## V2          -0.200528   0.135312 -1.4820   0.1385
## V3          -0.182814   0.129691 -1.4096   0.1588
## V4          -0.036111   0.130543 -0.2766   0.7821
## V5           0.154095   0.143897  1.0709   0.2844
## ---
## Signif. codes:  0 '***' 0.001 '**' 0.01 '*' 0.05 '.' 0.1 ' ' 1
```

Assessing Model Fit and Heterogeneity in Causal Forests

To ensure that our causal forest model effectively captures treatment effect heterogeneity, we can assess its goodness of fit using various diagnostic methods. These include calibration tests, subgroup comparisons, and rank-based assessments.

The function `test_calibration` in the `grf` package helps evaluate the fit of the causal forest. This function performs a regression of the true treatment effect (from held-out data) on the predicted treatment effects of the forest. It returns two key coefficients:

- **`mean.forest.prediction`**: A coefficient close to 1 suggests that the forest's mean treatment effect prediction is accurate.
- **`differential.forest.prediction`**: A coefficient close to 1 suggests that the forest captures treatment effect heterogeneity correctly.

Below is an example of running the calibration test:

```
# Perform calibration test
calibration_results <- test_calibration(tau.forest)
calibration_results
```

```
##
## Best linear fit using forest predictions (on held-out data)
## as well as the mean forest prediction as regressors, along
## with one-sided heteroskedasticity-robust (HC3) SEs:
##
##                                Estimate Std. Error t value    Pr(>t)
## mean.forest.prediction         1.002792   0.111722  8.9758 < 2.2e-16 ***
## differential.forest.prediction 1.156974   0.080232 14.4203 < 2.2e-16 ***
## ---
## Signif. codes:  0 '***' 0.001 '**' 0.01 '*' 0.05 '.' 0.1 ' ' 1
```

A well-calibrated forest should yield estimates for both coefficients near 1. If the estimates deviate significantly, it may indicate issues such as insufficient data, inadequate feature selection, or overfitting.

Let's discuss the second method. A heuristic way to test for heterogeneity is to split observations into two groups: high and low estimated CATE. We then estimate the ATE separately for each subgroup.

```
# Predict treatment effects
tau.hat <- predict(tau.forest)$predictions

# Define high and low CATE groups
high.effect <- tau.hat > median(tau.hat)

# Estimate ATE for high and low effect groups
ate.high <- average_treatment_effect(tau.forest, subset = high.effect)
ate.low <- average_treatment_effect(tau.forest, subset = !high.effect)

# Compute 95% confidence interval for difference in ATE
ate_diff_ci <- ate.high["estimate"] - ate.low["estimate"] +
  c(-1, 1) * qnorm(0.975) * sqrt(ate.high["std.err"]^2 + ate.low["std.err"]^2)

# Display results
print(ate_diff_ci)
```

```
## [1] 2.768666 3.796747
```

If there is significant heterogeneity, we expect a meaningful difference in ATE between the high and low CATE groups. A larger confidence interval suggests greater uncertainty in this heterogeneity estimate.

The most recent and recommended method for evaluating treatment effect heterogeneity is the rank-weighted average treatment effect (RATE). Proposed by Yadlowsky et al. (2022), RATE measures how well a CATE estimator ranks units based on estimated treatment benefits. Similar to the area under the curve (AUC) metric, a higher RATE indicates better discrimination of treatment effects across subgroups.

The RATE is closely linked to the treatment overlap curve (TOC), which plots the difference in expected outcomes between treated and control units as a function of the fraction of the population treated. Specifically, TOC at a fraction q represents the incremental benefit of treating the top q fraction of units with the largest estimated CATEs compared to the overall ATE. The area under the TOC (AUTOC) provides a single summary measure of heterogeneity: a higher AUTOC indicates substantial variation in treatment effects, while a lower AUTOC suggests more homogeneous effects. Other RATE metrics, such as the Qini coefficient, offer alternative ways of weighting the area under the TOC.

To estimate RATE reliably, it is standard practice to split the data into separate training and evaluation sets. The training set is used to fit the CATE estimator, while the evaluation set assesses the RATE. This approach reduces overfitting but may yield different results depending on the random split, highlighting a potential limitation.

The TOC plot visually represents RATE by showing how treatment effects vary across quantiles of predicted CATEs. A steep TOC suggests significant heterogeneity, while a flat TOC implies more uniform effects. Additionally, a 95% confidence interval for the AUTOC provides a measure of the uncertainty around the RATE estimate. Narrow intervals suggest precise estimates, while wider intervals indicate greater uncertainty in the heterogeneity captured by the model. Check https://grf-labs.github.io/grf/articles/rate.html for more information.

```
# Split data into training and evaluation sets
train <- sample(1:N, N / 2)
train.forest <- causal_forest(X[train, ], Y[train], W[train])
eval.forest <- causal_forest(X[-train, ], Y[-train], W[-train])

# Estimate treatment effects using the training forest for the evaluation set
tau.hat.eval <- predict(train.forest, X[-train, ])$predictions

# Calculate rank-based average treatment effect (RATE)
rate.cate <- rank_average_treatment_effect(eval.forest,
              tau.hat.eval,target = c("AUTOC"))
rate.cate
```

```
##  estimate  std.err            target
##  1.662433 0.194141 priorities | AUTOC
```

The following code plots the TOC to illustrate how treatment effects vary across different quantiles of predicted CATEs. It also computes a 95% confidence interval for the AUTOC, summarizing the extent of heterogeneity captured by the causal forest.

```
# Plot the Treatment Overlap Curve (TOC)
plot(rate.cate, main = "Treatment Overlap Curve (TOC)", xlab = "Quantiles",
     ylab = "Average Treatment Effect")
```

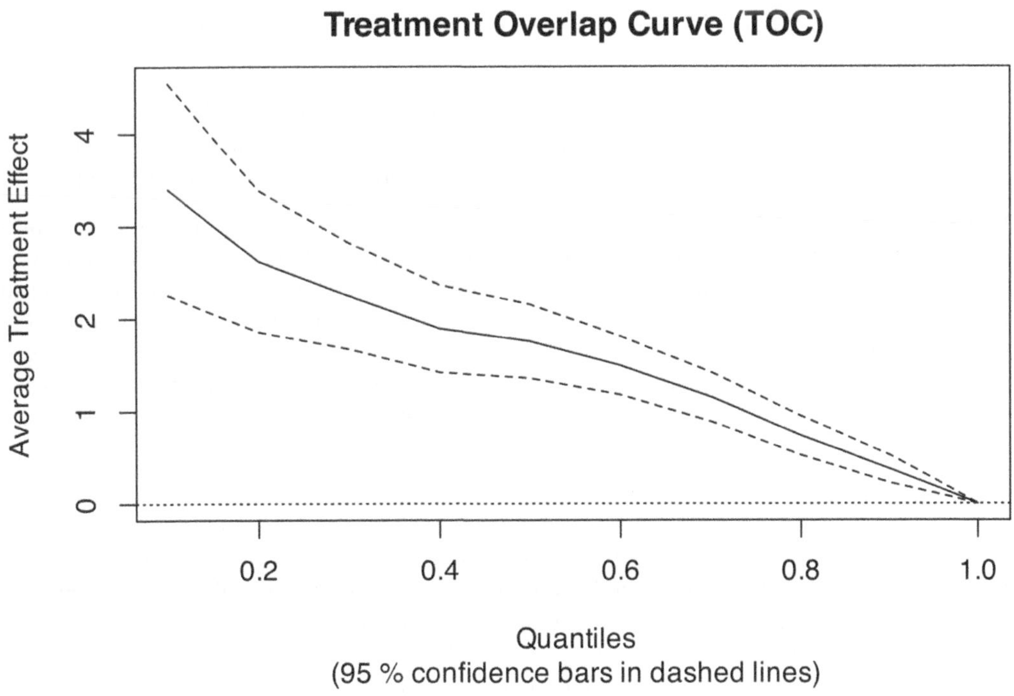

```
# Compute a 95% Confidence Interval for the Area Under the TOC (AUTOC)
autoc_ci <- average_treatment_effect(eval.forest, target.sample = "all")
cat("95% CI for AUTOC:", autoc_ci[1], "+/-", 1.96 * autoc_ci[2], "\n")
```

```
## 95% CI for AUTOC: -1.054098 +/- 0.3980702
```

In summary, RATE and AUTOC provide robust and interpretable metrics for assessing treatment effect heterogeneity, complementing the CATE estimates by offering insights into how effectively the model ranks treatment benefits across different subgroups.

By leveraging these diagnostic tools, we can validate that our causal forest model effectively estimates and captures heterogeneity in treatment effects, ensuring robustness in policy recommendations and applied economic analyses.

We have not covered causal survival forests, quantile forests, or instrumental forests here. However, the underlying idea is similar, even though different models may use distinct splitting criteria and process data differently. Additionally, GRF can be used for policy learning via optimal decision trees, helping to determine treatment assignment rules that maximize expected outcomes. The `policytree` package complements GRF by providing tools for learning optimal treatment policies from heterogeneous treatment effects. For more details on these models and their implementation, refer to the GRF package documentation and the official website.

24.3 Application of Causal Forests in Economics, Health, and Social Sciences

In the past five years, causal forest has strengthened causal inference in economics, social sciences, and health research. While traditional methods remain relevant, it offers a powerful tool for estimating heterogeneous treatment effects. Here, we review its latest applications.

Fuller & Rametta (2024) apply causal forests in political science to improve causal inference using doubly robust machine learning. Their analysis demonstrates the benefits of cross-validation within a random forest framework for estimating treatment effects in observational studies, providing an alternative to traditional econometric methods. Rehill (2025) reviews 133 papers implementing the GRF package in social sciences, assessing assumptions, strengths, and limitations. His findings indicate a rising preference for machine learning-based approaches in estimating heterogeneous treatment effects in policy and behavioral sciences.

Lechner & Mareckova (2025) compare selection-on-observables approaches, including *MCF*, *GRF*, and *double machine learning (DML)*, offering guidance on selecting appropriate methods for estimating heterogeneous treatment effects in applied econometrics. Jin, Naghi & Pick (2019) utilize GRF to examine educational interventions in Romania, Brazil, and Burkina Faso, identifying individual-level variations in treatment effects. Their findings underscore the synergy between machine learning and traditional econometric models in educational policy evaluation. Crudu (2023) applies GRF to labor economics, estimating the long-term impact of completing education during adverse labor market conditions using European cohort data from 1960 to 1990. His study highlights how causal forests uncover heterogeneity in economic outcomes. Busshoff et al. (2022) introduce the *MCF* package, validating it across epidemiology, medicine, and labor economics by replicating well-known studies. They demonstrate its ability to refine treatment effect estimates and uncover nuanced effect heterogeneity.

Health economics research increasingly employs causal forests for policy evaluation and treatment heterogeneity analysis. Shah (2023) examines Indonesia's national health insurance scheme using causal forests, highlighting their role in improving policy impact evaluations. Kreif, DiazOrdaz & Moreno-Serra (2022) compare GRF with traditional econometric models in health policy evaluations, confirming its effectiveness in estimating treatment heterogeneity in large-scale interventions. Sverdrup, Petukhova & Wager (2024) extend GRF applications to psychiatric research, showcasing its utility in estimating heterogeneous treatment effects for mental health interventions. Shirvaikar, Lin & Holmes (2023) modify causal forests to analyze risk heterogeneity in public health research, introducing a novel node-splitting procedure targeting *relative risk*. Their results indicate that *relative risk causal forests* can reveal previously undetected sources of heterogeneity in health outcomes.

Environmental and healthcare economics research also leverage causal forests. Zhang, Chen & Liu (2024) construct an *Environmental Quality Index* using GRF to assess how environmental factors influence medical expenditures in China. Their study demonstrates the importance of accounting for treatment heterogeneity in environmental health research. Davis & Heller (2020) analyze a youth summer jobs program using causal forests, showing that the program significantly reduces violent crime for high-risk youths but has negligible effects for lower-risk individuals. Their work highlights the value of causal forests in identifying treatment heterogeneity for targeted policy design.

24.4 Technical: Optimizing Causal Forests and Derivations

Causal trees and forests can be computationally intensive, especially with large datasets and many variables, due to the need to evaluate all possible splits at each node. As data size and complexity grow, this process becomes increasingly slow, making it essential to adopt strategies that balance speed and accuracy. For applied economists, focusing on techniques that can be directly implemented within existing software like the `grf` package in R is the most practical approach.

Key methods to speed up computation include subsampling and feature subsetting, both supported by `grf` through options like the `mtry` parameter. These techniques reduce the number of variables and observations evaluated at each split, significantly cutting down processing time. Setting early stopping criteria based on minimum information gain or using the `cp` parameter in R's `rpart` package also helps by preventing the model from wasting time on splits with minimal impact. Additionally, using approximate splitting criteria, such as evaluating splits only at quantile points, further reduces computation time without a substantial loss in accuracy.

The `grf` package also supports parallel processing, allowing it to use multiple cores by default if available. This capability enables efficient handling of large datasets without requiring manual configuration. For more technical optimizations like memory mapping and sparse matrices, which often require knowledge of computer architecture, data scientists or engineers usually take the lead. While these advanced strategies can further speed up computation, applied economists benefit most from methods that are easy to implement with existing package options. Combining these practical techniques allows for efficient and robust causal models suited for large-scale economic and social science applications.

In Python, the primary package for causal forests is `econml`, which includes an implementation of honest trees and GRF-based models. The `causalml` package also supports causal forests, offering built-in tools for treatment effect estimation and policy learning. The `mcf` (Modified Causal Forest) package is another option for improved effect heterogeneity estimation. Researchers can also leverage `sklearn` for preprocessing and hyperparameter tuning alongside these causal inference tools. Users can call Python or R scripts within Stata using `stata_call` or `python` integration.

24.4.1 Equivalence of Step 4 Estimators in the `grf` Algorithm for the Causal Forest

In the causal forest, we estimate the CATE, $\tau(x) = E[Y(1) - Y(0)|X = x]$, using a moment equation that accounts for local heterogeneity in treatment effects. The estimation relies on a weighted local regression approach, where the weights $\alpha_i(x)$ are derived from the forest structure.

The initial moment equation for $\tau(x)$ expresses the conditional expectation of the residualized outcome $\tilde{Y}_i$ as a function of the residualized treatment $\tilde{W}_i$:

$$E[\tilde{Y}_i|X_i = x] = \tau(x) \cdot E[\tilde{W}_i|X_i = x] + \text{constant} \tag{24.23}$$

Since $\tilde{W}_i$ and $\tilde{Y}_i$ are residuals (mean-zero given X), the constant term vanishes locally. The best linear predictor of $\tilde{Y}_i$ on $\tilde{W}_i$ within the neighborhood defined by the forest is weighted by $\alpha_i(x)$, leading to the estimator:

$$\hat{\tau}(x) = \frac{\sum_{i=1}^{n} \alpha_i(x)\tilde{Y}_i\tilde{W}_i}{\sum_{i=1}^{n} \alpha_i(x)\tilde{W}_i^2} \tag{24.24}$$

where: - $\alpha_i(x) = \frac{1}{B}\sum_{b=1}^{B} \frac{I(X_i \in L_b(x))}{|L_b(x)|}$ represents the weight for each training observation, determined by how often it appears in the same leaf as x across the B trees in the forest. - $\tilde{Y}_i = Y_i - \hat{m}^{(-i)}(X_i)$ is the OOB residualized outcome, adjusting for potential confounding. - $\tilde{W}_i = W_i - \hat{m}^{(-i)}(X_i)$ is the OOB residualized treatment, ensuring robustness to treatment heterogeneity.

Substitute the expression of $\alpha_i(x)$ into the equation of $\hat{\tau}(x)$ so:

$$\hat{\tau}(x) = \frac{\sum_{i=1}^{n}\left(\frac{1}{B}\sum_{b=1}^{B}\frac{I(X_i \in L_b(x))}{|L_b(x)|}\tilde{Y}i\tilde{W}i\right)}{\sum_{i=1}^{n}\left(\frac{1}{B}\sum_{b=1}^{B}\frac{I(X_i \in L_b(x))}{|L_b(x)|}\tilde{W}i^2\right)} \tag{24.25}$$

Rearrange the summations in the denominator and numerator:

$$\hat{\tau}(x) = \frac{\frac{1}{B}\sum_{b=1}^{B}\sum_{i=1}^{n}\frac{I(X_i \in L_b(x))}{|L_b(x)|}\tilde{Y}i\tilde{W}i}{\frac{1}{B}\sum_{b=1}^{B}\sum_{i=1}^{n}\frac{I(X_i \in L_b(x))}{|L_b(x)|}\tilde{W}i^2} \tag{24.26}$$

To express the estimator in terms of leaf-level aggregates, we introduce several averages computed within each leaf $L_b(x)$ of tree b. These averages summarize how the OOB residualized outcomes and treatments behave in each tree's local neighborhood. There averages are the leaf level precomputed sufficient statistics is step 3 if `grf` algorithm.

Leaf-Level Mean Outcome (Residualized)

$$\bar{Y}_b(x) = \frac{1}{|L_b(x)|} \sum_{i=1}^{n} I(X_i \in L_b(x)) \tilde{Y}_i \tag{24.27}$$

This is the average residualized outcome $\tilde{Y}_i$ among all units that fall into leaf $L_b(x)$.This captures how much the observed outcome deviates from the expected value given covariates within the leaf.

Leaf-Level Mean Treatment (Residualized)

$$\bar{W}_b(x) = \frac{1}{|L_b(x)|} \sum_{i=1}^{n} I(X_i \in L_b(x)) \tilde{W}_i \tag{24.28}$$

The average of residualized treatment $\tilde{W}_i$ within the leaf. This term represents how much the treatment assignment varies around its expected value within the leaf.

Leaf-Level Mean Treatment Squared

$$\bar{W}_b^2(x) = \frac{1}{|L_b(x)|} \sum_{i=1}^{n} I(X_i \in L_b(x)) \tilde{W}_i^2 \tag{24.29}$$

The average squared value of $\tilde{W}_i$ within the leaf.

Leaf Weighting Term

$$\bar{w}_b(x) = \frac{1}{|L_b(x)|} \sum_{i=1}^{n} I(X_i \in L_b(x)) \tag{24.30}$$

The mean weight applied to observations in leaf $L_b(x)$, often reducing to 1 if weights are uniform. This term normalizes sums, ensuring proper weighting when aggregating across trees.

Leaf-Level Mean Product of Residualized Outcome and Treatment

$$Y\bar{W}_b(x) = \frac{1}{|L_b(x)|} \sum_{i=1}^{n} I(X_i \in L_b(x)) \tilde{Y}_i \tilde{W}_i \tag{24.31}$$

The average product of the residualized outcome and residualized treatment within the leaf.

Express $\hat{\tau}(x)$ in Terms of Leaf Averages

Now, we rewrite the original estimator of $\hat{\tau}$ with α using these leaf-level summaries (precomputed statistics). This transformation leverages the structure of the causal forest, allowing us to compute the unit level treatment effect estimate more efficiently.

$$\hat{\tau}(x) = \frac{\frac{1}{B}\sum_{b=1}^{B} \left(Y\bar{W}_b(x)\bar{w}_b(x) - \bar{Y}_b(x)\bar{W}_b(x) \right)}{\frac{1}{B}\sum_{b=1}^{B} \left(\bar{W}_b^2(x)\bar{w}_b(x) - \bar{W}_b(x)\bar{W}_b(x) \right)} \tag{24.32}$$

Thus, the estimator $\hat{\tau}(x)$ is essentially a weighted ratio of the covariance between treatment and outcome residuals to the variance of the treatment residuals, computed across all trees. This provides an estimate of the conditional treatment effect at $X = x$, leveraging the structure of the causal forest.

Notes

1. This approach assumes random treatment assignment, ensuring that within each leaf, the outcome difference provides an unbiased estimate of the treatment effect. If the outcome Y_i is binary, the treatment effect is estimated using differences in probabilities rather than means:

$$\hat{\tau}(R_m) = \hat{p}_T - \hat{p}_C = \left(\frac{1}{N_T}\sum_{i\in T} Y_i\right) - \left(\frac{1}{N_C}\sum_{i\in C} Y_i\right)$$

Alternatively, one can use the log-odds ratio for a more scale-invariant measure:

$$\hat{\tau}_{\text{logit}}(R_m) = \log\left(\frac{\hat{p}_T}{1-\hat{p}_T}\right) - \log\left(\frac{\hat{p}_C}{1-\hat{p}_C}\right)$$

This transformation stabilizes estimates when probabilities are close to 0 or 1.

2. An additional parameter α allows adjusting the balance between MSE and variance terms in the EMSE criterion added 'causaltree' package :

$$-\widehat{\text{EMSE}}_\tau(S^{\text{tr}}, S^{\text{est}}, \Pi, \alpha) = \alpha\cdot\frac{1}{N^{\text{tr}}}\sum_{i\in S^{\text{tr}}} \hat{\tau}^2(X_i; S^{\text{tr}}, \Pi) - (1-\alpha)\cdot\left(\frac{1}{N^{\text{tr}}} + \frac{1}{N^{\text{est}}}\right)\cdot\sum_{\ell\in\Pi}\left(\frac{S^2_{\text{treat}}(\ell)}{p} + \frac{S^2_{\text{control}}(\ell)}{1-p}\right)$$

Where $\alpha \in (0,1)$ is a tuning parameter that controls the emphasis on MSE versus variance. A higher α places more emphasis on maximizing differences in treatment effects, while a lower α emphasizes minimizing variance.

3.

$$\hat{\tau}^{ATE}_{AIPW} = \frac{1}{N}\sum_{i=1}^{N}\left[\frac{W_i(Y_i - \hat{m}_1(X_i))}{\hat{e}(X_i)} - \frac{(1-W_i)(Y_i - \hat{m}_0(X_i))}{1-\hat{e}(X_i)} + \hat{m}_1(X_i) - \hat{m}_0(X_i)\right]$$

, same equation in Ch19 but different notation

4. The final average treatment effect estimate for a given sample is estimated with doubly robust, AIPW, method as:

$$\hat{\tau} = \frac{1}{N}\sum_{i=1}^{N}\hat{\Gamma}_i$$

where

$$\hat{\Gamma}_i = \hat{\tau}^{(-i)}(X_i) + \frac{W_i - \hat{e}^{(-i)}(X_i)}{\hat{e}^{(-i)}(X_i)(1-\hat{e}^{(-i)}(X_i))}(Y_i - \hat{\mu}(X_i, W_i))$$

with

$$\hat{\mu}(X_i, W_i) = \hat{m}^{(-i)}(X_i) + (W_i - \hat{e}^{(-i)}(X_i))\hat{\tau}^{(-i)}(X_i)$$

Here, $\hat{\tau}^{(-i)}(X_i)$ is the initial treatment effect estimate for each observation (point prediction of treatment effect), obtained using ρ_i as splitting criteria for honest trees and precomputed statistics. The term $\hat{e}^{(-i)}(X_i)$ is the estimated propensity score for observation i, excluding observation i from the estimation. The function $\hat{\mu}(X_i, W_i)$ represents the estimated conditional mean outcome, while the residual $Y_i - \hat{\mu}(X_i, W_i)$ captures deviations from this expected value. The term $\frac{W_i - \hat{e}^{(-i)}(X_i)}{\hat{e}^{(-i)}(X_i)(1-\hat{e}^{(-i)}(X_i))}$ serves as the debiasing weight, ensuring consistency and efficiency in estimation.

5. In equation (20) from main GRF 2019 paper the pseudo-outcome is specialized for causal forests to target the estimation of heterogeneous treatment effects. It is given by

$$\rho_i = \xi^\top A_P^{-1}(W_i - \bar{W}_P)\Big(Y_i - \bar{Y}_P - (W_i - \bar{W}_P)\hat{\beta}_P\Big)$$

where the elements are as follows: ξ is a vector that "picks out" or selects the component of the coefficient vector β that you are most interested in. In many cases—like when estimating the average treatment effect—ξ might simply be 1, meaning you are interested in the full effect. A_P is the local second-moment (or covariance) matrix for the treatment variable W in the parent node P. It is defined as

$$A_P = \frac{1}{\left|\{i : X_i \in P\}\right|} \sum_{i:X_i \in P} (W_i - \bar{W}_P)^{\otimes 2}$$

with $\bar{W}_P$ being the average of W_i in node P and $(W_i - \bar{W}_P)^{\otimes 2}$ denoting the outer product $(W_i - \bar{W}_P)(W_i - \bar{W}_P)^\top$.

In this formulation, the term $Y_i - \bar{Y}_P - (W_i - \bar{W}_P)\hat{\beta}_P$ represents the residualized outcome after removing the effect captured by the regression of Y on W in the parent node. Multiplying by $(W_i - \bar{W}_P)$ and then pre-multiplying by $\xi^\top A_P^{-1}$ serves to adjust (or standardize) this residual by the local variation in W, effectively constructing an influence function–like quantity that is used to guide the splits. Specifically, using out-of-bag residuals in causal forest, the program calculates the pseudo-outcomes ρ_i for all observations in the parent node one time only, and inserting $\hat{Y}_i = Y_i - \hat{m}^{-i}(X_i)$ and $\hat{W}_i = W_i - \hat{e}^{-i}(X_i)$ to the main equation we get the causal forest pseudo-outcomes ρ_i for each observation i in the parent node P.

6. https://grf-labs.github.io/grf/DEVELOPING.html

7. https://grf-labs.github.io/grf/articles/muhats.html

25

Meta Learners for Treatment Effects

In the Rubin causal framework, estimating treatment effects—particularly heterogeneous treatment effects—has traditionally relied on a range of econometric methods, including regression adjustment, propensity score methods, matching, inverse probability weighting, and instrumental variables. These approaches emphasize valid statistical inference, allowing researchers to establish confidence intervals, conduct hypothesis tests, and make policy-relevant conclusions with appropriate uncertainty quantification. However, in recent years, machine learning methods have introduced new ways to estimate treatment effects, focusing not just on inference but also on improving predictive performance. One prominent class of these methods is known as meta-learners.

Meta-learners provide a systematic way to estimate conditional average treatment effects (CATEs) by leveraging machine learning models as base learners. The name "meta-learner" arises from their structure: rather than using a single estimator, these methods decompose the problem into subcomponents—often estimating separate functions for different treatment groups and then combining these estimates to recover the treatment effect. This modular framework allows for greater flexibility, making meta-learners particularly useful in settings with high-dimensional covariates or complex treatment effect heterogeneity. In broad terms, S-learners use a single model for simplicity, T-learners employ two models to better capture heterogeneity, X-learners improve efficiency by balancing bias and variance, R-learners focus on residualized outcomes to control for confounding, and double learners improve robustness by orthogonalizing nuisance components. Additionally, meta-learners can incorporate deep learning architectures or ensembles of multiple models to further improve predictive accuracy, leveraging neural networks, boosting, or stacking techniques.

At first glance, meta-learners may appear similar to the traditional econometric methods used in the Rubin framework. Both approaches rely on estimating potential outcomes under different treatment assignments, and both can be used to estimate individualized treatment effects. However, their core objectives differ. Traditional methods in the Rubin framework prioritize inference—ensuring unbiased estimation, consistency, and well-defined uncertainty quantification. Meta-learners, in contrast, are primarily designed for prediction. The goal is to minimize mean squared error when estimating CATEs rather than focusing on statistical properties such as asymptotic normality or valid confidence intervals. Unlike traditional econometric models, meta-learners do not prioritize interpretability; instead, they are optimized for predictive performance. While some meta-learning methods can be adapted to provide inference, their primary strength lies in their ability to flexibly adapt to complex data structures, leveraging machine learning to estimate treatment effect heterogeneity more efficiently than standard parametric methods. Recent work, such as the R-Learner, has demonstrated that under certain conditions, meta-learners can achieve valid asymptotic properties, bridging the gap between machine learning and econometric inference.

Despite their focus on prediction rather than inference, meta-learners are an important topic for applied economists and social scientists. Even if the primary objective remains causal inference, understanding these methods can offer practical benefits. First,

DOI: 10.1201/9781003504795-25

meta-learners can serve as powerful tools for exploratory analysis, helping identify subpopulations where treatment effects may be particularly strong or weak. Second, in policy applications where targeting treatments efficiently is as important as establishing statistical significance, predictive models that rank individuals by treatment effect can be highly valuable. Finally, when combined with techniques that enable valid inference, such as double machine learning or conformal prediction, meta-learners can strike a balance between flexible modeling and statistical rigor.

Given their increasing adoption in applied research, this chapter focuses on meta-learners from the perspective of economists and social scientists. We build on the concepts covered in previous chapters, where we explored traditional methods for treatment effect estimation in the Rubin framework. Here, we introduce meta-learners as a complementary set of tools, discussing their strengths, limitations, and practical implementation. While these methods differ from standard approaches in their emphasis on prediction, they offer a new perspective on treatment effect heterogeneity that can improve empirical analysis. Understanding meta-learners allows researchers to leverage machine learning tools while maintaining a focus on causal questions, ensuring that the insights gained from these methods can be meaningfully integrated into applied research.

25.1 Meta-learners and CATE

Meta-learners are algorithms that harness machine learning to estimate CATE, the expected difference in outcomes between treated and control groups for individuals defined by specific covariates. This approach is particularly appealing as it accommodates complex data relationships without stringent parametric assumptions, aligning well with the practical challenges faced by applied researchers and practitioners dealing with real-world datasets. The flexibility of meta learners allows for capturing heterogeneous treatment effects, essential for personalized policy interventions.

In the Rubin causal model, potential outcomes for an individual i given treatment T_i are defined as:

$$Y_i(1), Y_i(0) \tag{25.1}$$

The individual treatment effect (ITE) is:

$$\tau_i = Y_i(1) - Y_i(0) \tag{25.2}$$

Since we only observe one of these potential outcomes, CATE estimation requires modeling assumptions and statistical methods as we discussed in previous chapters, and (undercertain conditions) CATE is estimated as:

$$\hat{\tau}(X) = \mathbb{E}[Y_i(1)|X_i = x] - \mathbb{E}[Y_i(0)|X_i = x] = \mathbb{E}[Y_i(1) - Y_i(0)|X_i = x] \tag{25.3}$$

Traditional methods such as regression adjustment, matching, inverse probability weighting, and propensity score attempt to recover these effects under identification assumptions (e.g., unconfoundedness or unbiasness). Meta-learners, instead of relying on fixed functional forms, learn patterns from data using flexible models such as random forests, neural networks, and boosting.

25.2 S-Learner (Single Model Approach)

The S-learner, or single learner, operates by training a single machine learning model to predict outcomes based on both the treatment indicator and covariates. The model is trained on the entire dataset using:

$$Y_i = f(X_i, W_i) + \varepsilon_i \tag{25.4}$$

where X_i represents covariates and $W_i \in \{0, 1\}$ is the treatment indicator. The S-learner treats the treatment variable W_i as if it were just another covariate in X_i. The estimated CATE for a given set of covariates x is derived as the difference in predictions when W_i is set to 1 (treated) versus 0 (control):

$$\hat{\tau}(x) = \hat{f}(x, 1) - \hat{f}(x, 0) \tag{25.5}$$

where $\hat{f}(x, w)$ is the predicted outcome from the model. By including the treatment as a feature, the model captures its effect alongside other covariates. This simplicity makes the S-learner computationally efficient, requiring only one model to be trained—a benefit for quick analyses. However, it tends to bias the treatment effect toward zero, especially when the treatment effect is weak relative to other covariates or when regularization is used. This occurs because the model might not sufficiently distinguish between treatment and control groups, potentially underestimating the true effect. For applied researchers, the S-learner is best suited when the relationship among treatment, covariates, and outcomes is relatively simple and when computational efficiency is prioritized.

25.3 T-Learner (Separate Models for Treatment and Control)

The T-learner, or two learner, employs two separate machine learning models: one for the treatment group and one for the control group. Each model predicts the outcome based on the covariates for its respective group. Observations in the control group ($W_i = 0$) are used to estimate the response under control, and observations in the treatment group ($W_i = 1$) are used to estimate the response under treatment. The T-learner fits separate models:

$$Y_i(1) = f_1(X_i) + \varepsilon_i, \quad Y_i(0) = f_0(X_i) + \varepsilon_i \tag{25.6}$$

The CATE is then estimated as the difference between these predictions:

$$\hat{\tau}(x) = \hat{f}_1(x) - \hat{f}_0(x) \tag{25.7}$$

where $\hat{f}_1(x)$ is the predicted outcome from the model trained on the treatment group ($W_i = 1$) and $\hat{f}_0(x)$ is from the model trained on the control group ($W_i = 0$). This approach allows for different outcome structures between treated and control units, avoiding the bias toward zero that can affect the S-learner. It is particularly effective when the outcome models for the two groups are distinct and capture group-specific patterns. However, the T-learner requires sufficient data in both groups to train separate models effectively; if one group has fewer observations, this may lead to overfitting or underfitting. Thus, the T-learner is ideal when there are clear differences in outcome models between groups and when sample sizes are balanced.

25.3.1 Simulation of S-Learner and T-Learner

Let's illustrate the S-learner and T-learner for estimating conditional average treatment effects. First, we generate a dataset with two covariates (X_1, X_2), assign treatment probabilistically based on a logistic function, and define a heterogeneous treatment effect that increases with X_1. The outcome for treated and control units is computed separately, adding random noise.

```
# Load required libraries
library(randomForest)
library(ggplot2)

# Set seed for reproducibility and define the sample size
set.seed(42)
n <- 1000

# Simulate covariates
X1 <- pmax(pmin(rnorm(n, mean = 2, sd = 1), 4), 0)
X2 <- pmax(pmin(rnorm(n, mean = 4, sd = 0.5), 6), 0)
df <- data.frame(X1, X2)

# Simulate treatment assignment based on a logistic model of X1 and X2
p <- plogis(0.5 * X1 - 0.25 * X2)  # true probability of treatment
df$W <- rbinom(n, 1, p)

# Define the true treatment effect as a function of X1 (heterogeneous effect)
tau <- 2 * X1  # treatment effect increases with X1

# Define potential outcomes for control and treated units
mu0 <- 1 + X1 + X2          # baseline outcome for control
mu1 <- mu0 + tau            # outcome for treated includes the treatment effect

# Generate observed outcomes (add noise to simulate randomness)
df$Y <- ifelse(df$W == 1, mu1, mu0) + rnorm(n)
```

In the S-learner, a single model (random forest) is trained using both covariates and the treatment indicator. CATE is estimated by predicting outcomes under treatment $(W = 1)$ and control $(W = 0)$ and taking the difference. This approach is computationally efficient but may suffer from bias when treatment effects are small.

In the T-learner, separate models are trained for treated and control units (random forest for treated, linear regression for controls). Each model predicts outcomes, and CATE is obtained by computing the difference between the predictions. This method is more flexible but requires sufficient data in both groups to avoid overfitting.

```
# S-Learner Simulation
# Train a single model on the entire dataset including the treatment indicator W.
s_learner_model <- randomForest(Y ~ X1 + X2 + W, data = df)

# Predict potential outcomes for each unit under treatment (W = 1) and control.
df$Yhat_1 <- predict(s_learner_model, newdata = transform(df, W = 1))
df$Yhat_0 <- predict(s_learner_model, newdata = transform(df, W = 0))

# Estimate the CATE as the difference between predicted outcomes
df$s_learner_tau_hat <- df$Yhat_1 - df$Yhat_0

# T-Learner Simulation
# Subset the data into treated and control groups
treated_data <- subset(df, W == 1)
control_data <- subset(df, W == 0)

# Train separate models for the treated and control groups using X1 and X2
t_model <- randomForest(Y ~ X1 + X2, data = treated_data)
c_model <- lm(Y ~ X1 + X2, data = control_data)


# Predict outcomes for all observations using both models
df$Yhat_treated <- predict(t_model, newdata = df)
df$Yhat_control <- predict(c_model, newdata = df)

# Estimate the CATE as the difference in predictions from the two models
df$t_learner_tau_hat <- df$Yhat_treated - df$Yhat_control

# For comparison, Print average treatment effect estimates for each method
df$tau_true <- tau
cat("Average True Treatment Effect:", mean(df$tau_true), "\n")

## Average True Treatment Effect: 3.949584

cat("Average S-Learner Tau Hat:", mean(df$s_learner_tau_hat), "\n")

## Average S-Learner Tau Hat: 4.031725

cat("Average T-Learner Tau Hat:", mean(df$t_learner_tau_hat), "\n")

## Average T-Learner Tau Hat: 4.004821

# Visualization for S-Learner and T-Learner with white background
ggplot(df, aes(x = tau_true)) +
  geom_point(aes(y = s_learner_tau_hat, color = "S-Learner"), alpha = 0.5) +
  geom_point(aes(y = t_learner_tau_hat, color = "T-Learner"), alpha = 0.5) +
  geom_abline(slope = 1, intercept = 0, linetype = "dashed") +
  labs(title = "Estimated vs True Treatment Effects: S-Learner vs T-Learner",
```

```
    x = "True Treatment Effect",
    y = "Estimated Treatment Effect",
    color = "Method") +
  scale_color_manual(values = c("S-Learner" = "gray", "T-Learner" = "black")) +
  theme_minimal()
```

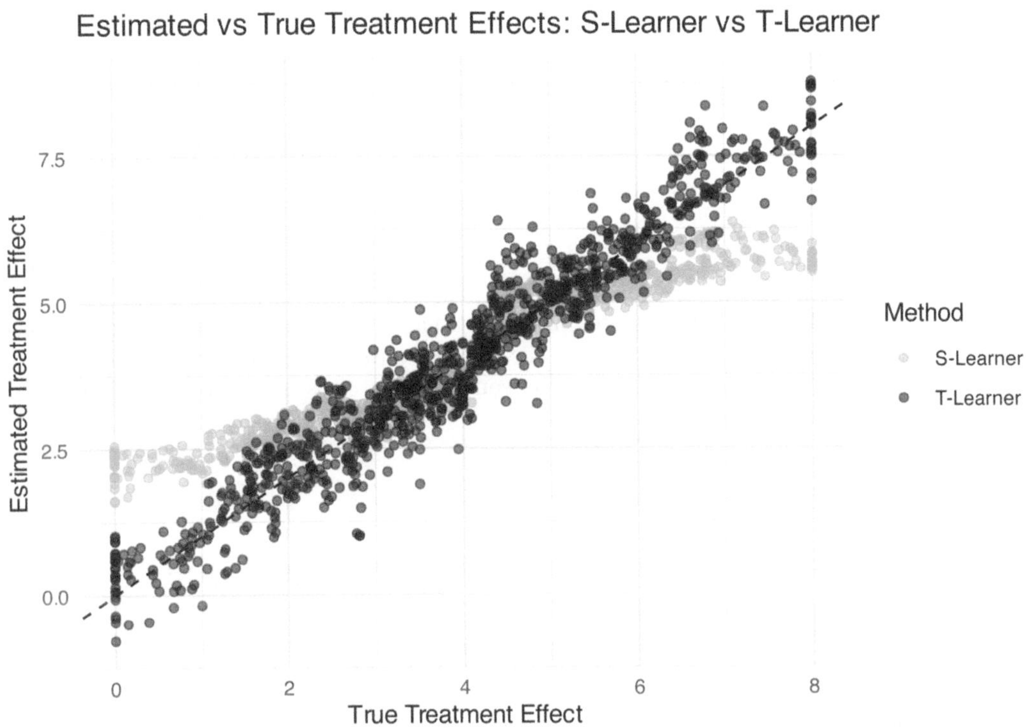

Finally, we compare the average estimated treatment effects and visualize the estimated vs. true CATE to assess accuracy.

25.4 X-Learner (Cross-Fitting)

The X-learner, a more advanced meta learner, seeks to balance the bias and variance trade-off by combining elements of both S and T learners. It begins by estimating the propensity score, the probability of receiving treatment given the covariates, which helps in weighting the observations. Then, it estimates outcome models for both treatment and control groups, similar to the T-learner, but uses these estimates to create a combined estimator for the CATE that minimizes certain loss functions.

The exact formula of CATE estimate for the X-learner is multilevel, involving both the outcome models and the propensity score, but the intuition is to leverage the separate modeling of the T-learner while adjusting with the single model approach of S-learner, weighted by the propensity score. This method is designed to perform well across various scenarios, offering robustness in estimation. However, its complexity makes it more challenging to implement and understand, requiring careful tuning of the machine learning models.

We start by estimating the outcome functions for the treated and control groups. Let

$$\hat{m}_1(x) = E[Y \mid X = x, W = 1] \quad \text{and} \quad \hat{m}_0(x) = E[Y \mid X = x, W = 0] \tag{25.8}$$

which are the models for the treated and control outcomes, respectively.

For each treated unit (where $W = 1$), we impute its treatment effect (pseudo-effects from Künzel et al. (2019)) as:

$$\hat{D}^1 = Y - \hat{m}_0(x) \tag{25.9}$$

and for each control unit (where $W = 0$), the imputed effect is:

$$\hat{D}^0 = \hat{m}_1(x) - Y \tag{25.10}$$

These imputed differences serve as preliminary estimates of the individual treatment effects. Next, we fit two regression models: one on the treated units to estimate a function $\hat{\tau}_1(x)$ that predicts $\hat{D}^1$ based on x, and another on the control units to estimate $\hat{\tau}_0(x)$ predicting $\hat{D}^0$.

Finally, using the estimated propensity score:

$$e(x) = P(W = 1 \mid X = x) \tag{25.11}$$

we combine the two estimates to obtain the final CATE for a new observation with covariates x as:

$$\hat{\tau}(x) = e(x)\,\hat{\tau}_0(x) + \big(1 - e(x)\big)\,\hat{\tau}_1(x) \tag{25.12}$$

This step-by-step process—estimating outcome models, computing imputed treatment effects, fitting separate regression models, and finally combining them—forms the basis of the X-learner approach for estimating CATE.

For practitioners, the X-learner is particularly useful when dealing with complex data where a balanced approach to bias and variance is needed, and when a good estimate of the propensity score is available, such as in marketing studies aiming to optimize campaign targeting with heterogeneous customer responses.

25.4.1 Simulation of X-Learner

Let's illustrate the X-learner approach in R with clear explanations at each step. First, run R-snippet from S and T-learner simulation above to generate simulated data again. Next, we estimate a propensity score (the probability of treatment) using logistic regression on the covariates. This propensity score helps adjust for selection bias when estimating treatment effects.

In step one of the X-learner, we build two outcome models. For treated units, we fit a random forest model predicting Y from X1 and X2 using only the treated subset, and for control units, we fit a linear model on the controls. Using these models, we impute the counterfactual outcomes: for treated observations we predict the outcome they would have had if they had not been treated, and for control observations we predict the outcome if they had been treated. The imputed treatment effects are calculated as the difference between the observed outcome and the predicted counterfactual.

Next, we fit two models on these imputed treatment effects as a function of the covariates. Here we use a linear regression model for the imputed effects from the treated group and a random forest model for those from the control group. Finally, we predict the treatment effect for every observation by combining the two estimates using the propensity score as a weight. The plot at the end helps visualize how closely the estimated treatment effects match the true ones.

Below is the complete R code implementing this simulation:

```
# Run the data R-snippet from S and T-learner before running this code
# Estimate the propensity score using logistic regression
ps_model <- glm(W ~ X1 + X2, family = binomial, data = df)
df$ps <- predict(ps_model, type = "response")

# ---------------------- Step 1: Outcome Models ----------------------
# For treated units, we fit a random forest model to predict Y using X1 and X2.
treated_data <- subset(df, W == 1)
rf_treated <- randomForest(Y ~ X1 + X2, data = treated_data)

# For control units, we use a linear regression model to predict Y.
control_data <- subset(df, W == 0)
lm_control <- lm(Y ~ X1 + X2, data = control_data)

# ----------------- Step 2: Impute Counterfactual Outcomes -----
# For treated units: imputed treatment effect is the observed outcome minus
# the predicted control outcome.
df$D1 <- NA  # imputed effect for treated units
df$D1[df$W == 1] <- df$Y[df$W == 1] - predict(lm_control,
                        newdata = df[df$W == 1,])

# For control units: imputed treatment effect is the predicted
# treated outcome minus the observed outcome.
df$D0 <- NA  # imputed effect for control units
df$D0[df$W == 0] <- predict(rf_treated,
          newdata = df[df$W == 0,]) - df$Y[df$W == 0]

# ----------- Step 3: Fit Models on Imputed Treatment Effects ----
# For the treated group imputed effects, we fit a linear reg. model.
model_tau1 <- lm(D1 ~ X1 + X2, data = df[df$W == 1,])

# For the control group imputed effects, we fit a random forest model.
rf_tau0 <- randomForest(D0 ~ X1 + X2, data = df[df$W == 0,])

# --------------- Step 4: Combine Estimates for Final CATE ---------------
# Predict the imputed treatment effect for every obs. from both models.
df$tau1_hat <- predict(model_tau1, newdata = df)  # from treated group
df$tau0_hat <- predict(rf_tau0, newdata = df)       # from control group
```

```
# Combine the two estimates using the propensity score as weights.
# This gives the final CATE estimate for each observation.
df$tau_hat <- df$ps * df$tau0_hat + (1 - df$ps) * df$tau1_hat

# For comparison, store the true treatment effect.
df$tau_true <- tau
cat("Average tau_hat:", mean(df$tau_hat), "\n")
```

```
## Average tau_hat: 3.952427
```

```
# ---------------------------- Visualization ----------------------------
# Plot the estimated treatment effects versus the true treatment effects.
ggplot(df, aes(x = tau_true, y = tau_hat)) +
  geom_point(alpha = 0.5) +
  geom_abline(slope = 1, intercept = 0, linetype = "dashed") +
  labs(title = "X-learner: Estimated vs True Treatment Effects",
       x = "True Treatment Effect",
       y = "Estimated Treatment Effect")+
  theme_minimal()
```

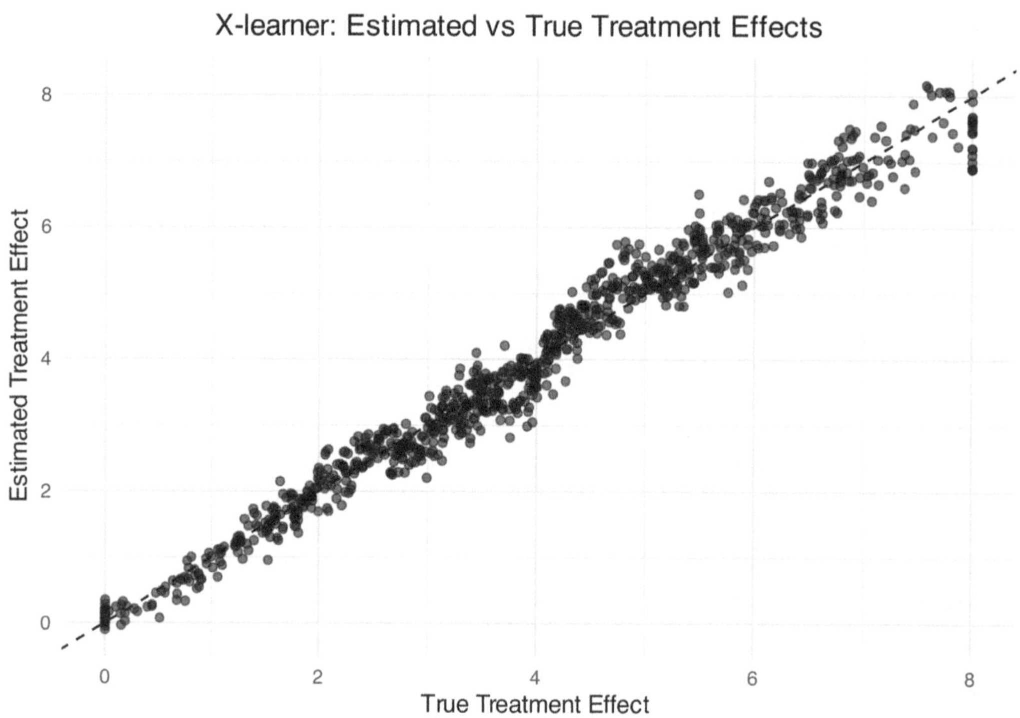

25.5 R-Learner (Residualized Estimation)

The R-Learner is a two-step approach for estimating heterogeneous treatment effects (CATE) while controlling for confounding (Nie & Wager, 2021). Key idea of R-learner is residualize everything first, then regress the residuals on each other. To estimate the R-learner using cross-fitting, we introduce sample splitting, ensuring that the residualization step and the final estimation step are performed on separate data folds. This helps reduce overfitting and improves generalization. The cross-fitted R-Learner follows these steps:

Split the Data into K Folds: Divide the dataset into K non-overlapping folds. Common choices include $K = 2$ (double machine learning) or $K = 5$.

Cross-Fitted Residualization: - For each fold k, train the propensity score model $\hat{e}^{(-k)}(X)$ and the outcome model $\hat{m}^{(-k)}(X)$ using all other folds except k. - Compute the residualized outcome and treatment for observations in fold k:

$$\tilde{Y}^{(k)} = Y - \hat{m}^{(-k)}(X), \quad \tilde{W}^{(k)} = W - \hat{e}^{(-k)}(X) \tag{25.13}$$

Estimate Treatment Effects: - Using the residualized data from all folds, estimate the CATE function:

$$\hat{\tau}(X) = \arg\min_{\tau} \sum_i (\tilde{Y}_i - \tau(X)\tilde{W}_i)^2 + \lambda R(\tau) \tag{25.14}$$

- This regression step can be performed using any flexible machine learning model (e.g., penalized regression, boosting, or neural networks).

Repeat for Each Fold and Aggregate: Aggregate the CATE estimates across folds to obtain the final cross-fitted estimate of $\hat{\tau}(X)$.

By cross-fitting, we ensure that the residualization step does not use the same data as the final estimation, reducing overfitting and improving robustness in finite samples. We can rewrite the estimation equation in a familiar form that we used to estimate unit-level CATEs from the causal forest chapter. If $\hat{m}^{-i}(X_i)$ and $\hat{e}^{-i}(X_i)$ represent cross-fitted estimates—meaning they are trained on all observations except i (i.e., leave-one-out or general cross-fitting)—then the equation holds:

$$Y_i - \hat{m}^{-i}(X_i) = (W_i - \hat{e}^{-i}(X_i))\tau(X_i) + \epsilon_i \tag{25.15}$$

This formulation follows the R-Learner's residualization step, ensuring that the estimation of $\tau(X_i)$ is based on residuals computed out-of-sample, thereby reducing bias. By transforming the problem into a standard regression of residualized outcomes on residualized treatment, it allows for the use of flexible machine learning methods to estimate $\tau(X_i)$. The choice of method depends on the complexity of treatment effect heterogeneity. Simple cases may be handled with linear regression, while more complex relationships require nonparametric methods such as random forests, gradient boosting, or neural networks.

The R-Learner and double machine learning (DML) may look very similar. Both approaches rely on cross-fitting and residualization to mitigate confounding. However, they differ in their primary objectives. DML is designed for estimating sample ATE or parametric CATE in high-dimensional settings while ensuring Neyman-orthogonality, which improves inference validity by reducing bias from regularized machine learning estimators. In contrast, the

R-Learner is more flexible and explicitly tailored for nonparametric CATE estimation. While DML often employs penalized regression techniques like LASSO or Ridge, R-Learner can accommodate a broader range of machine learning estimators to model heterogeneous treatment effects more effectively. Thus, although the two methods share similarities, the R-Learner provides a more general framework for capturing treatment effect heterogeneity, while DML prioritizes high-dimensional inference with robustness to regularization bias.

25.6 DR-Learner (Doubly Robust Learner)

Doubly robust (DR) Learners are a class of meta-learners that extend the augmented inverse probability weighting (AIPW) framework to improve estimation of treatment effects. They combine outcome modeling and propensity score weighting in a way that ensures consistency even if one of the models is misspecified. This doubly robust property makes DR-learners particularly attractive in observational studies, where modeling assumptions are often uncertain. Even though there are various DR-Learners, we focus on the one most widely used in machine learning models for estimating CATEs in R and Python (e.g., `grf`, `econml`, `causalML`), as Kennedy (2022) establishes that it achieves the semiparametric efficiency bound for CATE estimation under standard conditions, meaning it is statistically optimal in minimizing variance while remaining robust to model misspecification.

Definition and Estimation Given an observational dataset $\{(X_i, W_i, Y_i)\}_{i=1}^N$, where $W_i \in \{0, 1\}$ is the treatment indicator, Y_i is the observed outcome, X_i is the vector of covariates, the DR-Learner estimates the CATE, $\tau(X) = E[Y(1) - Y(0)|X]$, in three steps:

Step 1A: Estimating Outcome Models

Estimate the conditional outcome regression functions separately for treated and control groups:

$$\hat{m}_1(X) = E[Y|X, W = 1], \quad \hat{m}_0(X) = E[Y|X, W = 0] \tag{25.16}$$

These can be estimated using any flexible machine learning method such as random forests, neural networks, or gradient boosting.

Step 1B: Estimating the Propensity Score

Estimate the propensity score, $e(X) = P(W = 1|X)$, using a classification model:

$$\hat{e}(X) = P(W = 1|X) \tag{25.17}$$

This is typically done using logistic regression or machine learning models such as gradient boosting or random forests.

Step 2: Constructing the DR-Adjusted Residuals

Compute pseudo-outcomes, which correct for selection bias:

$$Y_i^{DR} = \frac{W - \hat{e}(X)}{\hat{e}(X)(1 - \hat{e}(X))} \Big(Y - \hat{f}_W(X)\Big) + \hat{f}_1(X) - \hat{f}_0(X) \tag{25.18}$$

These pseudo-outcomes adjust the observed data using both outcome models and propensity scores to remove confounding effects.

Step 3: Final Estimation of $\tau(X)$

CATE estimation can be done using any machine learning regression model (e.g., random forests, boosting, neural networks). Train a machine learning model $\hat{\tau}(X)$ to predict the pseudo-outcomes:

$$\hat{\tau}(X) = \arg\min_{\tau} \sum_{i=1}^{N} \left(Y_i^{DR} - \tau(X_i)\right)^2 \tag{25.19}$$

This is a classic doubly robust, AIPW, estimation where both the regression-based outcome model $\hat{m}(X)$ and inverse probability weighting (IPW) are used to correct for potential confounding factors. We do not present a simulation here, as it is already covered in the Doubly Robust, AIPW chapter. Y_i^{DR} is the mathematically identical AIPW equation, which we covered in doubly robust chapter, just expressed in different notation but equivalent forms. The DR-Learner formulation rewrites doubly robust pseudo-outcome using a single term, $(W - e(X))/(e(X)(1 - e(X)))$, which captures the deviation of treatment assignment from the predicted probability. The AIPW formulation explicitly separates the treatment and control contributions using indicator functions W and $1 - W$.

$$Y_i^{DR} = Y_i^{AIPW} = \frac{W_i(Y_i - \hat{m}_1(X_i))}{\hat{e}(X_i)} - \frac{(1 - W_i)(Y_i - \hat{m}_0(X_i))}{1 - \hat{e}(X_i)} + \hat{m}_1(X_i) - \hat{m}_0(X_i) \tag{25.20}$$

Step 4: Cross-fitting (Optional but highly recommended)

To further improve robustness and reduce overfitting, cross-fitting is applied. Cross-fitting improves the reliability of treatment effect estimation by addressing overfitting and bias. By ensuring that nuisance models for propensity scores and outcome regressions are trained on a separate data subset, it prevents overfitting and improves generalization. This separation also reduces bias that arises when the same sample is used for both training and estimation. Additionally, cross-fitting stabilizes estimates, particularly when using flexible machine learning models prone to overfitting. As a result, it is widely applied in DML and DR-Learners to improve robustness in causal inference. This involves repeating steps 1 and 2, but swapping the roles of the training and test samples.

Partition the Data:

- Split the dataset into two independent subsamples: D_1^n and D_2^n.
- Estimate the propensity score $\hat{e}(X)$ and outcome models $\hat{m}_0(X), \hat{m}_1(X)$ using D_1^n.
- Compute pseudo-outcomes using the test sample D_2^n.

Swap the Samples:

- Now, use D_2^n for training and D_1^n for testing.
- Re-estimate the propensity score $\hat{e}(X)$ and outcome models $\hat{m}_0(X), \hat{m}_1(X)$ on D_2^n.
- Compute pseudo-outcomes for observations in D_1^n.

Final Estimation:

- Take the average of the two resulting estimates as the final treatment effect estimate:

$$\hat{\tau}(X) = \frac{1}{2}\left(\hat{\tau}^{(1)}(X) + \hat{\tau}^{(2)}(X)\right) \tag{25.21}$$

where $\hat{\tau}^{(1)}(X)$ and $\hat{\tau}^{(2)}(X)$ are the treatment effect estimates from the two folds.

Generalization to K-Fold Cross-Fitting:

- Instead of using two folds, the process can be extended to K-fold cross-fitting.
- The dataset is split into K folds, and the above process is repeated across all folds.
- The final estimate is obtained by averaging over all folds.

The R-Learner estimates treatment effects by first residualizing outcomes and treatment, making it effective when outcome–treatment relationships are smooth. However, if either the outcome model or the propensity score model is misspecified, those errors propagate into the final treatment effect estimate, potentially leading to bias. In contrast, the DR-Learner applies AIPW, ensuring robustness if either model is correctly specified. This makes DR-Learner more reliable in cases where there is uncertainty about the correct functional form of the outcome or treatment assignment models.

When choosing between these methods, the decision depends on the structure of the data and the reliability of the underlying models. The R-Learner is preferable when one is confident that the outcome and treatment relationships can be well captured through residual regression, making it a useful choice for relatively smooth and stable data. On the other hand, the DR-Learner is the better option when robustness is a priority, particularly in situations where confounding must be addressed even when one of the nuisance models is misspecified.

25.7 CATE vs. ITE

Estimating the impact of an intervention at the individual level is a key challenge in causal inference. While the individual treatment effect (ITE) is the true difference in potential outcomes for a specific individual, it is fundamentally unobservable because we never see both treated and untreated outcomes for the same person. Instead, researchers estimate the CATE, which represents the expected treatment effect for individuals with the same observed characteristics.

Meta-learners are designed to estimate CATE by leveraging machine learning models. When covariates fully explain treatment effect heterogeneity, CATE can serve as a close approximation for ITE. However, if unobserved factors influence treatment effects, CATE remains an average effect over a group rather than an individual-level estimate. This distinction is critical in applications like personalized medicine or policy targeting, where decision-making relies on precise individual predictions. While meta-learners provide powerful tools for estimating heterogeneous effects, their reliability in approximating ITE depends on the richness of available covariates and the assumption of no unobserved confounders.

Alaa, Ahmad, and van der Laan (2023) introduce conformal meta-learners, an extension of traditional meta-learning frameworks that integrates conformal prediction to provide valid uncertainty quantification for ITE estimation. Unlike standard meta-learners, which primarily focus on point estimates of CATE, this approach ensures that the predicted treatment effects come with reliable confidence intervals, maintaining finite-sample validity regardless of the underlying machine learning model. By leveraging conformal inference, the method adapts to distributional complexities and provides robust uncertainty estimates, making it especially useful in high-stakes decision-making scenarios like personalized medicine, policy interventions, and targeted marketing.

The framework builds upon existing meta-learners (e.g., T-, S-, X-, and R-Learners) but incorporates a calibration step using a holdout dataset to generate prediction intervals that maintain a pre-specified confidence level. This makes the approach model-agnostic while improving robustness against heteroskedasticity and distributional shifts. The paper demonstrates that conformal meta-learners can significantly improve the reliability of treatment effect estimates, particularly in settings where standard meta-learners may struggle with uncertainty quantification. Researchers and practitioners looking for causal inference methods with formal guarantees on prediction reliability may find this framework particularly useful. Their GitHub repository is (https://github.com/AlaaLab/conformal-metalearners)).

For a comprehensive overview of meta-learners, Kunzel et al.(2019) provide a detailed review covering the S-, T-, and X-learners, along with their applications in translating supervised learning methods into CATE estimators. Their discussion offers a clear and accessible introduction to these techniques, making it a useful resource for researchers interested in applying machine learning to causal inference. Additionally, Salditt et al. (2024) offer a tutorial introduction to heterogeneous treatment effect estimation with meta-learners, including detailed explanations and simulations with code, which can help practitioners implement these methods in empirical research.

For readers interested in more advanced methods, Nie & Wager (2021) introduce the R-Learner, a powerful approach that optimizes residualized learning for CATE estimation. This method is particularly effective in high-dimensional settings where balancing covariates is crucial. Additionally, Kennedy (2022) discusses improvements in DR Learning, a method designed to improve causal effect estimation by combining outcome modeling and propensity score weighting. His work explores optimal approaches for reducing bias and variance in heterogeneous treatment effect estimation.

26

Difference in Differences and DML-DiD

This chapter returns to one of the most common strategies for addressing selection on unobservables: difference-in-differences (DiD). After introducing instrumental variables in Chapter 22 and developing a range of tools for estimating heterogeneous treatment effects—including causal forests and meta-learners in Chapters 23 to 25—we now revisit settings where unobserved factors may bias treatment assignment, but where panel data and policy variation over time provide opportunities for credible causal identification. The chapter begins with fixed effect models, classical DiD and two-way fixed effects (TWFE) models, extends to event study designs and cohort-specific treatment effects, and highlights the limitations of traditional DiD estimators under treatment heterogeneity or staggered adoption.

We then introduce the double machine learning for DiD (DML-DiD) estimator, which combines Neyman-orthogonal scores with machine learning to address high-dimensional confounding while preserving valid inference. The chapter includes an extended discussion on how to implement DML-DiD in practice, followed by a detailed simulation. We also cover how to extend DML-DiD to more realistic settings involving multiple time periods and staggered treatment adoption, as well as a discussion on handling time-varying continuous treatments within this framework. Throughout the chapter, we highlight recent advances, including papers, software packages, and resources released within the last few months, ensuring readers are equipped with both the foundational intuition and up-to-date tools needed to navigate and apply recent DiD methods. In the next chapter, we continue exploring solutions to selection on unobservables through synthetic control and regression discontinuity designs.

26.1 Panel Data, Fixed Effects, and Causal Inference

Until now, we have focused on models based on cross-sectional data, where observations come from different units at a single point in time. However, many empirical questions involve data that track the same units over multiple periods. This introduces panel data (also called longitudinal data), which consists of repeated observations on the same units—such as individuals, firms, or regions—over time. Unlike purely cross-sectional data, panel data allows us to account for time-invariant characteristics that could otherwise introduce bias in traditional regression models. By following the same entities over time, we can control for unobserved heterogeneity that does not change over time, improving the reliability of our estimates.

Panel data is widely used in labor economics, health economics, and firm-level studies where individual or firm-specific factors remain constant over time. For example, estimating the impact of job training on wages using cross-sectional data may suffer from omitted variable

DOI: 10.1201/9781003504795-26

bias due to unobserved ability differences. Panel data allows us to control for such factors by tracking individuals over time, improving causal inference.

Panel data methods provide a powerful framework for addressing unobserved confounders that do not vary over time. Unlike standard cross-sectional approaches that rely on selection on observables, panel techniques allow us to control for unobserved heterogeneity by leveraging within-unit variation. The key idea is that while some confounders may remain unmeasured, as long as they are constant for each unit over time, they can be removed through appropriate transformations. A key assumption in panel models is strict exogeneity, requiring that $E(\varepsilon_{it}|X_{i1}, ..., X_{iT}, \alpha_i) = 0$. However, in many cases—such as policy interventions or endogenous regressors—this assumption is violated, leading to biased estimates.

In Chapter 2, we examined how to estimate the effect of X on Y within a regression framework. The coefficient β_1 captures this effect, but in a simple cross-sectional setting, it primarily reflects association rather than causation. To obtain an unbiased estimate of β_1, we assume a randomly drawn sample and model the relationship as:

$$y_i = \beta_0 + \beta_1 x_{i1} + \ldots + \beta_k x_{ik} + \epsilon_i, \tag{26.1}$$

where the error term ϵ_i satisfies $E(\epsilon_i) = 0 \quad \text{and} \quad \text{Cov}(x_{ij}, \epsilon_i) = 0 \quad \text{for all } j$. These conditions ensure unbiased and consistent coefficient estimates. However, as discussed, we cannot include all covariates related to Y due to various limitations, including unobservability, which introduces bias in the coefficient estimates. If we have panel data, incorporating fixed effects allows us to control for all time-invariant unit-specific factors, improving the reliability of our estimates.

A typical panel data regression model extends the standard cross-sectional framework to incorporate both unit-specific and time-specific effects:

$$y_{it} = \alpha_i + \beta X_{it} + \gamma_t + \varepsilon_{it}, \tag{26.2}$$

where y_{it} is the outcome for unit i at time t, X_{it} represents explanatory variables, α_i captures unit-specific effects (fixed or random), γ_t accounts for time-fixed effects, controlling for common shocks (e.g., macroeconomic trends), ε_{it} is the error term.

There are several approaches to estimating the effect of X on Y in panel data settings, each with different assumptions and trade-offs.[1]

One of the most common methods is the fixed effects (FE) model, which accounts for unobserved heterogeneity by allowing each unit to have its own intercept (α_i). This approach eliminates time-invariant confounders by focusing on within-unit variation over time. The basic transformation in a fixed effects model removes α_i by demeaning or differencing:

$$(y_{it} - \bar{y}_i) = \beta(X_{it} - \bar{X}_i) + (\varepsilon_{it} - \bar{\varepsilon}_i) \tag{26.3}$$

where $\bar{y}_i$ and $\bar{X}_i$ are the averages of y_{it} and X_{it} over time for each unit i. By subtracting these averages, any time-invariant factors (including unobserved confounders) are removed. This method is particularly useful when selection into treatment is correlated with individual characteristics that do not change over time. FE estimation provides unbiased estimates of β under the assumption that omitted variables are time-invariant. However, it does not address time-varying unobserved confounders, which can still bias estimates. Additionally, because it removes between-unit variation, it is less efficient than alternative models when within-unit variation is small.

While panel methods help control for time-invariant unobserved confounders, they still rely on the assumption that within-unit variation in X_{it} is exogenous. That is, we assume:

$$E(\varepsilon_{it}|X_{it}, \alpha_i) = 0 \tag{26.4}$$

However, this assumption is often unrealistic, especially when policy changes, treatment assignments, or external shocks are not randomly assigned. Standard panel models do not directly account for unobserved, time-varying confounders, requiring more sophisticated approaches to causal inference.
In earlier discussions on causal inference, we introduced the Rubin causal model, which emphasizes the importance of a well-defined treatment and control group for identifying causal effects. Using panel data techniques, we can improve causal estimation by controlling for individual-specific and time-specific confounders (both observable and unobservable), leading to more credible estimates of treatment effects.

In Chapter 18, we discussed how controlling for observed covariates (X) allows us to estimate the average treatment effect (ATE) under the assumption of selection on observables. A basic linear regression model with treatment serves as the baseline approach:

$$Y_i = \alpha + \delta D_i + X_i^\top \beta + \varepsilon_i \tag{26.5}$$

where δ represents the difference between treated and untreated units after adjusting for X_i. If treatment effects are homogeneous, δ captures a constant ATE. However, in many cases, treatment effects may vary across individuals based on observable characteristics, leading to the CATE:

$$CATE(X) = E[Y(1) - Y(0)|X] \tag{26.6}$$

This formulation accounts for treatment effect heterogeneity, recognizing that policy impacts or interventions may differ across units.

Incorporating panel data further refines this estimation by allowing us to control for time-invariant confounders. The baseline treatment model can be extended to a panel data setting as follows:

$$Y_{it} = \alpha_i + \delta D_{it} + X_{it}^\top \beta + \gamma_t + \varepsilon_{it} \tag{26.7}$$

where α_i captures unit-specific effects, and γ_t accounts for time-fixed effects. This specification improves estimation by using within-unit variation in treatment while controlling for common time shocks.

While panel models mitigate bias from time-invariant unobserved heterogeneity, they still require strong assumptions. In particular, strict exogeneity of X_{it} is crucial:

$$E(\varepsilon_{it}|X_{i1}, ..., X_{iT}, \alpha_i) = 0 \tag{26.8}$$

Violations of this assumption—such as policy endogeneity or time-varying unobserved confounders—can still lead to biased estimates.

A natural experiment arises when external events or policy changes create variation in exposure to an intervention that mimics random assignment. We discussed several examples in Chapter 22. In such cases, treatment is not assigned by the researcher but occurs due to circumstances outside their control (i.e., exogenous shock), often allowing for meaningful comparisons between treated and control groups. By using techniques such as DiD and two-way fixed effects (TWFE). These methods leverage variation in treatment timing to account for confounding factors, improving causal inference beyond standard panel models.

26.2 Difference-in-Differences

DiD is a widely used tool in empirical research to estimate the causal effect of a treatment or intervention when randomized experiments are not feasible. The core idea is simple: compare how outcomes change over time in a treated group versus a control group. The key identifying assumption is known as the parallel trends assumption—that, in the absence of treatment, the treated and control groups would have experienced the same average change in outcomes over time. This assumption allows us to attribute any additional change in the treated group, beyond what is observed in the control group, to the effect of the treatment.

To build intuition, we begin with the simplest 2×2 setup involving a binary treatment. This setting includes two groups—treated and untreated—and two time periods—before and after treatment. One group is exposed to the treatment in the second period, while the other remains untreated throughout. We observe the same outcome variable across both groups and time periods. The goal is to estimate the causal effect of the treatment by comparing changes in outcomes over time between the two groups.

This setup is often summarized in a 2×2 table, where we examine average outcomes across groups and time:

TABLE 26.1 Group Mean Outcomes by Treatment and Time

	Pre-Treatment ($T = 0$)	Post-Treatment ($T = 1$)
Treated ($D = 1$)	$E[Y_i(0) \mid D = 1]$	$E[Y_i(1) \mid D = 1]$
Untreated ($D = 0$)	$E[Y_i(0) \mid D = 0]$	$E[Y_i(1) \mid D = 0]$

Suppose we want to evaluate whether a job training program improves wages. We measure average wages for two groups—those who enroll in the program (treated) and those who don't (control)—before and after the program starts.

Imagine the treated group's average wage rises from \$2600 to \$3200, while the control group's average wage increases from \$2100 to \$2500 over the same time. Both groups see wage growth, which could be due to factors like inflation or general economic trends. However, the treated group's increase is larger. The DiD method isolates the portion of this extra growth that can be attributed to the program. Specifically, we subtract the control group's wage change (\$400) from the treated group's wage change (\$600), yielding a DiD estimate of \$200. This \$200 reflects the program's estimated effect on wages, assuming that, without treatment, both groups would have experienced similar wage trends.

This logic can be expressed formally with the DiD formula:

$$\delta^{ATT} = (E[Y_{post}|D = 1] - E[Y_{pre}|D = 1]) - (E[Y_{post}|D = 0] - E[Y_{pre}|D = 0]) \quad (26.9)$$

where $D = 1$ indicates treatment and $D = 0$ indicates control, and "pre" and "post" refer to before and after treatment. The first term measures the actual change in the treated group, while the second captures the expected change without treatment (approximated using the control group's change). In the sections that follow, we will show how to estimate this using regression, extend the model to include covariates and continuous treatments, and finally consider the TWFE framework where treatment timing or intensity may vary. At each stage, we will highlight the assumptions required and potential limitations.

DiD can also be estimated using a simple regression model. The regression includes a group indicator (D_i), a time indicator (T_t), and an interaction term ($D_i \times T_t$), which captures the treatment effect. The model is:

$$Y_{it} = \alpha + \beta_1 D_i + \beta_2 T_t + \delta(D_i \times T_t) + \varepsilon_{it} \tag{26.10}$$

Here, δ is the coefficient on the interaction term and represents the DiD estimate. It tells us how much more the treated group's outcome changed relative to the control group's change. Using the wage example, this regression would recover the same \$200 effect if estimated properly.

TABLE 26.2 Expected Outcomes from DiD Regression Model

	Pre-Treatment (T=0)	**Post-Treatment (T=1)**	**Difference (Post - Pre)**
Treated (D=1)	$\alpha + \beta_1$	$\alpha + \beta_1 + \beta_2 + \delta$	$\beta_2 + \delta$
Untreated (D=0)	α	$\alpha + \beta_2$	β_2
Difference (T - U)	β_1	$\beta_1 + \delta$	**DiD** $= \delta$

The table presents the expected outcomes implied by the regression model for both treated and untreated groups, before and after treatment. For the untreated group, the average outcome increases from α in the pre-treatment period to $\alpha + \beta_2$ in the post-treatment period, capturing the baseline time trend. For the treated group, the outcome rises from $\alpha + \beta_1$ to $\alpha + \beta_1 + \beta_2 + \delta$, reflecting both the time trend and the treatment effect. The change over time for the control group is β_2, while for the treated group it is $\beta_2 + \delta$. The DiD estimate δ isolates the additional change in the treated group relative to the control, which is attributed to the treatment.

Similarly, we can interpret the DiD estimate by comparing the treated and control groups at each point in time—essentially a cross-sectional comparison. Prior to treatment, the treated group has an expected outcome of $\alpha + \beta_1$, while the control group's is α, yielding a difference of β_1. This term captures any existing differences between the groups, which may reflect both observable and unobservable characteristics unrelated to the treatment. In randomized experiments, careful design aims to minimize this baseline gap by ensuring that the groups are as similar as possible before the intervention.

Following treatment, the expected outcome for the treated group becomes $\alpha + \beta_1 + \beta_2 + \delta$, while for the control group it is $\alpha + \beta_2$, producing a difference of $\beta_1 + \delta$. Taking the difference between these two cross-sectional gaps—$(\beta_1 + \delta) - \beta_1$—removes the pre-existing difference and isolates the causal impact of the treatment, δ. This group-by-time comparison provides an alternative but equivalent way to understand the DiD estimate obtained from the regression.

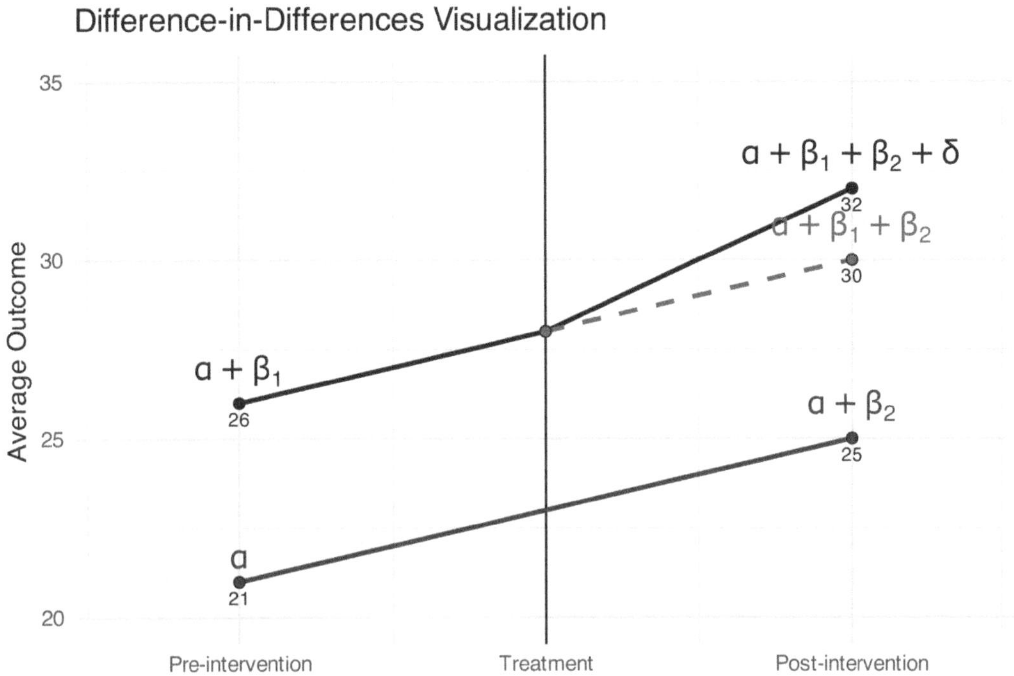

The figure above illustrates the DiD framework and the key elements of the identification strategy. The solid red and blue lines show the observed trends for the control and treated groups, respectively. Prior to the intervention, both groups follow similar trajectories, consistent with the parallel trends assumption. The vertical black line marks the onset of treatment. After this point, the treated group's outcome increases more sharply than the control group's, suggesting a treatment effect. The dashed green line depicts the counterfactual—what the treated group's outcome would have been in the absence of treatment, assuming they had continued along the same path as the control group. The vertical distance between the treated group's observed outcome and this counterfactual captures the DiD estimate.

One key feature illustrated in the figure is the constant difference in outcome before treatment. This is depicted by the parallel movement of the treated and control groups before the intervention, reinforcing the parallel trends assumption required for DiD to yield unbiased estimates. The difference between the treated and counterfactual post-treatment values represents the estimated treatment effect (δ).

Additionally, the inclusion of a pre-treatment period extending further back in time (if data for multiple pre-treatment periods is available) provides further reassurance that the parallel trends assumption holds. Extending the visual representation in the pre-treatment period highlights the stable gap between the treated and control groups, reinforcing the idea that any deviation after treatment is likely due to the intervention rather than pre-existing differences.

From the cross-sectional perspective, the treated group starts at β_1 above the control group; after treatment, the gap grows to $\beta_1 + \delta$, again implying that δ captures the treatment effect. Now, suppose we only observed post-treatment data. In that case, we could compare outcomes between treated and untreated individuals and use methods like regression adjustment or matching under the selection-on-observables framework. These methods rely on the assumption that, after conditioning on observed covariates, the treated and control groups are comparable. Any difference in outcomes could then be attributed to the treatment. However, if there are unobserved differences between groups that also affect outcomes, the estimated treatment effect would be biased.

This is where the strength of the DiD approach becomes clear. Because it uses both pre- and post-treatment data, DiD accounts not only for observed covariates (which can be included in the regression, as discussed below) but also for any time-invariant unobserved differences between the groups. These may include observable or unobservable factors and characteristics that do not change over time, all of which are differenced out by this method. As long as the parallel trends assumption holds, DiD effectively eliminates this potential bias. That's a major advantage of the method—it protects against omitted variable bias due to unobserved, time-invariant group differences and common shocks affecting all groups equally over time, which purely cross-sectional methods cannot address.

26.2.1 Controlling for Additional Covariates in DiD

The basic DiD model estimates the causal effect of a treatment by comparing the changes in outcomes between treated and control groups over time. A fundamental assumption of this approach is the parallel trends assumption, which states that in the absence of treatment, the outcome variable would have followed the same trajectory in both groups. While this assumption enables identification, the precision of the estimated treatment effect can be improved by controlling for additional covariates that influence the outcome variable.

A more general DiD specification that includes time-varying covariates is given by:

$$Y_{it} = \alpha + \beta_1 D_i + \beta_2 T_t + \delta(D_i \times T_t) + \beta X_{it} + \varepsilon_{it}, \tag{26.11}$$

where X_{it} is a vector of time-varying covariates that may affect Y_{it}, such as demographic characteristics, economic conditions, or firm-specific factors. β represents the coefficients associated with these covariates, capturing their independent influence on the outcome. δ is the DiD estimate, which measures the treatment effect after controlling for covariates.

Including X_{it} in the model serves several purposes. First, it reduces residual variance, thereby improving the efficiency of the estimate for γ. Second, it helps account for potential confounding if observable factors differ systematically between treated and control groups. However, it is important to note that adding covariates does not alter the identification strategy—the parallel trends assumption must still hold after conditioning on X_{it}. If unobserved factors simultaneously influence both the treatment decision and the outcome, the estimate may still suffer from omitted variable bias.

Assessing the Parallel Trends Assumption

Since the credibility of DiD depends on the parallel trends assumption, researchers should conduct robustness checks to validate this assumption. Several diagnostic tools can help assess whether the assumption is reasonable:

- **Pre-trends analysis:** This involves testing whether the treated and control groups exhibit similar trends in the outcome variable before treatment. If the groups show systematic differences in trends before the intervention, the estimated treatment effect may be biased.
- **Placebo tests:** A falsification test where treatment is artificially assigned at an earlier period when no intervention actually occurred. If a significant treatment effect is detected, it suggests that unobserved factors may be driving the results rather than the actual treatment.

- **Sensitivity analyses:** Researchers can test the robustness of their findings by modifying the control group, adjusting covariates, or varying the sample period to see if the estimated effect remains consistent.

If the parallel trends assumption appears violated, alternative approaches such as synthetic control methods—which construct a weighted combination of control units to better match the pre-treatment trajectory of the treated group—or interactive fixed effects models, which allow for heterogeneous trends across units, may be more appropriate.

26.2.2 Two-Way Fixed Effects with Covariates

The TWFE model is a standard extension of the basic DiD framework, particularly well-suited for panel data where units are observed over multiple time periods. It generalizes the simple two-group, two-period DiD design to settings with many units and multiple time periods, enabling researchers to estimate causal effects while controlling for both unit-specific and time-specific sources of unobserved heterogeneity.

The basic TWFE specification, when including time-varying covariates, takes the following form:

$$Y_{it} = \alpha_i + \gamma_t + \gamma^{TWFE}(D_i \times T_t) + \beta X_{it} + \varepsilon_{it} \tag{26.12}$$

where Y_{it} is the outcome for unit i at time t, α_i represents unit fixed effects, accounting for time-invariant unobserved heterogeneity across units (e.g., baseline productivity, geographic characteristics), γ_t captures time fixed effects, which absorb shocks common to all units at time t (e.g., macroeconomic trends, national policies), $D_i \times T_t$ is the interaction term indicating whether unit i is treated in period t, γ^{TWFE} is the coefficient of interest and represents the estimated treatment effect, X_{it} is a vector of time-varying covariates, and β denotes the associated coefficients, ε_{it} is the idiosyncratic error term.

This model is attractive for several reasons. First, by controlling for both α_i and γ_t, it eliminates confounding from any unobserved factors that are constant within a unit over time or that affect all units equally in a given period. This is particularly important in observational settings where unobserved heterogeneity is a primary threat to identification. For example, regional differences in institutional quality or persistent individual characteristics like ability can be absorbed by α_i, while national-level events or policies influencing all units are captured by γ_t.

Second, the inclusion of X_{it} allows the researcher to control for observed time-varying covariates that may affect the outcome. These might include income, education, employment status, household size, or any other variable that varies over time and could confound the relationship between treatment and outcome. Including these covariates can improve estimation precision and may also make the parallel trends assumption more plausible by adjusting for compositional changes over time.

It is worth emphasizing, however, that controlling for X_{it} does not change the identifying assumptions of the model. The validity of the TWFE estimate still depends on the assumption that, conditional on fixed effects and X_{it}, the treated and untreated units would have followed parallel trends in the absence of treatment. If this conditional parallel trends assumption is violated, the TWFE estimate may be biased even if relevant covariates are included.

One useful feature of the TWFE framework is that it absorbs all time-invariant covariates, whether observed or unobserved. That means researchers do not need to include static characteristics such as gender or ethnicity directly in the model, as their effects are captured by the unit fixed effects. However, the model only leverages within-unit variation over time, and any variable that does not change across time for a given unit will not contribute to identification.

Despite these advantages, the TWFE estimator has important limitations. In particular, the model implicitly assumes that the treatment effect is constant across all treated units and time periods. In many empirical settings, this assumption is unlikely to hold. For example, the impact of a training program might differ between early and late adopters, or the effect of a tax policy might change as the economy evolves. When treatment effects are heterogeneous across units or over time, and especially when treatment is adopted at different times by different units (i.e., staggered adoption), the TWFE estimator can produce biased and misleading results.

This issue arises because, under staggered timing, TWFE estimators use already-treated units as controls for later-treated units. If the treatment effect varies by treatment timing or duration, this comparison is problematic. In some cases, it can even reverse the sign of the estimated effect or generate estimates that reflect a non-convex, weighted average of various treatment effects that lacks a clear interpretation.

Due to these concerns, researchers increasingly turn to alternative methods when treatment timing varies or treatment effects are expected to differ across cohorts. These alternatives—such as group-time average treatment effect estimators or event study designs—allow for more flexible modeling of treatment dynamics and avoid the contamination issues associated with staggered adoption.

In summary, the TWFE model with covariates is a powerful and widely used method for estimating causal effects using panel data. It effectively controls for time-invariant unit characteristics and period-specific shocks, and it improves precision through the inclusion of time-varying covariates. However, its validity rests on strong assumptions—particularly, that treatment effects are homogeneous and that treated and control groups would have followed parallel trends. When these assumptions are not met, especially in the presence of staggered treatment timing, alternative strategies should be considered to ensure credible and interpretable results.

The treatment effect γ^{TWFE} is identified from within-unit variation over time. This means the model compares changes in outcomes within the same unit before and after treatment and contrasts that change with the corresponding change in units that did not receive treatment. Any covariates that do not vary over time for a given unit (e.g., race, gender, baseline location) are absorbed by the fixed effects and do not need to be included explicitly in the model.

26.2.3 Event Study Designs

Event study designs are a natural and powerful extension of the DiD framework, especially well-suited to settings where treatment is not assigned at a single point in time or where treatment effects are expected to evolve. While traditional DiD models estimate a single average treatment effect, this approach is often too restrictive in real-world applications where treatments occur in staggered fashion (i.e., different units are treated at different times), or where the effects of the intervention vary over time or across units.

Event study designs address these limitations by allowing the estimation of dynamic treatment effects—that is, the causal effects of treatment at different time periods before and after the treatment event. Instead of summarizing treatment impact with one number, event study methods trace out how the outcome variable responds over time relative to the moment of treatment. This is especially useful in applications like evaluating labor market reforms, education policies, health interventions, or any gradual policy rollout, where understanding timing and dynamics of effects is essential.

The standard event study regression model is specified as:

$$Y_{it} = \alpha_i + \gamma_t + \sum_{k \neq -1} \beta_k D_{it}^k + \varepsilon_{it} \tag{26.13}$$

where Y_{it} is the outcome of interest for unit i at time t, α_i are unit (individual, firm, region) fixed effects that control for time-invariant differences across units, γ_t are time fixed effects that absorb aggregate shocks or trends common to all units in each period, D_{it}^k is a dummy variable equal to 1 if unit i is k periods away from treatment at time t (e.g., 2 years before or 1 year after), β_k captures the effect of being k periods relative to treatment (the event time), the reference period, often $k = -1$, is omitted so all estimated β_k coefficients are interpreted relative to it, ε_{it} is an idiosyncratic error term.

This structure provides a sequence of estimated treatment effects over time, allowing for a dynamic view of causal impact. For example, one can estimate the effect one year before treatment (β_{-1}), at the time of treatment (β_0), and at various lags after treatment ($\beta_1, \beta_2, \ldots$).

Why Event Study Designs Matter

Event studies offer several advantages over conventional DiD models:

1. *They uncover dynamics*: Rather than assuming an immediate or constant treatment effect, event studies reveal when the treatment effect materializes, how it builds over time, and whether it fades out or persists.

2. *They test the parallel trends assumption*: A key assumption in DiD is that treated and control groups would have followed similar trends in the absence of treatment. Event study estimates before treatment (i.e., leads) help assess this. If the pre-treatment coefficients β_k for $k < 0$ are flat and close to zero, it supports the idea of parallel pre-trends. If they are statistically or economically significant, it raises concerns about potential bias.

3. *They reveal anticipation or spillover effects*: If outcomes change before treatment is implemented, this could indicate anticipation of the policy (e.g., individuals adjusting behavior in advance) or leakage/spillovers from treated to untreated units. Both challenge the internal validity of standard DiD models and can be visualized and tested using event study estimates.

4. *They are interpretable and visual*: Event study coefficients are typically plotted with confidence intervals against event time, making it easy to communicate patterns over time. Such plots are widely used in applied research to present treatment dynamics and validate identification assumptions.

Practical Considerations

To estimate an event study model, researchers must define treatment timing at the unit level (i.e., when each treated unit receives the intervention). Untreated units are included as control observations across all time periods, and treated units contribute to estimation based on their event time. When using balanced panels, every unit contributes information for the same number of leads and lags relative to treatment. In unbalanced or staggered panels, care is needed to ensure appropriate interpretation and avoid composition bias—where the sample of units contributing to each event time changes over time.

One practical challenge is the limited support for certain event time dummies, particularly at long leads or lags, where few units remain in the sample. In such cases, estimates for those periods may be imprecise. Researchers sometimes bin far-out event times (e.g., grouping all $k < -4$ into a single lead) to address this.

Another important issue is normalization: typically, one event time (often $k = -1$) is omitted from the regression and serves as the reference period. The interpretation of all estimated coefficients is therefore relative to this omitted baseline, and results can be sensitive to this choice if pre-trends are present.

Event Studies in the Presence of Staggered Treatment Timing

In recent years, growing attention has been paid to how event study models interact with staggered adoption—where units are treated in different periods. When treatment effects are heterogeneous (i.e., vary across units or over time), standard TWFE event studies can produce biased or misleading estimates. This is because units treated earlier may inadvertently act as controls for units treated later, contaminating the counterfactual.

Several recent contributions address this problem. Alternative estimators such as those proposed by Callaway and Sant'Anna (2021), Sun and Abraham (2021), and de Chaisemartin and D'Haultfœuille (2020) adjust for the fact that units treated at different times contribute unequally to identification. These estimators often involve:

- Computing group-time average treatment effects, estimating the effect separately for each cohort and time period,
- Aggregating effects across cohorts only when appropriate, with attention to weighting and dynamic structure,
- Providing event-study-style plots that are robust to heterogeneity and contamination.

A more recent development by de Chaisemartin and D'Haultfœuille (2024) further extends this framework to accommodate non-binary, non-absorbing, and dynamic treatment settings. In many empirical applications, treatment intensity varies across units and over time—consider tax rates, pollution exposure, or credit availability—where the treatment is not a simple "on/off" switch. Moreover, treatment may be non-absorbing, meaning that once treated, units are not necessarily always treated thereafter; they can exit treatment or move to different treatment levels. This violates the assumptions underpinning many common event-study estimators.

Their proposed estimator, $\text{DID}_{g,\ell}$, captures the effect of being exposed to a higher (or lower) treatment level for ℓ periods, relative to a group's baseline treatment level. It is constructed by comparing changes in outcomes for "switcher" groups to those of comparable groups that have not yet changed treatment and share the same initial treatment level. Crucially, this estimator maintains the core logic of classical DiD: it computes differences in outcomes over time, between treated and untreated groups, and then takes differences of those differences. At its core, $\text{DID}_{g,\ell}$ is a linear estimator. Although it is designed to handle far more complex settings—including time-varying, continuous, and lagged treatments—it still relies on linear combinations of group-time average outcomes, just like the standard DID.

In addition to this baseline estimator, the authors propose normalized estimators that scale treatment effects by the magnitude of the treatment change, making them interpretable as dose-response slopes. They also define a cost-benefit parameter that aggregates the total impact of treatment changes over time, allowing policy analysts to evaluate whether observed treatment trajectories led to welfare gains relative to a counterfactual of no change.

These estimators are implemented in the Stata command `did_multiplegt_dyn` and the companion R package `didMultiplegt`, both of which are user-friendly and designed for applied researchers. Full documentation, including detailed examples, simulation programs, and replication materials, is available at the authors' GitHub page: https://github.com/chaisemartinPackages/did_multiplegt_dyn. The site provides guidance on how to structure data, apply the estimators, interpret output, and generate event-study plots robust to treatment heterogeneity and lagged dynamics. Whether using Stata or R, users can follow well-documented code templates to implement actual-versus-status-quo estimators, normalized treatment effects, or dynamic policy evaluation parameters. In the last section of this chapter, we will also revisit time-varying continuous treatments and introduce estimators that implement the DML method, allowing for the incorporation of various machine learning techniques.

Best Practices

To apply event study methods effectively, researchers should:

- Clearly define the treatment event and ensure that timing is measured consistently across units;
- Choose an appropriate window of event time for estimation and decide whether to bin tails;
- Always include unit and time fixed effects, unless compelling theoretical reasons suggest otherwise;
- Examine pre-treatment coefficients carefully for signs of violation of the parallel trends assumption;
- Report robustness checks using alternative specifications (e.g., excluding certain leads or using alternative control groups);
- Consider newer methods if treatment is staggered or heterogeneous, to ensure valid causal interpretation.

In summary, event study designs represent a flexible, informative, and transparent approach to estimating treatment effects over time. They improve the standard DiD framework by allowing for rich temporal dynamics, testing identification assumptions, and improving credibility in the presence of complex policy rollouts. When used carefully, they serve as an essential tool for applied researchers seeking to understand not just whether an intervention worked—but when, how, and for whom it had its effects.

26.2.4 Heterogeneous Treatment Effects in DiD

A core assumption in traditional DiD models—especially those estimated via TWFE—is that the treatment effect is homogeneous, meaning it is constant across units and time. However, this assumption rarely holds in practice. In many empirical settings, the impact of a policy or intervention varies depending on when a unit is treated, how long it has been exposed, and who the unit is. These differences are especially pronounced in settings with staggered treatment adoption, where units receive treatment at different points in time rather than all at once.

When treatment effects are heterogeneous, TWFE DiD models can become problematic. Although these models are appealing for their simplicity and ability to control for unit and time fixed effects, they implicitly construct comparisons not just between treated and untreated units but also between earlier- and later-treated units. In such cases, units that have already been treated can serve as controls for units that are just about to be treated. If the treatment effect is not constant over time or across units, this practice generates invalid counterfactuals and can lead to biased or misleading estimates.

The bias stems from the way the TWFE estimator weights the comparisons. In settings with staggered treatment and varying treatment effects, the TWFE estimator often produces a weighted average of many different treatment effects, including some with negative weights. This means that the estimated coefficient is not simply an average of all individual treatment effects, but rather a mixture of effects that are aggregated in potentially nontransparent and nonrepresentative ways. As a result, the TWFE coefficient may not reflect the actual treatment effect for any identifiable group, and in some cases, it may even have the wrong sign.

These concerns have led to a broader rethinking of how to estimate treatment effects when timing and heterogeneity matter. Rather than relying on a single, pooled treatment effect, recent approaches emphasize the need to allow treatment effects to vary flexibly over time and across groups. This involves estimating what are often called group-time average treatment effects, where effects are computed separately for each cohort of treated units (i.e., those treated in the same time period) and each post-treatment period. Such disaggregated estimates are then aggregated across cohorts using transparent and policy-relevant weighting schemes.

Importantly, the identification of these group-time effects requires appropriate control groups—typically, units that have not yet been treated by that time or those that are never treated. By focusing only on valid comparisons (e.g., treated versus not-yet-treated or never-treated), these methods avoid the contamination issues that plague TWFE estimators in staggered settings.

Another useful approach involves dynamic DiD designs, where treatment effects are modeled as a function of "event time"—the number of periods relative to when a unit received treatment. This allows researchers to estimate how treatment effects evolve over time (e.g., short-run vs. long-run impacts) and whether effects accumulate or dissipate. These designs also allow for checking the parallel trends assumption by examining pre-treatment dynamics.

In applied work, accounting for heterogeneity is not just a technical refinement—it can meaningfully change the interpretation of results. If, for example, early adopters of a policy benefit more (or less) than late adopters, or if the treatment effect grows over time, averaging across these differences can obscure key policy insights. For policymakers, understanding when and for whom an intervention is effective is often as important as knowing whether it works on average.

In sum, when treatment timing varies and heterogeneous effects are likely, relying on traditional TWFE DiD estimators may be insufficient and even misleading. Instead, researchers should adopt frameworks that allow for cohort-specific, time-varying, and event-time-dependent treatment effects.

26.2.5 Where to Go Next: Tools, Resources, and Recent Advances

The treatment of heterogeneous effects in DiD settings is one of the most active areas of causal inference research today. While this chapter has provided the conceptual foundation and highlighted key challenges, our goal is not to offer a full econometric treatment. Instead, the purpose here is to build foundational intuition and introduce the types of challenges that arise when treatment effects vary across units or evolve over time, especially in settings with staggered adoption. Rather than detailing every methodological advance, our aim is to provide a clear framework for recognizing these issues and to point you to the most practical, readable, and up-to-date resources for further exploration. If you are interested in applying more robust DiD methods—particularly those that integrate modern causal inference and machine learning techniques—several resources stand out:

- For a broader theoretical and practical overview, the recent survey by Dmitry Arkhangelsky and Guido Imbens (*Causal Models for Longitudinal and Panel Data: A Survey*, 2024) is an excellent reference. It discusses a wide range of models, including DiD extensions, factor models, synthetic control methods, and interactive fixed effects, with a strong emphasis on empirical implementation and practical guidance for applied researchers.
  The published version appears in *The Econometrics Journal*: https://doi.org/10.1093/ectj/utae014
- A valuable application and review of these newer methods can be found in the large-scale reanalysis study by Chiu, Lan, Liu, and Xu (2025) titled *Causal Panel Analysis under Parallel Trends: Lessons from A Large Reanalysis Study*. This paper replicates 49 published articles that used TWFE estimators, reanalyzing them with multiple modern, heterogeneity-robust approaches. Their results underscore how sensitive conclusions can be to modeling assumptions—and how necessary it is to test for violations of parallel trends and to account for heterogeneous effects.
  You can access the paper here: https://doi.org/10.48550/arXiv.2309.15983
- Finally, for a hands-on, design-oriented perspective, the Practitioner's Guide by Baker, Callaway, Cunningham, Goodman-Bacon, and Sant'Anna (*Difference-in-Differences Designs: A Practitioner's Guide*, 2025) provides a well-organized roadmap through the current DiD landscape. The paper offers practical discussion of design choices, covariate adjustment, multiple periods, and weighting strategies across different DiD estimators—including advice on navigating real-world complications. The working paper is available here: https://arxiv.org/pdf/2503.13323 and wonderful web page is here https://psantanna.com/did-resources/[2]
- A comprehensive and highly accessible resource for applied researchers is the fortcoming book by Clément de Chaisemartin and Xavier D'Haultfœuille titled *Credible Answers to Hard Questions: Differences-in-Differences for Natural Experiments* (2025). This work introduces modern DiD estimators suited for complex natural experiments where randomized trials are infeasible. The book provides intuitive explanations, detailed practical guidance, and implementation-ready insights for navigating heterogeneity, dynamic treatment effects, and violations of standard assumptions. It is particularly useful for researchers aiming to obtain credible causal estimates in messy, real-world settings. The most recent version is available at authors webpage or SSRN: https://ssrn.com/abstract=4487202.
- For readers seeking an up-to-date and practical overview of the most recent developments in Difference-in-Differences methods across R, Stata, and Julia, a highly recommended resource is Asjad Naqvi's website: https://asjadnaqvi.github.io/DiD/. This site offers a curated and hands-on guide to modern DiD packages, including detailed explanations, code examples, and comparisons across platforms. It covers tools such as `did`, `did2s`, `bacondecomp`, and others, and provides working implementations, tutorials, and links to the underlying methods. For researchers looking to move beyond traditional TWFE and explore robust, modern DiD estimators, this site's Resources section is especially valuable—it

compiles recent papers, lecture notes, workshops, and videos that provide both conceptual background and practical guidance on current best practices in causal panel analysis.

- The `fect` R package developed by Yiqing Xu and colleagues provides a user-friendly implementation of factor-based models and interactive fixed effects for causal panel data analysis. This package is particularly useful when treatment timing is staggered and unobserved time-varying confounders may be present. You can explore the package and documentation here: https://yiqingxu.org/packages/fect/.

The goal of this book is to provide a machine-learning-oriented approach to causal inference in panel settings, with an emphasis on flexible methods that go beyond the standard TWFE model. We do not attempt to fully cover the expanding list of heterogeneity-robust DiD estimators; rather, our goal in this chapter is to build foundational intuition and to introduce the types of challenges that arise when treatment effects vary or when treatment is staggered. Readers seeking to apply the most recent heterogeneity-robust DiD estimators today should view this chapter as a conceptual bridge and use the references above as their starting point for practical implementation.

26.3 Double Machine Learning for DiD Estimation

This section introduces a recent, machine-learning-based approach to estimating causal effects in panel settings with treatment variation across time and units. We focus on the DML-DiD estimator, which blends the robustness of Neyman-orthogonal score functions with the flexibility of machine learning to handle high-dimensional covariates. We begin with the foundational ideas from semiparametric DiD estimators like Abadie (2005), then show how these ideas break down when naively combined with machine learning. We build intuition for Chang's (2020) DMLDiD estimator, which resolves these issues using orthogonal scores and cross-fitting. A full simulation demonstrates how to implement the estimator step by step, including model selection and trimming. We then extend the framework to multiple time periods and staggered treatment designs, following Callaway and Sant'Anna (2021), and show how to apply DML-DiD in practice to estimate dynamic group-time treatment effects. This section prepares readers to apply DML in realistic policy settings—where treatment is introduced gradually and treatment effects may vary over time and across units—while maintaining valid inference even with complex, high-dimensional data.

This section focuses on the estimator proposed in *Double/Debiased Machine Learning for Difference-in-Differences Models* by Chang (2020), which introduces the DMLDiD estimator. This estimator allows researchers to use ML methods in high-dimensional settings while still obtaining valid causal inferences using the DiD approach.

The paper addresses a common challenge in causal inference: how to incorporate high-dimensional covariates and ML methods into DiD models without sacrificing the validity of the estimates. Traditional and semiparametric DiD estimators work well under low-dimensional settings and with classical nonparametric methods (like kernels), but they can break down when ML methods are used to estimate first-stage nuisance parameters (like propensity scores). This is due to potential overfitting and slow convergence rates.

A previous study is Abadie (2005), which proposes a semiparametric DiD estimator for the average treatment effect on the treated (ATT). The identification of ATT relies on conditional parallel trends and overlap assumptions. Under these assumptions, Abadie shows that ATT can be identified using the following expression:

$$\theta_0 = \mathbb{E}\left[(Y_i(1) - Y_i(0)) \cdot \frac{D_i - P(D_i = 1 \mid X_i)}{P(D_i = 1)(1 - P(D_i = 1 \mid X_i))}\right] \tag{26.14}$$

In practice, the semiparametric estimator is implemented by estimating the propensity score $\hat{g}(X_i) = \hat{P}(D_i = 1 \mid X_i)$ and then computing the sample analog:

$$\hat{\theta} = \frac{1}{N}\sum_{i=1}^{N}(Y_i(1) - Y_i(0)) \cdot \frac{D_i - \hat{g}(X_i)}{\hat{p}(1 - \hat{g}(X_i))} \tag{26.15}$$

where $\hat{p}$ is the estimator of treated share. As we estimate ATT, inverse probability weighting is applied only to the control group, so treated units are not weighted by the propensity score.Samples with covariates similar to the treated group receive larger weights because they contribute more credible counterfactual information.

Samples in the control group that have a distribution of covariates closer to that of the treated units are given higher weights because they provide more relevant counterfactual information. This estimator performs well when $\hat{g}(X)$ is estimated using classical nonparametric methods like kernels or series in low-dimensional settings but is sensitive to misspecification or overfitting when ML is used to estimate the propensity score.

Chang builds on Abadie's (2005) semiparametric DiD estimator by designing a new estimator that is Neyman orthogonal, meaning it is robust to small errors in first-stage ML estimates.[3]

To estimate ATT, Chang combines this orthogonal score (see footnote) with the cross-fitting algorithm from Chernozhukov et al. (2018). The dataset is partitioned into K folds, and for each fold I_k, the nuisance functions—the propensity score $\hat{g}_k(X)$ and the outcome model $\hat{\ell}_{1k}(X)$—are estimated using only the data from the complementary sample I_k^c. The key insight is that these functions are estimated separately from the fold in which the final treatment effect is computed, which guards against overfitting and improves robustness. The intermediate fold-specific estimator is:

$$\tilde{\theta}_k = \frac{1}{n}\sum_{i \in I_k} \frac{D_i - \hat{g}_k(X_i)}{\hat{p}_k(1 - \hat{g}_k(X_i))} \cdot \left((Y_i(1) - Y_i(0)) - \hat{\ell}_{1k}(X_i)\right) \tag{26.16}$$

where $\hat{p}_k = \frac{1}{n}\sum_{i \in I_k^c} D_i$. The final DMLDiD estimator aggregates across folds: $\hat{\theta} = \frac{1}{K}\sum_{k=1}^{K} \tilde{\theta}_k$.

The component $\hat{\ell}_{1k}(X_i)$ estimates the counterfactual pre-post outcome difference for unit i had they not been treated. It is learned using only the control group data and models the expected evolution of outcomes conditional on X, i.e., $\mathbb{E}[Y(1) - Y(0) \mid X, D = 0]$. The DMLDiD estimator thus subtracts this predicted counterfactual change from the observed change in treated units. This isolates the treatment effect while controlling for differential trends due to observed characteristics.

As in Abadie (2005), the untreated units are reweighted by the inverse of the estimated propensity score to ensure comparability with the treated group. Both the propensity score $\hat{g}_k(X)$ and the treatment share $\hat{p}_k$ are estimated via cross-fitting, further preventing overfitting. Samples that closely resemble the treated group based on covariates receive larger weights, reinforcing their role as credible counterfactuals in the estimation.

A Monte Carlo simulation shows that while plugging ML into Abadie's estimator creates bias, the proposed DMLDiD corrects for this and remains centered at the true treatment effect. Chang applies DMLDiD to Sequeira's (2016) dataset on tariff reductions and bribe payments between South Africa and Mozambique. While Sequeira's traditional DiD model estimated a significant drop in corruption from tariff cuts, Chang finds an even stronger effect using DML-DiD. However, DMLDiD provides more robust estimates with smaller standard errors, suggesting that the original estimates may have understated the average treatment effect due to functional form misspecification and omitted heterogeneity.

26.3.1 Step-by-Step Implementation

Estimate the average treatment effect on the treated (ATT) in a DiD setup using panel data. This approach combines flexible ML methods to estimate nuisance functions—propensity scores and outcome regressions—while preserving valid inference through Neyman-orthogonal scores.

Step 0: Inputs

Start with panel data: - Y_{i0}: Pre-treatment outcome
- Y_{i1}: Post-treatment outcome
- $D_i \in \{0, 1\}$: Treatment indicator (treated only in post-period)
- X_i: Covariate vector (possibly high-dimensional)

Step 1: Compute Outcome Differences

For each unit:

$$\Delta Y_i = Y_{i1} - Y_{i0}$$

Step 2: Cross-Fitting Setup

Split the data into K folds (e.g., $K = 5$).
Each observation is assigned to one fold for test prediction; models are trained on the remaining $K - 1$ folds.

Step 3: Estimate Nuisance Functions with ML (Fold-by-Fold)

For each fold $k \in \{1, \ldots, K\}$: - Use fold k as the test set. - Use remaining folds as the training set.

3.1 Estimate Treatment Share Compute the average treatment probability in the training data:

$$\hat{p} = \mathbb{E}[D] = \frac{1}{n_{\text{train}}} \sum_{i \in \text{train}} D_i$$

This value is used to normalize the orthogonal score. Compute and store it separately for each fold.

3.2 Estimate Propensity Score

$$\hat{g}(X) = \mathbb{P}(D = 1 \mid X)$$

Estimate this using a classification model fit on the training data:

```
fit <- cv.glmnet(X_train, D_train, family = "binomial")
g_hat <- predict(fit, newx = X_test, type = "response")
```

Alternative ML methods: - Logistic regression (`glm`) - Random forests (`randomForest`) - Gradient boosting (`xgboost`) - Neural nets (`keras`)

3.3 Estimate Outcome Model for Controls

$$\hat{\ell}(X) = \mathbb{E}[\Delta Y \mid D = 0, X]$$

Train this model using **only untreated units** in the training data. Predict on all test observations:

```
lasso_model <- cv.glmnet(X_train[D_train == 0, ],
                         DeltaY_train[D_train == 0])
ell_hat <- predict(lasso_model, newx = X_test)
```

Alternative ML methods: Ridge, Lasso, forests, boosting, etc.

Step 4: Model Evaluation and Trimming (Optional)

Before trimming, evaluate model performance:

```
rmse_d <- sqrt(mean((D - g_hat)^2))
rmse_y <- sqrt(mean((DeltaY[D == 0] - ell_hat[D == 0])^2))
```

Trimming is optional but helps avoid instability when $\hat{g}(X)$ is close to 0 or 1:

```
trim_idx <- g_hat > 0.05 & g_hat < 0.95
g_hat    <- g_hat[trim_idx]
DeltaY   <- DeltaY[trim_idx]
ell_hat  <- ell_hat[trim_idx]
D        <- D[trim_idx]
```

Step 5: Compute Neyman-Orthogonal Score (Chang, 2020)

Using the estimated nuisance functions and $\hat{p}$ from each fold, compute the orthogonal score for each observation i:

$$\psi_i = \frac{D_i - \hat{g}(X_i)}{\hat{p}(1 - \hat{g}(X_i))} \cdot \left(\Delta Y_i - \hat{\ell}(X_i)\right)$$

This score is orthogonal to first-stage estimation error, which ensures robustness even when ML is used.

```
p_hat <- mean(D_train)  # estimated on training data
psi <- ((D - g_hat) / (p_hat * (1 - g_hat))) * (DeltaY - ell_hat)
att_meanpsi <- mean(psi)
```

Step 6: Estimate ATT

Normalize by the reweighted treatment indicator:

$$\text{Normalization factor} = \frac{1}{n}\sum_{i=1}^{n}\frac{D_i}{\hat{p}}$$

In most cases—where $\hat{p}$ is constant across units—this factor is ≈ 1 and can be skipped. But it ensures correct scaling with cross-fitting.

$$\hat{\theta}_{\text{DMLDiD}} = \begin{cases} \frac{1}{n}\sum_{i=1}^{n}\psi_i & \text{(no normalization)} \\ \frac{\text{mean}(\psi_i)}{\text{Normalization factor}} & \text{(with normalization)} \end{cases}$$

```
att_normfactor <- mean(D / p_hat)
att <- att_meanpsi / att_normfactor
```

Step 7: Estimate Standard Error (Plug-in Estimator)

The standard error is calculated from the variance of ψ_i:

$$\text{SE}(\hat{\theta}) = \sqrt{\frac{1}{n}\sum_{i=1}^{n}\psi_i^2}$$

This follows from Neyman orthogonality, which ensures $\mathbb{E}[\psi_i] = \theta$ and $\text{Var}(\hat{\theta}) = \frac{1}{n}\mathbb{E}[\psi_i^2]$.

```
phihat <- ((D - g_hat) / (p_hat * (1 - g_hat))) * (DeltaY - ell_hat) / att_normfactor
se <- sqrt(mean(phihat^2) / length(phihat))  # Plug-in standard error
```

26.3.2 Double Machine Learning DiD Simulation

Let's now demonstrate a simulation that follows the step-by-step procedure outlined in the previous section. While we have already discussed the intuition and mechanics of each component in detail, this simulation puts them into practice using synthetic panel data. The code reflects the core logic of the DML-DiD estimator: it uses cross-fitting to estimate nuisance functions, compares multiple candidate models based on out-of-fold RMSE, applies trimming to avoid extreme weights, and computes a Neyman-orthogonal score to obtain a bias-robust ATT estimate. The procedure evaluates multiple candidate models for both the propensity score (e.g., Logit, Random Forest) and the outcome model for controls (e.g., Lasso, XGBoost).

There is currently no ready-to-use R package that implements this full DML-DiD pipeline with candidate model selection, trimming, and normalization. This simulation is designed to be modular and transparent so that readers can understand each part of the process and adapt or improve it for their own applications. For example, you can easily extend the code to include more candidate ML methods, adjust cross-fitting settings, or apply it to real-world panel datasets. The code also improves upon and corrects the original version by Chang (2020) by implementing a normalized version of the estimator and introducing a more flexible structure for candidate model comparison and selection. The code can serve as both a learning tool and a template for empirical implementation.

```r
# --- Load necessary libraries ---
library(randomForest); library(xgboost); library(caret); library(glmnet)

# --- Step 0: Simulate and Prepare Data ---
set.seed(123)
n <- 1000
data <- data.frame(X1 = rnorm(n), X2 = rnorm(n), X3 = rnorm(n))
data$D <- rbinom(n, 1, 1/(1 + exp(-0.5*data$X1 + 0.3*data$X2 - 0.1*data$X3)))
data$Y0 <- 2 + 0.5*data$X1 + 0.3*data$X2 + rnorm(n)
data$Y1 <- data$Y0 + 1 + 2*data$D + rnorm(n)
# --- Step 1: Compute Outcome Differences as DeltaY ---
data$DeltaY <- data$Y1 - data$Y0
# --- Step 2: Cross-Fitting Setup ---
K <- 5  # Number of folds
folds <- createFolds(data$D, k = K, list = TRUE, returnTrain = FALSE)
# Vectors to store candidate predictions and observed outcomes across all folds
g_hat_logit_all <- g_hat_rf_all <- ell_hat_lasso_all <- ell_hat_xgb_all <- rep(NA, n)
D_obs <- DeltaY_obs <- p_hat_all <- rep(NA, n)

# --- Loop over Folds ---
for(k in 1:K) {
  test_idx <- folds[[k]]
  train_idx <- setdiff(1:n, test_idx)

  train_data <- data[train_idx, ]
  test_data  <- data[test_idx, ]

  # --- 3.1 Estimate Treatment Share p_hat---
  p_hat_all[test_idx] <-mean(train_data$D)

  # --- 3.2 Estimate Propensity Score ---
  #Trained on training data, predict on test data
  #Gives the estimated probability g_hat(X_i) for each test unit using
  # Candidate 1: Logit model using glm
  logit_model <- glm(D ~ X1 + X2 + X3, family = binomial(link = "logit"),
                     data = train_data)
  g_hat_logit_all[test_idx] <- predict(logit_model, newdata = test_data,
                                       type = "response")

  # Candidate 2: Random Forest model
  rf_model <- randomForest(as.factor(D) ~ X1 + X2 + X3, data = train_data)
  g_hat_rf_all[test_idx] <- predict(rf_model, newdata = test_data,
                                    type = "prob")[,2]

#Store out-of-fold predictions and observed D_i for later comparison and scoring
  D_obs[test_idx] <- test_data$D

  # --- 3.3 Estimate Outcome Model l_hat(X) for Controls ---
  # Candidate model trained on training data using only control units (D == 0)
```

```
  control_train <- train_data[train_data$D == 0, ]

  # Candidate 1: Lasso model using cv.glmnet
  X_train_ctrl <- as.matrix(control_train[, c("X1", "X2", "X3")])
  y_train_ctrl <- control_train$DeltaY
  lasso_model <- cv.glmnet(X_train_ctrl, y_train_ctrl)
  #Prediction is made for all test units, regardless of their treatment status.
  ell_hat_lasso_all[test_idx] <- predict(lasso_model,
        newx = as.matrix(test_data[, c("X1", "X2", "X3")]), s = "lambda.min")

  # Candidate 2: XGBoost model
  X_train_ctrl_xgb <- as.matrix(control_train[, c("X1", "X2", "X3")])
  params <- list(objective = "reg:squarederror", max_depth = 3,
                 eta = 0.1, subsample = 0.8)
  xgb_model <- xgboost(data = X_train_ctrl_xgb, label = y_train_ctrl,
                       params = params, nrounds = 50, verbose = 0)
  ell_hat_xgb_all[test_idx] <- predict(xgb_model,
                       newdata = as.matrix(test_data[, c("X1", "X2", "X3")]))
  #Store predictions of l_hat and observed \deltaY_i for use in orthogonal score
  DeltaY_obs[test_idx] <- test_data$DeltaY
}

# --- Overall Candidate Selection and Trimming (after the loop) ---
#Calculate RMSE for each candidate using the full test-set predictions
#Select the candidate with the lowest RMSE
# Propensity Score Selection
rmse_prop <- c(
  sqrt(mean((D_obs - g_hat_logit_all)^2, na.rm = TRUE)),  # Model 1
  sqrt(mean((D_obs - g_hat_rf_all)^2, na.rm = TRUE))      # Model 2
)
best_prop_idx <- which.min(rmse_prop)
g_hat_final <- list(g_hat_logit_all, g_hat_rf_all)[[best_prop_idx]]
cat("Best (lowest rmse) propensity score is model:", best_prop_idx, "\n")
```

```
## Best (lowest rmse) propensity score is model: 1
```

```
# Outcome Model Selection
control_idx <- which(D_obs == 0)
rmse_outcome <- c(
  sqrt(mean((DeltaY_obs[control_idx] - ell_hat_lasso_all[control_idx])^2,
          na.rm = TRUE)),  # Model 1
  sqrt(mean((DeltaY_obs[control_idx] - ell_hat_xgb_all[control_idx])^2,
          na.rm = TRUE))     # Model 2
)
best_outcome_idx <- which.min(rmse_outcome)
ell_hat_final <- list(ell_hat_lasso_all, ell_hat_xgb_all)[[best_outcome_idx]]
# 1=Lasso, 2=XGBoost
cat("Best (lowest rmse) outcome is model:", best_outcome_idx, "\n")
```

```
## Best (lowest rmse) outcome is model: 1
```

```
# Trimming: Remove extreme propensity scores (applied on the overall predictions)
trim_idx_final <- (g_hat_final > 0.05 & g_hat_final < 0.95)
g_hat_final <- g_hat_final[trim_idx_final]
p_hat_final <- p_hat_all[trim_idx_final]  # corresponding p_hat values
D_obs_final <- D_obs[trim_idx_final]
DeltaY_final <- DeltaY_obs[trim_idx_final]
ell_hat_final <- ell_hat_final[trim_idx_final]

# --- Step 5: Compute Neyman-Orthogonal Score ---
# All components used are predictions from earlier steps
# on test data to compute the orthogonal score
psi_final <- ((D_obs_final - g_hat_final) / (p_hat_final *
          (1 - g_hat_final))) * (DeltaY_final - ell_hat_final)

# --- Step 6: Estimate ATT with Normalization ---
# Normalize by the average of D_obs_final / p_hat_final
att_normfactor <- mean(D_obs_final / p_hat_final, na.rm = TRUE)

# Final ATT is the normalized mean of the scores
ATT_est <- mean(psi_final, na.rm = TRUE) / att_normfactor

# --- Step 7: Estimate Standard Error (Plug-in Estimator) ---
SE_ATT <- sqrt(var(psi_final, na.rm = TRUE) / sum(!is.na(psi_final))) /
 att_normfactor

# --- Output the Results ---
cat("Estimated ATT (DMLDiD):", round(ATT_est, 3), "\n")
```

```
## Estimated ATT (DMLDiD): 1.966
```

```
cat("Standard Error:", round(SE_ATT, 3), "\n")
```

```
## Standard Error: 0.091
```

26.3.3 Extension of DML-DiD to Multiple Time Periods

The DML-DiD estimator, originally developed for two-period treatment-control settings, can be naturally extended to staggered treatment adoption and multiple time periods, following the framework proposed by Callaway and Sant'Anna (2021, *REStat*). Their method estimates group-time average treatment effects, ATT(g,t), where g is the period when a unit is first treated and t is any post-treatment period. For each (g,t) pair, treated units are compared to those not yet treated by time t. This structure supports flexible estimation of treatment effects that vary across groups and over time.

They estimate ATT(g,t) using doubly robust (DR) estimators that combine inverse probability weighting and outcome regression.[4]
This approach is consistent if either model is correctly specified and corrects for both covariate imbalance and potential model misspecification. As we covered doubly robust estimation

in earlier chapters, readers should find it straightforward to apply the same principles in the DiD setting for estimating ATT(g,t).

The group-time effects can be aggregated using weights reflecting the relative size of each group-time cell:

$$\text{ATT}^{\text{overall}} = \sum_{(g,t)} w_{g,t} \cdot \text{ATT}(g,t), \quad w_{g,t} = \frac{N_{g,t}}{\sum_{g',t'} N_{g',t'}} \tag{26.17}$$

The choice of weights depends on the type of overall treatment effect desired: weights can reflect group size ("simple"), time since treatment ("dynamic"), cohort-specific shares ("group"), or the distribution of observations across calendar time ("calendar"). Each weighting scheme targets a different policy-relevant estimand. Their R package `did` and Stata implementation provide built-in functions to estimate and aggregate these effects. The method and code are well-documented at https://bcallaway11.github.io/did/index.html, which offers detailed examples and practical guidance.

The same logic can be applied to extend DML-DiD to staggered settings. For each group g (first treated in period g), define the control group as those not yet treated by t. Then estimate ATT(g,t) using the DML procedure—leveraging machine learning for flexible nuisance estimation—and aggregate these using weights as in Callaway and Sant'Anna.

This extension also naturally accommodates heterogeneous treatment effects. By allowing treatment effects to vary with covariates—either by adding interaction terms in the outcome model or using machine learning models (e.g., random forests or gradient boosting) that capture non-linearities—researchers can estimate subgroup-specific effects or model how impacts vary across the covariate distribution. This makes the DML-DiD framework particularly well-suited for complex, high-dimensional, and staggered treatment settings where standard DiD methods fall short.

Applying DML-DiD in Staggered Policy Settings

Suppose a new education policy is introduced across school districts over a ten-year period. Each district adopts the policy in a different year—or not at all. We observe annual panel data on student outcomes (e.g., test scores), treatment timing, and baseline covariates (e.g., income, teacher-student ratios, demographics).

This setting is ideal for a staggered adoption DiD design. We can use the DML-DiD framework to estimate the group-time average treatment effects, ATT(g,t), while flexibly adjusting for covariates.

To apply the method in practice, you can follow these steps (or modify based on the policy structure):

1. **Organize the data**: Ensure that for each district, you have a balanced panel over 10 years with outcomes Y_{it}, treatment indicators D_{it}, and baseline covariates X_i.
2. **Identify first treatment years**: For each treated unit, define the year they first receive treatment as g. For untreated units, label them as "never treated."
3. **Loop over event years**: For each year $t = g, g+1, \ldots, 10$, estimate $\text{ATT}(\text{g}, \text{t})$.
    - Treated group: Units first treated in year g, observed at time t.
    - Control group: Units not yet treated by year t (or never treated).

4. **Use year 1 (pre-treatment) as the baseline** for identifying conditional trends. Estimate treatment effects in years 2 to 10. For dynamic effects, define event time as $t - g$ and estimate ATT for each event year (e.g., year 0, 1, 2,... after treatment).

5. **Estimate nuisance functions** using cross-fitting:

   - Propensity score: $m(X) = \mathbb{E}[D = 1 \mid X]$
   - Outcome model for controls: $g(0, X) = \mathbb{E}[\Delta Y \mid D = 0, X]$, where $\Delta Y = Y_t - Y_1$
   - Use flexible ML methods (e.g., random forests, gradient boosting) to estimate these components.

6. **Apply the DML estimator** for each (g,t) pair using the normalized Neyman-orthogonal score.

7. **Repeat steps 3–6 for each treated cohort** (e.g., those treated in year 2, year 3, ..., year 9) and each post-treatment period. In this way, you compute ATT(g,t) for all groups and time periods.

8. **Aggregate** the ATT(g,t) estimates using appropriate weights:

$$\text{ATT}^{\text{overall}} = \sum_{(g,t)} w_{g,t} \cdot \text{ATT}(g, t)$$

   Weights can reflect group size ("simple"), time since treatment ("dynamic"), cohort shares ("group"), or calendar time ("calendar").

9. **Conduct placebo checks**: Use a pre-treatment year (e.g., year 1) as the reference and test for a false treatment effect in an earlier period (e.g., year 0), or use a placebo group that should not be affected by the treatment (e.g., a non–school-age population). These checks help assess the plausibility of the conditional parallel trends assumption, as is standard in Difference-in-Differences analyses.

10. **Visualize results**: Plot the estimated ATT(g,t) over time to assess the treatment effect dynamics. This can help identify whether effects are immediate, gradual, or delayed.

26.4 Time-varying Continuous Treatments

In many empirical settings, treatment is continuous—examples include pollution exposure, tax rates, or infrastructure intensity—requiring generalizations of the standard difference-in-differences (DiD) framework. Despite its widespread use in empirical research, the theoretical foundation for continuous DiD remains relatively underdeveloped, particularly in comparison to the extensive body of literature on DiD with binary or discrete treatments. Recently, a few recent studies have begun to address this gap. We will discuss the most related ones in this section.

Callaway, Goodman-Bacon, and Sant'Anna (2024) show that two-way fixed effects (TWFE) models like

$$Y_{it} = \alpha_i + \lambda_t + \beta^{TWFE}(D_i \cdot Post_t) + \varepsilon_{it} \tag{26.18}$$

can yield invalid or uninterpretable estimates when treatment effects vary by dose, due to irregular weighting and non-monotonic aggregation. To address this, they propose a sieve-based nonparametric estimator using B-spline basis functions to approximate $d \mapsto \mathbb{E}[\Delta Y \mid D = d]$, then subtracting $\mathbb{E}[\Delta Y \mid D = 0]$ to estimate dose-specific treatment effects:

$$\widehat{ATT}(d) = \widehat{\mathbb{E}}[\Delta Y \mid D = d] - \widehat{\mathbb{E}}[\Delta Y \mid D = 0]. \tag{26.19}$$

This method achieves minimax-optimal convergence rates and supports uniform inference. Their simulations confirm the estimator performs well across a range of treatment heterogeneity and sample sizes, with minimal bias and good coverage when the spline dimension is chosen via their proposed algorithm (drawing on the method of Chen, Christensen, and Kankanala, 2023).

In their empirical analysis, Callaway, et.al.(2024) reanalyze data from Acemoglu and Finkelstein (2007) on the effects of Medicare hospital spending using their continuous DiD framework. They show that standard two-way fixed effects estimates substantially differ from their nonparametric dose-specific treatment effects, highlighting that the TWFE model understates both the magnitude and heterogeneity of the true policy impact. Their approach reveals a more flexible and interpretable dose-response relationship between Medicare spending intensity and outcomes.

However, covariates are not incorporated into their main estimator. While their supplementary appendix discusses a conditional version of strong parallel trends (SE.3) that could identify conditional ATTs, no estimation procedure is proposed. This gap—between theoretical identification and practical implementation—points naturally to recent work using DML, which allows for robust covariate adjustment even in high-dimensional settings and offers an avenue for extending continuous treatment DiD to richer, more flexible empirical environments.

Zhang (2025) extends the difference-in-differences framework to accommodate continuous treatments by incorporating covariates nonparametrically into both identification and estimation, adapting ideas from Abadie (2005). The ATT is identified through a two-step process: first, estimating nuisance parameters such as the conditional density of treatment; second, substituting these into a weighted average to estimate the causal parameter. While machine learning offers flexibility in estimating these nuisance components, it can introduce substantial bias, especially in high-dimensional settings. To address this, Zhang adopts the double/debiased machine learning (DML) framework of Chernozhukov et al. (2018), using orthogonalization and cross-fitting to reduce overfitting and bias. As an application, the paper re-examines Acemoglu and Finkelstein's (2008) study of the 1983 Medicare PPS reform, where the share of Medicare inpatients serves as a continuous treatment variable. Zhang finds heterogeneous effects of the reform across hospitals, in contrast to the original linear estimates, and compares these findings to those in Callaway et al. (2024). Further details and replication resources are available on the author's webpage: https://lucaszz-econ.github.io/notes/Continuous_DiD.html.

To estimate the average treatment effect on the treated (ATT, or ATET as denoted in the follow-up paper) in settings with time-varying continuous treatments, we follow the methodology developed by Haddad, Huber, and Zhang (2024). In their working paper, "Difference-in-Differences with Time-varying Continuous Treatments using Double/Debiased Machine Learning", the authors build on Zhang (2025)'s two-period continuous DiD model and extend it in several important directions. Specifically, they allow (i) comparisons between two non-zero treatment doses, (ii) more than two time periods with varying treatment doses across time (as in staggered adoption settings), and (iii) a flexible treatment assignment mechanism where the treatment dose depends on both time and covariates. These extensions support settings where treatment intensity may vary across both units and time.

The framework applies to both panel data and repeated cross-section data. For implementation, the authors' `causalweight` package handles these estimators directly. If users are primarily interested in applying the method, functions like `didcontDMLpanel` and `didcontDML` offer convenient, general-purpose tools requiring only a minimal set of user inputs. While the paper presents several conditional and average ATET estimators, and introduces multiple notation-heavy expressions, the core logic remains consistent: use kernel weights and machine learning-based nuisance parameter estimation to compare treated and reweighted control units.

The conditional ATET of treatment dose $d > 0$ in period $t > 0$, given covariates X, is identified as:

$$\begin{aligned}\mathbb{E}[Y_t(d) - Y_0(0) \mid D = d, X] &= \mathbb{E}[Y_t \mid D = d, X] - \mathbb{E}[Y_0 \mid D = d, X] \\ &\quad - \{\mathbb{E}[Y_t \mid D = 0, X] - \mathbb{E}[Y_0 \mid D = 0, X]\}\end{aligned} \tag{26.20}$$

Averaging over the distribution of X among the treated in period t yields the ATET for that period. Importantly, the proposed estimator allows for dynamic treatment effects, meaning past treatment doses D_{t-1} can influence current outcomes. Still, the approach identifies the marginal effect of a change in treatment at time t, net of historical exposure.

The estimator is based on observed outcome differences $\Delta Y = Y_t - Y_c$, comparing units across treatment intensities using kernel weights centered around specified treatment values. Let w_i^1 and w_i^0 denote the normalized kernel weights for unit i around the treated and control levels. Estimation proceeds by predicting the potential outcome under control $\hat{\mu}_0(X_i) = \mathbb{E}[Y_i(D = 0) \mid X_i]$. That is similar to $\hat{\ell}(X_i)$ in the previous section. In addition, generalized propensity scores (GPS)—defined as conditional treatment densities—are estimated instead of binary propensity scores. For each treatment level d, $\rho(d, X_i)$ captures the likelihood of unit i receiving dose d given covariates X_i. Separate models are used to estimate $\hat{p}_1(X_i)$ and $\hat{p}_0(X_i)$ for the treated and control levels, respectively [5].

All nuisance functions—outcome predictions and GPS—are estimated using k-fold cross-fitting to reduce overfitting. Letting $\Delta y_i = Y_{it} - Y_{i,t-1}$, the kernel weights $\alpha_i = w_i^1$ for treated units and reweighted control weights $\beta_i = w_i^0 \cdot \frac{\hat{p}_1(X_i)}{\hat{p}_0(X_i)}$ yield the following normalized ATT estimator:

$$\hat{\tau}_{ATT} = \frac{1}{\sum \alpha_i} \sum \alpha_i \left(\Delta y_i - \hat{\mu}_0(X_i)\right) - \frac{1}{\sum \beta_i} \sum \beta_i \left(\Delta y_i - \hat{\mu}_0(X_i)\right) \tag{26.21}$$

This expression gives the difference in residualized outcomes between kernel weighted treated units and a kernel and GPS-reweighted control group [6]:

The `causalweight` package provides a user-friendly interface for implementing these methods in R. Package link:10.32614/CRAN.package.causalweight[7]

Internally, the function computes kernel weights, estimates nuisance parameters using [SuperLearner] (https://cran.r-project.org/web/packages/SuperLearner/index.html), performs cross-fitting, and calculates the ATET using the DML framework. Optional trimming and clustered standard errors are supported. Altogether, this method offers a robust and flexible framework for causal inference with continuous, time-varying treatments in observational panel and cross-section data.

Below is an example using panel data:

```
# Install the package from GitHub (if not already installed)
# devtools::install_github("fhernanb/causalweight")

library(causalweight)

# Load your panel dataset
# Assume df is a data.frame with columns:
# y0, y1: outcome at t0 and t1
# d: treatment in period t1 (continuous)
# x1, x2, ...: covariates

# Construct outcome difference; no missing values.
df$ydiff <- df$y1 - df$y0

# Estimate ATET using didcontDMLpanel
results <- didcontDMLpanel(
  ydiff     = df$ydiff,
  d         = df$d,
  t         = rep(1, nrow(df)),         # time variable (set to 1 for all in 2-period
                                        # panel)
  dtreat    = 1,                        # treatment level
  dcontrol  = 0,                        # control level
  controls  = df[, c("x1", "x2")],      # covariates
  MLmethod  = "lasso",                  # or "randomforest", etc.
  k         = 3                         # number of cross-fitting folds
)

# View the results
cat("ATET estimate:", round(results$ATET, 3),
    "\nStandard Error:", round(results$se, 3),
    "\nP-value:", round(results$pval, 4))
```

For repeated cross-sections, use `didcontDML()` and provide outcome and time variables separately, along with other parameters. The function handles cross-fitting, kernel weighting, and GPS estimation internally—offering an easy interface to apply the advanced estimators described in the paper.

For Python users, the same estimator can be implemented using the `DoubleML` package, which supports Difference-in-Differences models as part of its flexible API. The implementation is described in detail at https://docs.doubleml.org/stable/guide/models.html#difference-in-differences-models-did. It allows users to specify outcome, treatment, and time variables, and seamlessly integrates cross-fitting and orthogonal score computation for valid inference using machine learning–based nuisance estimation.

Before we end this section, we want to introduce another estimator that can be used in this ongoing stream of methodological developments. There is another recent paper by Hettinger et al. (2025) that introduces a multiply robust estimator for causal effect curves in DiD designs with continuous exposures. They define an analogous estimand called the Average Dose Effect on the Treated (ADT), given by:

$$\Psi(\delta) = \mathbb{E}[Y_1^{(1,\delta)} - Y_1^{(0,\emptyset)} \mid A = 1] \tag{26.22}$$

This expression represents the average causal effect on the treated group if all treated units had received exposure level δ (same as ATT in this section). In Hettinger et al. (2025), A indicates treatment status (1 = treated, 0 = control), while D represents a continuous exposure or dosage level received by treated units. The goal is to estimate how treatment effects vary with D, extending standard ATT to a dose-response setting.

The authors propose an estimator that is multiply robust—consistent if at least one model from each of two pairs (treatment/exposure models and outcome models) is correctly specified. They build the estimator using a decomposition of the efficient influence function (EIF) of a related stochastic estimand and then recover $\Psi(\delta)$ through a regression step. Specifically, they define nuisance functions such as: - $\pi_A(X) = P(A = 1 \mid X)$ - $\pi_D(\delta \mid X, A = 1) =$ density of dose given covariates among the treated, - $\mu_{1,\Delta}(D, X) = \mathbb{E}[Y_1 - Y_0 \mid A = 1, D, X]$ - $\mu_{0,\Delta}(X) = \mathbb{E}[Y_1 - Y_0 \mid A = 0, X]$

Then they construct two key components:

- a dose-specific effect estimator using inverse probability weighting and outcome regression among the treated group,
- a counterfactual trend estimator using both treated and control groups.

The estimated causal effect curve is obtained by regressing a pseudo-outcome on dose level using a flexible nonparametric smoother (local linear regression). This structure is similar to doubly robust estimators you may already know (Abadie, 2005; Sant'Anna and Zhao, 2020) and connects naturally to the DML-DiD framework. They also provide standard errors and confidence intervals using two approaches: a sandwich variance estimator and a nonparametric weighted bootstrap. Both methods are applicable to their multiply robust estimator and enable valid inference for the estimated effect curves.

Although the paper is notation-heavy, the intuition aligns with familiar concepts from semiparametric efficiency theory and DiD with confounder adjustment. Importantly, the authors provide a full implementation, including code and examples, at: https://github.com/garyhettinger/DiD-continuous. We do not go into further detail here to avoid lengthening this section, but we flag this work given its relevance, flexibility, and recency—it was published just two weeks before finalizing this chapter.

In this section, we introduced the DML-DiD estimator, demonstrated its implementation with a simulation, and showed how it can be extended to staggered treatment settings using group-time ATT estimation. With these tools in place, we now turn to Synthetic Control methods as another strategy for causal identification.

Notes

1. Random Effects (RE) Models: The random effects (RE) model assumes that unit-specific effects α_i are uncorrelated with explanatory variables ($E(\alpha_i|X_{it}) = 0$). If true, RE estimation is more efficient than fixed effects, as it retains between-unit variation. However, if $\text{Cov}(\alpha_i, X_{it}) \neq 0$, RE estimates are biased, making fixed effects preferable when unobserved heterogeneity is a concern.

Dynamic Panel Models: When past values of the dependent variable influence future outcomes, dynamic panel models introduce lagged dependent variables:

$$y_{it} = \rho y_{i,t-1} + \beta X_{it} + \alpha_i + \gamma_t + \varepsilon_{it}. \tag{26.23}$$

OLS or fixed effects estimation in this case leads to bias when T is small, as $y_{i,t-1}$ is correlated with the error term. Methods like Generalized Method of Moments (GMM) address this bias but add complexity and require strong assumptions. In panel settings, especially with long T, serial correlation in errors ($E(\varepsilon_{it}, \varepsilon_{i,t-1}) \neq 0$) is common, requiring robust standard errors or techniques like generalized least squares (GLS) for correction. These topics are covered in standard econometrics texts, while this section focuses on estimating causal/treatment effects.

2. DiD%20Resources

3. The orthogonal score function for the repeated outcomes case, shown as Equation (3.1) in Chang (2020), is:

$$\begin{aligned}\psi_1(W, \theta_0, p_0, \eta_{10}) = & \frac{Y(1) - Y(0)}{P(D=1)} \cdot \frac{D - P(D=1 \mid X)}{1 - P(D=1 \mid X)} - \theta_0 \\ & - \frac{D - P(D=1 \mid X)}{P(D=1)(1 - P(D=1 \mid X))} \cdot \mathbb{E}[Y(1) - Y(0) \mid X, D=0]\end{aligned}$$

This score function is constructed to be Neyman orthogonal, meaning its first-order derivative with respect to the nuisance components is zero. That is, small estimation errors in the first-stage nuisance parameters—specifically the propensity score $g_0(X) = P(D = 1 \mid X)$ and the outcome regression function $\ell_{10}(X) = \mathbb{E}[Y(1) - Y(0) \mid X, D = 0]$—do not affect the leading term of the estimator's distribution. This orthogonality enables valid inference even when these nuisance functions are estimated using flexible, high-dimensional ML methods.

4. The doubly robust estimator for ATT(g,t) is

$$\widehat{\text{ATT}}^{\text{DR}}(g,t) = \frac{1}{n_g} \sum_{i:G_g=1} \left(Y_{it} - \widehat{\mu}_0(X_i, t)\right) - \sum_{i:C=1} \frac{\widehat{p}_g(X_i)}{1 - \widehat{p}_g(X_i)} \left(Y_{it} - \widehat{\mu}_0(X_i, t)\right),$$

where Y_{it} is the observed outcome, $\widehat{\mu}_0(X_i, t)$ is the predicted outcome for untreated units based on covariates X_i, and $\widehat{p}_g(X_i)$ is the estimated propensity score for being in group g (treated at time g) versus the control group. The first term estimates the outcome for treated units relative to their counterfactuals, while the second term adjusts for differences using reweighted outcomes among controls.

5. When treatment D is continuous (e.g., hours of training, dosage, subsidy amount), we replace probabilities with conditional densities. The generalized propensity score is: $\rho(d, X) = f_{D|X}(d \mid X)$. That is, the conditional density of receiving treatment level d given covariates X. In continuous treatment settings, units at different treatment levels may systematically differ in covariates. GPS helps reweight comparisons, mimicking randomized assignment and enabling causal estimation—much like binary propensity scores do in standard settings.

6. The exact panel data estimator appears and used in step 4 in DML algorithm in Haddad et al. (2024), and is written as:

$$\begin{aligned}\hat{\Delta}^h_{d_t, d'_t, t} = \sum_{i=1}^{n} \Bigg[& \underbrace{\frac{\omega(D_{i,t}; d_t, h) \cdot [Y_{i,t} - Y_{i,t-1} - \hat{m}_{d'_t}(t, t-1, \mathbf{D}_{i,t-1}, X_i)]}{\sum_{i=1}^{n} \omega(D_{i,t}; d_t, h)}}_{\text{Treated part}} \\ & \underbrace{- \frac{\omega(D_{i,t}; d'_t, h) \cdot \hat{\rho}_{d_t}(\mathbf{D}_{i,t-1}, X_i) \cdot [Y_{i,t} - Y_{i,t-1} - \hat{m}_{d'_t}(t, t-1, \mathbf{D}_{i,t-1}, X_i)] / \hat{\rho}_{d'_t}(\mathbf{D}_{i,t-1}, X_i)}{\sum_{i=1}^{n} \omega(D_{i,t}; d'_t, h) \cdot \hat{\rho}_{d_t}(\mathbf{D}_{i,t-1}, X_i) / \hat{\rho}_{d'_t}(\mathbf{D}_{i,t-1}, X_i)}}_{\text{Reweighted control part}} \Bigg]\end{aligned}$$

where $\omega(D_{i,t}; d, h)$ are kernel weights; $\hat{m}_{d'_t}$ is the predicted outcome under control; $\hat{\rho}_{d_t}, \hat{\rho}_{d'_t}$ are generalized propensity scores (GPS), i.e., conditional treatment densities.

7. https://doi.org/10.32614/CRAN.package.causalweight

27

Synthetic DiD and Regression Discontinuity

This chapter continues our focus on causal inference under selection on unobservables and serves as the final chapter in this sequence. In the previous chapter, we examined difference-in-differences (DiD) and its extensions—including double machine learning for DiD—as strategies for identifying treatment effects when unobserved confounding may bias estimates. Building on that foundation, we now expand the causal inference toolkit to include methods that construct more flexible and data-driven counterfactuals.

We begin with a class of techniques that estimate counterfactual outcomes by explicitly combining information from untreated units in observational panel data settings. These methods—starting with the original synthetic control method (SCM) and extending to generalized SCM (GSCM), augmented SCM (ASCM), and synthetic DiD (SDiD)—offer alternatives to traditional approaches like DiD when standard assumptions (e.g., parallel trends) may not hold. At their core, these approaches aim to improve causal inference by flexibly weighting control units and/or time periods to approximate the untreated outcome trajectory for treated units. They have become increasingly popular for evaluating policies, shocks, and interventions in cases where randomized experiments are infeasible and only a limited number of treated units are available. This section introduces each method in turn, highlighting their assumptions, estimation strategies, and practical applications.

The chapter concludes with regression discontinuity designs (RDD), a powerful quasi-experimental method that exploits sharp or fuzzy thresholds in treatment assignment. RDD provides credible causal estimates near the cutoff point under minimal assumptions, making it one of the most widely used non-experimental designs. We review both sharp and fuzzy RDD frameworks, discuss practical estimation issues like bandwidth selection and functional form, and provide simulation-based examples to reinforce key ideas. We also highlight recent extensions that integrate machine learning tools for covariate adjustment and variance reduction. By the end of this chapter, readers will have a comprehensive understanding of advanced, design-based methods that address selection on unobservables using both structural and flexible, data-adaptive techniques. Throughout the chapter, we revisit tools and concepts from earlier parts of the book—including machine learning estimators, panel data structure, and local estimation—to show how they integrate into these advanced designs for causal inference.

27.1 The Synthetic Control Method

In empirical economics and related fields, identifying credible causal effects of interventions often faces a major challenge: the absence of a valid counterfactual. While randomized experiments are ideal for causal inference, they are rarely feasible in large-scale policy contexts. The SCM, introduced by Abadie and Gardeazabal (2003) and extended by Abadie,

DOI: 10.1201/9781003504795-27

Diamond, and Hainmueller (2010, 2015), offers a transparent, data-driven approach for such situations. SCM constructs a synthetic control unit—a weighted average of untreated units—that approximates the trajectory the treated unit would have followed in the absence of intervention.

SCM is particularly suited to comparative case studies where a single unit, such as a country or region, is uniquely exposed to a treatment, and no individual control unit alone serves as a good counterfactual. Instead of selecting a comparator arbitrarily, SCM determines weights on multiple control units so that the synthetic control matches the treated unit's pre-intervention characteristics, including lagged outcomes and relevant covariates. A good pre-treatment fit implies that any post-treatment divergence between the treated unit and its synthetic control can be interpreted as the causal effect.

Unlike difference-in-differences methods, SCM does not rely on a parallel trends assumption. It constructs the counterfactual directly, minimizing pre-treatment discrepancies. Because SCM typically focuses on a single treated unit, traditional inference methods are not applicable. Instead, SCM uses placebo tests: the method is applied to each control unit as if it were treated, generating a distribution of placebo effects. If the observed treatment effect is unusually large relative to this distribution, it supports a causal interpretation.

SCM results are commonly shown through plots comparing the treated and synthetic control paths over time, or by displaying the gap between them. These visuals, especially when pre-treatment fit is tight and post-treatment divergence is clear, provide intuitive and persuasive evidence. The post-/pre-treatment MSPE ratio is often used to assess the strength of the estimated effect.

What distinguishes SCM is that it constructs rather than assumes a counterfactual. The synthetic control is explicitly designed to match the treated unit before the intervention using a convex combination of control units. This makes SCM especially useful when there is only one treated unit and no randomized assignment.

Estimation Procedure (Simplified): The SCM is applied in panel data settings where we observe a treated unit and several untreated units over time. Suppose we have a balanced panel with $J+1$ units and T periods. Unit 1 is exposed to a treatment beginning at time T_0+1, while the remaining J units are potential controls. Let Y_{it}^N denote the potential outcome for unit i at time t without treatment, and Y_{it}^I the potential outcome under treatment. For $t \leq T_0$, we observe $Y_{it} = Y_{it}^N$ for all units, and for $t > T_0$, $Y_{1t} = Y_{1t}^I$. The treatment effect is defined as:

$$\alpha_{1t} = Y_{1t}^I - Y_{1t}^N, \quad t > T_0 \tag{27.1}$$

Our goal is to estimate \(Y_{1t}N\)—the counterfactual outcome the treated unit would have experienced without treatment—so we can recover the treatment effect \(α_{1t}\). SCM approximates this counterfactual by constructing a weighted average of the outcomes of the \(J\) control units:

$$\hat{Y}_{1t}^N = \sum_{j=2}^{J+1} w_j Y_{jt}, \quad t = 1, \ldots, T \tag{27.2}$$

The weights w_j are non-negative and sum to one, making the synthetic control a convex combination of controls. This ensures interpretability and avoids extrapolation.[1] The estimated treatment effect is then:

$$\hat{\alpha}_{1t} = Y_{1t} - \hat{Y}_{1t}^N \tag{27.3}$$

For SCM to be credible, several conditions must be met. First, the treated unit's characteristics must lie within the convex hull of the controls—otherwise, it cannot be closely approximated. Second, the method assumes no anticipation, so treatment does not influence outcomes before it begins. Third, there should be no interference, meaning the treatment affects only the treated unit.

Practical Implementation

One of the most widely cited and instructive applications of the SCM is the study by Abadie and Gardeazabal (2003), which examined the economic impact of terrorism in the Basque Country. The `Synth` R package, developed to operationalize this method, includes this very dataset as a built-in example, allowing users to replicate and explore the methodology in detail. This makes the Basque Country case a compelling entry point for learning SCM.

The core idea is to estimate the causal impact of terrorism by comparing the actual trajectory of per capita GDP in the Basque Country with that of a synthetic control—constructed from a weighted combination of other Spanish regions that were unaffected by terrorism. By aligning the synthetic control with the Basque Country's pre-treatment trajectory using education, investment, and demographic indicators, the method allows for a meaningful comparison after terrorism began to escalate around 1970.

```
library(Synth)
data(basque) # Load the data
# Set time parameters
pre_period <- 1960:1969
post_period <- 1970:1997
# Prepare input
dataprep.out <- dataprep(
  foo = basque,
  predictors = c("school.illit", "school.prim", "school.med",
                 "school.high", "invest", "popdens"),
  predictors.op = "mean",
  time.predictors.prior = pre_period,
  special.predictors = list(
    list("gdpcap", 1960:1969, "mean"),
    list("popdens", 1969, "mean")
  ),
  dependent = "gdpcap",
  unit.variable = "regionno",
  time.variable = "year",
  treatment.identifier = 17,
  controls.identifier = setdiff(unique(basque$regionno), 17),
  time.optimize.ssr = pre_period,
  time.plot = 1955:1997
)
synth.out <- synth(dataprep.out)


# Plot treated vs synthetic
path.plot(synth.res = synth.out, dataprep.res = dataprep.out,
          Main = "Basque Country: Treated vs Synthetic GDP per Capita",
          Ylab = "GDP per Capita", Xlab = "Year")
```

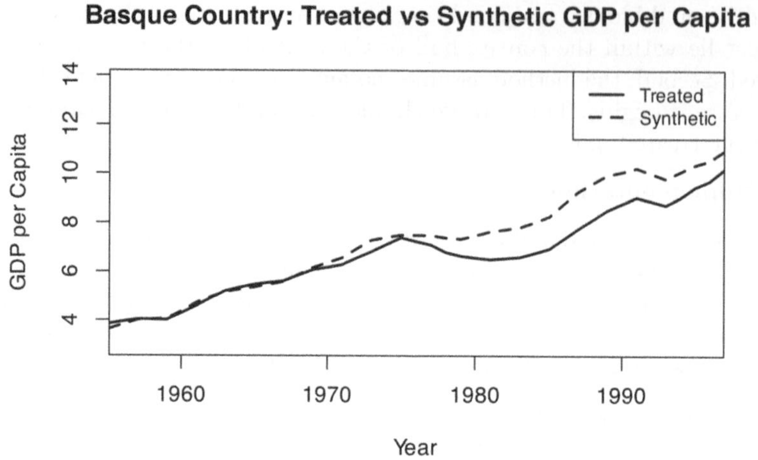


```
# Show the treatment effect (gap)
gaps.plot(synth.res = synth.out, dataprep.res = dataprep.out,
        Main = "Gap (Basque - Synthetic)", Ylab = "GDP Difference")
```

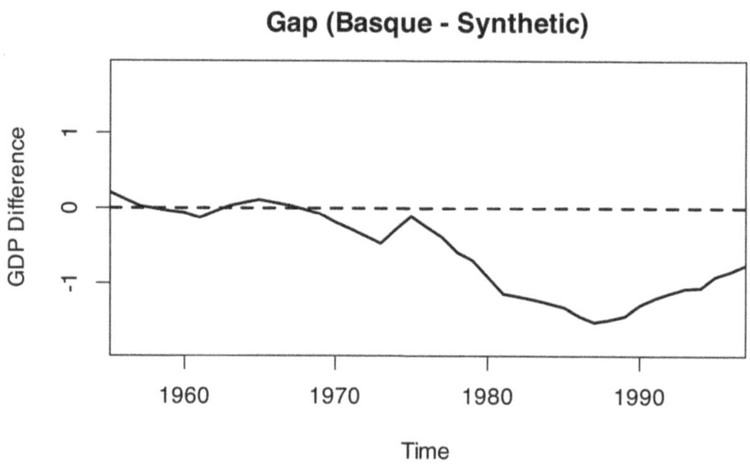


```
rm(list = ls())
```

The output typically includes two central plots. The first shows the paths of actual and synthetic GDP per capita over time, which ideally align closely before the intervention. The second plot displays the gap between the two series (i.e., the estimated treatment effect), revealing a substantial decline in the Basque Country's GDP relative to its synthetic counterpart after terrorism intensified. The tight pre-treatment match provides visual credibility, while the post-treatment divergence illustrates the estimated causal impact.

This example serves as an effective demonstration of how SCM constructs a counterfactual, assesses treatment effects, and visualizes the results. It also illustrates the core strengths of the method: its transparency, its minimal reliance on functional form assumptions, and its ability to handle one-time, region-specific interventions.

For readers interested in implementing SCM or further exploring its capabilities, the sources in the footnote provide comprehensive guidance and reproducible code.[2] While this chapter continues with more recent developments in this method, the materials listed in the footnote form a clear and practical foundation for learning the method and applying it to new contexts.

The SCM has been applied in a range of studies evaluating the effects of policy changes, shocks, and interventions when only a single treated unit is available. Abadie and Gardeazabal (2003) used SCM to quantify the economic impact of terrorism in the Basque Country, finding a significant reduction in per capita GDP relative to a synthetic Spain without terrorism. In a policy context, Abadie, Diamond, and Hainmueller (2010) analyzed the effect of California's tobacco control program, showing that smoking declined substantially compared to the synthetic control. Cavallo et al. (2013) examined the macroeconomic consequences of natural disasters, showing that major events like earthquakes can have large and persistent negative impacts. Bohn et al. (2014) applied SCM to study the impact of immigration enforcement, concluding that stricter enforcement policies did not reduce violent crime. In the health sector, Kreif et al. (2016) evaluated English health system reforms and found evidence of improved performance in treated areas relative to synthetic comparisons. Building on the foundations of SCM, researchers have developed extensions such as the generalized SCM (GSCM) to better accommodate settings with multiple treated units, unobserved heterogeneity, and time-varying confounders.

27.1.1 Generalized Synthetic Control Method

Building on the foundations of SCM, the GSCM extends the approach to settings involving multiple treated units, unobserved heterogeneity, and time-varying confounders. Developed by Xu (2017) and implemented in the `gsynth` R package, GSCM accommodates more complex treatment structures than the original SCM framework. For readers interested in deeper exploration of the method and its implementation, we refer to the original papers and documentation.[3]

Unlike classical SCM, which constructs a weighted average of control units to match the treated unit in the pre-treatment period, GSCM employs a matrix completion framework with interactive fixed effects. Potential outcomes are modeled as a combination of latent factors along with unit- and time-specific effects, allowing the method to flexibly account for unobserved heterogeneity and time-varying trends. The estimation algorithm iteratively recovers these latent components and estimates treatment effects by comparing observed outcomes with model-predicted counterfactuals. GSCM supports multiple treated units, staggered treatment adoption, dynamic effects, and provides tools for cross-validation to determine the number of factors. It also includes options for computing standard errors and conducting inference.

```
library(gsynth)
data(gsynth)
out <- gsynth(Y ~ D + X1 + X2, data = simdata,
              index = c("id", "time"),
              force = "two-way",
              CV = TRUE, r = c(0, 5),
              se = TRUE, inference = "parametric",
              nboots = 1000, parallel = FALSE)
```

```
# Plot average treatment effect
plot(out)
```

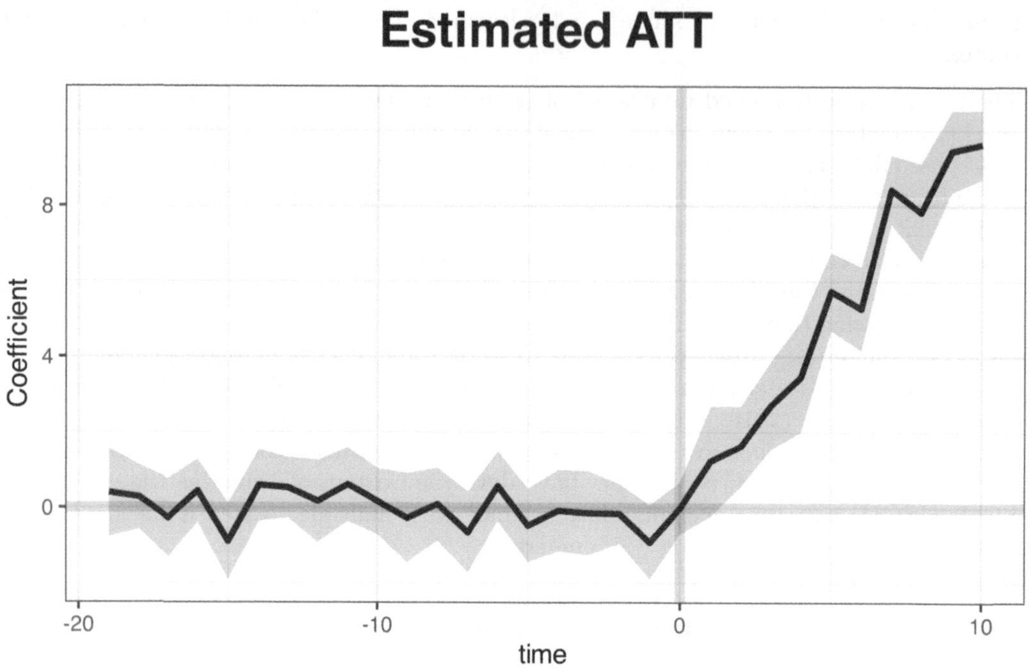


```r
# Plot treatment gap
plot(out, type = "gap", ylim = c(-3, 12), xlab = "Period",
     main = "Simulation GSynth Plot")
```

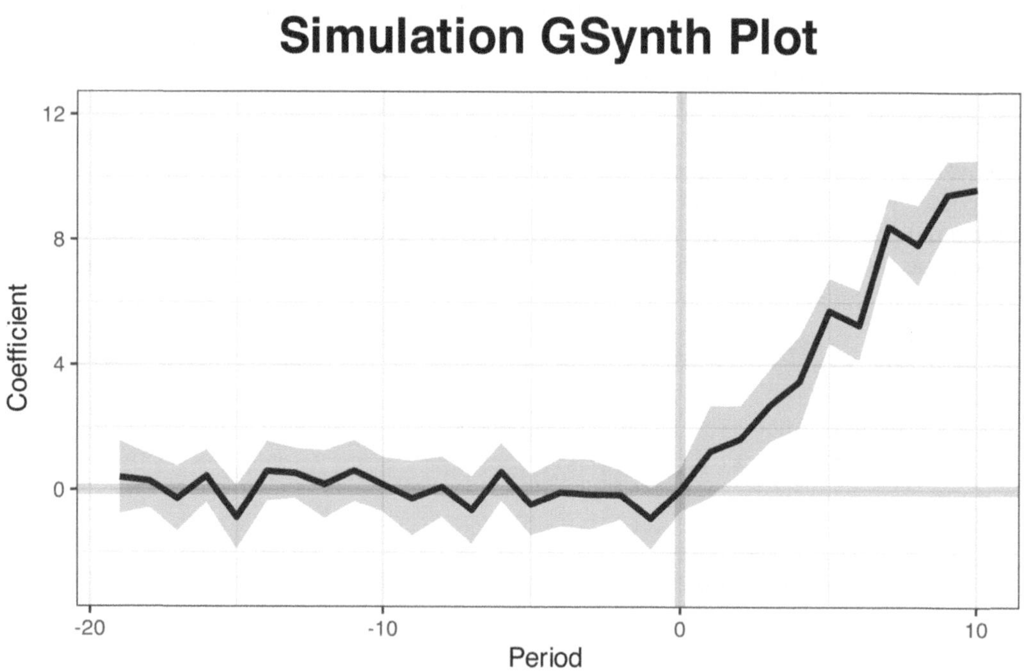

In this simulation, the `simdata` dataset includes 50 units observed over 30 periods, with five units treated from period 21 onward. Covariates `X1` and `X2`, along with a treatment indicator `D`, help explain the outcome variable `Y`. The model uses a two-way fixed effects specification and cross-validation to select the number of latent factors. Once estimated, the results are visualized using built-in plotting tools. The first plot displays the estimated average treatment effects over time, while the second shows the gap between observed and predicted outcomes for the treated units. These visualizations allow researchers to assess both the dynamics and the significance of the treatment effect.

GSCM is particularly powerful when the number of treated units is large or treatment timing varies across units. Its ability to model unobserved heterogeneity and account for complex temporal dynamics makes it a natural and robust extension of SCM for modern panel data applications.

A natural progression from GSCM is the augmented SCM (ASCM), which combines the original SCM approach with outcome modeling to improve pre-treatment fit and reduce extrapolation bias.

27.1.2 Augmented Synthetic Control Method

The ASCM builds on the original SCM framework by incorporating outcome modeling to improve pre-treatment fit and reduce potential extrapolation bias. Introduced by Ben-Michael, Feller, and Rothstein (2021), ASCM combines SCM with a parametric or semi-parametric model—often ridge regression—to estimate counterfactual outcomes more robustly, particularly when the treated unit lies outside the convex hull of the control units. This addresses a key limitation of SCM: its inability to match the treated unit well when no convex combination of controls approximates its pre-treatment behavior.

Let Y_{it} be the observed outcome for unit i at time t, and suppose unit 1 is treated beginning at time $T_0 + 1$. The goal is to estimate the counterfactual outcome Y_{1t}^N for $t > T_0$. As in SCM, ASCM constructs a weighted combination of control units:

$$\hat{Y}_{1t}^N = \sum_{j=2}^{J+1} w_j^{\text{ascm}} Y_{jt} \tag{27.4}$$

where $w_j^{\text{ascm}} = w_j^{\text{scm}} + w_j^{\text{adj}}$. Here, w_j^{scm} are the SCM weights that minimize the pre-treatment discrepancy between the treated unit and the synthetic control, and w_j^{adj} are adjustment terms obtained by fitting a penalized regression model to the residuals.[4] Unlike SCM, which imposes $w_j \geq 0$ and $\sum w_j = 1$, ASCM relaxes these constraints, allowing for negative weights and weights greater than one. This flexibility enables better prediction accuracy, though it comes with a trade-off in terms of interpretability. The final estimate of the treatment effect is:

$$\hat{\alpha}_{1t} = Y_{1t} - \hat{Y}_{1t}^N, \quad t > T_0 \tag{27.5}$$

ASCM supports inference through placebo-based standard errors and confidence intervals. By modeling pre-treatment residuals and incorporating them into post-treatment counterfactual prediction, ASCM yields more accurate and robust effect estimates, especially when SCM fails to achieve a close pre-treatment match.

The method is implemented in the `augsynth` R package (see footnote), which allows users to fit models with ridge augmentation, visualize results, and conduct sensitivity analysis.[5]

ASCM thus represents a powerful generalization of SCM, retaining its transparency while gaining flexibility through model-based augmentation. It is especially useful in applications where pre-treatment imbalance remains even after SCM weighting.

ASCM can be implemented in two key modes: `singlesynth` and `multisynth`. The `singlesynth` approach applies when there is a single treated unit, and the counterfactual is constructed using a weighted combination of control units with ridge-augmented adjustments. In contrast, `multisynth` extends ASCM to settings with staggered treated units. The `augsynth` package supports both approaches, allowing users to flexibly choose based on the number and structure of treated units in their application.

To illustrate the use of the ASCM, we consider the effects of the 2012 personal income tax cuts in Kansas, using the built-in `kansas` dataset from the `augsynth` package. The outcome of interest is log GSP per capita (`lngdpcapita`) across U.S. states, observed quarterly from 1990 to 2016. Kansas enacted substantial tax cuts in the second quarter of 2012, and we aim to estimate the causal impact of this policy on its economic performance.

We first estimate a standard synthetic control model (not shown), which uses only pre-treatment outcomes to construct a counterfactual. We then apply ASCM by augmenting the synthetic control with a ridge regression model:

```
## Install devtools if noy already installed
##install.packages("devtools", repos='http://cran.us.r-project.org')
##library(devtools)
## Install augsynth from github
##devtools::install_github("ebenmichael/augsynth", dependencies = TRUE)
library(augsynth)
data(kansas)

#syn <- augsynth(lngdpcapita ~ treated, fips, year_qtr, kansas,
#                progfunc = "None", scm = TRUE)
# plot(syn)
asyn <- augsynth(lngdpcapita ~ treated, fips, year_qtr, kansas,
                progfunc = "Ridge", scm = TRUE)
```

```
## One outcome and one treatment time found. Running single_augsynth.
```

```
plot(asyn)
```

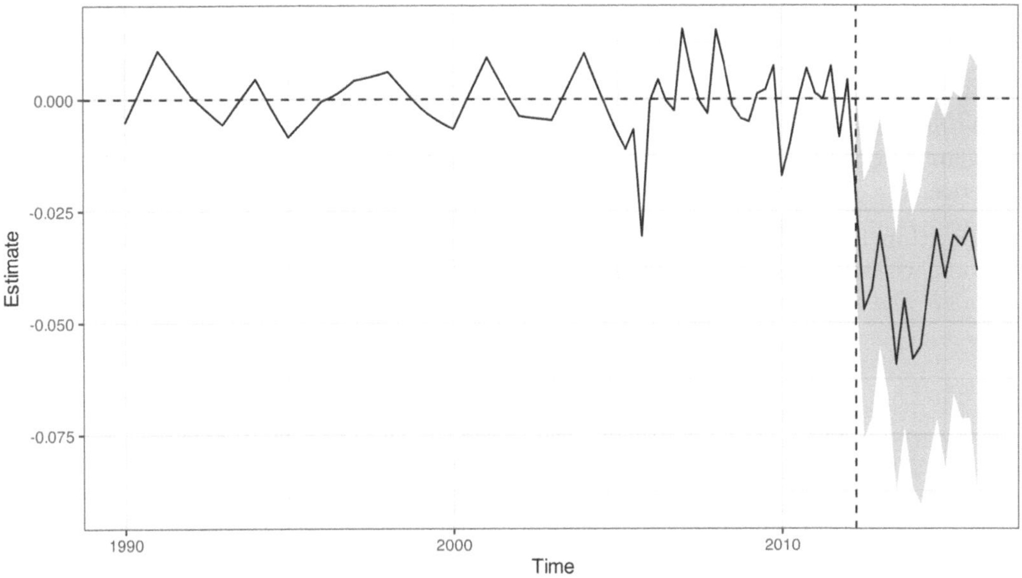

ASCM improves pre-treatment fit by adjusting residual differences between Kansas and its synthetic control, allowing for limited extrapolation beyond the convex hull of controls, regulated by a ridge penalty term. In this example, it results in clearer post-treatment effects and narrower confidence intervals, underscoring the method's advantages when traditional SCM fit is imperfect.

For staggered treatment adoption across units—where different units receive the intervention at different times—the `multisynth` function in the `augsynth` package can be used. One illustrative application comes from Paglayan (2018), who examined the effects of states implementing mandatory collective bargaining agreements for public sector unions. The implementation and replication materials are available on the package's GitHub page, linked in the earlier footnote.

A further recent development enhances SCM by incorporating multiple outcomes to improve estimation precision and robustness. In Ben-Michael, Feller, and Sun (2025), the authors propose a framework that jointly models several related outcomes to strengthen causal inference. The method is demonstrated through simulation and a re-analysis of the Flint water crisis's impact on student achievement. See footnote for replication materials.[6]

27.2 Synthetic Difference-in-Differences

As discussed earlier in this chapter, DiD and SC are widely used methods for estimating treatment effects, each with its own strengths and limitations. DiD relies on the assumption of parallel trends and applies equal weight to all units and time periods, while SC constructs a synthetic version of the treated unit using optimally chosen weights but ignores fixed effects and time variation in contributions. The SDiD estimator brings these two approaches together within a single framework. To overcome these challenges, **synthetic difference-in-differences (SDiD)**—proposed by Arkhangelsky et al. (2021)—combines the strengths of both approaches in a unified framework. SDiD improves robustness, enhances pre-treatment fit, and reduces extrapolation bias.

SDiD achieves this by applying weights across both units and time periods. Unit weights are learned from the pre-treatment period, similar to SC, while time weights—similar in spirit to DiD—are learned across both pre- and post-treatment periods. This structure allows SDiD to adapt flexibly to heterogeneous settings, achieving improved estimation performance in practice.

Estimation Procedure (Simplified): Assume a balanced panel with N units and T time periods, with the treatment beginning at period $T_0 + 1$. For simplicity, assume one treated unit. Let the observed outcome be modeled as:

$$Y\_\{it\} = \mu + \alpha i + \beta t + W\{it\}\tau + \varepsilon\{it\} \tag{27.6}$$

whereμ: overall mean, α_i: unit-specific fixed effects, β_t: time-specific fixed effects, W_{it}: binary treatment indicator, τ: treatment effect, ε_{it}: idiosyncratic error

The SDiD estimator seeks parameters $\hat{\tau}, \hat{\mu}, \hat{\alpha}, \hat{\beta}$ by minimizing a doubly-weighted least squares criterion:

$$(\hat{\tau}^{\text{sdid}}, \hat{\mu}, \hat{\alpha}, \hat{\beta}) = \arg \min_{\tau,\mu,\alpha,\beta} \sum_{i=1}^{N} \sum_{t=1}^{T} (Y_{it} - \mu - \alpha_i - \beta_t - W_{it}\tau)^2 \cdot \hat{\omega}_i^{\text{sdid}} \cdot \hat{\lambda}_t^{\text{sdid}} \tag{27.7}$$

where $\hat{\omega}_i^{\text{sdid}}$ and $\hat{\lambda}_t^{\text{sdid}}$ are estimated weights over units and time periods respectively.[7] The structure of the SDiD estimator makes clear how it generalizes both DiD and SC. If we remove both sets of weights—setting all unit and time weights equal—the SDiD objective reduces to a standard two-way fixed effects regression, which corresponds exactly to the traditional DiD estimator. This highlights that DiD assumes all units and time periods contribute equally, without differential weighting. On the other hand, if we omit unit fixed effects and time weights from the SDiD formulation—retaining only unit weights estimated to match pre-treatment outcomes—we recover the SC estimator. In this sense, DiD and SC are both nested within the SDiD framework, which enhances flexibility by allowing weighted contributions from both units and time periods while adjusting for unit and time-specific heterogeneity.

To illustrate the method, consider estimating the effect of California Proposition 99, a tobacco tax law passed in 1988, on cigarette consumption. The `synthdid` R package provides a built-in dataset and implementation tools:

```
devtools::install_github("synth-inference/synthdid")
library(synthdid)

# Estimate the effect of California Proposition 99 on cigarette consumption
data('california_prop99')
setup = panel.matrices(california_prop99)
tau.hat = synthdid_estimate(setup$Y, setup$N0, setup$T0)
se = sqrt(vcov(tau.hat, method='placebo'))
sprintf('point estimate: %1.2f', tau.hat)
```

```
## [1] "point estimate: -15.60"
```

```
sprintf('95%% CI (%1.2f, %1.2f)', tau.hat - 1.96 * se, tau.hat + 1.96 * se)
```

```
## [1] "95% CI (-33.16, 1.95)"
```

```
plot(tau.hat)
```

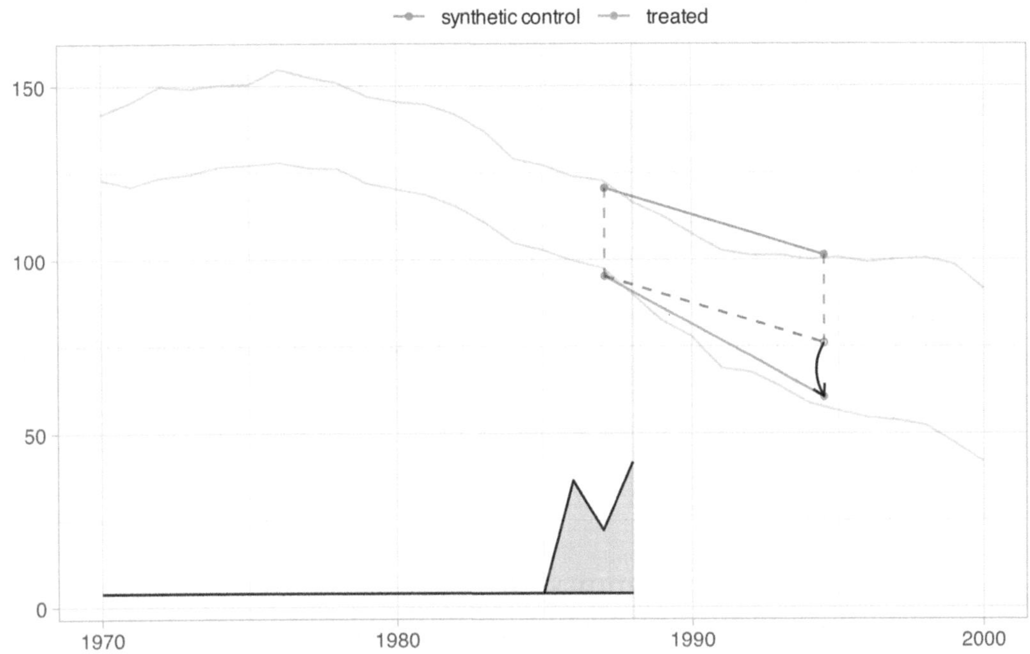


```
tau.sc   = sc_estimate(setup$Y, setup$N0, setup$T0)
tau.did  = did_estimate(setup$Y, setup$N0, setup$T0)
estimates = list(tau.did, tau.sc, tau.hat)
names(estimates) = c('Diff-in-Diff', 'Synthetic Control', 'Synthetic Diff-in-Diff')
print(unlist(estimates))
```

```
##           Diff-in-Diff      Synthetic Control Synthetic Diff-in-Diff
##              -27.34911              -19.61966              -15.60383
```

This comparison highlights how SDiD often yields estimates (the black arrow in the figure is estimated ATT) that are more conservative than DiD (when parallel trends fail) and more stable than SC (when matching is imperfect). The SDiD plot reveals a noticeable divergence post-1988, reflecting the estimated treatment effect of the law on smoking rates. The comparison shows that SDiD typically yields estimates that are more conservative than DiD (which may overstate effects under non-parallel trends) and more stable than SC (which may be sensitive to pre-treatment imbalance).

The method is readily accessible through the `synthdid` R package, which is supported by comprehensive theoretical documentation and clear implementation examples. A valuable companion to the R tools is the official documentation site https://synth-inference.github.io/synthdid/, which includes tutorials and replication materials.

For users working in Stata, Clarke et al. (2024) provide a clear and accessible exposition of the SDiD methodology. Their paper covers both single and staggered treatment adoption settings and introduces the `sdid` command for implementation in Stata. Even for researchers who do not use Stata, the paper is highly recommended for its lucid explanation of the method, its assumptions, inference strategies, and for the well-developed examples it provides.

In addition, Ciccia (2024) proposes an event-study extension to the SDiD framework. This work shows how SDiD estimators can be decomposed into dynamic treatment effects in both simple and staggered designs. These estimators can be implemented using the `sdid_event` package in Stata.

For Python users, there are also open-source implementations of SDiD under development and available at https://github.com/MasaAsami/pysynthdid and https://github.com/d2cml-ai/synthdid.py.

Taken together, the methods discussed in this section—SCM and its extensions—provide powerful tools for constructing credible counterfactuals and estimating treatment effects in observational data. These approaches are especially well-suited to situations involving a limited number of treated units and no clear natural control. However, when treatment assignment varies over time or across units in more structured ways, such as around a known threshold or cutoff, alternative strategies like regression discontinuity design (RDD) may offer a better fit. We turn next to RDD to examine its conceptual foundation, estimation strategy, and empirical use.

27.3 Regression Discontinuity Designs

RDD is a powerful quasi-experimental framework that exploits a sharp rule: treatment is assigned based on whether a continuous running variable crosses a known cutoff. This rule creates a discontinuity with no overlap—units just above the threshold receive treatment, and those just below do not. RDD is widely used for estimating causal effects in settings where randomization is infeasible but treatment follows a strict threshold policy. When properly implemented, RDD yields internally valid causal estimates for units near the cutoff, often providing insights comparable to those from randomized controlled trials (RCTs).

The canonical setup involves a policy or program assigned based on whether an individual's score on a running variable crosses a threshold—for example, students with test scores above 75 receiving a scholarship. Those just above and just below the cutoff are likely similar in observable and unobservable characteristics. Therefore, any systematic difference in outcomes such as college attendance or earnings can be attributed to the treatment.

The figure illustrates a sharp regression discontinuity design using a simulated example where students receive a scholarship if their test score exceeds 75. The vertical dashed line marks the cutoff, while the shaded region shows the estimation bandwidth h on either side. The two regression lines are fit separately to the left and right of the threshold, and the bold vertical line represents the estimated treatment effect—the jump in expected outcome at the cutoff. The design assumes individuals near the threshold are otherwise similar, justifying the causal interpretation of the observed discontinuity.

The key idea is that if individuals near the cutoff are otherwise comparable, then any observed difference in outcomes can be attributed to the treatment. At the threshold, the probability of treatment changes abruptly. For identification to be valid, the assignment

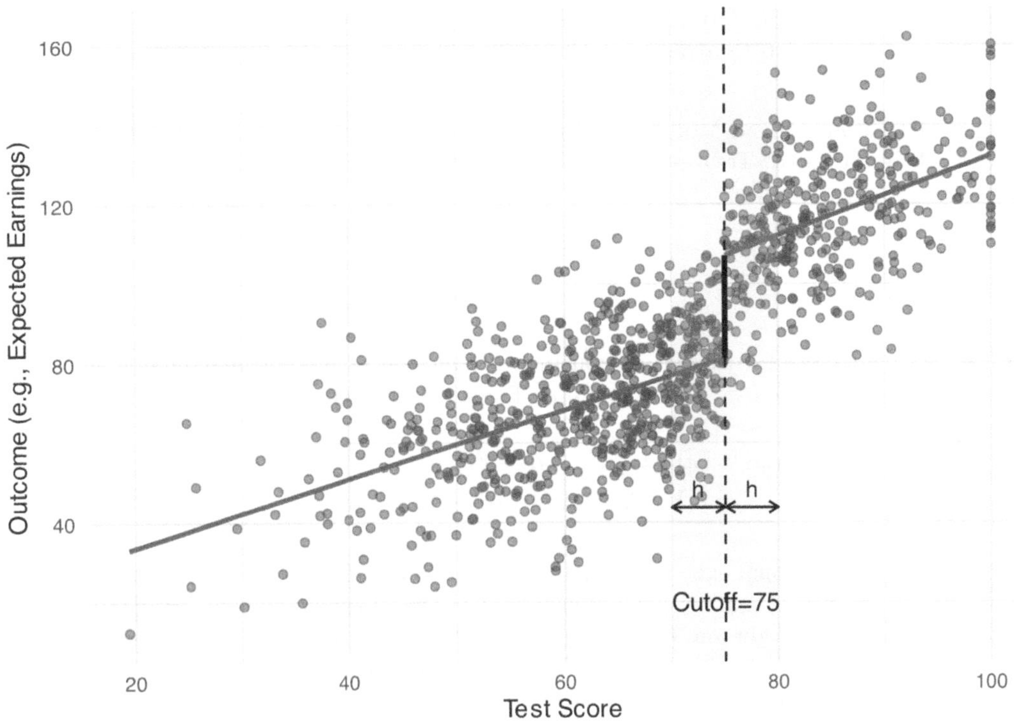

FIGURE 27.1 Simulated sharp RDD with scholarship cutoff.

rule does not need to be random—it only needs to be precisely defined, transparent, and immune to manipulation. This local comparison relies on minimal assumptions but critically depends on individuals being unable to sort around the cutoff. If units can sort around the threshold, internal validity is compromised. McCrary's (2008) density test is commonly used to detect such manipulation. While density tests can help detect sorting, they cannot rule out precise manipulation based on unobserved factors. Ultimately, the assumption functions as an exclusion restriction and must be supported by institutional knowledge.

When "*no manipulation/sorting with precision*" assumption holds, the assignment is effectively random in a neighborhood around the cutoff, identifying a Local Average Treatment Effect (LATE) that may not generalize beyond that range. This does not require X to be independent of Y, but only that around the cutoff, the value of X is as good as random for identification purposes. This identification strategy hinges on the assumption that, in the absence of treatment, potential outcomes would evolve smoothly with the running variable across the threshold.

RDD comes in two forms: sharp and fuzzy. In a sharp RDD, treatment is deterministically assigned—everyone above the cutoff is treated, and no one below is. Estimation involves comparing the conditional expectation of the outcome just above and just below the threshold. In contrast, a fuzzy RDD arises when the probability of treatment jumps at the cutoff, but treatment is not strictly assigned. This occurs when, for example, a policy is loosely enforced or individuals can opt out. In such cases, the cutoff indicator serves as an instrument for actual treatment and the design resembles an instrumental variables (IV) setup. The treatment effect is estimated as the ratio of the discontinuity in the outcome to the discontinuity in treatment probability, identifying a LATE for compliers. To summarize, sharp RDD assumes treatment is deterministically assigned at the cutoff, while fuzzy RDD involves a discontinuous jump in treatment probability, requiring instrumental variable methods to recover causal effects.

Under the Rubin Causal Model, each unit i has two potential outcomes: $Y_i(1)$ if treated and $Y_i(0)$ if untreated. Treatment D_i is assigned based on whether the running variable X_i exceeds a known cutoff c:

$$D_i = \begin{cases} 1, & \text{if } X_i \geq c, \\ 0, & \text{if } X_i < c. \end{cases} \tag{27.8}$$

Here, D_i is a deterministic and discontinuous function of X_i—no matter how close to c, treatment jumps from 0 to 1 at the cutoff. The treatment effect at the threshold is:

$$\begin{aligned} \tau_{\text{LATE}} &= \mathbb{E}[Y_i(1) \mid X_i = c] - \mathbb{E}[Y_i(0) \mid X_i = c] \\ &= \mathbb{E}[Y_i \mid X_i = c, D_i = 1] - \mathbb{E}[Y_i \mid X_i = c, D_i = 0] \end{aligned} \tag{27.9}$$

Assuming the conditional expectations of the potential outcomes are continuous in X around the cutoff c, we can estimate the treatment effect at the threshold by comparing the observed outcomes just to the right and just to the left of c.[8] Because treatment jumps discontinuously at c, the average outcome should also jump if the treatment has an effect. This jump reflects the causal effect at the cutoff and is estimated as:

$$\hat{\tau} = \lim_{x \downarrow c} \mathbb{E}[Y \mid X = x] - \lim_{x \uparrow c} \mathbb{E}[Y \mid X = x] \tag{27.10}$$

Here, the first term represents the expected outcome for units just above the cutoff (treated), and the second term for units just below the cutoff (untreated). The difference between these two conditional expectations isolates the causal effect of the treatment, assuming no other factor changes discontinuously at c. This nonparametric approach does not require specifying a functional form but relies on local comparisons near the threshold.

In a fuzzy RDD, treatment is not perfectly determined by the cutoff but the probability of receiving treatment increases discontinuously at $X = c$. This setup still allows identification of a local average treatment effect (LATE) at the threshold, but it requires adjusting for the fact that not all individuals above the cutoff are treated, and some below may also receive treatment.

Formally, the treatment assignment probability satisfies:

$$P(D_i = 1 \mid X_i) = \begin{cases} p_0, & X_i < c, \\ p_1, & X_i \geq c, \end{cases} \quad \text{with } p_1 > p_0 \tag{27.11}$$

The intuition is that the cutoff introduces a jump in the likelihood of treatment. If potential outcomes are continuous in X at c, then the treatment effect can be recovered by scaling the observed discontinuity in the outcome by the discontinuity in treatment probability. This is analogous to an IV approach, where the cutoff indicator serves as an instrument for actual treatment receipt.[9]

Estimation and Practical Implementation

Implementing regression discontinuity designs requires a series of coordinated estimation and validation decisions, including the choice of estimation method (parametric or nonparametric), functional form, bandwidth selection, specification of the local neighborhood around the cutoff, and diagnostic checks such as visual inspections, balance tests, and falsification exercises to assess validity and robustness. Researchers face several implementation choices—such as whether to use parametric or nonparametric methods, how to select bandwidths, and which functional forms or kernels to use—each involving trade-offs between bias and variance.

For estimation, researchers often use a parametric regression model where the outcome varies smoothly with the running variable, except for a discontinuity at the cutoff. A commonly used sharp RDD specification is:

$$Y_i = \alpha + \tau D_i + f(X_i - c) + \varepsilon_i \tag{27.12}$$

Here, Y_i is the observed outcome, D_i is the treatment indicator defined by whether $X_i \geq c$, and $f(X_i - c)$ is a smooth function of the running variable. The coefficient τ captures the size of the jump in the outcome at the threshold—interpreted as the treatment effect for units right at the cutoff. Including the smooth function helps control for any underlying trend in the running variable, isolating the discontinuity due to treatment.

Local polynomial regression is the workhorse method for RDD estimation. It approximates the regression function separately on either side of the cutoff using polynomials, allowing both the slope and intercept to differ at the threshold. A common starting point is a local linear specification—a degree-1 polynomial—estimated using the running variable centered at the cutoff. That is, defining $\tilde{X}_i = X_i - c$, the model becomes:

$$Y_i = \alpha + \tau D_i + \beta \tilde{X}_i + \gamma (D_i \cdot \tilde{X}_i) + \varepsilon_i \tag{27.13}$$

Here, $\tilde{X}_i$ is negative to the left of the cutoff, zero at the cutoff, and positive to the right. The treatment indicator D_i equals 1 if $X_i \geq c$. This setup fits separate lines on either side of the threshold, which may differ in both intercept (capturing the treatment effect) and slope.

Because there is no overlap in treatment assignment, treatment effects are estimated by extrapolating from either side of the cutoff. If the functional form is mis-specified, this extrapolation can introduce bias—highlighting the importance of flexible, data-driven methods focused on observations near the threshold. Although higher-order polynomials offer more flexibility, they can overfit and behave erratically near the cutoff. As a result, researchers are generally advised to use local linear or quadratic models only, as recommended by Gelman and Imbens (2019). Standard errors are typically heteroskedasticity-robust due to outcome variability near the cutoff. Recent work also discourages clustering standard errors by the running variable in this context (Calonico et al., 2017).[10]

A crucial design decision is bandwidth selection, which directly affects the bias–variance trade-off. A smaller bandwidth focuses on observations closest to the cutoff, increasing comparability and reducing bias but inflating variance due to a smaller sample size. A larger bandwidth provides more data and lowers variance but risks bias by including less comparable observations. Several data-driven algorithms address this trade-off. The Imbens–Kalyanaraman (IK) method offers a rule-of-thumb optimized for mean squared error, while the Calonico–Cattaneo–Titiunik (CCT) method introduces bias correction for robustness. These selectors help target a window around the cutoff where the functional form is most stable, minimizing extrapolation risk. In practice, researchers often assess sensitivity by plotting the estimated treatment effect $\hat{\tau}$ across different bandwidths to visualize how estimates change with window size.

Another way of approximating $f(X_i - c) = f(\tilde{X}_i)$ is to use a nonparametric kernel. Kernel-based methods estimate local averages by assigning weights that decline with distance from the cutoff, focusing inference on the neighborhood where treatment status changes. The triangular kernel is especially popular—not only because it emphasizes nearby observations, but also because it minimizes mean squared error in boundary estimation settings like RDD.[11] These methods are flexible and avoid imposing global functional form assumptions, but still require bandwidth selection and involve the same bias–variance trade-off considerations as local polynomial regressions.

Functional form decisions, particularly the choice of polynomial order, also influence RDD estimates. Higher-order polynomials offer more flexibility but risk overfitting and can produce erratic behavior near the threshold, especially when sensitive to outliers. In contrast, local linear or local quadratic models are generally preferred for balancing fit and interpretability. Researchers often inspect fitted curves and residuals to assess model adequacy and determine whether additional flexibility is needed.

Careful graphical analysis is a central component of RDD diagnostics. The typical plot shows the outcome variable on the vertical axis and the running variable on the horizontal, with fitted curves on each side of the cutoff. Beyond offering intuitive evidence, these plots also help guide bandwidth and functional form choices by revealing curvature or irregularities in the running variable–outcome relationship. A clear jump in the outcome at the threshold provides intuitive evidence of a treatment effect. However, these plots can be sensitive to binning choices and bandwidth, so visual interpretations must be approached with caution.

Validating the RD design requires ruling out manipulation around the threshold. If individuals can precisely sort around the cutoff, the identifying assumption fails. The McCrary (2008) test detects any discontinuity in the density of the running variable, implemented in R via `DCdensity()`. Researchers also assess whether predetermined covariates remain smooth across the cutoff. If a covariate shows a discontinuous jump at the threshold, it may signal the presence of a confounder rather than a treatment effect—raising concerns that the observed outcome discontinuity could be driven by imbalance rather than the intervention itself. Balance checks can include visual and regression-based assessments of other baseline characteristics around the cutoff.[12] Placebo tests are also useful—researchers examine whether estimated treatment effects arise at other values of the running variable where no policy change occurred. Finding significant jumps away from the actual cutoff would cast doubt on the design's validity.

27.3.1 RDD Simulation

In this simulation we generate data which we used the earlier RDD example: a scholarship awarded to students whose test scores exceed a cutoff of 75. As shown in the earlier figure, the design creates a sharp discontinuity in treatment assignment, allowing causal inference based on local comparisons near the threshold.

While the initial plot is constructed manually using `ggplot2`, in this simulation we use the `rdrobust` package to formally implement and estimate the RDD. The command `rdplot(y, x, c = 0)` automatically generates a data-driven RD plot using local linear regression, where x is the running variable centered at the cutoff and y is the simulated outcome. This function bin-scatters the data and overlays separate polynomial fits on each side of the cutoff, providing a quick and informative way to visualize the discontinuity and inspect model adequacy.

We estimate the treatment effect using `rdrobust()`. It applies local polynomial regression with robust bias correction and inference procedures developed by Calonico, Cattaneo, and Titiunik (2014). The call to `rdrobust(y, x, c = 0, p = 1, kernel = "triangular", h = ...)` estimates the treatment effect at the cutoff using a local linear regression (p = 1) with a triangular kernel, which weights observations closer to the cutoff more heavily. The bandwidth h is selected using the CCT optimal bandwidth selector via the `rdbwselect()` function with the `bwselect = "mserd"` option. The output includes both the conventional (uncorrected) and robust bias-corrected estimates of the treatment effect, along with corresponding standard errors. The local quadratic version `p = 2` is estimated in the same way, offering a check on the sensitivity of results to polynomial order. As discussed earlier, local linear models are generally preferred over higher-order polynomials due to their favorable bias–variance trade-off and stability near the cutoff.

```
library(rdrobust)

# Simulate test scores between 0 and 100
set.seed(42)  # For reproducibility
score <- rnorm(1000, mean = 70, sd = 15)
score <- pmin(pmax(score, 0), 100)  # Restrict scores between 0 and 100

# Scholarship assigned if score > 75
cutoff <- 75
treated <- ifelse(score > cutoff, 1, 0)
y <- 20 + 30 * treated + 0.8 * score + rnorm(1000, 0, 15)  # Potential outcome
x <- score - cutoff  # Center the running variable at the cutoff

# Bandwidth selection using CCT method
bw_cct <- rdbwselect(y, x, kernel = "triangular", bwselect = "mserd")
print(bw_cct)
```

```
## Call: rdbwselect
##
## Number of Obs.                 1000
## BW type                       mserd
## Kernel                   Triangular
## VCE method                       NN
##
## Number of Obs.                  647          353
## Order est. (p)                    1            1
## Order bias  (q)                   2            2
## Unique Obs.                     647          333
```

```
# Visual RD plot (uses local linear by default)
rdplot(y, x, c = 0, kernel = "triangular", p = 1)
```

```
# Local linear regression (p = 1) with triangular kernel
out <- rdrobust(y, x, c = 0, p = 1, kernel = "triangular", h = bw_cct$bws[1,1])
out$Estimate
```

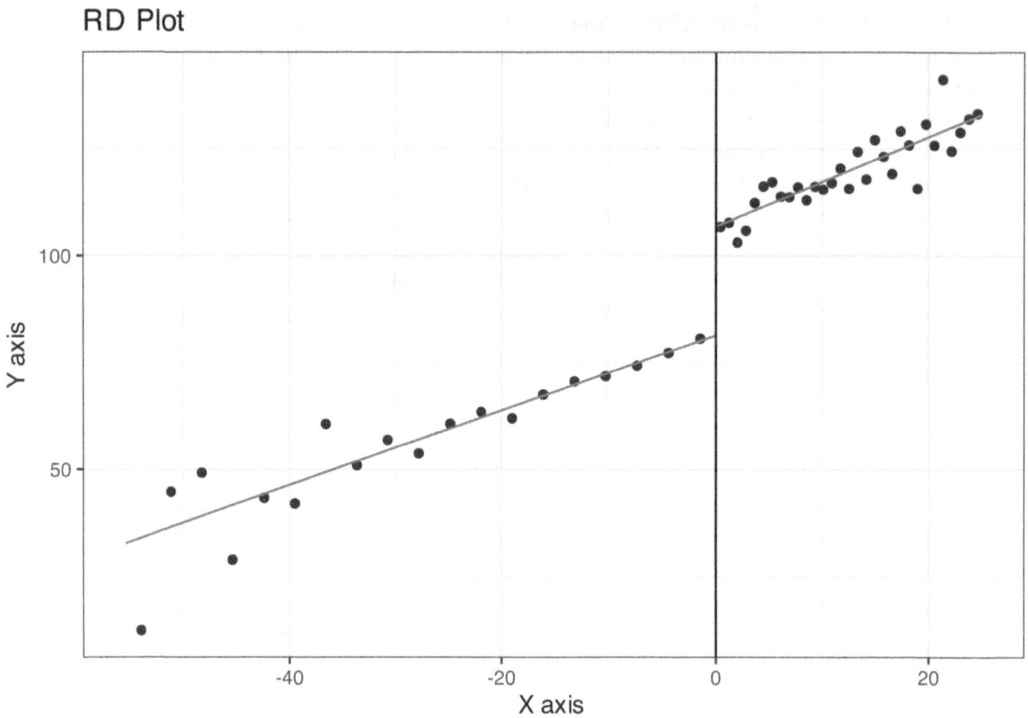

FIGURE 27.2 RD Plot with triangular kernel and CCT bandwidth (local linear fit).

```
##        tau.us   tau.bc    se.us    se.rb
## [1,] 20.82346 24.46106 3.442885 4.352538
```

```
# Local quadratic regression (p = 2) with triangular kernel
out2 <- rdrobust(y, x, c = 0, p = 2, kernel = "triangular", h = bw_cct$bws[1,1])
out2$Estimate
```

```
##        tau.us   tau.bc    se.us   se.rb
## [1,] 24.46106 29.94459 4.352538 5.20353
```

The second figure, produced using `rdplotdensity()` from the `rddensity` package, serves a different but crucial purpose. It implements the McCrary (2008) density test, which assesses the possibility that units have manipulated their position relative to the cutoff (e.g., by artificially boosting scores to qualify for the scholarship). The underlying assumption of RDD is that the running variable cannot be precisely manipulated; otherwise, the groups just above and below the threshold would not be comparable. `rddensity()` estimates the density of the running variable on either side of the cutoff and tests for a discontinuous jump. A significant result would suggest sorting or manipulation, which would undermine the causal interpretation of the estimated treatment effect. In our simulated data, which was generated without manipulation, the test is expected to yield a non-significant p-value, consistent with the design assumption.

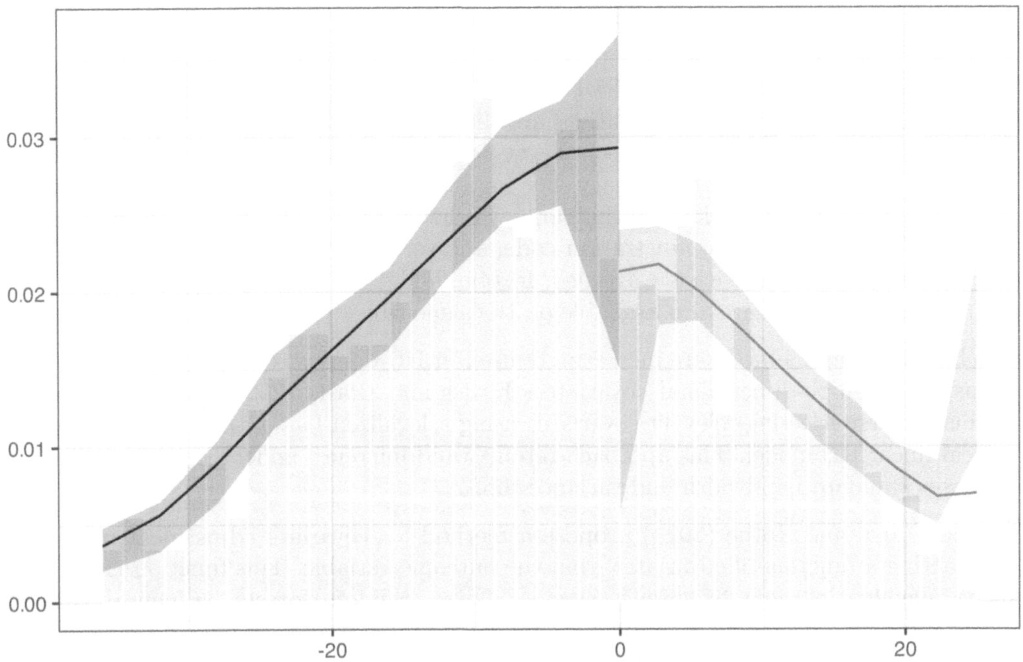

Together, these commands (`rdplot`, `rdrobust`, `rdbwselect`, and `rddensity`) form a robust pipeline for implementing, estimating, and validating an RDD. Researchers can modify key elements of this simulation to suit their empirical application: for instance, changing the cutoff value, using different kernels (e.g., uniform or Epanechnikov), trying alternative bandwidth selection methods (like IK), or experimenting with other polynomial orders. For more complex settings involving covariates or fuzzy designs (where treatment probability jumps but assignment is not deterministic), `rdrobust` also supports extensions such as instrumental variables estimation.

27.3.2 Credibility and Extensions in RDD Applications

While RDD designs offer high internal validity, they identify effects only for units near the cutoff. In sharp RDD, this corresponds to the average treatment effect at the threshold; in fuzzy RDD, it reflects the LATE for compliers—those whose treatment status is influenced by crossing the cutoff. These are not estimates of the treatment effect for everyone in the sample, but only for units close to the threshold—the "local" in LATE. This is a trade-off: while we gain credible identification, we give up broader generalizability. In many cases, however, this local group is of direct policy interest—for example, students just on the margin of qualifying for a scholarship or patients just eligible for an intervention. RDD also requires large samples near the cutoff to obtain precise estimates and depends on a well-defined assignment rule. Thin density near the threshold can limit power and precision, making RDD less suitable when few observations fall within the effective estimation window. Despite these limitations, RDD remains one of the most credible nonexperimental methods for causal inference, especially in evaluating policy interventions.

This section a very brief explanation of RDD. There are also extensions of RDD that go beyond the standard single-threshold, continuous running variable setup. These include RD designs with discrete running variables, multiple cutoffs and multiple scores (`rdmulti`), geographic or spatial RD (`SpatialRDD`),[13] and methods addressing issues like covariate adjustment, clustered data, and inference under weak identification. Calonico et al. (2025) introduced estimation and inference procedures, along with the `rdhte` package, for analyzing heterogeneous treatment effects in RDD designs. Hsu and Shen (2024) extend RD model to a dynamic framework, where observations are eligible for repeated RD events (see replication package from the reference doi link). These topics introduce additional complexity and are not covered in this brief summary. For implementation, the most comprehensive and up-to-date resource is rdpackages.github.io,[14] which hosts the main packages used in practice, along with replication files, illustration code, and software in R, Stata, and Python. One of the best recent overviews is Cattaneo, Idrobo, and Titiunik (2024). Additional related volumes and resources are also available on the same site.

Machine learning methods are increasingly used in RD designs to improve precision when adjusting for high-dimensional covariates. Kreiss and Rothe (2024) propose a two-step estimator: first selecting relevant covariates using a localized Lasso, then including them in a local linear RD model. This approach retains valid inference under approximate sparsity and uses standard bandwidth and variance tools.

Noack, Olma, and Rothe (2024) propose a method for covariate adjustment in RDD by subtracting a function of covariates from the outcome variable. This function is estimated using machine learning techniques such as Lasso, random forests, or neural networks. Their approach can handle high-dimensional covariates and aims to reduce the variance of treatment effect estimates. Simulation studies demonstrate that this method often yields shorter confidence intervals compared to traditional RD estimators, particularly when covariates are informative. The method is implemented in the`DoubleML` package in python.

To conclude, the regression discontinuity design provides a robust framework for causal inference under relatively mild assumptions. Its power lies in its simplicity and transparency: when a threshold-based rule determines treatment, and individuals cannot manipulate their position relative to the cutoff, RD can uncover the causal effect with minimal structural assumptions. The challenge, as always, is in the details: careful choice of bandwidth, correct functional form, attention to manipulation, and thoughtful interpretation of results are all essential to making RD work in practice.

Notes

1. To determine the optimal weights w_j, SCM relies on matching the treated unit to a weighted average of the control units based on observed characteristics and lagged outcomes. To determine the optimal weights, let $\mathbf{X}_1$ be a $k \times 1$ vector of pre-treatment predictors for the treated unit, and $\mathbf{X}_0$ a $k \times J$ matrix of the same variables for the control units. The weights $\mathbf{W}$ are chosen to minimize the weighted distance:

$$\|\mathbf{X}_1 - \mathbf{X}_0\mathbf{W}\|_V = \sqrt{(\mathbf{X}_1 - \mathbf{X}_0\mathbf{W})'V(\mathbf{X}_1 - \mathbf{X}_0\mathbf{W})}$$

where V is a positive semi-definite matrix that sets the relative importance of the predictors. Typically, V is selected to minimize the mean squared prediction error (MSPE) in the pre-treatment period. The optimization problem becomes:

$$\min_{\mathbf{W}} \|\mathbf{X}_1 - \mathbf{X}_0\mathbf{W}\|_V \quad \text{subject to } w_j \geq 0, \sum w_j = 1$$

2. Carlos Mendez's R Tutorial on Synthetic Control: https://carlos-mendez.quarto.pub/r-synthetic-control-tutorial/. CRAN page for the Synth Package: https://cran.r-project.org/web/packages/Synth/index.html. Abadie, Diamond and Hainmueller (2011). These resources offer detailed documentation and examples using

the default dataset provided in the package. They cover data preparation, model estimation, visualization, and interpretation.

3. See Xu (2017) and Xu & Liu (2022). These sources offer a comprehensive and clear overview of GSCM, including theoretical foundations, estimation details, and further applications. Note that the "gsynth" package has recently been retired and integrated into the 'fect' package. The updated implementation and full documentation are available at https://yiqingxu.org/packages/fect/04-gsynth.html.

4. Formally, the ASCM estimator can be understood as solving the following optimization problem:

$$\min_{\beta,\mathbf{w}} \sum_{t=1}^{T_0} \left(Y_{1t} - \mathbf{X}_t'\beta - \sum_{j=2}^{J+1} w_j Y_{jt} \right)^2 + \lambda \|\beta\|_2^2$$

where $\mathbf{X}_t$ is a vector of time-varying covariates or latent components, β is a vector of regression coefficients, and λ is a ridge penalty that controls the extent of regularization. The ridge penalty shrinks the coefficients to avoid overfitting and manages the degree of extrapolation beyond the convex hull.

5. See Ben-Michael, E., Feller, A., & Rothstein, J. (2021). "The Augmented Synthetic Control Method." *Journal of the American Statistical Association*, 116(536), 1789–1803. The method is available through the 'augsynth' R package: https://github.com/ebenmichael/augsynth.

6. Replication materials are available via Harvard Dataverse: Sun, Liyang; Ben-Michael, Eli; Feller, Avi, 2025. "Replication data for: Using Multiple Outcomes to Improve the Synthetic Control Method." https://doi.org/10.7910/DVN/KBVB9F. V1

7. Please see Algorithm 1 and 2, and Equations 4–6 in Arkhangelsky et al. (2021) for the formal estimation procedures. The weights are derived by solving the following optimization problems:

Unit Weights:

$$(\hat{\omega}_0, \hat{\omega}^{\text{sdid}}) = \arg \min_{\omega \in \mathbb{R}_+, \omega \in \Omega} \sum_{t=1}^{T_{\text{pre}}} \left(\omega_0 + \sum_{i=1}^{N_{\text{co}}} \omega_i Y_{it} - \frac{1}{N_{\text{tr}}} \sum_{i=N_{\text{co}}+1}^{N} Y_{it} \right)^2 + \zeta T_{\text{pre}} \|\omega\|_2^2$$

Time Weights:

$$(\hat{\lambda}_0, \hat{\lambda}^{\text{sdid}}) = \arg \min_{\lambda \in \mathbb{R}, \lambda \in \Lambda} \sum_{i=1}^{N_{\text{co}}} \left(\lambda_0 + \sum_{t=1}^{T_{\text{pre}}} \lambda_t Y_{it} - \frac{1}{T_{post}} \sum_{t=T_{\text{pre}}+1}^{T} Y_{it} \right)^2$$

Here, ζ is a ridge regularization parameter that prevents overfitting.

8. Note that in sharp RDD, there is complete lack of overlap: $P(D_i = 1|X_i < c) = 0$ and $P(D_i = 1|X_i \geq c) = 1$, so traditional identification via common support does not apply.

9. The treatment effect at the cutoff is estimated as:

$$\tau = \frac{\lim_{x \downarrow c} \mathbb{E}[Y \mid X = x] - \lim_{x \uparrow c} \mathbb{E}[Y \mid X = x]}{\lim_{x \downarrow c} \mathbb{E}[D \mid X = x] - \lim_{x \uparrow c} \mathbb{E}[D \mid X = x]}$$

This Wald-type estimator isolates the causal effect for compliers—those whose treatment status is influenced by crossing the threshold. A corresponding parametric specification using Local Two-Stage Least Squares (2SLS) is: First stage (treatment model):

$$D_i = \pi_0 + \pi_1 \mathbf{1}(X_i \geq c) + g(X_i - c) + \nu_i$$

Second stage (outcome model):

$$Y_i = \alpha + \tau \hat{D}_i + f(X_i - c) + \varepsilon_i$$

Here, $\mathbf{1}(X_i \geq c)$ is the instrument, $g(\cdot)$ and $f(\cdot)$ are smooth functions of the running variable, and $\hat{D}_i$ is the fitted value from the first stage. The second-stage coefficient τ gives the LATE at the cutoff for compliers.

10. Control variables are usually omitted in sharp RDD because the design itself—if valid—ensures treated and control groups are comparable near the cutoff. Including controls can improve precision but may imply concern about residual imbalance. Their inclusion is more relevant in fuzzy designs, where treatment is not strictly assigned by the cutoff. See Calonico et al. (2019) for methods incorporating covariates.

11. The triangular kernel assigns weights linearly decreasing with distance from the cutoff:

$$K(u) = (1 - |u|) \cdot \mathbf{1}(|u| \leq 1)$$

where $u = \frac{X_i - c}{h}$, c is the cutoff, and h is the bandwidth.

12. For example, researchers may examine visual plots of $E[W_i|\tilde{X}_i, D_i]$, where W_i represents a predetermined covariate and $\tilde{X}_i = X_i - c$, as well as local regressions of the form:

$$E[W_i|\tilde{X}_i, D_i] = \alpha + \beta_0 \tilde{X}_i + \beta_1 (\tilde{X}_i \cdot D_i) + \tau_W D_i.$$

A statistically significant τ_W would indicate a discontinuity in covariates at the threshold, suggesting possible imbalance.

13. For more details, see the package vignette at https://cran.r-project.org/web/packages/SpatialRDD/vignettes/spatialrdd_vignette.html.

14. https://rdpackages.github.io/

28

Time Series Forecasting

Time series forecasting is a task that involves using a model to predict future values of a time series based on its past values. The data consists of sequences of values that are recorded at regular intervals over a period of time, such as daily stock prices or monthly weather data. Time series forecasting can be approached using a variety of machine learning techniques, including linear regression, decision trees, and neural networks.

One key difference between time series forecasting and other types of machine learning tasks is the presence of temporal dependencies in the data. In time series data, the value at a particular time point is often influenced by the values that came before it, which means that the order in which the data points are presented is important. This can make time series forecasting more challenging, as the model must take into account the relationships between past and future values in order to make accurate predictions.

One of the most accessible and comprehensive source on forecasting using R is *Forecasting: Principles and Practice* (https://otexts.com/fpp3/) by Rob J. Hyndman and George Athanasopoulos (2021). The book now has the 3rd edition that uses the `tsibble` and `fable` packages rather than the `forecast` package. This brings a better integration to the tidyverse collection of packages. A move from FPP2 to FPP3 brings a move from `forecast` to `fable`. The main difference is that `fable` is designed for `tsibble` objects and `forecast` is designed for `ts` objects. There is a paper by Wang et al. (2020b) which is an excellent source for learning `tsibble` in more details.

In this section, we will use the `tsibble` and `fable` packages along with the `fpp3` package and cover five main topics: applications with Autoregressive Integrated Moving Average (ARIMA) models, grid search for ARIMA, time series embedding, forecasting with random forests, and artificial neural network applications, recurrent neural network (RNN) and long short-term memory (LSTM). The time series analysis and forecasting is a very deep and complex subject, which is beyond the scope of this book to cover in detail. FPP3 is free and very accessible even for those without a strong background on time-series forecasting. Therefore, this section assumes that some major concepts, like stationarity, time series decomposition, and exponential smoothing, are already understood by further readings of FPP3.

28.1 ARIMA: A Statistical Model for Time Series Forecasting

The **ARIMA** model is a fundamental statistical approach for time series forecasting. As a linear parametric model, ARIMA captures temporal dependencies such as seasonality and autocorrelation, making it useful for analyzing and predicting time-dependent data.

DOI: 10.1201/9781003504795-28

The model consists of three key components:

- **Autoregressive (AR) Component**: Captures relationships between the current value and its past values.
- **Integrated (I) Component**: Represents the degree of differencing applied to the data to achieve stationarity—where the mean, variance, and covariance remain constant over time.
- **Moving Average (MA) Component**: Models dependencies between the current and past forecast errors. The MA component of an ARIMA model is used to capture the short-term fluctuations in data that are not captured by the AR component. For example, if the time series data exhibits random noise or sudden spikes, the MA component can help to smooth out these fluctuations and improve the forecast accuracy.

An ARIMA model is expressed as ARIMA(p, d, q), where:

- p: Order of the autoregressive component (number of lagged values used).
- d: Degree of differencing needed to make the series stationary.
- q: Order of the moving average component (number of past errors included).

Selecting the optimal values for $(p, d,)$ and (q) depends on the characteristics of the time series. Before applying ARIMA, the data must be preprocessed to remove trends and seasonality, ensuring stationarity. Once the model is fitted, it generates forecasts based on learned patterns.

The mathematical foundation of the ARIMA model is based on the concept of AR and MA processes. An AR process is a type of stochastic process in which the current value of a time series depends on a linear combination of past values of the series. An AR process can be represented mathematically as:

$$X_t = c + \sum_{i=1}^{p} (\phi_i X_{t-i}) + \epsilon_t, \tag{28.1}$$

where X_t is the value of the time series at time t, c is a constant, ϕ_i is the AR coefficient for lag i, and ϵ_t is white noise (a sequence of random variables with a mean of zero and a constant variance).

A moving average process is a type of stochastic process in which the current value of a time series depends on a linear combination of past errors or residuals (the difference between the actual value and the forecasted value). A moving average process can be represented mathematically as:

$$X_t = c + \sum_{i=1}^{q} (\theta_i \epsilon_{t-i}) + \epsilon_t, \tag{28.2}$$

where θ_i is the moving average coefficient for lag i, and ϵ_t is again white noise.

The ARIMA model, which is a combination of AR and MA processes, can be represented mathematically as:

$$X_t = c + \sum_{i=1}^{p}(\phi_i X_{t-i}) + \sum_{i=1}^{q}(\theta_i \epsilon_{t-i}) + \epsilon_t \tag{28.3}$$

It is possible to write any stationary AR(p) model as an MA(∞) model by using repeated substitution. Here is the example for an AR(1) model without a constant:

$$\begin{aligned}
X_t &= \phi_1 X_{t-1} + \epsilon_t \quad \text{and} \quad X_{t-1} = \phi_1 X_{t-2} + \epsilon_{t-1} \\
X_t &= \phi_1 (\phi_1 X_{t-2} + \epsilon_{t-1}) + \epsilon_t \\
X_t &= \phi_1^2 X_{t-2} + \phi_1 \epsilon_{t-1} + \epsilon_t \\
X_t &= \phi_1^3 X_{t-3} + \phi_1^2 \epsilon_{t-2} + \phi_1 \epsilon_{t-1} + \epsilon_t \\
&\vdots
\end{aligned} \tag{28.4}$$

With $-1 < \phi_1 < 1$, the value of ϕ_1^k will get smaller as k gets bigger. Therefore, AR(1) becomes an MA (∞) process:

$$X_t = \epsilon_t + \phi_1 \epsilon_{t-1} + \phi_1^2 \epsilon_{t-2} + \phi_1^3 \epsilon_{t-3} + \cdots \tag{28.5}$$

The parameters of the ARIMA model (c, ϕ_i, θ_i) are estimated using maximum likelihood estimation (MLE), which involves finding the values of the parameters that maximize the likelihood of the observed data given the model. Once the model has been fit to the data, it can be used to make point forecasts (predictions for a specific time point) or interval forecasts (predictions with a range of possible values).

Some common methods for selecting p and q include in the ARIMA(p,d,q):

- Autocorrelation function (ACF) plot, which shows the correlations between the time series data and lagged versions of itself. A high positive autocorrelation at a lag of p suggests that p may be a good value for p in ARIMA(p,d,q).
- Partial autocorrelation function (PACF) plot, which shows the correlations between the time series data and lagged versions of itself, after accounting for the correlations at all lower lags. A high positive autocorrelation at a lag of q suggests the value for q in ARIMA(p,d,q).
- There are several statistical measures that can be used to compare the goodness of fit of different ARIMA models, such as Akaike's Information Criterion (AIC) and the Bayesian Information Criterion (BIC). These measures can be used to select the model with the lowest value, which is generally considered to be the best model.

It is important to note that determining the values of p and q is an iterative process, and we may need to try different values and evaluate the results in order to find the best fit for our data.

28.2 Hyndman–Khandakar Algorithm

The Hyndman–Khandakar algorithm (Hyndman and Khandakar, 2008a) combines several steps for modeling (and estimation) of the ARIMA model: unit root tests, minimization of the AICc, and MLE to obtain an ARIMA model. The arguments to `ARIMA()` in the `fable` package provide for many variations for modeling ARIMA. The modeling procedure to a set of (non-seasonal) time series data for ARIMA is defined in FPP3 as follows:

1. Plot the data to identify any outliers.
2. If the data shows variation that increases or decreases with the level of the series, transform the data (Box-Cox transformation) to stabilize the variance.
3. Check if the data are non-stationary. And, make them stationary, if they are not.
4. Start with an ARIMA $(p, d, 0)$ or ARIMA $(0, d, q)$ depending of what ACF/PACF indicates.
5. Try your chosen model(s), and use the AICc to search for a better model.

However, after step 5, the residuals from the chosen model are supposed to be white noise. Otherwise, the model has to be modified. Once the residuals look like white noise, the ARIMA model is ready for forecasting.

We will show all these steps by using the epidemic curve of COVID-19 in Toronto covering 266 days between the March 1st and the November 21st of 2020. An epidemic curve (or epi curve) is a visual display of the onset of illness among cases associated with an outbreak. The data contain the first wave and the first part of the second wave. It is from Ontario Data Catalogue[1] sorted by `Episode Date`, which is the date when the first symptoms were started. Our data set also contains the mobility data is from Facebook, `all_day_bing_tiles_visited_relative_change`, which reflects positive or negative changes in movement relative to baseline.

28.3 TS Plots

Let's first load the data and convert it to `tsibble`.

```
library(tsibble)
library(fpp3)
library(gridExtra)

#setwd("/Users/mutlu/Dropbox/MLCausalChapman/bookdown-crc-master")
load("dftoronto.RData")
day <- seq.Date(
  from = as.Date("2020/03/01"),
  to = as.Date("2020/11/21"),
  by = 1
)

tdata <- tibble(Day = day,
                mob = data$mob,
                cases = data$cases)

toronto <- tdata %>%
  as_tsibble(index = Day)

toronto
```

```
## # A tsibble: 266 x 3 [1D]
##    Day              mob cases
##    <date>         <dbl> <dbl>
##  1 2020-03-01 -0.0172       4
##  2 2020-03-02 -0.0320       6
##  3 2020-03-03 -0.0119      10
##  4 2020-03-04  0.0186       7
##  5 2020-03-05  0.0223       7
##  6 2020-03-06 -0.00626     10
##  7 2020-03-07  0.0261       8
##  8 2020-03-08  0.0273      10
##  9 2020-03-09 -0.0158      18
## 10 2020-03-10 -0.0521      29
## # i 256 more rows
```

Note the `[1D]` in the header indicating daily data. Dealing with daily and sub-daily data with `ts` class is not an easy process. The `tsibble` class handles such data with no problem. More details on `tsibbles` can be found at "Tidy time series data using tsibbles".[2]

Although there are better plotting option cosmetically, we will stick to what `fpp3` simply offers:

```
a <- toronto %>% autoplot(mob, col = 'gray20') +
  labs(
    title = "Mobility Index",
    subtitle = "Toronto 2020",
    x = "Days",
    y = "Index"
  )
b <- toronto %>% autoplot(cases, col = 'black') +
  labs(
    title = "Covid-19 Cases",
    subtitle = "Toronto 2020",
    x = "Days",
    y = "Cases"
  )

require(gridExtra)
grid.arrange(b, a, ncol = 2)
```

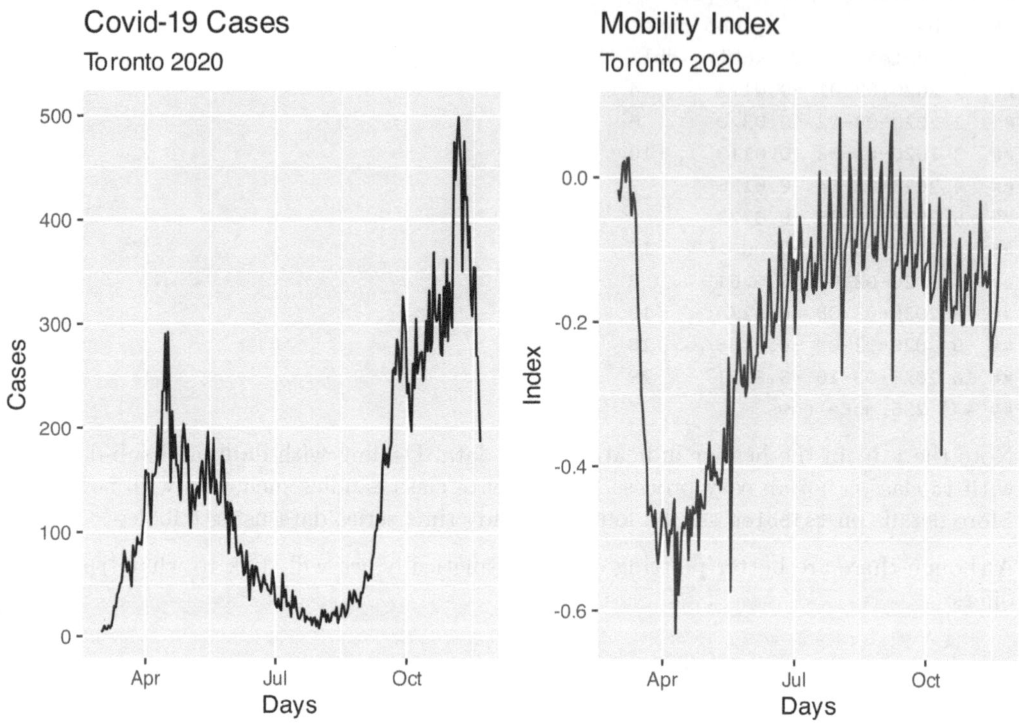

28.4 Box–Cox transformation

We would like to make the size of the variation about the same across the whole series. A proper variance-stabilizing transformation makes the forecasting model simpler and better. For example, Proietti and Lutkepohl (2012) find that the Box–Cox transformation produces forecasts which are significantly better than the untransformed data at the one-step-ahead horizon.

In time series analysis, stabilizing variance is crucial for building effective forecasting models. Many real-world time series exhibit heteroscedasticity, where the magnitude of fluctuations varies over time. This variability can make it challenging to model trends and seasonality accurately. A proper variance-stabilizing transformation helps simplify the modeling process, improves statistical properties, and enhances forecast accuracy.

One widely used transformation for this purpose is the Box–Cox transformation, introduced by George Box and David Cox (1964). This technique applies a power transformation to the data, making variance more uniform across the series and improving the assumptions required for many statistical and machine learning models.

For a given time series X_t, the Box–Cox transformation is defined as:

$$X_t^{(\lambda)} = \begin{cases} \frac{X_t^{\lambda}-1}{\lambda}, & \lambda \neq 0 \\ \ln(X_t), & \lambda = 0 \end{cases}$$

where:

- X_t represents the original time series data.
- λ is the transformation parameter that determines the nature of the transformation.
- If $\lambda = 1$, the data remains unchanged (no transformation).
- If $\lambda = 0$, the transformation reduces to a natural logarithm, which is useful for handling exponential growth.

The optimal value of λ is typically estimated from the data using MLE to achieve the best variance stabilization and normality.

Why Use the Box–Cox Transformation?

1. Stabilizing Variance: Many time series exhibit increasing variability over time. Applying Box–Cox reduces this heteroscedasticity, leading to better model performance.

2. Improving Normality: Many forecasting models, including ARIMA, assume normally distributed residuals. Transforming the data can help meet this assumption.

3. Enhancing Forecast Accuracy: Proietti and Lütkepohl (2012) found that forecasts based on Box–Cox-transformed data significantly outperform forecasts from untransformed data, particularly for short-term predictions (e.g., one-step-ahead forecasts).

4. Linearizing Relationships: In cases where relationships between variables are non-linear, applying Box–Cox can make them more linear, improving interpretability and the effectiveness of regression-based models.

Here are some common scenarios where the Box–Cox transformation is beneficial:

- When the time series exhibits a strong trend with increasing or decreasing variance.

- When residual diagnostics from an initial ARIMA model suggest non-constant variance (heteroscedasticity).

- When the distribution of the time series data is skewed or non-normal, affecting model assumptions.

- When using log transformations, but some values in the data are zero or negative (since log transformations are undefined for non-positive values).

```
lmbd <- toronto %>%
  features(cases, features = guerrero) %>%
  pull(lambda_guerrero)

toronto %>%
  autoplot(box_cox(cases, lambda = lmbd), col = "black") +
  labs(y = "",
       title = latex2exp::TeX(paste0(
         "Cases - Transformed with $\\lambda$ = ",
```

```
    round(lmbd, 2)
  )))
```

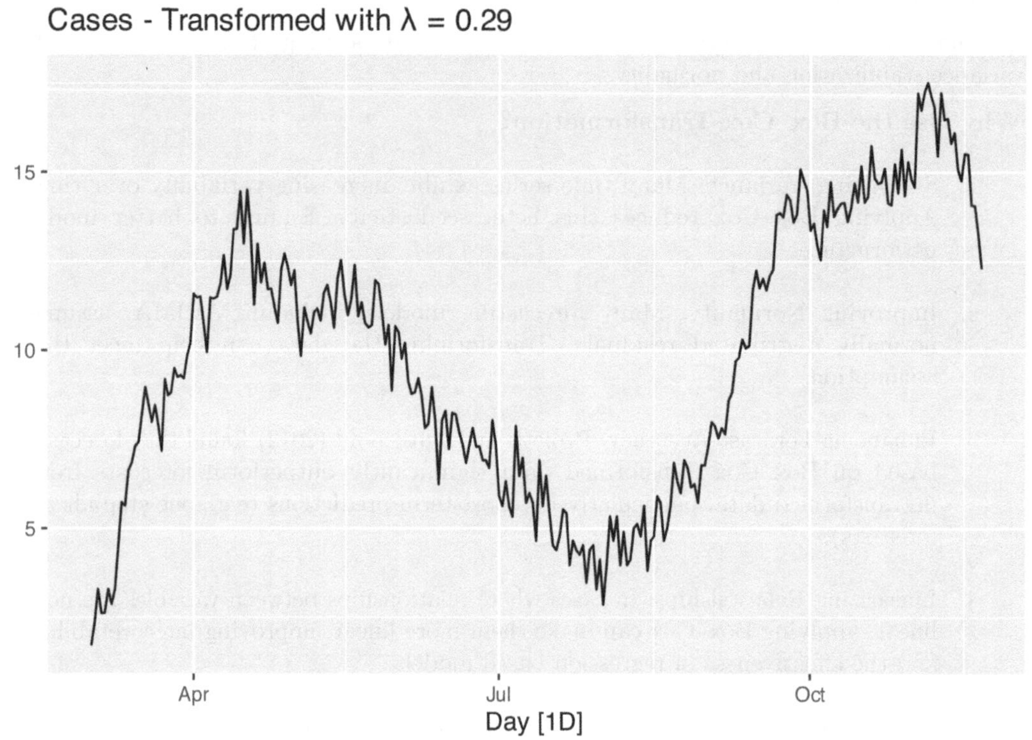

The option guerrero computes the optimal λ value for a Box–Cox transformation using the Guerrero[3] method (Guerrero, 1993).

Note that, since the number of tests performed in a given day changes the numbers of cases, we should use "positivity rates," which is the percentage of positive results in all COVID-19 tests given any day, instead of case numbers. We ignore this problem for now.

28.5 Stationarity

A time series is considered stationary if its statistical properties—mean, variance, and covariance—remain constant over time. This means that shifting the time axis does not alter the underlying distribution of the data. Stationarity is a fundamental assumption in many time series forecasting methods, as models built on non-stationary data may fail to generalize future trends and patterns effectively.

Many forecasting models, particularly ARIMA and other linear time series models, assume stationarity because they rely on past values to predict future ones. If the statistical properties of the data change over time, the model's parameters may become unreliable, leading to poor forecast accuracy.

Non-stationary data often exhibit trends, seasonality, or time-dependent variance, which complicates forecasting by making historical patterns less meaningful for future predictions. Converting a time series to a stationary form improves model stability and predictive performance.

To determine whether a time series is stationary, analysts use statistical tests such as:

1. Dickey–Fuller test (ADF Test): This tests for the presence of a unit root, which indicates non-stationarity. A significant test result suggests that the series is stationary.

2. Kwiatkowski–Phillips–Schmidt–Shin (KPSS) test: This test checks for stationarity by examining whether the variance of the series changes over time. A significant result here suggests that the series is *not* stationary.

A combination of both tests can provide a clearer picture:
- If the ADF test suggests stationarity but the KPSS test does not, the series may still require differencing.
- If both tests indicate non-stationarity, the data must be transformed before modeling.

If a time series is found to be non-stationary, transformation techniques can be applied to stabilize it before modeling:

1. Differencing: The most common technique, differencing involves subtracting consecutive values in the series to remove trends.
   - First-order differencing: $Y_t' = Y_t - Y_{t-1}$ removes linear trends.
   - Second-order differencing: $Y_t'' = Y_t' - Y_{t-1}'$ removes quadratic trends.
   - Seasonal differencing: Used when the data exhibits seasonal patterns.
2. Logarithmic Transformation: Taking the natural logarithm of the data can stabilize variance, particularly in exponential growth patterns.
3. Box–Cox Transformation: A more flexible transformation that stabilizes variance while making the data more normally distributed.
4. Decomposition: Separating a time series into trend, seasonal, and residual components allows for targeted transformations.

Ensuring stationarity is a critical step in time series analysis and forecasting. By using statistical tests to diagnose stationarity and applying appropriate transformations like differencing or Box-Cox, analysts can enhance model reliability and improve forecasting accuracy. Whether working with financial markets, climate data, or economic indicators, stationarity remains a foundational concept for robust time series modeling.

Let's first formally test all these series and see what we get:

```
# number of first differences
toronto %>%
  features(cases, unitroot_ndiffs)
```

```
## # A tibble: 1 x 0
```

```
# Formal KPSS test on level
toronto %>%
  features(cases, unitroot_kpss)
```

```
## # A tibble: 1 x 0
```

```
# Formal KPSS test on the first difference
toronto %>%
  mutate(diffcases = difference(cases)) %>%
  features(diffcases, unitroot_kpss)
```

```
## # A tibble: 1 x 0
```

It seems that the first difference can make the `cases` series stationary. The null in this test suggests that the series are stationary, and the p-value indicates that the null is rejected. So, it seems that the test after first differencing gives us a green light! However, ACF's are telling us that seasonal differencing would be needed:

```
level <- toronto %>% ACF(cases) %>%
  autoplot() + labs(subtitle = "Covid-19 Cases") +
    theme_bw() + scale_color_grey()

fdiff <- toronto %>% ACF(difference(cases)) %>%
  autoplot() + labs(subtitle = "First-difference") +
    theme_bw() + scale_color_grey()

diffbc <- toronto %>% ACF(difference(box_cox(cases, lmbd))) %>%
  autoplot() + labs(subtitle = "First-difference Box-Cox") +
    theme_bw() + scale_color_grey()

ddiff <-
  toronto %>% ACF(difference(difference(box_cox(cases, lmbd)))) %>%
  autoplot() + labs(subtitle = "Double-difference Box-Cox") +
    theme_bw() + scale_color_grey()

require(gridExtra)
grid.arrange(level, fdiff, diffbc, ddiff, ncol = 2, nrow = 2)
```

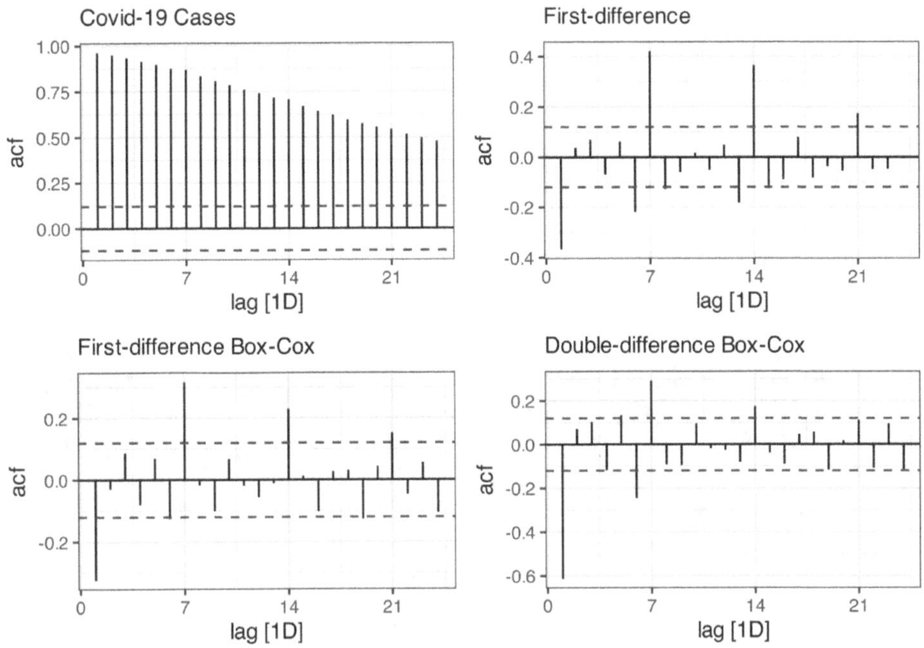

From ACF's, there seems to be a weekly seasonal pattern at 7, 14, and 21, which are Sundays. We know that reported COVID-19 cases on Sundays tend to be lower than the rest of the week at least during the first wave.

We can also test if we need seasonal differencing:

```
toronto %>%
  features(cases, unitroot_nsdiffs)
```

```
## # A tibble: 1 x 1
##   nsdiffs
##     <int>
## 1       0
```

```
# with Box-Cox
toronto %>%
  features(box_cox(cases, lmbd), unitroot_nsdiffs)
```

```
## # A tibble: 1 x 1
##   nsdiffs
##     <int>
## 1       0
```

The feature `unitroot_nsdiffs` returns 0 for both original and transformed series indicating no seasonal difference is required. We will stick to this "advice" because of two reasons. First, an unnecessary differencing would create more problems than a solution. Second, we can also modify ARIMA to incorporate seasonallity in the data, which we will see shortly.

Yet, out of curiosity, let's remove the "seemingly" weekly seasonality and see what happens to ACF's. Since, the order of differencing is not important, we first applied the seasonal differencing then applied the first difference:

```
toronto %>%
  gg_tsdisplay(difference(box_cox(cases, lmbd), 7) %>% difference(),
               plot_type = 'partial',
               lag = 36) +
  labs(title = "Seasonal & first differenced", y = "")
```

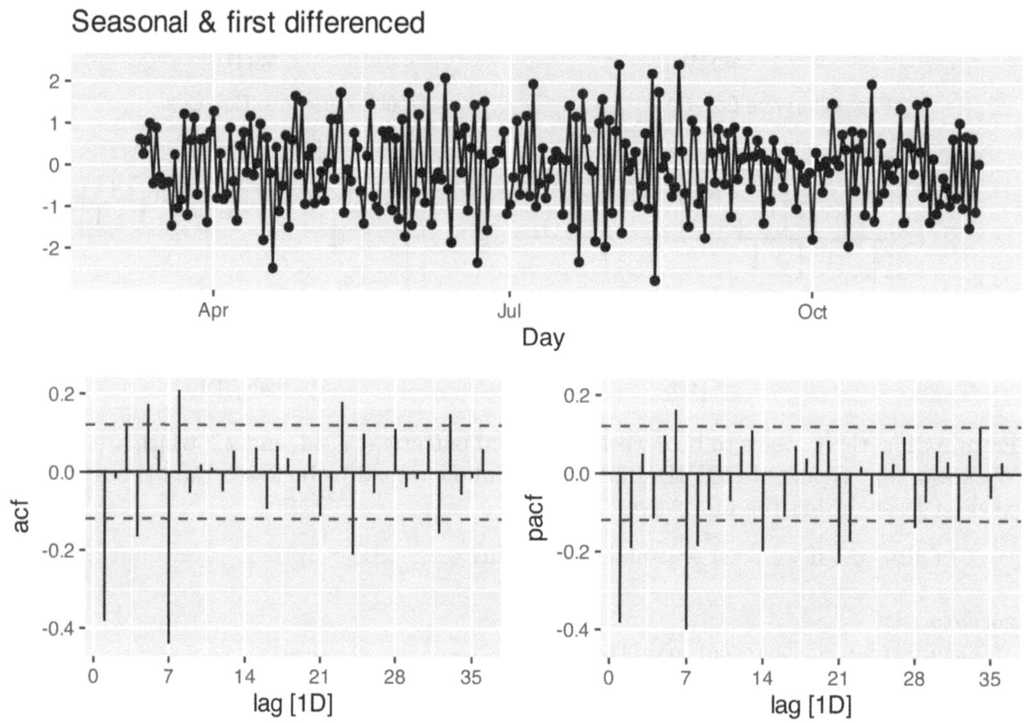

We can calculate the strength of the trend (T) and seasonality (S) in the time series, $y_t = T_t + S_t + R_t$, by

$$F_{Trend} = \max\left(0, 1 - \frac{\text{Var}\,(R_t)}{\text{Var}\,(T_t + R_t)}\right),$$
$$F_{Seasonality} = \max\left(0, 1 - \frac{\text{Var}\,(R_t)}{\text{Var}\,(S_t + R_t)}\right),$$

where R_t is the remainder component:

```
t <- toronto %>% features(cases, feat_stl)
t(t[1:2])
```

```
##                           [,1]
## trend_strength         0.9843102
## seasonal_strength_week 0.5142436
```

Relative to F_{Trend}, the seasonality is not robust in the data. So, our decision is to go with a simple first-differencing with Box–Cox transformation. However, we will look at the final predictive performance if the transformation provides any benefit.

28.6 Modeling ARIMA

In his post, "Forecasting COVID-19",[4] Rob J Hyndman makes the following comment in March 2020:

> (...) the COVID-19 pandemic, it is easy to see why forecasting its effect is difficult. While we have a good understanding of how it works in terms of person-to-person infections, we have limited and misleading data. The current numbers of confirmed cases are known to be vastly underestimated due to the limited testing available. There are almost certainly many more cases of COVID-19 that have not been diagnosed than those that have. Also, the level of under-estimation varies enormously between countries. In a country like South Korea with a lot of testing, the numbers of confirmed cases are going to be closer to the numbers of actual cases than in the US where there has been much less testing. So we simply cannot easily model the spread of the pandemic using the data that is available.

> The second problem is that the forecasts of COVID-19 can affect the thing we are trying to forecast because governments are reacting, some better than others. A simple model using the available data will be misleading unless it can incorporate the various steps being taken to slow transmission.

> In summary, fitting simple models to the available data is pointless, misleading and dangerous.

With our selection of the data, we do not intent to create another debate on forecasting COVID-19. There are hundreds of different forecasting models currently operational in a hub, The COVID-19 Forecast Hub,[5] that can be used live. We will start with an automated algorithm `ARIMA()` that will allow a seasonal parameters:

$$\text{ARIMA } (p, d, q) \times (P, D, Q)S$$

The first term is the non-seasonal part of ARIMA with $p =$ AR order, $d =$ non-seasonal differencing, $q =$ MA order. The secon term is seasonal part of the model with $P =$ seasonal

AR order, D = seasonal differencing, Q = seasonal MA order, and S = seasonal pattern, which defines the number of time periods until the pattern repeats again.

In our case, low values tend always to occur in some particular days, Sundays. Therefore, we may think that S = 7 is the span of the periodic seasonal behavior in our data. We can think of a seasonal first order autoregressive model, AR(1), would use X_{t-7} to predict X_t. Likewise, a seasonal second order autoregressive model would use X_{t-7} and X_{t-14} to predict X_t. A seasonal first order MA(1) model would use ϵ_{t-7} as a predictor. A seasonal second order MA(2) model would use ϵ_{t-7} and ϵ_{t-14}.

Let's use our data first-differenced and transformed:

```
toronto <- toronto %>%
  mutate(boxcases = box_cox(cases, lambda = lmbd))

toronto %>%
  gg_tsdisplay(difference(boxcases), plot_type='partial')
```

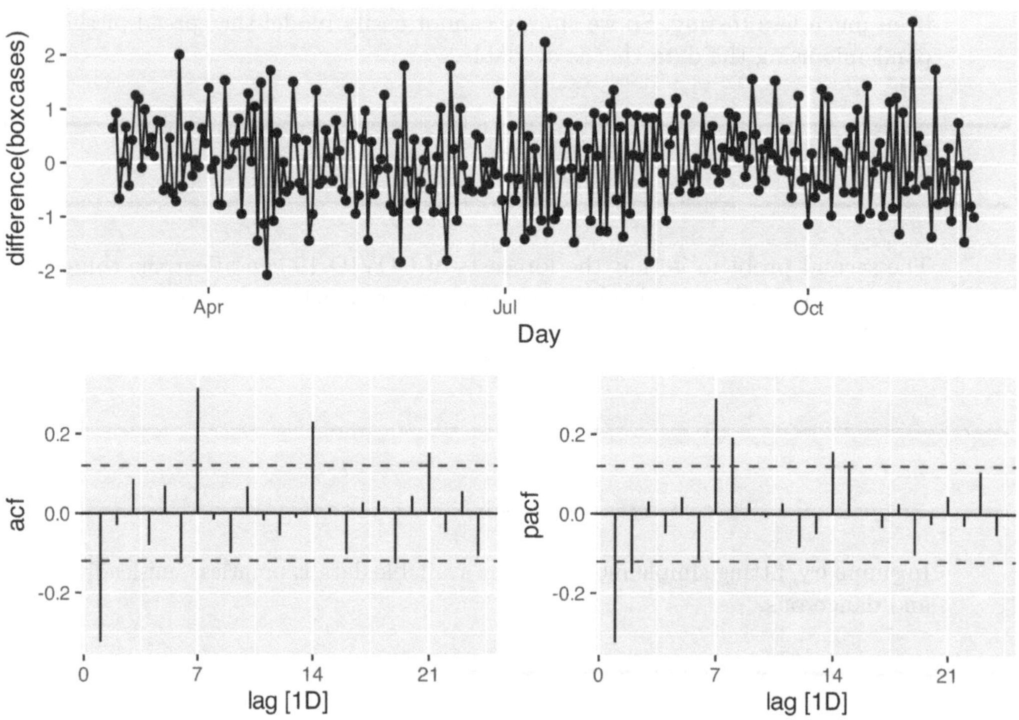

We look at the spikes and decays in ACF and PACF: a exponential decay in ACF is observed at seasonal spikes of 7, 14, and 21 as well as two spikes at 7 and 14 in PACF indicate seasonal AR(2). We will also add non-seasonal AR(2) due to 2 spikes in PACF at days 1 and 2. Here are our initial models:

$$\text{ARIMA}(2,1,0)(2,1,0)_7 \ \text{ARIMA}(0,1,2)(0,1,3)_7$$

```
covfit <- toronto %>%
  model(
    AR2 = ARIMA(boxcases ~ pdq(2, 1, 0) + PDQ(3, 1, 0)),
    MA3 = ARIMA(boxcases ~ pdq(0, 1, 2) + PDQ(0, 1, 3)),
    auto = ARIMA(boxcases, stepwise = FALSE, approx = FALSE)
  )

t(cbind(
  "AR2" = covfit$AR2,
  "MA3" = covfit$MA3,
  "auto" = covfit$auto
))
```

```
##      [,1]
## AR2  ARIMA(2,1,0)(3,1,0)[7]
## MA3  ARIMA(0,1,2)(0,1,3)[7]
## auto NULL model
```

```
glance(covfit) %>% arrange(AICc) %>% dplyr::select(.model:BIC)
```

```
## # A tibble: 2 x 6
##   .model sigma2 log_lik   AIC  AICc   BIC
##   <chr>   <dbl>   <dbl> <dbl> <dbl> <dbl>
## 1 MA3     0.468   -277.  567.  567.  588.
## 2 AR2     0.534   -285.  582.  583.  604.
```

```
covfit %>% dplyr::select(MA3) %>% report()
```

```
## Series: boxcases
## Model: ARIMA(0,1,2)(0,1,3)[7]
##
## Coefficients:
##           ma1     ma2     sma1     sma2     sma3
##       -0.4340  0.1330  -0.8617  -0.0573  -0.0809
## s.e.   0.0648  0.0612   0.0827   0.0733   0.0600
##
## sigma^2 estimated as 0.4684:  log likelihood=-277.29
## AIC=566.58   AICc=566.92   BIC=587.9
```

The `ARIMA()` function uses `unitroot_nsdiffs()` to determine D when it is not specified. Earlier, we run this function that suggested no seasonal differencing.

All other parameters are determined by minimizing the AICc (Akaike's Information Criterion with a correction for finite sample sizes), which is similar to Akaike's Information Criterion (AIC), but it includes a correction factor to account for the fact that the sample size may be small relative to the number of parameters in the model. This correction helps to reduce the bias in the AIC estimate and make it more accurate for small sample sizes. When the sample size is large, AIC and AICc are nearly equivalent and either one can be used.

Although AICc values across the models are not comparable (for "auto", as it has no seasonal differencing), it seems that our manually constructed ARIMA, ARIMA$(0,1,2)(0,1,3)_7$ could also be an option. This brings the possibility of a grid search to our attention.

Before that, however, let's check their residuals:

```
rbind(
  augment(covfit) %>%
    filter(.model == "auto") %>%
    features(.innov, ljung_box, lag = 24, dof = 5),
  augment(covfit) %>%
    filter(.model == "MA3") %>%
    features(.innov, ljung_box, lag = 24, dof = 5),
  augment(covfit) %>%
    filter(.model == "AR2") %>%
    features(.innov, ljung_box, lag = 24, dof = 5)
)
```

```
## # A tibble: 3 x 3
##   .model lb_stat lb_pvalue
##   <chr>    <dbl>     <dbl>
## 1 auto      NA      NA
## 2 MA3       27.3     0.0971
## 3 AR2       21.1     0.331
```

```
covfit %>%dplyr::select(MA3) %>% gg_tsresiduals(lag=36)
```

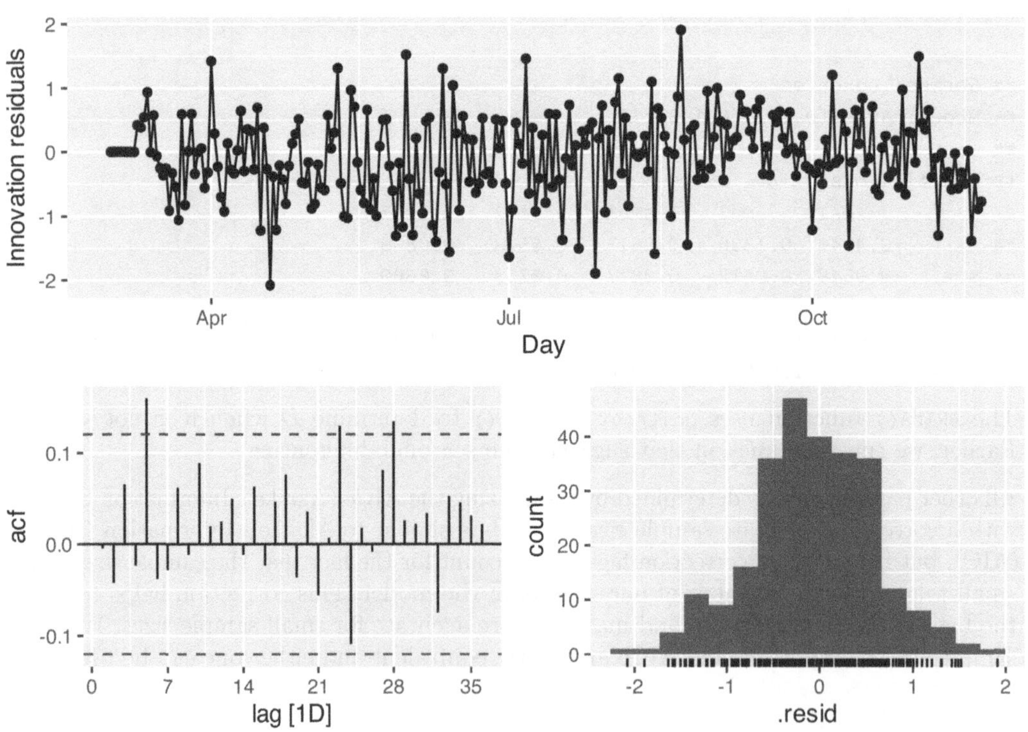

There are several significant spikes in the ACF. But, the model passes the Ljung–Box test at the 5 percent significance level.

Meanwhile, a model without white noise errors can still be used for forecasting, but the prediction intervals may not be accurate due to the correlated residuals. Sometimes, we cannot find a model that passes this test. In practice, we may have to look at the trade-off between prediction accuracy and reliable confidence intervals. If the difference is too high, we may chose the best model with the highest prediction accuracy.

Before looking at a cross-validation approach for model selection in ARIMA modeling, let use our model to predict a week ahead (2020-11-22 to 2020-11-28):

```
fc <- covfit %>%
  forecast(h = 7)
fc
```

```
## # A fable: 21 x 4 [1D]
## # Key:     .model [3]
##    .model Day            boxcases .mean
##    <chr>  <date>           <dist> <dbl>
##  1 AR2    2020-11-22 N(12, 0.53)  12.1
##  2 AR2    2020-11-23 N(13, 0.68)  13.3
##  3 AR2    2020-11-24 N(13, 0.87)  12.8
##  4 AR2    2020-11-25  N(13, 1.1)  12.8
##  5 AR2    2020-11-26  N(12, 1.3)  12.2
##  6 AR2    2020-11-27  N(12, 1.5)  12.3
##  7 AR2    2020-11-28  N(12, 1.7)  11.5
##  8 MA3    2020-11-22 N(12, 0.48)  12.4
##  9 MA3    2020-11-23 N(13, 0.63)  13.2
## 10 MA3    2020-11-24 N(13, 0.87)  13.1
## # i 11 more rows
```

```
library(gridExtra)

# Produce a 7-day ahead forecast once
fc <- covfit %>% forecast(h = 7)

# 1) Full forecast (no intervals)
fc %>%
  autoplot(toronto, level = c()) +
  scale_color_manual(values = c("black", "gray30", "gray60")) +
  scale_fill_manual(values = c("gray90", "gray70", "gray50")) +
  xlab("Days") +
  ylab("Transformed Cases with Box-Cox") +
  theme_minimal()
```

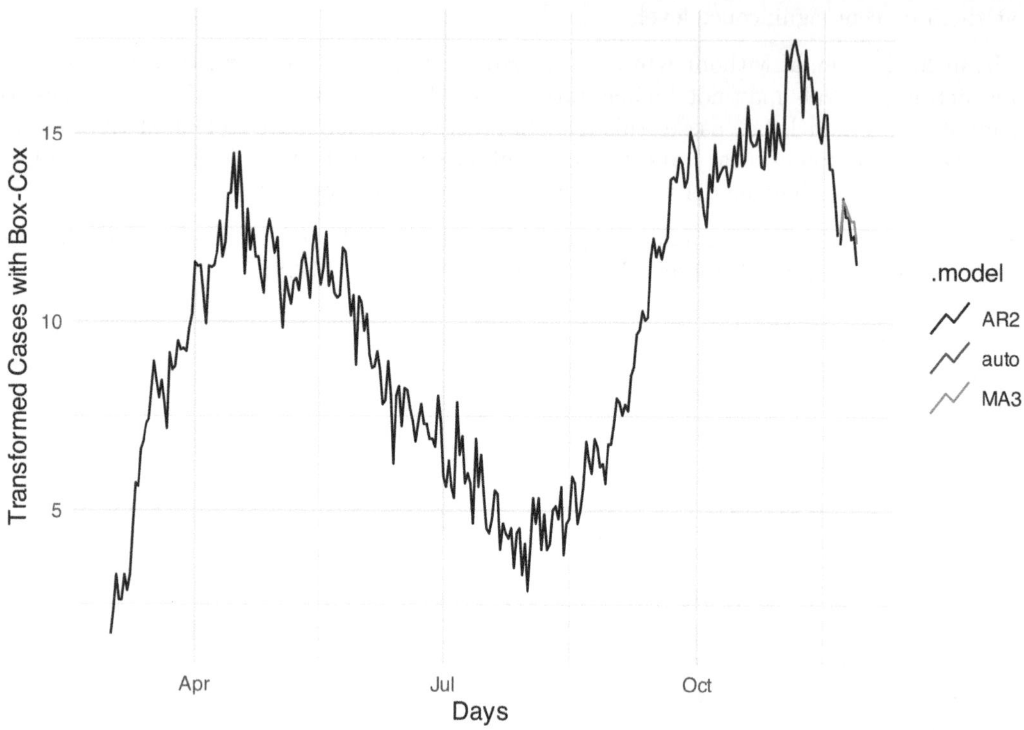


```
# 2) Individual-model plots (no intervals, no legends, grayscale)
a <- fc %>%
  filter(.model == "auto") %>%
  autoplot(toronto, level = c()) +
  aes(color = ".model") +  # force color mapping
  scale_color_manual(values = c("auto" = "black")) +
  scale_fill_manual(values = c("auto" = "gray70")) +
  labs(title = "COVID-19 Forecasting - Auto",
       y = "Box-Cox Transformed Cases") +
  theme_minimal() +
  theme(legend.position = "none")


b <- fc %>%
  filter(.model == "MA3") %>%
  autoplot(toronto, level = c()) +
  aes(color = ".model") +  # ensure color is mapped
  scale_color_manual(values = c("MA3" = "black")) +
  scale_fill_manual(values = c("MA3" = "gray70")) +
  labs(title = "COVID-19 Forecasting - MA3",
       y = "Box-Cox Transformed Cases") +
  theme_minimal() +
  theme(legend.position = "none")
```

```
# Arrange side-by-side
gridExtra::grid.arrange(a, b, ncol = 2)
```

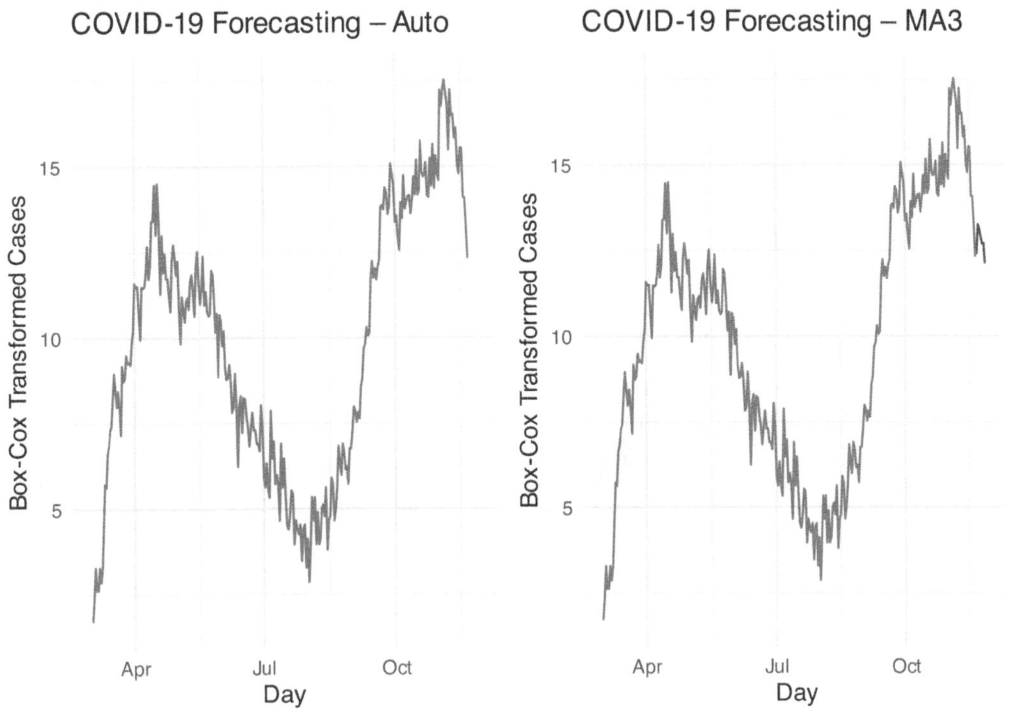

We have predicted values for coming 7 days but we do not have realized values. Hence, we cannot compare these models in terms of their accuracy. We can look at the forecast accuracy of these models by using a training set containing all data up to 2020-11-14. When we forecast the remaining seven days in the data, we can calculate the prediction accuracy.

```
train <- toronto %>%
  filter_index( ~ "2020-11-14")

fit <- train %>%
  model(
    AR2 = ARIMA(boxcases ~ pdq(2, 1, 0) + PDQ(3, 1, 0)),
    MA3 = ARIMA(boxcases ~ pdq(0, 1, 2) + PDQ(0, 1, 3)),
    auto = ARIMA(boxcases, stepwise = FALSE, approx = FALSE)
  ) %>%
  mutate(mixed = (auto + AR2 + MA3) / 3)
```

Although mixing several different ARIMA models does not make sense, we can have an ensemble forecast mixing several different time series models in addition ARIMA modeling. A nice discussion can be found in this post[6] at Stackoverflow.

And, now the accuracy measures:

```
fc <- fit %>% forecast(h = 7)
fc %>%
  autoplot(toronto, level = NULL)
```

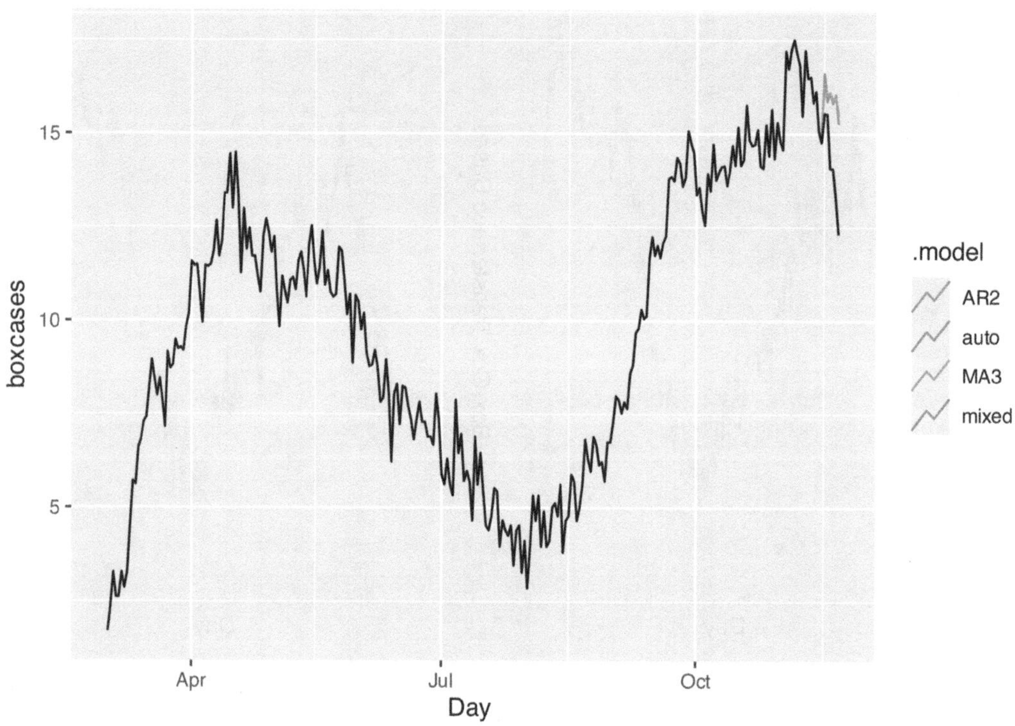


```
accuracy(fc, toronto)
```

```
## # A tibble: 4 x 10
##   .model .type     ME   RMSE    MAE   MPE  MAPE   MASE  RMSSE   ACF1
##   <chr>  <chr>  <dbl>  <dbl>  <dbl> <dbl> <dbl>  <dbl>  <dbl>  <dbl>
## 1 AR2    Test   -1.57   1.88   1.57 -11.6  11.6   1.35   1.30  0.359
## 2 MA3    Test   -1.61   1.91   1.61 -11.9  11.9   1.38   1.32  0.501
## 3 auto   Test  NaN    NaN    NaN    NaN   NaN   NaN    NaN    NA
## 4 mixed  Test  NaN    NaN    NaN    NaN   NaN   NaN    NaN    NA
```

In all measures, the model "auto" (ARIMA with the Hyndman-Khandakar algorithm) is better than others.

Finally, it is always good to check ARIMA (or any time series forecasting) against the base benchmark (lm is the horizantal line in the figure below).

```
bfit <- train %>%
  model(ave = MEAN(boxcases),
        lm = TSLM(boxcases ~ trend() + season()))
```

```
bfc <- bfit %>% forecast(h = 7)

bfc %>%
  autoplot(toronto, level = NULL)
```

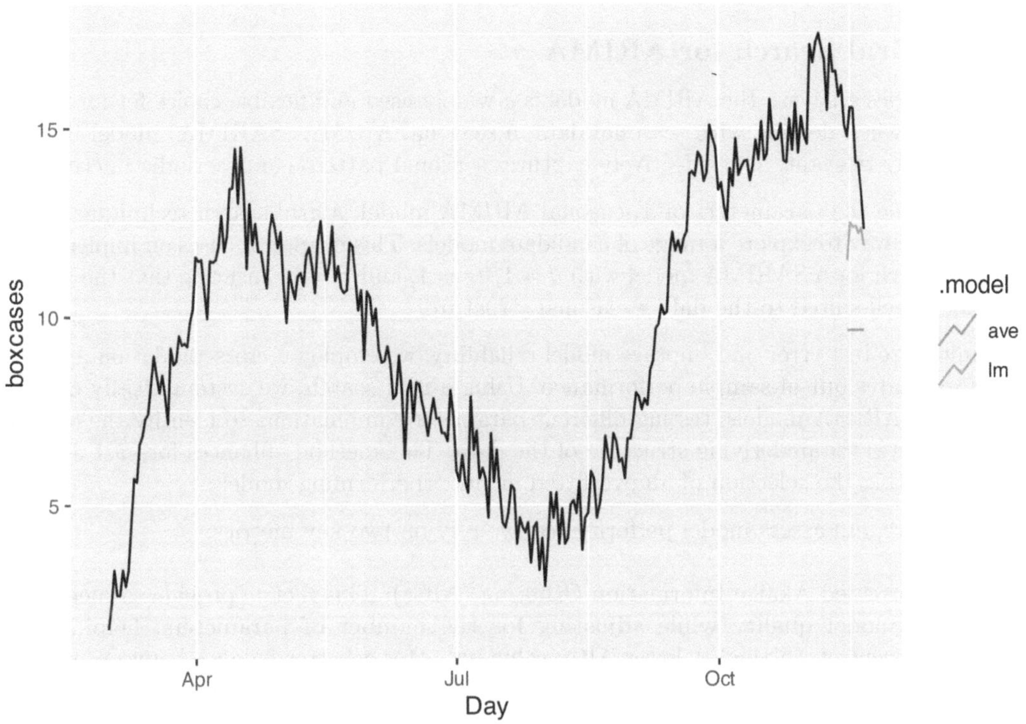


```
accuracy(bfc, toronto)
```

```
## # A tibble: 2 x 10
##   .model .type    ME  RMSE   MAE   MPE  MAPE  MASE RMSSE  ACF1
##   <chr>  <chr> <dbl> <dbl> <dbl> <dbl> <dbl> <dbl> <dbl> <dbl>
## 1 ave    Test   4.59  4.72  4.59  31.8  31.8  3.94  3.26 0.507
## 2 lm     Test   2.07  2.32  2.07  14.1  14.1  1.77  1.60 0.516
```

The results shows our ARIMA model is doing much better job relative to a time-series linear model or a simple average.

As we discussed earlier in this book, there are basically two ways to select a best fitting predictive model: ex-post and ex-ante tools to penalize the overfitting. With AIC and BIC measures, we can indirectly estimate the test (out-of-sample) error by making an adjustment to the training (in-sample) error to account for the bias due to overfitting. Therefore, these methods are ex-post tools to penalize the overfitting. The Hyndman-Khandakar algorithm uses this ex-post approach by selecting the best predictive ARIMA model with minimum AICc among alternatives.

We can directly estimate the test error (out-sample) and choose the model that minimizes it. Instead of selecting a model with AICc, we can do it by tuning the parameters of ARIMA with a cross-validation approach so that the tuned model achieves the highest predictive accuracy.

28.7 Grid Search for ARIMA

In time series analysis, the ARIMA model is a widely used and flexible choice for forecasting. However, when dealing with seasonal data, a seasonal ARIMA (SARIMA) model becomes particularly relevant, as it effectively captures seasonal patterns and periodic fluctuations.

To fine-tune the parameters of a seasonal ARIMA model, a grid search technique offers a systematic way to explore a range of candidate models. This chapter focuses on implementing a grid search for a SARIMA model with $d = 1$, $D = 1$, and $S = 7$, ensuring that the selected model is well-suited to the data's seasonal structure.

To minimize test error and improve model reliability, we employ a cross-validation approach that evaluates out-of-sample performance. Using a grid search, we systematically examine multiple SARIMA models, testing different parameter combinations to identify the one that best captures the underlying structure of the data. This method enhances forecast accuracy by preventing the selection of an overfitted or underperforming model.

To compare and assess model performance, we rely on two key metrics:

1. Corrected Akaike Information Criterion (AICc): This metric provides a measure of model quality while adjusting for the number of parameters, helping to prevent overfitting. A lower AICc value indicates a better balance between model complexity and goodness-of-fit.

2. Root Mean Squared Error (RMSE): RMSE quantifies the deviation between observed and predicted values, offering an intuitive measure of forecast accuracy. A lower RMSE suggests a model that produces more precise predictions.

By integrating grid search, cross-validation, and rigorous model evaluation, we ensure that the selected SARIMA model not only captures seasonal trends effectively but also delivers robust and reliable forecasts.

```
#In-sample grid-search
p <- 0:3
q <- 0:3
P <- 0:3
Q <- 0:2

comb <- as.matrix(expand.grid(p, q, P, Q))

# We remove the unstable grids
comb <- as.data.frame(comb[-1,])
ind <- which(comb$Var1 == 0 & comb$Var2 == 0, arr.ind = TRUE)
```

```
comb <- comb[-ind,]
row.names(comb) <- NULL
colnames(comb) <- c("p", "q", "P", "Q")

aicc <- c()
RMSE <- c()

for (k in 1:nrow(comb)) {
  tryCatch({
    fit <- toronto %>%
      model(ARIMA(boxcases ~ 0 + pdq(comb[k, 1], 1, comb[k, 2])
                  + PDQ(comb[k, 3], 1, comb[k, 4])))
    wtf <- fit %>% glance
    res <- fit %>% residuals()
    aicc[k] <- wtf$AICc
    RMSE[k] <- sqrt(mean((res$.resid) ^ 2))
  }, error = function(e) {
  })
}

cbind(comb[which.min(aicc), ], "AICc" = min(aicc, na.rm = TRUE))
```

```
##    p q P Q     AICc
## 75 3 3 0 1 558.7746
```

```
cbind(comb[which.min(RMSE), ], "RMSE" = min(RMSE, na.rm = TRUE))
```

```
##     p q P Q      RMSE
## 120 3 3 3 1 0.6478857
```

Although we initially set the ARIMA model without a constant, we can extend the grid search to include models with a constant term. Additionally, we can incorporate a Ljung-Box test (ljung_box) to assess the adequacy of each model's residuals. By evaluating both criteria—minimizing the AICc and passing the Ljung–Box test—we can refine our model selection process for improved accuracy.

However, a full grid search may not always be necessary. The Hyndman–Khandakar algorithm offers an efficient alternative for automatic ARIMA model selection, optimizing the model based on the minimum AICc. While this method does not perform a Ljung-Box test for each model, it effectively automates the selection process.

In practice, we can take a different approach: instead of checking for residual autocorrelation via the Ljung–Box test, we can implement a grid search with cross-validation, selecting the model that yields the minimum out-of-sample prediction error. This ensures that the chosen model generalizes well to new data, focusing on predictive accuracy rather than residual diagnostics.

Here is a simple example demonstrating how this selection process can be applied:

```
#In-sample grid-search
p <- 0:3
```

```
q <- 0:3
P <- 0:3
Q <- 0:2

comb <- as.matrix(expand.grid(p, q, P, Q))

# We remove the unstable grids
comb <- as.data.frame(comb[-1,])
ind <- which(comb$Var1 == 0 & comb$Var2 == 0, arr.ind = TRUE)
comb <- comb[-ind, ]
row.names(comb) <- NULL
colnames(comb) <- c("p", "q", "P", "Q")

train <- toronto %>%
  filter_index( ~ "2020-11-14")

RMSE <- c()

for (k in 1:nrow(comb)) {
  tryCatch({
    amk <- train %>%
      model(ARIMA(boxcases ~ 0 + pdq(comb[k, 1], 1, comb[k, 2])
                  + PDQ(comb[k, 3], 1, comb[k, 4]))) %>%
      forecast(h = 7) %>%
      accuracy(toronto)
    RMSE[k] <- amk$RMSE
  }, error = function(e) {
  })
}

cbind(comb[which.min(RMSE), ], "RMSE" = min(RMSE, na.rm = TRUE))
```

```
##    p q P Q      RMSE
## 12 0 3 0 0 0.7937723
```

```
g <- which.min(RMSE)
toronto %>%
  model(ARIMA(boxcases ~ 0 + pdq(comb[g, 1], 1, comb[g, 2])
              + PDQ(comb[g, 3], 1, comb[g, 4]))) %>%
  forecast(h = 7) %>%
  autoplot(toronto, level = NULL)
```

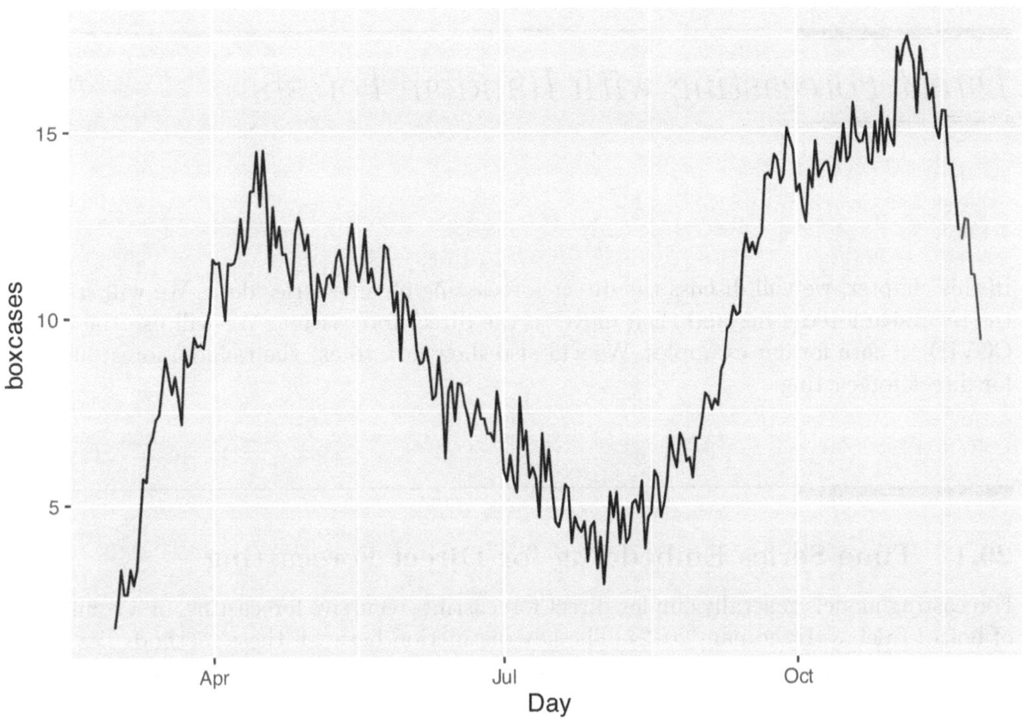

We will not apply h-step-ahead rolling-window cross-validations for ARIMA, which can be found in the post, Time series cross-validation using fable,[7] by Hyndman (2021). However, when we have multiple competing models, we may not want to compare their predictive accuracy by looking at their error rates using only few out-of-sample observations. If we use rolling windows or continuously expanding windows, we can effectively create a large number of days tested within the data.

Notes

1. https://data.ontario.ca/en/dataset?groups=2019-novel-coronavirus#byPHU
2. https://robjhyndman.com/hyndsight/tsibbles/
3. https://onlinelibrary.wiley.com/doi/10.1002/for.3980120104
4. https://robjhyndman.com/hyndsight/forecasting-covid19/#how-can-we-forecast-covid-19
5. https://covid19forecasthub.org
6. https://stackoverflow.com/questions/70183054/time-series-forecasting-using-fable-in-r-determining-most-optimum-combination-o
7. https://robjhyndman.com/hyndsight/tscv-fable/

29

Direct Forecasting with Random Forests

In this chapter, we will discuss the direct forecasting of time series data. We will start with the recursive forecasting and then move to the direct forecasting. We will use the Toronto COVID-19 data for our examples. We will also show how to use the random forest algorithm for direct forecasting.

29.1 Time Series Embedding for Direct Forecasting

Forecasting models generally employ direct forecasting, recursive forecasting, or a combination of both (Taieb & Hyndman, 2012). The key distinction between these methods lies in their approach to prediction accuracy and forecast variance, which in turn affects their suitability for different forecasting tasks.

Before exploring time series embedding for direct forecasting, let's first review the two primary forecasting strategies.

1. **Recursive Forecasting**:
   - Uses a single parametric model, such as AR(1), to generate predictions step by step.
   - The forecast for time $t+1$ is first estimated, then used as an input for predicting $t+2$, and so on.
   - This iterative process can accumulate errors, leading to increasing uncertainty as the forecast horizon expands.
   - It is particularly effective when the underlying time series follows a linear structure, but struggles when dealing with nonlinear dependencies.
2. **Direct Forecasting**:
   - Instead of iterating over previous predictions, direct forecasting builds separate models for each time step within the forecast horizon.
   - This approach allows for nonparametric methods, such as machine learning algorithms, which can capture complex and nonlinear relationships in the data.
   - However, direct forecasting often exhibits higher variance, especially as the forecast horizon increases, due to the need for estimating multiple models independently.

In multi-step forecasting, a common recursive approach involves using a single model—such as AR(1)—and iteratively substituting its own predictions to generate future values. Given a forecast horizon of h, the model is applied recursively to estimate:

DOI: 10.1201/9781003504795-29

$$
\begin{aligned}
\hat{y}_{t+1} &= f(y_t, y_{t-1}, \dots) \\
\hat{y}_{t+2} &= f(\hat{y}_{t+1}, y_t, \dots) \\
&\vdots \\
\hat{y}_{t+h} &= f(\hat{y}_{t+h-1}, \dots)
\end{aligned}
\tag{29.1}
$$

where $\hat{y}_{t+h}$ is the forecasted value at time $t + h$, and f represents the forecasting model. The recursive nature of this approach can lead to error accumulation, particularly for longer forecast horizons.

Direct forecasting, in contrast, requires restructuring the dataset to create individual models for each future time step. This technique, known as time series embedding, transforms the original time series into a supervised learning problem, where each target variable corresponds to a specific future time step. The process involves:

- Creating lagged features from past observations.
- Training separate models for each step within the forecast horizon.
- Ensuring that data preprocessing aligns with the expected forecast structure.

By embedding time series data effectively, direct forecasting can provide more flexible and adaptive predictions, particularly in cases where relationships between past and future values are highly nonlinear.

Choosing between recursive and direct forecasting depends on the trade-off between bias and variance. Recursive forecasting leverages a single model but risks error accumulation over longer horizons. Direct forecasting reduces error propagation but requires estimating multiple models, increasing variance.

In practice, hybrid approaches combining both methods are often employed to balance predictive accuracy and robustness. This chapter explores time series embedding in depth, demonstrating how to effectively restructure data to implement direct forecasting models.

Let's start with a simple AR(1) model to illustrate recursive multi-step forecasting:

$$x_{t+1} = \alpha_0 + \phi_1 x_t + \epsilon_t \tag{29.2}$$

To generate forecasts for the next three periods, we recursively substitute predicted values into the model:

$$
\begin{aligned}
\hat{x}_{t+1} &= \hat{\alpha}_0 + \hat{\phi}_1 x_t \\
\hat{x}_{t+2} &= \hat{\alpha}_0 + \hat{\phi}_1 \hat{x}_{t+1} \\
\hat{x}_{t+3} &= \hat{\alpha}_0 + \hat{\phi}_1 \hat{x}_{t+2}
\end{aligned}
\tag{29.3}
$$

Applying iterative substitution gives us the forecasts expressed in terms of x_t:

$$
\begin{aligned}
\hat{x}_{t+1} &= \hat{\alpha}_0 + \hat{\phi}_1 x_t \quad \text{(1st period)} \\
\hat{x}_{t+2} &= \hat{\alpha}_0 + \hat{\phi}_1 \hat{\alpha}_0 + \hat{\phi}_1^2 x_t \quad \text{(2nd period)} \\
\hat{x}_{t+3} &= \hat{\alpha}_0 + \hat{\phi}_1 \hat{\alpha}_0 + \hat{\phi}_1^2 \hat{\alpha}_0 + \hat{\phi}_1^3 x_t \quad \text{(3rd period)}
\end{aligned}
\tag{29.4}
$$

More generally, the h-step recursive forecast is:

$$\hat{x}_{t+h} = \hat{\alpha}_0 \sum_{i=1}^{h} \hat{\phi}_1^{i-1} + \hat{\phi}_1^h x_t \tag{29.5}$$

This approach uses the same estimated model for every step ahead; only the inputs change. Thus, a single model suffices to generate forecasts for any horizon.

Alternatively, the direct forecasting strategy fits a separate model for each forecast horizon. For example, with AR(1)-type direct forecasts for three periods ahead:

$$\begin{aligned} x_{t+1} &= \alpha_0 + \alpha_1 x_t + \epsilon_t \\ x_{t+2} &= \beta_0 + \beta_1 x_t + \epsilon_t \\ x_{t+3} &= \omega_0 + \omega_1 x_t + \epsilon_t \end{aligned} \tag{29.6}$$

Each model is estimated independently, and forecasts are given by:

$$\begin{aligned} \hat{x}_{t+1} &= \hat{\alpha}_0 + \hat{\alpha}_1 x_t \quad \text{(1st period)} \\ \hat{x}_{t+2} &= \hat{\beta}_0 + \hat{\beta}_1 x_t \quad \text{(2nd period)} \\ \hat{x}_{t+3} &= \hat{\omega}_0 + \hat{\omega}_1 x_t \quad \text{(3rd period)} \end{aligned} \tag{29.7}$$

Direct forecasting avoids compounding forecast errors across horizons, but requires estimating and maintaining multiple models—one for each step ahead.

29.2 VAR for Recursive Forecasting

The limitations of recursive multi-step forecasting become more apparent in multivariate settings. Consider the bivariate case where y_t depends not only on its own lag but also on another variable x_t:

$$y_{t+1} = \beta_0 + \beta_1 y_t + \beta_2 x_t + \epsilon_t \tag{29.8}$$

To forecast two steps ahead, we substitute forecasted values:

$$\hat{y}_{t+2} = \hat{\beta}_0 + \hat{\beta}_1 \hat{y}_{t+1} + \hat{\beta}_2 \hat{x}_{t+1} \tag{29.9}$$

This requires not only forecasting $\hat{y}_{t+1}$ but also estimating $\hat{x}_{t+1}$. A natural solution is to use a vector autoregressive (VAR) model.

A VAR model consists of a system of equations—one per variable—each including a constant and lags of **all** variables in the system. For a VAR(1), we might specify:

$$y_t = c_1 + \beta_1 y_{t-1} + \beta_2 x_{t-1} + \varepsilon_t \tag{29.10}$$

$$x_t = c_2 + \phi_1 x_{t-1} + \phi_2 y_{t-1} + e_t \tag{29.11}$$

Each equation is estimated separately by ordinary least squares (OLS), provided the variables are stationary. Multi-step forecasts are then computed recursively: one-step-ahead forecasts feed into the next step, just as in the univariate case. The VAR framework handles these dependencies internally and provides consistent forecasts for all variables in the system.

To determine the appropriate number of lags, model selection criteria such as the Bayesian Information Criterion (BIC) are typically used.

Let's have our COVID-19 data and include the mobility to our forecasting model.

```
library(tsibble)
library(fpp3)

load("dftoronto.RData")
day <- seq.Date(
  from = as.Date("2020/03/01"),
  to = as.Date("2020/11/21"),
  by = 1
)

tdata <- tibble(Day = day,
                mob = data$mob,
                cases = data$cases)

toronto <- tdata %>%
  as_tsibble(index = Day)

toronto
```

```
## # A tsibble: 266 x 3 [1D]
##    Day             mob cases
##    <date>        <dbl> <dbl>
##  1 2020-03-01 -0.0172      4
##  2 2020-03-02 -0.0320      6
##  3 2020-03-03 -0.0119     10
##  4 2020-03-04  0.0186      7
##  5 2020-03-05  0.0223      7
##  6 2020-03-06 -0.00626    10
##  7 2020-03-07  0.0261      8
##  8 2020-03-08  0.0273     10
##  9 2020-03-09 -0.0158     18
## 10 2020-03-10 -0.0521     29
## # i 256 more rows
```

We will estimate the recursive forecasts for 1 to 14 days ahead.

```
# We need make series stationary
trdf <- toronto %>%
  mutate(diffcases = difference(cases),
         diffmob = difference(mob))

# VAR with BIC
fit <- trdf[-1, ] %>%
```

```
  model(VAR(vars(diffcases, diffmob), ic = "bic"))
glance(fit)
```

```
## # A tibble: 1 x 6
##   .model                                    sigma2     log_lik   AIC  AICc   BIC
##   <chr>                                     <list>       <dbl> <dbl> <dbl> <dbl>
## 1 "VAR(vars(diffcases, diffmob), ic = \"bi~ <dbl[...]>   -
854. 1755. 1760. 1841.
```

```
fit %>% report()
```

```
## Series: diffcases, diffmob
## Model: VAR(5)
##
## Coefficients for diffcases:
##       lag(diffcases,1)  lag(diffmob,1)  lag(diffcases,2)  lag(diffmob,2)
##                -0.4074       -105.6524           -0.0703         11.0374
## s.e.            0.0639         28.3643            0.0695         29.9761
##       lag(diffcases,3)  lag(diffmob,3)  lag(diffcases,4)  lag(diffmob,4)
##                 0.0528         10.8093           -0.0123         -4.8989
## s.e.            0.0701         31.8601            0.0713         30.0019
##       lag(diffcases,5)  lag(diffmob,5)
##                 0.0227          6.1099
## s.e.            0.0640         29.2678
##
## Coefficients for diffmob:
##       lag(diffcases,1)  lag(diffmob,1)  lag(diffcases,2)  lag(diffmob,2)
##                  0e+00          -0.314             0e+00         -0.4688
## s.e.             1e-04           0.057             1e-04          0.0603
##       lag(diffcases,3)  lag(diffmob,3)  lag(diffcases,4)  lag(diffmob,4)
##                  1e-04         -0.2931            -1e-04         -0.2664
## s.e.             1e-04          0.0641             1e-04          0.0603
##       lag(diffcases,5)  lag(diffmob,5)
##                  3e-04         -0.4059
## s.e.             1e-04          0.0588
##
## Residual covariance matrix:
##           diffcases diffmob
## diffcases  811.6771 -0.1648
## diffmob     -0.1648  0.0033
##
## log likelihood = -853.64
## AIC = 1755.28    AICc = 1760.38  BIC = 1840.73
```

```
library(tsibble); library(fpp3)

# fit is your VAR model; now forecast 14 days ahead
fc14 <- fit %>% forecast(h = 14)
```

```
# plot only the point forecasts (no intervals, no legend)
fc14 %>%
  autoplot(trdf[-c(1:200), ], level = c()) +
  labs(title = "VAR recursive forecasts (1-14 days ahead)",
       x = "Date",
       y = "First differences") +
  theme(legend.position = "none")
```

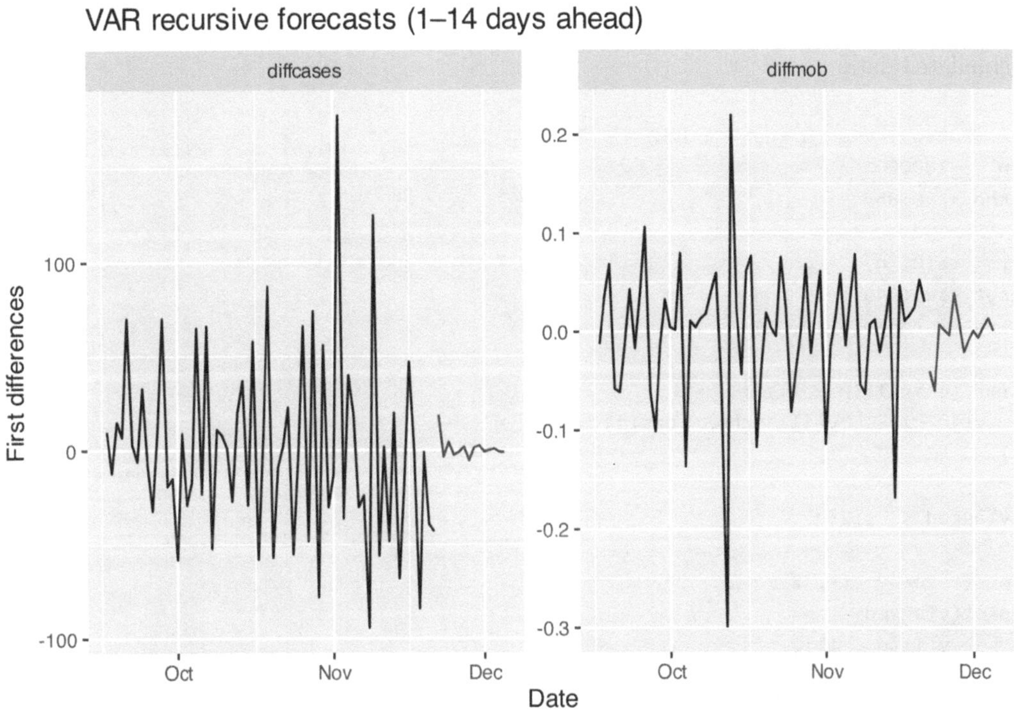

We should have transformed both series by the Box-Cox transformation, but we ignored it above.

29.3 Embedding for Direct Forecast

For direct forecasting, we need to rearrange the data in a way that we can estimate 7 models for forecasting ahead each day of 7 days. We will use `embed()` function to show what we mean with rearranging data for AR(3), for example:

```
Y <- 1:10
Y <- embed(Y, 3)
colnames(Y) = c("Y(t)", "Y(t-1)", "Y(t-2)")
Y
```

```
##       Y(t) Y(t-1) Y(t-2)
## [1,]    3      2      1
## [2,]    4      3      2
## [3,]    5      4      3
## [4,]    6      5      4
## [5,]    7      6      5
## [6,]    8      7      6
## [7,]    9      8      7
## [8,]   10      9      8
```

Now, the key point is that there is no a temporal dependence between each row so that shuffling this data after re-structuring it admissible. Let's have an AR(1) example on this simulated data:

```
# Stationary data rho < 1 but = 0.85
n <- 10000
rho <- 0.85

y <- c(0, n)
set.seed(345)
eps <- rnorm(n, 0, 1)

for (j in 1:(n - 1)) {
  y[j + 1] <- y[j] * rho + eps[j]
}

ylagged <- y[2:n]

par(mfrow = c(1, 2))
plot(ylagged,
     y[1:(n - 1)],
     col = "gray40",
     ylab = "y",
     xlab = "y(t-1)")
plot(y[1:500],
     type = "l",
     col = "gray20",
     ylab = "y",
     xlab = "t"
)
```

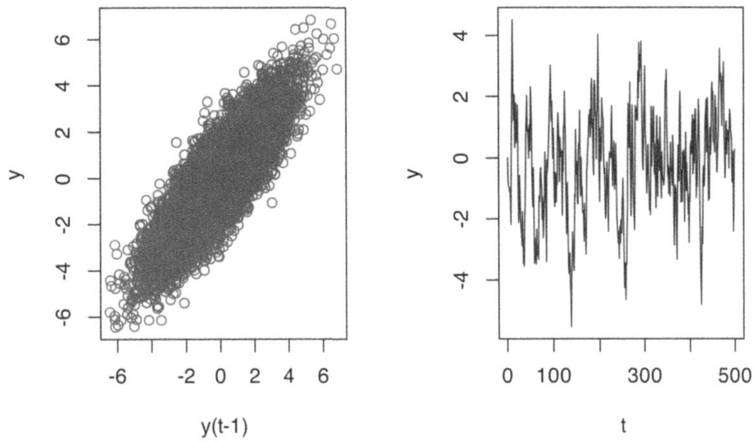

We will use an AR(1) estimation with OLS after embedding:

```
head(y)
```

```
## [1]  0.0000000 -0.7849082 -0.9466863 -0.9661413 -1.1118166 -1.0125757
```

```
y_em <- embed(y, 2)
colnames(y_em) <- c("yt", "yt_1")
head(y_em)
```

```
##              yt       yt_1
## [1,] -0.7849082  0.0000000
## [2,] -0.9466863 -0.7849082
## [3,] -0.9661413 -0.9466863
## [4,] -1.1118166 -0.9661413
## [5,] -1.0125757 -1.1118166
## [6,] -1.4942098 -1.0125757
```

And estimation of AR(1) with OLS:

```
y_em <- as.data.frame(y_em)
ar1 <- lm(yt ~ yt_1 - 1, y_em)
ar1
```

```
##
## Call:
## lm(formula = yt ~ yt_1 - 1, data = y_em)
##
## Coefficients:
##   yt_1
## 0.8496
```

Now, let's shuffle `y_em`:

```
# Shuffle
ind <- sample(nrow(y_em), nrow(y_em), replace = FALSE)
y_em_sh <- y_em[ind, ]

ar1 <- lm(yt ~ yt_1 - 1, y_em_sh)
ar1
```

```
##
## Call:
## lm(formula = yt ~ yt_1 - 1, data = y_em_sh)
##
## Coefficients:
##   yt_1
## 0.8496
```

This application shows the temporal independence across the observations in the rearranged data given that model (AR) is correctly specified. This is important because we can use conventional machine learning applications on time series data, like random forests, which we see in the next chapter.

This re-arrangement can also be applied to multivariate datasets:

```
tsdf <- matrix(c(1:10, 21:30), nrow = 10)
colnames(tsdf) <- c("Y", "X")
first <- embed(tsdf, 3)
colnames(first) <- c("y(t)","x(t)","y(t-1)","x(t-1)", "y(t-2)", "x(t-2)")
head(first)
```

```
##      y(t) x(t) y(t-1) x(t-1) y(t-2) x(t-2)
## [1,]    3   23      2     22      1     21
## [2,]    4   24      3     23      2     22
## [3,]    5   25      4     24      3     23
## [4,]    6   26      5     25      4     24
## [5,]    7   27      6     26      5     25
## [6,]    8   28      7     27      6     26
```

Now, we need to have three models for three forecasting horizons. Here are these models:

$$
\begin{aligned}
\hat{y}_{t+1} &= \hat{\alpha}_0 + \hat{\alpha}_1 y_t + \hat{\alpha}_2 y_{t-1} + \hat{\alpha}_3 x_t + \hat{\alpha}_4 x_{t-1} + \hat{\alpha}_5 x_{t-2} \quad & 1^{st} \text{ Period} \\
\hat{y}_{t+2} &= \hat{\beta}_0 + \hat{\beta}_1 y_t + \hat{\beta}_2 y_{t-1} + \hat{\beta}_3 x_t + \hat{\beta}_4 x_{t-1} + \hat{\beta}_5 x_{t-2} \quad & 2^{nd} \text{ Period} \\
\hat{y}_{t+3} &= \hat{\omega}_0 + \hat{\omega}_1 y_t + \hat{\omega}_2 y_{t-1} + \hat{\omega}_3 x_t + \hat{\omega}_4 x_{t-1} + \hat{\omega}_5 x_{t-2} \quad & 3^{rd} \text{ Period}
\end{aligned}
$$

Each one of these models requires a different rearrangement in the data. Here are the required arrangement for each model:

```
##      y(t) x(t) y(t-1) x(t-1) y(t-2) x(t-2)
## [1,]    3   23      2     22      1     21
## [2,]    4   24      3     23      2     22
## [3,]    5   25      4     24      3     23
## [4,]    6   26      5     25      4     24
## [5,]    7   27      6     26      5     25
```

```
## [6,]    8   28      7      27      6      26

##       y(t) x(t-1) y(t-2) x(t-2) y(t-3) x(t-3)
## [1,]    4     23      2     22      1     21
## [2,]    5     24      3     23      2     22
## [3,]    6     25      4     24      3     23
## [4,]    7     26      5     25      4     24
## [5,]    8     27      6     26      5     25
## [6,]    9     28      7     27      6     26

##       y(t) x(t-2) y(t-3) x(t-3) y(t-4) x(t-4)
## [1,]    5     23      2     22      1     21
## [2,]    6     24      3     23      2     22
## [3,]    7     25      4     24      3     23
## [4,]    8     26      5     25      4     24
## [5,]    9     27      6     26      5     25
## [6,]   10     28      7     27      6     26
```

We already rearranged the data for the first model. If we remove the first row in `y(t)` and the last row in the remaining set, we can get the data for the second model:

```
cbind(first[-1,1], first[-nrow(first),-1])
```

```
##      [,1] [,2] [,3] [,4] [,5] [,6]
## [1,]    4   23    2   22    1   21
## [2,]    5   24    3   23    2   22
## [3,]    6   25    4   24    3   23
## [4,]    7   26    5   25    4   24
## [5,]    8   27    6   26    5   25
## [6,]    9   28    7   27    6   26
## [7,]   10   29    8   28    7   27
```

We will use our COVID-19 data and a simple linear regression as an example of direct forecasting:

```
# Preparing data
df <- data.frame(dcases = trdf$diffcases, dmob = trdf$diffmob)
df <- df[complete.cases(df),]
rownames(df) <- NULL
df <- as.matrix(df)
head(df)
```

```
##      dcases     dmob
## [1,]      2 -0.01480
## [2,]      4  0.02013
## [3,]     -3  0.03049
## [4,]      0  0.00367
## [5,]      3 -0.02854
## [6,]     -2  0.03232
```

Now we need to decide on two parameters: the window size, that is, how many lags will be included in each row; and how many days we will forecast. The next section will use more advance functions for re-arranging the data and apply the direct forecasting with random forests. For now, let's use a 3-day window and a 3-day forecast horizon:

```
h = 3
w = 3
fh <- c() # storage for forecast

# Start with first
dt <- embed(df, w)
y <- dt[, 1]
X <- dt[, -1]

for (i in 1:h) {
  fit <- lm(y ~ X)
  l <- length(fit$fitted.values)
  fh[i] <- fit$fitted.values[l]
  y <- y[-1]
  X <- X[-nrow(X), ]
}

fh
```

```
## [1]  10.288416 -11.587090   0.302522
```

```
plot(1:266, trdf$diffcases, col = "gray40", type = "l")
lines(267:269, fh, col = "gray20")
```

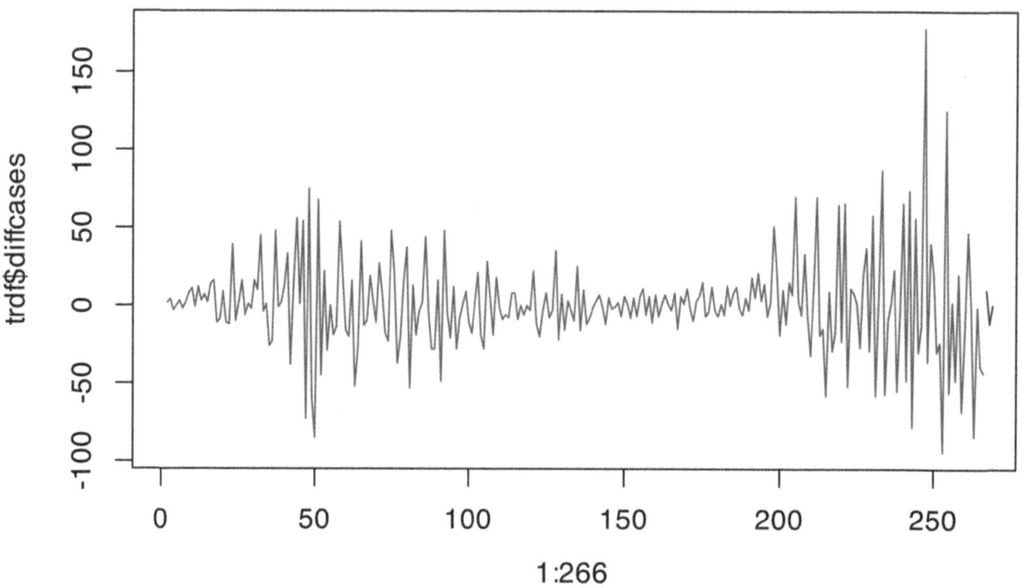

We haven't used training and test sets above. If we apply a proper splitting, we can even set the window size as our hyperparameter to minimize the forecast error:

```
# We set the last 7 days as our test set
train <- df[1:258,]
test <- df[-c(1:258),]

h = 7
w <- 3:14 # a grid for window size

fh <- matrix(0, length(w), h)
rownames(fh) <- w
colnames(fh) <- 1:7

for (s in 1:length(w)) {
  dt <- embed(train, w[s])
  y <- dt[, 1]
  X <- dt[, -1]
  for (i in 1:h) {
    fit <- lm(y ~ X)
    fh[s, i] <- last(fit$fitted.values)
    y <- y[-1]
    X <- X[-nrow(X), ]
  }
}

fh
```

```
##              1         2           3          4          5           6
## 3    -4.292862 -6.288479   5.2727764  10.692206  22.133103  -0.5252184
## 4    -5.014668 -1.626752   8.2861736  23.982849   4.611554  -0.2773355
## 5    -1.125996  1.634917  20.7212780   6.767507   5.115816  -0.5577792
## 6     1.533541 14.584416   5.6832803   8.066816   4.937718  -6.8419291
## 7    13.228621  1.612629   7.3973443   7.980486  -1.484987  -5.3696924
## 8     2.812780  3.308271   7.6799879   1.589578  -1.265470  -9.6077196
## 9    -5.430448  1.811491   0.7675925   1.698785  -7.123733 -16.9647249
## 10   -5.488847 -4.382922   0.8842250  -4.199708 -14.615359 -13.8839491
## 11 -11.104866 -4.133680  -5.3274242 -11.510596 -11.935885 -18.5728995
## 12 -11.656935 -8.289153 -11.9044832  -9.515252 -16.534428 -16.8239307
## 13 -18.314269 -13.292359 -9.2157517 -14.330746 -15.341226 -13.0680709
## 14 -23.661938 -10.963027 -13.9621680 -12.855445 -11.683527 -12.3975126
##            7
## 3  -19.79742
## 4  -19.62517
## 5  -26.29534
## 6  -23.77712
## 7  -20.07199
## 8  -27.04771
## 9  -25.44710
## 10 -30.22356
## 11 -29.91304
## 12 -25.62393
## 13 -25.15019
## 14 -27.72488
```

Rows in `fh` show the 7-day forecast for each window size. We can see which window size is the best:

```
rmspe <- c()

for (i in 1:nrow(fh)) {
  rmspe[i] <- sqrt(mean((fh[i, ] - test) ^ 2))
}

rmspe
```

```
##  [1] 33.45400 35.28827 31.67333 29.69115 31.57618 28.99568 28.53882 28.70796
##  [9] 27.16182 28.59872 28.77714 28.99870
```

```
which.min(rmspe)
```

```
## [1] 9
```

We used the last 7 days in our data as our test set. A natural question would be whether we could shuffle the data and use any 7 days as our test set? The answer is yes, because we do not need to follow a temporal order in the data after rearranging it with embedding. This is important because we can add a bootstrapping loop to our grid search above and get better tuning for finding the best window size.

We incorporate all these ideas with our random forest application in the next section.

29.4 Random Forest with Time Series

Now, we turn our attention to the application of embedding techniques for direct forecasting using random forests. The choice to employ random forests is driven by its inherent advantages, such as not requiring explicit tuning through grid search. Nevertheless, in practice, we can still optimize the model by searching for the optimal number of trees and the number of variables randomly sampled as candidates at each split.

By harnessing the power of random forests in conjunction with embedding methods, we can achieve more accurate and reliable time series forecasts. This chapter elucidates the process of integrating these two techniques, demonstrating how to effectively leverage random forests for direct forecasting in time series analysis.

Let's get our COVID-19 data:

```
library(tsibble)
library(fpp3)

load("toronto2.rds")
day <- seq.Date(
  from = as.Date("2020/03/01"),
  to = as.Date("2020/11/21"),
```

```
  by = 1
)

tdata <- tibble(Day = day, data[, -1])
toronto2 <- tdata %>%
  as_tsibble(index = Day)
toronto2
```

```
## # A tsibble: 266 x 8 [1D]
##    Day        cases      mob delay  male   age  temp   hum
##    <date>     <dbl>    <dbl> <dbl> <dbl> <dbl> <dbl> <dbl>
##  1 2020-03-01     4 -0.0172  36.8  0.75   55    -4.2   65.5
##  2 2020-03-02     6 -0.0320   8.5  1      45     3.8   84
##  3 2020-03-03    10 -0.0119  15    0.7    54     2.3   90
##  4 2020-03-04     7  0.0186  25.7  0.286  50     3.35  71
##  5 2020-03-05     7  0.0223  21    0.429  48.6   1.2   63.5
##  6 2020-03-06    10 -0.00626 13.1  0.5    36     0.04  75
##  7 2020-03-07     8  0.0261  10.4  0.5    46.2 -1.65   54
##  8 2020-03-08    10  0.0273  11.6  0.9    50     6.3   56
##  9 2020-03-09    18 -0.0158   8.89 0.611  35.6 12.5    55
## 10 2020-03-10    29 -0.0521   9.69 0.448  41.7   5.15  79
## # i 256 more rows
```

As before, the data contain the first wave and the initial part of the second wave in Toronto for 2020. It is from Ontario Data Catalogue[1] sorted by episode dates (`Day`), which is the date when the first symptoms were started. The mobility data is from Facebook, `all_day_bing_tiles_visited_relative_change`, which is reflects positive or negative change in movement relative to baseline. The other variables related to tests are `delay`, which is the time between test results and the episode date, the gender distribution of people is given by `male`, `age` shows the average age among tested people any given day. The last two variables, `temp` and `hum`, show the daily maximum day temperature and the average outdoor humidity during the day, respectively.

Except for `age` all other variables are non-stationary. We will take their first difference and make the series stationary before we proceed.

```
df <- toronto2 %>%
  mutate(
    dcases = difference(cases),
    dmob = difference(mob),
    ddelay = difference(delay),
    dmale = difference(male),
    dtemp = difference(temp),
    dhum = difference(hum)
  )

dft <- df[, -c(2:5, 7, 8)] #removing levels
dft <- dft[-1, c(1, 3:7, 2)] # reordering the columns
```

First, we will use a univariate setting for a single-window forecasting, which is the last 7 days.

29.4.1 Univariate

We will not have a grid search on the random forest algorithm, which could be added to the following script:

```
library(randomForest)

h = 7
w <- 3:21 # a grid for window size

fh <- matrix(0, length(w), h)
rownames(fh) <- w
colnames(fh) <- 1:h

for (s in 1:length(w)) {
  dt <- as.data.frame(embed(as.matrix(dft[, 2]), w[s]))
  test_ind = nrow(dt) - (h)
  train <- dt[1:test_ind,]
  test <- dt[-c(1:test_ind),]
  y <- train[, 1]
  X <- train[, -1]

  for (i in 1:h) {
    fit <- randomForest(X, y)
    fh[s,] <- predict(fit, test[, -1])
    y <- y[-1]
    X <- X[-nrow(X),]
  }
}

fh
```

```
##               1          2        3          4         5           6         7
## 3  -15.880433333   9.121167 19.74952  -6.609067 -5.722785  15.3516667 -2.087685
## 4   -9.544133333  -6.810200 23.81880  -4.161529 -1.214533   0.5277333  5.875767
## 5   -2.893600000  -6.885967 32.69350  -8.538333 -2.237667  -5.7720778 11.219167
## 6    0.070366667  -8.057467 22.55983 -12.094967 14.929700 -11.2058667 17.507433
## 7   -2.908866667 -15.868133 30.45953 -13.805133 19.191067 -23.7922000 14.290767
## 8    7.417066667 -22.705167 39.25673  -9.230933 13.765733 -15.4590667 27.734367
## 9   -3.620766667 -33.568367 68.65190 -20.472133 12.023133 -25.9268333 19.220533
## 10  -5.063400000 -33.309933 68.74610 -23.970333 14.040067 -25.0332000 12.531533
## 11  -0.009233333 -31.799500 60.88173 -23.562800 19.142833 -19.2536000 13.502400
## 12   0.805100000 -28.058533 55.59913 -26.090367 13.274300 -14.7545000 10.550400
## 13  -2.226166667 -30.789467 59.84473 -25.233567 15.489200 -11.8885333 15.370867
## 14  -1.639966667 -29.185000 58.61637 -21.431733 14.705067  -9.6757333 13.182233
## 15  -5.260733333 -33.169933 48.44817 -21.982467 14.593733 -12.0262667  9.309300
## 16  -5.429300000 -37.097867 61.44953 -21.993267 11.335233 -11.8728667 10.263967
## 17 -10.734700000 -33.021967 67.64857 -25.555300 16.965033 -10.5087667 10.809200
## 18  -6.696733333 -32.629567 59.85407 -23.227433 13.093067 -10.5785667 11.307200
## 19 -11.170300000 -36.225467 54.98873 -25.408567 16.683433 -14.8820000 12.992233
## 20  -3.524700000 -33.719100 61.33873 -23.823233 13.677867 -12.6076333 16.435500
```

```
## 21  -5.546933333 -31.462433 53.45857 -22.071567  9.086700 -13.1858667 15.337400
```

We can now see root mean square prediction error (RMSPE) for each row (window size):

```
actual <- test[, 1]
rmspe <- c()

for (i in 1:nrow(fh)) {
  rmspe[i] <- sqrt(mean((fh[i,] - actual) ^ 2))
}

rmspe
```

```
##  [1] 42.27364 44.57501 44.73248 44.35358 44.75067 51.39679 53.18043 51.53383
##  [9] 50.71377 48.11244 50.52535 50.43200 48.67944 51.68634 51.56775 50.45623
## [17] 50.10037 51.65413 49.68998
```

```
which.min(rmspe)
```

```
## [1] 1
```

And, if we plot several series of our forecast with different window sizes:

```
plot(
  actual,
  type = "l",
  col = "black",
  ylim = c(-80, 50),
  ylab = "Actual (red) vs. Forecasts",
  xlab = "Last 7 days",
  main = "7-Day Foerecasts",
  lwd = 3
)
lines(fh[1,], type = "l", col = "gray20")
lines(fh[2,], type = "l", col = "gray40")
lines(fh[5,], type = "l", col = "gray80")
lines(fh[12,], type = "l", col = "black")
legend(
  "bottomright",
  title = "Lags",
  legend = c("3-day", "4-day", "7-day", "14-day"),
  col = c("gray20", "gray40", "gray80"),
  lty = c(1, 1, 1, 1, 1),
  bty = "o",
  cex = 0.75
)
```

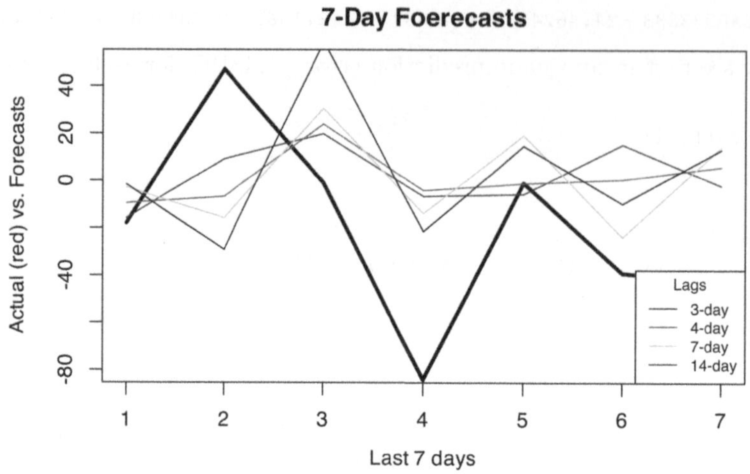

As the window size gets larger, the forecast becomes increasingly smooth missing the short term dynamics. Another observation is that, although "blue" (3-day window) has the minimum RMSPE, it is not able to capture ups and downs relative to 7-day or 14-day windows.

29.4.2 Multivariate

Can we increase the prediction accuracy with additional predictors?

```
library(randomForest)

h = 7
w <- 3:14 # a grid for window size

fh <- matrix(0, length(w), h)
rownames(fh) <- w
colnames(fh) <- 1:h

for (s in 1:length(w)) {
  dt <- as.data.frame(embed(as.matrix(dft[, -1]), w[s]))
  test_ind = nrow(dt) - (h)
  train <- dt[1:test_ind,]
  test <- dt[-c(1:test_ind),]
  y <- train[, 1]
  X <- train[, -1]

  for (i in 1:h) {
    fit <- randomForest(X, y)
    fh[s,] <- predict(fit, test[, -1])
    y <- y[-1]
    X <- X[-nrow(X),]
  }
}
```

```
fh
```

```
##            1           2        3          4          5          6         7
## 3  -17.788533  -0.3141000 16.53653  -9.326433 -18.102767  10.969800 -3.163967
## 4  -22.697100  -3.4509000 15.49413  -8.524767 -12.633133   2.173933 -3.292467
## 5  -13.800000  -0.7464333 16.56037 -11.723300  -7.062600 -13.184300  2.813033
## 6  -11.665500   1.9710333 20.05657 -10.684600   3.292100  -9.363967  6.406867
## 7   -9.373733  -5.8512667 23.13793 -17.310433   6.720633 -16.444167 11.981633
## 8  -10.233033 -12.3322000 22.61410  -9.796867   5.102200 -14.007967 12.177500
## 9  -12.145600 -22.5280667 45.52763 -22.078400   7.680600 -21.096567  8.656400
## 10 -11.179767 -19.4309000 47.87777 -23.726333   8.190267 -22.027567 12.756467
## 11  -8.145300 -18.9786333 49.77097 -26.724600  14.152900 -21.407233 11.465667
## 12 -10.947600 -18.4935667 47.21440 -23.789367   9.581900 -19.870133  8.059733
## 13  -9.108533 -16.9865000 40.43617 -22.952300   9.804567 -19.631100 10.097000
## 14 -10.174000 -20.3914333 43.31103 -26.960767  12.482533 -19.625367  5.761600
```

```
actual <- test[, 1]
rmspe <- c()

for (i in 1:nrow(fh)) {
  rmspe[i] <- sqrt(mean((fh[i, ] - actual) ^ 2))
}

rmspe
```

```
##  [1] 42.25558 41.30589 38.97195 40.04286 40.51307 43.67209 44.61230 44.66197
##  [9] 44.42342 43.71028 42.97556 42.68597
```

```
which.min(rmspe)
```

```
## [1] 3
```

```
plot(
  actual,
  type = "l",
  col = "black",
  ylim = c(-80, 50),
  ylab = "Actual (red) vs. Forecasts",
  xlab = "Last 7 days",
  main = "7-Day Foerecasts",
  lwd = 3
)
lines(fh[1,], type = "l", col = "gray20")
lines(fh[2,], type = "l", col = "gray40")
lines(fh[5,], type = "l", col = "gray80")
lines(fh[12,], type = "l", col = "black")
legend(
  "bottomright",
```

```
  title = "Lags",
  legend = c("3-day", "4-day", "7-day", "14-day"),
  col = c("gray20", "gray40", "gray80", "black"),
  lty = c(1, 1, 1, 1, 1),
  bty = "o",
  cex = 0.75
)
```

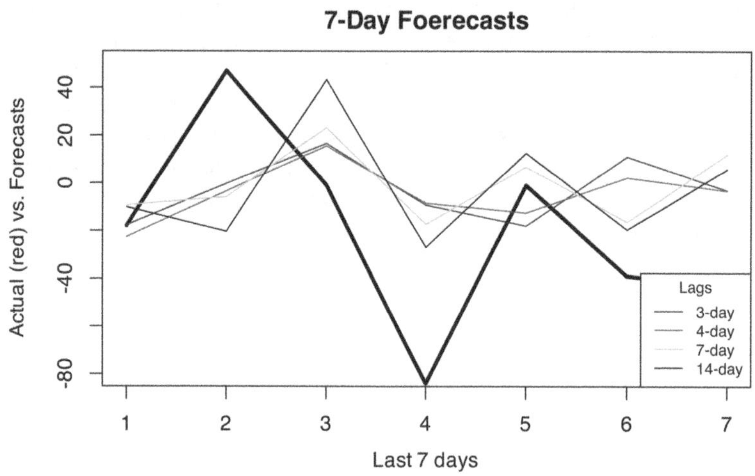

It seems that additional predictors do increase the accuracy. Again, relative to the best model (5-day window) our 7-day window correctly captures most ups and downs in the forecast. Now, a visual inspection shows that all RMSPE's are lower than the univariate forecasts. We would conclude that this is because of the new predictors, specially mobility, temperature, and humidity. As a side note, we need to test if those differences are statistical significant or not (i.e. Diebold–Mariano test).

29.5 Rolling and Expanding Windows

A 7-day window is not enough for a reliable judgment on the forecast accuracy. One way to deal with this issue is to use rolling or expanding windows to predict the next h days. The following example shows a 1-day-ahead forecast with varying lags for embedding.

```
library(randomForest)

l = 3:10 # lags for embedding
ws = 150 # size of each rolling window
rmspe <- c()

all_fh <- vector(mode = "list", length = length(l))
all_y <-  vector(mode = "list", length = length(l))

for (s in 1:length(l)) {
```

```
  dt <- as.data.frame(embed(as.matrix(dft[,-1]), l[s]))
  nwin <- nrow(dt) - ws #number of windows
  fh <- c()
  y <- c()

  for (i in 1:nwin) {
    train <- dt[i:(ws + i - 1),] # each loop, window moves one day forward
    test <- dt[(ws + i),]

    set.seed(i + s)
    fit <- randomForest(train[,-1], train[, 1])
    fh[i] <- predict(fit, test[,-1])
    y[i] <- test[, 1] # to use later for plotting
  }
  all_y[[s]] <- y
  all_fh[[s]] <- fh
  err <- test[, 1] - fh
  rmspe[s] <- sqrt(mean(err ^ 2))
}

rmspe
```

```
## [1] 45.00315 45.02472 45.31378 44.83595 45.81351 47.01531 47.14842 46.94298
```

```
bst <- which.min(rmspe)
l[bst] # Winning lag in embedding
```

```
## [1] 6
```

To adjust the application above to an expanding-window forecast, we just need to change `dt[i:(ws + i - 1), ]` to `dt[1:(ws + i - 1), ]` in the script.

Now, we can plot the results:

```
par(mfrow = c(1, 2))
plot(
  all_y[[bst]],
  type = "l",
  col = "gray50",
  ylab = "Actual (light gray) vs Predicted (black)",
  xlab = "Days",
  main = "1-Day-Ahead"
)
lines(all_fh[[bst]], col = "gray10")
plot(
  all_y[[bst]][60:110],
  type = "o",
  col = "gray50",
  ylab = "Actual (light gray) vs Predicted (black)",
  xlab = "Days",
```

```
  main = "Last 50 Days"
)
lines(all_fh[[bst]][60:110], col = "black")
```

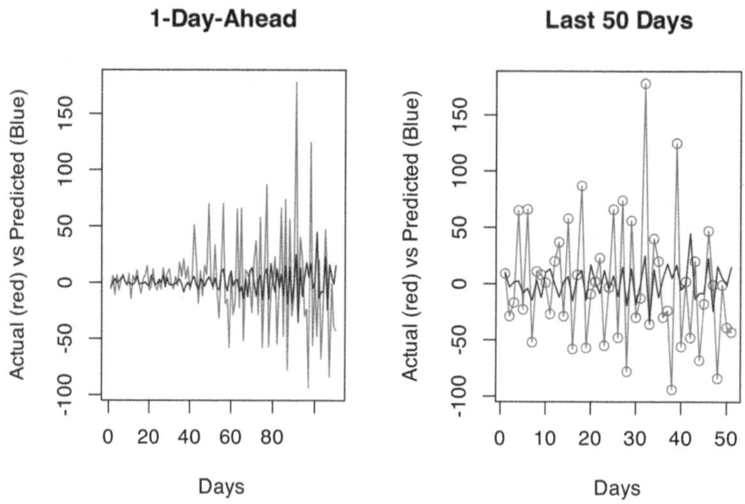

Getting the predicted values back to originals can be achieved by:

$$y_{t+1} = y_t + z_{t+1}$$
$$y_{t+2} = y_{t+1} + z_{t+2} = y_t + z_{t+1} + z_{t+2}$$

```
set.seed(321)
y <- rnorm(10)
z <- diff(y)     # first differences
back <- cumsum(c(y[1], z))
cbind(y, back)
```

```
##                y       back
##  [1,]  1.7049032  1.7049032
##  [2,] -0.7120386 -0.7120386
##  [3,] -0.2779849 -0.2779849
##  [4,] -0.1196490 -0.1196490
##  [5,] -0.1239606 -0.1239606
##  [6,]  0.2681838  0.2681838
##  [7,]  0.7268415  0.7268415
##  [8,]  0.2331354  0.2331354
##  [9,]  0.3391139  0.3391139
## [10,] -0.5519147 -0.5519147
```

Since our algorithm predicts the changes in observations, a simple sum would do the job for back transformation. For example, as a starting point, our algorithm predicts the change in Y from day 156 to 157 (window size 150 plus the best lag window, 6). When we add this predicted change to the actual Y at 156, it will give us the back-transformed forecast at day 157.

```
y <- df$cases

# The first forecast is at ws (150) + l[best] (6) + 1, which is 157
# The first actual Y should start a day earlier
# removing all Y's until ws+l[bst]

y_a_day_before <- y[-c(1:(ws + l[bst] - 1))]

# This adds predicted changes to observed values a day earlier
back_forecast <- head(y_a_day_before,-1) + all_fh[[bst]]

# Actual Y's in the test set starting at ws (150) + l[best] (6) + 1, which is 157
ytest <- y[-c(1:(ws + l[bst]))]

plot(
  ytest,
  type = "l",
  col = "gray30",
  ylab = "Actual Y (Blue) vs Forecast (light gray)",
  xlab = "Days",
  main = "Back-transformed Forecast"
)
lines(back_forecast, type = "l", col = "gray80")
```

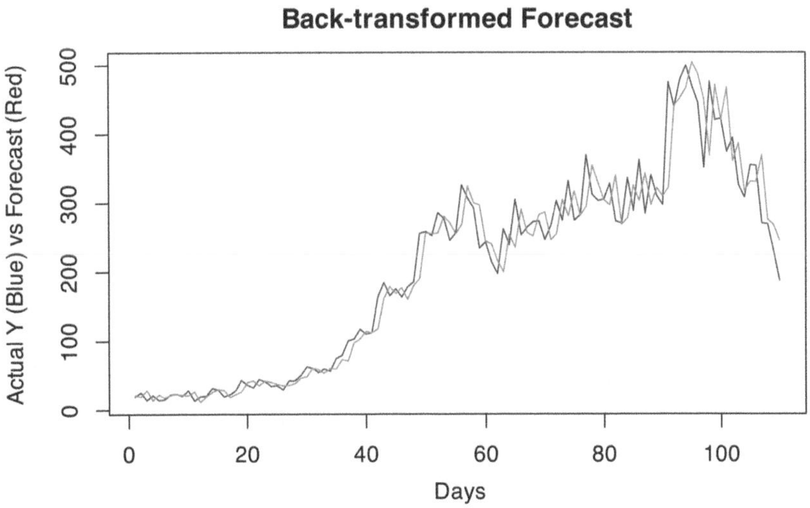

It seems that, for most days, our algorithm simply forecasts the next day by using the value from the day before. If we change our algorithm to a 7-day-ahead forecast, this would be different. This is also a common problem when the predictive model has a poor forecasting power. Again, this is not due to our algorithm, but forecasting an epi curve with imperfect test data is almost impossible job, as we highlighted earlier.

In practice, however, there are several ways that we can improve the scripts above. For example, we can consider the (rolling or expanding) window size as a hyperparameter. We can also have an explicit training for the random forest algorithm. We can have an ensemble forecasting by adding other predictive algorithms to the script, like boosting. Further, we can develop a base forecast that would give us a benchmark to see how much our algorithm improves against that base. Lastly, we could apply a transformation to the data in order to stabilize the variance in all variables.

Note

1. https://data.ontario.ca/en/dataset?groups=2019-novel-coronavirus#byPHU

30

Neural Networks and Deep Learning

Artificial neural networks (ANNs) are a type of machine learning model that are inspired by the structure and function of the human brain. They consist of interconnected units called artificial neurons or nodes which are organized into layers. The concept of ANNs dates back to the 1940s, when Warren McCulloch and Walter Pitts (1943) proposed a model of the neuron as a simple threshold logic unit. In the 1950s and 1960s, researchers began developing more complex models of neurons and exploring the use of neural networks for tasks such as pattern recognition and machine translation. However, these early efforts were largely unsuccessful due to the limited computational power at that time.

It wasn't until the 1980s and 1990s that significant progress was made in the development of ANNs, thanks to advances in computer technology and the availability of larger and more diverse datasets. In 1986[1], Geoffrey Hinton and his team developed the backpropagation algorithm, which revolutionized the field by allowing neural networks to be trained more efficiently and accurately. Since then, ANNs have been applied to a wide range of tasks, including image and speech recognition, natural language processing, and even playing games like chess and Go. They have also been used in a variety of fields, including finance, healthcare, and transportation. Today, ANNs are an important tool in the field of machine learning and continue to be an active area of research and development.

There have been many influential works accomplished in the field of ANNs over the years. Here are a few examples of some of the most important and influential works in the history of ANNs:

- Perceptrons[2] by Frank Rosenblatt (1958): This paper introduced the concept of the perceptron, which is a type of ANN that can be trained to recognize patterns in data. The perceptron became a foundational concept in the field of machine learning and was a key catalyst for the development of more advanced ANNs.
- Backpropagation[3] by Rumelhart, Hinton, and Williams (1986): This paper introduced the backpropagation algorithm, which is a method for training ANNs that allows them to learn and adapt over time. The backpropagation algorithm is still widely used today and has been a key factor in the success of ANNs in many applications.
- LeNet-5[4] by Yann LeCun et al. (1998): This paper described the development of LeNet-5, an ANN designed for recognizing handwritten digits. LeNet-5 was one of the first successful applications of ANNs in the field of image recognition and set the stage for many subsequent advances in this area.
- Deep Learning[5] by Yann LeCun, Yoshua Bengio, and Geoffrey Hinton (2015): This paper provided a comprehensive review of the field of deep learning, which is a type of ANN that uses many layers of interconnected neurons to process data. It has had a major impact on the development of deep learning and has helped to drive many of the recent advances in the field.

DOI: 10.1201/9781003504795-30

30.1 Neural Network: The Idea

Both support vector machines (SVM) and neural networks employ some kind of data transformation that moves them into a higher dimensional space. What the kernel function does for the SVM, the hidden layers do for neural networks.

Let's start with a predictive model with a single input (covariate). The simplest model could be a linear model:

$$y \approx \alpha + \beta x \tag{30.1}$$

where α is the intercept and β is the slope. The model is linear in the parameters, α and β. The model is linear in the covariate x as well. The model can be written as a weighted sum of fixed components: $\alpha + \beta x$. This is a linear model with fixed components.

Since this model could be a quite restrictive, we can have a more flexible one by a polynomial regression:

$$y \approx \alpha + \beta_1 x + \beta_2 x^2 + \beta_3 x^3 + \ldots = \alpha + \sum_{m=1}^{M} \beta_m x^m \tag{30.2}$$

The polynomial regression is based on fixed components, or bases: $x, x^2, x^3, \ldots, x^M$. The artificial neural net replaces these fixed components with adjustable ones or bases: $f(\alpha_1 + \delta_1 x)$, $f(\alpha_2 + \delta_2 x), \ldots, f(\alpha_M + \delta_M x)$. We can see the first simple ANN as nonlinear functions of linear combinations:

$$\begin{aligned} y &\approx \alpha + \beta_1 f(\alpha_1 + \delta_1 x) + \beta_2 f(\alpha_2 + \delta_2 x) + \beta_3 f(\alpha_3 + \delta_3 x) + \ldots \\ &= \alpha + \sum_{m=1}^{M} \beta_m f(\alpha_m + \delta_m x) \end{aligned} \tag{30.3}$$

where α is the intercept, β_m are the weights, and $f(.)$ is a nonlinear function. The model is linear in the parameters, α, β_m, and δ_m. The model is nonlinear in the covariate x. The model can be written as a weighted sum of flexible components: $\alpha + \sum_{m=1}^{M} \beta_m f(\alpha_m + \delta_m x)$.

where $f(.)$ is an **activation** function—a fixed nonlinear function. Common examples of activation functions are

- The **logistic** (or sigmoid) function: $f(x) = \frac{1}{1+e^{-x}}$;
- The **hyperbolic tangent** function: $f(x) = \tanh(x) = \frac{e^x - e^{-x}}{e^x + e^{-x}}$;
- The rectified linear unit (**ReLU**): $f(x) = \max(0, x)$;

The full list of activation functions can be found at Wikipedia[6].

Let us consider a realistic (simulated) sample:

```
n <- 200
set.seed(1)
x <- sort(runif(n))
y <- sin(12*(x + 0.2))/(x + 0.2) + rnorm(n)/2
df <- data.frame(y, x)
plot(x, y, main="Simulated data",  col= "grey")
```

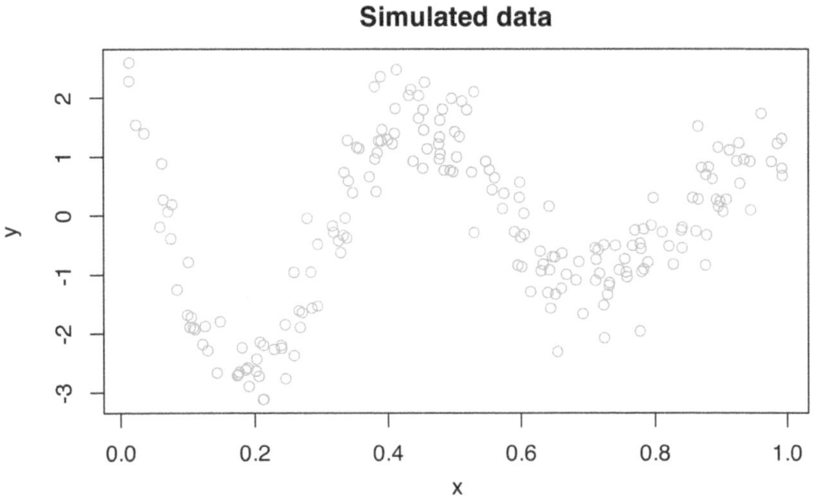

We can fit a polynomial regression with $M = 3$:

```
ols <- lm(y ~ x + I(x^2) + I(x^3))
plot(x, y, main="Polynomial: M = 3",  col= "grey")
lines(x, predict(ols), col="blue", lwd = 3)
```

```
ols <- lm(y ~ x + I(x^2) + I(x^3))
plot(x, y, main="Polynomial: M = 3",  col= "grey")
lines(x, predict(ols), col="black", lwd = 3)
```

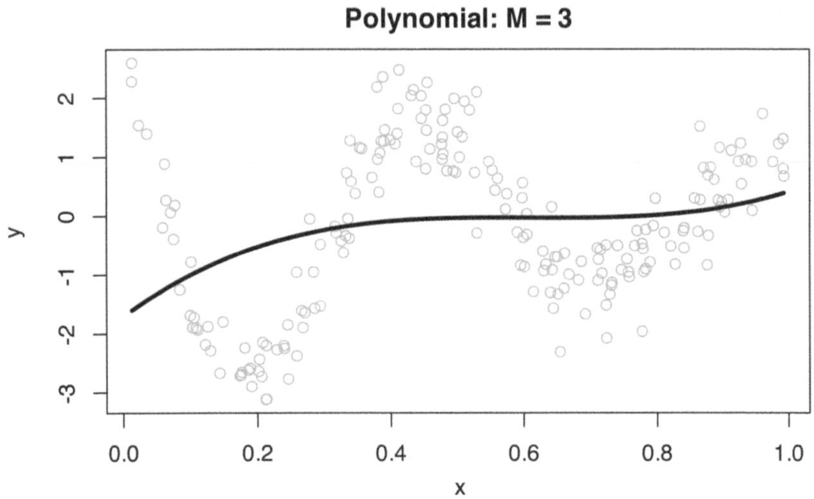

Now, we can think of the line as weighted sum of fixed components: $\alpha_1 + \beta_1 x + \beta_2 x^2 + \beta_3 x^3$.

```
# Parts
first <- ols$coefficients[2]*x
second <- ols$coefficients[3]*x^2
third <- ols$coefficients[4]*x^3
```

```
yhat <- ols$coefficients[1] + first + second + third

# Plots
par(mfrow=c(1,4), oma = c(0,0,2,0))
plot(x, first, ylab = "y", col = "gray60", main = "x")
plot(x, second, ylab = "y", col = "gray40", main = expression(x^2))
plot(x, third, ylab = "y", col = "gray20", main = expression(x^3))
plot(x, y, ylab="y", col = "grey",
     main = expression(y == alpha + beta[1]*x + beta[2]*x^2 + beta[3]*x^3))
lines(x, yhat, col = "gray10", lwd = 3)
mtext("Fixed Components",
      outer=TRUE, cex = 1.5, col="black")
```

```
# Parts
first <- ols$coefficients[2]*x
second <- ols$coefficients[3]*x^2
third <- ols$coefficients[4]*x^3
yhat <- ols$coefficients[1] + first + second + third

# Plots
par(mfrow=c(1,4), oma = c(0,0,2,0))
plot(x, first, ylab = "y", col = "gray60", main = "x")
plot(x, second, ylab = "y", col = "gray40", main = expression(x^2))
plot(x, third, ylab = "y", col = "gray20", main = expression(x^3))
plot(x, y, ylab="y", col = "grey",
     main = expression(y == alpha + beta[1]*x + beta[2]*x^2 + beta[3]*x^3))
lines(x, yhat, col = "gray10", lwd = 3)
mtext("Fixed Components",
      outer=TRUE, cex = 1.5, col="black")
```

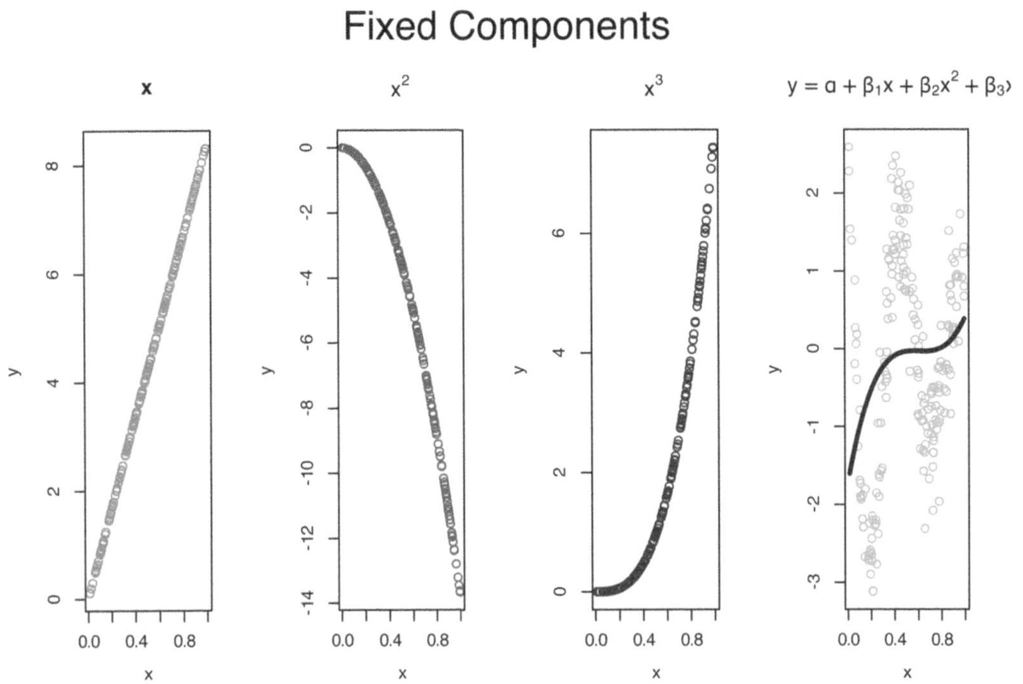

The artificial neural net replaces the fixed components in the polynomial regression with adjustable ones, $f(\alpha_1 + \delta_1 x)$, $f(\alpha_2 + \delta_2 x)$, $\ldots, f(\alpha_M + \delta_M x)$ that are more flexible. They are adjustable with tunable internal parameters. They can express several shapes, not just one (fixed) shape. Hence, adjustable components enable to capture complex models with fewer components (smaller M).

Let's replace those fixed components x, x^2, x^3 in our polynomial regression with $f(\alpha_1 + \delta_1 x)$, $f(\alpha_2 + \delta_2 x)$, $f(\alpha_3 + \delta_3 x)$.

The following code demonstrates the ability of a simple ANN with arbitrary parameters to capture the underlying signal relative to a third-degree polynomial regression model. It defines an ANN function with sigmoid activation functions for three nodes (M = 3), arbitrary parameters `a`, `b`, `beta`, and an intercept (`int`). For each node, the code calculates the weighted input (`z`) using `a` and `b`, and then applies the sigmoid activation function to obtain the output (`sig`). The output is then multiplied by the corresponding beta value. The final output (`yhat`) is calculated as the sum of the intercept and the weighted outputs from all three nodes.

```
a = c(1.5, 9, 3)
b = c(-20,-14,-8)
beta = c(15, 25,-40)
int = 3

ann <- function(a, b, beta, int) {
  #1st sigmoid
  a1 = a[1]
  b1 = b[1]
  z1 = a1 + b1 * x
```

```
  sig1 = 1 / (1 + exp(-z1))

  f1 <- sig1

  #2nd sigmoid
  a2 = a[2]
  b2 = b[2]
  z2 = a2 + b2 * x
  sig2 = 1 / (1 + exp(-z2))

  f2 <- sig2

  #3rd sigmoid
  a3 = a[3]
  b3 = b[3]
  z3 = a3 + b3 * x
  sig3 = 1 / (1 + exp(-z3))

  f3 <- sig3

  yhat = int + beta[1] * f1 + beta[2] * f2 + beta[3] * f3
  return(yhat)
}

yhat <- ann(a, b, beta, int)

plot(x, y, main = "ANN: M = 3", ylim = c(-5, 15))
lines(x, yhat, col = "gray40", lwd = 3)
```

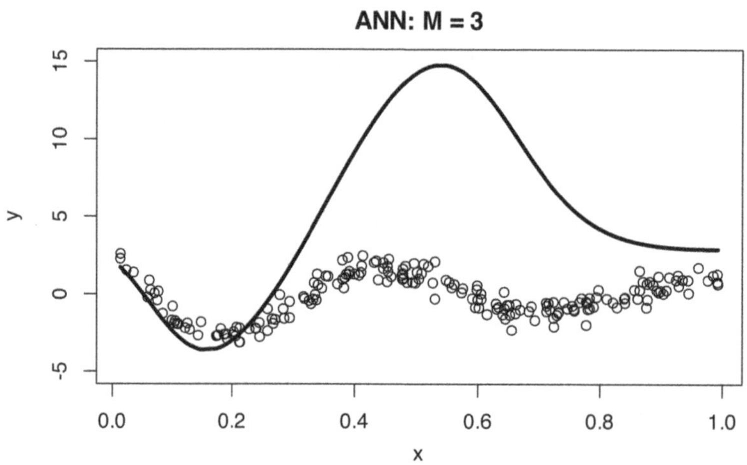

The resulting plot demonstrates the ANN's ability to fit the data compared to the third-degree polynomial regression model. The next section, backpropagation, will show us how to get the correct parameters for our simple ANN. For now, let's obstain them with `neuralnet`.

```
library(neuralnet)

set.seed(2)
# fit a single-hidden-layer net with 3 units
nn <- neuralnet(y ~ x, data = df, hidden = 3, threshold = 0.05)

# explicitly call neuralnet::compute()
pred <- neuralnet::compute(nn, data.frame(x = df$x))
yhat <- pred$net.result

# scatter the raw data...
plot(df$x, df$y,
     main = "Neural Networks: M = 3",
     xlab = "x", ylab = "y",
     pch = 16)

# ...and overlay the fitted curve
ord <- order(df$x)
lines(df$x[ord], yhat[ord], col = "gray40", lwd = 3)
```

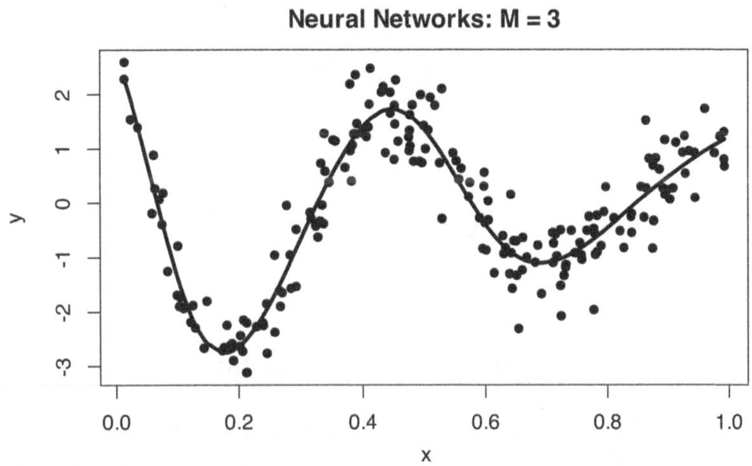

Why did neural networks perform better than polynomial regression in the previous example? Again, adjustable components enable to capture complex models. Let's delve little deeper. Here is the weight structure of

$$y \approx \alpha + \sum_{m=1}^{3} \beta_m f\left(\alpha_m + \delta_m x\right) = \alpha + \beta_1 f\left(\alpha_1 + \delta_1 x\right) + \beta_2 f\left(\alpha_2 + \delta_2 x\right) + \beta_3 f\left(\alpha_3 + \delta_3 x\right) \quad (30.4)$$

```
nn$weights
```

```
## [[1]]
## [[1]][[1]]
##              [,1]       [,2]       [,3]
## [1,]    1.26253    6.59977   2.504890
```

```
## [2,] -18.95937 -12.24665 -5.700564
##
## [[1]][[2]]
##             [,1]
## [1,]    2.407654
## [2,]   13.032092
## [3,]   19.923742
## [4,] -32.173264
```

```
plot(nn, rep = "best",
     col.axis = "black",
     col.border = "black",
     col.hidden = "gray80",
     col.intercept = "black",
     col.out = "black",
     col.synapse = "black")
```

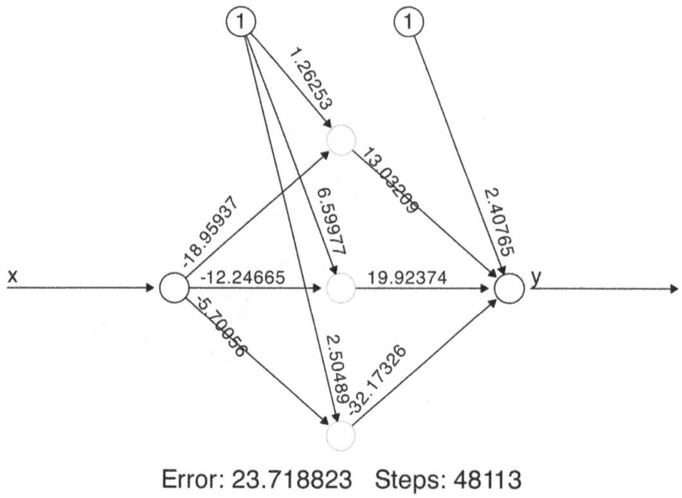

- Error: It represents the final error of the best-performing neural network after the training process has concluded. It's a snapshot of how well (or not) your model has learned the underlying pattern in our data.
- Steps: The 48113 is the number of steps (iterations) the algorithm took to reach its current state, under the constraints we provided (e.g., learning rate, threshold, etc.). It's a testament to how hard your network worked to get to its final form.

We used sigmoid (logistic) activation functions

$$\text{Node 1:} \quad f_1(x) = \frac{1}{1 + e^{-(1.26253 - 18.95937x)}}$$

$$\text{Node 2:} \quad f_2(x) = \frac{1}{1 + e^{-(6.599773 - 12.24665x)}}$$

$$\text{Node 3:} \quad f_3(x) = \frac{1}{1 + e^{-(2.504890 - 5.700564x)}}$$

We can calculate the value of each activation function by using our data, x:

```
X <- cbind(1, x)

# to 1st Node
n1 <- nn$weights[[1]][[1]][,1]
f1 <- nn$act.fct(X%*%n1)

# to 2nd Node
n2 <- nn$weights[[1]][[1]][,2]
f2 <- nn$act.fct(X%*%n2)

# to 3rd Node
n3 <- nn$weights[[1]][[1]][,3]
f3 <- nn$act.fct(X%*%n3)

par(mfrow=c(1,3), oma = c(0,0,2,0))
plot(x, f1, col = "gray20", main = expression(f(alpha[1] + beta[1]*x)))
plot(x, f2, col = "gray40", main = expression(f(alpha[2] + beta[2]*x)))
plot(x, f3, col = "gray60", main = expression(f(alpha[3] + beta[3]*x)))
mtext("Flexible Components",
      outer=TRUE, cex = 1.5, col="black")
```


```
X <- cbind(1, x)

# to 1st Node
n1 <- nn$weights[[1]][[1]][,1]
f1 <- nn$act.fct(X%*%n1)

# to 2nd Node
n2 <- nn$weights[[1]][[1]][,2]
f2 <- nn$act.fct(X%*%n2)

# to 3rd Node
n3 <- nn$weights[[1]][[1]][,3]
f3 <- nn$act.fct(X%*%n3)

par(mfrow=c(1,3), oma = c(0,0,2,0))
plot(x, f1, col = "gray20", main = expression(f(alpha[1] + beta[1]*x)))
plot(x, f2, col = "gray40", main = expression(f(alpha[2] + beta[2]*x)))
plot(x, f3, col = "gray60", main = expression(f(alpha[3] + beta[3]*x)))
mtext("Flexible Components",
      outer=TRUE, cex = 1.5, col="black")
```

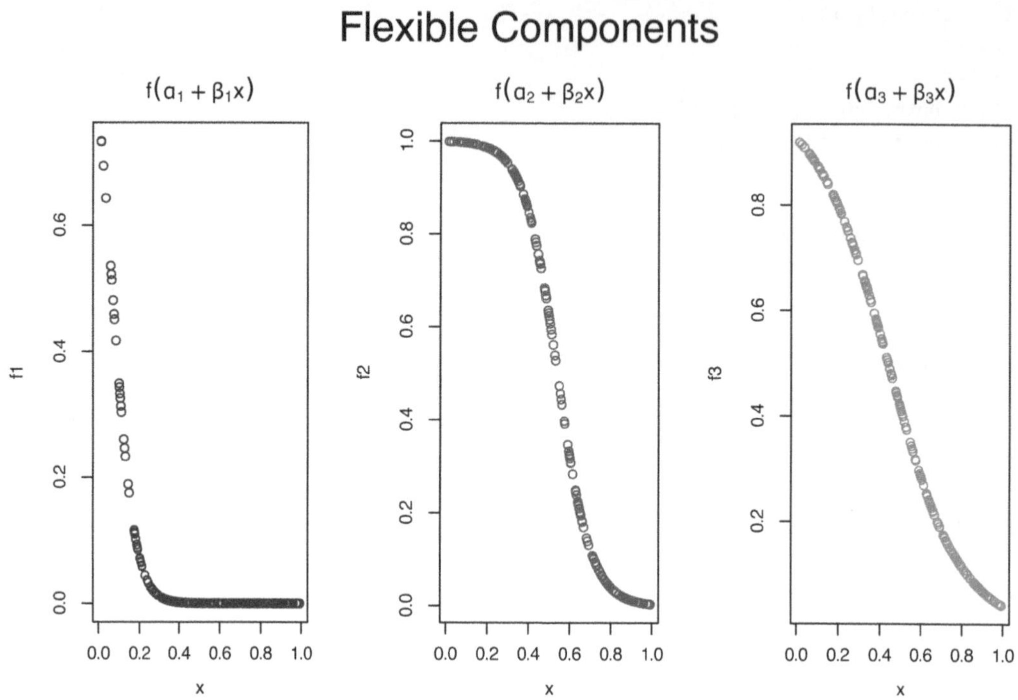

Now we will go from these nodes to the "sink":

$$\frac{1}{1+e^{-(1.26253-18.95937x)}} \times 13.032092$$
$$\frac{1}{1+e^{-(6.599773-12.24665x)}} \times 19.923742$$
$$\frac{1}{1+e^{-(2.504890-5.700564x)}} \times (-32.173264)$$

And the final output (including the bias term) as:

$$\begin{aligned}\text{Output}(x) = 2.407654 &+ \frac{1}{1+e^{-(1.26253-18.95937x)}} \times 13.032092\\ &+ \frac{1}{1+e^{-(6.599773-12.24665x)}} \times 19.923742\\ &+ \frac{1}{1+e^{-(2.504890-5.700564x)}} \times (-32.173264)\end{aligned}$$

Here are the results:

```
# From Nodes to sink (Y)
f12 <- f1*nn$weights[[1]][[2]][2]
f22 <- f2*nn$weights[[1]][[2]][3]
f23 <- f3*nn$weights[[1]][[2]][4]
```

```
## Results
yhat <- nn$weights[[1]][[2]][1] + f12 + f22 + f23
plot(x, y, main="ANN: M = 3")
lines(x, yhat, col="gray40", lwd = 3)
```

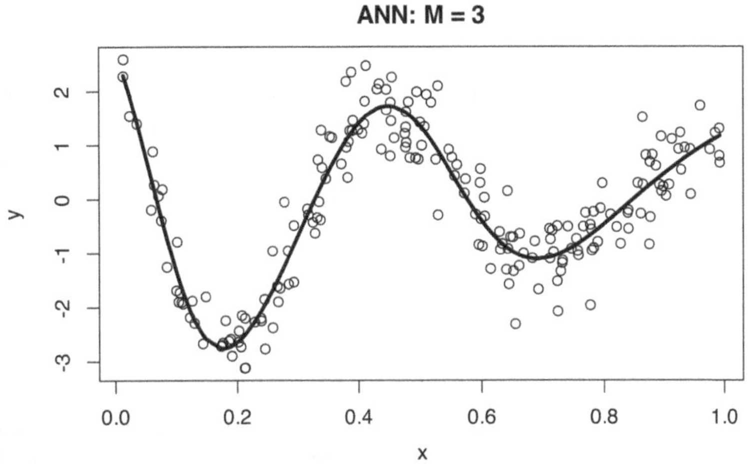

Backpropagation

In 1986, Rumelhart et al.[7] found a way to train neural networks with the backpropagation algorithm. Today, we would call it a gradient descent using reverse-mode autodiff. Backpropagation is an algorithm used to train neural networks by adjusting the weights and biases of the network to minimize the cost function. Suppose we have a simple neural network as follows:

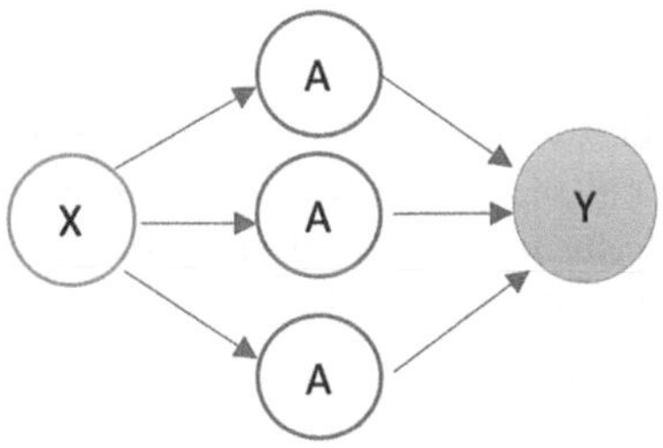

The first layer is the source layer (with X). The second layer is called as hidden layer with three "neurons" each of which has an activation function (A). The last layer is the "sink" or output layer. First, let's define a loss function, MSPE:

$$\text{MSPE} = \frac{1}{n}\sum_{i=1}^{n}(y_i - \hat{y})^2 \tag{30.5}$$

And we want to solve:

$$\omega^\star = \operatorname{argmin}\left\{\frac{1}{n}\sum_{i=1}^{n}(y_i - \hat{y})^2\right\} \tag{30.6}$$

where ω is the vector of weights and biases in the network. The goal of the backpropagation algorithm is to find the optimal weights and biases that minimize the error between the predicted output $\hat{y}$ and the actual output y.

To compute the gradient of the error with respect to a weight w or a bias b, we use the chain rule:

$$\frac{\partial \text{MSPE}}{\partial w} = \frac{\partial \text{MSPE}}{\partial \hat{y}} \frac{\partial \hat{y}}{\partial z} \frac{\partial z}{\partial w} \tag{30.7}$$

where z is the weighted sum of inputs to the neuron, and $\hat{y}$ is the output of the neuron. The first term, $\frac{\partial \text{MSPE}}{\partial \hat{y}}$, is the derivative of the error with respect to the output of the neuron. The second term, $\frac{\partial \hat{y}}{\partial z}$, is the derivative of the output with respect to the weighted sum of inputs. The third term, $\frac{\partial z}{\partial w}$, is the derivative of the weighted sum of inputs with respect to the weight or bias.

Remember,

$$\hat{y} = f(x) = \frac{1}{1 + e^{-z}} = \frac{1}{1 + e^{-(\alpha + wx)}} \tag{30.8}$$

where $f(.)$ is the activation function, x is the input to the neuron, and α is the bias. The output of the neuron is a function of the weighted sum of inputs, z, which is given by $z = \alpha + wx$.

By repeating this process for each weight and bias in the network, we can calculate the error gradient and use it to adjust the weights and biases in order to minimize the error of the neural network. This can be done by using gradient descent (See Chapter 32), which is an iterative method for optimizing a differentiable objective function, typically by minimizing it.

However, with a multilayer ANN, we use stochastic gradient descent (SGD), which is a faster method iteratively minimizing the loss function by taking small steps in the opposite direction of the gradient of the function at the current position. The gradient is calculated using a randomly selected subset of the data, rather than the entire dataset, which is why it is called "stochastic." One of the main advantages of SGD is that it can be implemented very efficiently and can handle large data sets very well.

We will use a very simple hand-coded gradient descent algorithm here (instead of `keras`) to solve our ANN problem as an example:

```
n <- 200
set.seed(1)
x <- sort(runif(n))
Y <- sin(12*(x + 0.2))/(x + 0.2) + rnorm(n)/2
df <- data.frame(Y, x)
plot(x, Y, main="Simulated data",  col= "grey")

# starting points
set.seed(234)
alpha = runif(1); beta1 = runif(1); beta2 = runif(1); beta3 = runif(1)
a1 = runif(1); a2 = runif(1); a3 = runif(1)
b1 = runif(1); b2 = runif(1); b3 = runif(1)

# initial prediction
```

```
z1 = a1 + b1*x; z2 = a2 + b2*x; z3 = a3 + b3*x
sig1 = 1/(1 + exp(-z1)); sig2 = 1/(1 + exp(-z2)); sig3 = 1/(1 + exp(-z3))
yhat <- alpha + beta1 * sig1 + beta2 * sig2 + beta3 * sig3
MSE <- sum((Y - yhat) ^ 2) / n
prev_MSE <- MSE
iterations = 0

while (TRUE) {
  # gradients
  alpha_new <- alpha - (0.01 / n) * sum((Y - yhat) * (-1))
  beta1_new <- beta1 - (0.01 / n) * sum((Y - yhat) * sig1 * (-1))
  beta2_new <- beta2 - (0.01 / n) * sum((Y - yhat) * sig2 * (-1))
  beta3_new <- beta3 - (0.01 / n) * sum((Y - yhat) * sig3 * (-1))
  a1_new <- a1 - (0.01 / n) * sum((Y - yhat) * (-beta1 * exp(-z1) / ((1 + exp(-z1))^2)))
  a2_new <- a2 - (0.01 / n) * sum((Y - yhat) * (-beta2 * exp(-z2) / ((1 + exp(-z2))^2)))
  a3_new <- a3 - (0.01 / n) * sum((Y - yhat) * (-beta3 * exp(-z3) / ((1 + exp(-z3))^2)))
  b1_new <- b1 - (0.01 / n) * sum((Y - yhat) * (-beta1 * x * exp(-z1) / ((1 + exp(-z1))^2)))
  b2_new <- b2 - (0.01 / n) * sum((Y - yhat) * (-beta2 * x * exp(-z2) / ((1 + exp(-z2))^2)))
  b3_new <- b3 - (0.01 / n) * sum((Y - yhat) * (-beta3 * x * exp(-z3) / ((1 + exp(-z3))^2)))

  # update
  alpha <- alpha_new; beta1 <- beta1_new; beta2 <- beta2_new; beta3 <- beta3_new
  a1 <- a1_new; a2 <- a2_new; a3 <- a3_new
  b1 <- b1_new; b2 <- b2_new; b3 <- b3_new

  # new prediction
  z1 = a1 + b1*x; z2 = a2 + b2*x; z3 = a3 + b3*x
  sig1 = 1 / (1 + exp(-z1)); sig2 = 1 / (1 + exp(-z2)); sig3 = 1 / (1 + exp(-z3))
  yhat <- alpha + beta1 * sig1 + beta2 * sig2 + beta3 * sig3
  MSE_new <- sum((Y - yhat)^2) / n

  # convergence check
  if (abs(prev_MSE - MSE_new) < 1e-6) break
  prev_MSE <- MSE_new
  iterations <- iterations + 1
  if (iterations > 50000) {
    warning("Did not converge in 50,000 iterations.")
    break
  }
}

# plot final fit
plot(x, Y, main="Simulated data", col="grey")
lines(x, yhat, col="gray40", lwd = 3)
```

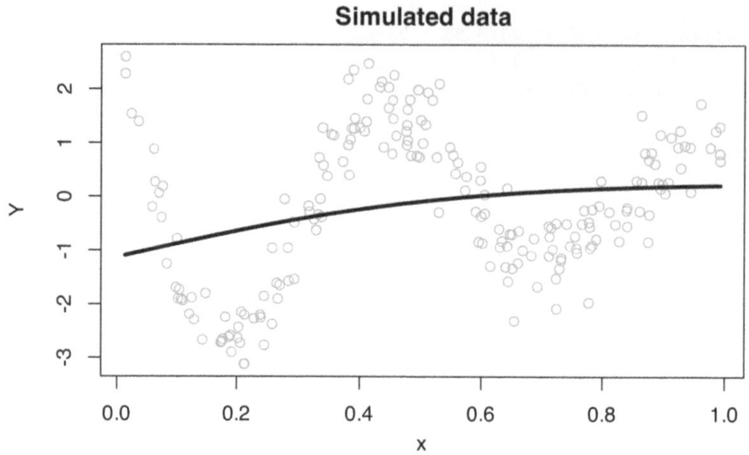

This plain gradient algorithm is very simple, identifying only ten parameters. In a regular network, the number of parameters is usually very large and it gets even larger in deep neural networks. Therefore, finding the most efficient (fast) backpropagation method is an active research area in the field of artificial intelligence. One of the recent developments in this area is Tensors and Tensorflow that we will talk in Section VII. More details on gradient descent algorithms are provided in Chapter 32.

30.2 Neural Network: More inputs

With a set of covariates $X = (1, x_1, x_2, \ldots, x_k)$, we have

$$
\begin{aligned}
y &\approx \alpha + \sum_{m=1}^{M} \beta_m f\left(\alpha_m + \mathbf{X}\delta_m\right) \\
&= \alpha + \beta_1 f\left(\alpha_1 + \delta_{11}x_{1i} + \delta_{12}x_{2i} + \cdots + \delta_{1k}x_{ki}\right) \\
&\quad + \beta_2 f\left(\alpha_2 + \delta_{21}x_{1i} + \delta_{22}x_{2i} + \cdots + \delta_{2k}x_{ki}\right) \\
&\quad \vdots \\
&\quad + \beta_M f\left(\alpha_M + \delta_{M1}x_{1i} + \delta_{M2}x_{2i} + \cdots + \delta_{Mk}x_{ki}\right)
\end{aligned}
$$

where $f(.)$ is a nonlinear function (e.g., sigmoid, tanh, etc.). The δ_{mk} are the weights of the k-th input to the m-th neuron. The β_m are the weights of the m-th neuron to the output. The α_m are the biases of the m-th neuron. The α is the bias of the output.

By adding nonlinear functions of linear combinations with $M > 1$, we have seen that we can capture nonlinearity. With multiple features, we can now capture interaction effects and, hence, obtain a more flexible model. This can be seen in blue and orange arrows in the following picture:

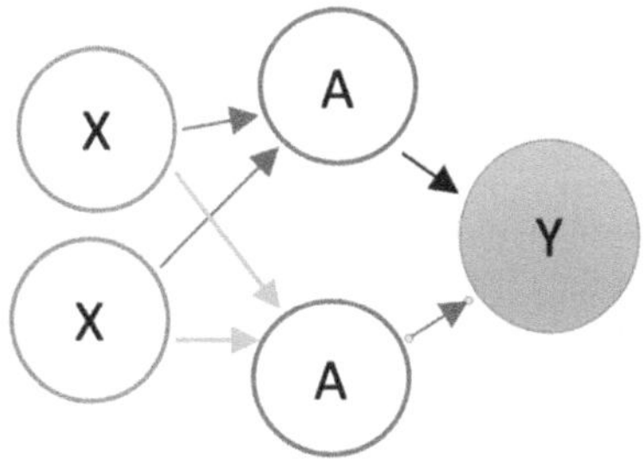

Let's have an application using a Mincer equation and the data (SPS 1985 - cross-section data originating from the May 1985 Current Population Survey by the US Census Bureau) from the `AER` package.

Before we start, there are few important pre-processing steps to complete. First, ANN are inefficient when the data are not scaled. The reason is backpropagation. Since ANN use gradient descent, the different scales in features will cause different step sizes. Scaling the data before feeding it to the model enables the steps in gradient descent updated at the same rate for all the features. Second, indicator predictors should be included in the input matrix by dummy-coding. Finally, the formula for the model needs to be constructed to initialize the algorithm. Let's see all of these pre-processing steps below:

```
library(AER)
data("CPS1985")

df <- CPS1985

# Scaling and Dummy coding
df[,sapply(df, is.numeric)] <- scale((df[, sapply(df, is.numeric)]))
ddf <- model.matrix(~.-1, data= df, contrasts.arg =
                       lapply(df[,sapply(df, is.factor)],
                              contrasts, contrasts = FALSE))

ddf <- as.data.frame(ddf)

# formula to pass on ANN and lm()
w.ind <- which(colnames(ddf)=="wage")
frmnn <- as.formula(paste("wage~",
                   paste(colnames(ddf[-w.ind]), collapse='+')))
frmln <- as.formula(paste("wage~", "I(experience^2)", "+",
                   paste(colnames(ddf[-w.ind]), collapse = "+")))

# Bootstrapping loops instead of CV
mse.test <- matrix(0, 10, 5)
for(i in 1:10){

  set.seed(i+1)
  trainid <- unique(sample(nrow(ddf), nrow(ddf), replace = TRUE))
  train <- ddf[trainid,]
  test <- ddf[-trainid,]
```

```
  # Models
  fit.lm <- lm(frmln, data = train)
  fit.nn <- neuralnet(frmnn, data = train, hidden = 1,threshold=0.05,
                      linear.output = FALSE)
  fit.nn2 <- neuralnet(frmnn, data = train, hidden = 2,threshold=0.05)
  fit.nn3 <- neuralnet(frmnn, data = train, hidden = 3,threshold=0.05)
  fit.nn4 <- neuralnet(frmnn, data = train, hidden = 3,threshold=0.05,
                       act.fct = "tanh", linear.output = FALSE)

  # Prediction errors
  mse.test[i,1] <- mean((test$wage - predict(fit.lm, test))^2)
  mse.test[i,2] <- mean((test$wage - predict(fit.nn, test))^2)
  mse.test[i,3] <- mean((test$wage - predict(fit.nn2, test))^2)
  mse.test[i,4] <- mean((test$wage - predict(fit.nn3, test))^2)
  mse.test[i,5] <- mean((test$wage - predict(fit.nn4, test))^2)
}
colMeans(mse.test)
```

```
## [1] 0.7296417 0.8919442 0.9038211 1.0403616 0.8926576
```

This experiment alone shows that a **linear** Mincer equation (with `I(expreince^2)`) is a much better predictor than ANN. As the complexity of ANN rises with more neurons, the likelihood that ANN overfits goes up, which is the case in our experiment. In general, linear regression may be a good choice for simple, low-dimensional datasets with a strong linear relationship between the variables, while ANNs may be better suited for more complex, high-dimensional datasets with nonlinear relationships between variables.

Overfitting can be a concern when using ANNs for prediction tasks. Overfitting occurs when a model is overly complex and has too many parameters relative to the size of the training data, which results in fitting the noise in the training data rather than the underlying pattern. As a result, the model may perform well on the training data but poorly on new, unseen data. One way to mitigate overfitting in ANNs is to use techniques such as regularization, which imposes constraints on the model to prevent it from becoming too complex. Another approach is to use techniques such as early stopping, which involves interrupting the training process when the model starts to overfit the training data.

30.3 Deep Learning

Deep neural network (DNN), or deep learning, is an ANN that has two or more hidden layers. Even greater flexibility is achieved via composition of activation functions:

$$y \approx \alpha + \sum_{m=1}^{M} \beta_m f\left(\alpha_m^{(1)} + \underbrace{\sum_{p=1}^{P} f\left(\alpha_p^{(2)} + \mathbf{X}\delta_p^{(2)}\right)}_{\text{it replaces } \mathbf{X}} \delta_m^{(1)}\right) \tag{30.9}$$

Here, $\mathbf{X}$ is the input vector, and $f(\cdot)$ is a nonlinear activation function such as the sigmoid or hyperbolic tangent. The inner sum represents the output of the second hidden layer, where $\alpha_p^{(2)}$ are the biases and $\delta_p^{(2)}$ are the weights applied to the inputs. These outputs then become the effective inputs to the first hidden layer, which applies another transformation using its own weights $\delta_m^{(1)}$ and biases $\alpha_m^{(1)}$. Each transformed value is then scaled by an output-layer weight β_m, and the sum of these values is added to the output-layer bias α to produce the final prediction y. The indices P and M refer to the number of neurons in the second and first hidden layers, respectively. This nesting of nonlinear transformations is what gives deep networks their expressive power.

Before having an application, we should note the number of available packages that offer ANN implementations in R and with Python. For example, CRAN hosts more than 80 packages related to neural network modeling. Above, we just saw one example with `neuralnet`. The work by Mahdi et al, 2021[8] surveys and ranks these packages for their accuracy, reliability, and ease-of-use.

```
mse.test <- c()
for(i in 1:10){

  set.seed(i+1)
  trainid <- unique(sample(nrow(ddf), nrow(ddf), replace = TRUE))
  train <- ddf[trainid,]
  test <- ddf[-trainid,]

  # Models
  fit.nn22 <- neuralnet(frmnn, data = train, hidden = c(3,3),
                        threshold = 0.05)
  mse.test[i] <- mean((test$wage - predict(fit.nn22, test))^2)
}

mean(mse.test)
```

```
## [1] 1.211114
```

The overfitting gets worse with an increased complexity! Here is the plot for our DNN:

```
plot(fit.nn22, rep = "best", col.axis = "black", col.border = "black",
     col.hidden = "gray80", col.intercept = "black", col.out = "black",
     col.synapse = "black")
```

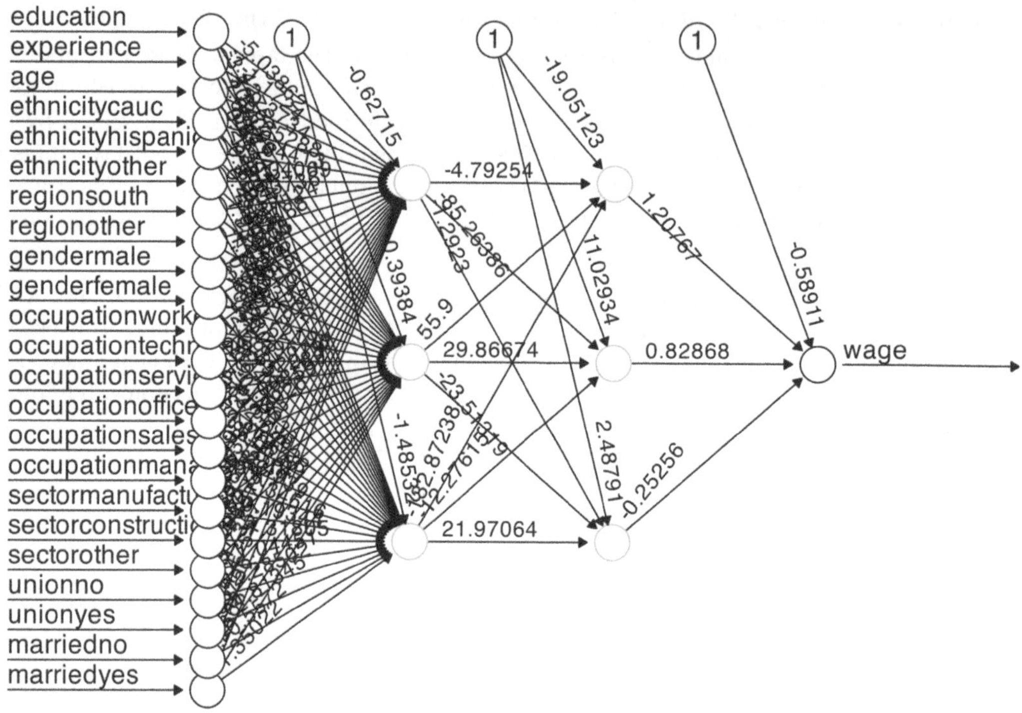

A better plot could be obtained by using the `NeuralNetTools` package:

```
library(NeuralNetTools)
plotnet(fit.nn22, alpha = 1, circle_col = "gray80", border_col = "black", line_col =
  "black")
```

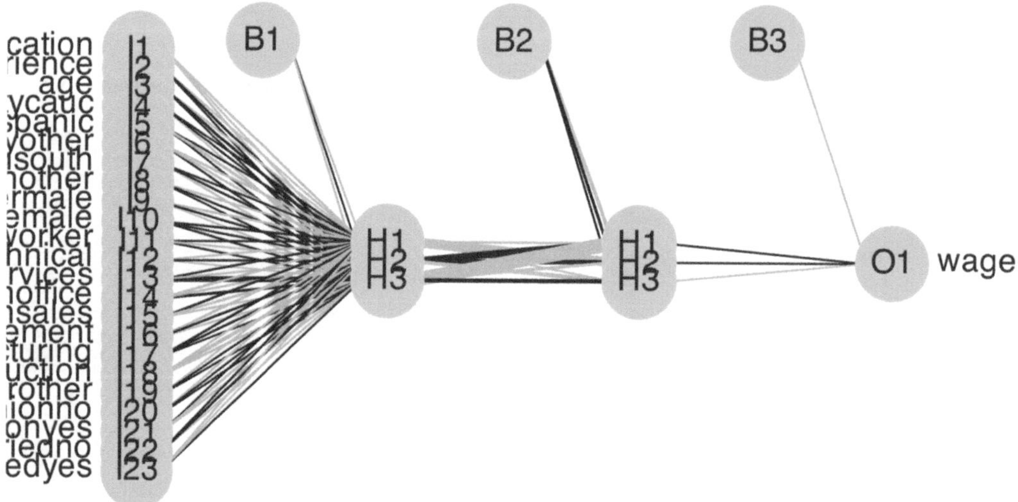

DNNs can model complex nonlinear relationships. With very complex problems, such as detecting hundreds of types of objects in high-resolution images, we need to train deeper

NNs, perhaps with 10 layers or more each with hundreds of neurons. Therefore, training a fully-connected DNN is a very slow process facing a severe risk of overfitting with millions of parameters. Moreover, gradients problems make lower layers very hard to train. A solution to these problems came with a different NN architect such as convolutional NNs (CNN or ConvNets) and recurrent NNs (RNN)

Moreover, the interpretability of an ANN, which is known to be a "blackbox" method, can be an issue regardless of the complexity of the network. However, it is generally easier to understand the decisions made by a simple ANN than by a more complex one.

A simple ANN might have only a few layers and a relatively small number of neurons, making it easier to understand how the input data is processed and how the final output is produced. However, even a simple ANN can still be a black box in the sense that the specific calculations and decisions made by the individual neurons within the network are not fully visible or understood. On the other hand, a more complex ANN with many layers and a large number of neurons can be more difficult to interpret, as the internal workings of the network are more complex and harder to understand. In these cases, it can be more challenging to understand how the ANN is making its decisions or to identify any biases or errors in its output. Overall, the interpretability of an ANN depends on the complexity of the network and the specific task it is being used for. Simple ANNs may be more interpretable, but even they can be considered black boxes to some extent.

Here are a few resources that provide information about the interpretability of ANNs:

- Interpretable Machine Learning[9] by Christoph Molnar (2021) is a online book that provides an overview of interpretability in machine learning, including techniques for interpreting ANNs.
- Interpretability of Deep Neural Networks[10] by Chakraborty et al. (2017) is a survey paper that discusses the interpretability of deep neural networks and presents an overview of the various techniques and approaches that have been developed to improve their interpretability.

Before concluding this section we apply DNN to a classification problem using the same data that we had earlier.

```
library(ISLR)
df <- Carseats
str(df)
```

```
## 'data.frame':	400 obs. of  11 variables:
##  $ Sales      : num  9.5 11.22 10.06 7.4 4.15 ...
##  $ CompPrice  : num  138 111 113 117 141 124 115 136 132 132 ...
##  $ Income     : num  73 48 35 100 64 113 105 81 110 113 ...
##  $ Advertising: num  11 16 10 4 3 13 0 15 0 0 ...
##  $ Population : num  276 260 269 466 340 501 45 425 108 131 ...
##  $ Price      : num  120 83 80 97 128 72 108 120 124 124 ...
##  $ ShelveLoc  : Factor w/ 3 levels "Bad","Good","Medium": 1 2 3 3 1 1 3 2 3 3 ...
##  $ Age        : num  42 65 59 55 38 78 71 67 76 76 ...
##  $ Education  : num  17 10 12 14 13 16 15 10 10 17 ...
##  $ Urban      : Factor w/ 2 levels "No","Yes": 2 2 2 2 2 1 2 2 1 1 ...
##  $ US         : Factor w/ 2 levels "No","Yes": 2 2 2 1 2 1 2 1 2 ...
```

```
#Change SALES to a factor variable
df$Sales <- as.factor(ifelse(Carseats$Sales <= 8, 0, 1))
dff <- df[, -1]

# Scaling and Dummy coding
dff[,sapply(dff, is.numeric)] <- scale((dff[, sapply(dff, is.numeric)]))
ddf <- model.matrix(~.-1, data= dff, contrasts.arg =
                    lapply(dff[, sapply(dff, is.factor)],
                           contrasts, contrasts = FALSE))

ddf <- data.frame(Sales = df$Sales, ddf)

# Formula
w.ind <- which(colnames(ddf) == "Sales")
frm <- as.formula(paste("Sales~", paste(colnames(ddf[-w.ind]),
                                        collapse = '+')))


library(ROCR)

n <- 10
AUC1 <- c()
AUC2 <- c()

for (i in 1:n) {
  set.seed(i)
  ind <- unique(sample(nrow(ddf), nrow(ddf), replace = TRUE))
  train <- ddf[ind, ]
  test <- ddf[-ind, ]

  # Models
  fit.ln <- glm(frm, data = train, family = binomial(link = "logit"))
  fit.dnn <- neuralnet(frm, data = train, hidden = 2, threshold = 0.05,
                       linear.output = FALSE, err.fct = "ce")

  #Predictions
  phat.ln <- predict(fit.ln, test, type = "response")
  phat.dnn <- predict(fit.dnn, test, type = "repsonse")[,2]

  #AUC for predicting Y = 1
  pred_rocr1 <- ROCR::prediction(phat.ln, test$Sales)
  auc_ROCR1 <- ROCR::performance(pred_rocr1, measure = "auc")
  AUC1[i] <- auc_ROCR1@y.values[[1]]

  pred_rocr2 <- ROCR::prediction(phat.dnn, test$Sales)
  auc_ROCR2 <- ROCR::performance(pred_rocr2, measure = "auc")
  AUC2[i] <- auc_ROCR2@y.values[[1]]
}

(c(mean(AUC1), mean(AUC2)))
```

```
## [1] 0.9471081 0.9186785
```

Again the results are not very convincing to use DNN in this example.

We now take on a more complex classification task using the Red Wine dataset from Kaggle (Cortez et al., 2009), available at Wine Quality Data Set[11]. The dataset contains 1,599 observations and 12 variables. Our goal is to predict wine quality (rated from 3 to 8) based on 11 physicochemical attributes such as acidity, residual sugar, and sulfur dioxide.

```
dfr <- read.csv("wineQualityReds.csv", header = TRUE)
dfr <- dfr[, -1] # removing the index
table(dfr$quality)
```

```
##
##   3   4   5   6   7   8
##  10  53 681 638 199  18
```

```
# Let's remove the outlier qualities:
indo <- which(dfr$quality == "3" | dfr$quality == "8")
dfr <- dfr[-indo, ]
dfr$quality <- as.factor(dfr$quality)
table(dfr$quality)
```

```
##
##   4   5   6   7
##  53 681 638 199
```

Then scale the data and get the formula,

```
# Scaling and Dummy coding
dfr[, sapply(dfr, is.numeric)] <-
  scale((dfr[, sapply(dfr, is.numeric)]))

ddf <- model.matrix( ~ quality - 1, data = dfr)
w.ind <- which(colnames(dfr) == "quality")
dfr <- dfr[, -w.ind] # removing 'quality`
df <- cbind(ddf, dfr)

frm <- as.formula(paste(
  paste(colnames(ddf), collapse = '+'),
  "~",
  paste(colnames(dfr), collapse = '+')
))
frm
```

```
## quality4 + quality5 + quality6 + quality7 ~ fixed.acidity + volatile.acidity +
##     citric.acid + residual.sugar + chlorides + free.sulfur.dioxide +
##     total.sulfur.dioxide + density + pH + sulphates + alcohol
```

And, our simple DNN application

```
ind <- sample(nrow(df), nrow(df) * .7)
train <- df[ind, ]
test <- df[-ind, ]

fit.nn <-
  neuralnet(
    frm,
    data = train,
    hidden = c(3, 2),
    threshold = 0.05,
    linear.output = FALSE,
    err.fct = "ce"
  )

plot(fit.nn, rep = "best", col.axis = "black", col.border = "black",
     col.hidden = "gray80", col.intercept = "black",
     col.out = "black", col.synapse = "black")
```

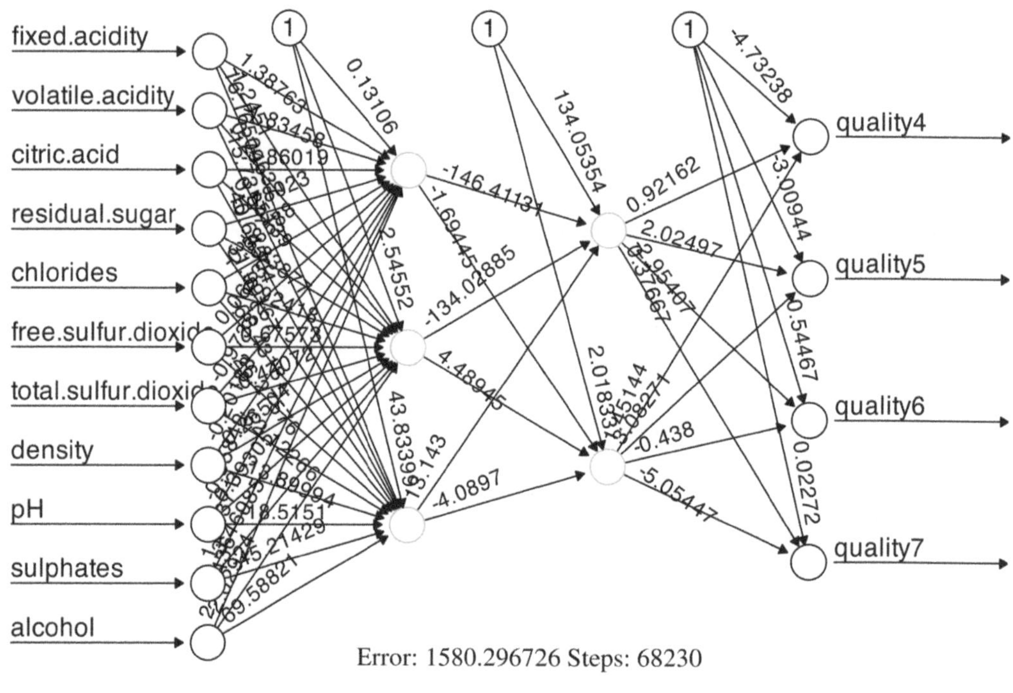

And our prediction:

```
library(utiml)

phat <- predict(fit.nn, test)
head(phat)
```

```
##           [,1]      [,2]      [,3]        [,4]
## 11 0.04931942 0.6859165 0.2825146 0.007453298
```

```
## 14 0.02015208 0.2300976 0.5716617 0.050629158
## 16 0.08538068 0.8882124 0.0550048 0.009824072
## 17 0.03592572 0.5136539 0.5273030 0.006685037
## 22 0.03616818 0.5173671 0.5263112 0.006541895
## 27 0.03092853 0.4318134 0.5389403 0.011412135
```

```
# Assigning label by selecting the highest phat
label.hat <- t(apply(phat, 1, function(x) as.numeric(x == max(x))))
head(label.hat)
```

```
##    [,1] [,2] [,3] [,4]
## 11    0    1    0    0
## 14    0    0    1    0
## 16    0    1    0    0
## 17    0    0    1    0
## 22    0    0    1    0
## 27    0    0    1    0
```

```
# Confusion Table
pred <- apply(phat, 1, which.max)
fck <- colnames(test)[1:4]
predicted <- fck[pred]

act<- apply(test[,1:4], 1, which.max)
actual <- fck[act]

table(predicted, actual)
```

```
##            actual
## predicted   quality4 quality5 quality6 quality7
##   quality5         4      107       32        3
##   quality6         7       78      160       51
##   quality7         1        0       11       18
```

This is just an example and the results are not reflecting a trained model.

Advance DNN applications with a proper training require a longer time and more capable machines. We can do a grid search on different number of hidden layers and neurons. However, a large datasets and more complex DNNs need better applications, like, `Keras` that uses GPU with capable operating systems allowing a much better efficiency in training. So far, we have used the `neuralnet` package. There are several packages in R that are also capable of implementing and training ANN. The most suitable one for our needs will depend on our specific requirements and preferences. For a powerful and flexible package for building and training ANNs, `neuralnet` or `deepnet` may be good options. When we just need a simple and easy-to-use package for training feedforward networks and making predictions, `nnet` may be another good choice. If we want a general-purpose package that can handle a wide range of machine learning tasks, including ANNs, `caret` would be a good option.

DNNs are neural networks with many layers, which can be difficult to train because of the large number of parameters that need to be optimized. This can make the training process computationally intensive and prone to overfitting. CNNs, on the other hand, are specifically designed to process data that has a grid-like structure, such as an image. One key aspect of

CNNs is that they use convolutional layers, which apply a set of filters to the input data and produce a set of transformed feature maps. These filters are able to detect specific features in the input data, such as edges, corners, or textures, and are able to share these features across the input data. This means that the number of parameters in a CNN is typically much smaller than in a DNN, which makes the model easier to train and less prone to overfitting. Overall, CNNs are well-suited for tasks such as image classification, object detection and, speech recognition. We will not cover the details of CNN here. There are several packages available in R for working with CNNs.

Finally, in a DNN, "dropout" and "regularization" are techniques used to prevent overfitting. Dropout is a regularization technique that randomly drops out, or removes, a certain percentage of neurons from the network during training. This has the effect of reducing the complexity of the model, as it can't rely on any one neuron or group of neurons to make predictions. Regularization is a general term that refers to any method used to prevent overfitting in a machine learning model. There are many types of regularization techniques, which add a penalty term to the parameters of the the activation functions.

30.4 Deep Learning with `keras` package

The `keras` package in R provides an interface to the Keras deep learning library, which is a high-level neural networks API written in Python. Keras allows you to build and train deep learning models with minimal code and provides a wide range of pre-built layers, activation functions, optimizers, and loss functions to choose from. The `keras` package in R provides an interface to the Keras deep learning library, which is a high-level neural networks API written in Python.

To use Keras with GPU acceleration with M processors, you'll need to set up TensorFlow to work with Apple's Metal Performance Shaders (MPS) framework. The M series Apple computers have powerful Apple Silicon GPUs that can be leveraged for machine learning tasks. Here's how to set it up:

1. Install TensorFlow with MPS support:

```
pip install tensorflow-macos
pip install tensorflow-metal
```

2. For R users specifically, you'll need to set up the proper environment:
   - Install R and RStudio
   - Install the `keras` and `tensorflow` R packages

```
install.packages("keras")
install.packages("tensorflow")
```

3. Initialize and configure Keras to use the MPS backend:

```
library(keras)
library(tensorflow)

# Install TensorFlow for R with MPS support
install_tensorflow(extra_packages = c("tensorflow-metal"))

# Configure to use Metal
use_backend("tensorflow")
```

4. Verify GPU availability:

```
library(tensorflow)
tf$config$list_physical_devices('GPU')
```

This should show your Apple GPU device if everything is set up correctly.

When building Keras models, your code will automatically utilize the GPU acceleration provided by the Metal framework when available.

Keep in mind that while the Metal GPU acceleration provides good performance, some TensorFlow operations may not be fully optimized for Apple Silicon yet. The M chips should provide excellent performance for most deep learning tasks, especially with the properly configured environment.

The following program trains a deep neural network regression model using the CPS1985 dataset (from the AER package) to predict log wages based on demographic and job characteristics. The code below trains a DNN with two hidden layers, each containing 64 neurons. The model is trained on the training data and evaluated on the test data using the mean squared error Error (MSE) and (MAE) as metrics. The results show the test loss and test MAE for the model. More details can be found here TensorFlow for R[12].

```
library(keras)
library(tensorflow)
library(AER)

# Check for GPU availability (if any)
print(tf$config$list_physical_devices("GPU"))
```

```
## [[1]]
## PhysicalDevice(name='/physical_device:GPU:0', device_type='GPU')
```

```
# Clear any previous Keras session
keras::k_clear_session()

# Load the dataset from the AER package
data("CPS1985")
df <- CPS1985
```

```r
# Scale numeric features
numeric_vars <- sapply(df, is.numeric)
df[, numeric_vars] <- scale(df[, numeric_vars])

# Create dummy variables for factor variables using model.matrix
ddf <- model.matrix(~ . - 1, data = df,
contrasts.arg = lapply(df[, sapply(df, is.factor)],contrasts,contrasts = FALSE))
ddf <- as.data.frame(ddf)

# Split the data into training (80%) and testing (20%) sets
set.seed(123)  # for reproducibility
train_indices <- sample(1:nrow(ddf), size=round(0.8 * nrow(ddf)),replace =FALSE)
train <- ddf[train_indices, ]
test <- ddf[-train_indices, ]

# Prepare predictor matrices and target vectors. Here 'wage' is the target.
train_x <- as.matrix(train[, -which(colnames(train) == "wage")])
train_y <- matrix(train$wage, ncol = 1)
test_x <- as.matrix(test[, -which(colnames(test) == "wage")])
test_y <- matrix(test$wage, ncol = 1)

# Define the input layer using the number of predictors
inputs <- layer_input(shape = ncol(train_x))

# Build the model using the functional API
outputs <- inputs %>%
  layer_dense(units = 64, activation = "relu") %>%
  layer_dense(units = 64, activation = "relu") %>%
  layer_dense(units = 1)  # Output layer for regression

# Create the model by specifying inputs and outputs
model <- keras_model(inputs = inputs, outputs = outputs)

# Compile the model with MSE loss and MAE as metric, using the RMSprop optimizer
model$compile(
  loss = "mean_squared_error",
  optimizer = optimizer_rmsprop(),
  metrics = list("mean_absolute_error")
)

# Fit the model. Setting verbose = 0 will suppress progress output.
history <- model$fit(
  x = train_x,
  y = train_y,
  epochs = as.integer(100),
  batch_size = as.integer(128),
  validation_split = 0.2,
  verbose = 0  # change to 1 if you wish to see training progress
)
```

```
# Evaluate the model on the test set
score <- model$evaluate(
  x = test_x,
  y = test_y,
  verbose = 0
)

cat("Test loss:", score[[1]], "\n")

## Test loss: 0.5787563

cat("Test mean absolute error:", score[[2]], "\n")

## Test mean absolute error: 0.5951508
```

Test Loss: This is the value of the loss function for our test dataset. In a regression problem like yours, where you might be using MSE as the loss function, this represents the average of the squares of the differences between the predicted and actual values across all observations in the test set. It's a measure of how well the model is able to predict the target variable on unseen data. A lower test loss indicates better model performance.

Test MAE (Mean Absolute Error): MAE is another metric used to evaluate the performance of a regression model. It's calculated as the average of the absolute differences between the predicted and actual values across all observations in the test set. Unlike the MSE, the MAE gives a linear score, meaning all individual differences are weighted equally in the average. It's particularly useful because it's easy to interpret; for instance, an MAE of 5 would mean that, on average, the predicted values differ from the actual values by 5 units of the target variable.

30.4.1 Setting up a Keras Model

Step 1: Defining the Model

In Keras, a model is essentially a graph of layers. The most common type of model is a stack of layers called the Sequential model, which allows you to create models layer-by-layer in a linear fashion.

- Sequential Model: This is a linear stack of layers. You start by creating an instance of Sequential and then add layers to it. Each layer has weights that correspond to the connections between units (neurons) in this layer and the next layer.
- Layers: The building blocks of models. Layers extract representations out of the data fed into them. For most deep learning models, dense (fully connected) layers are used. The first Dense layer we add plays a special role. While technically every Dense layer is just a layer of neurons fully connected to the previous layer, the first Dense layer you add to the model also implicitly serves as the input layer. This is because it needs to know the shape of the input data it will be receiving, which you specify with the input_shape parameter.
- Activation Functions: These are applied to the output of a layer to introduce non-linearity into the model, allowing it to learn complex patterns in the data. Common activation functions include ReLU, Sigmoid, and Tanh.

- Input Shape: This specifies the shape of the input data. The input shape is the number of features in your dataset. For example, if you have 10 features, the input shape would be 10.

Step 2: Compiling the Model

After defining the model, you need to compile it. This involves specifying the loss function, optimizer, and metrics to be used during training.

- Loss Function: This is the objective that the model will try to minimize during training. For regression problems, Mean Squared Error (MSE) is commonly used.
- Optimizer: This is the algorithm used to update the weights of the model during training. Common optimizers include Stochastic Gradient Descent (SGD), Adam, and RMSprop.
- Metrics: These are used to evaluate the performance of the model. For regression problems, Mean Absolute Error (MAE) is a common metric.

Step 3: Training the Model

Once the model is compiled, you can train it on your training data. This involves passing the input data and target labels to the model and specifying the number of epochs (iterations over the entire dataset) and batch size (number of samples per gradient update).

- Epochs: One epoch is when an entire dataset is passed forward and backward through the neural network once.
- Batch Size: The number of training examples utilized in one iteration.
- Validation Split: The fraction of the training data to be used as validation data. The model will set apart this fraction of the training data, won't train on it, and will evaluate the loss and any model metrics on this data at the end of each epoch.

Step 4: Evaluating the Model

After training the model, you can evaluate its performance on the test data. This involves passing the test data to the model and calculating the loss and metrics specified during compilation.

30.4.2 Keras with Python

Lets do the same thing with Python. First, we need to save the `ddf` dataframe to a CSV file so that we can load it in Python.

```
write.csv(ddf, "ddf.csv", row.names = FALSE)
```

Now, we can run the following Python code to train a deep neural network on the same data and evaluate its performance on the test set.

```
from tensorflow.keras import backend as K
from tensorflow.keras.models import Sequential
from tensorflow.keras.layers import Dense
from tensorflow.keras.optimizers import RMSprop
import numpy as np
import pandas as pd

# Clearing the session
```

```
K.clear_session()

# Loading the ddf dataframe from the CSV file
ddf = pd.read_csv("ddf.csv")

# Let's split the data into training and testing sets
np.random.seed(42)  # For reproducibility
sample_indices = np.random.choice(ddf.index, size = len(ddf), replace = True)
train = ddf.loc[sample_indices]
out_of_bag_indices = np.setdiff1d(ddf.index, sample_indices, assume_unique = False)
test = ddf.loc[out_of_bag_indices]

# Preparing the data
train_x = train.drop(columns=['wage']).values
train_y = train['wage'].values
test_x = test.drop(columns=['wage']).values
test_y = test['wage'].values

# Define the model
model = Sequential([
    Dense(64, activation='relu', input_shape=(train_x.shape[1],)),
    Dense(64, activation='relu'),
    Dense(1)
])

# Compile the model
model.compile(optimizer = RMSprop(), loss = 'mean_squared_error',
              metrics = ['mean_absolute_error'])

# Fit the model
history = model.fit(
    train_x,
    train_y,
    epochs=100,
    batch_size=128,
    validation_split=0.2,
    verbose=0
)

# Evaluate the model
score = model.evaluate(test_x, test_y, verbose=0)

print('Test loss:', score[0])

## Test loss: 0.6580724716186523

print('Test mean absolute error:', score[1])

## Test mean absolute error: 0.6071147322654724
```

30.4.3 Tuning a Keras Model

Tuning a Keras model involves experimenting with various configurations of the model's architecture, the compilation settings, and the training process to find the best performance. Here are some common ways to tune a simple Keras model:

30.4.3.1 Adjusting Layer Configurations:

- Number of Units: Vary the number of neurons in the Dense layers. More neurons can capture more complex patterns but can also lead to overfitting.
- Activation Functions: Try different activation functions (e.g., relu, tanh, sigmoid) to see how they affect model performance.

30.4.3.2 Experimenting with Optimizers:

- Different optimizers (e.g., adam, sgd, rmsprop) have different optimization algorithms and hyperparameters (like learning rate). Switching optimizers or adjusting their parameters (e.g., learning rate) can significantly impact model performance.
- Varying the Loss Functions: Depending on the problem type (regression vs. classification), different loss functions might be more appropriate. Even within a problem type, experimenting with different loss functions can yield better results.

30.4.3.3 Regularization Techniques:

To combat overfitting, try adding regularization methods such as L1 or L2 regularization, dropout layers, or batch normalization. These techniques can help the model generalize better to unseen data. L1 and L2 regularization add a penalty term to the loss function based on the weights, while dropout randomly sets a fraction of input units to zero during training to prevent overfitting. Dropout is a regularization technique that helps prevent overfitting by randomly setting a fraction of input units to zero during training. This forces the model to learn more robust features and reduces the risk of overfitting.

30.4.3.4 Model Training Adjustments:

- Epochs and Batch Size: The number of epochs and the batch size can affect how well your model learns. Too few epochs might underfit, while too many can overfit. The batch size can influence the stability and speed of the learning process. Experiment with different values to find the optimal combination.
- Validation Split: Adjusting the proportion of data used for validation can provide a better estimate of model performance on unseen data. A common choice is 20% of the training data.
- Early Stopping: Implement early stopping to halt training when the validation performance stops improving. This prevents overfitting and saves time by avoiding unnecessary epochs.

30.4.3.5 Hyperparameter Tuning Tools:

Consider using tools like Keras Tuner, Hyperopt, or scikit-learn's `GridSearchCV` or `RandomizedSearchCV` for systematic hyperparameter tuning. These tools can help you search through a large hyperparameter space efficiently and find the best configuration for your model.

In the following example, L1 and L2 regularization, dropout, and batch normalization are added to combat overfitting. An early stopping callback halts training when the validation loss hasn't improved for a specified number of epochs (patience), preventing overfitting and saving computational resources.

```
from tensorflow.keras.layers import Input, Dense, Dropout, BatchNormalization
from tensorflow.keras.models import Sequential
from tensorflow.keras.optimizers.legacy import RMSprop
from tensorflow.keras.regularizers import l1_l2
from tensorflow.keras.callbacks import EarlyStopping

model = Sequential([
    Input(shape=(train_x.shape[1],)),
    Dense(64, activation='relu', kernel_regularizer=l1_l2(l1=0.01, l2=0.01)),
    Dropout(0.5),
    BatchNormalization(),
    Dense(64, activation='relu'),
    Dense(1)
])

model.compile(optimizer='adam', loss='mean_squared_error',
              metrics=['mean_absolute_error'])

early_stopping = EarlyStopping(monitor='val_loss', patience=10)

history = model.fit(train_x, train_y, epochs=100, batch_size=128,
          validation_split=0.2, callbacks=[early_stopping], verbose=0)
```

In this example, L1 and L2 regularization are added to the first hidden layer to penalize large weights, dropout is used to prevent overfitting, and batch normalization is applied to stabilize training. The early stopping callback monitors the validation loss and stops training if it doesn't improve for 10 epochs. These techniques can help improve the model's generalization performance and prevent overfitting.

```
score = model.evaluate(test_x, test_y, verbose=0)

print('Test loss:', score[0])
```

```
## Test loss: 1.243335247039795
```

```
print('Test mean absolute error:', score[1])
```

```
## Test mean absolute error: 0.6312177777290344
```

30.4.4 Visualizing Model Performance

```
import pydot
from tensorflow.keras.utils import plot_model

plot_model(model, to_file='png/model_plot.png',
           show_shapes=True, show_layer_names=True)
```

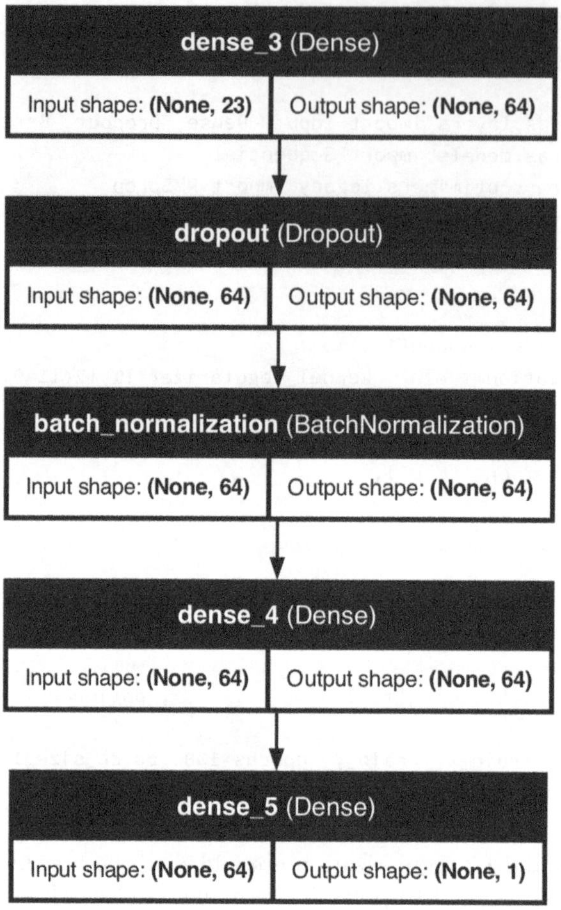

The `plot_model` function generates a visual representation of the model architecture, showing the connections between layers and the flow of data through the network. This can be useful for understanding the model's structure and sharing it with others.

And we can plot the training history to visualize the model's performance during training:

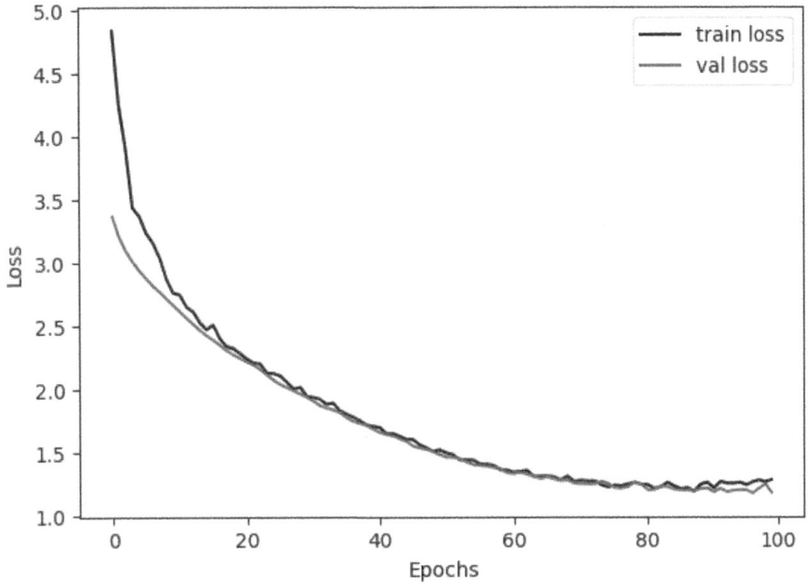

30.4.5 Grid Search with Keras Tuner

Step 1: Import Libraries: First, you import necessary libraries and modules. This includes Keras Tuner itself, TensorFlow's Keras API for building neural network models, and specific classes for model layers and the sequential model type.

Step 2: Define the Model-Building Function: The `build_model` function is where you define the architecture of your neural network. This function takes a `HyperParameters` object (`hp`) as its argument, which is used to define the space of hyperparameters you want to explore.

- Sequential Model: You start by initializing a Sequential model, indicating you're building a model layer by layer in sequence.
- Dense Layers: You then add Dense layers, which are fully connected neural network layers. The number of units (neurons) in these layers is what you're tuning, so instead of hardcoding a number, you use `hp.Int('units', min_value=32, max_value=512, step=32)` to define a range of values to explore during the tuning process.
- Compilation: Finally, you compile the model, specifying the optimizer, loss function, and metrics to monitor.

Step 3: Initialize the Tuner: You instantiate a `RandomSearch` tuner object. This object will manage the tuning process, including:

- Running trials (each trial tests a unique set of hyperparameters).
- Evaluating model performance for each set of hyperparameters.
- Selecting the next set of hyperparameters to test.

Parameters for the tuner include:

- The model-building function (`build_model`).
- The optimization objective (`val_loss`).

- The maximum number of trials (max_trials) and executions per trial (executions_per_trial).
- Directories for storing tuning results (directory and project_name).

Step 4: Perform the Search: Using the tuner.search() method, you start the hyperparameter search process. This method requires your training data (train_x, train_y), the number of epochs for training each model, and the validation split. The tuner will train different models using varying hyperparameters within the defined space and evaluate their performance.

Step 5: Get the Best Model and Hyperparameters: After the search concludes, you retrieve the best-performing model and its hyperparameters using tuner.get_best_models() and tuner.get_best_hyperparameters().

```
import keras_tuner as kt
from keras_tuner import RandomSearch
from keras_tuner.engine.hyperparameters import HyperParameters
from tensorflow.keras.models import Sequential
from tensorflow.keras.layers import Dense

def build_model(hp):
    model = Sequential()
    # Define the hyperparameter space for 'units'
    units = hp.Int('units', min_value=32, max_value=512, step=32)

    model.add(Dense(units=units, activation='relu', input_shape=(train_x.shape[1],)))
    model.add(Dense(units=units, activation='relu'))  # Reuse the 'units' variable
    model.add(Dense(1))
    model.compile(optimizer='adam', loss='mean_squared_error',
                  metrics=['mean_absolute_error'])
    return model

tuner = RandomSearch(
    build_model,
    objective='val_loss',
    max_trials=5,
    executions_per_trial=3,
    directory='my_dir',
    project_name='helloworld'
)

tuner.search(train_x, train_y, epochs=100, validation_split=0.2, verbose=0)

best_model = tuner.get_best_models(num_models=1)[0]
best_hyperparameters = tuner.get_best_hyperparameters(num_trials=1)[0]

# best_model.summary()

score = best_model.evaluate(test_x, test_y, verbose=0)

print('Test loss:', score[0])
```

```
## Test loss: 0.6608307957649231
```

```r
print('Test mean absolute error:', score[1])
```

```
## Test mean absolute error: 0.6022828817367554
```

30.4.6 Grid Search with R

Hyperparameter tuning in R, especially for Keras models, can be approached in several ways, but it does not directly support `keras_tuner` like Python. However, you can use a combination of custom loops or grid search strategies to achieve a similar outcome. Below is an R version that approximates the hyperparameter tuning process described in your Python script. This example uses a manual grid search approach:

```r
library(keras)
library(dplyr)

# Clear the Keras session
keras::k_clear_session()

train_x <- as.matrix(ddf %>% select(-wage))
train_y <- matrix(ddf$wage, ncol = 1)

# Define the model-building function accepting hyperparameters
build_model <- function(units) {
  # Clear session for each model
  keras::k_clear_session()

  # Define model using functional API instead of sequential
  inputs <- layer_input(shape = ncol(train_x))
  outputs <- inputs %>%
    layer_dense(units = units, activation = 'relu') %>%
    layer_dense(units = units, activation = 'relu') %>%
    layer_dense(units = 1)

  # Create model
  model <- keras_model(inputs = inputs, outputs = outputs)

  # Compile model
  model$compile(
    optimizer = 'adam',
    loss = 'mean_squared_error',
    metrics = list('mean_absolute_error')
  )

  return(model)
}

# Manual hyperparameter tuning setup
units_grid <- seq(32, 512, by = 32)
```

```
val_losses <- numeric(length(units_grid))
names(val_losses) <- as.character(units_grid)

# Grid search loop
for (units in units_grid) {
  cat("Training with units =", units, "\n")
  model <- build_model(units)

  history <- model$fit(
    x = train_x,
    y = train_y,
    epochs = as.integer(100),
    batch_size = as.integer(32),
    validation_split = 0.2,
    verbose = 0
  )

  val_losses[as.character(units)] <- min(history$history$val_loss)
}
```

```
## Training with units = 32
## Training with units = 64
## Training with units = 96
## Training with units = 128
## Training with units = 160
## Training with units = 192
## Training with units = 224
## Training with units = 256
## Training with units = 288
## Training with units = 320
## Training with units = 352
## Training with units = 384
## Training with units = 416
## Training with units = 448
## Training with units = 480
## Training with units = 512
```

```
# Find the best performing model
best_units <- as.numeric(names(which.min(val_losses)))
cat("Best units:", best_units, "\n")
```

```
## Best units: 224
```

```
# Build and evaluate the best model
best_model <- build_model(best_units)
best_model$fit(
  x = train_x,
  y = train_y,
  epochs = as.integer(100),
```

```
  batch_size = as.integer(32),
  verbose = 0
)

## <keras.src.callbacks.history.History object at 0x3a84b9280>

# Evaluate on test data
score <- best_model$evaluate(
  x = test_x,
  y = test_y,
  verbose = 0
)

cat('Test loss:', score[[1]], '\n')

## Test loss: 0.9601009

cat('Test mean absolute error:', score[[2]], '\n')

## Test mean absolute error: 0.7689717
```

In this R script, we define a `build_model` function that creates a Keras model with a specified number of units in the Dense layers. We then set up a manual grid search loop to iterate over different values of `units` and train models with those hyperparameters. The best-performing model is selected based on the validation loss, and its performance is evaluated on the test set.

30.5 Clasification with keras Package

Applying deep learning to classification problems introduces some differences in model architecture, training process, and evaluation compared to regression problems. Here's an overview of these differences and some key considerations:

30.5.1 Model Outcome and Activation Function

Regression models typically have a single output unit without an activation function (or with a linear activation function), predicting a continuous quantity. For binary classification, there's usually a single output unit with a sigmoid activation function. The sigmoid function maps the output to a probability score between 0 and 1, indicating the likelihood of the input belonging to the positive class.

For multi-class classification, the output layer has one unit per class and uses the softmax activation function. The softmax function converts the model outputs to probabilities that sum to 1, with each value representing the probability that the input belongs to one of the classes.

Regression models commonly use MSE or mean absolute error (MAE) as the loss function. For binary classification, the binary cross-entropy loss function is standard.

For multi-class classification, categorical cross-entropy (or softmax cross-entropy) loss is typically used. This loss function compares the predicted class probabilities with the true class labels and penalizes the model based on the difference between them.

Cross-entropy loss is a measure of the difference between two probability distributions: the predicted distribution (output of the model) and the true distribution (one-hot encoded labels). The goal is to minimize this difference during training. The cross-entropy loss is defined as:

$$\text{Cross-Entropy Loss} = -\sum_i y_i \log(p_i) \tag{30.10}$$

where y_i is the true class label (0 or 1 for binary classification, one-hot encoded for multi-class classification), and p_i is the predicted probability for class i. For a binary classification, we have two possible outcomes: the true label y is either 0 or 1 , and the predicted probability (from the neural network) is p. The binary cross-entropy loss for a single observation is:

$$L(y,p) = -[y\log(p) + (1-y)\log(1-p)] \tag{30.11}$$

- If the true label $y = 1$, the loss becomes $L(1,p) = -\log(p)$. As p (the predicted probability of the positive class) decreases, $-\log(p)$ increases, meaning the loss increases when the model's prediction is further from the actual class.
- If the true label $y = 0$, the loss becomes $L(0,p) = -\log(1-p)$. As p increases, $-\log(1-p)$ increases, meaning the loss increases when the model's prediction is incorrectly confident in the positive class.

This loss function effectively penalizes predictions that are confident and wrong, encouraging the model to produce probabilities close to the actual labels. And the total loss is

$$J = \sum_{i=1}^{n} y_i \log p(x_i) + (1-y_i)\log(1-p(x_i)), \tag{30.12}$$

where n is the number of observations, y_i is the true label (0 or 1), and $p(x_i)$ is the predicted probability for observation i. For each observation,

$$p(\mathbf{x_i}) = \frac{e^{\beta_0+\beta x_i}}{1+e^{\beta_0+\beta x_i}} \tag{30.13}$$

where β_0 is the intercept and β is the coefficient vector. The sigmoid function maps the linear combination of features to a probability between 0 and 1.

30.5.2 Probabilistic Model and Thresholding

Neural networks can be considered probabilistic in the context of classification. The output layer (using sigmoid or softmax) provides probabilities indicating the likelihood of each class. Although the raw output of the network provides these probabilities, making a final class decision often requires thresholding

As before, we cannot use a common threshold, like 0.5, for the sigmoid output. We need to rely on AUC. In multi-class classification, the class with the highest probability (from the softmax output) is typically selected as the model's prediction.

Unlike regression, where you might use MSE or MAE, classification models are often evaluated using accuracy, precision, recall, F1 score. We use, however, a complex metrics like ROC-AUC or confusion matrices, depending on the specific requirements of the problem.

30.5.3 **keras** Package

First the preprocessing

```
library(keras)
library(tensorflow)
library(caret) # for dummyVars
library(dplyr) # for data manipulation

# Clearing the workspace
rm(list = ls())

# Clearing the Keras session
keras::k_clear_session()

# Load the data
dfr <- read.csv("wineQualityReds.csv", header = TRUE)
dfr <- dfr[, -1] # removing the index column
# Ensure quality is treated as numeric, not string
indo <- which(dfr$quality == 3 | dfr$quality == 8)
dfr <- dfr[-indo, ]

# Convert quality to a factor for one-hot encoding
dfr$quality <- as.factor(dfr$quality)

# One-hot encode the quality variable
quality_one_hot <- model.matrix(~ quality - 1, data = dfr)

# Prepare the data excluding the original quality column
df_transformed <- dfr[, -which(names(dfr) == "quality")]

# Splitting the dataset into training and test sets
set.seed(123) # For reproducibility
# Randomly sample 80% of the data
indexes <- sample(1:nrow(dfr), size = 0.8 * nrow(dfr), replace = FALSE)

train_x <- df_transformed[indexes, ]
test_x <- df_transformed[-indexes, ]
train_y <- quality_one_hot[indexes, ]
test_y <- quality_one_hot[-indexes, ]

# Define the input layer using the number of predictors
inputs <- layer_input(shape = ncol(train_x))

# There are 4 classes after your preprocessing
outputs<- inputs %>%
```

```
  layer_dense(units = 32, activation = 'relu') %>%
  layer_dense(units = 32, activation = 'relu') %>%
  # Update the following to match the number of classes
  layer_dense(units = 4, activation = 'softmax')

# Create the model by specifying inputs and outputs
model <- keras_model(inputs = inputs, outputs = outputs)

# Compile the model for multi-class classification
model$compile(
  loss = 'categorical_crossentropy', # Correct loss function for multi-class
  optimizer = 'adam', # Accuracy is a common metric for classification
  metrics = list(metric_binary_accuracy(), tf$keras$metrics$AUC(name = "auc"))
)

# Fit the model
history <- model$fit(
  x = as.matrix(train_x),
  y = train_y,
  epochs = as.integer(100),
  batch_size = as.integer(128),
  validation_split = 0.2,
  verbose = 0
)

# Predictions as probabilities
predictions <- model$predict(as.matrix(test_x), verbose = 0)
head(predictions)
```

```
##              [,1]      [,2]      [,3]        [,4]
## [1,] 0.033331841 0.6760932 0.2657034 0.024871647
## [2,] 0.023892730 0.7402552 0.2192302 0.016621800
## [3,] 0.003671232 0.8908125 0.1035926 0.001923616
## [4,] 0.020731958 0.6742111 0.2771741 0.027882790
## [5,] 0.049422119 0.4999054 0.3798771 0.070795409
## [6,] 0.058197383 0.3829152 0.4227718 0.136115640
```

```
# Evaluate the model on the test set
test_x_matrix <- as.matrix(test_x)

model$evaluate(test_x_matrix, test_y, verbose = 0)
```

```
## [1] 1.0048255 0.7793651 0.8267927
```

Now AUC

```
library(pROC)

# Initialize an empty vector to store AUC for each class
```

```
aucs <- c()

# Calculate AUC for each class
for (i in 1:ncol(test_y)) {
  roc_res <- roc(response = test_y[, i], predictor = predictions[, i])
  aucs[i] <- auc(roc_res)
}

# Calculate mean AUC across all classes
mean_auc <- mean(aucs)

print(aucs)

## [1] 0.7073284 0.7681047 0.6265222 0.7883951

print(mean_auc)

## [1] 0.7225876
```

As you can see test and OOB AUC's are close!

In Keras, when compiling the model, the metrics parameter is used to specify the metrics that will be evaluated by the model during training and testing. However, directly measuring AUC during training as a metric within the compile function wasn't originally supported in earlier versions of Keras. This limitation was primarily due to AUC being a more complex calculation that requires the entire dataset's predictions to compute accurately, as opposed to metrics like accuracy, which can be calculated batch-wise.

However, with updates to TensorFlow and Keras, you now have the ability to include AUC as a metric during model compilation if you are using a version of TensorFlow that supports it (TensorFlow 2.2.0 and later).

30.6 PyTorch and Keras

PyTorch and Keras are both popular open-source libraries for deep learning, but they have different design philosophies, ecosystems, and usage patterns. Here's a brief overview of both:

PyTorch

Developed by Facebook's AI Research lab (FAIR).

- Dynamic Computation Graphs: PyTorch uses dynamic computation graphs (also known as define-by-run paradigm), which means the graph is built on the fly as operations are executed. This feature provides a more intuitive approach for dynamic models and iterative processes, making debugging easier and more interactive.
- Pythonic Nature: It seamlessly integrates with the Python data science stack (e.g., NumPy, SciPy) and supports native Python debugging tools. Its API and model definitions are very Pythonic, so it feels more like writing standard Python code.

- Strong GPU Acceleration: PyTorch provides strong support for CUDA and cuDNN, allowing efficient and easy GPU computation.
- Research and Flexibility: Initially favored more by researchers due to its flexibility and dynamic graph feature, which is ideal for experimental projects and rapid prototyping.

Keras

Developed by François Chollet - initially as an independent project, now part of TensorFlow under the TensorFlow umbrella managed by Google.

- High-Level API: Keras is designed as a high-level neural networks library. It runs on top of other deep learning libraries like TensorFlow, Theano, or Microsoft Cognitive Toolkit (CNTK). However, since TensorFlow 2.0, Keras has been integrated more deeply into TensorFlow as tf.keras.
- Ease of Use: It focuses on being user-friendly, modular, and extensible. It provides high-level building blocks for developing deep learning models with a few lines of code, making it accessible for beginners.
- Static Computation Graphs (with TensorFlow backend): TensorFlow (and thereby Keras, when used with TensorFlow as the backend) primarily uses static computation graphs, meaning the model's structure is defined and compiled before it's run. This approach can lead to faster execution and easier deployment of models but might be less intuitive for dynamic model adjustments.
- Wide Adoption in Industry: Due to its simplicity and TensorFlow's strong support, Keras has seen wide adoption in the industry for developing and deploying production models.

Differences at a Glance

PyTorch offers a more dynamic and intuitive approach (dynamic computation graphs) that's particularly suited for research and projects where model architectures change frequently. Keras, especially as part of TensorFlow, focuses on ease of use and quick deployment, making it an excellent choice for beginners and applied deep learning projects.

Both have strong communities. PyTorch is often cited as the go-to for academic research and projects requiring dynamic models, while Keras (through TensorFlow) has a broad adoption in industry with extensive support for production deployment. Keras is generally considered easier to learn for beginners due to its high-level API, while PyTorch, with its Pythonic design and dynamic nature, is also very approachable, especially for those who have a solid grounding in Python.

Choosing between PyTorch and Keras (TensorFlow) often comes down to personal preference, project requirements, and specific features or functionalities that one might need for their deep learning tasks.

Is R Competitive in Neural Networks?

R is a programming language and environment primarily used for statistical computing and graphics. It's heavily used in academia, research, and industry for statistical analysis, data visualization, and data science.

When it comes to neural networks and deep learning, R is not as prominently featured as Python. The Python ecosystem, with libraries like TensorFlow, Keras, and PyTorch, has become the de facto standard for deep learning research and development due to its extensive libraries, community support, and flexibility.

However, this does not mean R is not competitive or incapable of deep learning tasks. R has packages such as `keras` and `tensorflow` that provide interfaces to Keras and TensorFlow, respectively, allowing R users to build and train deep learning models within the R environment. There are also other packages and tools for machine learning and advanced statistical modeling.

The choice between R and Python often comes down to personal preference, specific project requirements, and the ecosystem one is more comfortable with. For users deeply embedded in the R ecosystem, especially those working in statistics, bioinformatics, or fields where R has strong domain-specific libraries, continuing in R for deep learning tasks might be preferable. For new projects or users starting from scratch, especially with a focus on deep learning, Python might offer a more direct path to a wide range of tools and community support.

Notes

1. https://www.iro.umontreal.ca/~vincentp/ift3395/lectures/backprop_old.pdf
2. https://psycnet.apa.org/record/1959-09865-001
3. https://www.iro.umontreal.ca/~vincentp/ift3395/lectures/backprop_old.pdf
4. http://yann.lecun.com/exdb/publis/pdf/lecun-01a.pdf
5. https://pubmed.ncbi.nlm.nih.gov/26017442/
6. https://en.wikipedia.org/wiki/Activation_function
7. https://www.iro.umontreal.ca/~vincentp/ift3395/lectures/backprop_old.pdf
8. https://www.inmodelia.com/exemples/2021-0103-RJournal-SM-AV-CD-PK-JN.pdf
9. https://christophm.github.io/interpretable-ml-book/
10. https://ieeexplore.ieee.org/document/8397411
11. https://archive.ics.uci.edu/ml/datasets/wine+quality
12. https://tensorflow.rstudio.com/guides/

31

Matrix Decomposition and Applications

Matrix decomposition, or matrix factorization, plays a central role in modern data analysis and machine learning. It refers to the process of breaking down a complex matrix into simpler, constituent components that uncover hidden patterns, simplify computation, and enable dimensionality reduction. This chapter systematically explores a suite of decomposition techniques and their powerful applications.

We begin with **eigenvectors and eigenvalues**, the foundational concepts behind many decomposition methods. These mathematical constructs reveal how matrices stretch or compress space along particular directions. They form the basis of **eigenvalue decomposition (EVD)**, which allows certain square matrices to be represented in terms of their eigenvectors and eigenvalues, unlocking deeper insights into their structure and behavior.

We then turn to **singular value decomposition (SVD)**, a more general technique that can be applied to any real or complex matrix, whether square or rectangular. SVD breaks a matrix into orthogonal components, exposing the latent features that drive its structure. This section connects SVD with familiar applications such as recommender systems and image compression.

Following this, we explore **rank (r) approximations**, which involve reconstructing a matrix using only its top singular values and associated vectors. This provides a powerful way to reduce dimensionality, eliminate noise, and compress data without significant information loss.

Next, we examine the **Moore–Penrose inverse**, a generalized matrix inverse especially useful for non-square or singular matrices. Computed via SVD, the pseudoinverse is essential in solving **least-squares problems**, particularly in linear regression and under- or over-determined systems.

Building on these tools, we introduce **principal component analysis (PCA)**, a widely-used method for dimensionality reduction. PCA reorients data along the directions of greatest variance using the eigenvectors of the covariance matrix or through SVD. This transformation reduces the number of variables while preserving the essential structure of the dataset, making PCA a staple in exploratory data analysis and preprocessing for machine learning models.

Finally, we explore **factor analysis (FA)**, a technique closely related to PCA but with a different conceptual foundation. FA models observed data as linear combinations of latent variables (factors) plus error terms. Unlike PCA, which focuses purely on variance, FA aims to uncover the underlying constructs that explain correlations among variables. It is especially useful in social sciences, psychometrics, and behavioral research, where latent structures such as personality traits or intelligence are of primary interest.

Together, these matrix decomposition techniques form a cohesive framework for understanding high-dimensional data. They simplify complex relationships, reveal structure, and provide the foundation for a wide range of analytical and machine learning tools.

DOI: 10.1201/9781003504795-31

31.1 Eigenvectors and Eigenvalues

In this section, we introduce the fundamental concepts of eigenvectors and eigenvalues, which are central to understanding matrix decomposition. These mathematical constructs offer deep insights into the internal structure of matrices and are foundational to numerous applications in mathematics, physics, and data science. From describing linear transformations to powering algorithms like PCA, eigenvectors and eigenvalues play a critical role.

At a high level, eigenvectors and eigenvalues arise in the context of matrix multiplication. While most vectors change both magnitude and direction when multiplied by a matrix $\mathbf{A}$, certain special vectors—called eigenvectors—only change in magnitude, not in direction. In other words, when a matrix $\mathbf{A}$ acts on an eigenvector $\mathbf{v}$, the result is a scaled version of the same vector:

$$\mathbf{A}\mathbf{v} = \lambda\mathbf{v} \tag{31.1}$$

Here, $\mathbf{v}$ is a non-zero vector and λ is a scalar known as the eigenvalue associated with that eigenvector. This simple yet powerful equation captures the essence of how a matrix transforms space: eigenvectors point in invariant directions, and eigenvalues tell us how much those directions are stretched or compressed.

These concepts are especially important in machine learning. For example, PCA involves computing the eigenvectors and eigenvalues of a data set's covariance matrix. The eigenvectors (also called principal components in this context) identify the directions of maximum variance in the data, while the eigenvalues indicate the amount of variance captured by each component. By selecting the top eigenvectors, we can project the data into a lower-dimensional space while preserving its most important patterns.

In the next section, we explore how these ideas underpin various matrix decomposition techniques, beginning with **EVD**. To obtain the eigenvalues and eigenvectors of a matrix $\mathbf{A}$, we start from the fundamental condition ~(31.1).

This can be rewritten as:

$$\mathbf{A}\mathbf{v} - \lambda\mathbf{I}\mathbf{v} = 0 \tag{31.2}$$

which simplifies to:

$$(\mathbf{A} - \lambda\mathbf{I})\mathbf{v} = 0 \tag{31.3}$$

Here, $\mathbf{I}$ is the identity matrix. For this equation to have a **non-trivial solution** (i.e., $\mathbf{v} \neq 0$), the matrix $\mathbf{A} - \lambda\mathbf{I}$ must be **non-invertible**. Otherwise, we would get:

$$\mathbf{v} = (\mathbf{A} - \lambda\mathbf{I})^{-1} \cdot 0 = 0 \tag{31.4}$$

which is the trivial solution. A matrix is non-invertible if and only if its determinant is zero. Therefore, to find the eigenvalues λ, we solve the **characteristic equation**:

$$\det(\mathbf{A} - \lambda\mathbf{I}) = 0 \tag{31.5}$$

This equation gives all values of λ for which a non-zero eigenvector $\mathbf{v}$ satisfies equation (31.1). Each solution λ corresponds to one or more eigenvectors that satisfy the original eigenvalue problem.

We start with a square matrix, **A**, like

$$A = \begin{bmatrix} 1 & 2 \\ 3 & -4 \end{bmatrix}$$

$$\begin{aligned} \det(\mathbf{A} - \lambda\mathbf{I}) &= \begin{vmatrix} 1-\lambda & 2 \\ 3 & -4-\lambda \end{vmatrix} = (1-\lambda)(-4-\lambda) - 6 \\ &= -4 - \lambda + 4\lambda + \lambda^2 - 6 \\ &= \lambda^2 + 3\lambda - 10 \\ &= (\lambda - 2)(\lambda + 5) = 0 \\ &\therefore \lambda_1 = 2, \lambda_2 = -5 \end{aligned}$$

We have two eigenvalues. We now need to consider each eigenvalue indivudally

$$\begin{aligned} \lambda_1 &= 2 \\ (A1 - \lambda I)\mathbf{v} &= 0 \\ \begin{bmatrix} 1-\lambda_1 & 2 \\ 3 & -4-\lambda_1 \end{bmatrix} \begin{bmatrix} v_1 \\ v_2 \end{bmatrix} &= \begin{bmatrix} 0 \\ 0 \end{bmatrix} \\ \begin{bmatrix} -1 & 2 \\ 3 & -6 \end{bmatrix} \begin{bmatrix} v_1 \\ v_2 \end{bmatrix} &= \begin{bmatrix} 0 \\ 0 \end{bmatrix} \end{aligned}$$

Hence,

$$\begin{aligned} -v_1 + 2v_2 &= 0 \\ 3v_1 - 6v_2 &= 0 \\ v_1 = 2, v_2 &= 1 \end{aligned}$$

And for $\lambda_2 = -5$,

$$\begin{aligned} \begin{bmatrix} 1-\lambda_2 & 2 \\ 3 & -4-\lambda_2 \end{bmatrix} \begin{bmatrix} v_1 \\ v_2 \end{bmatrix} &= \begin{bmatrix} 0 \\ 0 \end{bmatrix} \\ \begin{bmatrix} 6 & 2 \\ 3 & 1 \end{bmatrix} \begin{bmatrix} v_1 \\ v_2 \end{bmatrix} &= \begin{bmatrix} 0 \\ 0 \end{bmatrix} \end{aligned}$$

Hence,

$$\begin{aligned} 6v_1 + 2v_2 &= 0 \\ 3v_1 + v_2 &= 0 \\ v_1 = -1, v_2 &= 3 \end{aligned}$$

We have two eigenvalues:

$$\begin{aligned} \lambda_1 &= 2 \\ \lambda_2 &= -5 \end{aligned}$$

And two corresponding eigenvectors:

$$\begin{bmatrix} 2 \\ 1 \end{bmatrix}, \begin{bmatrix} -1 \\ 3 \end{bmatrix}$$

for $\lambda_1 = 2$

$$\begin{bmatrix} 1 & 2 \\ 3 & -4 \end{bmatrix} \begin{bmatrix} 2 \\ 1 \end{bmatrix} = \begin{bmatrix} 2+2 \\ 6-4 \end{bmatrix} = \begin{bmatrix} 4 \\ 2 \end{bmatrix} = 2 \begin{bmatrix} 2 \\ 1 \end{bmatrix}$$

Let's see the solution in R:

```
A <- matrix(c(1, 3, 2, -4), 2, 2)
eigen(A)
```

```
## eigen() decomposition
## $values
## [1] -5  2
##
## $vectors
##            [,1]      [,2]
## [1,] -0.3162278 0.8944272
## [2,]  0.9486833 0.4472136
```

The eigenvectors are typically normalized by dividing by its length $\sqrt{v'v}$, which is 5 in our case for $\lambda_1 = 2$.

```
# For the ev (2, 1), for lambda
c(2, 1) / sqrt(5)
```

```
## [1] 0.8944272 0.4472136
```

There some nice properties that we can observe in this application.

```
# Sum of eigenvalues = sum of diagonal terms of A (Trace of A)
ev <- eigen(A)$values
sum(ev) == sum(diag(A))
```

```
## [1] TRUE
```

```
# Product of eigenvalues = determinant of A
round(prod(ev), 4) == round(det(A), 4)
```

```
## [1] TRUE
```

```
# Diagonal matrix D has eigenvalues = diagonal elements
D <- matrix(c(2, 0, 0, 5), 2, 2)
eigen(D)$values == sort(diag(D), decreasing = TRUE)
```

```
## [1] TRUE TRUE
```

We can see that if one of the eigenvalues is zero for a matrix, the determinant of the matrix will be zero. We willl return to this issue in SVD.

Let's finish this section with diagonalization and eigendecomposition.

Suppose we have m linearly independent eigenvectors ($\mathbf{v_i}$ is eigenvector i in a column vector in $\mathbf{V}$) of $\mathbf{A}$.

$$\mathbf{AV} = \mathbf{A}\left[\mathbf{v_1}\mathbf{v_2}\cdots\mathbf{v_m}\right] = \left[\mathbf{Av_1}\mathbf{Av_2}\ldots\mathbf{Av_m}\right] = \left[\begin{array}{cccc} \lambda_1\mathbf{v_1} & \lambda_2\mathbf{v_2} & \ldots & \lambda_m\mathbf{v_m} \end{array}\right]$$

because

$$\mathbf{Av} = \lambda\mathbf{v}$$

$$\mathbf{AV} = \left[\mathbf{v_1}\mathbf{v_2}\cdots\mathbf{v_m}\right]\left[\begin{array}{cccc} \lambda_1 & 0 & \cdots & 0 \\ 0 & \lambda_2 & \cdots & 0 \\ \vdots & \vdots & \ddots & \vdots \\ 0 & 0 & \cdots & \lambda_m \end{array}\right] = \mathbf{V}\Lambda$$

So that

$$\mathbf{AV} = \mathbf{V\Lambda}$$

Hence,

$$\mathbf{A} = \mathbf{V}\Lambda\mathbf{V}^{-1}$$

Eigendecomposition (a.k.a. spectral decomposition) decomposes a matrix $\mathbf{A}$ into a multiplication of a matrix of eigenvectors $\mathbf{V}$ and a diagonal matrix of eigenvalues $\mathbf{\Lambda}$.

This can only be done if a matrix is diagonalizable. In fact, the definition of a diagonalizable matrix $\mathbf{A} \in \mathbb{R}^{n\times n}$ is that it can be eigendecomposed into n eigenvectors, so that $\mathbf{V}^{-1}\mathbf{AV} = \Lambda$.

$$\begin{aligned} \mathbf{A}^2 &= \left(\mathbf{V}\Lambda\mathbf{V}^{-1}\right)\left(\mathbf{V}\Lambda\mathbf{V}^{-1}\right) \\ &= \mathbf{V}\Lambda\mathbf{I}\Lambda\mathbf{V}^{-1} \\ &= \mathbf{V}\Lambda^2\mathbf{V}^{-1} \end{aligned}$$

in general

$$\mathbf{A}^k = \mathbf{V}\Lambda^k\mathbf{V}^{-1}$$

Example:

```
A = matrix(sample(1:100, 9), 3, 3)
A
```

```
##      [,1] [,2] [,3]
## [1,]   76   25   69
## [2,]   95   51   36
## [3,]   79   31   37
```

```
eigen(A)
```

```
## eigen() decomposition
## $values
## [1] 165.01306 -14.65182  13.63876
##
## $vectors
##            [,1]       [,2]         [,3]
## [1,] -0.5749122 -0.5983678  0.362770047
## [2,] -0.6400025  0.5425789 -0.931827660
## [3,] -0.5097772  0.5895457  0.009752211
```

```
V = eigen(A)$vectors
Lam = diag(eigen(A)$values)
# Prove that AV = VLam
round(A %*% V, 4) == round(V %*% Lam, 4)
```

```
##      [,1] [,2] [,3]
## [1,] TRUE TRUE TRUE
## [2,] TRUE TRUE TRUE
## [3,] TRUE TRUE TRUE
```

```
# And decomposition
A == round(V %*% Lam %*% solve(V), 4)
```

```
##      [,1] [,2] [,3]
## [1,] TRUE TRUE TRUE
## [2,] TRUE TRUE TRUE
## [3,] TRUE TRUE TRUE
```

And, matrix inverse with eigendecomposition:

$$\mathbf{A}^{-1} = \mathbf{V}\Lambda^{-1}\mathbf{V}^{-1}$$

Example:

```
A = matrix(sample(1:100, 9), 3, 3)
A
```

```
##      [,1] [,2] [,3]
## [1,]   94   86   88
## [2,]   96   12   60
## [3,]   32   85   38
```

```
V = eigen(A)$vectors
Lam = diag(eigen(A)$values)

# Inverse of A
solve(A)
```

```
##              [,1]         [,2]        [,3]
## [1,] -0.04684096  0.042483660  0.04139434
## [2,] -0.01742919  0.007625272  0.02832244
## [3,]  0.07843137 -0.052832244 -0.07189542
```

```
# And
V %*% solve(Lam) %*% solve(V)
```

```
##              [,1]         [,2]        [,3]
## [1,] -0.04684096  0.042483660  0.04139434
## [2,] -0.01742919  0.007625272  0.02832244
## [3,]  0.07843137 -0.052832244 -0.07189542
```

The inverse of $\mathbf{\Lambda}$ is just the inverse of each diagonal element (the eigenvalues). But, this can only be done if a matrix is diagonalizable. So if $\mathbf{A}$ is not $n \times n$, then we can use $\mathbf{A}'\mathbf{A}$ or $\mathbf{A}\mathbf{A}'$, both symmetric now.

Example:

$$\mathbf{A} = \begin{pmatrix} 1 & 2 \\ 2 & 4 \end{pmatrix}$$

As $\det(\mathbf{A}) = 0$, $\mathbf{A}$ is singular and its inverse is undefined. In other words, since $\det(\mathbf{A})$ equals the product of the eigenvalues λ_j of A, the matrix $\mathbf{A}$ has an eigenvalue which is zero.

To see this, consider the spectral (eigen) decomposition of A :

$$\mathbf{A} = \sum_{j=1}^{p} \theta_j \mathbf{v}_j \mathbf{v}_j^\top$$

where $\mathbf{v}_\mathrm{j}$ is the eigenvector belonging to θ_j.

The inverse of $\mathbf{A}$ is then:

$$\mathbf{A}^{-1} = \sum_{j=1}^{p} \theta_j^{-1} \mathbf{v}_j \mathbf{v}_j^\top$$

A has eigenvalues 5 and 0. The inverse of A via the spectral decomposition is then undefined:

$$\mathbf{A}^{-1} = \frac{1}{5} \mathbf{v}_1 \mathbf{v}_1^\top + \frac{1}{0} \mathbf{v}_1 \mathbf{v}_1^\top$$

While EVD is a foundational tool, it is limited to square matrices—matrices with the same number of rows and columns. However, in practice, especially in data analysis and machine learning, we often work with rectangular matrices, where the number of features and observations differ. This is where SVD becomes invaluable. SVD generalizes the concept of eigendecomposition to any matrix, whether square or rectangular, making it one of the most widely used matrix factorization techniques in applied mathematics and machine learning.

31.2 Single Value Decomposition

While **EVD** is a foundational tool, it is limited to **square matrices**—matrices with the same number of rows and columns. However, in practice, especially in data analysis and machine learning, we often work with **rectangular matrices**, where the number of features and observations differ. This is where **SVD** becomes invaluable. SVD generalizes the concept of eigendecomposition to any matrix, whether square or rectangular, making it one of the most widely used matrix factorization techniques in applied mathematics and machine learning.

This section delves into **SVD**, a powerful and flexible matrix factorization technique with broad applications in data analysis, dimensionality reduction, and recommender systems. SVD reveals the latent structure of data by decomposing any real (or complex) matrix into three interpretable components.

Unlike **eigendecomposition**, which is only defined for **square matrices**, SVD can be applied to **rectangular matrices**, making it much more useful in real-world settings where data matrices often have more rows than columns, or vice versa. Given an $m \times n$ matrix $\mathbf{A}$, the SVD of $\mathbf{A}$ is given by:

$$\mathbf{A} = \mathbf{U}\mathbf{\Sigma}\mathbf{V}^T$$

Here:

- $\mathbf{U}$ is an $m \times m$ **orthogonal matrix** whose columns are the **left singular vectors**,
- $\mathbf{A}$ is an $m \times n$ **diagonal matrix** with non-negative values called **singular values**, sorted in descending order,
- $\mathbf{V}^T$ is the transpose of an $n \times n$ **orthogonal matrix** whose columns are the **right singular vectors**.

The singular values in $\mathbf{\Sigma}$ represent the importance (or strength) of each corresponding singular vector. By retaining only the top k singular values and their associated vectors, we can construct a **low-rank approximation** of $\mathbf{A}$ that captures the most important structure in the data. This is particularly useful for:

- **Dimensionality reduction** (e.g., Latent Semantic Analysis in NLP),
- **Data compression** (e.g., image compression),
- **Recommender systems** (e.g., collaborative filtering with user-item matrices).

SVD provides not only a compact representation of data but also deep insight into its intrinsic geometry. In the next section, we will explore how this decomposition helps in approximating complex data sets efficiently using lower-dimensional representations.

For any matrix $\mathbf{A}$, both $\mathbf{A}^\top\mathbf{A}$ and $\mathbf{A}\mathbf{A}^\top$ are symmetric. Therefore, they have n and m **orthogonal* eigenvectors, respectively. The proof is simple:

Suppose we have a 2×2 symmetric matrix, $\mathbf{A}$, with two distinct eigenvalues (λ_1, λ_2) and two corresponding eigenvectors ($\mathbf{v}_1$ and $\mathbf{v}_1$). Following the rule,

$$\mathbf{A}\mathbf{v}_1 = \lambda_1\mathbf{v}_1$$
$$\mathbf{A}\mathbf{v}_2 = \lambda_2\mathbf{v}_2$$

Let's multiply (inner product) the first one with $\mathbf{v}_2^\top$:

$$\mathbf{v}_2^\top \mathbf{A}\mathbf{v}_1 = \lambda_1 \mathbf{v}_2^\top \mathbf{v}_1$$

And, the second one with $\mathbf{v}_1^\top$

$$\mathbf{v}_1^\top \mathbf{A}\mathbf{v}_2 = \lambda_2 \mathbf{v}_1^\top \mathbf{v}_2$$

If we take the transpose of both side of $\mathbf{v}_2^\top \mathbf{A}\mathbf{v}_1 = \lambda_1 \mathbf{v}_2^\top \mathbf{v}_1$, it will be:

$$\mathbf{v}_1^\top \mathbf{A}\mathbf{v}_2 = \lambda_1 \mathbf{v}_1^\top \mathbf{v}_2$$

And, subtract these last two:

$$\begin{array}{r} \mathbf{v}_1^\top \mathbf{A}\mathbf{v}_2 = \lambda_2 \mathbf{v}_1^\top \mathbf{v}_2 \\ \mathbf{v}_1^\top \mathbf{A}\mathbf{v}_2 = \lambda_1 \mathbf{v}_1^\top \mathbf{v}_2 \\ \hline 0 = (\lambda_2 - \lambda_1)\, \mathbf{v}_1^\top \mathbf{v}_2 \end{array}$$

Since , λ_1 and λ_2 are distinct, $\lambda_2 - \lambda_1$ cannot be zero. Therefore, $\mathbf{v}_1^\top \mathbf{v}_2 = 0$. The dot products of two vectors can be expressed geometrically,

$$\begin{aligned} a \cdot b &= \|a\| \|b\| \cos(\theta), \\ \cos(\theta) &= \tfrac{a \cdot b}{\|a\| \|b\|} \end{aligned}$$

Hence, $\cos(\theta)$ has to be zero for $\mathbf{v}_1^\top \mathbf{v}_2 = 0$. Since $\cos(90) = 0$, the two vectors are orthogonal.

We start with the following eigendecomposition for $\mathbf{A}^\top\mathbf{A}$ and $\mathbf{A}\mathbf{A}^\top$:

$$\begin{aligned} \mathbf{A}^\top \mathbf{A} &= \mathbf{V}\mathbf{D}\mathbf{V}^\top \\ \mathbf{A}\mathbf{A}^\top &= \mathbf{U}\mathbf{D}'\mathbf{U}^\top \end{aligned}$$

where $\mathbf{V}$ is an $n \times n$ **orthogonal** matrix consisting of the eigenvectors of $\mathbf{A}\mathbf{A}^\top$ and, $\mathbf{D}$ is an $n \times n$ diagonal matrix with the eigenvalues of $\mathbf{A}^\top\mathbf{A}$ on the diagonal. The same decomposition for $\mathbf{A}\mathbf{A}^\top$, now $\mathbf{U}$ is an $m \times m$ **orthogonal** matrix consisting of the eigenvectors of $\mathbf{A}\mathbf{A}^\top$, and $\mathbf{D}'$ is an $m \times m$ diagonal matrix with the eigenvalues of $\mathbf{A}\mathbf{A}^\top$ on the diagonal.

It turns out that $\mathbf{D}$ and $\mathbf{D}'$ have the same non-zero diagonal entries except that the order might be different.

We can write SVD for any real $m \times n$ matrix as:

$$\mathbf{A} = \mathbf{U}\mathbf{\Sigma}\mathbf{V}^\top$$

where $\mathbf{U}$ is an $m \times m$ orthogonal matrix whose columns are the eigenvectors of $\mathbf{A}\mathbf{A}^\top$, $\mathbf{V}$ is an $n \times n$ orthogonal matrix whose columns are the eigenvectors of $\mathbf{A}^\top\mathbf{A}$, and $\mathbf{\Sigma}$ is an $m \times n$ diagonal matrix of the form:

$$\mathbf{\Sigma} = \begin{pmatrix} \sigma_1 & & \\ & \ddots & \\ & & \sigma_n \\ 0 & 0 & 0 \\ 0 & 0 & 0 \end{pmatrix}$$

with $\sigma_1 \geq \sigma_2 \geq \cdots \geq \sigma_n > 0$. The number of non-zero singular values is equal to the rank of rank($\mathbf{A}$). In $\mathbf{\Sigma}$ above, $\sigma_1, \ldots, \sigma_n$ are the square roots of the eigenvalues of $\mathbf{A}^\top \mathbf{A}$. They are called the **singular values** of $\mathbf{A}$.

One important point is that, although $\mathbf{U}$ in $\mathbf{U\Sigma V}^\top$ is $m \times m$, when it is multiplied by $\mathbf{\Sigma}$, it reduces to $n \times n$ due to zeros in $\mathbf{\Sigma}$. Hence, we can actually select only those in $\mathbf{U}$ that are not going to be zeroed out due to that multiplication. When we take only $n \times n$ from $\mathbf{U}$ matrix, it is called "Economy SVD", $\hat{\mathbf{U}}\hat{\mathbf{\Sigma}}\mathbf{V}^\top$, where all matrices will be $n \times n$.

The singular value decomposition is very useful when our basic goal is to "solve" the system $\mathbf{A}x = b$ for all matrices $\mathbf{A}$ and vectors b with a numerically stable algorithm. Some important applications of the SVD include computing the pseudoinverse, matrix approximation, and determining the rank, range, and null space of a matrix. We will see some of them in the following chapters.

Here is an example:

```
set.seed(104)
A <- matrix(sample(100, 12), 3, 4)
A
```

```
##      [,1] [,2] [,3] [,4]
## [1,]   77   24   32   78
## [2,]   67   61   39   96
## [3,]   34   94   42   28
```

```
svda <- svd(A)
svda
```

```
## $d
## [1] 199.83933  70.03623  16.09872
##
## $u
##           [,1]       [,2]       [,3]
## [1,] 0.5515235  0.5259321  0.6474699
## [2,] 0.6841400  0.1588989 -0.7118312
## [3,] 0.4772571 -0.8355517  0.2721747
##
## $v
##           [,1]       [,2]       [,3]
## [1,] 0.5230774  0.3246068  0.7091515
## [2,] 0.4995577 -0.8028224 -0.1427447
## [3,] 0.3221338 -0.1722864  0.2726277
## [4,] 0.6107880  0.4694933 -0.6343518
```

```
# Singular values = sqrt(eigenvalues of t(A)%*%A))
ev <- eigen(t(A) %*% A)$values
round(sqrt(ev), 5)
```

```
## [1] 199.83933  70.03623  16.09872   0.00000
```

Note that this "Economy SVD" is using only the non-zero eigenvalues and their respective eigenvectors.

```
Ar <- svda$u %*% diag(svda$d) %*% t(svda$v)
Ar
```

```
##      [,1] [,2] [,3] [,4]
## [1,]   77   24   32   78
## [2,]   67   61   39   96
## [3,]   34   94   42   28
```

The true strength of SVD lies not only in its ability to decompose any matrix, but also in how it enables us to create low-rank approximations of complex data. By retaining only the most significant singular values and their corresponding singular vectors, we can reconstruct a simplified version of the original matrix that preserves its most important features while discarding noise or redundancy. This leads us directly to one of the most powerful applications of SVD: Rank(r) approximations.

31.3 Rank(r) Approximations

One of the most valuable applications of **SVD** is the ability to construct **rank approximations**, also known as **matrix approximations**. In this section, we explore the concept of **Rank(r) approximations**, a critical technique for reducing the complexity of large matrices while maintaining their essential structure and information.

The idea is simple yet powerful: instead of keeping all singular values from the SVD, we keep only the top r largest ones, along with their associated singular vectors. This produces a new matrix of rank r—a compressed version of the original—that closely approximates it in terms of structure and content. Rank approximations are widely used in machine learning, signal processing, and data compression, where balancing accuracy and efficiency is essential.

We can write $\mathbf{A} = \mathbf{U}\mathbf{\Sigma}\mathbf{V}^\top$ as:

$$= \sigma_1 u_1 v_1^\top + \sigma_2 u_2 v_2^\top + \ldots + \sigma_n u_n v_n^\top + 0.$$

Each term in this equation is a Rank(1) matrix: u_1 is $n \times 1$ column vector and v_1 is $1 \times n$ row vector. Since these are the only orthogonal entries in the resulting matrix, the first term with σ_1 is a Rank(1) $n \times n$ matrix. All other terms have the same dimension. Since σ's are ordered, the first term is the carries the most information. So, Rank(1) approximation is taking only the first term and ignoring the others. Here is a simple example:

```
#rank-one approximation
A <- matrix(c(1, 5, 4, 2), 2 , 2)
A
```

```
##      [,1] [,2]
## [1,]    1    4
## [2,]    5    2
```

```
v1 <- matrix(eigen(t(A) %*% (A))$vector[, 1], 1, 2)
sigma <- sqrt(eigen(t(A) %*% (A))$values[1])
```

```
u1 <- matrix(eigen(A %*% t(A))$vector[, 1], 2, 1)

# Rank(1) approximation of A
Atilde <- sigma * u1 %*% v1
Atilde
```

```
##           [,1]      [,2]
## [1,] -2.560369 -2.069843
## [2,] -4.001625 -3.234977
```

And, Rank(2) approximation can be obtained by adding the first 2 terms. As we add more terms, we can get the full information in the data. But often times, we truncate the ranks at r by removing the terms with small $sigma$. This is also called noise reduction.

There are many examples on the Internet for real image compression, but we apply rank approximation to a heatmap from our own work. The heatmap shows moving-window partial correlations between daily positivity rates (Covid-19) and mobility restrictions for different time delays (days, "lags").

```
comt <- readRDS("comt.rds")

heatmap(
  comt,
  Colv = NA,
  Rowv = NA,
   col = gray.colors(256, start = 1, end = 0),
  main = "Heatmap - Original",
  xlab = "Lags",
  ylab = "Starting days of 7-day rolling windows"
)

# Rank(2) with SVD
fck <- svd(comt)
r = 2
comt.re <-
  as.matrix(fck$u[, 1:r]) %*% diag(fck$d)[1:r, 1:r] %*% t(fck$v[, 1:r])

heatmap(
  comt.re,
  Colv = NA,
  Rowv = NA,
  main = "Heatmap Matrix - Rank(2) Approx",
  xlab = "Lags",
  ylab = "Startting days of 7-day rolling windows"
)
```

```
comt <- readRDS("comt.rds")

heatmap(
  comt,
```

```
  Colv = NA,
  Rowv = NA,
  col = gray.colors(256, start = 1, end = 0),
  main = "Heatmap - Original",
  xlab = "Lags",
  ylab = "Starting days of 7-day rolling windows"
)
```

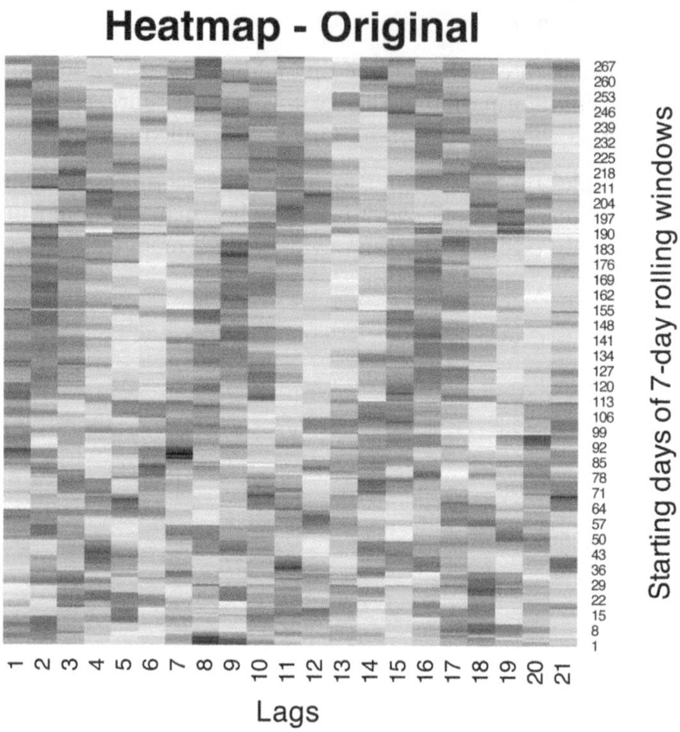


```
# Rank(2) with SVD
fck <- svd(comt)
r = 2
comt.re <-
  as.matrix(fck$u[, 1:r]) %*% diag(fck$d)[1:r, 1:r] %*% t(fck$v[, 1:r])

heatmap(
  comt.re,
  Colv = NA,
  Rowv = NA,
  col = gray.colors(256, start = 1, end = 0),
  main = "Heatmap Matrix - Rank(2) Approx",
  xlab = "Lags",
  ylab = "Startting days of 7-day rolling windows"
)
```

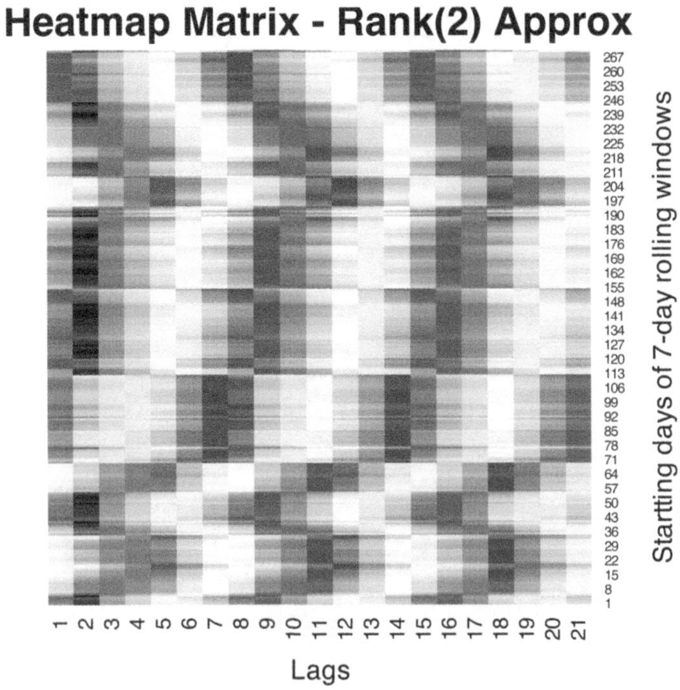

This Rank(2) approximation reduces the noise in the moving-window partial correlations so that we can see the clear trend about the delay in the effect of mobility restrictions on the spread.

We change the order of correlations in the original heatmap, and make it row-wise correlations:

```
#XX' and X'X SVD
wtf <- comt %*% t(comt)
fck <- svd(wtf)
r = 2
comt.re2 <-
  as.matrix(fck$u[, 1:r]) %*% diag(fck$d)[1:r, 1:r] %*% t(fck$v[, 1:r])

heatmap(
  comt.re2,
  Colv = NA,
  Rowv = NA,
  col = gray.colors(256, start = 1, end = 0),
  main = "Row Corr. - Rank(2)",
  xlab = "Startting days of 7-day rolling windows",
  ylab = "Startting days of 7-day rolling windows"
)
```

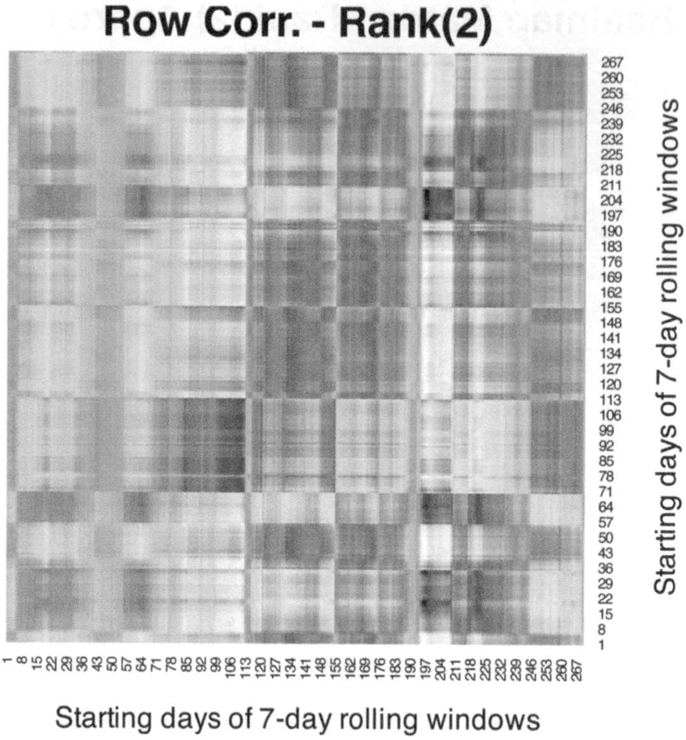

This is now worse than the original heatmap we had ealier. When we apply a Rank(2) approximation, however, we have a very clear picture:

```
wtf <- t(comt) %*% comt
fck <- svd(wtf)
r = 2
comt.re3 <-
  as.matrix(fck$u[, 1:r]) %*% diag(fck$d)[1:r, 1:r] %*% t(fck$v[, 1:r])

heatmap(
  comt.re3,
  Colv = NA,
  Rowv = NA,
  col = gray.colors(256, start = 1, end = 0),
  main = "Column Corr. - Rank(2)",
  xlab = "Lags",
  ylab = "Lags"
)
```

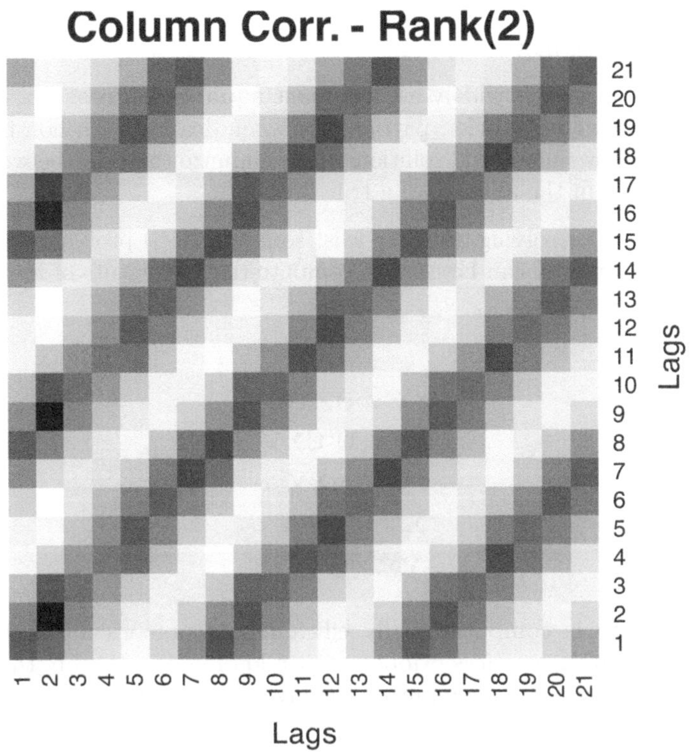

Low-rank approximations not only help us simplify data but also lead to efficient and stable solutions for problems involving large or ill-conditioned matrices. One such problem is solving systems of linear equations when the matrix is not square or not invertible in the usual sense. In these cases, we turn to a generalized notion of the matrix inverse—known as the Moore–Penrose inverse—which is closely tied to the SVD framework and provides a powerful tool for obtaining least-squares solutions.

31.4 Moore–Penrose Inverse

In this section, we introduce the **Moore–Penrose inverse**, also known as the **pseudoinverse**, a generalized matrix inverse that plays a central role in solving linear systems, especially when the system is **underdetermined** or **overdetermined**. Unlike the standard inverse, which is only defined for square, full-rank matrices, the Moore–Penrose inverse can be computed for **any matrix**, whether square or rectangular, full-rank or rank-deficient.

Given a matrix $\mathbf{A}$, its Moore–Penrose inverse, denoted $\mathbf{A}^+$, is defined as the unique matrix that satisfies four specific conditions (known as the Penrose conditions). When $\mathbf{A}$ is decomposed using SVD as:

$$\mathbf{A} = \mathbf{U}\mathbf{\Sigma}\mathbf{V}^T$$

the pseudoinverse is given by:

$$\mathbf{A}^+ = \mathbf{V}\mathbf{\Sigma}^+\mathbf{U}^T$$

where $\mathbf{\Sigma}^+$ is formed by taking the reciprocal of each non-zero singular value in $\mathbf{\Sigma}$ and transposing the resulting matrix.

The Moore–Penrose inverse is widely used in **least squares regression**, **dimensionality reduction**, and **neural networks**, particularly when dealing with non-invertible or non-square matrices. It allows us to find solutions that minimize the error between observed and predicted values, even in the absence of a true inverse.

The SVD can be used for solving ordinary least squares (OLS) problems. In particular, the SVD of the design matrix $\mathbf{X}$ can be used to compute the coefficients of the linear regression model. Here are the steps:

$$\begin{aligned}
\mathbf{y} &= \mathbf{X}\beta \\
\mathbf{y} &= \mathbf{U\Sigma V}'\beta \\
\mathbf{U}'\mathbf{y} &= \mathbf{U}'\mathbf{U\Sigma V}'\beta \\
\mathbf{U}'\mathbf{y} &= \mathbf{\Sigma V}'\beta \\
\mathbf{\Sigma}^{-1}\mathbf{U}'\mathbf{y} &= \mathbf{V}'\beta \\
\mathbf{V\Sigma}^{-1}\mathbf{U}'\mathbf{y} &= \beta
\end{aligned}$$

This formula for beta is computationally efficient and numerically stable, even for ill-conditioned or singular $\mathbf{X}$ matrices. Moreover, it allows us to compute the solution to the OLS problem without explicitly computing the inverse of $\mathbf{X}^T\mathbf{X}$.

Menawhile, the term

$$\mathbf{V\Sigma}^{-1}\mathbf{U}' = \mathbf{M}^+$$

is called **"generalized inverse" or the Moore–Penrose Pseudoinverse**.

If $\mathbf{X}$ has full column rank, then the pseudoinverse is also the unique solution to the OLS problem. However, if $\mathbf{X}$ does not have full column rank, then its pseudoinverse may not exist or may not be unique. In this case, the OLS estimator obtained using the pseudoinverse will be a best linear unbiased estimator (BLUE), but it will not be the unique solution to the OLS problem.

To be more specific, the OLS estimator obtained using the pseudoinverse will minimize the sum of squared residuals subject to the constraint that the coefficients are unbiased, i.e., they have zero expected value. However, there may be other linear unbiased estimators that achieve the same minimum sum of squared residuals. These alternative estimators will differ from the OLS estimator obtained using the pseudoinverse in the values they assign to the coefficients.

In practice, the use of the pseudoinverse to estimate the OLS coefficients when $\mathbf{X}$ does not have full column rank can lead to numerical instability, especially if the singular values of $\mathbf{X}$ are very small. In such cases, it may be more appropriate to use regularization techniques such as ridge or Lasso regression to obtain stable and interpretable estimates. These methods penalize the size of the coefficients and can be used to obtain sparse or "shrunken" estimates, which can be particularly useful in high-dimensional settings where there are more predictors than observations.

Here are some application of SVD and pseudoinverse.

```
library(MASS)

##Simple SVD and generalized inverse
A <- matrix(c(1, 1, 1, 1, 1, 1, 1, 1, 1, 1, 1, 1, 0, 0, 0, 0, 0,
              0, 0, 0, 0, 1, 1, 1, 0, 0, 0, 0, 0, 0, 0, 0, 0, 1, 1, 1), 9, 4)

a.svd <- svd(A)
ds <- diag(1 / a.svd$d[1:3])
u <- a.svd$u
v <- a.svd$v
us <- as.matrix(u[, 1:3])
vs <- as.matrix(v[, 1:3])
(a.ginv <- vs %*% ds %*% t(us))
```

```
##             [,1]        [,2]        [,3]        [,4]        [,5]        [,6]
## [1,]  0.08333333  0.08333333  0.08333333  0.08333333  0.08333333  0.08333333
## [2,]  0.25000000  0.25000000  0.25000000 -0.08333333 -0.08333333 -0.08333333
## [3,] -0.08333333 -0.08333333 -0.08333333  0.25000000  0.25000000  0.25000000
## [4,] -0.08333333 -0.08333333 -0.08333333 -0.08333333 -0.08333333 -0.08333333
##             [,7]        [,8]        [,9]
## [1,]  0.08333333  0.08333333  0.08333333
## [2,] -0.08333333 -0.08333333 -0.08333333
## [3,] -0.08333333 -0.08333333 -0.08333333
## [4,]  0.25000000  0.25000000  0.25000000
```

```
ginv(A)
```

```
##             [,1]        [,2]        [,3]        [,4]        [,5]        [,6]
## [1,]  0.08333333  0.08333333  0.08333333  0.08333333  0.08333333  0.08333333
## [2,]  0.25000000  0.25000000  0.25000000 -0.08333333 -0.08333333 -0.08333333
## [3,] -0.08333333 -0.08333333 -0.08333333  0.25000000  0.25000000  0.25000000
## [4,] -0.08333333 -0.08333333 -0.08333333 -0.08333333 -0.08333333 -0.08333333
##             [,7]        [,8]        [,9]
## [1,]  0.08333333  0.08333333  0.08333333
## [2,] -0.08333333 -0.08333333 -0.08333333
## [3,] -0.08333333 -0.08333333 -0.08333333
## [4,]  0.25000000  0.25000000  0.25000000
```

We can use SVD for solving a regular OLS on simulated data:

```
#Simulated DGP
x1 <- rep(1, 20)
x2 <- rnorm(20)
x3 <- rnorm(20)
u <- matrix(rnorm(20, mean = 0, sd = 1), nrow = 20, ncol = 1)
X <- cbind(x1, x2, x3)
beta <- matrix(c(0.5, 1.5, 2), nrow = 3, ncol = 1)
Y <- X %*% beta + u
```

```
#OLS
betahat_OLS <- solve(t(X) %*% X) %*% t(X) %*% Y
betahat_OLS
```

```
##         [,1]
## x1 0.6310514
## x2 1.5498699
## x3 1.7014166
```

```
#SVD
X.svd <- svd(X)
ds <- diag(1 / X.svd$d)
u <- X.svd$u
v <- X.svd$v
us <- as.matrix(u)
vs <- as.matrix(v)
X.ginv_mine <- vs %*% ds %*% t(us)

# Compare
X.ginv <- ginv(X)
round((X.ginv_mine - X.ginv), 4)
```

```
##      [,1] [,2] [,3] [,4] [,5] [,6] [,7] [,8] [,9] [,10] [,11] [,12] [,13] [,14]
## [1,]    0    0    0    0    0    0    0    0    0     0     0     0     0     0
## [2,]    0    0    0    0    0    0    0    0    0     0     0     0     0     0
## [3,]    0    0    0    0    0    0    0    0    0     0     0     0     0     0
##      [,15] [,16] [,17] [,18] [,19] [,20]
## [1,]     0     0     0     0     0     0
## [2,]     0     0     0     0     0     0
## [3,]     0     0     0     0     0     0
```

```
# Now OLS
betahat_ginv <- X.ginv %*% Y
betahat_ginv
```

```
##           [,1]
## [1,] 0.6310514
## [2,] 1.5498699
## [3,] 1.7014166
```

```
betahat_OLS
```

```
##         [,1]
## x1 0.6310514
## x2 1.5498699
## x3 1.7014166
```

The Moore–Penrose inverse is one example of how matrix decomposition techniques can be used to solve complex problems involving data structure and linear systems. Another key application—perhaps the most widely used dimensionality reduction technique in data science—is PCA. Rooted in the mathematics of eigendecomposition and closely related to SVD, PCA allows us to simplify high-dimensional datasets while preserving the most important patterns in the data.

31.5 Principle Component Analysis

In this section, we explore **PCA**, a foundational technique in data analysis and machine learning for **dimensionality reduction** and **feature extraction**. PCA transforms a high-dimensional dataset into a lower-dimensional space in a way that retains as much of the original variance (i.e., information) as possible.

The core idea behind PCA is to identify the directions—called **principal components**—along which the data varies the most. These directions are obtained by computing the **eigenvectors** and **eigenvalues** of the **covariance matrix** of the data. The eigenvectors determine the new coordinate system (axes), and the eigenvalues indicate how much variance is captured along each axis.

Mathematically, PCA can be understood as follows:

1. Center the data (subtract the mean).
2. Compute the covariance matrix.
3. Perform eigendecomposition on the covariance matrix.
4. Select the top k eigenvectors corresponding to the largest eigenvalues.
5. Project the data onto the new k-dimensional subspace formed by these eigenvectors.

Alternatively, PCA can be performed using **SVD**, which often provides a more stable and computationally efficient solution, especially for large datasets.

PCA is extensively used in applications such as **data visualization**, **noise reduction**, **compression**, and **preprocessing** before applying other machine learning models. By reducing dimensionality, PCA helps us simplify complex datasets, uncover latent structure, and reduce computational costs—all while minimizing information loss.

In short, **PCA is a eigenvalue decomposition of a covariance matrix** (of data matrix $\mathbf{X}$). Since a covariance matrix is a square symmetric matrix, we can apply the eigenvalue decomposition, which reveals the unique orthogonal directions (variances) in the data so that their orthogonal linear combinations maximize the total variance.

The goal is here a dimension reduction of the data matrix. Hence by selecting a few loading, we can reduce the dimension of the data but capture a substantial variation in the data at the same time.

Principal components are the ordered (orthogonal) lines (vectors) that best account for the maximum variance in the data by their magnitude. To get the (unique) variances (direction and the magnitude) in data, we first obtain the mean-centered covariance matrix.

When we use the covariance matrix of the data, we can use eigenvalue decomposition to identify the unique variation (eigenvectors) and their relative magnitudes (eigenvalues) in the data. Here is a simple procedure:

1. $\mathbf{X}$ is the data matrix,
2. $\mathbf{B}$ is the mean-centered data matrix,
3. $\mathbf{C}$ is the covariance matrix ($\mathbf{B}^T\mathbf{B}$). Note that, if $\mathbf{B}$ is scaled, i.e. "z-scored", $\mathbf{B}^T\mathbf{B}$ gives correlation matrix. We will have more information on covariance and correlation in Chapter 32.
4. The eigenvectors and values of $\mathbf{C}$ by $\mathbf{C} = \mathbf{V}\mathbf{D}\mathbf{V}^\top$. Thus, $\mathbf{V}$ contains the eigenvectors (loadings) and $\mathbf{D}$ contains eigenvalues.
5. Using $\mathbf{V}$, the transformation of $\mathbf{B}$ with $\mathbf{B}\mathbf{V}$ maps the data of p variables to a new space of p variables which are uncorrelated over the dataset. $\mathbf{T} = \mathbf{B}\mathbf{V}$ is called the **principle component or score matrix**.
6. Since SVD of $\mathbf{B} = \mathbf{U}\Sigma\mathbf{V}^\top$, we can also get $\mathbf{B}\mathbf{V} = \mathbf{T} = \mathbf{U}\boldsymbol{\Sigma}$. Hence the principle components are $\mathbf{T} = \mathbf{B}\mathbf{V} = \mathbf{U}\boldsymbol{\Sigma}$.
7. However, not all the principal components need to be kept. Keeping only the first r principal components, produced by using only the first r eigenvectors, gives the truncated transformation $\mathbf{T}_r = \mathbf{B}\mathbf{V}_r$. Obviously you choose those with higher variance in each directions by the order of eigenvalues.
8. We can use $\frac{\lambda_k}{\sum_{i=1} \lambda_k}$ to identify r. Or cumulatively, we can see how much variation could be captured by r number of λs, which gives us an idea how many principle components to keep:

$$\frac{\sum_{i=1}^{r} \lambda_k}{\sum_{i=1}^{n} \lambda_k}$$

We use the `factorextra` package and the `decathlon2` data for an example.

```
library("factoextra")
data(decathlon2)

X <- as.matrix(decathlon2[, 1:10])
head(X)
```

```
##            X100m Long.jump Shot.put High.jump X400m X110m.hurdle Discus
## SEBRLE     11.04      7.58    14.83      2.07 49.81        14.69  43.75
## CLAY       10.76      7.40    14.26      1.86 49.37        14.05  50.72
## BERNARD    11.02      7.23    14.25      1.92 48.93        14.99  40.87
## YURKOV     11.34      7.09    15.19      2.10 50.42        15.31  46.26
## ZSIVOCZKY  11.13      7.30    13.48      2.01 48.62        14.17  45.67
## McMULLEN   10.83      7.31    13.76      2.13 49.91        14.38  44.41
##            Pole.vault Javeline X1500m
## SEBRLE           5.02    63.19  291.7
## CLAY             4.92    60.15  301.5
## BERNARD          5.32    62.77  280.1
## YURKOV           4.72    63.44  276.4
## ZSIVOCZKY        4.42    55.37  268.0
## McMULLEN         4.42    56.37  285.1
```

```
n <- nrow(X)
B <- scale(X, center = TRUE)
```

```
C <- t(B) %*% B / (n - 1)
head(C)
```

```
##                    X100m  Long.jump   Shot.put  High.jump      X400m
## X100m          1.0000000 -0.7377932 -0.3703180 -0.3146495  0.5703453
## Long.jump     -0.7377932  1.0000000  0.3737847  0.2682078 -0.5036687
## Shot.put      -0.3703180  0.3737847  1.0000000  0.5747998 -0.2073588
## High.jump     -0.3146495  0.2682078  0.5747998  1.0000000 -0.2616603
## X400m          0.5703453 -0.5036687 -0.2073588 -0.2616603  1.0000000
## X110m.hurdle   0.6699790 -0.5521158 -0.2701634 -0.2022579  0.5970140
##               X110m.hurdle     Discus  Pole.vault    Javeline      X1500m
## X100m            0.6699790 -0.3893760  0.01156433 -0.26635476 -0.17805307
## Long.jump       -0.5521158  0.3287652  0.07982045  0.28806781  0.17332597
## Shot.put        -0.2701634  0.7225179 -0.06837068  0.47558572  0.00959628
## High.jump       -0.2022579  0.4210187 -0.55129583  0.21051789 -0.15699017
## X400m            0.5970140 -0.2545326  0.11156898  0.02350554  0.18346035
## X110m.hurdle     1.0000000 -0.4213608  0.12118697  0.09655757 -0.10331329
```

```
#Check it
head(cov(B))
```

```
##                    X100m  Long.jump   Shot.put  High.jump      X400m
## X100m          1.0000000 -0.7377932 -0.3703180 -0.3146495  0.5703453
## Long.jump     -0.7377932  1.0000000  0.3737847  0.2682078 -0.5036687
## Shot.put      -0.3703180  0.3737847  1.0000000  0.5747998 -0.2073588
## High.jump     -0.3146495  0.2682078  0.5747998  1.0000000 -0.2616603
## X400m          0.5703453 -0.5036687 -0.2073588 -0.2616603  1.0000000
## X110m.hurdle   0.6699790 -0.5521158 -0.2701634 -0.2022579  0.5970140
##               X110m.hurdle     Discus  Pole.vault    Javeline      X1500m
## X100m            0.6699790 -0.3893760  0.01156433 -0.26635476 -0.17805307
## Long.jump       -0.5521158  0.3287652  0.07982045  0.28806781  0.17332597
## Shot.put        -0.2701634  0.7225179 -0.06837068  0.47558572  0.00959628
## High.jump       -0.2022579  0.4210187 -0.55129583  0.21051789 -0.15699017
## X400m            0.5970140 -0.2545326  0.11156898  0.02350554  0.18346035
## X110m.hurdle     1.0000000 -0.4213608  0.12118697  0.09655757 -0.10331329
```

Eigenvalues and vectors:

```
#Eigens
evalues <- eigen(C)$values
evalues
```

```
##  [1] 3.7499727 1.7451681 1.5178280 1.0322001 0.6178387 0.4282908 0.3259103
##  [8] 0.2793827 0.1911128 0.1122959
```

```
evectors <- eigen(C)$vectors
evectors #Ordered
```

```
##                [,1]        [,2]          [,3]         [,4]        [,5]         [,6]
```

```
##  [1,]  0.42290657 -0.2594748 -0.081870461 -0.09974877  0.2796419 -0.16023494
##  [2,] -0.39189495  0.2887806  0.005082180  0.18250903 -0.3355025 -0.07384658
##  [3,] -0.36926619 -0.2135552 -0.384621732 -0.03553644  0.3544877 -0.32207320
##  [4,] -0.31422571 -0.4627797 -0.003738604 -0.07012348 -0.3824125 -0.52738027
##  [5,]  0.33248297 -0.1123521 -0.418635317 -0.26554389 -0.2534755  0.23884715
##  [6,]  0.36995919 -0.2252392 -0.338027983  0.15726889 -0.2048540 -0.26249611
##  [7,] -0.37020078 -0.1547241 -0.219417086 -0.39137188  0.4319091  0.28217086
##  [8,]  0.11433982  0.5583051 -0.327177839  0.24759476  0.3340758 -0.43606610
##  [9,] -0.18341259 -0.0745854 -0.564474643  0.47792535 -0.1697426  0.42368592
## [10,] -0.03599937  0.4300522 -0.286328973 -0.64220377 -0.3227349 -0.10850981
##              [,7]        [,8]        [,9]       [,10]
##  [1,]  0.03227949 -0.35266427  0.71190625 -0.03272397
##  [2,] -0.24902853 -0.72986071  0.12801382 -0.02395904
##  [3,] -0.23059438  0.01767069 -0.07184807  0.61708920
##  [4,] -0.03992994  0.25003572  0.14583529 -0.41523052
##  [5,] -0.69014364  0.01543618 -0.13706918 -0.12016951
##  [6,]  0.42797378 -0.36415520 -0.49550598  0.03514180
##  [7,]  0.18416631 -0.26865454 -0.18621144 -0.48037792
##  [8,] -0.12654370  0.16086549 -0.02983660 -0.40290423
##  [9,]  0.23324548  0.19922452  0.33300936 -0.02100398
## [10,]  0.34406521  0.09752169  0.19899138  0.18954698
```

Now with `prcomp()`. First, eigenvalues:

```
# With `prcomp()`
Xpca <- prcomp(X, scale = TRUE)
#Eigenvalues
Xpca$sdev
```

```
##  [1] 1.9364846 1.3210481 1.2320016 1.0159725 0.7860272 0.6544393 0.5708855
##  [8] 0.5285666 0.4371645 0.3351059
```

They are the square root of the eigenvalues that we calculated before and they are ordered.#

```
sqrt(evalues)
```

And, the "loadings" (Eigenvectors):

```
#Eigenvectors
Xpca$rotation # 10x10
```

```
##                     PC1        PC2          PC3         PC4        PC5
## X100m       -0.42290657 -0.2594748  0.081870461  0.09974877 -0.2796419
## Long.jump    0.39189495  0.2887806 -0.005082180 -0.18250903  0.3355025
## Shot.put     0.36926619 -0.2135552  0.384621732  0.03553644 -0.3544877
## High.jump    0.31422571 -0.4627797  0.003738604  0.07012348  0.3824125
## X400m       -0.33248297 -0.1123521  0.418635317  0.26554389  0.2534755
## X110m.hurdle -0.36995919 -0.2252392  0.338027983 -0.15726889  0.2048540
## Discus       0.37020078 -0.1547241  0.219417086  0.39137188 -0.4319091
## Pole.vault  -0.11433982  0.5583051  0.327177839 -0.24759476 -0.3340758
## Javeline     0.18341259 -0.0745854  0.564474643 -0.47792535  0.1697426
```

```
## X1500m         0.03599937  0.4300522  0.286328973  0.64220377  0.3227349
##                      PC6         PC7         PC8         PC9        PC10
## X100m          0.16023494 -0.03227949 -0.35266427  0.71190625  0.03272397
## Long.jump      0.07384658  0.24902853 -0.72986071  0.12801382  0.02395904
## Shot.put       0.32207320  0.23059438  0.01767069 -0.07184807 -0.61708920
## High.jump      0.52738027  0.03992994  0.25003572  0.14583529  0.41523052
## X400m         -0.23884715  0.69014364  0.01543618 -0.13706918  0.12016951
## X110m.hurdle   0.26249611 -0.42797378 -0.36415520 -0.49550598 -0.03514180
## Discus        -0.28217086 -0.18416631 -0.26865454 -0.18621144  0.48037792
## Pole.vault     0.43606610  0.12654370  0.16086549 -0.02983660  0.40290423
## Javeline      -0.42368592 -0.23324548  0.19922452  0.33300936  0.02100398
## X1500m         0.10850981 -0.34406521  0.09752169  0.19899138 -0.18954698
```

```
loadings <- Xpca$rotation
```

The signs of eigenvectors are flipped and opposites of what we calculated with `eigen()` above. This is because the definition of an eigenbasis is ambiguous of sign. There are multiple discussions about the sign reversals in eignevectores.

Let's visualize the order:

```
plot(Xpca$sdev) # Eigenvalues
fviz_eig(Xpca) # Cumulative with "factoextra"

# Or
var <- (Xpca$sdev) ^ 2
var_perc <- var / sum(var) * 100

barplot(
  var_perc,
  xlab = 'PC',
  ylab = 'Percent Variance',
  names.arg = 1:length(var_perc),
  las = 1,
  ylim = c(0, max(var_perc)),
  col = 'lightgreen'
)

abline(h = mean(var_perc), col = 'red')
```

```
plot(Xpca$sdev) # Eigenvalues
```

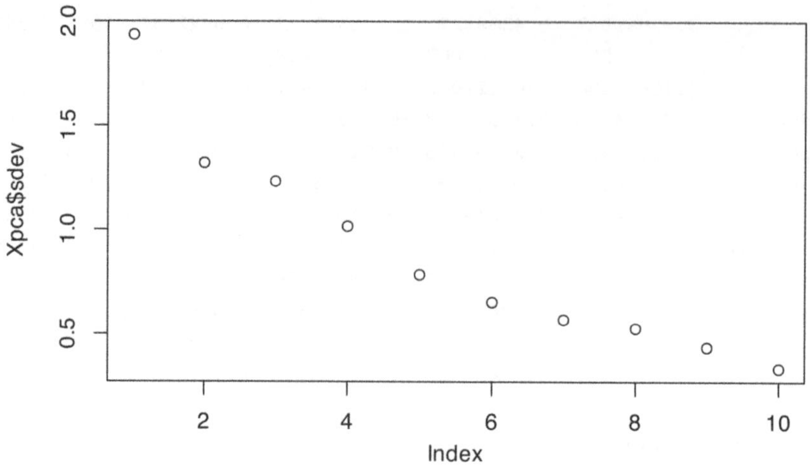


```
fviz_eig(Xpca, barfill = "gray70", barcolor = "black") + theme_minimal()
```

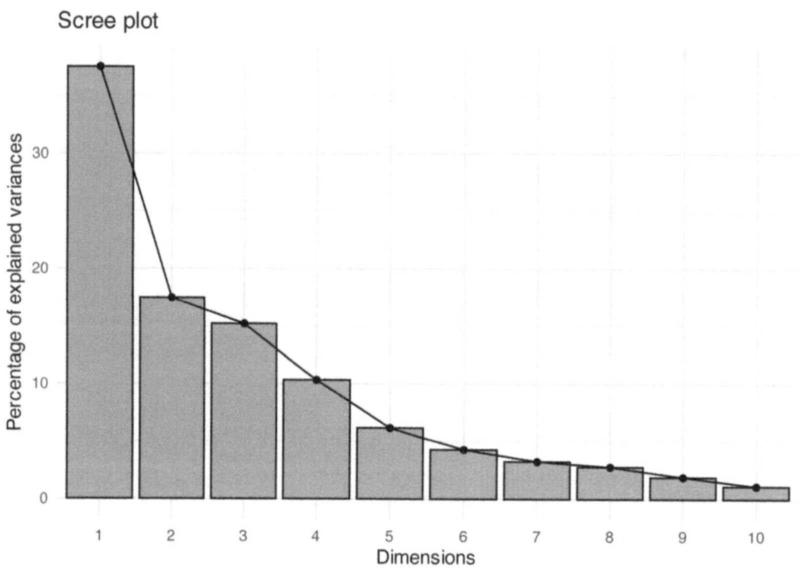


```
# Or
var <- (Xpca$sdev) ^ 2
var_perc <- var / sum(var) * 100

barplot(
  var_perc,
  xlab = 'PC',
  ylab = 'Percent Variance',
  names.arg = 1:length(var_perc),
  las = 1,
  ylim = c(0, max(var_perc)),
  col = 'gray'
)

abline(h = mean(var_perc), col = 'gray60')
```

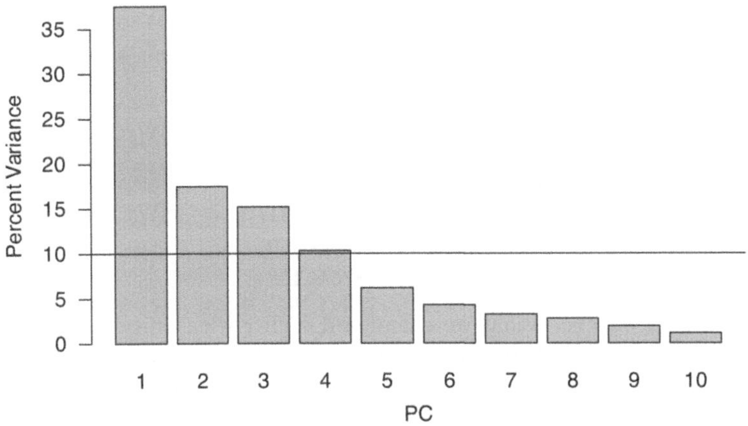

Since we have ten variables, if each variable contributed equally, they would each contribute 10% to the total variance (red line). This criterion suggests we should also include principal component 4 (but barely) in our interpretation.

And principle component scores $\mathbf{T} = \mathbf{XV}$ (a.k.a score matrix) with `prcomp()`:

```
pc <- scale(X) %*% Xpca$rotation
head(pc)
```

```
##                  PC1        PC2        PC3         PC4         PC5        PC6
## SEBRLE     0.2727622  0.5264068  1.5556058  0.10384438  1.05453531  0.7177257
## CLAY       0.8879389  2.0551314  0.8249697  1.81612193 -0.40100595 -1.5039874
## BERNARD   -1.3466138  1.3229149  0.9439501 -1.46516144 -0.17925232  0.5996203
## YURKOV    -0.9108536 -2.2390912  1.9063730  0.09501304  0.18735823  0.3754439
## ZSIVOCZKY -0.1018764 -1.0694498 -2.0596722  0.07056229 -0.03232182 -0.9321431
## McMULLEN   0.2353742 -0.9215376 -0.8028425  1.17942532  1.79598700 -0.3241881
##                   PC7         PC8         PC9        PC10
## SEBRLE    -0.04935537 -0.02990462  0.63079187  0.07728655
## CLAY      -0.75968352  0.06536612 -0.05920672  0.15812336
## BERNARD   -0.75032098  0.49570997 -0.07483747 -0.03288604
## YURKOV    -0.29565551 -0.09332310  0.06769776  0.13791531
## ZSIVOCZKY -0.30752133 -0.29476740  0.48055837  0.44234659
## McMULLEN   0.02896393  0.53358562 -0.05116850  0.37610188
```

```
dim(pc)
```

```
## [1] 27 10
```

```
# which is also given by `prcomp()`
head(Xpca$x)
```

```
##                  PC1        PC2        PC3         PC4         PC5        PC6
## SEBRLE     0.2727622  0.5264068  1.5556058  0.10384438  1.05453531  0.7177257
## CLAY       0.8879389  2.0551314  0.8249697  1.81612193 -0.40100595 -1.5039874
## BERNARD   -1.3466138  1.3229149  0.9439501 -1.46516144 -0.17925232  0.5996203
## YURKOV    -0.9108536 -2.2390912  1.9063730  0.09501304  0.18735823  0.3754439
```

```
## ZSIVOCZKY -0.1018764 -1.0694498 -2.0596722  0.07056229 -0.03232182 -0.9321431
## McMULLEN   0.2353742 -0.9215376 -0.8028425  1.17942532  1.79598700 -0.3241881
##                  PC7         PC8         PC9        PC10
## SEBRLE    -0.04935537 -0.02990462  0.63079187  0.07728655
## CLAY      -0.75968352  0.06536612 -0.05920672  0.15812336
## BERNARD   -0.75032098  0.49570997 -0.07483747 -0.03288604
## YURKOV    -0.29565551 -0.09332310  0.06769776  0.13791531
## ZSIVOCZKY -0.30752133 -0.29476740  0.48055837  0.44234659
## McMULLEN   0.02896393  0.53358562 -0.05116850  0.37610188
```

Now you can think that if we use `evectors` that we calculated earlier with filliped signs, the data would be different. It's similar to multiply the entire data with -1. So the data would not change in a sense that that captures the variation between observations and variables. That's why the sign of eigenvalues are arbitraray.

Now, with SVD:

```
# With SVD
Xsvd <- svd(scale(X))
pc_2 <- Xsvd$u %*% diag(Xsvd$d)
dim(pc_2)
```

```
## [1] 27 10
```

```
head(pc_2)
```

```
##            [,1]       [,2]       [,3]        [,4]        [,5]       [,6]
## [1,]  0.2727622  0.5264068  1.5556058  0.10384438  1.05453531  0.7177257
## [2,]  0.8879389  2.0551314  0.8249697  1.81612193 -0.40100595 -1.5039874
## [3,] -1.3466138  1.3229149  0.9439501 -1.46516144 -0.17925232  0.5996203
## [4,] -0.9108536 -2.2390912  1.9063730  0.09501304  0.18735823  0.3754439
## [5,] -0.1018764 -1.0694498 -2.0596722  0.07056229 -0.03232182 -0.9321431
## [6,]  0.2353742 -0.9215376 -0.8028425  1.17942532  1.79598700 -0.3241881
##             [,7]        [,8]        [,9]       [,10]
## [1,] -0.04935537 -0.02990462  0.63079187  0.07728655
## [2,] -0.75968352  0.06536612 -0.05920672  0.15812336
## [3,] -0.75032098  0.49570997 -0.07483747 -0.03288604
## [4,] -0.29565551 -0.09332310  0.06769776  0.13791531
## [5,] -0.30752133 -0.29476740  0.48055837  0.44234659
## [6,]  0.02896393  0.53358562 -0.05116850  0.37610188
```

Here we can reduce the dimensionality by selecting only 4 PC (the first 4 PC's are above the average, which explain more than 80% of the variation in the data - see the graph above)

```
reduced <- pc[, 1:4]
dim(reduced)
```

```
## [1] 27  4
```

```
head(reduced)
```

```
##                   PC1        PC2        PC3         PC4
## SEBRLE      0.2727622  0.5264068  1.5556058  0.10384438
## CLAY        0.8879389  2.0551314  0.8249697  1.81612193
## BERNARD    -1.3466138  1.3229149  0.9439501 -1.46516144
## YURKOV     -0.9108536 -2.2390912  1.9063730  0.09501304
## ZSIVOCZKY -0.1018764 -1.0694498 -2.0596722  0.07056229
## McMULLEN    0.2353742 -0.9215376 -0.8028425  1.17942532
```

The individual columns of $\mathbf{T}$ successively inherit the maximum possible variance from $\mathbf{X}$, with each coefficient vector in $\mathbf{V}$ constrained to be a unit vector. In $\mathbf{T} = \mathbf{XV}$, $\mathbf{V}$ is a $p \times p$ matrix of weights whose columns are the eigenvectors of $\mathbf{X}^\top\mathbf{X}$. The columns of $\mathbf{V}$ multiplied by the square root of corresponding eigenvalues, that is, eigenvectors scaled up by the variances, are called loadings in PCA and factor analysis.

Note that if we make a singular value decomposition for a covariance matrix,

$$\begin{aligned}\mathbf{X}^T\mathbf{X} &= \mathbf{V}\boldsymbol{\Sigma}^\top\mathbf{U}^\top\mathbf{U}\boldsymbol{\Sigma}\mathbf{V}^\top \\ &= \mathbf{V}\boldsymbol{\Sigma}^\top\boldsymbol{\Sigma}\mathbf{V}^\top \\ &= \mathbf{V}\hat{\boldsymbol{\Sigma}}^2\mathbf{V}^\top\end{aligned}$$

where $\hat{\boldsymbol{\Sigma}}$ is the square diagonal matrix with the singular values of $\mathbf{X}$ and the excess zeros are chopped off so that it satisfies $\hat{\boldsymbol{\Sigma}}^2 = \boldsymbol{\Sigma}^\top\boldsymbol{\Sigma}$.

Comparison with the eigenvector factorization of $\mathbf{X}^\top\mathbf{X}$ establishes that the right singular vectors $\mathbf{V}$ of $\mathbf{X}$ are equivalent to the eigenvectors of $\mathbf{X}^\top\mathbf{X}$, while the singular values $\sigma_{(k)}$ of $\mathbf{X}$ are equal to the square-root of the eigenvalues $\lambda_{(k)}$ of $\mathbf{X}^\top\mathbf{X}$.

While PCA focuses on capturing the directions of maximum variance in the observed data, it does not attempt to model any underlying latent causes. In contrast, FA takes a more probabilistic approach. It assumes that the observed variables are influenced by a smaller number of unobserved latent factors, along with some measurement error. This shift in perspective—from explaining variance to modeling hidden structure—marks a key distinction between PCA and FA.

31.6 Factor Analysis

FA is a statistical technique used to model the underlying relationships between observed variables and a smaller set of **unobserved latent variables**, called **factors**. Unlike PCA, which is purely algebraic and focused on maximizing explained variance, FA is based on a **generative model** of the data. It assumes that the observed data can be expressed as a linear combination of latent factors plus some noise.

Mathematically, the FA model is:

$$\mathbf{x} = \boldsymbol{\Lambda}\mathbf{f} + \epsilon$$

where: $\mathbf{x}$ is the observed data vector, $\mathbf{f}$ is a vector of latent factors (typically fewer in number than the observed variables), - $\boldsymbol{\Lambda}$ is the **factor loading matrix** that relates the factors to the observed variables, ϵ is the vector of **unique errors**, assumed to be uncorrelated with the factors.

The goal of factor analysis is to estimate the loading matrix $\mathbf{\Lambda}$ and understand the structure of relationships among variables by identifying which latent factors drive them. FA is particularly useful in **psychometrics**, **survey analysis**, **finance**, and **social sciences**, where researchers seek to uncover hidden constructs like intelligence, satisfaction, or market forces from observed indicators.

Unlike PCA, which treats all variance as meaningful, FA separates **shared variance** (explained by common factors) from **unique variance** (attributable to measurement error or other sources), making it more suitable when latent structure is of primary interest.

FA and PCA both involve reducing the dimensionality of a dataset, but they are not the same. PCA is a mathematical technique that transforms a dataset of possibly correlated variables into a smaller set of uncorrelated variables known as principal components. The principal components are linear combinations of the original variables, and each principal component accounts for as much of the variation in the data as possible.

The main difference between FA and PCA lies in their objectives. PCA aims to reduce the number of variables by identifying the most important components, while factor analysis aims to identify the underlying factors that explain the correlations among the variables. Therefore, PCA is more commonly used for data reduction or data compression, while factor analysis is more commonly used for exploring the relationships among variables.

As shown below, a factor model can be represented by as a series of multiple regressions, where each X_i $(i = 1, \cdots, p)$ is a function of m number of unobservable common factors f_i:

$$\begin{aligned}
X_1 &= \mu_1 + \beta_{11} f_1 + \beta_{12} f_2 + \cdots + \beta_{1m} f_m + \epsilon_1 \\
X_2 &= \mu_2 + \beta_{21} f_1 + \beta_{22} f_2 + \cdots + \beta_{2m} f_m + \epsilon_2 \\
&\vdots \\
X_p &= \mu_p + \beta_{p1} f_1 + \beta_{p2} f_2 + \cdots + \beta_{pm} f_m + \epsilon_p
\end{aligned}$$

where $\mathrm{E}(X_i) = \mu_i$, ϵ_i are called the **specific factors**. The coefficients, β_{ij}, are the factor **loadings**. We can expressed all of them in a matrix notation.

$$\mathbf{X} = \boldsymbol{\mu} + \mathbf{Lf} + \boldsymbol{\epsilon} \tag{31.6}$$

where

$$\mathbf{L} = \begin{pmatrix} \beta_{11} & \beta_{12} & \cdots & \beta_{1m} \\ \beta_{21} & \beta_{22} & \cdots & \beta_{2m} \\ \vdots & \vdots & & \vdots \\ \beta_{p1} & \beta_{p2} & \cdots & \beta_{pm} \end{pmatrix}$$

There are multiple assumptions:

- $E(\epsilon_i) = 0$ and $\operatorname{var}(\epsilon_i) = \psi_i$, which is called as "specific variance",
- $E(f_i) = 0$ and $\operatorname{var}(f_i) = 1$,
- $\operatorname{cov}(f_i, f_j) = 0$ for $i \neq j$,
- $\operatorname{cov}(\epsilon_i, \epsilon_j) = 0$ for $i \neq j$,
- $\operatorname{cov}(\epsilon_i, f_j) = 0$,

Given these assumptions, the variance of X_i can be expressed as:

$$\text{var}(X_i) = \sigma_i^2 = \sum_{j=1}^{m} \beta_{ij}^2 + \psi_i$$

There are two sources of the variance in X_i: $\sum_{j=1}^{m} \beta_{ij}^2$, which is called the **Communality** for variable i, and **specific variance**, ψ_i.

Moreover,

- $\text{cov}(X_i, X_j) = \sigma_{ij} = \sum_{k=1}^{m} \beta_{ik}\beta_{jk}$,
- $\text{cov}(X_i, f_j) = \beta_{ij}$

The factor model for our variance-covariance matrix of **X** can then be expressed as:

$$\text{var-cov}(\mathbf{X}) = \Sigma = \mathbf{L}\mathbf{L}' + \mathbf{\Psi}$$

which is the sum of the shared variance with another variable, $\mathbf{L}\mathbf{L}'$ (the common variance or **communality**) and the unique variance, $\mathbf{\Psi}$, inherent to each variable (**specific variance**)

We need to look at $\mathbf{L}\mathbf{L}'$, where $\mathbf{L}$ is the $p \times m$ matrix of loadings. In general, we want to have $m \ll p$.

The ith diagonal element of $\mathbf{L}\mathbf{L}'$, the sum of the squared loadings, is called the ith communality. The communality values represent the percent of variability explained by the common factors. The sizes of the communalities and/or the specific variances can be used to evaluate the goodness of fit.

To estimate factor loadings with PCA, we first calculate the principal components of the data, and then compute the factor loadings using the eigenvectors of the correlation matrix of the standardized data. When PCA is used, the matrix of estimated factor loadings, **L**, is given by:

$$\hat{\mathbf{L}} = \left[\sqrt{\hat{\lambda}_1}\hat{\mathbf{v}}_1 \quad \sqrt{\hat{\lambda}_2}\hat{\mathbf{v}}_2 \quad \ldots \sqrt{\hat{\lambda}_m}\hat{\mathbf{v}}_m \right]$$

where

$$\hat{\beta}_{ij} = \hat{\mathbf{v}}_{ij}\sqrt{\hat{\lambda}_j}$$

where i is the index of the original variable, j is the index of the principal component, eigenvector (i, j) is the i-th component of the j-th eigenvector of the correlation matrix, eigenvalue (j) is the j-th eigenvalue of the correlation matrix.

This method tries to find values of the loadings that bring the estimate of the total communality close to the total of the observed variances. The covariances are ignored. Remember, the communality is the part of the variance of the variable that is explained by the factors. So a larger communality means a more successful factor model in explaining the variable.

Let's have an example. The data set is called `bfi` and comes from the `psych` package.

The data includes 25 self-reported personality items from the International Personality Item Pool, gender, education level, and age for 2,800 subjects. The personality items are split into five categories: Agreeableness (A), Conscientiousness (C), Extraversion (E), Neuroticism

(N), Openness (O). Each item was answered on a six point scale: 1 Very Inaccurate to 6 Very Accurate.

```
library(psych)
library(GPArotation)
data("bfi")
str(bfi)
```

```
## 'data.frame':    2800 obs. of  28 variables:
##  $ A1       : int  2 2 5 4 2 6 2 4 4 2 ...
##  $ A2       : int  4 4 4 4 3 6 5 3 3 5 ...
##  $ A3       : int  3 5 5 6 3 5 5 1 6 6 ...
##  $ A4       : int  4 2 4 5 4 6 3 5 3 6 ...
##  $ A5       : int  4 5 4 5 5 5 5 1 3 5 ...
##  $ C1       : int  2 5 4 4 4 6 5 3 6 6 ...
##  $ C2       : int  3 4 5 4 4 6 4 2 6 5 ...
##  $ C3       : int  3 4 4 3 5 6 4 4 3 6 ...
##  $ C4       : int  4 3 2 5 3 1 2 2 4 2 ...
##  $ C5       : int  4 4 5 5 2 3 3 4 5 1 ...
##  $ E1       : int  3 1 2 5 2 2 4 3 5 2 ...
##  $ E2       : int  3 1 4 3 2 1 3 6 3 2 ...
##  $ E3       : int  3 6 4 4 5 6 4 4 NA 4 ...
##  $ E4       : int  4 4 4 4 4 5 5 2 4 5 ...
##  $ E5       : int  4 3 5 4 5 6 5 1 3 5 ...
##  $ N1       : int  3 3 4 2 2 3 1 6 5 5 ...
##  $ N2       : int  4 3 5 5 3 5 2 3 5 5 ...
##  $ N3       : int  2 3 4 2 4 2 2 2 2 5 ...
##  $ N4       : int  2 5 2 4 4 2 1 6 3 2 ...
##  $ N5       : int  3 5 3 1 3 3 1 4 3 4 ...
##  $ O1       : int  3 4 4 3 3 4 5 3 6 5 ...
##  $ O2       : int  6 2 2 3 3 3 2 2 6 1 ...
##  $ O3       : int  3 4 5 4 4 5 5 4 6 5 ...
##  $ O4       : int  4 3 5 3 3 6 6 5 6 5 ...
##  $ O5       : int  3 3 2 5 3 1 1 3 1 2 ...
##  $ gender   : int  1 2 2 2 1 2 1 1 1 2 ...
##  $ education: int  NA NA NA NA NA 3 NA 2 1 NA ...
##  $ age      : int  16 18 17 17 17 21 18 19 19 17 ...
```

To get rid of missing observations and the last three variables,

```
df <- bfi[complete.cases(bfi[, 1:25]), 1:25]
```

The first decision that we need make is the number of factors that we will need to extract. For $p = 25$, the variance-covariance matrix Σ contains

$$\frac{p(p+1)}{2} = \frac{25 \times 26}{2} = 325$$

unique elements or entries. With m factors, the number of parameters in the factor model would be

$$p(m+1) = 25(m+1)$$

Taking $m = 5$, we have 150 parameters in the factor model. How do we choose m? Although it is common to look at the results of the principal components analysis, often in social sciences, the underlying theory within the field of study indicates how many factors to expect.

```
scree(df)
```

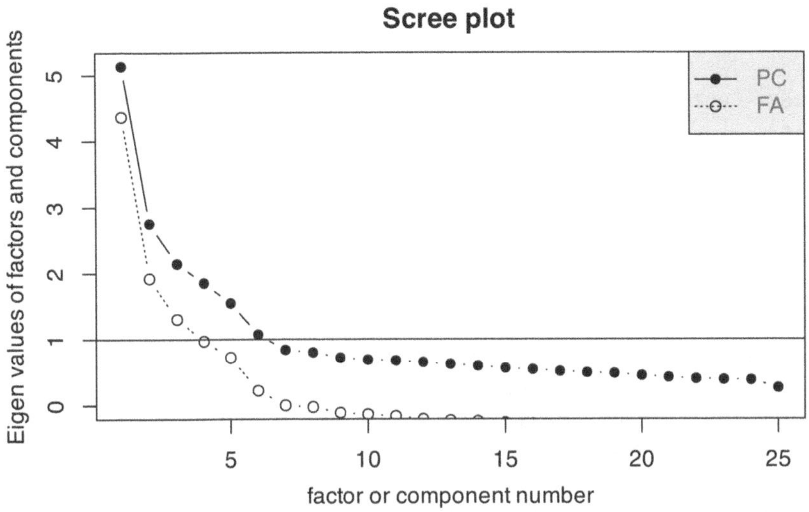

Let's use the `factanal()` function of the build-in `stats` package, which performs maximum likelihood estimation.

```
pa.out <- factanal(df, factors = 5)
pa.out
```

```
##
## Call:
## factanal(x = df, factors = 5)
##
## Uniquenesses:
##    A1    A2    A3    A4    A5    C1    C2    C3    C4    C5    E1    E2    E3
## 0.830 0.576 0.466 0.691 0.512 0.660 0.569 0.677 0.510 0.557 0.634 0.454 0.558
##    E4    E5    N1    N2    N3    N4    N5    O1    O2    O3    O4    O5
## 0.468 0.592 0.271 0.337 0.478 0.507 0.664 0.675 0.744 0.518 0.752 0.726
##
## Loadings:
##    Factor1 Factor2 Factor3 Factor4 Factor5
## A1  0.104                  -0.393
## A2          0.191   0.144   0.601
## A3          0.280   0.110   0.662
## A4          0.181   0.234   0.454  -0.109
## A5 -0.124   0.351           0.580
## C1                  0.533           0.221
## C2                  0.624   0.127   0.140
## C3                  0.554   0.122
```

```
## C4  0.218          -0.653
## C5  0.272  -0.190  -0.573
## E1         -0.587           -0.120
## E2  0.233  -0.674  -0.106   -0.151
## E3          0.490            0.315   0.313
## E4 -0.121   0.613            0.363
## E5          0.491   0.310   0.120   0.234
## N1  0.816                   -0.214
## N2  0.787                   -0.202
## N3  0.714
## N4  0.562  -0.367  -0.192
## N5  0.518  -0.187            0.106  -0.137
## O1          0.182   0.103           0.524
## O2  0.163          -0.113   0.102  -0.454
## O3          0.276            0.153   0.614
## O4  0.207  -0.220            0.144   0.368
## O5                                  -0.512
##
##                Factor1 Factor2 Factor3 Factor4 Factor5
## SS loadings      2.687   2.320   2.034   1.978   1.557
## Proportion Var   0.107   0.093   0.081   0.079   0.062
## Cumulative Var   0.107   0.200   0.282   0.361   0.423
##
## Test of the hypothesis that 5 factors are sufficient.
## The chi square statistic is 1490.59 on 185 degrees of freedom.
## The p-value is 1.22e-202
```

The first chunk provides the "uniqueness" (specific variance) for each variable, which range from 0 to 1 . The uniqueness explains the proportion of variability, which cannot be explained by a linear combination of the factors. That's why it's referred to as noise. This is the $\hat{\Psi}$ in the equation above. A high uniqueness for a variable implies that the factors are not the main source of its variance.

The next section reports the loadings ranging from -1 to 1. This is the $\hat{\mathbf{L}}$ in the equation (31.2) above. Variables with a high loading are well explained by the factor. Note that R does not print loadings less than 0.1.

The communalities for the ith variable are computed by taking the sum of the squared loadings for that variable. This is expressed below:

$$\hat{h}_i^2 = \sum_{j=1}^{m} \hat{\beta}_{ij}^2$$

A well-fit factor model has low values for uniqueness and high values for communality. One way to calculate the communality is to subtract the uniquenesses from 1.

```
apply(pa.out$loadings ^ 2, 1, sum) # communality
```

```
##        A1        A2        A3        A4        A5        C1        C2        C3
## 0.1703640 0.4237506 0.5337657 0.3088959 0.4881042 0.3401202 0.4313729 0.3227542
##        C4        C5        E1        E2        E3        E4        E5        N1
## 0.4900773 0.4427531 0.3659303 0.5459794 0.4422484 0.5319941 0.4079732 0.7294156
```

```
##        N2        N3        N4        N5        O1        O2        O3        O4
## 0.6630751 0.5222584 0.4932099 0.3356293 0.3253527 0.2558864 0.4815981 0.2484000
##        O5
## 0.2740596
```

```
1 - apply(pa.out$loadings ^ 2, 1, sum) # uniqueness
```

```
##        A1        A2        A3        A4        A5        C1        C2        C3
## 0.8296360 0.5762494 0.4662343 0.6911041 0.5118958 0.6598798 0.5686271 0.6772458
##        C4        C5        E1        E2        E3        E4        E5        N1
## 0.5099227 0.5572469 0.6340697 0.4540206 0.5577516 0.4680059 0.5920268 0.2705844
##        N2        N3        N4        N5        O1        O2        O3        O4
## 0.3369249 0.4777416 0.5067901 0.6643707 0.6746473 0.7441136 0.5184019 0.7516000
##        O5
## 0.7259404
```

The table under the loadings reports the proportion of variance explained by each factor. `Proportion Var` shows the proportion of variance explained by each factor. The row `Cumulative Var` is the cumulative `Proportion Var`. Finally, the row `SS loadings` reports the sum of squared loadings. This can be used to determine a factor worth keeping (Kaiser Rule).

The last section of the output reports a significance test: The null hypothesis is that the number of factors in the model is sufficient to capture the full dimensionality of the data set. Hence, in our example, we fitted not an appropriate model.

Finally, we may compare estimated correlation matrix, $\hat{\Sigma}$ and the observed correlation matrix:

```
Lambda <- pa.out$loadings
Psi <- diag(pa.out$uniquenesses)
Sigma_hat <- Lambda %*% t(Lambda) + Psi
```

Let's check the differences:

```
round(head(cor(df)) - head(Sigma_hat), 2)
```

```
##       A1    A2    A3    A4    A5    C1    C2    C3    C4    C5    E1    E2
## A1  0.00 -0.12 -0.03  0.01  0.04  0.05  0.06  0.03  0.09  0.00  0.08  0.03
## A2 -0.12  0.00  0.03  0.02 -0.03 -0.04 -0.05  0.03 -0.03  0.02 -0.04 -0.01
## A3 -0.03  0.03  0.00  0.02  0.02 -0.02 -0.02 -0.02 -0.01 -0.01  0.03  0.01
## A4  0.01  0.02  0.02  0.00 -0.02 -0.04  0.04 -0.06  0.01 -0.04  0.01  0.01
## A5  0.04 -0.03  0.02 -0.02  0.00  0.02 -0.01  0.01  0.00  0.00  0.03  0.03
## C1  0.05 -0.04 -0.02 -0.04  0.02  0.00  0.07  0.01  0.01  0.05  0.01  0.01
##       E3    E4    E5    N1    N2    N3    N4    N5    O1    O2    O3    O4
## A1  0.07  0.05  0.01 -0.01 -0.02  0.02  0.01  0.00  0.06  0.06  0.02 -0.02
## A2 -0.06 -0.04  0.07  0.00  0.04 -0.03 -0.01 -0.01 -0.01 -0.01 -0.03  0.01
## A3  0.01 -0.03 -0.01  0.02  0.01 -0.01 -0.02 -0.05  0.00 -0.02  0.00 -0.03
## A4 -0.01  0.01 -0.02  0.02 -0.02  0.01 -0.02  0.00  0.02 -0.02  0.00 -0.02
## A5  0.03  0.04 -0.01  0.00  0.00 -0.01 -0.01  0.00  0.00  0.00  0.00  0.00
## C1 -0.02  0.06  0.02 -0.02 -0.01  0.02  0.01  0.01 -0.01  0.02  0.00  0.04
##       O5
```

```
## A1  0.07
## A2 -0.05
## A3 -0.01
## A4 -0.01
## A5  0.00
## C1  0.02
```

This matrix is also called as the **residual matrix**.

In this chapter, we explored the foundational role of matrix decomposition in the broader context of dimension reduction and data analysis. We began with the concepts of eigenvectors and eigenvalues, which provide critical insight into how matrices transform space and underlie many decomposition techniques.

We then examined SVD, a versatile tool that extends decomposition to rectangular matrices and serves as the basis for a range of applications—including Rank(r) approximations, which allow us to represent complex data using lower-dimensional structures. These approximations play a vital role in data compression, noise reduction, and algorithm efficiency.

From there, we introduced the Moore–Penrose inverse, a generalized matrix inverse that enables us to find least-squares solutions to linear systems even when a traditional inverse does not exist. This concept is particularly useful in overdetermined or underdetermined systems, and its computation via SVD highlights the deep connection between decomposition and problem-solving in linear algebra.

Finally, we explored two major applications of decomposition in machine learning: PCA, which identifies the directions of greatest variance in data for dimensionality reduction, and FA, which models data as driven by latent, unobserved variables. While PCA is primarily focused on variance and orthogonal transformations, FA emphasizes the discovery of hidden factors that account for observed relationships.

Taken together, these tools provide a powerful framework for simplifying data, uncovering structure, and improving the performance and interpretability of machine learning models. As we move forward, these techniques will serve as a foundation for more advanced topics such as supervised learning, neural networks, and probabilistic modeling—where the ability to manage high-dimensional inputs is both essential and transformative.

32

Optimization Algorithms: Basics

This final chapter concludes the book by shifting focus from causal estimation and predictive modeling to the foundational engine behind nearly all model estimation procedures: optimization. While previous chapters have emphasized designing models and choosing estimators, successful application of these tools relies on algorithms that solve optimization problems efficiently, accurately, and at scale. Whether estimating coefficients in a regression model, optimizing causal forest splits, or solving orthogonal score functions in double machine learning, optimization drives the computational engine behind the inference.

According to Wikipedia[1]:

> An optimization algorithm is a procedure which is executed iteratively by comparing various solutions until an optimum or a satisfactory solution is found. Optimization algorithms help us to minimize or maximize an objective function $F(x)$ with respect to the internal parameters of a model mapping a set of predictors (X) to target values(Y).

We explore this topic from both conceptual and practical perspectives. The chapter begins by introducing the core logic of optimization: finding the best possible solution given an objective function and a feasible set. We then present a range of optimization algorithms commonly used in machine learning, econometrics, and operations research—highlighting when and why each class of method is appropriate. We also distinguish between constrained and unconstrained optimization problems, a distinction that arises frequently in estimation tasks such as regularization or resource allocation. These algorithms include zero-order (derivative-free) approaches like brute-force search and grid search; first-order methods such as gradient descent and its variants (stochastic, mini-batch, adaptive); and second-order techniques like Newton–Raphson, which leverage curvature information for faster convergence.

Through a series of step-by-step simulations, we demonstrate how these algorithms operate in practice. We begin with simple one-dimensional examples to illustrate key ideas like learning rate, convergence, and local vs. global optima. We then build toward multivariable and constrained optimization problems relevant to econometric estimation, including maximum likelihood and regularization. Along the way, we show how optimization concepts tie directly to earlier chapters—for example, how gradient descent underlies LASSO, or how Newton's method appears in maximum likelihood estimation for logistic regression.

DOI: 10.1201/9781003504795-32

Optimization methods are broadly categorized as follows:

- **Zero-Order Methods:** Also known as derivative-free algorithms, these rely only on the value of the objective function (not gradients). They are ideal when the function is a black box, noisy, or non-differentiable. Examples are random search, grid search, Nelder–Mead.
- **First-Order Methods:** These methods use the gradient (first derivative) of the objective function to guide the search for an optimum. They are widely used in machine learning. Examples are gradient descent, stochastic gradient descent (SGD), Adam, RMSProp.
- **Second-Order Methods:** These utilize both the gradient and the Hessian (second derivative matrix) to adjust the search direction. They typically offer faster convergence, especially for convex problems, but are more computationally demanding. Examples: Newton's method, Quasi-Newton (BFGS).

Throughout the chapter, we emphasize both intuition and implementation. Code snippets, numerical illustrations, and links to optimization packages in R are provided to help readers move seamlessly from mathematical formulation to real-world application. By the end of this chapter—and the book—readers will have developed not only a strong conceptual foundation in modeling and inference, but also the practical skills needed to implement, evaluate, and improve those models through modern optimization techniques.

Optimization is a vast and evolving field, with hundreds of algorithms developed for different objective functions, data structures, and computational environments. Rather than attempt an exhaustive survey, this chapter focuses on building strong foundational understanding and introduces the most widely used optimization methods that underpin the models and estimators presented throughout the book. From LASSO and logistic regression to causal forests and double machine learning, these tools rely on optimization routines—often buried inside software packages—that solve problems efficiently behind the scenes. By working through step-by-step examples, we aim to make these inner mechanics visible, helping readers understand not just what the algorithms do, but how they do it.

32.1 Brute-Force Optimization

Brute-force optimization refers to a class of methods that search over a predefined grid or range of values to find the one that yields the best outcome according to a specified objective function. These methods do not use gradients or curvature information and are especially useful when the function is not differentiable or has an unknown form. For instance, in earlier chapters we used grid search to tune hyperparameters in k-nearest neighbors and penalized regression, and to select the bandwidth selection in nonparametric methods. While computationally inefficient for large problems, brute-force methods are intuitive and transparent, making them ideal for illustrating core ideas in optimization.

Let's look at a simplified example about optimal retirement two plan and solve it with a zero-order algorithm.

Suppose that there are two groups of workers who are planning for their retirement at the age of 65. Both consider spending 40,000 dollars each year for the rest of their lives after retirement. On average, people in both groups expect to live 20 more years after retirement with some uncertainty. The people in the first group (A) have the following risk profile: 85% chance to live 20 years and 15% chance to live 30 years. The same risk profile for the people in the second group (B) is: 99% for 20 years and 1% for 30 years. Suppose that in each group, their utility (objective) function is $U = C^{0.5}$.

What's the maximum premium (lump-sum payment) that a person in each group would be willing to pay for a life-time annuity of 40K?

Without a pension plan, people in each group have the following utilities:

```
#For people in group A
U_A = 0.85*sqrt(40000*20) + 0.15*sqrt(10*0)
U_A
```

```
## [1] 760.2631
```

```
#For people in group B
U_B = 0.99*sqrt(40000*20) + 0.01*sqrt(10*0)
U_B
```

```
## [1] 885.4829
```

For example, they would not pay 200,000 dollars to cover their retirement because that would make them worse than their current situation (without a pension plan).

```
#For people in group A
U_A = 0.85*sqrt(40000*20 - 200000) + 0.15*sqrt(40000*10 - 200000)
U_A
```

```
## [1] 725.4892
```

```
#For people in group B
U_B = 0.99*sqrt(40000*20 - 200000) + 0.01*sqrt(40000*10 - 200000)
U_B
```

```
## [1] 771.3228
```

Hence, the payment they would be willing to make for reduction in uncertainty during their retirement should not make them worse off. Or more technically, their utility should not be lower than their current utility levels. Therefore `Pmax`, the maximum premium that a person would be willing to pay, can be found by minimizing the following **cost function** for people, for example, in Group A:

$$f(P_{\max}) = p\sqrt{40000 \times 20 \text{ years} - P_{\max}} + (1-p)\sqrt{40000 \times 10 \text{ years} - P_{\max}} - p\sqrt{40000 \times 20 \text{ years}}$$

Where p is the probability of living 20 years. The cost function is a function of `Pmax` and we want to find the value of `Pmax` that makes the cost function equal to zero. This is a simple optimization problem, but it can be solved with different algorithms. We will start with a brute-force method and then move to more sophisticated ones.

Here is the iteration to solve for `Pmax` for people in Group A. We created a cost function, `costf`, that we try to minimize. Change the parameters to play with it. The same algorithm can be used to find `Pmax` for people in Group B.

```
library(stats)

p <- 0.85
w1 <- 800000
w2 <- 400000

converged = F
iterations = 0
maxiter <- 600000
learnrate <- 0.5
Pmax <- 10000

while(converged == FALSE){
  costf <- p*sqrt(w1 - Pmax) + (1 - p)*sqrt(w2 - Pmax) - p*sqrt(w1)
  if(costf > 0){
    Pmax <- Pmax + learnrate
    iterations = iterations + 1

    if(iterations > maxiter) {
      print("It cannot converge before finding the optimal Pmax")
      break
    }
    converged = FALSE
  }else{
    converged = TRUE
    print(paste("Maximum Premium:",
                Pmax, "achieved with",
                iterations, "iterations"))
  }
}
```

```
## [1] "Maximum Premium: 150043 achieved with 280086 iterations"
```

```
#let's verify it by `uniroot()` which finds the roots for f(x) = 0
costf <- function(x){
  p * sqrt(800000 - x) +
    (1-p) * sqrt(400000 - x) -
    p*sqrt(800000)
  }

paste("Unitroot for f(x) = 0 is ",
      uniroot(costf, c(10000, 200000))$root)
```

```
## [1] "Unitroot for f(x) = 0 is  150042.524874307"
```

There are better functions that we could use for this purpose, but this example works well for our experiment.

There several of important parameters in our algorithm. The first one is the starting `Pmax`, which can be set up manually. If the starting value is too low, iteration could not converge.

If it's too high, it can give us an error. Another issue is that our iteration does not know if the learning rate should increase or decrease when the starting value is too high or too low. This can be done with additional lines of code, but we will not address it here.

This situation leads us to the learning rate: the incremental change in the value of the parameter. This parameter should be conditioned on the value of cost function. If the cost function for a given `Pmax` is negative, for example, the learning rate should be negative. Second, the number of maximum iterations must be set properly, otherwise the algorithm may not converge or take too long to converge. In the next section, we will address these issues with a smarter algorithm.

There are other types of approaches. For example, the algorithm may create a grid of `Pmax` and then try all the possible values to see which one approximately makes the cost function minimum.

32.2 Derivative-based Methods

Derivative-based methods are a class of optimization algorithms that utilize the first and/or second derivatives of the objective function to guide the search for an optimum. These methods are particularly effective for smooth, continuous functions where derivatives can be computed or approximated. They are widely used in machine learning, econometrics, and operations research due to their efficiency and ability to converge quickly to local optima.

One of the most widely known derivative-based methods is the **Newton–Raphson** algorithm. If we assume that the objective function is differentiable and has a unique minimum (or maximum), we can develop an iterative procedure that searches for the point in parameter space where the derivative equals zero. This is the basis for many estimation techniques, including maximum likelihood estimation.

In fact, derivative-based optimization is embedded in many of the methods we've encountered throughout the book. For example, logistic regression estimates—used in causal inference and classification—are typically computed using Newton-Raphson or its variant, Fisher Scoring, as implemented in the `glm()` function in R. Similarly, in double machine learning frameworks, internal steps such as estimating propensity scores or outcome models often rely on these solvers. Although the optimization process is handled by software packages, understanding these underlying mechanics reveals what happens behind the scenes during model fitting. There are other related methods, such as Fisher Scoring[2] and Iteratively Reweighted Least Squares,[3] that we will not cover in detail here.

First, let's take a closer look at the Newton-Raphson method. This is a well-known extension of your calculus class about derivatives in High School. The method is very simple and used to find the roots of $f(x) = 0$ by iterations. In first-year computer science courses, this method is used to teach loop algorithms that calculate the value of, for example, $e^{0.71}$ or $\sqrt{12}$. It is a simple iteration that converges in a few steps.

$$x_{n+1} = x_n - \frac{f(x_n)}{f'(x_n)} \tag{32.1}$$

To understand it, let's look at the function $y = f(x)$ shown in the following graph:

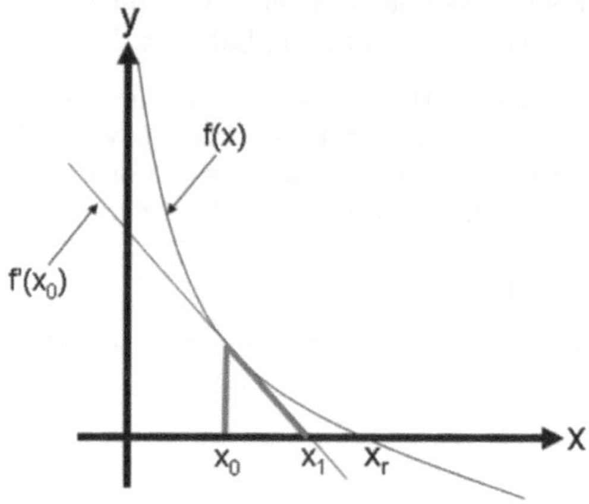

It has a zero at $x = x_r$, which is not known. To find it, we start with x_0 as an initial estimate of X_r. The tangent line to the graph at the point $(x_0, f(x_0))$ has the point x_1 at which the tangent crosses the x-axis. The slope of this line can be defined as

$$\frac{y - f(x_0)}{x - x_0} = f'(x_0) \tag{32.2}$$

Hence,

$$y - f(x_0) = f'(x_0)(x - x_0) \tag{32.3}$$

At the point where the tangent line cross the x-axis, $y = 0$ and $x = x_1$. Hence solving the equation for x_1, we get

$$x_1 = x_0 - \frac{f(x_0)}{f'(x_0)} \tag{32.4}$$

And the second approximations:

$$x_2 = x_1 - \frac{f(x_1)}{f'(x_1)} \tag{32.5}$$

And with multiple iterations one can find the solution. Here is the example:

```
newton <- function(f, x0, tol = 1e-5, n = 1000) {
  require(numDeriv) # Package for computing f'(x)

for (i in 1:n) {
    dx <- genD(func = f, x = x0)$D[1] # First-order derivative f'(x0)
    x1 <- x0 - (f(x0) / dx) # Calculate next value x1
    if (abs(x1 - x0) < tol) {
      res <- paste("Root approximation is ", x1, " in ", i, " iterations")
      return(res)
    }
    # If Newton-Raphson has not yet reached convergence set x1 as x0 and continue
    x0 <- x1
  }
```

```
    print('Too many iterations in method')
}

func2 <- function(x) {
  x^15 - 2
}
newton(func2, 1)

## [1] "Root approximation is  1.04729412282063  in  5  iterations"

#Check it
paste("Calculator result: ", 2^(1/15))

## [1] "Calculator result:  1.04729412282063"
```

Newton's method is often used to solve two different, but related, problems:

1. Finding x such that $f(x) = 0$ (try to solve our insurance problem with this method)
2. Finding x that $g'(x) = 0$, or find x that minimizes/maximizes $g(x)$.

The relation between these two problems is obvious when we define $f(x) = g'(x)$. Hence, for the second problem, the Newton-Raphson method becomes:

$$x_{n+1} = x_n - \frac{g'(x_n)}{g''(x_n)} \tag{32.6}$$

Connection between these two problems are defined in a post[4] very nicely.

Let's pretend that we are interested in determining the parameters of a random variable $X \sim N(\mu, \sigma^2)$. Here is the log-likelihood function for X:

$$\log(\mathcal{L}(\mu, \sigma)) = -\frac{n}{2}\log\left(2\pi\sigma^2\right) - \frac{1}{2\sigma^2}\sum_{i=1}^{n}(x_i - \mu)^2 \tag{32.7}$$

We use `dnorm()` which calculates the pdf of a normal variable. First let's have the data and the log-likelihood:

```
# Let's create a sample of normal variables
set.seed(2019)
X <- rnorm(100, 0, 1)

# And the log-likelihood of this function.
# Remember likelihood function would be prod(dnorm()) with log=F
normalLL <- function(prmt){
  sum(dnorm(X, mean = prmt[1], sd = prmt[2], log = TRUE))
}

# Let's try several parameters
normalLL(prmt = c(1,1.5))
```

```
## [1] -176.078
```

```
normalLL(prmt = c(2,1))
```

```
## [1] -347.4119
```

```
normalLL(prmt = c(mean(X),sd(X)))
```

```
## [1] -131.4619
```

As you can see, the last one is the best. And we can verify it because we had created X with 0 mean and 1 sd, approximately. Now we will use the Newton–Raphson method to calculate those parameters that minimize the negative log-likelihood.

First, let's build a function that estimates the slope of the function (first-derivative) numerically at any arbitrary point in parameter space for mean and sd, separately. Don't forget, **the log-likelihood is a function of parameters (mean and sd) not X**.

```
# First partial (numerical) derivative w.r.t. mean
firstM <- function(p1, p2, change = 0.0001){
  prmt <- c(p1, p2)
  high <- normalLL(prmt + c(change,0))
  low <- normalLL(prmt - c(change,0))
  slope <- (high-low)/(change*2)
  return(slope)
}
firstM(mean(X), sd(X))
```

```
## [1] 1.421085e-10
```

```
# First partial (numerical) derivative w.r.t. sd
firstSD <- function(p1, p2, change = 0.0001){
  prmt <- c(p1, p2)
  high <- normalLL(prmt + c(0, change))
  low <- normalLL(prmt - c(0, change))
  slope <- (high-low)/(change*2)
  return(slope)
}

firstSD(mean(X), sd(X))
```

```
## [1] -1.104417
```

```
#Verify them with the grad()
library(numDeriv)
f <- function(x) {
    a <- x[1]; b <- x[2]
    sum(dnorm(X, mean = a, sd = b, log = TRUE))
    }
grad(f,c(mean(X),sd(X)))[1]
```

```
## [1] -1.367073e-12
```

```
grad(f,c(mean(X),sd(X)))[2]
```

```
## [1] -1.104419
```

```
# Or better
round(jacobian(f,c(mean(X),sd(X))), 4) #First derivatives
```

```
##      [,1]    [,2]
## [1,]    0 -1.1044
```

```
round(hessian(f,c(mean(X),sd(X))), 4) #Second derivatives
```

```
##           [,1]     [,2]
## [1,] -121.9741    0.000
## [2,]    0.0000 -240.289
```

Let's try them now in the Newton-Raphson method.

$$x_{n+1} = x_n - \frac{g'(x_n)}{g''(x_n)} \tag{32.8}$$

Similar to the first one, we can also develop a function that calculates the second derivatives. However, instead of using our own functions, let's use `grad()` and `hessian()` from the `numDeriv` package.

```
set.seed(2019)
X <- rnorm(100, 2, 2)

NR <- function(f, x0, y0, tol = 1e-5, n = 1000) {
  for (i in 1:n) {
    dx <- grad(f,c(x0, y0))[1] # First-order derivative f'(x0)
    ddx <- hessian(f,c(x0, y0))[1,1] # Second-order derivative f''(x0)
    x1 <- x0 - (dx / ddx) # Calculate next value x1
      if (abs(x1 - x0) < tol) {
        res <- paste("The mean approximation is ", x1, " in ", i, " iterations")
        return(res)
      }
    # If Newton-Raphson has not yet reached convergence set x1 as x0 and continue
    x0 <- x1
  }
  print('Too many iterations in method')
}

func <- function(x) {
    a <- x[1]; b <- x[2]
    sum(dnorm(X, mean = a, sd = b, log = TRUE))
    }
```

```
NR(func, -3, 1.5)
```

```
## [1] "The mean approximation is  1.85333200301108  in  2  iterations"
```

```
#Let;s verify it
mean(X)
```

```
## [1] 1.853332
```

Finding sd is left to the practice questions. But the way to do it should be obvious. Use our approximation of the mean (1.853332) as a fixed parameter in the function and run the same algorithm for finding sd. **When the power of computers and the genius of mathematics intercepts, beautiful magics happen.**

32.3 ML Estimation with Logistic Regression

The PDF of Bernoulli distribution is

$$P(Y=y)=p^y(1-p)^{1-y}$$

It's likelihood

$$\begin{aligned}
L(\boldsymbol{\beta} \mid \mathbf{y} ; \mathbf{x}) &= L\left(\beta_0, \beta_1 \mid \left(y_1, \ldots, y_n\right) ; \left(x_1, \ldots, x_n\right)\right) \\
&= \prod_{i=1}^n p_i^{y_i}\left(1-p_i\right)^{1-y_i}
\end{aligned}$$

And log-likelihood

$$\begin{aligned}
\ell(\boldsymbol{\beta} \mid \mathbf{y} ; \mathbf{x}) &= \log \left(\prod_{i=1}^n p_i^{y_i}\left(1-p_i\right)^{1-y_i}\right) \\
&= \sum_{i=1}^n\left(\log \left(p_i^{y_i}\right)+\log \left(1-p_i\right)^{1-y_i}\right) \\
&= \sum_{i=1}^n y_i\left(\log \left(p_i\right)+\left(1-y_i\right) \log \left(1-p_i\right)\right)
\end{aligned}$$

where

$$\begin{aligned}
\mathrm{L}\left(p_i\right) &= \log \left(\frac{p_i}{1-p_i}\right) \\
&= \beta_0+\beta_1 x_1
\end{aligned}$$

So,

$$p_i=\frac{\exp \left(\beta_0+x_1 \beta_1\right)}{1+\exp \left(\beta_0+x_1 \beta_1\right)}$$

First partial derivative with respect to β_0

$$\begin{aligned}\frac{\partial p_i}{\partial \beta_0} &= \frac{\exp(\beta_0 + x_1\beta_1)}{(1+\exp(\beta_0 + x_1\beta_1))^2} \\ &= p_i(1-p_i)\end{aligned}$$

And

$$\begin{aligned}\frac{\partial p_i}{\partial \beta_1} &= \frac{x_1\exp(\beta_0 + x_1\beta_1)}{(1+\exp(\beta_0 + x_1\beta_1))^2} \\ &= x_1 p_i(1-p_i)\end{aligned}$$

Newton-Raphson's equation is

$$\boldsymbol{\beta}^{(t+1)} = \boldsymbol{\beta}^{(t)} - \left(\boldsymbol{H}^{(t)}\right)^{-1}\boldsymbol{u}^{(t)},$$

where

$$\boldsymbol{\beta}^{(t)} = \begin{bmatrix} \beta_0^{(t)} \\ \beta_1^{(t)} \end{bmatrix}$$

$$\boldsymbol{u}^{(t)} = \begin{bmatrix} u_0^{(t)} \\ u_1^{(t)} \end{bmatrix} = \begin{bmatrix} \frac{\partial \ell(\beta^{(t)}|y;x)}{\partial \beta_0} \\ \frac{\partial \ell(\beta^{(t)}|y;x)}{\partial \beta_1} \end{bmatrix} = \begin{bmatrix} \sum_{i=1}^n \left(y_i - p_i^{(t)}\right) \\ \sum_{i=1}^n x_i\left(y_i - p_i^{(t)}\right) \end{bmatrix}$$

where,

$$p_i^{(t)} = \frac{\exp\left(\beta_0^{(t)} + x_1\beta_1^{(t)}\right)}{1+\exp\left(\beta_0^{(t)} + x_1\beta_1^{(t)}\right)}$$

$\boldsymbol{H}^{(t)}$ can be considered as Jacobian matrix of $\boldsymbol{u}(\cdot)$,

$$\boldsymbol{H}^{(t)} = \begin{bmatrix} \frac{\partial u_0^{(t)}}{\partial \beta_0} & \frac{\partial u_0^{(t)}}{\partial \beta_1} \\ \frac{\partial u_1^{(t)}}{\partial \beta_0} & \frac{\partial u_1^{(t)}}{\partial \beta_1} \end{bmatrix}$$

Let's simulate data and solve it the Newton-Raphson's method described above.

```
rm(list=ls())

#Simulating data
set.seed(1)
n <- 500
X = rnorm(n) # this is our x
z = -2 + 3 * X

#Prob. is defined by logistic function
p = 1 / (1 + exp(-z))

#Bernoulli is the special case of the binomial distribution with size = 1
```

```
y = rbinom(n, size = 1, prob = p)

#And we create our data
df <-  data.frame(y, X)
head(df)
```

```
##   y          X
## 1 0 -0.6264538
## 2 0  0.1836433
## 3 0 -0.8356286
## 4 0  1.5952808
## 5 0  0.3295078
## 6 0 -0.8204684
```

```
logis <- glm(y ~ X, data = df, family = binomial)
summary(logis)
```

```
##
## Call:
## glm(formula = y ~ X, family = binomial, data = df)
##
## Coefficients:
##             Estimate Std. Error z value Pr(>|z|)
## (Intercept)  -1.8253     0.1867  -9.776   <2e-16 ***
## X             2.7809     0.2615  10.635   <2e-16 ***
## ---
## Signif. codes:  0 '***' 0.001 '**' 0.01 '*' 0.05 '.' 0.1 ' ' 1
##
## (Dispersion parameter for binomial family taken to be 1)
##
##     Null deviance: 605.69  on 499  degrees of freedom
## Residual deviance: 328.13  on 498  degrees of freedom
## AIC: 332.13
##
## Number of Fisher Scoring iterations: 6
```

```
library(numDeriv)

func_u <- function(b) {
  c(sum(df$y - exp(b[1] + b[2] * df$X)/ (1 + exp(b[1] + b[2] * df$X))),
    sum(df$X * (df$y - exp(b[1] + b[2] * df$X)/ (1 + exp(b[1] + b[2] * df$X)))))
}

# Starting points
delta <- matrix(1:2, nrow = 2) # starting delta container (with any number > 0)
b <- array(c(-2,3))

while(abs(sum(delta)) > 0.0001){
  B <- b #current b
```

```
  b <- as.matrix(b) - solve(jacobian(func_u, x = b)) %*% func_u(b) #new b
  delta <- b - as.matrix(B)
}
b
```

```
##              [,1]
## [1,] -1.825347
## [2,]  2.780929
```

32.4 Gradient Descent Algorithm

While Newton-Raphson uses both first and second derivatives to refine estimates, many optimization problems—especially in high-dimensional settings—are solved using first-order methods that rely only on gradient information. **Gradient descent** is one of the most widely used algorithms in this class. It forms the backbone of many machine learning and econometric procedures, including training neural networks, fitting regularized regressions, and solving loss minimization problems in double machine learning.

Gradient descent works by iteratively updating parameter values in the direction of the negative gradient, gradually moving toward a local minimum of the objective function. Its simplicity, scalability, and compatibility with noisy or incomplete data make it ideal for large-scale problems where second-order methods may be computationally expensive or infeasible.

Let's begin with a simple regression problem to illustrate how gradient descent operates in practice. The cost function in OLS is the residual sum of squares, $\text{RSS} = \sum_{i=1}^{n}\left(\widehat{e}_i\right)^2 = \sum_{i=1}^{n}\left(y_i - \hat{y}\right)^2 = \sum_{i=1}^{n}\left(y_i - \left(b_1 + b_2 x_i\right)\right)^2$, which is a convex function. Our objective to find b_1 and b_2 that minimize RSS. How can we find those parameters to minimize a cost function if we don't know much about it? The trick is to start with some point and move a bit (locally) in the direction that reduces the value of the cost function. In general, this search process for finding the minimizing point has two components: the direction and the step size. The direction tells us which direction we move next, and the step size determines how far we move in that direction. For example, the iterative search for b_2 of gradient descent can be described by the following recursive rule:

$$b_2^{(k+1)} = b_2^{(k)} - lr \nabla RSS^k$$

Here, lr is *learning rate* and ∇RSS^k is the slope of RSS at step k. Hence, $lr \nabla RSS^k$ is the total step size at step k. Note that, as we move from either directions towards b_2^*, ∇RSS^k gets smaller. In fact, it becomes zero at b_2^*. Therefore, ∇RSS^k helps iterations find the proper adjustment in each step in terms of direction and magnitude. Since RSS is a convex function, it's easy to see how sign of ∇RSS^k will direct the arbitrary $b_2^{''}$ towards the optimal b_2.

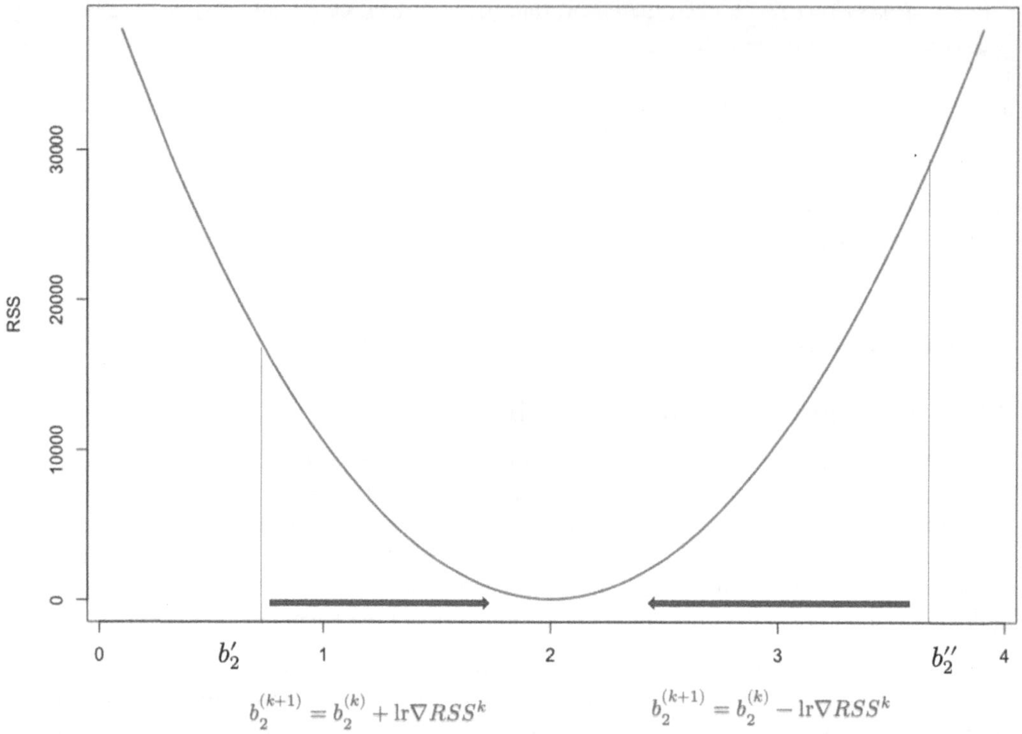

Since first-order approximation at $b_2^{''}$ is good only for small Δb_2, a small $lr > 0$ is needed to o make Δb_2 small in magnitude. Moreover, when a high learning rate used it leads to "overshooting" past the local minima and may result in diverging algorithm.

Below, we first use a simple linear regression function on simulated data and estimate its parameters with `lm()`. Let's simulate a sample with our DGM.

```
set.seed(1001)
N <- 100
int <- rep(1, N)
x1 <- rnorm(N, mean = 10, sd = 2)
Y <- rnorm(N, 2*x1 + int, 1)
model <- lm(Y ~ x1)
b <- coef(model)
b
```

```
## (Intercept)          x1
##    1.209597    1.979643
```

```
plot(x1, Y, col = "black", pch = 20)
abline(b)
```

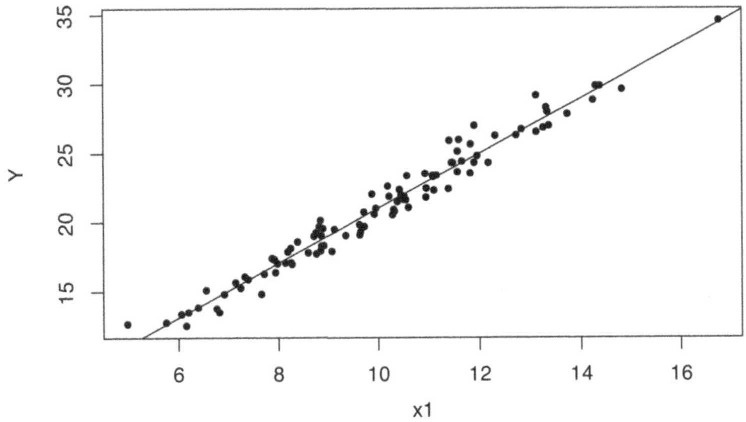

The cost function that we want to minimize is

$$y_i = 1 + 2x_i + \epsilon_i RSS = \sum \epsilon_i^2 = \sum (y_i - 1 - 2x_i)^2$$

And, its plot for a range of coefficients is already shown earlier.

32.4.1 One-variable

Below, we create a function, `grdescent`, to show how sensitive gradient descent algorithms would be to different calibrations:

```
grdescent <- function(x, y, lr, maxiter) {
  #starting points
  set.seed(234)
  b <- runif(1, 0, 1)
  c <- runif(1, 0, 1)
  n <- length(x)

  #function
  yhat <- c + b * x

  #gradient
  MSE <- sum((y - yhat) ^ 2) / n
  converged = F
  iterations = 0

  #while loop
  while (converged == F) {
    b_new <- b - ((lr * (1 / n)) * (sum((y - yhat) * x * (-1))))
    c_new <- c - ((lr * (1 / n)) * (sum(y - yhat) * (-1)))
    b <- b_new
    c <- c_new
    yhat <- b * x + c

    MSE_new <- sum((y - yhat) ^ 2) / n
    MSE <- c(MSE, MSE_new)
```

```
    d = tail(abs(diff(MSE)), 1)

    if (round(d, 12) == 0) {
      converged = T
      return(paste("Iterations: ",
                   iterations, "Intercept: ",
                   c, "Slope: ", b))
    }
    iterations = iterations + 1
    if (iterations > maxiter) {
      converged = T
      return(paste("Max. iter. reached, ", "Intercept:",
                   c, "Slope:", b))
    }
  }
}
```

Note that the key part in this algorithm is `b_new <- b + (learnrate * (1 / n)) * sum((y - yhat) * x*(-1)`. The first b that is picked randomly by `b <- runif(1, 0, 1)` is adjusted by `learnrate * (1 / n) * (sum((y - yhat) * -x))`.

Furthermore, `sum((y - yhat) * x)` is the first order condition of the cost function (RSS - Residual Sum of Squares) for the slope coefficient. The cost function (RSS) is a convex function where the minimum can be achieved by the optimal b. It is a linear Taylor approximation of RSS at b that provides the **steepest** descent, that is just a simple adjustment for identifying the direction of the adjustment of b until the minimum RSS is reached.

Now we will see if this function will give us the same intercept and slope coefficients already calculated with `lm()` above.

```
grdescent(x1, Y, 0.01, 100000)
```

```
## [1] "Iterations:  16388 Intercept:  1.20949474169039 Slope:  1.97965284879392"
```

This is good. But, if start a very low number with a small learning rate, then we need more iteration

```
grdescent(x1, Y, 0.005, 1000000)
```

```
## [1] "Iterations:  31363 Intercept:  1.20945256472045 Slope:  1.97965686098386"
```

The main question is how to choose the learning rate. This remains an active research area, and a full answer lies beyond the scope of this chapter. As a general guideline, it is best to start with a small value and tune it during the training process.

32.4.2 Adjustable *lr* and SGD

An adjustable learning rate has several advantages over a fixed learning rate in gradient-based optimization algorithms like stochastic gradient descent:

- Faster convergence: An adjustable learning rate can help the algorithm converge faster by starting with a larger learning rate, allowing the model to make bigger steps in the initial phase. This can help escape local minima or saddle points more quickly and reach the vicinity of the global minimum.
- Improved precision: As the learning rate decreases over time, the algorithm takes smaller steps, allowing for more precise updates to the model parameters. This can help the model fine-tune its parameters and potentially achieve a lower loss value compared to a fixed learning rate.
- Prevent oscillations: A fixed learning rate might cause oscillations around the optimal solution, whereas an adjustable learning rate can help dampen these oscillations by gradually reducing the step size. This can result in a more stable convergence.
- Adaptive to problem complexity: Some optimization problems might require different learning rates for different stages of the optimization process. An adjustable learning rate can adapt to the problem's complexity, allowing the model to learn at a more suitable pace for each stage.
- Robustness: An adjustable learning rate can make the optimization algorithm more robust to the choice of the initial learning rate. Even if the initial learning rate is not perfect, the algorithm can adapt over time and still reach a reasonable solution.

Although an adjustable learning rate can lead to faster convergence, improved precision, and better overall performance in gradient-based optimization algorithms, it also introduces additional hyperparameters (e.g., decay rate, annealing schedule) that need to be tuned for optimal performance.

Below, we made some changes to earlier gradient descent to make it stochastic with an adaptive learning rate. In this modified code, we have implemented the following changes:

- Shuffled the data points using the `sample()` function.
- Iterated over the data points in mini-batches of size 1 (`batch_size = 1`). This makes it **stochastic gradient descent**.
- Re-calculated the gradients and updated the weights for each mini-batch.

This should give us a simple stochastic gradient descent implementation for our linear regression problem. To implement an adjustable learning rate, we can use a learning rate scheduler or a learning rate annealing method. The following example shows how to use a simple exponential learning rate annealing method, which will decrease the learning rate over time:

```
# Set the seed
set.seed(1001)

# Generate data
N <- 100
int <- rep(1, N)
x1 <- rnorm(N, mean = 10, sd = 2)
Y <- rnorm(N, 2*x1 + int, 1)
model <- lm(Y ~ x1)
b <- coef(model)
b
```

```
## (Intercept)          x1
##    1.209597    1.979643
```

```
# Starting points
set.seed(234)
b <- runif(1, 0, 1)
c <- runif(1, 0, 1)
n <- length(x1)

# Parameters
initial_learning_rate <- 0.01
decay_rate <- 0.99999
batch_size <- 1
max_iterations <- 300000
tolerance <- 1e-12

# Function
yhat <- c + b * x1

# Gradient
MSE <- sum((Y - yhat) ^ 2) / n
converged = F
iterations = 0
num_batches <- ceiling(n / batch_size)

# While loop
while (converged == F) {
  # Shuffle data points
  indices <- sample(n, n)

  for (i in seq(1, n, by = batch_size)) {
    idx <- indices[i:min(i + batch_size - 1, n)]
    x_batch <- x1[idx]
    y_batch <- Y[idx]

    yhat_batch <- c + b * x_batch

    learning_rate <- initial_learning_rate * decay_rate^iterations

    b_new <- b - learning_rate * ((1 / length(idx)) *
                                  sum((y_batch - yhat_batch) *
                                        x_batch * (-1)))
    c_new <- c - learning_rate * ((1 / length(idx)) *
                                  sum(y_batch - yhat_batch) * (-1))

    b <- b_new
    c <- c_new
  }

  yhat <- b * x1 + c
  MSE_new <- sum((Y - yhat) ^ 2) / n
  d = abs(MSE_new - tail(MSE, 1))
```

```
  if (d < tolerance) converged = T
  MSE <- c(MSE, MSE_new)

  iterations = iterations + 1
  if (iterations > max_iterations) converged = T
}

c(iterations, c, b)
```

```
## [1] 3.000010e+05 1.205426e+00 1.966007e+00
```

Stochastic gradient descent (SGD) tends to be faster than plain gradient descent (GD) when working with large datasets. The main reason for this is that SGD updates the model parameters more frequently, using only a random subset of data points (or even a single data point) in each update, while GD uses the entire dataset for each update.

The main advantage of using SGD over plain GD is related to the convergence speed and the ability to escape local minima. In SGD, the model parameters are updated after each mini-batch (in this case, a single data point), whereas in GD, the updates happen after going through the entire dataset. As a result, SGD can converge faster than GD because it performs more frequent updates, which can be especially beneficial when working with large datasets.

Moreover, SGD introduces randomness in the optimization process due to the random sampling of data points. This stochastic nature can help the algorithm to escape local minima and find a better (global) minimum. In the case of plain GD, the algorithm always follows the true gradient, which can cause it to get stuck in sharp, non-optimal minima.

However, there are some trade-offs when using SGD. The updates in SGD can be noisy because they are based on a random subset of data points, which can lead to fluctuations in the learning process. This can make the algorithm's convergence path look less smooth than in the case of plain GD. Further, SGD often requires more careful tuning of hyperparameters, such as the learning rate and batch size. In some cases, a learning rate schedule (decreasing the learning rate over time) can be used to improve convergence.

In summary, while SGD can offer faster convergence and better ability to escape local minima, it comes with the trade-off of noisier updates and may require more careful hyperparameter tuning. When working with large datasets, we can also consider using mini-batch GD, which is a compromise between GD and SGD. Mini-batch GD uses a small batch of data points to compute the gradient, rather than the entire dataset (GD) or a single data point (SGD). This can offer a good balance between computational efficiency and convergence properties.

In the SGD code above, the decay rate is a hyperparameter that controls the rate at which the learning rate decreases over time in an adjustable learning rate schedule. Choosing an appropriate decay rate depends on the specific problem, the model, and the optimization algorithm being used. In practice, the decay rate is often chosen empirically through experimentation or by using techniques such as cross-validation or grid search.

Here are some guidelines to help us choose an appropriate decay rate:

- A common starting point for the decay rate is 0.99, as it provides a relatively slow decay of the learning rate. However, this value might not be optimal for all problems, so you should treat it as a starting point and experiment with different values to see what works best for your specific problem.

- If the optimization problem is complex or has a highly non-convex loss surface, you might want to choose a smaller decay rate (e.g., 0.9 or 0.95) to allow for a faster reduction in the learning rate. This can help the model escape local minima or saddle points more quickly. On the other hand, if the problem is relatively simple, you might want to choose a larger decay rate (e.g., 0.995 or 0.999) to keep the learning rate higher for a longer period.
- Since there is no one-size-fits-all answer for the decay rate, it is essential to experiment with different values and observe how they affect the optimization process. You can use techniques such as cross-validation or grid search to systematically explore different decay rate values and choose the one that yields the best performance.
- It can be helpful to monitor the learning rate during training to ensure that it is decaying at an appropriate pace. If the learning rate is decreasing too quickly, it might result in slow convergence or getting stuck in local minima. If the learning rate is decreasing too slowly, it might cause oscillations around the optimal solution and prevent the model from converging.

Learning rate scheduler is a more general concept than the specific exponential decay method we demonstrated here. A learning rate scheduler is a technique used to adjust the learning rate during the training process according to a predefined schedule or rule. The exponential decay method is just one example of a learning rate scheduler.

There are various learning rate scheduler strategies:

- Exponential decay: The learning rate is multiplied by a fixed decay rate at each iteration or epoch, as demonstrated in the previous example.
- Step decay: The learning rate is reduced by a fixed factor at specific intervals, such as every N epochs. For example, the learning rate could be reduced by a factor of 0.5 every 10 epochs.
- Time-based decay: The learning rate is reduced according to a function of the elapsed training time or the number of iterations. For example, the learning rate could be reduced by a factor proportional to the inverse of the square root of the number of iterations.
- Cosine annealing: The learning rate is reduced following a cosine function, which allows for periodic "restarts" of the learning rate, helping the optimization process escape local minima or saddle points.
- Cyclic learning rates: The learning rate is varied cyclically within a predefined range, allowing the model to explore different areas of the loss surface more effectively.
- Adaptive learning rates: These learning rate schedulers adjust the learning rate based on the progress of the optimization process, such as the improvement in the loss function or the validation accuracy. Some well-known adaptive learning rate methods include AdaGrad, RMSprop, and Adam.

The choice of the learning rate scheduler depends on the specific problem, the model, and the optimization algorithm being used. It's essential to experiment with different learning rate schedulers and monitor the training progress to find the best strategy for a particular problem.

32.4.3 Multivariable

We will expand the gradient descent algorithms with an multivariable example using matrix algebra. First, the data and model simulation:

```
set.seed(1001)
N <- 100
int <- rep(1, N)
x1 <- rnorm(N, mean = 10, sd = 2)
x2 <- rnorm(N, mean = 5, sd = 1)
x3 <- rbinom(N, 1, 0.5)
x4 <- rbinom(N, 1, 0.5)
x5 <- rbinom(N, 1, 0.5)
x6 <- rnorm(N, 1, 0.25)
x7 <- rnorm(N, 1, 0.2)
x2x3 <- x2*x3
x4x5 <- x4*x5
x4x6 <- x5*x6
x3x7 <- x3*x7

Y <- rnorm(N, 2*x1 + -0.5*x2 - 1.75*x2x3 + 2*x4x5 - 3*x4x6 + 1.2*x3x7 + int, 1)
X <- cbind(int, x1, x2, x2x3, x4x5, x4x6, x3x7)
```

We can solve it with linear algebra manually:

```
betaOLS <- solve(t(X)%*%X)%*%t(X)%*%Y
print(betaOLS)
```

```
##             [,1]
## int    0.4953323
## x1     1.9559022
## x2    -0.3511182
## x2x3 -1.9112623
## x4x5  1.7424723
## x4x6 -2.8323934
## x3x7  2.1015442
```

We can also solve it with `lm()`

```
model1.lm <- lm(Y ~ X -1)
summary(model1.lm)
```

```
##
## Call:
## lm(formula = Y ~ X - 1)
##
## Residuals:
##      Min       1Q   Median       3Q      Max
## -2.84941 -0.45289 -0.09686  0.57679  2.07154
##
## Coefficients:
##       Estimate Std. Error t value Pr(>|t|)
## Xint   0.49533    0.75410   0.657  0.51290
## Xx1    1.95590    0.03868  50.571  < 2e-16 ***
## Xx2   -0.35112    0.12600  -2.787  0.00645 **
```

```
## Xx2x3 -1.91126    0.13358 -14.308  < 2e-16 ***
## Xx4x5  1.74247    0.24396   7.143 2.01e-10 ***
## Xx4x6 -2.83239    0.18831 -15.041  < 2e-16 ***
## Xx3x7  2.10154    0.64210   3.273  0.00149 **
## ---
## Signif. codes:  0 '***' 0.001 '**' 0.01 '*' 0.05 '.' 0.1 ' ' 1
##
## Residual standard error: 0.8448 on 93 degrees of freedom
## Multiple R-squared:  0.9972, Adjusted R-squared:  0.997
## F-statistic:  4677 on 7 and 93 DF,  p-value: < 2.2e-16
```

Now the function for gradient descent:

```
grdescentM <- function(x, y, lr, maxiter) {
  set.seed(123)
  b <- runif(ncol(x), 0, 1)
  yhat <- x%*%b
  e <- y - yhat
  RSS <- t(e)%*%e
  converged = F
  iterations = 0
  n <- length(y)

  while(converged == F) {
    b_new <- b - (lr*(1/n))*t(x)%*%(x%*%b - y)
    b <- b_new
    yhat <- x%*%b
    e <- y - yhat

    RSS_new <- t(e)%*%e
    RSS <- c(RSS, RSS_new)
    d = tail(abs(diff(RSS)), 1)

    if (round(d, 12) == 0) {
      converged = T
      return(b)
    }
    iterations = iterations + 1
    if(iterations > maxiter) {
      converged = T
      return(b)
   }
  }
}
```

```
grdescentM(X, Y, 0.01, 100000)
```

```
##           [,1]
## int   0.4953843
## x1    1.9559009
```

```
## x2    -0.3511257
## x2x3 -1.9112548
## x4x5  1.7424746
## x4x6 -2.8323944
## x3x7  2.1015069
```

32.5 Optimization with R

A good summary of tools for optimization in R given in this guide: Optimization and Mathematical Programming (https://cran.r-project.org/web/views/Optimization.html). There are many optimization methods, each of which would only be appropriate for specific cases. In choosing a numerical optimization method, we need to consider following points:

1. We need to know if it is a constrained or unconstrained problem. For example, the MLE method is an unconstrained problem. Most regularization problems, like LASSO or Ridge, are constraint optimization problems.

2. Do we know how the objective function is shaped a priori? MLE and OLS methods have well-known objective functions (RSS and Log-Likelihood). Maximization and minimization problems can be used in both cases by flipping the sign of the objective function.

3. Multivariate optimization problems are much harder than single-variable optimization problems. There is, however, a large set of available optimization methods for multivariate problems.

4. In multivariate cases, the critical point is whether the objective function has available gradients. If only the objective function is available without gradient or Hessian, the Nelder–Mead algorithm is the most common method for numerical optimization. If gradients are available, the best and most used method is the gradient descent algorithm. We have seen its application for OLS. This method can be applied to MLE as well. It is also called a Steepest Descent Algorithm. In general, the gradient descent method has three types: Batch Gradient Descent, SGD, and Mini-Batch Gradient Descent.

5. If the gradient and Hessian are available, we can also use the Newton–Raphson method. This is only possible if the dataset is not high-dimensional, as calculating the Hessian would otherwise be very expensive.

6. Usually the Nelder–Mead method is slower than the GD method. `Optim()` uses the Nelder–Mead method, but the optimization method can be chosen in its arguments.

The most detailed and advance source is Numerical Recipes (https://en.wikipedia.org/wiki/Numerical_Recipes), which uses `C++` and `R`.

Notes

1. https://en.wikipedia.org/wiki/Mathematical_optimization#Optimization_algorithms
2. https://en.wikipedia.org/wiki/Scoring_algorithm
3. https://en.wikipedia.org/wiki/Iteratively_reweighted_least_squares
4. https://stats.stackexchange.com/questions/376191/why-is-the-second-derivative-required-for-newtons-method-for-back-propagation, (Gulzar, 2018)

References

Abadie, A., & Gardeazabal, J. (2003). The economic costs of conflict: A case study of the Basque Country. *American Economic Review, 93*(1), 113–132.

Abadie, A., & Imbens, G. W. (2002). Simple and bias-corrected matching estimators for average treatment effects. *NBER Technical Working Paper No. 283.* https://doi.org/10.3386/t0283

Abadie, A., & Imbens, G. W. (2006). Large sample properties of matching estimators for average treatment effects. *Econometrica, 74*(1), 235–267. https://doi.org/10.1111/j.1468-0262.2006.00655.x

Abadie, A., & Imbens, G. W. (2008). On the failure of the bootstrap for matching estimators. *Econometrica, 76*(6), 1537–1557. https://doi.org/10.3982/ECTA6474

Abadie, A., & Imbens, G. W. (2011). Bias-corrected matching estimators for average treatment effects. *Journal of Business & Economic Statistics, 29*(1), 1–11. https://doi.org/10.1198/jbes.2009.07333

Abadie, A., & Imbens, G. W. (2016). Matching on the estimated propensity score. *Econometrica, 84*, 781–807. https://doi.org/10.3982/ECTA11293

Abadie, A., Diamond, A., & Hainmueller, J. (2010). Synthetic control methods for comparative case studies: Estimating the effect of California's tobacco control program. *Journal of the American Statistical Association, 105*(490), 493–505.

Abadie, A., Diamond, A., & Hainmueller, J. (2011). Synth: An R package for synthetic control methods in comparative case studies. *Journal of Statistical Software, 42*(13), 1–17. https://doi.org/10.18637/jss.v042.i13

Abadie, A., Diamond, A., & Hainmueller, J. (2015). Comparative politics and the synthetic control method. *American Journal of Political Science, 59*(2), 495–510.

Acemoglu, D., & Finkelstein, A. (2008). Input and technology choices in regulated industries: Evidence from the health care sector. *Journal of Political Economy, 116*, 837–880.

Acemoglu, D., Johnson, S., & Robinson, J. A. (2001). The colonial origins of comparative development: An empirical investigation. *American Economic Review, 91*(5), 1369–1401.

Aizer, A., & Doyle, J. J. (2015). Juvenile incarceration, human capital, and future crime: Evidence from randomly assigned judges. *Quarterly Journal of Economics, 130*(2), 759–803.

Amjad, M., Shah, D., & Shen, D. (2018). Robust synthetic control. *Journal of Machine Learning Research, 19*(22), 1–50.

Andrews, I., Stock, J. H., & Sun, Y. (2018). Weak instruments in instrumental variables regression: Theory and practice. *Annual Review of Economics, 10*, 683–706. Retrieved from https://scholar.harvard.edu/files/wirev_092218-_corrected_0.pdf

Angrist, J. D. (1990). Lifetime earnings and the Vietnam era draft lottery: Evidence from Social Security administrative records. *American Economic Review, 80*(3), 313–336.

Angrist, J. D., & Krueger, A. B. (1991). Does compulsory school attendance affect schooling and earnings? *Quarterly Journal of Economics, 106*(4), 979–1014.

Angrist, J. D., & Pischke, J.-S. (2009). *Mostly Harmless Econometrics: An Empiricist's Companion.* Princeton University Press.

Arcidiacono, P., & Ellickson, P. B. (2022). Practical Methods for Estimating Dynamic Discrete Choice Models. arXiv:1808.02569.

Arkhangelsky, D., & Imbens, G. (2024). Causal models for longitudinal and panel data: A survey. *The Econometrics Journal, 27*(3), C1–C61. https://doi.org/10.1093/ectj/utae014

Arkhangelsky, D., Athey, S., Hirshberg, D. A., Imbens, G. W., & Wager, S. (2021). Synthetic difference-in-differences. *American Economic Review, 111*(12), 4088–4118. https://doi.org/10.1257/aer.20190159

Athey, S., & Imbens, G. W. (2019). Machine Learning Methods That Economists Should Know About. *Annual Review of Economics*, 11, 685–725. https://doi.org/10.1146/annurev-economics-080217-053433

Athey, S., & Wager, S. (2019). Estimating treatment effects with causal forests: An application. *The Annals of Statistics, 47*(2), 1148–1178.

Athey, S., Tibshirani, J., & Wager, S. (2019). Generalized random forests. *Annals of Statistics, 47*(2), 1148–1178.

Austin, P. C. (2009). Balance diagnostics for comparing the distribution of baseline covariates between treatment groups in propensity-score matched samples. *Statistics in Medicine, 28*(25), 3083–3107. https://doi.org/10.1002/sim.3697

Austin, P. C., & Small, D. S. (2014). The use of bootstrapping when using propensity-score matching without replacement: A simulation study. *Statistics in Medicine, 33*(24), 4306–4319.

Autor, D. H., Dorn, D., & Hanson, G. H. (2013). The China syndrome: Local labor market effects of import competition in the United States. *American Economic Review, 103*(6), 2121–2168.

Banerjee, A., & Duflo, E. (2009). "The Experimental Approach to Development Economics." *Annual Review of Economics*, 1, 151–178.

Belloni, A., Chernozhukov, V., & Hansen, C. (2014a). Inference on treatment effects after selection among high-dimensional controls. *Review of Economic Studies, 81*(2), 608–650.

Belloni, A., Chernozhukov, V., & Hansen, C. (2014b). High-dimensional methods and inference on structural and treatment effects. *Journal of Economic Perspectives, 28*(2), 29–50.

Ben-Michael, E., Feller, A., & Rothstein, J. (2021). The augmented synthetic control method. *Journal of the American Statistical Association, 116*(536), 1789–1803. https://doi.org/10.1080/01621459.2021.1929245

Bergstra, J., & Bengio, Y. (2012). Random search for hyper-parameter optimization. *Journal of Machine Learning Research, 13*(Feb), 281–305.

Berrar, D. (2019). Cross-validation. In *Encyclopedia of Bioinformatics and Computational Biology* (pp. 542–545). Elsevier.

Berry, S., & Haile, P. (2021). *Foundations of Demand Estimation.* CFDP 2301, revised May 2022. https://cowles.yale.edu/research/cfdp-2301-foundations-demand-estimation

Bohn, S., Lofstrom, M., & Raphael, S. (2014). Did the 2007 Legal Arizona Workers Act reduce the state's unauthorized immigrant population? *Review of Economics and Statistics, 96*(2), 258–269.

Borjas, G. J. (2021). *Labor Economics* (8th ed.). McGraw-Hill Education.

Boshnjaku, A., Krasniqi, E., & Kamberi, F. (2025). The emerging need to integrate digital health literacy as a course into health-related and care-related profession curricula. *Frontiers in Public Health.* Retrieved from https://www.frontiersin.org/journals/public-health/articles/10.3389/fpubh.2025.1534139/full

Boulesteix, A.-L., Janitza, S., Kruppa, J., & König, I. R. (2012). Overview of random forest methodology and practical guidance with emphasis on computational biology and bioinformatics. *WIREs Data Mining and Knowledge Discovery, 2*(6), 493–507.

Box, G. E. P. (1976). "Science and Statistics." *Journal of the American Statistical Association,* 71(356), 791–799.

Box, G. E. P., & Jenkins, G. M. (1976). *Time Series Analysis: Forecasting and Control.* Holden-Day.

Breiman, L. (2001). Random forests. *Machine Learning, 45*(1), 5–32.

Breiman, L. (2001). Statistical modeling: The two cultures (with comments and a rejoinder by the author). *Statistical Science, 16*(3), 199–231. https://doi.org/10.1214/ss/1009213726

Busshoff, H., Bodory, H., & Lechner, M. (2022). High-resolution treatment effects estimation: Uncovering effect heterogeneities with the modified causal forest. *Entropy, 24*(8), 1039. https://doi.org/10.3390/e24081039

Bühlmann, P., & van de Geer, S. (2011). *Statistics for High-Dimensional Data: Methods, Theory and Applications.* Springer.

Callaway, B., & Sant'Anna, P. H. C. (2021). Difference-in-differences with multiple time periods. *Journal of Econometrics, 225*(2), 200–230.

Calonico, S., Cattaneo, M. D., & Titiunik, R. (2014). Robust nonparametric confidence intervals for regression-discontinuity designs. *Econometrica, 82*(6), 2295–2326.

Calonico, S., Cattaneo, M. D., & Titiunik, R. (2017). Robust nonparametric confidence intervals for regression-discontinuity designs. *Econometrica, 85*(1), 229–261.

Calonico, S., Cattaneo, M. D., Farrell, M. H., & Titiunik, R. (2019). Regression discontinuity designs using covariates. *Review of Economics and Statistics, 101*(3), 442–451.

Calonico, S., Cattaneo, M. D., Farrell, M. H., Palomba, F., & Titiunik, R. (2025). *Treatment effect heterogeneity in regression discontinuity designs* [Working paper]. *arXiv preprint.* https://arxiv.org/abs/2503.13696

Campbell, D. T., & Stanley, J. C. (1963). *Experimental and Quasi-Experimental Designs for Research.* Houghton Mifflin.

Card, D. (1990). The impact of the Mariel Boatlift on the Miami labor market. *Industrial and Labor Relations Review, 43*(2), 245–257.

Card, D. (1995). Using geographic variation in college proximity to estimate the return to schooling. In L. N. Christofides, E. K. Grant, & R. Swidinsky (Eds.), *Aspects of Labour Market Behaviour: Essays in Honour of John Vanderkamp* (pp. 201–222). University of Toronto Press.

Cartwright, N. (2007). *Hunting Causes and Using Them: Approaches in Philosophy and Economics.* Cambridge University Press.

Caruana, R., Lou, Y., Gehrke, J., Koch, P., Sturm, M., & Elhadad, N. (2006). Intelligible models for healthcare: Predicting pneumonia risk and hospital readmission. *Proceedings of the 12th ACM SIGKDD International Conference on Knowledge Discovery and Data Mining*, 1135–1144.

Cattaneo, M. D., Idrobo, N., & Titiunik, R. (2024). *A practical introduction to regression discontinuity designs: Extensions.* Cambridge Elements: Quantitative and Computational Methods for Social Science. Cambridge University Press.

Cavallo, E., Galiani, S., Noy, I., & Pantano, J. (2013). Catastrophic natural disasters and economic growth. *Review of Economics and Statistics, 95*(5), 1549–1561.

Chang, N. C. (2020). Double/debiased machine learning for difference-in-differences models. *Econometrics Journal, 23*, 177–191.

Chen, H., Tian, X., & Yu, Y. (2020). CausalML: Python package for causal inference with machine learning. *GitHub Repository.* Retrieved from https://github.com/uber/causalml

Chen, X., Chernozhukov, V., Fernández-Val, I., & Kostyshak, S. (2021). Debiased/double machine learning for instrumental variable quantile regressions. *arXiv preprint.* https://arxiv.org/abs/1909.12592

Chen, X., Christensen, T., & Kankanala, S. (2025). Adaptive estimation and uniform confidence bands for nonparametric structural functions and elasticities. *The Review of Economic Studies, 92*(1), 162–196. https://doi.org/10.1093/restud/rdae025

Chernozhukov, V., Chetverikov, D., Demirer, M., Duflo, E., Hansen, C., & Newey, W. (2018). Double/debiased machine learning for treatment and structural parameters. *The Econometrics Journal, 21*(1), C1–C68. https://academic.oup.com/ectj/article/21/1/C1/5056401

Chernozhukov, V., Hansen, C., & Spindler, M. (2015). Post-selection and post-regularization inference in linear models with many controls and instruments. *American Economic Review: Papers & Proceedings, 105*(5), 486–490.

Chiu, A., Lan, X., Liu, Z., & Xu, Y. (2025). Causal panel analysis under parallel trends: Lessons from a large reanalysis study. *American Political Science Review* (conditionally accepted). *arXiv preprint arXiv:2309.15983.* https://doi.org/10.48550/arXiv.2309.15983

Ciccia, D. (2024). A short note on event-study synthetic difference-in-differences estimators. *arXiv preprint arXiv:2407.09565.* https://doi.org/10.48550/arXiv.2407.09565

Clarke, D., Pailañir, D., Athey, S., & Imbens, G. (2024). On synthetic difference-in-differences and related estimation methods in Stata. *The Stata Journal, 24*(4), 557–598. https://doi.org/10.1177/1536867X241297914

Cochran, W. G. (1968). The effectiveness of adjustment by subclassification in removing bias in observational studies. *Biometrics, 24*(2), 295–313. https://doi.org/10.2307/2528036

Cochran, W. G., & Rubin, D. B. (1973). Controlling bias in observational studies: A review. *Sankhyā: The Indian Journal of Statistics, Series A, 35*(4), 417–446. Retrieved from https://www.jstor.org/stable/25049893

Cox, D. R. (2001). Statistical modeling: The two cultures [Comment]. Statistical Science, 16(3), 216–218.

Crudu, P. (2023). Long-term effects of early adverse labour market conditions: A causal machine learning approach. *SSRN Working Paper.* https://ssrn.com/abstract=4592117

Currie, J., & Gruber, J. (1996). Saving babies: The efficacy and cost of recent changes in the Medicaid eligibility of pregnant women. *Journal of Political Economy, 104*(6), 1263–1296.

Cutler, D. R., Edwards Jr., T. C., Beard, K. H., Cutler, A., Hess, K. T., Gibson, J., & Lawler, J. J. (2007). Random forests for classification in ecology. *Ecology, 88*(11), 2783–2792.

Daubechies, I., Defrise, M., & De Mol, C. (2004). An iterative thresholding algorithm for linear inverse problems with a sparsity constraint. *Communications on Pure and Applied Mathematics, 57*(11), 1413–1457.

Davies, N. M., Smith, G. D., & Windmeijer, F. (2023). Reading and conducting instrumental variable studies: A guide for researchers. *The BMJ, 387*, e078093.

Davis, J. M. V., & Heller, S. B. (2020). Rethinking the benefits of youth employment programs: The heterogeneous effects of summer jobs. *The Review of Economics and Statistics, 102*(4), 664–677. https://doi.org/10.1162/rest_a_00850

de Chaisemartin, C., & d'Haultfœuille, X. (2020). Two-way fixed effects estimators with heterogeneous treatment effects. *American Economic Review, 110*(9), 2964–2996.

de Chaisemartin, C., & d'Haultfœuille, X. (2023). Credible answers to hard questions: Differences-in-differences for natural experiments. *SSRN Working Paper.* https://ssrn.com/abstract=4487202 or https://doi.org/10.2139/ssrn.4487202

de Chaisemartin, C., & D'Haultfœuille, X. (2024). Difference-in-differences estimators of intertemporal treatment effects. *The Review of Economics and Statistics.* https://doi.org/10.1162/rest_a_01414

Deaton, A., & Cartwright, N. (2018). "Understanding and Misunderstanding Randomized Controlled Trials." *Social Science & Medicine*, 210, 2–21.

Diebold, F. X. (2015). *Forecasting in Economics, Business, Finance, and Beyond.* Princeton University Press.

Dinkelman, T. (2011). The effects of rural electrification on employment: New evidence from South Africa. *American Economic Review, 101*(7), 3078–3108.

Dobkin, C., Finkelstein, A., Kluender, R., & Notowidigdo, M. J. (2018). The economic consequences of hospital admissions. *American Economic Review, 108*(2), 308–352.

Dube, A., & Zipperer, B. (2022). Minimum wages and employment: A case study of the fast-food industry. *Quarterly Journal of Economics.*

Efron, B. (2001). Statistical modeling: The two cultures [Comment]. Statistical Science, 16(3), 218–219.

Efron, B. (2020). Prediction, Estimation, and Attribution. *Journal of the American Statistical Association*, 115(530), 636–655. https://doi.org/10.1080/01621459.2020.1762613

Efron, B., & Tibshirani, R. J. (1994). *An Introduction to the Bootstrap*. Chapman & Hall.

Feng, Z., Lu, Y., & Wang, R. (2022). Ranking the importance of demographic, socioeconomic, and underlying health factors on US COVID-19 deaths: A geographical random forest approach. *Health & Place, 75*, 102825. Retrieved from https://www.sciencedirect.com/science/article/pii/S1353829222000053

Freijeiro-González, L., Febrero-Bande, M., & González-Manteiga, W. (2022). A critical review of LASSO and its derivatives for variable selection under dependence among covariates. *International Statistical Review, 90*(1), 12469. https://doi.org/10.1111/insr.12469

Friedman, J., Hastie, T., & Tibshirani, R. (2007). Pathwise coordinate optimization. *The Annals of Applied Statistics, 1*(2), 302–332.

Friedman, J., Hastie, T., & Tibshirani, R. (2010). Regularization paths for generalized linear models via coordinate descent. *Journal of Statistical Software, 33*(1), 1–22.

Friedman, M. (1953). "The Methodology of Positive Economics." In *Essays in Positive Economics*, University of Chicago Press.

Fu, W. J. (1998). Penalized regressions: The bridge versus the lasso. *Journal of Computational and Graphical Statistics, 7*(3), 397–416.

Fuller, S., & Rametta, J. T. (2024). *Causal forest and doubly robust machine learning for political science*. OSF.

Funk, M. J., Westreich, D., Wiesen, C., Stürmer, T., Brookhart, M. A., & Davidian, M. (2011). Doubly robust estimation of causal effects. *American Journal of Epidemiology, 173*(7), 761–767. https://doi.org/10.1093/aje/kwq439

Gelman, A. (2021). Reflections on Breiman's two cultures of statistical modeling. *Observational Studies, 7*(1), 95–98. https://doi.org/10.1353/obs.2021.0025

Gelman, A., & Imbens, G. W. (2019). Why high-order polynomials should not be used in regression discontinuity designs. *Journal of Business & Economic Statistics, 37*(3), 447–456.

Glynn, A. N., & Quinn, K. M. (2010). An introduction to the augmented inverse propensity weighted estimator. *Political Analysis, 18*(1), 36–56.

Goldsmith-Pinkham, P., Hull, P., & Kolesár, M. (2020). Assessing the validity of using Bartik instruments for natural experiments. *NBER Working Paper No. 24426.*

Goodman-Bacon, A. (2021). Difference-in-differences with variation in treatment timing. *Journal of Econometrics, 225*(2), 254–277.

Grembi, V., Nannicini, T., & Troiano, U. (2016). Do fiscal rules matter? *American Economic Journal: Applied Economics, 8*(3), 1–30.

Gruber, S., & van der Laan, M. J. (2009). Targeted maximum likelihood estimation: A gentle introduction. *U.C. Berkeley Division of Biostatistics Working Paper Series*, Working Paper 252. Retrieved from https://biostats.bepress.com/ucbbiostat/paper252

Haddad, M., Huber, M., & Zhang, L. (2024). Difference-in-differences with time-varying continuous treatments using double/debiased machine learning. *arXiv preprint arXiv:2410.21105*. https://arxiv.org/abs/2410.21105

Hahn, J., Todd, P., & Van der Klaauw, W. (2001). Identification and estimation of treatment effects with a regression-discontinuity design. *Econometrica, 69*(1), 201–209.

Handayani, Y. (2025). Implementation of good governance in Indonesia in an effort to minimize corruption crime. *Journal of Health, Economics, and Social Sciences.* Retrieved from https://www.jurnal.unismuhpalu.ac.id/index.php/IJHESS/article/download/6878/4872

Hansen, B. E. (2022). Econometrics. Princeton University Press

Hartford, J., Lewis, G., Leyton-Brown, K., & Taddy, M. (2017). Deep IV: A flexible approach for counterfactual prediction. *Proceedings of the 34th International Conference on Machine Learning (ICML)*, 1414–1423. Retrieved from http://proceedings.mlr.press/v70/hartford17a.html

Hastie, T., Tibshirani, R., & Friedman, J. (2009). *The Elements of Statistical Learning: Data Mining, Inference, and Prediction* (2nd ed.). Springer.

Hettinger, G., Lee, Y., & Mitra, N. (2025). Multiply robust difference-in-differences estimation of causal effect curves for continuous exposures. *Biometrics, 81*(1), ujaf015. https://doi.org/10.1093/biomtc/ujaf015

Hoerl, A. E., & Kennard, R. W. (1970a). Ridge regression: Biased estimation for nonorthogonal problems. *Technometrics, 12*(1), 55–67. https://doi.org/10.1080/00401706.1970.10488634

Hoerl, A. E., & Kennard, R. W. (1970b). Ridge regression: Applications to nonorthogonal problems. *Technometrics, 12*(1), 69–82. https://doi.org/10.1080/00401706.1970.10488635

Hsu, Y., & Shen, S. (2024). Dynamic regression discontinuity under treatment effect heterogeneity. *Quantitative Economics, 15*(4), 1035–1064. https://doi.org/10.3982/QE2150

Huang, J., Ma, S., & Zhang, C. H. (2008). Adaptive lasso for sparse high-dimensional regression models. *Statistica Sinica*, 18, 1603–1618.

Hyndman, R. J., & Athanasopoulos, G. (2021). *Forecasting: Principles and Practice* (3rd ed.). OTexts.

Imbens, G. W., & Angrist, J. D. (1994). Identification and estimation of local average treatment effects. *Econometrica, 62*(2), 467–475.

Imbens, G. W., & Kalyanaraman, K. (2012). Optimal bandwidth choice for the regression discontinuity estimator. *The Review of Economic Studies, 79*(3), 933–959.

Imbens, G. W., & Rubin, D. B. (2015). *Causal Inference in Statistics, Social, and Biomedical Sciences.* Cambridge University Press.

Imbens, G. W., & Wooldridge, J. M. (2009). Recent developments in the econometrics of program evaluation. *Journal of Economic Literature, 47*(1), 5–86. https://doi.org/10.1257/jel.47.1.5

Javanmard, A., & Montanari, A. (2014). Confidence intervals and hypothesis testing for high-dimensional regression. *Journal of Machine Learning Research, 15*, 2869–2909.

Jiang, K., Xu, W., & Danowski, J. (2025). Network analysis of social media texts. *Frontiers in Research Metrics and Analytics.* Retrieved from https://www.frontiersin.org/journals/research-metrics-and-analytics/articles/10.3389/frma.2025.1558798/full

Jin, F. F., Naghi, A. A., & Pick, A. (2019). *Heterogeneous treatment effects of educational interventions by using random forests.* Erasmus University Thesis.

Kang, H., Guo, Z., Liu, Z., & Small, D. (2024). Identification and inference with invalid instruments. *arXiv preprint arXiv:2407.19558.* https://arxiv.org/abs/2407.19558

Kennedy, E. H. (2022). Towards optimal doubly robust estimation of heterogeneous causal effects. *arXiv preprint arXiv:2004.14497.* https://arxiv.org/abs/2004.14497

King, G., & Nielsen, R. (2019). Why propensity scores should not be used for matching. *Political Analysis, 27*(4), 435–454. https://doi.org/10.1017/pan.2019.11

Klemperer, P., & Meyer, M. (1986). Price competition vs. quantity competition: The role of uncertainty. *RAND Journal of Economics, 17*(4), 618–638.

Kreif, N., DiazOrdaz, K., & Moreno-Serra, R. (2022). Estimating heterogeneous policy impacts using causal machine learning: A case study of health insurance reform in Indonesia. *Health Services and Outcomes Research Methods, 22*, 192–227. https://doi.org/10.1007/s10742-021-00259-3

Kreif, N., Grieve, R., Hangartner, D., Turner, A. J., Nikolova, S., & Sutton, M. (2016). Examination of the synthetic control method for evaluating health policies with multiple treated units. *Health Economics, 25*(12), 1514–1528.

Kreiss, A., & Rothe, C. (2024). Inference in regression discontinuity designs with high-dimensional covariates. *The Econometrics Journal, 26*(2), 105–123. https://doi.org/10.1093/ectj/utac029

Kuhn, M., & Johnson, K. (2013). *Applied Predictive Modeling.* Springer.

Kunzel, S. R., Sekhon, J. S., Bickel, P. J., & Yu, B. (2019). Metalearners for estimating heterogeneous treatment effects using machine learning. *Proceedings of the National Academy of Sciences, 116*(10), 4156–4165. https://doi.org/10.1073/pnas.1804597116

Kwak, D., Liang, Y., Shi, X., & Tan, X. (2024). Comparing machine learning and advanced methods with traditional methods to generate weights in inverse probability of treatment weighting: The INFORM study. *Pragmatic and Observational Research, 15*, 173–183. https://doi.org/10.2147/POR.S466505

Lechner, M. (2002). Some practical issues in the evaluation of heterogeneous labour market programmes by matching methods. *Journal of the Royal Statistical Society: Series A (Statistics in Society), 165*(1), 59–82.

Lechner, M., & Mareckova, J. (2025). *Comprehensive causal machine learning. arXiv preprint arXiv:2405.10198.* https://doi.org/10.48550/arXiv.2405.10198

Lee, B. K., Lessler, J., & Stuart, E. A. (2010). Improving propensity score weighting using machine learning. *Statistics in Medicine, 29*(3), 337–346. https://doi.org/10.1002/sim.3782

Lesko, C. R., Henderson, N. C., & Varadhan, R. (2018). Considerations when assessing heterogeneity of treatment effect in patient-centered outcomes research. *Journal of Clinical Epidemiology, 100*, 22–31. https://doi.org/10.1016/j.jclinepi.2018.04.005

Leuven, E., & Sianesi, B. (2003). psmatch2: Stata module to perform full Mahalanobis and propensity score matching. *Statistical Software Components.*

Levis, B., Kennedy, C. J., & Keele, L. (2024). Nonparametric identification and efficient estimation of causal effects using instrumental variables. *Econometrics Journal,* forthcoming.

Lin, W. (2013). Agnostic notes on regression adjustments to experimental data: Reexamining Freedman's critique. *Annals of Applied Statistics, 7*, 295–318. Retrieved from https://arxiv.org/pdf/1208.2301

Lin, W., Ding, P., & Han, Z. (2023). Efficiency gains of matching over regression adjustment in causal inference. *Journal of the American Statistical Association.* https://doi.org/10.1080/01621459.2023.2166827

Love, T. E. (2002). Displaying covariate balance: Love plots. *Cleveland Clinic Biostatistics Workshop.* Retrieved from https://bioinformatics.ccf.org

Lucas, R. E. (1976). Econometric Policy Evaluation: A Critique. In K. Brunner & A. Meltzer (Eds.), *The Phillips Curve and Labor Markets.* Carnegie-Rochester Conference Series on Public Policy.

Lunceford, J. K., & Davidian, M. (2004). Augmented inverse probability weighting and the double robust estimator. *Biometrics, 60*(2), 353–361.

Ma, X., Karimpour, A., & Wu, Y.-J. (2023). A causal inference approach to eliminate the impacts of interfering factors on traffic performance evaluation. *arXiv preprint.* https://arxiv.org/abs/2308.03545

Ma, Y., & Wang, L. (2018). Robust inference using inverse probability weighting. *arXiv preprint arXiv:1810.11397.* Retrieved from https://arxiv.org/abs/1810.11397

Manski, C. F. (2019). *Patient Care under Uncertainty.* Princeton University Press.

McCrary, J. (2008). Manipulation of the running variable in the regression discontinuity. *Journal of Econometrics, 142*(2), 698–714.

McLean, C. (2025). The aquanaut: Still a tool for ocean science. *Frontiers in Marine Science.* Retrieved from https://www.frontiersin.org/journals/marine-science/articles/10.3389/fmars.2025.1527929/full

Miguel, E., Satyanath, S., & Sergenti, E. (2004). Economic shocks and civil conflict: An instrumental variables approach. *Journal of Political Economy, 112*(4), 725–753.

Mogstad, M., & Torgovitsky, A. (2018). Identification and extrapolation of causal effects with instrumental variables. *Annual Review of Economics, 10*, 577–613.

Mullainathan, S., & Spiess, J. (2017). Machine learning: An applied econometric approach. *Journal of Economic Perspectives, 31*(2), 87–106.

Nakamura, E., & Steinsson, J. (2018). Identification in macroeconomics. *Journal of Economic Perspectives, 32*(3), 59–86.

Nie, X., & Wager, S. (2021). Quasi-oracle estimation of heterogeneous treatment effects. *Biometrika, 108*(2), 299–319. https://doi.org/10.1093/biomet/asaa076

Noack, C., Olma, T., & Rothe, C. (2024). Flexible covariate adjustments in regression discontinuity designs. *arXiv preprint arXiv:2107.07942.* https://arxiv.org/abs/2107.07942

Nunn, N., & Qian, N. (2014). U.S. food aid and civil conflict. *American Economic Review, 104*(6), 1630–1666.

Observational Studies Special Issue: Commentaries on Breiman's Two Cultures Paper. (2021). *Observational Studies, 7*(1). https://muse.jhu.edu/issue/45147

Ogburn, E. L., & Shpitser, I. (2021). Causal modelling: The two cultures. *Observational Studies, 7*(1), 179–183. https://doi.org/10.1353/obs.2021.0006

Piot-Lepetit, I. (2025). Strategies of digitalization and sustainability in agrifood value chains. *Frontiers in Sustainable Food Systems.* Retrieved from https://www.frontiersin.org/journals/sustainable-food-systems/articles/10.3389/fsufs.2025.1565662/full

Prasad, A. M., Iverson, L. R., & Liaw, A. (2006). Newer classification and regression tree techniques: Bagging and random forests for ecological prediction. *Ecosystems, 9*(2), 181–199.

Puterman, E., Weiss, J., Hives, B. A., Gemmill, A., Karasek, D., Mendes, W. B., & Adler, N. (2020). Predicting mortality from 57 economic, behavioral, social, and psychological factors. *Proceedings of the National Academy of Sciences, 117*(15), 8456–8463. Retrieved from https://www.pnas.org/doi/abs/10.1073/pnas.1918455117

Rahayu, N. S., & Pradita, A. R. (2025). Optimizing the use of domestic products within the Ministry of State Apparatus and Bureaucratic Reforms. *International Journal of Health, Economics, and Social Sciences.* Retrieved from https://www.jurnal.unismuhpalu.ac.id/index.php/IJHESS/article/download/6383/4836

Rajagopal, D., & Subramanian, P. K. T. (2025). AI augmented edge and fog computing for Internet of Health Things (IoHT). *PeerJ Computer Science.* Retrieved from https://peerj.com/articles/cs-2431.pdf

Rehill, P. (2025). How do applied researchers use the causal forest? A methodological review. *International Statistical Review.* https://doi.org/10.1111/insr.12610

Robins, J. M., Rotnitzky, A., & Zhao, L. P. (1994). Estimation of regression coefficients when some predictors are not always observed. *Journal of the American Statistical Association, 89*(427), 846–866.

Rosenbaum, P. R., & Rubin, D. B. (1983). The central role of the propensity score in observational studies for causal effects. *Biometrika, 70*(1), 41–55. https://doi.org/10.1093/biomet/70.1.41

Rosenbaum, P. R., & Rubin, D. B. (1984). Reducing bias in observational studies using subclassification on the propensity score. *Journal of the American Statistical Association, 79*(387), 516–524. https://doi.org/10.2307/2288398

Rubin, D. B. (1978). Bayesian inference for causal effects: The role of randomization. *Annals of Statistics, 6*, 34–58.

Rubin, D. B. (1980). Randomization analysis of experimental data: The Fisher randomization test comment. *Journal of the American Statistical Association, 75*(371), 591–593.

Rubin, D. B. (2001). Using propensity scores to help design observational studies: Application to the tobacco litigation. *Health Services and Outcomes Research Methodology, 2*, 169–188. https://doi.org/10.1023/A:1020363010465

Sadorsky, P. (2021). A random forests approach to predicting clean energy stock prices. *Journal of Risk and Financial Management, 14*(2), 48. Retrieved from https://www.mdpi.com/1911-8074/14/2/48

Salditt, M., Eckes, T., & Nestler, S. (2024). A tutorial introduction to heterogeneous treatment effect estimation with meta-learners. *Administration and Policy in Mental Health and Mental Health Services Research, 51*, 650–673. https://doi.org/10.1007/s10488-023-01303-9

Schultz Lindenmeyer, G., & da Silva Torrent, H. (2024). Boosting and predictability of macroeconomic variables: Evidence from Brazil. *Computational Economics, 64*, 377–409. https://doi.org/10.1007/s10614-023-10421-3

Shah, V., Kreif, N., & Jones, A. M. (2023). Evaluating the heterogeneous impacts of Indonesia's national health insurance scheme using causal machine learning. In N. Hashimzade & M. Thornton (Eds.), *Handbook of Research Methods and Applications in Empirical Microeconomics.* Edward Elgar Publishing.

Shao, D., Soleymani, A., Quinzan, F., & Kwiatkowska, M. (2024). Learning decision policies with instrumental variables through double machine learning. *arXiv preprint.* https://arxiv.org/abs/2405.08498

Sharma, S., & Bangur, P. (2025). Circling to wellness: Health implications of transitioning to a circular economy. *Circular Economy and Sustainability.* Retrieved from https://link.springer.com/article/10.1007/s43615-025-00507-5

Shirvaikar, V., Lin, X., & Holmes, C. (2023). Targeting relative risk heterogeneity with causal forests. *arXiv preprint.* https://arxiv.org/abs/2309.15793

Shmueli, G. (2010). To Explain or To Predict? *Statistical Science*, 25(3), 289–310. https://dx.doi.org/10.2139/ssrn.1351252

Singer, J. D., & Willett, J. B. (2003). *Applied Longitudinal Data Analysis: Modeling Change and Event Occurrence.* New York: Oxford Academic. Online edition published September 1, 2009.

Staiger, D., & Stock, J. H. (1997). Instrumental variables regression with weak instruments. *Econometrica, 65*(3), 557–586. Retrieved from https://www.nber.org/system/files/working_papers/t0284/t0284.pdf

Stock, J. H., & Yogo, M. (2005). Testing for weak instruments in linear IV regression. In D. W. K. Andrews & J. H. Stock (Eds.), *Identification and Inference for Econometric Models: Essays in Honor of Thomas Rothenberg* (pp. 80–108). Cambridge University Press. Retrieved from https://scholar.harvard.edu/files/stock/files/testing_for_weak_instruments_in_linear_iv_regression.pdf

Stuart, E. A. (2010). Matching methods for causal inference: A review and a look forward. *Statistical Science, 25*(1), 1–21. https://doi.org/10.1214/09-STS313

Sverdrup, E., Petukhova, M., & Wager, S. (2024). Estimating treatment effect heterogeneity in psychiatry: A review and tutorial with causal forests. *arXiv preprint.* https://arxiv.org/abs/2409.01578

Słoczyński, T., Uysal, S. D., & Wooldridge, J. M. (2023). Covariate balancing and the equivalence of weighting and doubly robust estimators of average treatment effects. *arXiv preprint arXiv:2310.18563.* Retrieved from https://arxiv.org/abs/2310.18563

Tamba, W. P., Yanti, F., & Tamba, D. (2025). Jakarta waste management policy and the capacity crisis of Bantargebang TPST: An environmental justice review. *Renai Journal.* Retrieved from https://renai-journal.percik.or.id/index.php/renai/article/download/15/12

Tetlock, P. E., & Gardner, D. (2015). *Superforecasting: The Art and Science of Prediction.* Crown.

Thoplan, R. (2014). Random forests for poverty classification. *International Journal of Sciences: Basic and Applied.* Retrieved from https://www.researchgate.net/profile/Ruben-Thoplan/publication/264785074_Random_Forests_for_Poverty_Classification/links/53ef8d0a0cf2711e0c42f4b4/Random-Forests-for-Poverty-Classification.pdf

Tibshirani, R. J. (2013). The lasso problem and uniqueness. *arXiv preprint*, arXiv:1206.0313. https://doi.org/10.48550/arXiv.1206.0313

van de Geer, S., Bühlmann, P., & Zhou, S. (2011). The adaptive and the thresholded lasso for potentially misspecified models (and a lower bound for the lasso). *Electronic Journal of Statistics, 5*, 688–749. https://doi.org/10.1214/11-EJS624

van de Geer, S., Bühlmann, P., Ritov, Y., & Dezeure, R. (2014). On asymptotically optimal confidence regions and tests for high-dimensional models. *Annals of Statistics, 42*(3), 1166–1202. https://doi.org/10.1214/14-AOS1221

Varadhan, R., & Seeger, J. D. (2013). Estimation and reporting of heterogeneity of treatment effects. In P. Velentgas, N. A. Dreyer, P. Nourjah, et al. (Eds.), *Developing a Protocol for Observational Comparative Effectiveness Research: A User's Guide* (Chapter 3). Rockville, MD: Agency for Healthcare Research and Quality. Retrieved from https://www.ncbi.nlm.nih.gov/books/NBK126188/

Wager, S., & Athey, S. (2018). Estimation and Inference of Heterogeneous Treatment Effects using Random Forests. *Journal of the American Statistical Association*, 113(523), 1228–1242. arXiv:1712.09988.

Wang, C., Wang, S., Shi, F., & Wang, Z. (2018). Robust propensity score computation method based on machine learning with label-corrupted data. *arXiv preprint*. https://arxiv.org/abs/1801.03132

Wooldridge, J. M. (2010). *Econometric Analysis of Cross Section and Panel Data* (2nd ed.). MIT Press.

Xu, Y. (2017). Generalized synthetic control method. *Political Analysis, 25*(1), 57–76.

Xu, Y., & Liu, L. (2022). *gsynth: Generalized Synthetic Control Method.* R package version 1.2.1. https://yiqingxu.org/packages/gsynth/

Yadlowsky, S., Fleming, S., Shah, N., Brunskill, E., & Wager, S. (2021). Evaluating treatment prioritization rules via rank-weighted average treatment effects. *arXiv preprint arXiv:2111.07966.* https://arxiv.org/abs/2111.07966

Young, A. (2022). Channeling Fisher: Randomization tests and the statistical insignificance of seemingly significant experimental results. *Quarterly Journal of Economics, 137*(2), 611–661.

Yıldız, A. Y., & Kalayci, A. (2025). Gradient boosting decision trees on medical diagnosis over tabular data. *arXiv preprint.* https://doi.org/10.48550/arXiv.2410.03705

Zhang, C.-H., & Zhang, S. S. (2014). Confidence intervals for low-dimensional parameters in high-dimensional linear models. *Journal of the Royal Statistical Society: Series B, 76*(1), 217–242.

Zhang, L. Z. (2025). Continuous difference-in-differences with double/debiased machine learning. *arXiv preprint arXiv:2408.10509.* https://arxiv.org/abs/2408.10509

Zhang, Y., Chen, S., & Liu, D. (2024). A measurement study of the environmental quality and medical expenditures of elderly individuals: Causal inference based on machine learning. *Archives of Public Health, 82*, 195. https://doi.org/10.1186/s13690-024-01386-2

Zou, H. (2006). The adaptive lasso and its oracle properties. *Journal of the American Statistical Association, 101*(476), 1418–1429.

Zou, H., & Hastie, T. (2005). Regularization and variable selection via the elastic net. *Journal of the Royal Statistical Society: Series B (Statistical Methodology), 67*(2), 301–320. https://doi.org/10.1111/j.1467-9868.2005.00503.x

Index